Student Solutions Manual

Chemistry: The Molecular Science

FIFTH EDITION

John W. Moore
University of Wisconsin, Madison

Conrad L. Stanitski
Franklin and Marshall College

Prepared by

Judy L. Ozment
The Pennsylvania State University

CENGAGE Learning

Australia • Brazil • Mexico • Singapore • United Kingdom • United States

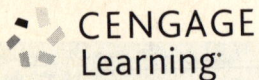

CENGAGE Learning

ISBN-13: 978-1-285-77865-5
ISBN-10: 1-285-77865-0

Cengage Learning
200 First Stamford Place, 4th Floor
Stamford, CT 06902
USA

Cengage Learning is a leading provider of customized learning solutions with office locations around the globe, including Singapore, the United Kingdom, Australia, Mexico, Brazil, and Japan. Locate your local office at: **www.cengage.com/global**.

Cengage Learning products are represented in Canada by Nelson Education, Ltd.

To learn more about Cengage Learning Solutions, visit **www.cengage.com**.

Purchase any of our products at your local college store or at our preferred online store **www.cengagebrain.com**.

Printed in the United States of America
1 2 3 4 5 6 7 18 17 16 15 14

Table of Contents

Solutions for Red-Numbered Questions for Review and Thought

Introduction

This solutions manual was written specifically to accompany the fifth edition of the textbook *Chemistry The Molecular Science*, by Moore and Stanitski. It presents detailed solutions for the some of the Questions for Review and Thought at the end of each chapter. If a question's number is bold and blue, then its solution will be in this book.

Using this Book

Many of these solutions are presented using the same format described in Chapter 1 and Appendix A. They use a four-stage process: *analyze* the problem, *plan* a solution, and *execute* the plan, then do a *reasonable result check* to show how is reasonable. Following these stages should help to make methods more readily applicable to similar problems, or the same kind problem in a different context. Some of the solutions are patterned after the methods shown in Problem-Solving Examples throughout the text.

It is important to try to answer a question for yourself, before looking at the solutions book. When you find it necessary to use this book, try first to use it to get hints or directions by reading the first part of the *Analyze* section to see if that information clarifies the problem. If you find that your interpretation or evaluation of the problem is routinely incorrect or incomplete, then you might seek help from your instructor, a teaching assistant, a learning center staff person, or a tutor in reading word problems for comprehension. Sometimes, the best help for some problems can be gained from math tutors, since they often have experience helping students specifically with reading word problems.

If you find that you usually analyze the problem in a similar fashion as described here, but still need help understanding how to plan a solution, then read how the *plan* is developed. This will give general step-by-step instructions for how the problem is solved. For numerical answers, it is at this point where you should try to estimate what answer you anticipate. Do you expect it to be a large or small number? What units and sign do you expect it to have? What significant figures will it have? The more you are able to frame an expectation for the answer, the less likely you are to make mistakes along the way.

Step-by-step solutions are shown in the *Execute* section. If you can *analyze* the questions, *plan* solutions, and *execute* the plans before actually looking here, you will start gaining confidence in your ability to learn how to answer the questions on your own.

Once you have the result for a question, take a moment to think about whether it makes sense. Often, reflection will help you confirm the correctness of an answer, or expose its flaws. For example, if you just determined that an atom of gold weighs four times more than the mass of the entire earth, the *Reasonable Result Check* stage of the problem-solving method might help you see that you made a mistake. It is wise to go back to the question and read it again, then ask yourself: Does this result answer this question? Is the result the right size and sign? Is it what you expected? It is important to check the units and the significant figures at this point, also.

A true sign that you are learning how to do chemistry problems is if you find yourself relying on this solutions book less and less. Set goals for yourself whenever you use this book to limit how often you consult it and how extensively. It is easy to such books as a crutch, and crutches keep you from walking on your own.

There are many ways to solve problems. Often times, the right answer can be derived in several ways that are equally valid. Your instructors may have different ways of describing how to work some of the problems solved here. All good methods have three things in common: (1) They always give the right answer. (2) They demonstrate how the answer was achieved. (3) And, they make sense. If any one of these three is missing, the method is flawed.

Cautions

RYou must resist the temptation to just read the solutions in this book without doing any work on your own. There are two very obvious reasons for this: (1) Recognition is easier than recall. It is far easier to look over a solution done correctly and believe that you understand it, than it is to look at the same question followed by a blank space (as would happen on a test, for example) and recall how to do it. (2) Most teachers will not let you use this book when they are testing you; hence, these solutions will not be there when you need to know how to do things.

All teachers using the textbook *Chemistry – The Molecular Science* also have a copy of the solutions in this book, so you must resist the temptation to copy work from this book into your assignments and call it your work. Besides being unethical, doing this defeats the purpose of an assignment designed to have you practice problem-solving. The instructor is asking you to do your own work, and to show what you have learned about the subject.

Learn How to Learn

Finally, keep in mind that you are learning basic chemistry and introductory physical science as a building block for other things. Those things are much more complicated. Always try to learn a subject with maximum flexibility. Whenever possible, look for how a solution can be generalized. Relying on rote memorization and narrowly-defined systems used to solve very specific types of problems may work to get you through this course, but it can be detrimental to your learning science or learning how to apply science to more complicated things, including life, the universe and everything. Sometimes, you will have to give up on preconceived notions to be able to learn more.

> "The man who grasps principles can successfully select his own methods. The man who tries methods, ignoring principles, is sure to have trouble."
>
> – Ralph Waldo Emerson

> "Imagination is more important than knowledge."
>
> – Albert Einstein

> "I know I have not found the answers to all of my questions. The answers I have found only serve to raise a whole set of new questions. In some ways I am as confused as ever, but I believe that I am confused on a higher level and about more important things."
>
> – unknown

Acknowledgements

I want to especially thank Karen Pesis, from the American River College in California, and Mr. Arya Kermansha, from Penn State University, for their patient and expert help in checking the accuracy of these solutions. I greatly appreciate the assistance and support of both of the book authors, Dr. Conrad Stantiski, and especially Dr. John Moore. I also want to express my thanks again to Leslie Kinsland, since much of what she helped me learn on the first edition of this book is still a part of these newer ventures.

Lastly, I gratefully acknowledge the fantastic support of my family, especially Karen and Emma Pesis, Lynda Webb, Susan Thompson, and Loretta Ozment, and good friends, especially Debra Lavagnino, Sara Shriner, Paul Bomboy, and Ann Schmiedekamp.

Chapter 1: The Nature of Chemistry
Solutions for Red-Numbered Questions for Review and Thought

Topical Questions

How Science is Done (Section 1-3)

9. *Result:* **(a) Qualitative (b) Quantitative and qualitative (c) Quantitative and qualitative (d) Qualitative**

Explanation:

(a) The details of the appearance of a substance (silvery-white) and information about the specific element it contains (sodium) are both **qualitative**.

(b) The temperature at which a solid melts (660 °C) is **quantitative** information. Information about what element it is (aluminum) is **qualitative**.

(c) The mass percentage of an element in the human body (about 23%) is **quantitative** information. Information about what element it is (carbon) is **qualitative**.

(d) The allotropic forms of an element (graphite, diamond, and fullerenes) and information about what element it is (carbon) are both **qualitative**.

Identifying Matter: Physical Properties (Section 1-4)

11. *Result/Explanation:* Bromine is a reddish-brown liquid. Sulfur is a chalky yellowish solid. They appear to have no property in common. The physical phase, shape, color, and appearance are different.

13. *Result:* **The solid will melt because your body temperature of 37°C is above the melting point of 29.76°C.**

Analyze and Plan: Many Americans only remember the human body temperature in the Fahrenheit scale. That is 98.6 °F. If that is the case, we can quickly apply the °F to °C conversion equation, so we can compare it to the melting point.

Execute:

$$^{\circ}\mathrm{C} = \frac{5}{9} \times \left(^{\circ}\mathrm{F} - 32\right) = \frac{5}{9} \times \left(98.6 - 32\right) = 37.0^{\circ}\mathrm{C}$$

If the sample melts at a temperature of 29.76 °C and your hand is 37 °C, the liquid will boil when exposed to the heat energy emitted by your hand when you hold the sample.

Measurements, Units, and Calculations (Section 1-5)

15. *Result:* **0.00283 kg, Ca and F**

Analyze and Plan: Given the mass of the crystal as 2.83 grams, find the mass in kilograms. Using the conversion factor between grams and kilograms, determine the mass in kilograms.

Execute:

$$2.83 \text{ g} \times \frac{1 \text{ kg}}{1000 \text{ g}} = 0.00283 \text{ kg}$$

The symbols for the elements in this crystal are Ca (calcium) and F (fluorine).

☑ *Reasonable Result Check:* Kilograms are larger units than grams, so the number of kilograms should be smaller than the number of grams.

17. *Result:* **No, 23.4 mi/hr < 25 mi/hr**

Analyze: Given the length of the track and the time to run its length, determine the miles per hour and compare to 25 mi/hr.

Plan: Divide the meters by the seconds, then use conversion factors between meters and centimeters, centimeters and inches, inches and feet, and feet and miles to determine the distance in miles. And use the conversion factor between second and hours to determine the time in hours.

Execute:

$$\frac{100\ \text{m}}{9.58\ \text{s}} \times \left(\frac{100\ \text{cm}}{1\ \text{m}}\right) \times \left(\frac{1\ \text{in}}{2.54\ \text{cm}}\right) \times \left(\frac{1\ \text{ft}}{12\ \text{in}}\right) \times \left(\frac{1\ \text{mi}}{5280\ \text{ft}}\right) \times \left(\frac{3600\ \text{s}}{1\ \text{hr}}\right) = 23.4\frac{\text{mi}}{\text{hr}}$$

No, this runner could not be arrested for exceeding the 25 mi/hr speed limit, since 23.4 mi/hr < 25 mi/hr.

☑ *Reasonable Result Check:* No one on foot could chase and catch up with a car going 25 mph.

19. *Result:* **(a) Four (b) Three (c) Four (d) Four (e) Three (f) Four**

Analyze: Given several measured quantities, determine the number of significant figures.

Plan: Use rules given in Section 1-5, summarized here: All non-zeros are significant. Zeros that precede (sit to the left of) non-zeros are never significant (e.g., 0.003). Zeros trapped between non-zeros are always significant (e.g., 3.003). Zeros that follow (sit to the right of non-zeros are (a) significant, if a decimal point is explicitly given (e.g., 3300.) OR (b) not significant, if a decimal point is not specified (e.g., 3300).

Execute:

(a) 1374 kg has **four** significant figures. The 1, 3, 7, and 4 digits are each significant.

(b) 0.00348 s has **three** significant figures. The 3, 4, and 8 digits are each significant. The zeros are all before the first non-zero-digit 3 and therefore they are not significant.

(c) 5.619 mm has **four** significant figures. The 5, 6, 1, and 9 digits are each significant.

(d) 2.475×10^{-3} cm has **four** significant figures. The 2, 4, 7, and 5 digits are each significant.

(e) 33.1 mL has **three** significant figures. The 3, 3, and 1 digits are each significant.

(f) 2300. m has **four** significant figures. The 2, 3, 0, and 0 digits are each significant.

☑ *Reasonable Result Check:* Only 2 answers have zeros in them. Those in (b) are to the left of the first non-zero digit, so none of the zeros there were significant. Those in (f) are in a number with a decimal point, so all of them were significant.

21. *Result:* **(a) 1.9 g/mL (b) 291.2 cm^3 (c) 0.0217 (d) 5.21 × 10^{-5}**

Analyze: Given numbers combined in calculations, determine the result with proper significant figures.

Plan: Perform the mathematical steps according to order of operations, applying the proper significant figures (addition and subtraction retains the least number of decimal places in the result; multiplication and division retain the least number of significant figures in the result). *Notice: if operations are combined that use different rules, we must stop and determine the intermediate result any time the rule switches.*

Execute:

(a)
$$\frac{4.850\ \text{g} - 2.34\ \text{g}}{1.3\ \text{mL}}$$

The numerator uses the subtraction rule. The first number has three decimal places (the 8, 5, and 0 are all decimal places -- digit that follow the decimal point to the right) and the second number has two decimal places (the 3 and the 4 are both decimal places), so the result of the subtraction has two decimal places.

$$\frac{2.51\ \text{g}}{1.3\ \text{mL}}$$

The ratio uses the division rule. The numerator has three significant figures and the denominator has two significant figures, so the answer will have two significant figures. Therefore, the answer is **1.9 g/mL.**

(b) $$V = {}^4/_3\ \pi r^3 = {}^4/_3 \times (3.1415926) \times (4.112\ \text{cm})^3$$

This whole calculation uses the multiplication rule, with four significant figures, limited by the measurement of r. The numerals 4 and 3 are exact, in this context. The value of π must be carried to *more than four* significant figures, such as 3.1415926… The answer comes out **291.2 cm^3**.

(c) $$(4.66 \times 10^{-3}) \times 4.666$$

This calculation uses the multiplication rule. The first number, 4.66×10^{-3}, has three significant figures and the second number, 4.666, has four significant figures, so the answer has three significant figures **0.0217**.

(d) $$\frac{0.003400}{65.2}$$

This calculation uses the division rule. The numerator has four significant figures and the denominator has three significant figures, so the answer has three significant figures 0.0000521 or **5.21×10^{-5}**.

☑ *Reasonable Result Check:* The significant figures, size, and units of the answers are appropriate.

23. *Result:* **Copper**

Analyze and Plan: We have the mass of the metal and some volume information. We need to determine the density. Use the initial and final volumes to find the volume of the metal piece, then use the mass and the volume to get the density. The metal piece displaces the water when it sinks, making the volume level in the graduated cylinder rise.

Execute: The metal piece volume is the difference between the starting volume and the final volume:

$$V_{\text{metal}} = V_{\text{final}} - V_{\text{initial}} = (37.2\ \text{mL}) - (25.4\ \text{mL}) = 11.8\ \text{mL}$$

$$d = \frac{m}{V} = \frac{105.5\ \text{g}}{11.8\ \text{mL}} = 8.94\ \frac{\text{g}}{\text{mL}}$$

According to Table 1.1, this is very close to the density of copper (d = 8.93 g/mL).

☑ *Reasonable Result Check:* The metal piece sinks, so the density of the metal piece must be higher than water. (Table 1.1 gives water density as 0.998 g/mL.)

25. *Result:* **Aluminum**

Analyze: We have the three linear dimensions of a regularly shaped piece of metal.

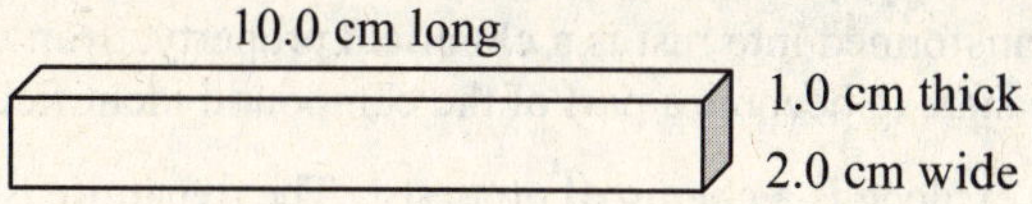

We also have its mass. We have a table of densities (Table 1.1). We need to determine the identity of the metal.

Plan: Use the three linear dimensions to find the volume of the metal piece. Use the volume and the mass to find the density. Use the table of densities to find the identity of the metal.

Execute: V = (thickness) × (width) × (length) = (1.0 cm) × (2.0 cm) × (10.0 cm) = 20. cm^3

Using conversion factors, find the volume in mL: $20.\ \text{cm}^3 \times \frac{1\ \text{mL}}{1\ \text{cm}^3} = 20.\ \text{mL}$

Find the density: $$d = \frac{m}{V} = \frac{54.0\ \text{g}}{20.\ \text{mL}} = 2.7\ \frac{\text{g}}{\text{mL}}$$

According to Table 1.1, the density that most closely matches this one is **aluminum** (d = 2.70 g/mL).

☑ *Reasonable Result Check:* The mass is larger than the volume. so d is larger than 1.

27. *Result:* **3.9 × 10³ g**

Analyze: We have the three linear dimensions of a regularly shaped sodium chloride crystal:

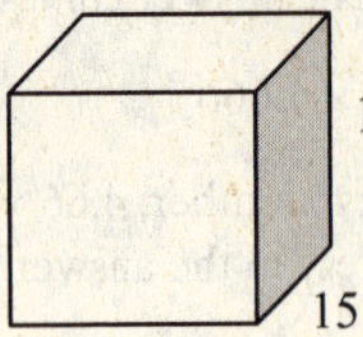

We have a table of densities (Table 1.1). We need to determine the mass of the crystal.

Plan and Execute: Use the three linear dimensions to find the volume of the crystal. Use the volume and density (Table 1.1) to find the mass.

$$V = (\text{thickness}) \times (\text{width}) \times (\text{length}) = (10.\ \text{cm}) \times (15\ \text{cm}) \times (12\ \text{cm}) = 1800\ \text{cm}^3 = 1.8 \times 10^3\ \text{cm}^3$$

Using conversion factors, find the volume in mL:

$$1.8 \times 10^3\ \text{cm}^3 \times \frac{1\ \text{mL}}{1\ \text{cm}^3} = 1.8 \times 10^3\ \text{mL}$$

Find the mass using the density: $$1.8 \times 10^3\ \text{mL} \times \frac{2.16\ \text{g}}{1\ \text{mL}} = 3.9 \times 10^3\ \text{g}$$

Notice: We're carrying two significant figures since the length data was only that precise.

☑ *Reasonable Result Check:* This crystal is pretty large. So, while the mass calculated is a large number, the volume is still about half the mass, consistent with a density around two.

Chemical Change and Chemical Properties (Section 1-6)

29. *Result:* **(a) Physical (b) Chemical (c) Chemical (d) Physical**

Explanation:

(a) The normal color of bromine is a **physical** property. Determining the color of a substance does not change its chemical form.

(b) The fact that iron can be transformed into rust is a **chemical** property. Iron undergoes a transformation, from its elemental metallic state to become a part of the compound identified as rust.

(c) The fact that dynamite can explode is a **chemical** property. The dynamite is chemically changed when it is observed to explode.

(d) Observing the shininess of aluminum does not change it, so this is a **physical** property. Melting aluminum does not change it to a different substance, though it does change its physical state. It is still aluminum, so melting at 660 °C is a **physical** property of aluminum.

31. *Result:* **(a) Chemical (b) Chemical (c) Physical**

Explanation:

(a) Bleaching clothes from purple to pink is a **chemical** change. The purple substance in the clothing reacts with the bleach to make a pink substance. The purple color cannot be brought back nor can the bleach.

(b) The burning of fuel in the space shuttle (hydrogen and oxygen) to form water and energy is a **chemical** change. The two elements react to form a compound.

(c) The ice cube melting in the lemonade is a **physical** change. The H_2O molecules do not change to a different form in the physical state change.

33. *Result:* **(a) Forcing a chemical reaction to occur (b) Causing work to be done (c) Causing work to be done (d) Forcing a chemical reaction to occur**

Explanation:

(a) The conversion of excess food into fat molecules is the body's way of storing energy for doing work later. So, this represents an outside source of energy (from the food we eat) **forcing a chemical reaction to occur** (the production of fat).

(b) Sodium reacts with water rather violently. It produces a lot of heat energy and **causes work to be done**.

(c) Sodium azide in an automobile's airbag decomposes causing the bag to inflate. This uses a chemical reaction to release energy and **cause work to be done** (inflation of the air bag).

(d) The process of hard-boiling an egg on your stove uses energy from the stove to **cause a chemical reaction to occur** (the coagulation of the white and yolk of the egg).

Classifying Matter: Substances and Mixtures (Section 1-7)

35. *Result/Explanation:* It is clear by visual inspection that the mixture is non-uniform (**heterogeneous**) at the macroscopic level. Iron could be separated from sand **using a magnet**, since iron is attracted to magnets and the sand is not.

37. *Result/Explanation:* Sometimes, it is necessary to try some tests to see if different parts of the mixture respond differently to physical separation techniques. It may require some experimentation, such as testing whether samples of the pure substances dissolve in water or are attracted to a magnet. Such information may also be available on the internet.

(a) Table salt dissolved in water can be separated by **evaporating the water**, which would leave the dry salt.

(b) Testing shows that iron filings are attracted to a magnet, but magnesium pieces are not. **Using a magnet**, the iron filings can be lifted out of the mixture, leaving behind the magnesium pieces.

(c) This mixture will have shiny silver metal pieces in a white crystalline powder, so the first thing we could try is use tweezers or forceps to pick out the shiny metal zinc pieces from the white sugar crystals. Solubility testing shows that sucrose dissolves in water, but zinc does not. Therefore, put the mixture in water, **dissolve the sucrose, separate the solution from the solid zinc** using a filter or a sieve, then **evaporate the water**.

Classifying Matter: Elements and Compounds (Section 1-8)

39. *Result/Explanation:*

(a) A blue powder turns white and loses mass. The loss of mass is most likely due to the creation of a gaseous product. That suggests that the original material was a **compound that decomposed** into the white substance (a compound or an element) and a gas (a compound or an element).

(b) If three different gases were formed, that suggests that the original material was a **compound that decomposed** into three compounds or elements.

41. *Result:* **(a) Heterogeneous mixture (b) Pure compound (c) Element (d) Homogeneous mixture**

Explanation:

(a) Chunky peanut butter is definitely a **heterogeneous mixture**. The uncrushed peanut chunks do not have the same properties as the smooth, sweetened part of the mixture.

(b) Distilled water is a **pure compound**. The distillation process removes other minerals and substances from water, leaving it just water.

(c) Platinum is an **element** with the symbol "Pt".

(d) Air is usually considered to be a **homogeneous mixture**. Now, sometimes air has enough variable properties to qualify as heterogeneous, such as near the tailpipe of a diesel truck; however, most of the time, the gases in a sample of air are sufficiently well mixed such that there is no visible difference in the properties of various regions of that air sample.

43. *Result:* **(a) No (b) Maybe**

Explanation:

(a) The black substance was both the source of the element that contributes to the red-orange substance and the source of the oxygen in the water.

(b) The red-orange substance may be a combination of two or more elements including possibly hydrogen or oxygen, or it maybe an elemental substance, since the water produced could account for the hydrogen and the oxygen in the products.

Nanoscale Theories and Models (Section 1-9)

45. *Result/Explanation:* Using Figure 1.16 to help define scales. The scale bar defines the image scale to be in nanometers. The images of electrons from the scanning tunneling microscope are at the **nanoscale**.

47. *Result/Explanation:* When we open a can of a carbonated beverage, the carbon dioxide gas expands rapidly as it rushes out of the can.

At the nanoscale, this can be explained as large number of **carbon dioxide molecules crowded into the unopened can**. When the can is opened, the molecules that were about to hit the can surface where the hole was made continue forward through the hole. A large number of the carbon dioxide particles that were contained within the can **escape** very quickly **through the** same **hole**.

49. *Result/Explanation:* The atoms in the solid sucrose molecules start off at a relatively low energy and compose a rather complex molecule. A significant amount of heat energy must be added to increase the motion of these atoms so that they are able to break free of the bonds that hold them together in the sugar molecule and to interact with each other to make the "caramelized" products.

The Atomic Theory (Section 1-10)

51. *Result/Explanation:* Conservation of mass is easy to see from the point of view of the atomic theory. A chemical change is described as the rearrangement of atoms. Because the atoms in the starting materials must all be accounted for in the substances produced, and because the mass of each atom does not change, there would be no change in the overall mass.

53. *Result/Explanation:* The law of multiple proportions says that if two compounds contain the same elements and samples of those two compounds both contain the same number of atoms of one element, then the ratio of the atoms of the other elements will be a small whole number.

Communicating Chemistry: Symbolism (Section 1-11)

55. *Result/Explanation:* Formula for each substance and nanoscale picture:

(a) Water H_2O

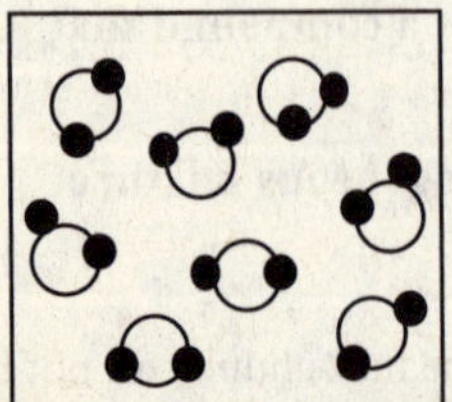

(b) Nitrogen N_2

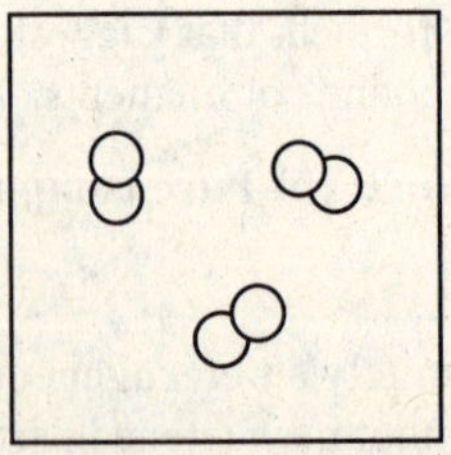

(c) Neon Ne

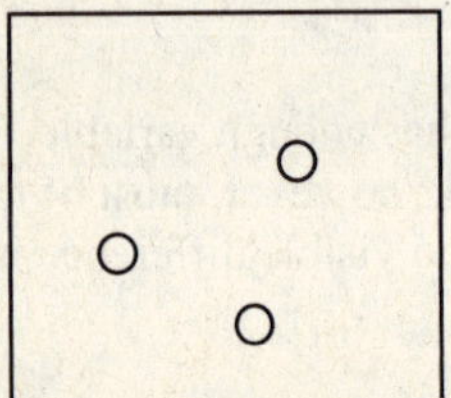

(d) Chlorine Cl_2

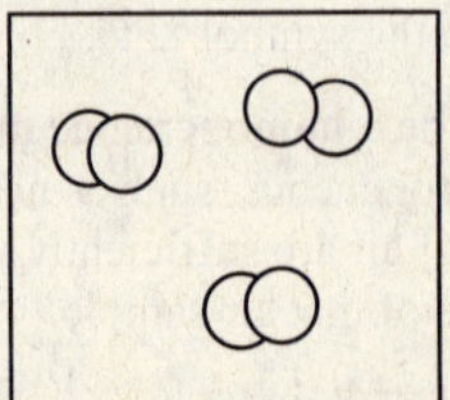

57. *Result/Explanation:* $S\ (s) + O_2\ (g) \longrightarrow SO_2\ (g)$

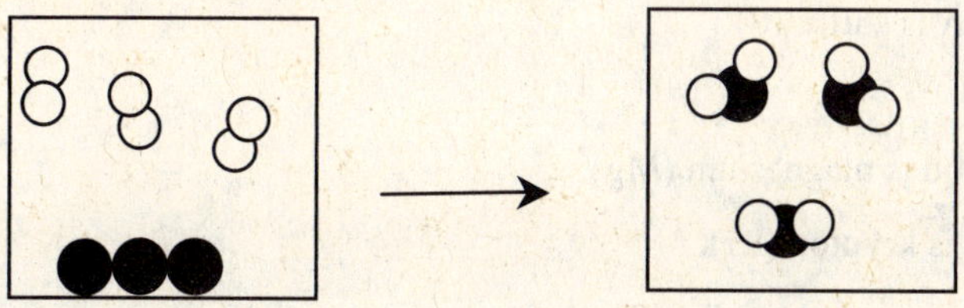

Sulfur solid and oxygen gas — Sulfur dioxide gas

59. *Result/Explanation:* $I_2\ (s) \longrightarrow I_2\ (g)$

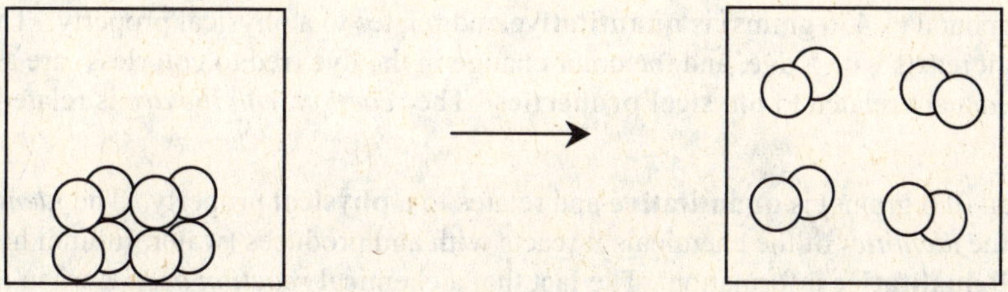

solid iodine — iodine gas

The Chemical Elements (Section 1-12)

61. *Result/Explanation:* Many pairs of responses are equally valid. Below are a few common examples. These lists are not comprehensive; many other answers are also right. The periodic table on the inside cover of the textbook is color coded to indicate metals, non-metals and metalloids.

(a) Common metallic elements: iron, Fe; gold, Au; lead, Pb; copper, Cu; aluminum, Al

(b) Common non-metallic elements: carbon, C; hydrogen, H; oxygen, O; nitrogen, N

(c) Metalloids: boron, B; silicon, Si; germanium, Ge; arsenic, As; antimony, Sb; tellurium, Te

(d) Elements that are diatomic molecules: nitrogen, N_2; oxygen, O_2; hydrogen, H_2; fluorine, F_2; chlorine, Cl_2; bromine, Br_2; iodine, I_2

The Periodic Table (Section 1-13)

65. *Result/Explanation:* There are currently six elements in Group 4A of the periodic table. They are non-metal: carbon (C), metalloids: silicon (Si) and germanium (Ge), and metals: tin (Sn), lead (Pb) and flerovium (Fl).

67. *Result:* **(a) I (b) In (c) Ir (d) Fe**

Analyze: Look up the four elements on Figure 1.26 in Section 1-13.

(a) I, iodine, is the halogen (because it is in Group 7A)

(b) In, indium, is a main group metal (because it is a metal found in an A group—Group 3A)

(c) Ir, iridium, is a transition metal (colored blue on Figure 1.26) in period 6. The period number 6 is given to the far left of the row in the periodic table next to the element Cs.

(d) Fe, iron, is a transition metal (colored blue on Figure 1.26) in period 4. The period number 4 is given to the far left of the row in the periodic table next to the element K.

69. *Result:* **(a) Mg (b) Na (c) C (d) S (e) I (f) Mg (g) Kr (h) O (i) Ge** *Notice: There are multiple Results to (a), (b) and (i) in this Question. The ones given here are only examples.*

Analyze: Use the periodic table and information given in Section 1-13.

(a) An element in Group 2A is magnesium (Mg).

(b) An element in the third period is sodium (Na).

(c) The element in the second period of Group 4A is carbon (C).

(d) The element in the third period in Group 6A is sulfur (S).

(e) The halogen in the fifth period is iodine (I).

(f) The alkaline earth element in the third period is magnesium (Mg).

(g) The noble gas element in the fourth period is krypton (Kr).

(h) The non-metal in Group 6A and the second period is oxygen (O).

(i) A metalloid in the fourth period is germanium (Ge).

General Questions

75. *Result/Explanation:*

(a) The mass of the compound (1.456 grams) is **quantitative** and relates to a physical property. The color (white), the fact that it reacts with a dye, and the color change in the dye (red to colorless) are all **qualitative**. The *colors* are related to **physical properties**. The *reaction with the dye* is related to a **chemical property**.

(b) The *mass* of the metal (0.6 grams) is **quantitative** and relates to a **physical** property. The *identity* of the metal (lithium) and the *identities* of the chemicals it reacts with and produces (water, lithium hydroxide, and hydrogen) are all **qualitative** information. The fact that a chemical *reaction occurs* when the metal is added to water is **qualitative** information and related to a **chemical** property.

77. *Result:* **Garden requires 3.0 ft^3, which is more than 1.45 ft^3 bag.**

Analyze and Plan: Given the linear dimensions of the plot and the depth expected, calculate the volume needed, then compare to the volume in the bag.

Execute:

$$6\text{ ft} \times 6\text{ ft} \times 1\text{ in} \times \frac{1\text{ ft}}{12\text{ in}} = 3.0\ \text{ft}^3$$

Garden requires 3.0 ft^3, which is more than 1.45 ft^3 bag.

☑ *Reasonable Result Check:* Perhaps the company meant to indicate ½" depth.

79. *Result:* **0.197 nm, 197 pm**

Analyze: A distance is given in angstroms (Å), which are defined. Determine the distance in nanometers and picometers.

Plan: Use the given relationship between angstroms and meters as a conversion factor to get from angstroms to meters. Then use the metric relationships between meters and the other two units to find the distance in nanometers and picometers.

Execute:

$$1.97\text{ Å} \times \frac{1\times10^{-10}\text{ m}}{1\text{ Å}} \times \frac{1\text{ nm}}{1\times10^{-9}\text{ m}} = 0.197\text{ nm}$$

$$1.97\text{ Å} \times \frac{1\times10^{-10}\text{ m}}{1\text{ Å}} \times \frac{1\text{ pm}}{1\times10^{-12}\text{ m}} = 197\text{ pm}$$

☑ *Reasonable Result Check:* The unit nanometer is larger than an angstrom, so the distance in nm should be smaller. The unit picometer is smaller than an angstrom, so the distance in pm should be larger.

81. *Result/Explanation:* If the density of solid calcium is almost twice that of solid potassium, but their masses are approximately the same size, then the volume must account for the difference. This suggests that the atoms of calcium are smaller than the atoms of potassium:

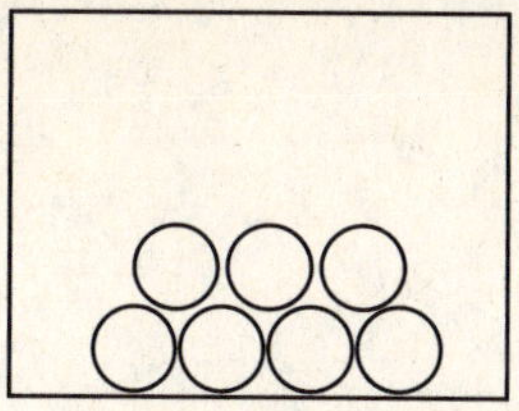

solid calcium
smaller atoms
closer packed
smaller volume

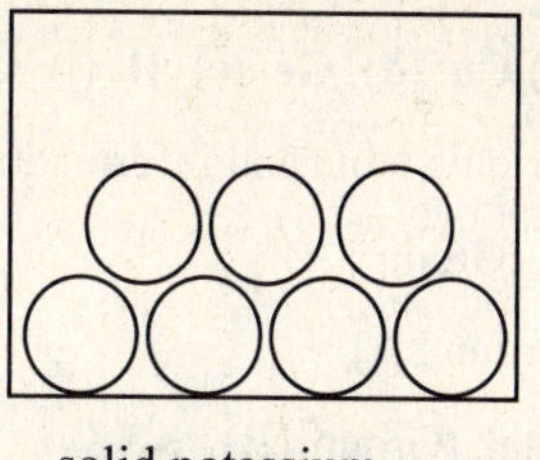

solid potassium
larger atoms
less closely packed
larger volume

83. *Result:* **508 m**

Analyze: We have the mass of a spool of aluminum wire with known diameter. Assuming the wire is a cylinder, find the length (ℓ) of wire in meters. Use the density to find the volume from the mass, then use the given volume equation and the known diameter to find the length.

$$10.0\text{ lb} \times \frac{453.59\text{ g}}{1\text{ lb}} \times \frac{1\text{ mL}}{2.70\text{ g}} \times \frac{1\text{ cm}^3}{1\text{ mL}} = 1680\text{ cm}^3$$

The radius is half the diameter. Determine the radius in centimeters:

$$R = \frac{1}{2} \times \left(0.0808\text{ in}\right) \times \frac{2.54\text{ cm}}{1\text{ in}} = 0.103\text{ cm}$$

Rearrange $V = \pi r^2 \ell$ to solve for ℓ, plug in the known values and convert to meters:

$$\ell = \frac{V}{\pi r^2} = \frac{1680\text{ cm}^3}{(3.14159) \times \left(0.103\text{ cm}\right)^2} \times \frac{1\text{ m}}{100\text{ cm}} = 508\text{ m}$$

☑ *Reasonable Result Check:* It makes sense that a quantity of wire that weighs 10 pound would be several hundred feet long.

85. *Result/Explanation:* The highest density materials will sink to the bottom, with increasingly less dense materials floating on top.

The solid material with the highest density is the Teflon plastic pieces ($d = 2.3\text{ g/cm}^3$), so those pieces will be found at the bottom of the graduated cylinder sitting in the liquid perfluorohexane, which has the highest density of the liquids ($d = 1.669\text{ g/cm}^3$).

Floating on the surface of the perfluorohexane will be the liquid water ($d = 1.00\text{ g/cm}^3$).

Floating on the water will be the pieces of HDPE plastic ($d = 0.97\text{ g/cm}^3$) and the liquid hexane ($d = 0.766\text{ g/cm}^3$).

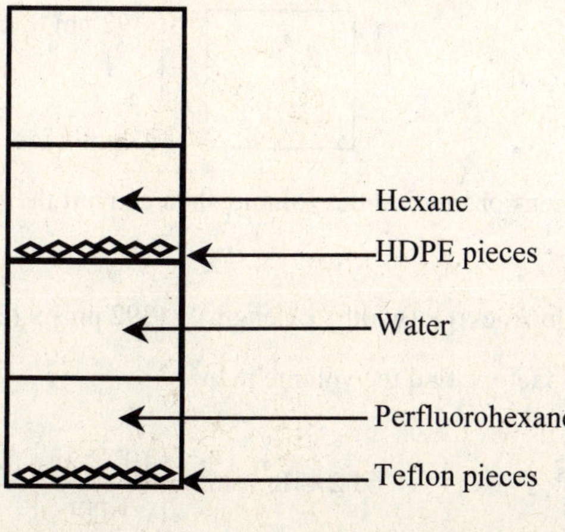

88. *Result:* **(a) K (b) Ar (c) Cu (d) Ge (e) H (f) Ca (g) Br (h) P**

Explanation: Use the periodic table and information given in Section 1-13.

(a) K is an alkali metal. (Group 1A)

(b) Ar is a noble gas. (Group 8A)

(c) Cu is a transition metal. (Group 1B)

(d) Ge is a metalloid. (Group 4A)

(e) H is a group 1 nonmetal.

(f) Ca is an alkaline earth metal. (Group 2A)

(g) Br is a halogen. (Group 7A)

(h) P is a nonmetal that is a solid. (Group 5A)

89. *Result/Explanation:* Look at the periodic table, given.

(a) A colorless gas is a non-metal. Those gases are found in the **lavender area**.

(b) A solid that is ductile and malleable are metals. Metals are found in the **gray and blue areas**.

(c) Non-metals and metalloids are poor electrical conductors. Solids with this characteristic are found in the **orange and lavender areas**.

91. *Result/Explanation:* Se and S have the greatest similarities in physical and chemical properties because they are both in the same periodic group (Group 6A).

93. *Result/Explanation:* A substance that can be broken down is not an element. A series of tests will result in a confirmation with one positive test. To prove that something is an element requires a battery of tests that all have negative results. A hypothesis that the substance is an element and cannot be broken down is more difficult to prove. (Section 1-3)

95. *Result:* **(a) Nickel, lead and magnesium (b) Titanium**

Explanation:

(a) According to the table of densities, a metal will float if the density is lower. That means that nickel, lead and magnesium will float on liquid mercury.

(b) The more different the densities, the smaller the fraction of the floating element will be below the surface. That means that titanium will float highest on mercury.

96. *Result:* **6.02×10^{-29} m^3**

Analyze and Plan: We have the length of the edge of a cube. Find the volume of the cube in m^3.

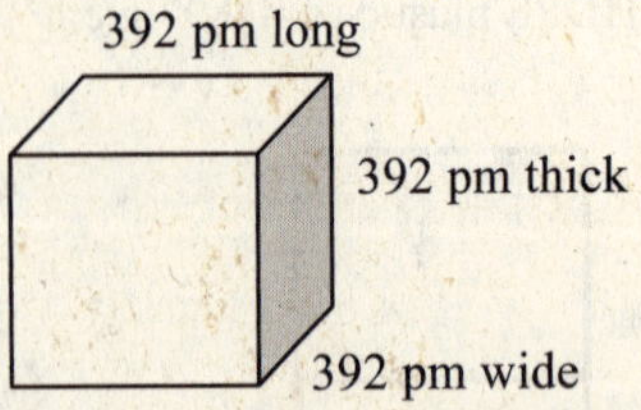

Use the linear dimensions to find the volume, then convert the volume to m^3 using metric conversions.

Execute:

$$V = (\text{thickness}) \times (\text{width}) \times (\text{length}) = (392 \text{ pm}) \times (392 \text{ pm}) \times (392 \text{ pm}) = 6.02 \times 10^7 \text{ pm}^3$$

Using conversion factors, find the volume in m^3:

$$6.02 \times 10^7 \text{ pm}^3 \times \left(\frac{10^{-12} \text{ m}}{1 \text{ pm}}\right)^3 = 6.02 \times 10^{-29} \text{ m}^3$$

☑ *Reasonable Result Check:* The cube is from the nanoscale, so it makes sense that it would be a very small volume using a macroscale unit of measure.

Applying Concepts

100. *Result:* **(a) Bromobenzene sample (b) Gold sample (c) Lead sample**

Explanation:

(a) (Table 1.1) density of butane = 0.579 g/mL; density of bromobenzene = 1.49 g/mL 1 mL butane weighs less than 1 mL of bromobenzene so, 20 mL butane weighs less than 20 mL of bromobenzene. The bromobenzene sample has a larger mass.

(b) (Table 1.1) density of benzene = 0.880 g/mL; density of gold = 19.32 g/mL There are 0.880 grams of benzene in 1 mL of benzene, so there are 8.80 grams of benzene in 10 mL of benzene. Since 1.0 mL of gold has a mass of 19.32 grams that means the gold sample has a larger mass.

(c) (Table 1.1) density of copper = 8.93 g/mL; density of lead = 11.34 g/mL Any volume of lead has a larger mass than the same volume of copper. That means the lead sample has a larger mass.

102. (a) *Result:* **2.7×10^2 mL ice (b) Deformed, overflowing, or broken**

Analyze: Use the volume of the bottle and the densities of water and ice to determine the volume of ice formed from a fixed amount of water.

Plan: Use the volume of the bottle and the density of water to determine the mass of water frozen, then calculate the volume of the ice.

Execute: At 25 °C, density of water is 0.997 g/mL.

At 0 °C, density of ice = 0.917 g/mL

$$250 \text{ mL water} \times \frac{0.997 \text{ g } H_2O(\ell)}{1 \text{ mL water}} \times \frac{1 \text{ g } H_2O(s)}{1 \text{ g } H_2O(\ell)} \times \frac{1 \text{ mL ice}}{0.917 \text{ g } H_2O(s)} = 2.7 \times 10^2 \text{ mL ice}$$

☑ *Reasonable Result Check:* The density of water is larger than the density of ice. It makes sense that the volume of the ice produced is larger than the volume of water.

(b) *Result/Explanation:* If the bottle is made of flexible plastic, it might be deformed and bulging if not cracked or broken. If the bottle is made of glass and the top came off, there would be ice (approximately 20 mL of it) oozing out of the top. Worst case scenario: if the bottle was glass and the top did not come off, it would be broken.

104. *Result/Explanation:*

(a) (Table 1.1) density of water = 0.998 g/mL, density of bromobenzene = 1.49 g/mL. Since water does not dissolve in bromobenzene, the lower density water will be the top layer of the immiscible layers.

(b) If poured slowly and carefully, the ethanol will float on top of the water and slowly dissolve in the water. Both ethanol and water will float on the bromobenzene.

(c) Stirring will speed up the ethanol dissolving with the water to make one phase. Assuming the new mixture has the average density of the original liquids, the water/ethanol layer (average density is 0.894 g/mL) will sit on top of the bromobenzene layer. (density is 1.49 g/mL)

106. *Result:* **Drawing (b)**

Explanation: The 90 °C mercury atoms will be a little bit further apart and moving somewhat more than the 10 °C mercury, though they still would be the same size atoms. The individual atoms in (c) are bigger – that doesn't happen. The individual atoms in (d) are smaller – that doesn't happen, either.

111. *Result:* **0.7527 g Ag, 0.2473 g Cl, 0.8854 g I**

Analyze and Plan: Write equations relating the mass of each atom to the sample masses, then use the relationship between the mass of an iodine atom to the mass of a chlorine atom to help calculate the mass of each element in the samples.

Execute:

$$1.0000\text{ g AgCl} = m_{Ag} + m_{Cl}$$

$$1.6381\text{ g AgI} = m_{Ag} + m_{I}$$

$$m_I = 3.580\ m_{Cl}$$

Subtract the first two equations to eliminate m_{Ag}:

$$1.6381\text{ g} - 1.0000\text{ g} = m_{Ag} + m_I - (m_{Ag} + m_{Cl}) = m_I - m_{Cl}$$

Substitute the third equation to eliminate m_I and solve for m_{Cl}:

$$0.6381\text{ g} = 3.580\ m_{Cl} - m_{Cl} = 2.580\ m_{Cl}$$

$$m_{Cl} = 0.2473\text{ g Cl}$$

Use the first equation to solve for m_{Ag}:

$$m_{Ag} = 1.0000\text{ g AgCl} - m_{Cl} = 1.0000\text{ g AgCl} - 0.2473\text{ g} = 0.7527\text{ g Ag}$$

Use the second equation to solve for m_I:

$$m_I = 1.6381\text{ g AgI} - m_{Ag} = 1.6381\text{ g AgI} - 0.7527\text{ g Ag} = 0.8854\text{ g I}$$

☑ *Reasonable Result Check:* The ratio of the calculated mass of I to the calculated mass of Cl (0.8854 g/0.2473 g) is 3.580, as indicated in the problem.

113. *Result:* **(a) 3×10^{22} molecules (b) Fraction = 3×10^{-20} (c) 300 molecules**

Analyze and Plan: (a) Use the time elapsed, breathing rate, one breath's volume, and molecules/mL to determine the number of molecules. (b) To determine the fraction, divide the molecules in speech by the molecules in the air (c) calculate the molecules in a single breath and multiply by the fraction.

Execute:

$$\text{(a)}\quad 10\text{ min} \times \left(\frac{20\text{ breaths}}{1\text{ min}}\right) \times \left(\frac{500\text{ mL}}{1\text{ breath}}\right) \times \left(\frac{2.5 \times 10^{19}\text{ molecules}}{1\text{ mL}}\right) = 2.5 \times 10^{24}\text{ molecules}$$

$$\simeq 3 \times 10^{22}\text{ molecules (1 sig fig)}$$

$$\text{(b)}\quad \text{fraction} = \frac{2.5 \times 10^{24}\text{ molecules}}{1.1 \times 10^{44}\text{ molecules}} = 2.3 \times 10^{-20} \simeq 3 \times 10^{-20}\text{ (1 sig fig)}$$

$$\text{(c)}\quad \frac{500\text{ mL}}{1\text{ breath}} \times \frac{2.5 \times 10^{19}\text{ molecules}}{1\text{ mL}} = 1.25 \times 10^{19}\ \frac{\text{molecules}}{\text{breath}}\text{ (1 sig fig)}$$

$$\left(1.25 \times 10^{22}\ \frac{\text{molecules}}{\text{breath}}\right) \times \left(2.3 \times 10^{-20}\right) = 280\text{ molecules} \simeq 300\text{ molecules (1 sig fig)}$$

☑ *Reasonable Result Check:* Considering the large number of molecules in the speech, it makes sense that there may be a small number of recycled molecules in a later breath, even though the speech molecules represent a very tiny fraction of all the air molecules.

115. *Result:* **7.056 g/cm³ ≠ 7.917 g/cm³ Densities are not the same, so they are not the same metal.**

Analyze and Plan: Calculate the volume of each cube and divide it into the mass to get the density. If the densities are not the same, then the metals are not the same.

Execute: Density = mass/volume = mass/(side length)^3

$$\text{Your cube: } \frac{16.23\text{ g}}{(1.32\text{ cm})^3} = 7.056\frac{\text{g}}{\text{cm}^3}$$

$$\text{Partner's cube: } \frac{24.64\text{ g}}{(1.46\text{ cm})^3} = 7.917\frac{\text{g}}{\text{cm}^3}$$

Because the two cubes do not have the same density, they are not made from the same metal.

☑ *Reasonable Result Check:* It makes sense that the instructor would give different unknowns.

More Challenging Questions

119. *Result:* **(a) Copper (b) 120 mL**

Analyze: Given the mass of a flask filled with a fixed volume of water, the mass of a flask filled with a given number of shots and the rest water, and the mass of the container with the same number of shots, (a) determine what pure metal the shots are made of and (b) determine the volume occupied by 500 shots.

Plan: Using the density of water, determine the mass of the water in the flask. From the combined mass of the flask + water, determine the mass of the flask. Use the combined mass of the flask, shots, plus water; the mass of the shots; and the density of water to determine the volume of water displaced by the shots. Subtract this volume from 100.0 mL to determine the volume of the water displaced by the shots. (a) Take a ratio of the mass of the shots to their volume to determine the density, then use Table 1.1 to determine what the metal is. (b) Use the volume of 20 shots to determine the volume of 500 shots.

Execute:

(a)

$$100.0\text{ mL water} \times \frac{0.998\text{ g water}}{1\text{ mL water}} = 99.8\text{ g water}$$

$$122.3\text{ g} - 99.8\text{ g} = 22.5\text{ g flask}$$

$$159.9\text{ g} - 42.3\text{ g} - 22.5\text{ g} = 95.1\text{ g water in the container with shots}$$

$$95.1\text{ g water} \times \frac{1\text{ mL water}}{0.998\text{ g water}} = 95.3\text{ mL water}$$

Volume of water displaced can be calculated by subtracting the volume of water with shots from the volume of water without shots:

$$\text{Volume of water displaced} = 100.0\text{ mL} - 95.3\text{ mL} = 4.7\text{ mL}$$

$$d = \frac{m}{V} = \frac{42.3\text{ g}}{4.7\text{ mL}} = 9.0\text{ g/mL}$$

The closest pure metal element listed in Table 1.1 to this value is **copper** (d = 8.93 g/mL)

(b)

$$500\text{ shots} \times \frac{4.7\text{ mL}}{20\text{ shots}} = 120\text{ mL}$$

☑ *Reasonable Result Check:* (a) The density of the metal is fairly close to copper. It would have been useful if the scientific observer would have reported the color of the metal to distinguish it from possibly being nickel (with d = 8.90 g/mL), instead. (b) When the number of shots increases, the volume increases.

122. *Result:* **(a) 32.1 g sulfur (b) 29.8 g zinc sulfide**

Analyze: Given the mass of one reactant and the mass of product of their combination, find the mass of the second reactant in the product, then determine the mass of product that could be formed from a different mass of reactant.

(a) *Plan:* The product is composed of two elements. The difference between the given masses must be the mass of the second element.

Execute: 97.5 g zinc sulfide – 65.4 g zinc = 32.1 g sulfur

(b) *Plan:* The elements combine in fixed ratios, so we can set up a zinc-to-product ratio to determine how the mass of product changes with a different mass of zinc.

Execute:

$$\frac{32.1\ \text{g zinc}}{65.4\ \text{g zinc sulfide}} = \frac{20.0\ \text{g zinc}}{x\ \text{g zinc sulfide}}$$

$$x = 29.8\ \text{g zinc sulfide}$$

☑ *Reasonable Result Check:* The sum of the reactant elements is the mass of the product, according to the conservation of mass. A smaller mass of one element will produce a smaller mass of product.

124. *Result:* **No, the samples contain variable percentages of iron**

Analyze: Given the mass of various portions of a sample known to contain only iron and sulfur and the mass of iron in each portion, determine if the sample is a compound of iron and sulfur, and explain the decision.

Plan and Execute:

If the sample is a compound, the elements will be combined in a fixed proportion, so we can calculate the percentage of iron in each portion and compare them.

$$\%\ \text{iron portion 1} = \frac{0.964\ \text{g iron}}{1.518\ \text{g portion}} \times 100\ \% = 63.5\ \%$$

$$\%\ \text{iron portion 2} = \frac{1.203\ \text{g iron}}{2.056\ \text{g portion}} \times 100\ \% = 58.51\ \%$$

$$\%\ \text{iron portion 3} = \frac{1.290\ \text{g iron}}{1.873\ \text{g portion}} \times 100\ \% = 69.87\ \%$$

The percentage of iron changes from portion to portion, so the sample is not composed of a single compound of iron and sulfur.

☑ *Reasonable Result Check:* The variable iron content proves that the sample cannot be a single compound.

126. *Result:* **3.1 L**

Analyze: Given the mass of one substance, the volume and density of a solution, and the mass of the solution after a reaction produces a gas with known density that escapes, determine the volume of the gas produced.

Plan and Execute: Calculate the total mass of the original mixture before the reaction occurs by adding the masses of the calcium carbonate and the hydrochloric acid solution:

$$\text{Total mass} = 12.6\ \text{g calcium carbonate} + 63.0\ \text{mL solution} \times \frac{1.096\ \text{g}}{1\ \text{mL}} = 81.6\ \text{g before reaction}$$

Use the conservation of mass to calculate the mass of escaped gas. Assuming that nothing else escaped the solution (such as water in the form of steam), the difference between the mass of the mixture before the reaction and the mass after must be the mass of the escaped gas:

$$\text{Mass gas} = 81.6\ \text{g before reaction} - 76.1\ \text{g after reaction} = 5.5\ \text{g gas}$$

Use the density of the gas to determine the volume from this mass:

$$5.5\ \text{g} \times \frac{1\ \text{L}}{1.798\ \text{g}} = 3.1\ \text{L}$$

☑ *Reasonable Result Check:* It makes sense that the solution's mass is smaller than the original mixture's mass because a gas escaped. The low density of the gas produces a large volume of gas.

Chapter 2: Chemical Compounds

Solutions for Red-Numbered Questions for Review and Thought

Topical Questions

Atomic Structure and Subatomic Particles (Section 2.1)

7. *Result/Explanation:* The masses and charges of the electron and proton are given in Section 2.1. Alpha particles are described as having two protons and two neutrons, so we double the charge of the proton to get the charge of the alpha particle. The alpha particle mass is not given in the textbook, but it is said to be the mass of one He^{2+} ion: mass of one He atom – 2(mass of electron)

$$\left(\frac{4.0026\text{ g}}{1\text{ mol}}\right)\times\left(\frac{1\text{ mole He atom}}{6.02214179\times10^{23}\text{ He atom}}\right)-2\times(9.1094\times10^{-28}\text{ g}) = 6.6447\times10^{-24}\text{ g/He}^{2+}$$

Name	Electric Charge (C)	Mass (g)	Deflected by Electric Field?
proton	1.6022×10^{-19}	1.6726×10^{-24}	**yes**
alpha particle	$\mathbf{3.2044\times10^{-19}}$	$\mathbf{6.6447\times10^{-24}}$	**yes**
electron	-1.6022×10^{-19}	$\mathbf{9.1094\times10^{-28}}$	**yes**

☑ *Reasonable Result Check:* The sum of two protons and two neutrons (6.6951×10^{-24} g) is slightly more than the mass of the alpha particle given at the National Institute of Standards and Technology web site: http://physics.nist.gov/cgi-bin/cuu/Value?mal .

9. *Result:* **40,000 cm**

Analyze: If the nucleus is scaled to a diameter of a golf ball (4 cm), determine the diameter of the atom.

Plan: Find the accepted relationship between the size of the nucleus and the size of the atom. Use size relationships to get the diameter of the "artificially large" atom.

Execute:

From Figure 2.4, nucleus diameter is approximately 10^{-14} m and an atom's diameter is approximately 10^{-10} m

Determine atom-diameter/nucleus-diameter ratio: $\frac{10^{-10}\text{ m}}{10^{-14}\text{ m}}=10^4$

So, the atom is about 10,000 times bigger than the nucleus. $10{,}000\times4\text{ cm}=40{,}000\text{ cm}$

☑ *Reasonable Result Check:* A much larger nucleus means a much larger atom with a large atomic diameter.

11. *Result:* (a) $^{67}_{34}Se$ (b) $^{72}_{36}Kr$ (c) $^{72}_{36}Kr$ (c) $^{67}_{34}Se$

Analyze and Plan: Given the symbol, $^{A}_{Z}S$, the number of neutrons is calculated with A – Z.

Execute:

For $^{67}_{34}Se$, the number of neutrons = 67 – 34 = 33. For $^{67}_{33}As$, the number of neutrons = 67 – 33 = 34.

For $^{67}_{35}Br$, the number of neutrons = 67 – 35 = 32. For $^{72}_{36}Kr$, the number of neutrons = 72 – 36 = 36.

(a) $^{67}_{34}Se$ contains 33 neutrons.

(b) $^{72}_{36}Kr$ contains the greatest number of neutrons

(c) $^{72}_{36}Kr$ contains equal number of protons and neutrons (36).

(d) Arsenic contains 33 protons. $^{67}_{34}Se$ contains 33 neutrons.

Tools of Chemistry (Section 2.2 and 2.3)

14. *Result/Explanation:* In Section 2.3, page 61, a "Tools of Chemistry" box explains the Mass Spectrometer. The species that is moving through a mass spectrometer during its operation are **ions** (usually +1 cations) that have been formed from the sample molecules by a bombarding electron beam.

16. *Result/Explanation:* The diatomic bromine molecule will be composed of: $^{79}Br–^{79}Br$, $^{79}Br–^{81}Br$, $^{81}Br–^{79}Br$, and $^{81}Br–^{81}Br$, causing peaks at mass numbers 158, 160, and 162.

Because the peak at 158 contains only bromine-79 and the relative abundance of that isotope is 50.69%, this peak has a relative abundance of (0.5069) × (0.5069)=0.2569, or 25.69%.

Because the peak at 162 contains only bromine-81 and the relative abundance of that isotope is 49.31%, this peak has a relative abundance of (0.4931) × (0.4931)=0.2569, or 24.31%.

Two isotopic combinations will have a peak at 160: $^{79}Br–^{81}Br$ and $^{81}Br–^{79}Br$. This peak has a relative abundance of 2 × (0.5069) × (0.4931)=0.4999, or 49.99%

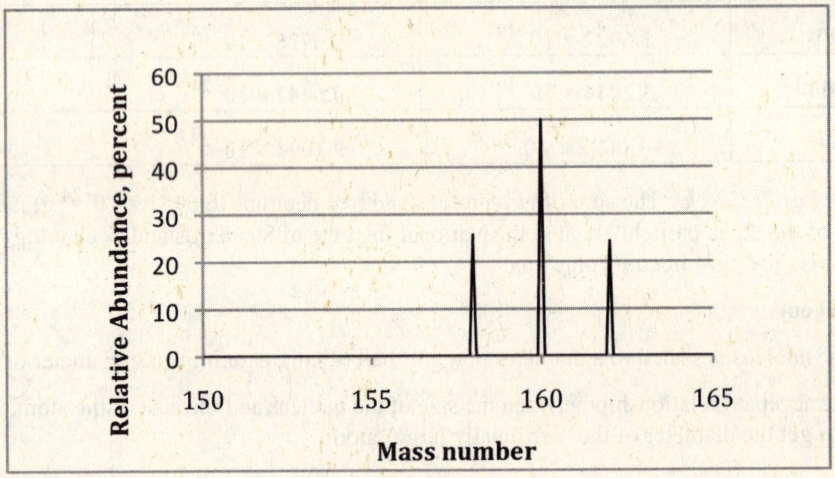

Isotopes (Section 2-3)

18. *Result:* **number of neutrons, by three**

Analyze, Plan, and Execute: Uranium-235 differs from uranium-238 in terms of the number of neutrons in the atoms. Uranium-235 has three (238 – 235) fewer neutrons uranium-237.

20. *Result:* **27 electrons, 27 protons, and 33 neutrons**

Analyze, Plan, and Execute: Given the identity of an element (cobalt) and the atom's mass number (60), find the number of electrons, protons, and neutrons in the atom.

Look up the symbol for cobalt and find that symbol on the periodic table. The periodic table gives the atomic number. The atomic number is the number of protons. The number of electrons is equal to the number of protons since the atom has no charge. The number of neutrons is the difference between the mass number and the atomic number.

The element cobalt has the symbol Co. On the periodic table, we find it listed with the atomic number 27. So, the atom has 27 protons, 27 electrons and (60 – 27 =) 33 neutrons.

☑ *Reasonable Result Check:* The number protons and electrons must be the same (27=27). The sum of the protons and neutrons is the mass number (27 + 33 = 60).

22. *Result:* **78.92 u/atom**

Analyze: Given the average atomic weight of an element and the percentage abundance of one isotope, determine the atomic weight of the only other isotope.

Plan: Using the fact that the sum of the percents must be 100%, determine the percent abundance of the second isotope. Knowing that the weighted average of the isotope masses must be equal to the reported atomic weight, set up a relationship between the known atomic mass and the various isotope masses using a variable to describe the second isotope's atomic weight.

Execute: We are told that the natural abundance of ^{81}Br is 49.31% and that there are only two isotopes. To calculate the percent abundance of the other isotope, subtract from 100%:

$$100.0\% - 49.31\% = 50.69\%$$

These percentages tell us that every 10000 atoms of bromine contain 4931 atoms of the ^{81}Br isotope and 5069 atoms of the other bromine isotope *(limited to 4 sig figs)*. The atomic weight for Br is given as 79.904 u/atom. The isotopic mass of ^{81}Br isotope is 80.916289 u/atom. Let X be the atomic mass of the other isotope of bromine.

$$\frac{4931 \text{ atoms } ^{81}\text{Br}}{10000 \text{ Br atoms}} \times \left(\frac{80.916289 \text{ amu}}{1 \text{ atom } ^{81}\text{Br}}\right) + \frac{5069 \text{ atoms other isotope}}{10000 \text{ Br atoms}} \times \left(X \frac{\text{amu}}{\text{atom}}\right) = 79.904 \frac{\text{amu}}{\text{Br atom}}$$

Solve for X

$$39.90 + 0.5069\,X = 79.904$$

$$X = 78.92 \text{ u/atom} \quad \textit{(limited to 4 sig figs)}$$

☑ *Reasonable Result Check:* Because the relative abundance is very close to 50% for each isotope, we expected the mass of the lighter isotope to be lower than the mass of the heavier isotope.

24. *Result:* **(a)** $^{23}_{11}\text{Na}$ **(b)** $^{39}_{18}\text{Ar}$ **(c)** $^{69}_{31}\text{Ga}$

Analyze: Given the identity of an element and the number of neutrons in the atom, determine the atomic symbol $^{A}_{Z}X$.

Plan: Look up the symbol for the element and find that symbol on the periodic table. The periodic table gives the atomic number (Z), which represents the number of protons. Add the number of neutrons to the number of protons to get the mass number (A).

Execute:

(a) The element sodium has the symbol Na. On the periodic table, we find it listed with the atomic number 11. The given number of neutrons is 12. So, (11 + 12 =) 23 is the mass number for this sodium atom. Its atomic symbol looks like this: $^{23}_{11}\text{Na}$.

(b) The element argon has the symbol Ar. On the periodic table, we find it listed with the atomic number 18. The given number of neutrons is 21. So, (18 + 21 =) 39 is the mass number for this argon atom. Its atomic symbol looks like this: $^{39}_{18}\text{Ar}$.

(c) The element gallium has the symbol Ga. On the periodic table, we find it listed with the atomic number 31. The given number of neutrons is 38. So, (31 + 38 =) 69 is the mass number for this gallium atom. Its atomic symbol looks like this: $^{69}_{31}\text{Ga}$.

☑ *Reasonable Result Check:* Mass number should be close to (but not exactly the same as) the atomic weight also given on the periodic table. Sodium's atomic weight (22.99) is close to the 23 mass number. Argon's atomic weight (39.95) is close to the 39 mass number. Gallium's atomic weight (69.72) is close to the 69 mass number.

26. *Result:* **See calculation below**

Analyze: Using the exact mass and the percent abundance of several isotopes of an element, determine the atomic weight.

Plan: Calculate the weighted average of the isotope masses.

Execute: Every 100 atoms of lithium contains 7.500 atoms of the ^{6}Li isotope and 92.50 atoms of the ^{7}Li isotope.

$$\frac{7.500\text{ atoms }^{6}\text{Li}}{100\text{ Li atoms}} \times \left(\frac{6.015121\text{ u}}{1\text{ atom }^{6}\text{Li}}\right) + \frac{92.50\text{ atoms }^{7}\text{Li}}{100\text{ Li atoms}} \times \left(\frac{7.016003\text{ u}}{1\text{ atom }^{7}\text{Li}}\right) = 6.941\text{ u/Li atom}$$

☑ *Reasonable Result Check:* The periodic table value for atomic weight is the same as calculated here.

28. *Result:* **60.12% ^{69}Ga, 39.88% ^{71}Ga**

Analyze: Using the exact mass of two isotopes and the atomic weight, determine the abundance of the isotopes.

Plan: Establish variables describing the isotope percentages. Set up two relationships between these variables. The sum of the percents must be 100%, and the weighted average of the isotope masses must be the reported atomic mass.

Execute: Set X% ^{69}Ga and Y% ^{71}Ga. This means: Every 100 atoms of gallium contain X atoms of the ^{69}Ga isotope and Y atoms of the ^{71}Ga isotope.

$$\frac{\text{X atoms }^{69}\text{Ga}}{100\text{ Ga atoms}} \times \left(\frac{68.9257\text{ u}}{1\text{ atom }^{69}\text{Ga}}\right) + \frac{\text{Y atoms }^{71}\text{Ga}}{100\text{ Ga atoms}} \times \left(\frac{70.9249\text{ u}}{1\text{ atom }^{71}\text{Ga}}\right) = 69.723\frac{\text{u}}{\text{Ga atom}}$$

And, X + Y = 100%. We now have two equations and two unknowns, so we can solve for X and Y algebraically. Solve the first equation for Y: Y = 100 – X. Plug that in for Y in the second equation. Then solve for X:

$$\frac{X}{100}\times(68.9257) + \frac{100-X}{100}\times(70.9249) = 69.723$$

$$0.689257X + 70.9249 - 0.709249X = 69.723$$

$$70.9249 - 69.723 = 0.709249X - 0.689257X = (0.709249 - 0.689257)X$$

$$1.202 = (0.019992)X$$

X = 60.12, so there is 60.12% ^{69}Ga

Now, plug the value of X in the first equation to get Y.

Y = 100 – X = 100 – 60.12 = 39.88, so there is 39.88% ^{69}Ga

Therefore the abundances for these isotopes are: 60.12% ^{69}Ga and 39.88% ^{71}Ga

☑ *Reasonable Result Check:* The periodic table value for the atomic weight is closer to 68.9257 than it is to 70.9249, so it makes sense that the percentage of ^{69}Ga is larger than ^{71}Ga. The sum of the two percentages is 100.00%.

Ions and Ionic Compounds (Section 2-4)

30. *Result:* **(a) Li^{+} (b) Sr^{2+} (c) Al^{3+} (d) Zn^{2+}**

Analyze and Plan: A general rule for the charge on a metal cation: the group number represents the number of electrons lost. Hence, the group number will be the cation's positive charge.

Execute:

(a) Lithium (Group 1A) Li^{+}

(b) Strontium (Group 2A) Sr^{2+}

(c) Aluminum (Group 3A) Al^{3+}

(d) Zinc (Group 2B) Zn^{2+}

32. *Result:* **(a) 2+ (b) 3– (c) 2+ or 3+ (d) 2–**

Analyze and Plan: A general rule for the charge on a monatomic metal cation: the group number represents the number of electrons lost. Hence, the group number will be the cation's positive charge. Transition metals often have a +2 charge. Some have +3 and +1 charged ions, as well. For nonmetal elements in Groups 5A-7A, the electrons gained by an atom to form a stable monatomic anion are calculated by subtracting the group number from 8. The difference between the group number and 8 is the negative charge of the anion.

Execute:

(a) Magnesium (Group 2A) has a 2+ charge. Mg^{2+}

(b) Phosphorus (Group 5A) $5 - 8 = -3$ P^{3-}

(c) Iron (a transition metal) has a 2+ or 3+ charge. Fe^{2+} or Fe^{3+}

(d) Selenium (Group 6A) $6 - 8 = -2$ Se^{2-}

34. *Result:* **CoO, Co_2O_3**

Analyze, Plan, and Execute: Cobalt ions are Co^{2+} and Co^{3+}. Oxide ion is O^{2-}. The two compounds containing cobalt and oxide are made from the neutral combination of the charged ions:

One Co^{2+} and one O^{2-} [net charge = +2 + (–2) = 0] CoO

Two Co^{3+} and three O^{2-} [net charge = 2(+3) + 3(–2) = 0] Co_2O_3

36. *Result:* **(c) and (d) are correct formulas. (a) $AlCl_3$, (b) NaF**

Analyze, Plan, and Execute:

(a) Aluminum ion (from Group 3A) is Al^{3+}. Chloride ion (from Group 7A) is Cl^-.

AlCl is **not** a neutral combination of these two ions. The correct formula would be $AlCl_3$.

[net charge = +3 + 3(–1) = 0]

(b) Sodium ion (Group 1A) is Na^+. Fluoride ion (from Group 7A) is F^-.

NaF_2 is **not** a neutral combination of these two ions. The correct formula would be NaF.

[net charge = +1 + (–1) = 0]

(c) Gallium ion (from Group 3A) is Ga^{3+}. Oxide ion (from Group 6A) is O^{2-}.

Ga_2O_3 **is** the correct neutral combination of these two ions.

[net charge = 2(+3) + 3(–2) = 0]

(d) Magnesium ion (from Group 2A) is Mg^{2+}. Sulfide ion (from Group 6A) is S^{2-}.

MgS **is** the correct neutral combination of these two ions.

[net charge = +2 + (–2) = 0]

38. *Result:* **(b), (c), and (e) are ionic, because the compounds contain metals and nonmetals together**

Analyze and Plan: To tell if a compound is ionic or not, look for metals and nonmetals together, or common cations and anions. If a compound contains only nonmetals or metalloids and nonmetals, it is likely not ionic.

Execute:

(a) CF_4 contains only nonmetals. Not ionic.

(b) $SrBr_2$ has a metal and nonmetal together. Ionic.

(c) $Co(NO_3)_3$ has a metal and nonmetals together. Ionic.

(d) SiO_2 contains a metalloid and a nonmetal. Not ionic.

(e) KCN has a metal and nonmetals together. Ionic.

(f) SCl_2 contains only nonmetals. Not ionic.

Naming Ions and Ionic Compounds (Section 2-5)

40. *Result:* **$BaSO_4$, barium ion, 2+, sulfate, 2–; $Mg(NO_3)_2$, magnesium ion, 2+, nitrate, 1–; $NaCH_3COO$, sodium ion, 1+, acetate, 1–**

Analyze, Plan, and Execute: Barium sulfate is $BaSO_4$. It contains a barium ion (Ba^{2+}), with a 2+ electrical charge, and a sulfate ion (SO_4^{2-}), with a 2– electrical charge. Magnesium nitrate is $Mg(NO_3)_2$. It contains a magnesium ion (Mg^{2+}), with a 2+ electrical charge, and two nitrate ions (NO_3^-), each with a 1– electrical charge. Sodium acetate is $NaCH_3COO$. It contains a sodium ion (Na^+), with a 1+ electrical charge, and an acetate ion (CH_3COO^-), with a 1– electrical charge. (Notice: Occasionally the Na^+ is written on the other end of the acetate formula like this CH_3COONa. It is done that way because the negative charge on acetate is on one of the oxygen atoms, so that's where the Na^+ cation will be attracted.)

42. *Result:* **(a) $Ni(NO_3)_2$ (b) $NaHCO_3$ (c) $LiClO$ (d) $Mg(ClO_3)_2$ (e) $CaSO_3$**

Analyze, Plan, and Execute:

(a) Nickel(II) ion is Ni^{2+}. Nitrate ion is NO_3^-. We use one Ni^{2+} and two NO_3^- to make neutral $Ni(NO_3)_2$.

(b) Sodium ion is Na^+. Bicarbonate ion is HCO_3^-. We use one Na^+ and one HCO_3^- to make neutral $NaHCO_3$.

(c) Lithium ion is Li^+. Hypochlorite ion is ClO^-. We use one Li^+ and one ClO^- to make neutral $LiClO$.

(d) Magnesium ion is Mg^{2+}. Chlorate ion is ClO_3^-. We use one Mg^{2+} and two ClO_3^- to make neutral $Mg(ClO_3)_2$.

(e) Calcium ion is Mg^{2+}. Sulfite ion is SO_3^{2-}. We use one Mg^{2+} and one SO_3^{2-} to make neutral $CaSO_3$.

44. *Result:* **(a) $(NH_4)_2CO_3$ (b) CaI_2 (c) $CuBr_2$ (d) $AlPO_4$**

Analyze and Plan: Make neutral combinations with the common ions involved.

Execute:

(a) Ammonium (NH_4^+) and carbonate (CO_3^{2-}) must be combined 2:1, to make $(NH_4)_2CO_3$.

(b) Calcium (Ca^{2+}) and iodide (I^-) must be combined 1:2, to make CaI_2.

(c) Copper(II) (Cu^{2+}) and bromide (Br^-) must be combined 1:2, to make $CuBr_2$.

(d) Aluminum (Al^{3+}) and phosphate (PO_4^{3-}) must be combined 1:1, to make $AlPO_4$.

46. *Result:* **(a) potassium sulfide (b) nickel(II) sulfate (c) ammonium phosphate (d) aluminum hydroxide (e) cobalt(III) sulfate**

Analyze and Plan: Give the name of the cation then the name of the anion.

Execute:

(a) K_2S contains cation K^+ called potassium and anion S^{2-} called sulfide, so it is potassium sulfide.

(b) $NiSO_4$ contains cation Ni^{2+} called nickel(II) and anion SO_4^{2-} called sulfate, so it is nickel(II) sulfate.

(c) $(NH_4)_3PO_4$ contains cation NH_4^+ called ammonium and anion PO_4^{3-} called phosphate, so it is ammonium phosphate.

(d) $Al(OH)_3$ contains cation Al^{3+} called aluminum and anion OH^- called hydroxide, so it is aluminum hydroxide.

(e) $Co_2(SO_4)_3$ contains cation Co^{3+} called cobalt(III) and anion SO_4^{2-} called sulfate, so it is cobalt(III) sulfate.

Ionic Compounds: Bonding and Properties (Section 2-6)

48. *Result:* **MgO; MgO has higher ionic charges and smaller ion sizes than NaCl**

Analyze, Plan, and Execute: Magnesium oxide is MgO, and it is composed of Mg^{2+} ions and O^{2-} ions. The relatively high melting temperature of MgO compared to NaCl (composed of Na^+ ions and Cl^- ions) is probably due to the larger ionic charges and smaller sizes of the ions. The large opposite charges sitting close together have very strong attractive forces between the ions. Melting requires that these attractive forces be overcome.

Molecular Compounds (Section 2-7)

50. *Result:* **(a) Ionic (b) Molecular (c) Molecular (d) Ionic**

Analyze and Plan: To tell if a compound is ionic or not, look at the formula for metals and nonmetals together, or common cations and anions. If a compound contains only nonmetals or metalloids and nonmetals, it is probably molecular. Ionic compounds have very high melting points (well above room temperature) and will conduct electricity when melted.

Execute:

(a) Rb_2O has a metal and a nonmetal together. **Ionic.**

(b) C_6H_{12} contains only nonmetals. **Molecular.**

(c) A compound that is a liquid at room temperature. **Molecular.**

(d) A compound that conducts electricity when molten. **Ionic.**

52. *Result/Explanation:*

(a) Structural

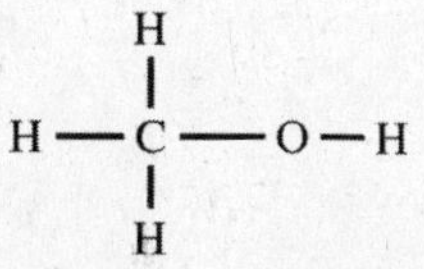

Molecular:

CH_4O

(b) Structural

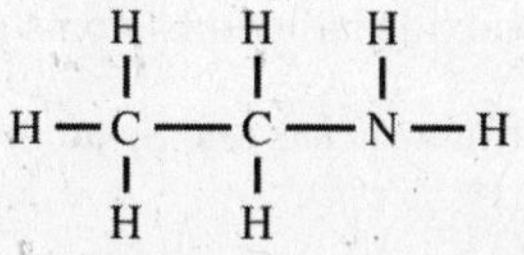

Molecular:

C_2H_7N

(c) Structural

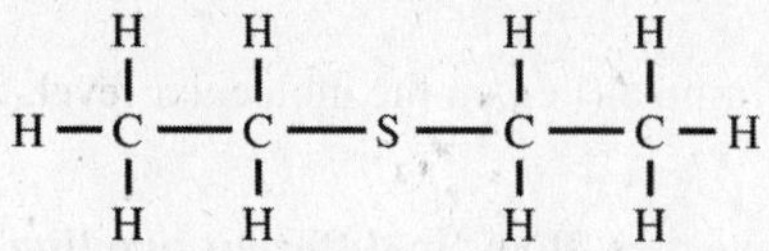

Molecular:

$C_4H_{10}S$

(d) Structural

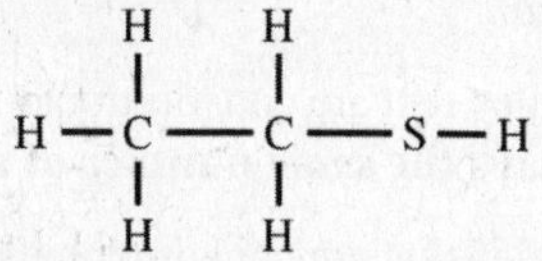

Molecular:

C_2H_6S

54. *Result/Explanation:*

(a) Heptane Molecular Formula: $\mathbf{C_7H_{16}}$

(b) Acrylonitrile Molecular Formula: $\mathbf{C_3H_3N}$

56. *Result:* **(a) 1 Ca, 2 C, 4 O (b) 8 C, 8 H (c) 2 N, 8 H, 1 S, 4 O (d) 1 Pt, 2 N, 6 H, 2 Cl (e) 4 K, 1 Fe, 6 C, 6 N**

Analyze and Plan: Keep in mind that atoms found inside parentheses that are followed by a subscript get multiplied by that subscript.

Execute:

(a) CaC_2O_4 contains one atom of calcium, two atoms of carbon, and four atoms of oxygen.

(b) $C_6H_5CHCH_2$ contains eight atoms of carbon and eight atoms of hydrogen.

(c) $(NH_4)_2SO_4$ contains two (1 × 2) atoms of nitrogen, eight (4 × 2) atoms of hydrogen, one atom of sulfur, and four atoms of oxygen.

(d) $Pt(NH_3)_2Cl_2$ contains one atom of platinum, two (1 × 2) atoms of nitrogen, six (3 × 2) atoms of hydrogen, and two atoms of chlorine.

(e) $K_4Fe(CN)_6$ contains four atoms of potassium, one atom of iron, six (1 × 6) atoms of carbon, and six (1 × 6) atoms of nitrogen.

Naming Binary Molecular Compounds (Section 2-8)

58. *Result/Explanation:* A general rule for naming binary compounds is to name the first element then take the first part of the name of the second element and add the ending -ide. Prefixes given in Table 2.6 are used to designate the number of a particular kind of atom, such as mono- for one, di- for two, tri- for three, etc.

(a) SO_2 is **sulfur dioxide**.

(b) CCl_4 is **carbon tetrachloride**.

(c) P_4S_{10} is **tetraphosphorus decasulfide**.

(d) SF_4 is **sulfur tetrafluoride**.

60. *Result/Explanation:* A general rule for applying the names of binary compounds to the formula is to list the symbol for first element named then the symbol for the second element. Use the prefixes described in Table 2.6 to learn the number of a particular kind of atom and use that number for the subscript on the symbol.

(a) nitrogen triiodide has an N atom and three I atoms: $\mathbf{NI_3}$.

(b) carbon disulfide has a C atom and two S atoms. $\mathbf{CS_2}$.

(c) dinitrogen tetraoxide has two N atoms and four O atoms: $\mathbf{N_2O_4}$.

(d) selenium hexafluoride has one Se atom and six F atoms: $\mathbf{SeF_6}$.

Organic Molecular Compounds (Section 2-9)

62. *Result/Explanation:* Carbon makes four bonds. A carbon atom in an alkane chain is bonded to at least one other C atom, so that leaves up to three remaining bonds that may each be to an H atom. So, in a noncyclic alkane other than methane, the maximum number of hydrogen atoms that can be bonded to one carbon atom is **three**.

64. *Result/Explanation:*

(a) Two molecules that are constitutional isomers have the same formula (i.e., on the molecular level, these molecules have **the same number of atoms of each kind**).

(b) Two molecules that are constitutional isomers of each other have their atoms in **different bonding arrangements**.

66. *Result/Explanation:* Noncyclic hydrocarbons have 2n + 2 hydrogen atoms, where n = number of carbon atoms. Eicosane has 20 carbon atoms, so it has 2(20) + 2 = **42 hydrogen atoms**.

Amount of Substance: The Mole (Section 2-10)

68. *Result:* $\mathbf{2 \times 10^8}$ **years**

Analyze: Determine how long it will take for all the people in the United States to count 1 mole of pennies if they spend eight hours a day every day counting.

Plan: Calculate the number of pennies each person has to count, then calculate how many days each person would spend counting their share.

Execute:

$$\frac{6.022 \times 10^{23} \text{ pennies}}{300{,}000{,}000 \text{ people}} = 2 \times 10^{15} \text{ pennies/person}$$

$$\frac{2 \times 10^{15} \text{ pennies}}{\text{person}} \times \frac{1 \text{ s}}{1 \text{ penny}} \times \frac{1 \text{ min.}}{60 \text{ s}} \times \frac{1 \text{ hour}}{60 \text{ min.}} \times \frac{1 \text{ day}}{8 \text{ hours}} \times \frac{1 \text{ year}}{365.25 \text{ days}} = 2 \times 10^{8} \text{ years}$$

Assuming that the population stays fixed over this period of time and that no one quits the job or dies without being replaced, it would take about 200 billion years for the people in the United States to count this one mole of pennies.

☑ *Reasonable Result Check:* The quantity of pennies in one mole is huge. It will take people a LONG time to count that many pennies.

Molar Mass (Section 2-11)

70. *Result:* **(a) 27 g B (b) 0.48 g O_2 (c) 6.98 × 10^{-2} g Fe (d) 2.61 × 10^3 g He**

Analyze: Determine mass in grams from given quantity in moles.

Plan: Look up the elements on the periodic table to get the atomic weight (with at least four significant figures). If necessary, calculate the molecular weight. Use that number for the molar mass (with units of grams per mole) as a conversion factor between moles and grams.

Notice: Whenever you use physical constants that you look up, it is important to carry more significant figures than the rest of the measured numbers, to prevent causing inappropriate round-off errors.

Execute:

(a) Boron (B) has atomic number 5 on the periodic table. Its atomic weight is 10.811 u/atom, so the molar mass is 10.811 g/mol.

$$2.5 \text{ mol B} \times \frac{10.811 \text{ g B}}{1 \text{ mol B}} = 27 \text{ g B}$$

(b) O_2 (diatomic molecular oxygen) is made with two atoms of the element with the atomic number 8 on the periodic table. Its atomic weight is 15.9994 u/atom; therefore, the molecular weight of O_2 is 2×15.9994 u/atom = 31.9988 u/molecule, and the molar mass is 31.9988 g/mol.

$$0.015 \text{ mol } O_2 \times \frac{31.9988 \text{ g } O_2}{1 \text{ mol } O_2} = 0.48 \text{ g } O_2$$

(c) Iron (Fe) has atomic number 26 on the periodic table. Its atomic weight is 55.845 u/atom, so the molar mass is 55.845 g/mol.

$$1.25 \times 10^{-3} \text{ mol Fe} \times \frac{55.845 \text{ g Fe}}{1 \text{ mol Fe}} = 6.98 \times 10^{-2} \text{ g Fe}$$

(d) Helium (He) has atomic number 2 on the periodic table. Its atomic weight is 4.0026 u/atom, so the molar mass is 4.0026 g/mol.

$$653 \text{ mol He} \times \frac{4.0026 \text{ g He}}{1 \text{ mol He}} = 2.61 \times 10^{3} \text{ g He}$$

☑ *Reasonable Result Check:* The mol units cancel when the factor is multiplied, leaving grams.

72. *Result:* **(a) 1.9998 mol Cu (b) 0.499 mol Ca (c) 0.6208 mol Al (d) 3.1 × 10^{-4} mol K (e) 2.1 × 10^{-5} mol Am**

Analyze: Determine the quantity in moles from given mass in grams.

Plan: Look up the elements on the periodic table to get the atomic weight. Use that number for the molar mass (with units of grams per mole) as a conversion factor between grams and moles.

Notice: Whenever you use physical constants that you look up, it is important to carry more significant figures than the rest of the measured numbers, to prevent causing inappropriate round-off errors.

Execute:

(a) Copper (Cu) has atomic number 29 on the periodic table. Its atomic weight is 63.546 u/atom, so the molar mass is 63.546 g/mol.

$$127.08 \text{ g Cu} \times \frac{1 \text{ mol Cu}}{63.546 \text{ g Cu}} = 1.9998 \text{ mol Cu}$$

(b) Calcium (Ca) has atomic number 20 on the periodic table. Its atomic weight is 40.078 u/atom, so the molar mass is 40.078 g/mol.

$$20.0 \text{ g Ca} \times \frac{1 \text{ mol Ca}}{40.078 \text{ g Ca}} = 0.499 \text{ mol Ca}$$

(c) Aluminum (Al) has atomic number 13 on the periodic table. Its atomic weight is 26.9815 u/atom, so the molar mass is 26.9815 g/mol.

$$16.75 \text{ g Al} \times \frac{1 \text{ mol Al}}{26.9815 \text{ g Al}} = 0.6208 \text{ mol Al}$$

(d) Potassium (K) has atomic number 19 on the periodic table. Its atomic weight is 39.0983 u/atom, so the molar mass is 39.0983 g/mol.

$$0.012 \text{ g K} \times \frac{1 \text{ mol K}}{39.0983 \text{ g K}} = 3.1 \times 10^{-4} \text{ mol K}$$

(e) Radioactive americium (Am) has atomic number 95 on the periodic table. The atomic weight given on the periodic table is the weight of its most stable isotope 243 u/atom, so the molar mass is 243 g/mol.

Convert milligrams into grams, first.

$$5.0 \text{ mg Am} \times \frac{1 \text{ g Am}}{1000 \text{ mg Am}} \times \frac{1 \text{ mol Am}}{243 \text{ g Am}} = 2.1 \times 10^{-5} \text{ mol Am}$$

☑ *Reasonable Result Check:* Notice that grams units cancel when the factor is multiplied, leaving moles.

74. *Result:* **4.131 × 10^{23} Cr atoms**

Analyze: Given a chromium sample with known mass, determine the number of atoms in it.

Plan: Start with the mass. Use the molar mass of chromium as a conversion factor between grams and moles. Then use Avogadro's number as a conversion factor between moles of chromium atoms and the actual number of chromium atoms.

Execute:

$$35.67 \text{ g Cr} \times \frac{1 \text{ mol Cr atoms}}{51.996 \text{ g Cr}} \times \frac{6.0221 \times 10^{23} \text{ Cr atoms}}{1 \text{ mol Cr atoms}} = 4.131 \times 10^{23} \text{ Cr atoms}$$

☑ *Reasonable Result Check:* A sample of chromium that a person can see and hold is macroscopic. It will contain a very large number of atoms.

76. *Result:* **(a) 12.63 g (b) 7.689 × 10^{20} molecules (c) 1.538 × 10^{21} N atoms (d) 14.4 g N**

Analyze: $C_{13}H_{10}N_2$ has 13 C atoms, 10 H atoms, and 2 N atoms.

Plan: Look up the elements on the periodic table to get their atomic weights. Combine those numbers for the molar mass (with units of grams per mole):

Execute: Carbon (C), with atomic number 6, has a molar mass of 12.0107 g/mol. Hydrogen (H), with atomic number 1, has a molar mass of 1.0079 g/mol. Nitrogen, with atomic number 7, has a molar mass of 14.0067 g/mol.

$$\left(\frac{13 \text{ mol C}}{1 \text{ mol comp}}\right) \times \left(\frac{12.0107 \text{g}}{1 \text{ mol C}}\right) + \left(\frac{10 \text{ mol H}}{1 \text{ mol comp}}\right) \times \left(\frac{1.0079 \text{ g}}{1 \text{ mol H}}\right) + \left(\frac{2 \text{ mol N}}{1 \text{ mol comp}}\right) \times \left(\frac{14.0067 \text{ g}}{1 \text{ mol N}}\right) = 194.2315 \text{ g/mol}$$

A shorthand version of this calculation looks like this:

$$13(12.0107 \text{ g/mol C}) + 10(1.0079 \text{ g/mol H}) + 2(14.0067 \text{ g/mol N}) = 194.2315 \text{ g/mol } C_{13}H_{10}N_2$$

(a) Use molar mass as a conversion factor between moles and grams.

$$0.06500 \text{ mol comp} \times \frac{194.2315 \text{ g comp}}{1 \text{ mol comp}} = 12.63 \text{ g comp}$$

(b) Use Avogadro's number to relate moles to molecules.

$$0.2480 \text{ g comp} \times \left(\frac{1 \text{ mol comp}}{194.2315 \text{ g comp}}\right) \times \left(\frac{6.022 \times 10^{23} \text{ molecules comp}}{1 \text{ mol comp}}\right) = 7.689 \times 10^{20} \text{ molecules comp}$$

(c) Use the chemical formula, $C_{13}H_{10}N_2$, to relate the atoms of N to molecules of compound.

$$7.689 \times 10^{20} \text{ molecules comp} \times \left(\frac{2 \text{ N atoms}}{1 \text{ molecule comp}}\right) = 1.538 \times 10^{21} \text{ N atoms}$$

(d) Calculate moles of compound using the molar mass, use the formula to relate moles of N atoms, then use the molar mass of N to calculate mass of nitrogen.

$$100. \text{ g comp} \times \left(\frac{1 \text{ mol comp}}{194.2315 \text{ g comp}}\right) \times \left(\frac{2 \text{ mol N}}{1 \text{ mol comp}}\right) \times \left(\frac{14.0067 \text{ g N}}{1 \text{ mol N}}\right) = 114.4 \text{ g N}$$

☑ *Reasonable Result Check:* Notice that several units cancel when the factors are multiplied.

78. *Result:* **(a) 41.7 pennies (b) 9.28 × 10^{-4} mol Cu (c) 5.59 × 10^{20} Cu atoms**

Analyze, Plan, and Execute:

(a) Use the percent Cu in a penny and the mass of one penny to calculate the number of pennies.

$$2.458 \text{ g Cu} \times \left(\frac{100 \text{ g penny}}{2.40 \text{ g Cu}}\right) \times \left(\frac{1 \text{ g penny}}{2.458 \text{ g penny}}\right) = 41.7 \text{ pennies}$$

(b) Calculate moles of copper using the molar mass of Cu, 63.546 g/mol.

$$2.458 \text{ g penny} \times \left(\frac{2.40 \text{ g Cu}}{100 \text{ g penny}}\right) \times \left(\frac{1 \text{ mol Cu}}{63.546 \text{ g Cu}}\right) = 9.28 \times 10^{-4} \text{ mol Cu}$$

(c) Use Avogadro's number to relate moles to atoms.

$$9.28 \times 10^{-4} \text{ mol Cu} \times \left(\frac{6.022 \times 10^{23} \text{ Cu atoms}}{1 \text{ mol Cu}}\right) = 5.59 \times 10^{20} \text{ Cu atoms}$$

☑ *Reasonable Result Check:* Notice that several units cancel when the factors are multiplied.

80. *Result:*

	CH_3OH	Carbon	Hydrogen	Oxygen
Amount of substance (mol)	**1 mol**	**1 mol**	**4 mol**	**1 mol**
No. of molecules or atoms	**6.022×10^{23} molecules**	**6.022×10^{23} atoms**	**2.409×10^{24} atoms**	**6.022×10^{23} atoms**
Molar mass (grams per mol methanol)	**32.0417 g/mol**	**12.0107 g/mol**	**4.0316 g/mol**	**15.9994 g/mol**

Analyze, Plan, and Execute: Consider a sample of 1 mol of methanol. The formula gives the mole ratio of each atom in one mole of methanol. Each mole of atoms is 6.022×10^{23} atoms. The molar masses can be determined by looking up each element on the period table, finding the atomic weight (which is also the grams/mol), and multiplying by the number of moles (see the solution to Question 77 for more details).

82. *Result:* **(a) 0.0312 mol (b) 0.0101 mol (c) 0.0125 mol (d) 0.00406 mol (e) 0.00599 mol**

Analyze: Determine the amount (in moles) in a given mass of a compound.

Plan: Use the formula and the periodic table to calculate the molar mass for the compound, then use the molar mass as a conversion factor between grams and moles.

Execute:

(a) Molar mass CH_3OH = (12.0107 g/mol C) + 4(1.0079 g/mol H)

+ (15.9994 g/mol O) = 32.0417 g/mol CH_3OH

$$1.00 \text{ g } CH_3OH \times \frac{1 \text{ mol } CH_3OH}{32.0417 \text{ g } CH_3OH} = 0.0312 \text{ mol } CH_3OH$$

(b) Molar mass Cl_2CO = 2(35.453 g/mol Cl) + (12.0107 g/mol C) + (15.9994 g/mol O) = 98.916 g/mol Cl_2CO

$$1.00 \text{ g } Cl_2CO \times \frac{1 \text{ mol } Cl_2CO}{98.916 \text{ g } Cl_2CO} = 0.0101 \text{ mol } Cl_2CO$$

(c) Ammonium nitrate is NH_4NO_3.
Molar mass NH_4NO_3 = 2(14.0067 g/mol N) + 4(1.0079 g/mol H)

+ 3(15.9994 g/mol O) = 80.0432 g/mol NH_4NO_3

$$1.00 \text{ g } NH_4NO_3 \times \frac{1 \text{ mol } NH_4NO_3}{80.0432 \text{ g } NH_4NO_3} = 0.0125 \text{ mol } NH_4NO_3$$

(d) Magnesium sulfate heptahydrate is $MgSO_4{\cdot}7\ H_2O$.

Molar mass $MgSO_4{\cdot}7\ H_2O$ = (24.305 g/mol Mg) + (32.065 g/mol S)

+ 11(15.9994 g/mol O) + 14(1.0079 g/mol H) = 246.474 g/mol $MgSO_4{\cdot}7\ H_2O$

$$1.00 \text{ g } MgSO_4 \cdot 7H_2O \times \frac{1 \text{ mol } MgSO_4 \cdot 7H_2O}{246.474 \text{ g } MgSO_4 \cdot 7H_2O} = 0.00406 \text{ mol } MgSO_4 \cdot 7H_2O$$

(e) Silver acetate is $AgC_2H_3O_2$.

Molar mass $AgC_2H_3O_2$ = (107.8682 g/mol Ag) + 2(12.0107 g/mol C)

+ 3(1.0079 g/mol H) + 2(15.9994 g/mol O) = 166.9121 g/mol $AgC_2H_3O_2$

$$1.00 \text{ g } AgC_2H_3O_2 \times \frac{1 \text{ mol } AgC_2H_3O_2}{166.9121 \text{ g } AgC_2H_3O_2} = 0.00599 \text{ mol } AgC_2H_3O_2$$

☑ *Reasonable Result Check:* The quantity in moles is always going to be smaller than the mass in grams.

84. *Result:* **(a) 151.1622 g/mol (b) 0.0352 mol (c) 25.1 g**

Analyze: Determine the molar mass of a compound and then determine the mass of a given number of moles and the number of moles in a given mass.

Plan: Use the formula and the periodic table to calculate the molar mass for the compound, then use the molar mass as a conversion factor between grams and moles.

Execute:

(a) Molar mass $C_8H_9O_2N$ = 8(12.0107 g/mol C) + 9(1.0079 g/mol H)

+ 2(15.9994 g/mol O) + (14.0067 g/mol N) = 151.1622 g/mol $C_8H_9O_2N$

(b) $$5.32 \text{ g } C_8H_9O_2N \times \frac{1 \text{ mol } C_8H_9O_2N}{151.1622 \text{ g } C_8H_9O_2N} = 0.0352 \text{ mol } C_8H_9O_2N$$

(c) $$0.166 \text{ mol } C_8H_9O_2N \times \frac{151.1622 \text{ g } C_8H_9O_2N}{1 \text{ mol } C_8H_9O_2N} = 25.1 \text{ g } C_8H_9O_2N$$

☑ *Reasonable Result Check:* The quantity in moles is always going to be smaller than the mass in grams.

Composition and Chemical Formulas (Section 2-12)

87. *Result:* **(a) 239.3 g/mol PbS, 86.60% Pb, 13.40% S (b) 30.0688 g/mol C_2H_6, 79.8881% C, 20.1119% H (c) 60.0518 g/mol CH_3COOH, 40.0011% C, 6.7135% H, 53.2854% O (d) 80.0432 g/mol NH_4NO_3, 34.9979% C, 5.0368% H, 59.9654% O**

Analyze: Given the formula of a compound, determine the molar mass, and the mass percent of each element.

Plan: Calculate the mass of each element in one mole of compound, while calculating the molar mass of the compound. Divide the calculated mass of the element by the molar mass of the compound and multiply by 100% to get mass percent. To get the last element's mass percent, subtract the other percentages from 100%.

Execute: (a)

$$\text{Mass of Pb per mole of PbS} = 207.2 \text{ g/mol Pb}$$

$$\text{Mass of S per mole of PbS} = 32.065 \text{ g/mol S}$$

$$\text{Molar mass PbS} = (207.2 \text{ g/mol Pb}) + (32.065 \text{ g/mol S}) = 239.3 \text{ g/mol PbS}$$

$$\%\text{ Pb} = \frac{\text{mass of Pb per mol PbS}}{\text{mass of PbS per mol PbS}} \times 100\% = \frac{207.2 \text{ g Pb}}{239.3 \text{ g PbS}} \times 100\% = 86.60\% \text{ Pb in PbS}$$

$$\%\text{ S} = 100\% - 86.60\% \text{ Pb} = 13.40\% \text{ S in PbS}$$

(b)

$$\text{Mass of C per mole of } C_2H_6 = 2(12.0107 \text{ g/mol C}) = 24.0214 \text{ g/mol C}$$

$$\text{Mass of H per mole of } C_2H_6 = 6(1.0079 \text{ g/mol H}) = 6.0474 \text{ g/mol H}$$

$$\text{Molar mass } C_2H_6 = (24.0214 \text{ g/mol C}) + (6.0474 \text{ g/mol H}) = 30.0688 \text{ g/mol } C_2H_6$$

$$\%\text{ C} = \frac{\text{mass of C / mol } C_2H_6}{\text{mass of } C_2H_6 \text{ / mol } C_2H_6} \times 100\% = \frac{24.0214 \text{ g C}}{30.0688 \text{ g } C_2H_6} \times 100\% = 79.8881\% \text{ C in } C_2H_6$$

$$\%\text{ H} = 100\% - 79.8881\% \text{ C} = 20.1119\% \text{ H in } C_2H_6$$

(c)

$$\text{Mass of C per mole of } CH_3COOH = 2(12.0107 \text{ g/mol C}) = 24.0214 \text{ g/mol C}$$

$$\text{Mass of H per mole of } CH_3COOH = 4(1.0079 \text{ g/mol H}) = 4.0316 \text{ g/mol H}$$

$$\text{Mass of O per mole of } CH_3COOH = 2(15.9994 \text{ g/mol O}) = 31.9988 \text{ g/mol O}$$

$$\text{Molar mass } CH_3COOH = (24.0214 \text{ g/mol C}) + (4.0316 \text{ g/mol H}) + (31.9988 \text{ g/mol O})$$

$$= 60.0518 \text{ g/mol } CH_3COOH$$

$$\%\text{C} = \frac{\text{mass of C / mol } CH_3COOH}{\text{mass of } CH_3COOH \text{ / mol } CH_3COOH} \times 100\%$$

$$= \frac{24.0214 \text{ g C}}{60.0518 \text{ g } CH_3COOH} \times 100\% = 40.0011\% \text{ C in } CH_3COOH$$

$$\%\text{ H} = \frac{\text{mass of H / mol } CH_3COOH}{\text{mass of } CH_3COOH \text{ / mol } CH_3COOH} \times 100\%$$

$$= \frac{4.0316 \text{ g H}}{60.0518 \text{ g } CH_3COOH} \times 100\% = 6.7135\% \text{ H in } CH_3COOH$$

$$\%\text{ O} = 100\% - 40.0011\% \text{ C} - 6.7135\% \text{ H} = 53.2854\% \text{ O in } CH_3COOH$$

(d) Mass of N per mole of NH_4NO_3 = 2(14.0067 g/mol N) = 28.0134 g/mol N

Mass of H per mole of NH_4NO_3 = 4(1.0079 g/mol H) = 4.0316 g/mol H

Mass of O per mole of NH_4NO_3 = 3(15.9994 g/mol O) = 47.9982 g/mol O

Molar mass NH_4NO_3 = (28.0134 g/mol N) + (4.0316 g/mol H)

+ (47.9982 g/mol O) = 80.0432 g/mol NH_4NO_3

$$\%\ \text{N} = \frac{\text{mass of N/mol NH}_4\text{NO}_3}{\text{mass of NH}_4\text{NO}_3/\text{mol NH}_4\text{NO}_3} \times 100\% = \frac{28.0134\ \text{g N}}{80.0432\ \text{g NH}_4\text{NO}_3} \times 100\%$$

$$= 34.9979\%\ \text{N in NH}_4\text{NO}_3$$

$$\%\ \text{H} = \frac{\text{mass of H/mol NH}_4\text{NO}_3}{\text{mass of NH}_4\text{NO}_3/\text{mol NH}_4\text{NO}_3} \times 100\% = \frac{4.0316\ \text{g H}}{80.0432\ \text{g NH}_4\text{NO}_3} \times 100\%$$

$$= 5.0368\%\ \text{H in NH}_4\text{NO}_3$$

$$\%\ \text{O} = 100\% - 34.9979\%\ \text{C} - 5.0368\%\ \text{H} = 59.9654\%\ \text{O in NH}_4\text{NO}_3$$

☑ *Reasonable Result Check:* Calculating the last element's mass percent using the formula gives the same answer as subtracting the other percentages from 100%.

89. *Result:* **245.745 g/mol, 25.858% Cu, 22.7992% N, 5.74197% H, 13.048% S, 32.5528% O**

Analyze: Given the formula of a compound, determine the molar mass, and the mass percent of each element

Plan: Calculate the mass of each element in one mole of compound, while calculating the molar mass of the compound (see full method on Question 68). Divide the calculated mass of the element by the molar mass of the compound and multiply by 100% to get mass percent.

Execute: The compound is $Cu(NH_3)_4SO_4{\cdot}H_2O$.

Mass of Cu per mole of compound = 63.546 g/mol Cu

Mass of N per mole of compound = 4(14.0067 g/mol N) = 56.0268 g/mol N

$Cu(NH_3)_4SO_4{\cdot}H_2O$ contains four NH_3 and one H_2O molecule so it has a total of (3×4 + 2) 14 H atoms.

Mass of H per mole of compound = 14(1.0079 g/mol H)= 14.1106 g/mol H

Mass of S per mole of compound = 32.065 g/mol S

$Cu(NH_3)_4SO_4{\cdot}H_2O$ contains one SO_4 and one H_2O molecule so it has a total of (4 + 1) 5 O atoms.

Mass of O per mole of compound = 5(15.9994 g/mol O) = 79.9970 g/mol O

Molar mass $Cu(NH_3)_4SO_4{\cdot}H_2O$ = (63.546 g/mol Cu) + (56.0268 g/mol N) + (14.1106 g/mol H)

+ (32.065 g/mol S) + (79.9970 g/mol O) = 245.745 g/mol $Cu(NH_3)_4SO_4{\cdot}H_2O$ (comp)

$$\%\ \text{element} = \frac{\text{mass of element/mol comp}}{\text{mass of comp/mol comp}} \times 100\%$$

$$\text{Mass percent Cu} = \frac{63.546\ \text{g Cu}}{245.745\ \text{g comp}} \times 100\% = 25.858\%\ \text{Cu in Cu(NH}_3)_4\text{SO}_4 \cdot \text{H}_2\text{O}$$

$$\text{Mass percent N} = \frac{56.0268\ \text{g N}}{245.745\ \text{g comp}} \times 100\% = 22.7992\%\ \text{N in Cu(NH}_3)_4\text{SO}_4 \cdot \text{H}_2\text{O}$$

$$\text{Mass percent H} = \frac{14.1106\ \text{g H}}{245.745\ \text{g comp}} \times 100\% = 5.74197\%\ \text{H in Co(NH}_3)_4\text{SO}_4 \cdot \text{H}_2\text{O}$$

$$\text{Mass percent S} = \frac{32.065 \text{ g S}}{245.745 \text{ g comp}} \times 100\% = 13.048\% \text{ S in Co(NH}_3)_4\text{SO}_4 \cdot \text{H}_2\text{O}$$

$$\text{Mass percent O} = \frac{79.9970 \text{ g O}}{245.745 \text{ g comp}} \times 100\% = 32.5528\% \text{ O in Co(NH}_3)_4\text{SO}_4 \cdot \text{H}_2\text{O}$$

☑ *Reasonable Result Check:* The sum of the percentages is 100%.

91. *Result:* **(a) $C_{10}H_{12}NO$ (b) $C_{20}H_{24}N_2O_2$**

Analyze: Given the percent by mass of elements in quinine and the molar mass determine the empirical formula and the molecular formula.

Plan and Execute:

(a) Choose a convenient sample of quinine, such as 100.00 g. Using the mass percents, determine the number of grams of C, H, N, and O in the sample. Convert these masses to moles, using the atomic weights. Set up a ratio, and simplify it by dividing each by the smallest number. Use the integers as the subscripts in the empirical formula

The compound is 74.04% C, 7.46% H, 8.64% N, and 9.86% O by mass. This means that a sample of 100.00 g has 74.04 g C, 7.46 g H, 8.64 g N, and 9.86 g O.

$$74.04 \text{ g C} \times \frac{1 \text{ mol C}}{12.0107 \text{ g C}} = 6.165 \text{ mol C} \qquad 7.46 \text{ g H} \times \frac{1 \text{ mol H}}{1.0079 \text{ g H}} = 7.40 \text{ mol H}$$

$$8.64 \text{ g N} \times \frac{1 \text{ mol N}}{14.0067 \text{ g N}} = 0.617 \text{ mol N} \qquad 9.86 \text{ g O} \times \frac{1 \text{ mol O}}{15.9994 \text{ g O}} = 0.616 \text{ mol O}$$

Mole Ratio 6.16 mol C : 7.40 mol H : 0.617 mol N : 0.616 mol O

Simplify by dividing each amount by 0.616 mol

Mole Ratio 10 C : 12 mol H : 1 mol N : 1 mol O

Therefore, the empirical formula is $C_{10}H_{12}NO$

(b) The molecular formula is $(C_{10}H_{12}NO)_n$. Determine the empirical formula molar mass. Then determine the value of n by dividing the molar mass of the compound by the molar mass of the empirical formula.

Molar mass of $C_{10}H_{12}NO$ = 10(12.0107 g/mol C) + 12(1.0079 g/mol H)

+ 15.9994 g/mol O + 14.0067 g/mol N = 162.2079 g/mol $C_{10}H_{12}NO$

$$n = \frac{324.41 \text{ g/ mol comp}}{162.2079 \text{ g/ mol emp formula}} = 2$$

So, the molecular formula is $(C_{10}H_{12}NO)_2$, or $C_{20}H_{24}N_2O_2$.

☑ *Reasonable Result Check:* The mole ratio is clearly a whole number relationship and the molar mass is very close to double the empirical formula mass.

94. *Result:* **One**

Analyze: Given the molar mass of a compound with an unknown formula and the mass percent of an element in that compound, determine the number of atoms of that element in the compound.

Plan Using a convenient sample size of compound, determine the mass of the element in that compound. Convert both masses to moles, then determine the mole ratio.

Execute: In 100.000 grams of the compound carbonic anhydrase (abbreviated as: c.a.) there are 0.218 g Zn.

$$100.000 \text{ g c.a.} \times \frac{1 \text{ mole c.a.}}{3.00 \times 10^4 \text{ g c.a.}} = 0.00333 \text{ mole c.a.}$$

$$0.218 \text{ g Zn} \times \frac{1 \text{ mole Zn}}{65.409 \text{ g Zn}} = 0.00333 \text{ mole Zn}$$

0.00333 mole Zn : 0.00333 mole c.a.

1 mole Zn : 1 mole c.a.

One Zn atom in every molecule of c.a.

☑ *Reasonable Result Check:* The contribution of the zinc mass to the mass of the molecule is very small, so it makes sense that the number of atoms of zinc is very small in this large molecule.

96. *Result:* **6**

Analyze: Given the percent by mass of an element in a compound and the compound's formula with an unknown subscript, determine the value of the unknown subscript.

Plan: Choose a convenient sample of Si_2H_x, such as 100.00 g. Using the percent by mass, determine the number of grams of Si and H in the sample. Use the molar mass of Si as a conversion factor to get the moles of Si. Use the molar mass of H as a conversion factor to get the moles of H. Set up a mole ratio to find the value of x.

Execute: The compound is 90.28% Si by mass. This means that 100.00 g of Si_2H_x contains 90.28 grams Si and the rest of the mass is from H.

$$\text{Mass of H in sample} = 100.00 \text{ g } Si_2H_x - 90.28 \text{ g Si} = 9.72 \text{ g H}$$

$$90.28 \text{ g Si} \times \frac{1 \text{ mol Si}}{28.0855 \text{ g Si}} = 3.214 \text{ mol Si}$$

$$9.72 \text{ g H} \times \frac{1 \text{ mol H}}{1.0079 \text{ g H}} = 9.64 \text{ mol H}$$

$$\text{Mole Ratio} = \frac{\text{moles of H in sample}}{\text{moles of Si in sample}} = \frac{9.64 \text{ mol H}}{3.214 \text{ mol Si}} = \frac{3 \text{ mol H}}{1 \text{ mol Si}} = \frac{6 \text{ mol H}}{2 \text{ mol Si}}$$

Therefore, the formula is Si_2H_6 and x = 6.

☑ *Reasonable Result Check:* The Mole Ratio is clearly a whole number relationship indicating a sensible number of hydrogen atoms in this molecule.

98. *Result:* $\mathbf{C_4H_8N_2O_2}$

Analyze: Given the empirical formula of a compound and the molar mass, determine the molecular formula.

Plan: Find the mass of 1 mol of the empirical formula. Divide the molar mass of the compound by the calculated empirical mass to get a whole number. Multiply all the subscripts in the empirical formula by this whole number.

Execute: The empirical formula is C_2H_4NO, the molecular formula is $(C_2H_4NO)_n$.

$$\text{Mass of 1 mol } C_2H_4NO = 2(12.0107 \text{ g/mol C}) + 4(1.0079 \text{ g/mol H})$$
$$+ 14.0067 \text{ g/mol N} + 15.9994 \text{ g/mol O} = 58.0591 \text{ g/mol } C_2H_4NO$$

$$n = \frac{\text{mass of 1 mol of molecule}}{\text{mass of 1 mol of } C_2H_4NO} = \frac{116.1 \text{ g}}{58.0591 \text{ g}} = 2.000 \approx 2$$

Molecular Formula is $(C_2H_4NO)_2 = C_4H_8N_2O_2$

☑ *Reasonable Result Check:* The molar mass is about 2 times larger than the mass of one mole of the empirical formula, so the molecular formula $C_4H_8N_2O_2$ makes sense.

General Questions

102. *Result:* **0.995 g Pt**

Analyze: Given the mass of a sample of cisplatin along with its percentage platinum, determine the grams of platinum.

Plan: Always start with the sample. Convert the mass of compound to mass of platinum using the percentage platinum as a conversion factor.

Execute: 100 grams of cisplatin contains 65.0 grams of Pt.

$$1.53 \text{ g cisplatin} \times \frac{65.0 \text{ g Pt}}{100 \text{ g cisplatin}} = 0.995 \text{ g Pt}$$

☑ *Reasonable Result Check:* The mass of Pt should be about $\frac{2}{3}$ of the mass of the compound cisplatin.

105. *Result:* **89 tons/yr**

Analyze and Plan: Start with the sample. Given the number of people in the city, use the volume of water each person needs per day, then calculate the total water needs for the day. Using the number of days in a year calculate the total volume of water used per year. Then using the mass of one gallon of water as a conversion factor, determine the grams of water. Convert the grams to tons. Then, using the fluoride concentration as a conversion factor, determine the number of tons of fluoride, and use the mass percentage of fluoride in sodium fluoride to determine the number of tons.

Execute:

$$150{,}000 \text{ people} \times \frac{175 \text{ gal water / person}}{1 \text{ day}} \times \frac{365 \text{ days}}{1 \text{ year}} \times \frac{8.34 \text{ lb water}}{1 \text{ gal water}} \times \frac{1 \text{ ton water}}{2000 \text{ lb water}}$$

$$\times \frac{1 \text{ ton fluoride}}{1{,}000{,}000 \text{ tons water}} \times \frac{100 \text{ tons sodium fluoride}}{45.0 \text{ tons flouride}} = 89 \frac{\text{tons sodium fluoride}}{\text{year}}$$

☑ *Reasonable Result Check:* The significant figures are limited to two by the 150,000 figure. The mass units are appropriately labeled. The units cancel appropriately to give tons per year. This is a large number of people using a large amount of water so the large quantity of sodium fluoride makes sense.

107. *Result:* **0.038 mol**

Analyze: Given the carat mass of a diamond and the relationship between carat and milligrams, determine how many moles of carbon are in the diamond.

Plan: Always start with the sample. Diamond is an allotropic form of pure carbon. Given the carats of the diamond, use the relationship between carats and milligrams as a conversion factor to determine milligrams of carbon. Then using metric relationships to determine grams of carbon, and the molar mass of carbon to determine the moles of carbon.

Execute:

$$2.3 \text{ carats C} \times \frac{200. \text{ mg C}}{1 \text{ carat C}} \times \frac{1 \text{ g C}}{1000 \text{ mg C}} \times \frac{1 \text{ mol C}}{12.01 \text{ g C}} = 0.038 \text{ mol C}$$

☑ *Reasonable Result Check:* A carat is 0.2 grams, so 2.3 carats is less than a half a gram of carbon. Since 12 grams of carbon represents a mole, half a gram should be a few hundredths of a mole.

109. *Result:* **(a) iodine monobromide (b) bromine trifluoride (c) diiodine hexachloride (d) chlorine pentafluoride (e) iodine heptafluoride**

Analyze and Plan: A general rule for applying the names of binary compounds to the formula is to list the symbol for first element named then the symbol for the second element. Use the prefixes described in Table 2.6 to learn the number of a particular kind of atom and use that number for the subscript on the symbol.

Execute:

(a) IBr has one I atom and one Br atom, so its name is **iodine monobromide.**

(b) BrF_3 has one Br atom and three F atoms, so its name is **bromine trifluoride.**

(c) I_2Cl_6 has two I atom and six Cl atoms, so its name is **diiodine hexachloride.**

(d) ClF_5 has one Cl atom and five F atoms, so its name is **chlorine pentafluoride.**

(e) IF_7 has one I atom and seven F atoms, so its name is **iodine heptafluoride.**

111. *Result:* **(a) 28.8515% N (b) 6.57 × 10^{20} molecules (c) 5.26 × 10^{21} C atoms (d) nine times greater**

Analyze: Given the formula caffeine, $C_8H_{10}N_4O_2$, determine the mass percent of nitrogen, molecules of caffeine, and C atoms. Compare caffeine content of beverages.

(a) *Plan:* Calculate the mass of N in one mole of caffeine, while calculating the molar mass of caffeine. Divide the calculated mass of N by the molar mass of caffeine and multiply by 100% to get percent.

Execute: Mass of N per mole of $C_8H_{10}N_4O_2$ = 2(14.0067g N) = 56.0268 g N

Mass of 1 mol $C_8H_{10}N_4O_2$ = 8(12.0107 g/mol C) + 10(1.0079 g/mol H)

+ 4(14.0067 g/mol N) + 2(15.9994 g/mol O) = 194.1902 g/mol $C_8H_{10}N_4O_2$

$$\text{Mass percent N} = \frac{56.0268 \text{ g N}}{194.1902 \text{ g caffeine}} \times 100\% = 28.8515\% \text{ N}$$

(b) *Plan:* Use molar mass and Avogadro's number.

$$\textit{Execute:}\ 212 \text{ mg caffeine} \times \left(\frac{1 \text{ g}}{1000 \text{ mg}}\right) \times \left(\frac{1 \text{ mol caffeine}}{194.1902 \text{ g caffeine}}\right) \times \left(\frac{6.022 \times 10^{23} \text{ caffeine molecules}}{1 \text{ mol caffeine}}\right)$$

$$= 6.57 \times 10^{20} \text{ caffeine molecules}$$

(c) *Plan:* Use the relationship between atoms of C in one molecule.

$$\textit{Execute:}\ 6.57 \times 10^{20} \text{ caffeine molecules} \times \left(\frac{8 \text{ C atoms}}{1 \text{ caffeine molecule}}\right) = 5.26 \times 10^{21} \text{ C atoms}$$

(d) *Plan:* The 1.93-oz 5-Hour Energy® drink has 212 mg caffeine. An 8-oz coffee has 100 mg caffeine. Calculate concentration (mg/oz) and compare.

$$\textit{Execute:}\ \text{Concentration in 5-Hour Energy® drink} = \frac{212 \text{ mg}}{1.93 \text{ oz}} = 109.8 \frac{\text{mg}}{\text{oz}}$$

$$\text{Concentration in coffee} = \frac{100 \text{ mg}}{8 \text{ oz}} = 12.5 \frac{\text{mg}}{\text{oz}} \text{ (1 sig fig)}$$

$$\text{Compare the concentrations, using a ratio: } \frac{109.8 \frac{\text{mg}}{\text{oz}}}{12.5 \frac{\text{mg}}{\text{oz}}} = 8.78 = 9$$

The 5-Hour Energy® drink has **nine times greater** caffeine concentration.

☑ *Reasonable Result Check:* This explains why people use energy drinks instead of coffee to keep awake.

113. *Result:* **(a) Ionic: (ii), (iv), and (vi) Molecular: (i), (v), (vii) No compound forms: (iii) (b) (i) BrCl, bromine monochloride; (ii) Li_2Te, lithium telluride; (iv) MgF_2, magnesium fluoride, (v) NF_3, nitrogen trifluoride; (vi) In_2S_3, indium sulfide; (vii) $SeBr_2$, selenium dibromide**

Analyze and Plan: Identify ionic, molecular or no compound. If a compound forms, determine the compound's formula and name.

Execute: (a) Ionic compound, molecular compound, or no compound. (b) Formula and name.

(i) Chlorine (Cl) and bromine (Br) are not likely to form an ionic compound, since they are both nonmetals in Group 7A. If a **molecular** compound formed, it would be covalent **BrCl, bromine monochloride**.

(ii) Lithium (Li) and tellurium (Te) might make an **ionic** compound. Lithium is a metal and tellurium is a metalloid. The likely compound contains ions Li^+ (Group 1A cation; charge is +1) and Te^{2-} (Group 6A anion; charge is –2). The compound's formula will be **Li_2Te, lithium telluride**.

(iii) Sodium (Na) and argon (Ar) are not likely to form an ionic compound, since argon is in Group 8A. Those elements a very unreactive and do not form ions at all. **No compound is expected to form**.

(iv) Magnesium (Mg) and fluorine (F) will make an **ionic** compound. Magnesium is a metal and fluorine is a nonmetal. The likely compound contains ions Mg^{2+} (Group 2A cation; charge is +2) and F^- (Group 7A anion; charge is –1). The compound's formula will be **MgF_2, magnesium fluoride**.

(v) Nitrogen (N) and bromine (Br) are not likely to form an ionic compound, since they are both nonmetals in Groups 5A and 7A, respectively. If a **molecular** compound formed, it would likely be covalent **NF_3, nitrogen trifluoride**.

(vi) Indium (In) and sulfur (S) will make an **ionic** compound. Indium is a metal and sulfur is a nonmetal. The likely compound contains ions In^{3+} (Group 3A cation; charge is +3) and S^{2-} (Group 6A anion; charge is –2). The compound's formula will be **In_2S_3, indium sulfide**.

(vii) Selenium (Se) and bromine (Br) are not likely to form an ionic compound, since they are both nonmetals in Groups 6A and 7A, respectively. If a **molecular** compound formed it would be covalent **$SeBr_2$, selenium dibromide**.

115. *Result:* **(a) 1.66×10^{-3} mol (b) 0.346 g**

Analyze: Given the number of tablets consumed, the mass of a compound in each tablet, and the formula of the compound, determine the moles of the compound consumed and the mass of one element consumed.

Plan: Always start with the sample—in this case, the number of tablets consumed. We assume that $C_7H_5BiO_4$ is the "active ingredient". Use the mass of $C_7H_5BiO_4$ per tablet to determine the mass of $C_7H_5BiO_4$ in the sample. Then use the molar mass of $C_7H_5BiO_4$ to determine the moles of $C_7H_5BiO_4$ in the sample. Then use the formula stoichiometry to get the moles of Bi in the sample. Then use the molar mass of Bi to get the grams of Bi in the sample.

Execute: The sample is composed of two tablets of Pepto-Bismol. Find grams of $C_7H_5BiO_4$:

$$2\text{ tablets} \times \frac{300.\text{ mg } C_7H_5BiO_4}{1\text{ tablet}} \times \frac{1\text{ g } C_7H_5BiO_4}{1000\text{ mg } C_7H_5BiO_4} = 0.600\text{ g } C_7H_5BiO_4$$

Molar mass of $C_7H_5BiO_4$= 7(12.0107 g/mol C) + 5(1.0079 g/mol H)

$$+ 208.9804\text{ g/mol Bi} + 4(15.9994\text{ g/mol O}) = 362.0924\text{ g/mol } C_7H_5BiO_4$$

(a) Find moles of $C_7H_5BiO_4$ in the sample

$$0.600\text{ g } C_7H_5BiO_4 \times \frac{1\text{ mol } C_7H_5BiO_4}{362.0924\text{ g } C_7H_5BiO_4} = 1.66 \times 10^{-3}\text{ mol } C_7H_5BiO_4$$

(b) Find grams of Bi in the sample

$$1.66 \times 10^{-3}\text{ mol } C_7H_5BiO_4 \times \frac{1\text{ mol Bi}}{1\text{ mol } C_7H_5BiO_4} \times \frac{208.9804\text{ g Bi}}{1\text{ mol Bi}} = 0.346\text{ g Bi}$$

☑ *Reasonable Result Check:* (a) The sample is somewhere between macroscale and microscale, so it makes sense that the number of moles is somewhat small. (b) The bismuth is almost 60% of the $C_7H_5BiO_4$ compound mass so it makes sense that the number of grams of Bi in the sample is about 60% of the mass of the compound in the sample.

Applying Concepts

117. *Analyze and Plan, and Execute:*

(a) A crystal of sodium chloride has alternating lattice of Na^+ and Cl^- ions.

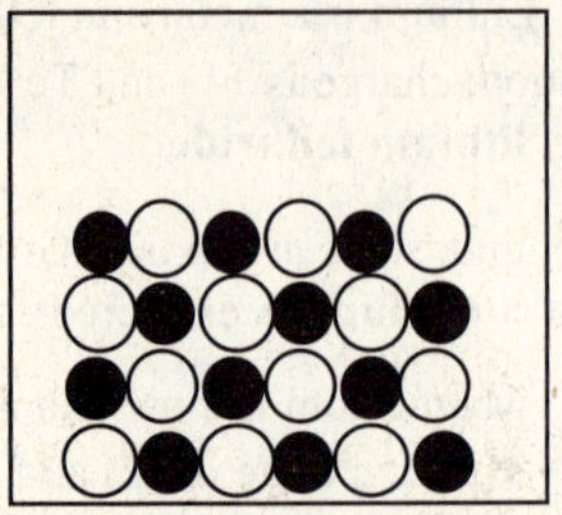

(b) The sodium chloride after it is melted has paired ions randomly distributed.

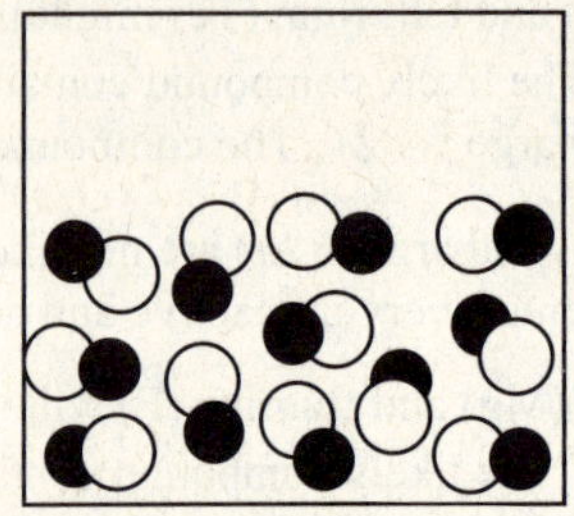

119. *Result:* **(a) not possible; mass number and atomic number wrong (b) possible (c) not possible; mass number wrong (d) not possible; mass number wrong (e) possible (f) not possible; mass number wrong**

Analyze and Plan: Check to see if the following statements are true: The atomic number must be the same as the number of protons and the mass number must be the sum of the number of protons and neutrons.

Execute:

(a) These values are **not possible**, since the atomic number (42) is not the same as the number of protons (19), and the sum of protons and neutrons (19+23=42) is not the same as the mass number (19).

(b) These values are **possible**. The atomic number (92) is the same as the number of protons (92), and the sum of protons and neutrons (92+143=235) is the same as the mass number (235).

(c) These values are **not possible**. The sum of protons and neutrons (131+79=210) is not the same as the mass number (53).

(d) These values are **not possible**. The sum of protons and neutrons (15+15=30) is not the same as the mass number (32).

(e) These values are **possible**. The atomic number (7) is the same as the number of protons (7), and the sum of protons and neutrons (7+7=14) is not the same as the mass number (14).

(f) These values are **not possible**. The sum of protons and neutrons (18+40=58) is not the same as the mass number (40).

121. *Result:* ^{39}**K**

Analyze, Plan, Execute: Potassium's atomic weight is 39.0983. The isotopes that contribute most to this mass are ^{39}K and ^{41}K, since the question tells us that ^{40}K has a very low abundance. Since the atomic mass is closer to 39 than 41, that confirms that the ^{39}**K isotope** is more abundant.

123. *Result:* **(a) 1 mol of Cl_2 (b) 1 mol of O_2 (c) one nitrogen molecule (d) 6.032×10^{23} molecules of F_2 (e) 20.3 grams of neon (f) 159.8 grams Br_2 (g) 9.6 grams of Li (h) 58.9 g Co (i) 6.022×10^{23} calcium atoms (j) Same**

Analyze and Plan: Molecules are made up of atoms. The unit mole is a convenient way of describing a large quantity of particles. It is also important to keep in mind that 1 mol of particles contains 6.022×10^{23} particles and the molar mass describes the mass of 1 mol of particles.

Execute:

(a) A sample containing **1 mol of Cl_2** has more atoms than a sample containing 1 mol Cl, since each molecule of Cl_2 contains two atoms of Cl.

(b) A sample containing **1 mol of O_2** contains 6.022×10^{23} molecules of O_2. This sample has many more atoms than a sample containing just 1 molecule of O_2.

(c) A sample containing **one nitrogen molecule** (N_2) has two atoms, which is more than one N atom.

(d) A sample containing **6.032×10^{23} molecules of F_2** contains more than 1 mol of F_2 molecules, which has only 6.022×10^{23} molecules of F_2.

(e) The molar mass of Ne is 20.18 g/mol, so a sample of **20.3 grams of neon** contains more than 1 mol of neon.

(f) The molar mass of bromine (Br_2) is 2(79.9 g/mol Br) = 159.8 g/mol Br_2, so the sample composed of **159.8 grams of bromine** contains 1 mol bromine, which equals 6.022×10^{23} molecules. This 159.8-gram sample has many more particles than a sample containing just 1 molecule of Br_2.

(g) The molar mass of Ag is 107.9 g/mol, so a sample of 107.9 grams of Ag contains 1 mol of Ag. The molar mass of Li is 6.9 g/mol, so a sample of **9.6 grams of Li** contains more than 1 mol of Li.

(h) The molar mass of Co is 58.9 g/mol, so the sample with **58.9 grams of Co** contains 1 mol of Co atoms. The molar mass of Cu is 63.55 g/mol, so the sample with 58.9 grams of Cu contains less than 1 mol of Cu atoms. Therefore the Co sample has more atoms than the Cu sample.

(i) The sample containing 6.022×10^{23} atoms of calcium, Ca, contains 1 mol of Ca atoms. The molar mass of Ca is 40.1 g/mol, so the sample composed of 1 gram of cobalt, Co, contains less than 1 mol of Co; thus the sample with **6.022×10^{23} calcium atoms** has more particles than a sample with 1 gram cobalt.

(j) Since chlorine atoms all weigh the same, so two samples with identical mass containing only chlorine must have the **same** number of Cl atoms. The molar mass of Cl_2 is twice that of Cl, but the number of molecule is half the number of atoms, if the samples are both 1 g.

126. *Result:* **(a) three (b) (i) and (iv), (ii) and (iii), (v) and (vi)**

Analyze, Plan, and Execute: There are three isomers given for C_3H_8O. Each of these isomers is represented by two of the structures. The following pairs are identical:

Isomer number one: Isomer number two: Isomer number three:

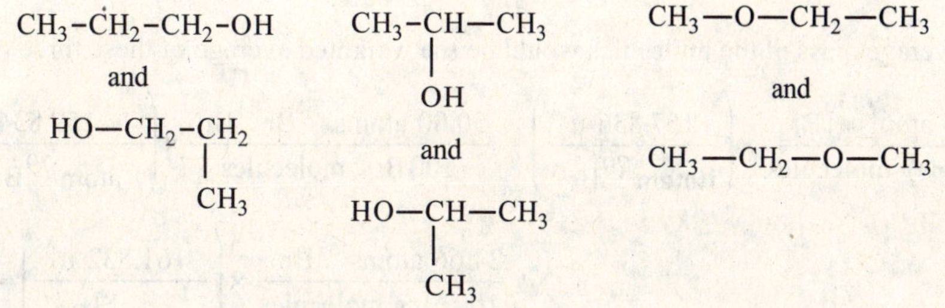

The first pair has an OH bonded to an end carbon: **(i) and (iv)**. The second pair has an OH bonded to the middle carbon: **(v) and (vi)**. The third pair has an O bonded between two carbon atoms: **(ii) and (iii)**.

127. *Result:* **Tl_2CO_3, Tl_2SO_4**

Analyze, Plan, and Execute: Thallium nitrate is $TlNO_3$. Since NO_3^- has a –1 charge, that means that thallium ion has a +1 charge, and is represented by Tl^+. The carbonate compound containing thallium will be a combination of Tl^+ and CO_3^{2-}, and the compound's formula will look like this: Tl_2CO_3. The sulfate compound containing thallium will be a combination of Tl^+ and SO_4^{2-}, and the compound's formula will look like this: Tl_2SO_4.

More Challenging Questions

130. *Result:* **(a) ^{79}Br–^{79}Br, ^{79}Br–^{81}Br, ^{81}Br–^{81}Br (b) 78.918 g/mol, 80.916 g/mol (c) 79.92 g/mol (d) 50.1% ^{79}Br, 49.9% ^{81}Br**

Analyze: Using the mass of several diatomic molecules of bromine that differ by iostopic composition and the relative heights of their spectral peaks on a mass spectrum, determine the identity of isotopes in the molecules, the mass of each isotope of bromine, the average atomic mass of bromine, and the abundance of the isotopes.

Plan: Identify which of the two isotopes contribute to each of the mass spectrum peaks. Then use that information to find that atomic mass of each isotope. Use the relative peak heights and the isotope masses to find the average molar mass of Br_2 and the average atomic mass of Br, then establish variables describing the isotope percentages. Set up two relationships between these variables. The sum of the percents must be 100%, and the weighted average of the molecular masses must be the reported atomic mass.

Execute:

(a) The peak representing the diatomic molecule with the lightest mass must be composed of two atoms of the lightest isotopes, $^{79}Br–^{79}Br$. The peak representing the diatomic molecule with the largest mass must be composed of two atoms of the heavier isotopes, $^{81}Br–^{81}Br$. The middle peak must represent molecules made with one of each, $^{79}Br–^{81}Br$.

(b) The mass of the molecule composed of two lightweight atoms is 157.836 g/mol, so each atom must weigh 0.5×(157.836 g/mol) = **78.918 g/mol**. The mass of the molecule composed of two heavy weight atoms is 161.832 g/mol, so each atom must weigh 0.5×(161.832 g/mol) = **80.916 g/mol**. This is verified by adding these two isotope masses to obtain the mass of the mixed-isotope peak: 78.918 g/mol + 80.916 g/mol = 159.834 g/mol

(c) The peak heights relate to the quantity of each isotope variation, so find the percentage abundance for each molecule by calculating the percentage of peak heights:

Total = 6.337 + 12.499 + 6.164 = 25.000

$$\%\ ^{79}Br–^{79}Br = \frac{6.337}{25.000} \times 100\ \% = 25.35\ \%$$

$$\%\ ^{79}Br–^{81}Br = \frac{12.499}{25.000} \times 100\ \% = 50.000\ \%$$

$$\%\ ^{81}Br–^{81}Br = \frac{6.164}{25.000} \times 100\ \% = 24.66\ \%$$

The average mass of the molecules would be the weighted average of these three isotopic variants:

$$\frac{25.35 \text{ atoms } ^{79}Br_2}{100 \text{ } Br_2 \text{ molecules}} \times \left(\frac{157.836 \text{ u}}{1 \text{ atom } ^{79}Br_2}\right) + \frac{50.00 \text{ atoms } ^{79}Br^{81}Br}{100\,Br_2 \text{ molecules}} \times \left(\frac{159.834 \text{ u}}{1 \text{ atom } ^{79}Br^{81}Br}\right) +$$

$$\frac{24.66 \text{ atoms } ^{81}Br_2}{100 \text{ } Br_2 \text{ molecules}} \times \left(\frac{161.832 \text{ u}}{1 \text{ atom } ^{81}Br_2}\right) = 159.84 \frac{\text{u}}{Br_2 \text{ molecules}}$$

The diatomic molecule mass gets divided by two, to determine the atomic mass: (159.84 g/mol Br_2)/2 = 79.92 g/mol Br.

(d) Define percent abundance in terms of two variables: X% ^{79}Br and Y% ^{81}Br. This means:

Every 100 atoms of bromine contains X atoms of the ^{79}Br isotope.

Every 100 atoms of bromine contains Y atoms of the ^{81}Br isotope.

$$X + Y = 100\%$$

$$\frac{X \text{ atoms } ^{79}Br}{100 \text{ Br atoms}} \times \left(\frac{78.918 \text{ u}}{1 \text{ atom } ^{79}Br}\right) + \frac{Y \text{ atoms } ^{91}Br}{100 \text{ Br atoms}} \times \left(\frac{80.916 \text{ u}}{1 \text{ atom } ^{91}Br}\right) = 79.92 \frac{\text{u}}{\text{Br atom}}$$

We now have two equations and two unknowns, so we can solve for X and Y algebraically. Solve the first equation for Y: Y = 100 – X. Plug that in for Y in the second equation. Then solve for X:

$$\frac{X}{100} \times (78.918) + \frac{100 - X}{100} \times (80.916) = 79.92$$

$$0.78918\ X + 80.916 - 0.80916\ X = 79.92$$

$$80.916 - 79.92 = 0.80916\text{ X} - 0.78918\text{ X}$$

$$1.00 = (0.01998)\text{X}$$

$$\text{X} = 50.1$$

Now, plug the value of X in the first equation to get Y. Y = 100 – X = 100 – 50.1 = 49.9

Therefore the abundance for these isotopes are: **50.1% ^{79}Br and 49.9% ^{81}Br.**

☑ *Reasonable Result Check:* The molar mass of Br from the periodic table is 79.904 g/mol, so the result from part (c) is close but slightly high. These isotopes are about equally abundant as indicated by the near 25%:50%:25% ratio of the isotopes in the three mass spectrum peaks. The % isotopic abundance values calculated in (d) are similar to but not the same as the percentages given in Question 16, 50.69% and 49.31%.

132. *Result:* **75.0% sulfate**

Analyze: Given a mixture of two sulfate compounds and the mass percent of one of the two compounds in the mixture, determine the mass percent of sulfate in the mixture.

Plan: Using a 100.0 gram sample of the mixture, determine the mass percent of the second compound, then calculate the mass of sulfate from each compound in the sample. Add these two masses and determine the mass percent of sulfate in the 100.0 g sample.

Execute:

100.0 g sample – 32.0 g $MgSO_4$ = 68.0 g $(NH_4)_2SO_4$

Molar mass $MgSO_4$ = 24.3050 g/mol Mg + 32.065 g/mol S + 4(15.9994 g/mol O) = 120.368 g/mol $MgSO_4$

Molar mass $(NH_4)_2SO_4$ = 2(14.0067 g/mol N) + 8(1.0079 g/mol H)

+ (32.065 g/mol S) + 4(15.9994 g/mol O) = 132.139 g/mol $(NH_4)_2SO_4$

Molar mass SO_4^{2-} = 32.065 g/mol S + 4(15.9994 g/mol O) = 96.063 g/mol SO_4^{2-}

Mass SO_4^{2-} from $MgSO_4$

$$= 32.0\text{ g MgSO}_4 \times \frac{1\text{ mol MgSO}_4}{120.368\text{ g MgSO}_4} \times \frac{1\text{ mol SO}_4^{2-}}{1\text{ mol MgSO}_4} \times \frac{96.063\text{ g SO}_4^{2-}}{1\text{ mol SO}_4^{2-}} = 25.54\text{ g SO}_4^{2-}$$

Mass SO_4^{2-} from $(NH_4)_2SO_4$

$$= 68.0\text{ g (NH}_4)_2\text{SO}_4 \times \frac{1\text{ mol (NH}_4)_2\text{SO}_4}{132.1392\text{ g (NH}_4)_2\text{SO}_4} \times \frac{1\text{ mol SO}_4^{2-}}{1\text{ mol (NH}_4)_2\text{SO}_4} \times \frac{96.063\text{ g SO}_4^{2-}}{1\text{ mol SO}_4^{2-}} = 49.43\text{ g SO}_4^{2-}$$

Total mass SO_4^{2-} = 25.54 g + 49.43 g = 74.97 g (round to one decimal place)

$$\%\text{ SO}_4^{2-} = \frac{74.97\text{ g SO}_4^{2-}}{100.0\text{ g sample}} \times 100\% = 75.0\%\text{ SO}_4^{2-}\text{ (3 sig figs)}$$

☑ *Reasonable Result Check:* The other atoms in these compounds are lightweight compared to sulfate, so it makes sense that a majority of the mass percent is sulfate.

134. *Result:* **6.73% ^{41}K and 0.01% ^{40}K**

Analyze: Given the average atomic weight of an element, the atomic weight of three isotopes, and the percentage abundance of one isotope, determine percentage abundance of the other two isotopes.

Plan: Knowing that the weighted average of the isotope masses must be equal to the reported atomic weight, set up a relationship between the known average atomic weight and the various isotope masses using a variable to describing the other two isotope's atomic weight.

Execute: We are told that potassium has an average atomic mass of 39.0983 u and has 93.2581% ^{39}K. This percentage/100 tell us the fraction of K with this isotopes mass. Let X be the percent of the ^{40}K isotope and Y be the percent of the ^{41}K isotope.

We know that the sum of the percents must be 100%.

$$100.0000\% = \%\,^{39}\text{K} + \%\,^{40}\text{K} + \%\,^{41}\text{K} = 93.2581\% + \text{X} + \text{Y}$$

We also know that the average atomic mass is the weighted average of the isotopic masses.

$$\frac{93.2581\ ^{39}\text{K atom}}{100\text{K atoms}} \times \left(\frac{38.963707\text{ u}}{1\text{ atom }^{39}\text{K}}\right) + \frac{\text{X}\ ^{40}\text{K atom}}{100\text{K atoms}} \times \left(\frac{39.963999\text{ u}}{1\text{ atom }^{40}\text{K}}\right) + \frac{\text{Y}\ ^{41}\text{K atom}}{100\text{K atoms}} \times \left(\frac{40.961825\text{ u}}{1\text{ atom }^{41}\text{K}}\right)$$

$$= 39.0983\ \frac{\text{u}}{\text{K atom}}$$

$$36.3368 + 0.39963999\text{X} + 0.40961825\text{Y} = 39.0983$$

Solve for the first equation for X in terms of Y:

$$\text{X} = 100.0000\% - 93.2581\% - \text{Y} = 6.7419 - \text{Y}$$

Plug this into the second equation:

$$36.3368 + 0.39963999(6.7419 - \text{Y}) + 0.40961825\text{Y} = 39.0983$$

Solve for Y:

$$2.6943 - 0.39963999\text{Y} + 0.40961825\text{Y} = 39.0983 - 36.3368$$

$$(0.40961825 - 0.39963999)\text{Y} = 39.0983 - 36.3368 - 2.6943$$

$$0.00997826\text{Y} = 0.0672$$

$$\text{Y} = \frac{0.0672}{0.00997826} = 6.73\%\ ^{41}\text{K}$$

Use Y to solve for X: $\text{X} = 100.0000\% - 93.2581\% - 6.73\% = 0.01\%\ ^{40}\text{K}$

So the percent abundance of the other two isotopes are **6.73% ^{41}K and 0.01% ^{40}K.**

☑ *Reasonable Result Check:* It makes sense that these two isotopes have lower percent abundance since the average atomic weight is closer to the isotopic mass of the lightest element.

136. *Result:* **71.6% Fe, 18.5% Cr, and 8.96% Ni**

Analyze: Given the mass of a stainless steel sample and the masses of oxide products produced from these samples, determine the mass percent of each metal in the sample.

Plan: Use molar masses to determine moles of products. Use chemical formula to relate moles of products to moles of metal. Use molar mass of metal to calculate the mass of the metal. Divide the mass of the metal by the mass of the sample and multiply by 100% to get mass percent for each metal.

Execute: Molar mass of Fe_2O_3 = 2(55.854 g/mol Fe) + 2(15.9994 g/mol O) = 159.688 g/mol Fe_2O_3

$$10.3\text{ g Fe}_2\text{O}_3 \times \left(\frac{1\text{ mol Fe}_2\text{O}_3}{159.688\text{ g Fe}_2\text{O}_3}\right) \times \left(\frac{2\text{ mol Fe}}{1\text{ mol Fe}_2\text{O}_3}\right) \times \left(\frac{55.854\text{ g Fe}}{1\text{ mol Fe}}\right) = 7.16\text{ g Fe}$$

$$\%\text{ Fe} = \frac{7.16\text{ g Fe}}{10.0\text{ g sample}} \times 100\% = 71.6\%\text{ Fe}$$

Molar mass of Cr_2O_3 = 2(51.9961 g/mol Cr) + 2(15.9994 g/mol O) = 151.9904 g/mol Cr_2O_3

$$2.71\text{ g Cr}_2\text{O}_3 \times \left(\frac{1\text{ mol Cr}_2\text{O}_3}{151.9904\text{ g Cr}_2\text{O}_3}\right) \times \left(\frac{2\text{ mol Cr}}{1\text{ mol Cr}_2\text{O}_3}\right) \times \left(\frac{51.9961\text{ g Cr}}{1\text{ mol Cr}}\right) = 1.85\text{ g Cr}$$

$$\% \text{ Cr} = \frac{1.85 \text{ g Cr}}{10.0 \text{ g sample}} \times 100\% = 18.5\% \text{ Cr}$$

Molar mass of NiO = 58.6934 g/mol Ni + 15.9994 g/mol O = 74.6928 g/mol NiO

$$1.14 \text{ g NiO} \times \left(\frac{1 \text{ mol NiO}}{74.6928 \text{ g NiO}}\right) \times \left(\frac{1 \text{ mol Ni}}{1 \text{ mol NiO}}\right) \times \left(\frac{58.6934 \text{ g Ni}}{1 \text{ mol Ni}}\right) = 0.896 \text{ g Ni}$$

$$\% \text{ Ni} = \frac{0.896 \text{ g Ni}}{10.0 \text{ g sample}} \times 100\% = 8.96\% \text{ Ni}$$

☑ *Reasonable Result Check:* The measured percentages from this sample match (in the first two sig figs) the expected percentages identified for the stainless steel used in the Gateway Arch in St. Louis, given at the beginning of the question: 71.6% ≈ 72.0%Fe, 18.5% ≈ 19.0%Cr, and 8.96% ≈ 9.0% Ni

140. *Result:* **I: K_2O II: KO_2 III: KO_3 IV: K_2O_2**

Analyze and Plan: Use mass percent K in these K_xO_y compounds to determine the percent O. Assume a 100.0 g sample, calculate the moles of K and O in the sample, set up a ratio, and find the simple whole number relationship for the empirical formula. Determine the empirical formula mass to help identify which compound has the given molar mass.

Execute:

I: 100.0% – 83.0% K = 17.0% O. A 100.0 g sample has 83.0 g K and 17.0 g O.

$$83.0 \text{ g K} \times \frac{1 \text{ mol K}}{39.0983 \text{ g K}} = 2.12 \text{ mol K} \qquad 17.0 \text{ g O} \times \frac{1 \text{ mol O}}{15.9994 \text{ g O}} = 1.06 \text{ mol O}$$

Mole ratio: 2.12 mol K : 1.06 mol O

Simplify by dividing by 1.06 mol: 2 mol K : 1 mol O

The empirical formula is: **K_2O**

The empirical formula mass for K_2O is 2(39.0983 g/mol K) + 15.9994 g/mol O = 94.1960 g/mol.

II: 100.0% – 55.0% K = 45.0% O. A 100.0 g sample has 55.0 g K and 45.0 g O.

$$55.0 \text{ g K} \times \frac{1 \text{ mol K}}{39.0983 \text{ g K}} = 1.41 \text{ mol K} \qquad 45.0 \text{ g O} \times \frac{1 \text{ mol O}}{15.9994 \text{ g O}} = 2.81 \text{ mol O}$$

Mole ratio: 1.41 mol K : 2.81 mol O

Simplify by dividing by 1.41 mol: 1 mol K : 2 mol O

The empirical formula is: **KO_2**

The empirical formula mass for KO_2 is 39.0983 g/mol K + 2(15.9994 g/mol O) = 71.0971 g/mol.

III: 100.0% – 44.9% K = 55.1% O. A 100.0 g sample has 44.9 g K and 55.1 g O.

$$44.9 \text{ g K} \times \frac{1 \text{ mol K}}{39.0983 \text{ g K}} = 1.15 \text{ mol K} \qquad 55.1 \text{ g O} \times \frac{1 \text{ mol O}}{15.9994 \text{ g O}} = 3.44 \text{ mol O}$$

Mole ratio: 1.15 mol K : 3.44 mol O

Simplify by dividing by 1.15 mol: 1 mol K : 3 mol O

The empirical formula is: **KO_3**

The empirical formula mass for KO_3 is 39.0983 g/mol K + 3(15.9994 g/mol O) = 87.0965 g/mol.

IV: 100.0% –71.0% K = 29.0% O. A 100.0 g sample has 71.0 g K and 29.0 g O.

$$71.0 \text{ g K} \times \frac{1 \text{ mol K}}{39.0983 \text{ g K}} = 1.82 \text{ mol K} \qquad 29.0 \text{ g O} \times \frac{1 \text{ mol O}}{15.9994 \text{ g O}} = 1.81 \text{ mol O}$$

Mole ratio: 1.82 mol K : 1.81 mol O

Simplify by dividing by 1.81 mol: 1 mol K : 1 mol O

The empirical formula is: **KO**

The empirical formula mass for KO is 39.0983 g/mol K + 15.9994 g/mol O = 55.098 g/mol.

One of the compounds is reported to have a molar mass of 110.2 g/mol. The first three empirical formula masses are not close to an integer multiple of this but, but looking at the last compound, $(KO)_n$:

$$n = \frac{110.2 \text{ g/mol}}{55.0977 \text{ g/mol}} = 2$$

Therefore, it is clear that compound IV must have a molecular formula of K_2O_2, potassium peroxide.

☑ *Reasonable Result Check:* All of the mole ratios were indisputable small integer ratios.

141. *Result:* **(a) I: XeF_4 II: XeF_2 III: XeF_6 (b) I: xenon tetrafluoride II: xenon difluoride III: xenon hexafluoride**

Analyze and Plan:

(a) Use the mass percent of Xe in compound II to determine the mass percent of F in compound II. Calculate moles of Xe and F, set up a ratio, and find the simple whole number relationship for the formula for compound II. Compound I has twice as many F atoms as Compound II (since it contains twice the mass of fluorine). Compound III has 1.5 as many F atoms as Compound I (since it contains 1.5 times the mass of fluorine contained in Compound I).

Execute:

II: 100.0% – 77.5% Xe = 22.5% F. A 100.0 g sample has 77.5 g Xe and 22.5 g F.

$$77.5 \text{ g Xe} \times \frac{1 \text{ mol Xe}}{131.293 \text{ g Xe}} = 0.590 \text{ mol Xe} \qquad 22.5 \text{ g F} \times \frac{1 \text{ mol F}}{18.9984 \text{ g F}} = 1.18 \text{ mol F}$$

Mole ratio: 0.590 mol Xe : 1.18 mol F

Simplify by dividing by 1.18 mol: 1 mol Xe : 2 mol F

The empirical formula is: $\mathbf{XeF_2}$.

I: Compound I contains twice the mass of fluorine in Compound II. That means there are twice as many atoms of F in the formula. 2 × 2 = 4. The empirical formula is: $\mathbf{XeF_4}$.

III: Compound III contains 1.5 times the mass of fluorine in Compound I. That means there are 1.5 times as many F atoms in the formula. 1.5 × 4 = 6. The empirical formula is: $\mathbf{XeF_6}$.

(b) *Explanation:* These are binary molecular compounds, so use the system described in Section 2-8.
I: The name of $\mathbf{XeF_4}$ is xenon tetrafluoride

II: The name of $\mathbf{XeF_2}$ is xenon difluoride

III: The name of $\mathbf{XeF_6}$ is xenon hexafluoride

☑ *Reasonable Result Check:* The mole ratio calculated for Compound II was very close to a small whole number. The mass of F in one mole of XeF_4 (75.9936 g) is twice the mass of F in one mole of XeF_2 (37.9968 g). The mass of F in one mole of XeF_6 (113.9904 g) is 1.5 times the mass in XeF_4 (75.9936 g).

145. *Result:* **45.0% $CaCl_2$**

Analyze and Plan: A mixture contains calcium chloride and sodium chloride. Use molar mass to calculate the moles of calcium in CaO, then relate moles of calcium to moles of $CaCl_2$, then use molar mass to calculate mass of $CaCl_2$.

Execute: Calcium chloride is $CaCl_2$ and sodium chloride is NaCl. The calcium in $CaCl_2$ is converted to Calcium carbonate, $CaCO_3$, which is then converted to calcium oxide, CaO. One mol $CaCl_2$ has 1 mol Ca atoms. One mol CaO has 1 mol Ca atoms.

Molar mass of $CaCl_2$ = 40.078 g/mol Ca + 2(35.453 g/mol Cl) = 110.984 g/mol $CaCl_2$

Molar mass of CaO = 40.078 g/mol Ca + 15.9994 g/mol O = 56.077 g/mol CaO

$$0.959 \text{ g CaO} \times \left(\frac{1 \text{ mol CaO}}{56.077 \text{ g CaO}}\right) \times \left(\frac{1 \text{ mol Ca}}{1 \text{ mol CaO}}\right) \times \left(\frac{1 \text{ mol CaCl}_2}{1 \text{ mol Ca}}\right) \times \left(\frac{110.984 \text{ g CaCl}_2}{1 \text{ mol CaCl}_2}\right) = 1.90 \text{ g CaCl}_2$$

$$\% \text{ CaCl}_2 \text{ in mix} = \frac{1.90 \text{ g CaCl}_2}{4.22 \text{ g mixture}} \times 100\% = 45.0\ \% \text{ CaCl}_2$$

☑ *Reasonable Result Check:* The mass of the $CaCl_2$ is smaller than the sample mass.

147. *Result:* **(a) –2 (b) Al_2X_3 (c) Se**

Analyze, Plan, and Execute:

(a) A group 6A element is likely to have a –2 charge, since: anion charge = (group number) – 8 = 6 – 8 = –2

(b) Aluminum ion, Al^{3+} combines with X^{2-} to form Al_2X_3.

(c) Given the percent by mass of an element in a compound and the compound's formula including an unknown element, determine the identity of the unknown element.

Choose a convenient sample of Al_2X_3, such as 100.00 g. Using the percent by mass, determine the number of grams of Al and X in the sample. Use the molar mass of Al as a conversion factor to get the moles of Al. Use the formula stoichiometry as a conversion factor to get the moles of X. Determine the molar mass by dividing the grams of X in the sample, by the moles of X in the sample. Using the periodic table, determine which Group 6A element has a molar mass nearest this value.

The compound is 18.55% Al by mass. This means that 100.00 g of Al_2X_3 contains 18.55 grams Al and the rest of the mass is from X.

Mass of X in sample = 100.00 g Al_2X_3 – 18.55 g Al = 81.45 g X

Formula stoichiometry: 1 mol of Al_2X_3 contains 2 mol of Al atoms.

$$18.55 \text{ g Al} \times \frac{1 \text{ mol Al}}{26.9815 \text{ g Al}} \times \frac{3 \text{ mol X}}{2 \text{ mol Al}} = 1.031 \text{ mol X}$$

$$\text{Molar Mass of X} = \frac{\text{mass of X in sample}}{\text{moles of X in sample}} = \frac{81.45 \text{ g X}}{1.031 \text{ mol X}} = 78.98 \text{ g/mol}$$

The periodic table indicates that X = Se (Z = 34, with atomic weight = 78.96 g/mol).

☑ *Reasonable Result Check:* The molar mass calculated is close to that of the Group 6A element, Se. Ions of the elements in Group 6A will typically have 2– charge, and the formula would be Al_2Se_3.

Chapter 3: Chemical Reactions
Solutions for Red-Numbered Questions for Review and Thought

Topical Questions

Chemical Equations (Section 3-1)

10. *Result:*

	NH_3	O_2	NO	H_2O
No. molecules	4	5	4	6
No. atoms	16	10	8	18
Amount of molecules (mol)	4	5	4	6
Mass (g)	68.121 6	159.994 0	120.024 4	108.091 2
Total mass of reactants (g)	228.1156		–	
Total mass of products (g)	–		228.1156	

Analyze: Given the equation for the reaction, find the related molecules, atoms, moles, masses of each of the reactants and products, and total masses of the reactants and the products.

Plan: The stoichiometric coefficients (the numbers in front of the molecules' formulas) in the balanced equation can be interpreted as the related number of molecules or the related number of moles of molecules. The molar mass of each molecule is multiplied by the number of moles of that molecule represented in the equation to find its mass in grams. Add the mass of NH_3 and O_2 for the total reactant mass. Add the mass of NO and H_2O for the total product mass.

Execute:

$$4\ NH_3(g) + 5\ O_2(g) \longrightarrow 4\ NO(g) + 6\ H_2O(g)$$

The stoichiometric coefficients are 4 for NH_3, 5 for O_2, 4 for NO and 6 for H_2O; these numbers represent relative numbers of molecules and relative amount of molecules (in moles).

	NH_3	O_2	NO	H_2O
No. molecules	4	5	4	6
No. atoms	4(1 N+ 3 H) = 16	5(2 O) = 10	4(1 N + 1 O) = 8	6(2 H + 1 O) = 18
Amount of molecules	4 mol	5 mol	4 mol	6 mol
Mass	4 mol × [1(14.0067g/mol N) + 3(1.0079g/mol H)] = 68.1216 g	5 mol × 2(15.9994g/mol O) = 159.9940 g	4 mol × (14.0067g/mol N + 15.9994g/mol O) = 120.0244 g	6 mol × [2(1.0079 g/mol H) + 15.9994g/mol O] = 108.0912 g
Total mass of reactants	68.1216 g + 159.9940 g = 228.1156 g		–	
Total mass of products	–		120.0244 g + 108.0912 g =228.1156 g	

☑ *Reasonable Result Check:* A properly balanced equation has the same numbers of atoms of each type (4 N, 12 H, and 10 O) in the products and reactants, so the totals on each side should be equal, also (16+10=8+18). The law of conservation of mass says that mass of the reactants must equal the mass of the products, 228.1156g.

12. *Result:* $\mathbf{A_2 + 2\ B \longrightarrow 2\ AB}$

Analyze: In the first box, three A_2 molecules are combined with six B atoms. After the reaction occurs, the box contains only six AB molecules. The balanced equation that can describe this reaction is:

$$3\ A_2 + 6\ B \longrightarrow 6\ AB$$

Plan and Execute: When we write balanced chemical equations, the stoichiometric coefficients ought to be the smallest whole numbers relating the reactants and products. Here, all of the coefficients are divisible by 3. Therefore, this chemical equation: $\mathbf{A_2 + 2\ B \longrightarrow 2\ AB}$ is the best description of the reaction.

15. *Result:* **(a) Box (a) (b) Box (c)**

Analyze, Plan and Execute:

(a) The reactants must all be diatomic molecules. The mole ratio from the balanced equation indicates that the reactants must have a 1:2 ratio. **Box (a)** has nine diatomic molecules; three of them are blue X_2 molecules and six of them are pink Y_2 molecules, for a ratio of 3:6 or 1:2.

(b) The products must be triatomic molecules, with the formula XY_2. That means each molecule must be composed of one blue (X) atom and two pink (Y) atoms. **Box (c)** has six molecules of XY_2 so this box represents the products

Balancing Chemical Equations (Section 3-2)

18. *Result:* $\mathbf{X_2 + Y_2 \longrightarrow XY_2 + X}$

Analyze, Plan, and Execute: In the first box, there are six diatomic molecules made with pink (X) atoms and six diatomic molecules made with purple (Y) atoms, so the reaction between X_2, and Y_2 is shown with a 1:1 mole ratio. The products are composed of six pink (X) atoms and six triatomic molecules made up of two purple atoms and one pink atom (XY_2), also in a 1:1 mole ratio.

$$6\ X_2 + 6\ Y_2 \longrightarrow 6\ XY_2 + 6\ X$$

Divide by six to get the smallest integer coefficient:

$$X_2 + Y_2 \longrightarrow XY_2 + X$$

20. *Result:* **(a)** $\mathbf{UO_2(s) + 4\ HF(\ell) \longrightarrow UF_4(s) + 2\ H_2O(\ell)}$ **(b)** $\mathbf{B_2O_3(s) + 6\ HF(\ell) \longrightarrow 2\ BF_3(s) + 3\ H_2O(\ell)}$ **(c)** $\mathbf{BF_3(g) + 3\ H_2O(\ell) \longrightarrow 3\ HF(\ell) + H_3BO_3(s)}$

Analyze: Balance the given equations.

Plan: When balancing equations, select a specific order in which the elements are balanced. Select first the atoms that are only in one product and one reactant, since they will be easier. Select next those elements that are in more than one reactant or product (such as H or O) and then those which are present in elemental form in either the reactants or products. Be systematic. If any of the coefficients end up fractional, multiply every coefficient by the same constant to eliminate the fraction.

Execute:

(a) $\_\ UO_2(s) + \_\ HF(\ell) \longrightarrow \_\ UF_4(s) + \_\ H_2O(\ell)$ Select order: U, F, H, O

$\underline{1}\ UO_2(s) + \_\ HF(\ell) \longrightarrow \underline{1}\ UF_4(s) + \_\ H_2O(\ell)$ 1 U

$\underline{1}\ UO_2(s) + \underline{4}\ HF(\ell) \longrightarrow \underline{1}\ UF_4(s) + \_\ H_2O(\ell)$ 4 F

$\underline{1}\ UO_2(s) + \underline{4}\ HF(\ell) \longrightarrow \underline{1}\ UF_4(s) + \underline{2}\ H_2O(\ell)$ 4 H and 2 O

(b) $\_\ B_2O_3(s) + \_\ HF(\ell) \longrightarrow \_\ BF_3(s) + \_\ H_2O(\ell)$ Select order: B, F, H, O

$\underline{1}\ B_2O_3(s) + \_\ HF(\ell) \longrightarrow \underline{2}\ BF_3(s) + \_\ H_2O(\ell)$ 2 B

$\underline{1}\ B_2O_3(s) + \underline{6}\ HF(\ell) \longrightarrow \underline{2}\ BF_3(s) + \_\ H_2O(\ell)$ 6 F

$\underline{1}\ B_2O_3(s) + \underline{6}\ HF(\ell) \longrightarrow \underline{2}\ BF_3(s) + \underline{3}\ H_2O(\ell)$ 6 H and 3 O

(c) $\_\ BF_3(g) + \_\ H_2O(\ell) \longrightarrow \_\ HF(g) + \_\ H_3BO_3(s)$ Select order: B, F, H, O

$\underline{1}\ BF_3(g) + \_\ H_2O(\ell) \longrightarrow \_\ HF(\ell) + \underline{1}\ H_3BO_3(s)$ 1 B

$\underline{1}\ BF_3(g) + \_\ H_2O(\ell) \longrightarrow \underline{3}\ HF(\ell) + \underline{1}\ H_3BO_3(s)$ 3 F

$\underline{1}\ BF_3(g) + \underline{3}\ H_2O(\ell) \longrightarrow \underline{3}\ HF(\ell) + \underline{1}\ H_3BO_3(s)$ 6 H and 3 O

☑ *Reasonable Result Check:* (a) 1 U, 2 O, 4 H & 4 F (b) 2 B, 3 O, 6 H & 6 F (c) 1 B, 3 F, 6 H & 3 O

22. *Result:* **(a) $H_2NCl(aq) + 2\ NH_3(g) \longrightarrow NH_4Cl(aq) + N_2H_4(aq)$**

(b) $(CH_3)_2N_2H_2(\ell) + 2\ N_2O_4(g) \longrightarrow 3\ N_2(g) + 4\ H_2O(g) + 2\ CO_2(g)$

(c) $CaC_2(s) + 2\ H_2O(\ell) \longrightarrow Ca(OH)_2(s) + C_2H_2(g)$

Analyze and Plan: Follow the method described in the solution for Question 20.

Execute:

(a) $\_\ H_2NCl(aq) + \_\ NH_3(g) \longrightarrow \_\ NH_4Cl(aq) + \_\ N_2H_4(aq)$ Select order: Cl, N, H

$\underline{1}\ H_2NCl(aq) + \_\ NH_3(g) \longrightarrow \underline{1}\ NH_4Cl(aq) + \_\ N_2H_4(aq)$ 1 Cl

$\underline{1}\ H_2NCl(aq) + \underline{2}\ NH_3(g) \longrightarrow \underline{1}\ NH_4Cl(aq) + \underline{1}\ N_2H_4(aq)$ 3 N and 8 H

(b) $\_\ (CH_3)_2N_2H_2(\ell) + \_\ N_2O_4(g) \longrightarrow \_\ N_2(g) + \_\ H_2O(\ell) + \_\ CO_2(g)$ Select order: H, C, O, N

$\underline{1}\ (CH_3)_2N_2H_2(\ell) + \_\ N_2O_4(g) \longrightarrow \_\ N_2(g) + \underline{4}\ H_2O(g) + \_\ CO_2(g)$ 8 H

$\underline{1}\ (CH_3)_2N_2H_2(\ell) + \_\ N_2O_4(g) \longrightarrow \_\ N_2(g) + \underline{4}\ H_2O(g) + \underline{2}\ CO_2(g)$ 2 C

$\underline{1}\ (CH_3)_2N_2H_2(\ell) + \underline{2}\ N_2O_4(g) \longrightarrow \_\ N_2(g) + \underline{4}\ H_2O(g) + \underline{2}\ CO_2(g)$ 8 O

$\underline{1}\ (CH_3)_2N_2H_2(\ell) + \underline{2}\ N_2O_4(g) \longrightarrow 3\ N_2(g) + \underline{4}\ H_2O(g) + \underline{2}\ CO_2(g)$ 6 N

(c) $\_\ CaC_2(s) + \_\ H_2O(\ell) \longrightarrow \_\ Ca(OH)_2(s) + \_\ C_2H_2(g)$ Select order: Ca, C, O, H

$\underline{1}\ CaC_2(s) + \_\ H_2O(\ell) \longrightarrow \underline{1}\ Ca(OH)_2(s) + \_\ C_2H_2(g)$ 1 Ca

$\underline{1}\ CaC_2(s) + \_\ H_2O(\ell) \longrightarrow \underline{1}\ Ca(OH)_2(s) + \underline{1}\ C_2H_2(g)$ 2 C

$\underline{1}\ CaC_2(s) + 2\ H_2O(\ell) \longrightarrow \underline{1}\ Ca(OH)_2(s) + \underline{1}\ C_2H_2(g)$ 2 O and 4 H

☑ *Reasonable Result Check:* (a) 8 H, 3 N & 1 Cl (b) 2 C, 8 H, 6 N & 8 O (c) 1 Ca, 2 C, 4 H & 2 O

24. *Result:* **(a) $C_6H_{12}O_6 + 6\ O_2 \longrightarrow 6\ CO_2 + 6\ H_2O$ (b) $C_5H_{12} + 8\ O_2 \longrightarrow 5\ CO_2 + 6\ H_2O$**

(c) $2\ C_7H_{14}O_2 + 19\ O_2 \longrightarrow 14\ CO_2 + 14\ H_2O$ (d) $C_2H_4O_2 + 2\ O_2 \longrightarrow 2\ CO_2 + 2\ H_2O$

Analyze: Balance the given combustion equations.

Plan: When balancing combustion equations, select a C, H, O as the order in which the elements are balanced. Be systematic. If the coefficient of O_2 ends up a fraction, multiply every coefficient by 2 to make all of the coefficients whole numbers.

Execute:

(a) $\_\ C_6H_{12}O_6 + \_\ O_2 \longrightarrow \_\ CO_2 + \_\ H_2O$

$\underline{1}\ C_6H_{12}O_6 + \_\ O_2 \longrightarrow \underline{6}\ CO_2 + \_\ H_2O$ 6 C

$\underline{1}\ C_6H_{12}O_6 + \_\ O_2 \longrightarrow \underline{6}\ CO_2 + \underline{6}\ H_2O$ 12 H

$\underline{1}\ C_6H_{12}O_6 + \underline{6}\ O_2 \longrightarrow \underline{6}\ CO_2 + \underline{6}\ H_2O$ 18 O

(b) $\_\ C_5H_{12} + \_\ O_2 \longrightarrow \_\ CO_2 + \_\ H_2O$

$\underline{1}\ C_5H_{12} + \_\ O_2 \longrightarrow \underline{5}\ CO_2 + \_\ H_2O$ 5 C

$\underline{1}\ C_5H_{12} + \_\ O_2 \longrightarrow \underline{5}\ CO_2 + \underline{6}\ H_2O$ 12 H

$\underline{1}\ C_5H_{12} + \underline{8}\ O_2 \longrightarrow \underline{5}\ CO_2 + \underline{6}\ H_2O$ 16 O

(c) $\_\ C_7H_{14} + \_\ O_2 \longrightarrow \_\ CO_2 + \_\ H_2O$

$\underline{1}\ C_7H_{14}O_2 + \_\ O_2 \longrightarrow \underline{7}\ CO_2 + \_\ H_2O$ 7 C

$\underline{1}\ C_7H_{14}O_2 + \_\ O_2 \longrightarrow \underline{7}\ CO_2 + \underline{7}\ H_2O$ 14 H

$\underline{1}\ C_7H_{14}O_2 + \frac{19}{2}\ O_2 \longrightarrow \underline{7}\ CO_2 + \underline{7}\ H_2O$ 21 O

$2\ C_7H_{14}O_2 + 19\ O_2 \longrightarrow 14\ CO_2 + 14\ H_2O$

(d) $\_\ C_2H_4O_2 + \_\ O_2 \longrightarrow \_\ CO_2 + \_\ H_2O$

$\underline{1}\ C_2H_4O_2 + \_\ O_2 \longrightarrow \underline{2}\ CO_2 + \_\ H_2O$ 2 C

$\underline{1}\ C_2H_4O_2 + \_\ O_2 \longrightarrow \underline{2}\ CO_2 + \underline{2}\ H_2O$ 4 H

$\underline{1}\ C_2H_4O_2 + \underline{2}\ O_2 \longrightarrow \underline{2}\ CO_2 + \underline{2}\ H_2O$ 6 O

✓ *Reasonable Result Check:* (a) 6 C, 12 H & 18 O, (b) 5 C, 12 H & 16 O, (c) 14 C, 28 H, 42 O (d) 2 C, 4 H, 6 O

Precipitation Reactions (Section 3-3)

25. *Result:* **(a) K^+ and OH^- (b) K^+ and SO_4^{2-} (c) Na^+ and NO_3^- (d) NH_4^+ and Cl^-**

Analyze: Identify the common ions present in the compounds.

Plan: Separate the cations and anions in each formula and give them appropriate charges.

Execute:

(a) The ions present in a solution of KOH are K^+ and OH^-.

(b) The ions present in a solution of K_2SO_4 are K^+ and SO_4^{2-}.

(c) The ions present in a solution of $NaNO_3$ are Na^+ and NO_3^-.

(d) The ions present in a solution of NH_4Cl are NH_4^+ and Cl^-.

27. *Result:* **(a) and (d)**

Analyze and Plan: Determine if the compound is an electrolyte, by determining if it is ionic.

Execute:

(a) NaCl is an ionic compound and will ionize to form Na^+ and Cl^-. When NaCl is dissolved in water, the resulting solution will conduct electricity.

(b) $CH_3CH_2CH_3$ is an organic hydrocarbon compound and will not ionize. When $CH_3CH_2CH_3$ is dissolved in water, the resulting solution will not conduct electricity.

(c) CH_3OH is an organic compound and will not ionize. When CH_3OH is dissolved in water, the resulting solution will not conduct electricity.

(d) $Ca(NO_3)_2$ is an ionic compound and will ionize to form Ca^{2+} and NO_3^-. When $Ca(NO_3)_2$ is dissolved in water, the resulting solution will conduct electricity.

29. *Result:* **(a) soluble, K^+ and HPO_4^{2-} (b) soluble, Na^+ and ClO^- (c) soluble, Mg^{2+} and Cl^- (d) soluble, Ca^{2+} and OH^- (e) soluble, Al^{3+} and Br^-**

Analyze: Given a compound's formula, determine whether it is water-soluble.

Plan: Identify the cation and the anion in the salt, then use Table 3.1.

Notice: Any rule that applies is sufficient to determine the compound's solubility. For example, if Rule 1 applies, there is no need to look for other rules related to the anion.

Notice: The question asks which ions are present, not how many. When answering this question, numbers that were part of the compound's ion mole ratio are not included.

Execute:

(a) Potassium hydrogen phosphate is K_2HPO_4.

Rule 1: "All ... group 1A ... compounds are soluble." Potassium is in group 1A, so potassium hydrogen phosphate is soluble. It ionizes to form potassium ion, K^+, and hydrogen phosphate ion, HPO_4^{2-}.

(b) Sodium hypochlorite is NaOCl.

Rule 1: "All ... group 1A ... compounds are soluble." Sodium is in group 1A, so sodium hypochlorite is soluble. It ionizes to form sodium ion, Na^+, and hypochlorite ion, ClO^-.

(c) Magnesium chloride is $MgCl_2$.

Rule 3: "All common... chlorides ... are soluble, except AgCl, Hg_2Cl_2, and $PbCl_2$..." Magnesium chloride is soluble. It ionizes to form magnesium ion, Mg^{2+}, and chloride ion, Cl^-.

(d) Calcium hydroxide is $Ca(OH)_2$.

Rule 10: " ... $Ca(OH)_2$ is slightly soluble." That means it dissolves only slightly, but the ions that will be formed are calcium ion, Ca^{2+}, and hydroxide ion, OH^-.

(e) Aluminum bromide is $AlBr_3$.

Rule 3: "All common... bromides ... are soluble, except AgBr, Hg_2Br_2, and $PbBr_2$..." Aluminum bromide is soluble. It ionizes to form aluminum ion, Al^{3+}, and bromide ion, Br^-.

31. *Result:* **Box (b)**

Analyze, Plan, and Execute: Calcium chloride has a 1:2 ion mole ratio in the formula: $CaCl_2$. It also ionizes completely when it dissolves. So, locate the box with twice as many pink ions (Cl^-) as blue ions (Ca^{2+}). **Box (b)** shows 16 pink ions and eight blue ions.

33. *Result:*

Complete ionic: $2\ K^+(aq) + CO_3^{2-}(aq) + Cu^{2+}(aq) + 2\ NO_3^-(aq) \longrightarrow CuCO_3(s) + 2\ K^+(aq) + 2\ NO_3^-(aq)$;

net ionic: $CO_3^{2-}(aq) + Cu^{2+}(aq) \longrightarrow CuCO_3(s)$; **precipitate is copper(II) carbonate.**

Analyze: Given the names of two reactants combined to form an unidentified precipitate, write complete and the net ionic equations for this reaction and name the precipitate.

Plan: Determine the formulas of the reactants, then determine the resulting products by rearranging the ions in the reactants to form new compounds. Balance the complete equation. Ionize the reactants and use the solubility rules to determine the actual physical state of the products. Ions that remain completely unchanged (same physical phase and same ionized form) are the spectator ions. Any ions that are the same on the product side as on the reactant side are eliminated to produce the net ionic equation. The insoluble product compound should then be named as the precipitate.

Notice: All soluble ionic compounds and strong acids and bases are ionized in the complete ionic equation. All solids, weak acids and bases, and molecular compounds remain un-ionized.

All ions have aqueous phase in the equations below.

Execute:

Potassium carbonate is K_2CO_3 and copper(II) nitrate is $Cu(NO_3)_2$.

The cross combinations are copper (II) carbonate, $CuCO_3$, which is insoluble (according to rule 9: All carbonates are insoluble…) and potassium nitrate, KNO_3, which is soluble (according to rule 1: All Group 1A… ion compounds are soluble.)

balanced overall equation: $K_2CO_3\text{(aq)} + Cu(NO_3)_2\text{(aq)} \longrightarrow CuCO_3\text{(s)} + 2\ KNO_3\text{(aq)}$

Complete ionic equation:

$$2\ K^+\text{(aq)} + CO_3^{2-}\text{(aq)} + Cu^{2+}\text{(aq)} + 2\ NO_3^-\text{(aq)} \longrightarrow CuCO_3\text{(s)} + 2\ K^+\text{(aq)} + 2\ NO_3^-\text{(aq)}$$

Eliminate spectator ions, K^+, potassium ions and NO_3^-, nitrate ions.

Net ionic equation: $CO_3^{2-}\text{(aq)} + Cu^{2+}\text{(aq)} \longrightarrow CuCO_3\text{(s)}$

The precipitate is copper(II) carbonate.

☑ *Reasonable Result Check:* One insoluble compound was formed, as indicated in the Question. All spectator ions have been identified and eliminated to make the net ionic equation. The atoms are all balanced and the charges are all balanced in each equation.

35. **(a)** $\mathbf{Ca(OH)_2(s) + CoCl_2(aq) \longrightarrow CaCl_2(aq) + Co(OH)_2(s)}$

complete ionic: $\mathbf{Ca(OH)_2(s) + Co^{2+}\text{(aq)} + 2\ Cl^-\text{(aq)} \longrightarrow Ca^{2+}\text{(aq)} + 2\ Cl^-\text{(aq)} + Co(OH)_2(s)}$

net ionic: $\mathbf{Ca(OH)_2(s) + Co^{2+}\text{(aq)} \longrightarrow Ca^{2+}\text{(aq)} + Co(OH)_2(s)}$

(b) $\mathbf{BaCl_2(aq) + Na_2CO_3(aq) \longrightarrow BaCO_3(s) + 2\ NaCl(aq)}$

complete ionic: $\mathbf{Ba^{2+}\text{(aq)} + 2\ Cl^-\text{(aq)} + 2\ Na^+\text{(aq)} + CO_3^{2-}\text{(aq)} \longrightarrow BaCO_3(s) + 2\ Na^+\text{(aq)} + 2\ Cl^-\text{(aq)}}$

net ionic: $\mathbf{Ba^{2+}\text{(aq)} + CO_3^{2-}\text{(aq)} \longrightarrow BaCO_3(s)}$

(c) $\mathbf{2\ Na_3PO_4(aq) + 3\ Ni(NO_3)_2(aq) \longrightarrow Ni_3(PO_4)_2(s) + 6\ NaNO_3(aq)}$; **complete ionic:**

$\mathbf{6\ Na^+\text{(aq)} + 2\ PO_4^{3-}\text{(aq)} + 3\ Ni^{2+}\text{(aq)} + 6\ NO_3^-\text{(aq)} \longrightarrow Ni_3(PO_4)_2(s) + 6\ Na^+\text{(aq)} + 6\ NO_3^-\text{(aq)}}$;

net ionic: $\mathbf{2\ PO_4^{3-} + 3\ Ni^{2+} \longrightarrow Ni_3(PO_4)_2(s)}$

Analyze and Plan: Given overall chemical equations, balance them, then write complete and net ionic equations. Use the method described in the solution to Question 33.

Execute:

(a) $Ca(OH)_2$ is slightly soluble (Table 3.1, Rule 10). $\mathbf{CoCl_2}$ is soluble (Table 3.1, Rule 3). $\mathbf{CaCl_2}$ is soluble (Table 3.1, Rule 2). $\mathbf{Co(OH)_2}$ is insoluble (Table 3.1, Rule 10).

Balanced equation: $\mathbf{Ca(OH)_2(s) + CoCl_2(aq) \longrightarrow CaCl_2(aq) + Co(OH)_2(s)}$

Complete ionic equation:

$$\mathbf{Ca(OH)_2(s) + Co^{2+}(aq) + 2\ Cl^-(aq) \longrightarrow Ca^{2+}(aq) + 2\ Cl^-(aq) + Co(OH)_2(s)}$$

Eliminate spectator ion, Cl^-

Net ionic equation: $\mathbf{Ca(OH)_2(s) + Co^{2+}\text{(aq)} \longrightarrow Ca^{2+}\text{(aq)} + Co(OH)_2(s)}$

(b) $BaCl_2$ is soluble (Table 3.1, Rule 3). Na_2CO_3 is soluble (Table 3.1, Rule 1).

$BaCO_3$ is insoluble (Table 3.1, Rule 9), NaCl is soluble (Table 3.1, Rule 1).

Balanced equation: $BaCl_2\text{(aq)} + Na_2CO_3\text{(aq)} \longrightarrow BaCO_3\text{(s)} + 2\ NaCl\text{(aq)}$

Complete ionic equation:

$$Ba^{2+}\text{(aq)} + 2\ Cl^-\text{(aq)} + 2\ Na^+\text{(aq)} + CO_3^{2-}\text{(aq)} \longrightarrow BaCO_3\text{(s)} + 2\ Na^+\text{(aq)} + 2\ Cl^-\text{(aq)}$$

Eliminate spectator ions, Cl^- and Na^+

Net ionic equation: $Ba^{2+}(aq) + CO_3^{2-}(aq) \longrightarrow BaCO_3(s)$

(c) Na_3PO_4 is soluble (Table 3.1, Rule 1). $Ni(NO_3)_2$ is soluble (Table 3.1, Rule 2). $Ni_3(PO_4)_2$ is insoluble (Table 3.1, Rule 8). $NaNO_3$ is soluble (Table 3.1, Rule 1).

$$Na_3PO_4(aq) + Ni(NO_3)_2(aq) \longrightarrow Ni_3(PO_4)_2(s) + NaNO_3(aq) \quad \text{unbalanced}$$

Select balancing order: Ni^{2+}, then PO_4^{3-}, then NO_3^-, and Na^+

Balanced equation: $2\ Na_3PO_4(aq) + 3\ Ni(NO_3)_2(aq) \longrightarrow Ni_3(PO_4)_2(s) + 6\ NaNO_3(aq)$

Complete ionic equation:

$$6\ Na^+(aq) + 2\ PO_4^{3-}(aq) + 3\ Ni^{2+}(aq) + 6\ NO_3^-(aq) \longrightarrow Ni_3(PO_4)_2(s) + 6\ Na^+(aq) + 6\ NO_3^-(aq)$$

Eliminate spectator ions, NO_3^- and Na^+

Net ionic equation: $2\ PO_4^{3-}(aq) + 3\ Ni^{2+}(aq) \longrightarrow Ni_3(PO_4)_2(s)$

☑ *Reasonable Result Check:* All of the balanced equations are balanced. The net ionic equations have no spectator ions. The atoms are all balanced and the charges are all balanced in each equation.

38. *Result:* **$Pb(NO_3)_2(aq) + 2\ KCl(aq) \longrightarrow PbCl_2(s) + 2\ KNO_3(aq)$; reactants: lead(II) nitrate and potassium chloride; products: lead(II) chloride and potassium nitrate**

Analyze: Given the reactants of a precipitation reaction, balance the equation.

Plan: Adapt the method described in the solution to Question 33.

Execute:

Lead(II) nitrate is $Pb(NO_3)_2$, a soluble compound (Table 3.1, Rule 2). Potassium chloride is KCl, a soluble compound (Table 3.1, Rule 1). The products of the reaction would be lead(II) chloride, $PbCl_2$, an insoluble compound (Table 3.1, Rule 3) and potassium nitrate, KNO_3, a soluble compound (Table 3.1, Rule 1).

$$Pb(NO_3)_2(aq) + 2\ KCl(aq) \longrightarrow PbCl_2(s) + 2\ KNO_3(aq) \quad \text{balanced overall equation}$$

☑ *Reasonable Result Check:* The equation is balanced. This precipitation reaction produces an insoluble solid.

Acid-Base Reactions (Section 3-4)

40. *Result:* **(a) base, strong, K^+ and OH^- (b) base, strong, Mg^{2+} and OH^- (c) acid, weak, H^+ and ClO^- (d) acid, strong, H^+ and Br^- (e) base, strong, Li^+ and OH^- (f) acid, weak; small amounts of H^+, HSO_3^-, and SO_3^{2-}**

Analyze: Given some chemical formulas, identify if they are acids or bases, identify whether they are weak or strong, and determine what ions produce when they dissolve in water.

Plan: Acids produce H^+ in aqueous solutions. Bases produce OH^- in aqueous solutions. In several of these cases, Table 3.1 or Table 3.2 can be used to determine whether a large or small amount of ions are produced, by determining if the compound is soluble and/or weak or strong. In some cases, it is not possible to look up that information in Chapter 3. It is almost always true that, if an acid is **not** one of the common strong acids listed in Table 3.2, it is a **weak** acid. That is the case with the weak acids in this Question. If the acid or base is strong, it will ionize. If the acid or base is weak it will not ionize to a great extent, remaining primarily in the molecular form.

Execute:

(a) KOH is a strong base. (Given in Table 3.2), producing K^+ and OH^- ions when dissolved in water.

(b) $Mg(OH)_2$ is an insoluble ionic compound (Table 3.1, Rule 10) so few ions are produced in water, though the $Mg(OH)_2$ that dissolves does ionize completely. So, practically, it may be considered weak, because the OH^- ion concentration will never be very large. Technically, it may be considered to be strong, because all the dissolved $Mg(OH)_2$ is ionized, producing Mg^{2+} and OH^- ions when dissolved in water.

(c) HClO is a weak acid. (It's not listed as a common strong acid in Table 3.2.). It will not ionize very much, remaining mostly in the HClO(aq) form. The small amount that does ionize will form H^+ and ClO^- ions.

(d) HBr is a strong acid (Given in Table 3.2), producing H^+ and Br^- ions when dissolved in water.

(e) LiOH is a strong base (Given in Table 3.2), producing Li^+ and OH^- ions when dissolved in water.

(f) H_2SO_3 is a weak acid (It's not listed as a common strong acid in Table 3.2.). It will not ionize very much, remaining mostly in the H_2SO_3(aq) form. The small amount that does ionize will form H^+, HSO_3^-, and SO_3^{2-} ions.

☑ *Reasonable Result Check:* All of the ions produced are common ions.

42. *Result:* **(a) HNO_2; NaOH; complete ionic: $HNO_2(aq) + Na^+(aq) + OH^-(aq) \longrightarrow H_2O(\ell) + Na^+(aq) + NO_2^-(aq)$; net ionic: $HNO_2(aq) + OH^-(aq) \longrightarrow H_2O(\ell) + NO_2^-(aq)$ (b) H_2SO_4; $Ca(OH)_2$; complete ionic & net ionic: $H^+(aq) + HSO_4^-(aq) + Ca(OH)_2(s) \longrightarrow 2\ H_2O(\ell) + CaSO_4(s)$ (c) HI; NaOH; complete ionic: $H^+(aq) + I^-(aq) + Na^+(aq) + OH^-(aq) \longrightarrow H_2O(\ell) + Na^+(aq) + I^-(aq)$; net ionic: $H^+(aq) + OH^-(aq) \longrightarrow H_2O(\ell)$ (d) H_3PO_4; $Mg(OH)_2$; complete ionic & net ionic: $2\ H_3PO_4(aq) + 3\ Mg(OH)_2(s) \longrightarrow 6\ H_2O(\ell) + Mg_3(PO_4)_2(s)$**

Analyze: Given some chemical formulas for salts, identify the acids and bases that would react to form them, then write the overall neutralization reaction both in complete and net ionic form.

Plan: The formation of a salt using neutralization comes from reacting an acid containing the salt's anion and a base containing the salt's cation. Set up the complete equation, then Adapt the methods described in previous Questions to find the complete and net ionic equations.

Execute:

(a) $NaNO_2$ is formed from the neutralization of HNO_2 (to supply the nitrite anion) and NaOH (to supply the sodium cation). HNO_2 is a weak acid. NaOH is a strong base. $NaNO_2$ is a soluble compound.

Balanced overall equation: $HNO_2(aq) + NaOH(aq) \longrightarrow H_2O(\ell) + NaNO_2(aq)$

Complete ionic equation: $HNO_2(aq) + Na^+(aq) + OH^-(aq) \longrightarrow H_2O(\ell) + Na^+(aq) + NO_2^-(aq)$

Eliminate spectator ion, Na^+

Net ionic equation: $HNO_2(aq) + OH^-(aq) \longrightarrow H_2O(\ell) + NO_2^-(aq)$

(b) $CaSO_4$ is formed from the neutralization of H_2SO_4 (to supply the sulfate anion) and $Ca(OH)_2$ (to supply the calcium cation). The first ionization of H_2SO_4 is strong, but HSO_4^- is a weak acid. The reactant $Ca(OH)_2$ is only slightly soluble and what dissolves is a strong base and will react with acid to make water. The product, $CaSO_4$, is insoluble.

Balanced overall equation: $H_2SO_4(aq) + Ca(OH)_2(s) \longrightarrow 2\ H_2O(\ell) + CaSO_4(s)$

Complete ionic and net ionic equations:

$$H^+(aq) + HSO_4^-(aq) + Ca(OH)_2(s) \longrightarrow 2\ H_2O(\ell) + CaSO_4(s)$$

(c) NaI is formed from the neutralization of HI (to supply the iodide anion) and NaOH (to supply the sodium cation). HI is a strong acid and NaOH is a strong base. The product, NaI, is soluble.

Balanced overall equation: $HI(aq) + NaOH(aq) \longrightarrow H_2O(\ell) + NaI(aq)$

Complete ionic equation: $H^+(aq) + I^-(aq) + Na^+(aq) + OH^-(aq) \longrightarrow H_2O(\ell) + Na^+(aq) + I^-(aq)$

Eliminate spectator ions, Na^+ and I^-

Net ionic equation: $H^+(aq) + OH^-(aq) \longrightarrow H_2O(\ell)$

(d) $Mg_3(PO_4)_2$ is formed from the neutralization of H_3PO_4 (to supply the phosphate anion) and $Mg(OH)_2$ (to supply the magnesium cation). H_3PO_4 is a weak acid. The reactant $Mg(OH)_2$ is insoluble but it can react with acid to make water. The product, $Mg_3(PO_4)_2$, is insoluble.

Balanced overall, complete ionic, and net ionic equation:

$$2\ H_3PO_4(aq) + 3\ Mg(OH)_2(s) \longrightarrow 6\ H_2O(\ell) + Mg_3(PO_4)_2(s)$$

☑ *Reasonable Result Check:* These acids and bases undergo neutralization to produce the appropriate salt and water. The net ionic equation does not always includes the whole salt, when one or both of the ions of the salt are found to be spectator ions.

44. *Result:* **(a) precipitation reaction; products are NaCl and MnS; $MnCl_2(aq) + Na_2S(aq) \longrightarrow 2\ NaCl(aq) + MnS(s)$ (b) precipitation reaction; products are NaCl and $ZnCO_3$; $Na_2CO_3(aq) + ZnCl_2(aq) \longrightarrow 2\ NaCl(aq) + ZnCO_3(s)$ (c) gas-forming reaction; products are $KClO_4$, H_2O and CO_2; $K_2CO_3(aq) + 2\ HClO_4(aq) \longrightarrow 2\ KClO_4(aq) + H_2O(\ell) + CO_2(g)$**

Analyze: Given the reactants of reactions, classify the reaction that occurs, identify the products, and balance the equations.

Plan: To classify these reactions, determine the exchange products and check their solubility and/or strength using Table 3.1 or Table 3.2. Remember that carbonate compounds reacting with acids produce CO_2 gas.

Execute:

(a) When $MnCl_2$ reacts with Na_2S, the exchange products are NaCl and MnS. Checking their solubility, we find that NaCl is soluble, but MnS is insoluble (Table 3.1, Rule 12). That makes this a **precipitation reaction**.

Balanced overall equation: $MnCl_2(aq) + Na_2S(aq) \longrightarrow 2\ NaCl(aq) + MnS(s)$

(b) When Na_2CO_3 reacts with $ZnCl_2$, the exchange products are NaCl and $ZnCO_3$. Checking the solubility, we see that NaCl is soluble, but $ZnCO_3$ is insoluble (Table 3.1, Rule 9), so this is a **precipitation reaction**.

Balanced overall equation: $Na_2CO_3(aq) + ZnCl_2(aq) \longrightarrow 2\ NaCl(aq) + ZnCO_3(s)$

(c) When K_2CO_3 reacts with $HClO_4$, the exchange products are $KClO_4$ and H_2CO_3, which decomposes into liquid H_2O and CO_2 gas. That makes this a **gas-forming reaction**. $KClO_4$ is soluble.

Balanced overall equation: $K_2CO_3(aq) + 2\ HClO_4(aq) \longrightarrow 2\ KClO_4(aq) + H_2O(\ell) + CO_2(g)$

☑ *Reasonable Result Check:* Precipitation reactions form insoluble compounds and gas-forming reactions produce gases. The atoms are all balanced and the charges are all balanced in each equation.

46. *Result:* **$MnCO_3(aq) + 2\ HCl(aq) \longrightarrow H_2O(\ell) + CO_2(g) + MnCl_2(aq)$**

Analyze: Given the reactants of a reaction, balance the equation.

Adapt the method described in the solution to Question 33.

Rhodochrosite is manganese(II) carbonate, $MnCO_3$, an insoluble compound (Table 3.1, Rule 9). Hydrochloric acid, HCl, is a strong acid (Table 3.2). The products of the reaction are manganese(II) chloride, $MnCl_2$, a soluble compound (Table 3.1, Rule 3) and carbonic acid, H_2CO_3, a weak acid (Table 3.2), which decomposes into liquid water and carbon dioxide gas in a gas-forming reaction.

$MnCO_3(aq) + 2\ HCl(aq) \longrightarrow H_2O(\ell) + CO_2(g) + MnCl_2(aq)$ balanced overall equation

☑ *Reasonable Result Check:* The equation is balanced. This reaction between a carbonate compound and a strong acid produces H_2CO_3, which decomposes into water and CO_2.

47. *Result:* **(a) strong electrolyte (b) weak electrolyte (c) strong electrolyte (d) strong electrolyte**

Strategy: Given formulas of compounds, determine if they are strong, weak or non- electrolytes.

Plan: Soluble ionic compounds; see Table 3.1 are strong electrolytes. Strong acids, and strong bases are strong electrolytes. Weak acids and bases are weak electrolytes; see Table 3.2. Molecular compounds that are not acids or bases are nonelectrolytes.

Execute:

(a) Na_2CO_3 is a soluble ionic compound, so this is a **strong electrolyte.**

(b) H_2CO_3 is a weak acid, so this is a **weak electrolyte**.

(c) HNO_3 is a strong acid, so this is a **strong electrolyte**.

(d) KOH is a strong base, so this is a **strong electrolyte**.

Oxidation-Reduction Reactions (Section 3-5)

49. *Result:* **(a) –1 (b) +1 (c) +3 (d) +5 (e) +7**

Analyze: Given several formulas, determine the oxidation state of the Cl atoms in each one.

Plan: Use the rules spelled out in Section 3-4c to determine the oxidation numbers of chlorine. Rule 3 says the sum of oxidation numbers in a neutral compound is 0, Rule 6(d) gives us the oxidation number of hydrogen as +1, and Rule 7(a) gives us the oxidation number for oxygen is –2. This information is used several times in this Question.

The term "oxidation number" is abbreviated below as "Ox. #". For example: Ox. # H = +1, Ox. # O = –2

Execute:

(a) HCl is a neutral compound, so: (Ox. #H) + (Ox. #Cl) = 0

(+ 1) + (Ox. #Cl) = 0 Ox. #Cl = **–1**

(b) HClO is a neutral compound, so: (Ox. #H) + (Ox. #Cl) + (Ox. #O) = 0

(+1) + (Ox. # Cl) + (–2) = 0 Ox. # Cl = **+1**

(c) $HClO_2$ is a neutral compound, so: (Ox. #H) + (Ox. #Cl) + 2(Ox. #O) = 0

(+1) + (Ox. # Cl) + 2(–2) = 0 Ox. # Cl = **+3**

(d) $HClO_3$ is a neutral compound, so: (Ox. #H) + (Ox. #Cl) + 3(Ox. #O) = 0

(+1) + (Ox. # Cl) + 3(–2) = 0 Ox. # Cl = **+5**

(e) $HClO_4$ is a neutral compound, so: (Ox. #H) + (Ox. #Cl) + 4(Ox. #O) = 0

(+1) + (Ox. # Cl) + 4(–2) Ox. # Cl = **+7**

☑ *Reasonable Result Check:* Because the oxygen atom carries a negative oxidation number, the chlorine atoms must ends up with positive oxidation numbers in all but part (a).

52. (a) precipitation (b) oxidation-reduction (c) acid-base neutralization

Analyze: Given several reactions, determine if they are oxidation-reduction reactions and classify the remaining reactions.

Plan: To determine if a reaction is an oxidation-reduction reaction, we need to see if any of the atoms change oxidation state. Oxidation-reduction reactions are ones in which the atoms change oxidation states from reactants to products. If no change in oxidation state is observed, then it is not an oxidation-reduction reaction.

Execute:

(a) The ionic compounds representing reactants and products in this reaction all contain Cd^{2+}, Cl^-, Na^+, and S^{2-} ions. Therefore, this is not an oxidation-reduction reaction. The formation of insoluble CdS classifies this reaction as a **precipitation reaction.**

(b) The reactants Ca and O_2 both have zero oxidation states. They are combined into an ionic compound, CaO, with ions Ca^{2+} and O^{2-} ions, which have +2 and –2 oxidation states, respective. Because the oxidation states changed from reactants to products, this is classified as an **oxidation-reduction reaction**.

(c) The ionic compounds representing reactants contain Ca^{2+}, OH^-, H^+, and Cl^- ions. The product ionic compound contains Ca^{2+} and Cl^-. The other product, water, has O in the –2 oxidation state and H in the +1 oxidation state. Therefore, this is not an oxidation-reduction reaction. The formation of water from the reaction of OH^- and H^+ classifies this reaction as an **acid-base neutralization** reaction.

☑ *Reasonable Result Check:* The reactions that are not oxidation-reduction reactions are readily classified as one of the other reactions we have studied in this Chapter.

55. *Result:* **Substances (b), (c), and (d)**

Explanation: (b) O_2, (c) HNO_3 and (d) MnO_4^- are good oxidizing agents. They are all capable of oxidizing other chemicals because they are readily reduced.

57. *Result:* **(a) Most: Groups 1A and 2A; Least: right side of the transition elements (b) No (c) Yes, Pb(s) + 2 $AgNO_3$(aq) ⟶ $Pb(NO_3)_2$(aq) + 2 Ag(s) (d) Al(s) > Pb(s) > Ag(s)**

Analyze, Plan, and Execute:

(a) The most reactive metals are in Groups 1A and 2A. (According to Table 3.4, some of the most reactive metals are: Li, K, Ba, Sr, Ca, and Na.) The metals on the right side of the transition elements are the least reactive. (According to Table 3.4, some of the least reactive metals are: Au, Pt, Pd, Ag, Hg, and Cu.)

(b) Table 3.4 shows that aluminum is more active than H_2. H_2 forms when Al is in acid. H_2 is more active than Ag, which is why it is not formed with Ag in acid, so Al is more active metal than Ag. This is confirmed in Table 3.4. A less reactive metal will not form a more reactive metal, so Ag will not react with Al^{3+} to form Al and Ag^+.

(c)
$$Pb + 2\,H^+ \longrightarrow Pb^{2+} + H_2$$

This happens slowly, so Pb is somewhat more active than H_2.

$$Al + 3\,Pb^{2+} \longrightarrow 2\,Al^{3+} + Pb$$

So, Al is more active than Pb.

In (b), we learned that Ag will not react in acid to make H_2. So, Pb is more reactive than Ag. Therefore Pb should react with Ag^+ to form Ag.

$$Pb + 2\,Ag^+ \longrightarrow Pb^{2+} + 2\,Ag$$

$$Pb(s) + 2\,AgNO_3(aq) \longrightarrow Pb(NO_3)_2(aq) + 2\,Ag(s)$$

(d) Summarizing the results of (a) to (c) above, the order of decreasing reactivity is: Al(s) > Pb(s) > Ag(s)

The Mole and Chemical Reactions (Section 3-6)

59. *Result:* **1.1 mol O_2, 35 g O_2, 1.0 × 10^2 g NO_2**

Analyze: Given a balanced chemical equation and the number of moles of a reactant, determine the moles and grams of another reactant needed and the grams of product produced.

Plan: Use the mole ratio in the balanced equation to convert the moles of one reactant to moles of the other reactant. Use molar mass to convert from moles to grams. Use the mole ratio in the equation to find the moles of product formed, then use the molar mass of the product to find the grams.

Execute: The balanced equation says: 2 mol NO react with 1 mol O_2.

$$2.2 \text{ mol NO} \times \frac{1 \text{ mol } O_2}{2 \text{ mol NO}} = 1.1 \text{ mol } O_2$$

$$1.1 \text{ mol } O_2 \times \frac{31.9988 \text{ g } O_2}{1 \text{ mol } O_2} = 35 \text{ g } O_2$$

The balanced equation says: 2 mol NO produces 2 mol NO_2.

$$2.2 \text{ mol NO} \times \frac{2 \text{ mol } NO_2}{2 \text{ mol NO}} \times \frac{46.0055 \text{ g } NO_2}{1 \text{ mol } NO_2} = 1.0 \times 10^2 \text{ g } NO_2 \text{ produced}$$

☑ *Reasonable Result Check:* Fewer moles of O_2 are needed than moles of NO. The mass of NO used is 2.2 mol × (30.0061 g/mol) = 66 g. The sum of the reactant masses (66 g + 35 g) is equal to the products mass (1.0 × 10^2 g), within known significant figures.

61. *Result:* The complete table looks like this:

$(NH_4)_2PtCl_6$	Pt	HCl
12.35 g	**5.428** g	**5.410** g
0.02782 mol	**0.02782** mol	**0.1484** mol

Analyze: Given a balanced chemical equation and the number of grams of a reactant, determine the moles of the reactant and the mass and moles of two products.

Plan: Use the molar mass of the reactant to find the quantity (in moles) of that substance. Then use the mole ratio in the balanced equation to convert moles of reactant to moles of each product formed, then use the molar mass of each product to find the grams.

Execute:

Molar mass of $(NH_4)_2PtCl_6$ = 2[(14.0067 g/mol N) + 4(1.0079 g/mol H)] + (195.078 g/mol Pt)+ 6(35.453 g/mol Cl) = 443.873 g/mol $(NH_4)_2PtCl_6$

$$12.35 \text{ g } (NH_4)_2PtCl_6 \times \frac{1 \text{ mol } (NH_4)_2PtCl_6}{443.873 \text{ g } (NH_4)_2PtCl_6} = 2.782 \times 10^{-2} \text{ mol } (NH_4)_2PtCl_6$$

The balanced equation says: 3 mol $(NH_4)_2PtCl_6$ react to form 3 mol Pt.

$$2.782 \times 10^{-2} \text{ mol } (NH_4)_2PtCl_6 \times \frac{1 \text{ mol Pt}}{1 \text{ mol } (NH_4)_2PtCl_6} = 2.782 \times 10^{-2} \text{ mol Pt}$$

$$2.782 \times 10^{-2} \text{ mol Pt} \times \frac{195.078 \text{ g Pt}}{1 \text{ mol Pt}} = 5.428 \text{ g Pt}$$

The balanced equation says: 3 mol $(NH_4)_2PtCl_6$ react to form 16 mol HCl.

$$2.782 \times 10^{-2} \text{ mol } (NH_4)_2PtCl_6 \times \frac{16 \text{ mol HCl}}{3 \text{ mol } (NH_4)_2PtCl_6} = 0.1484 \text{ mol HCl}$$

$$0.1484 \text{ mol HCl} \times \frac{36.461 \text{ mol HCl}}{1 \text{ mol HCl}} = 5.410 \text{ g HCl}$$

☑ *Reasonable Result Check:* The moles are smaller than the grams. This is appropriate. The moles of HCl are larger than the moles of $(NH_4)_2PtCl_6$ and Pt.

63. *Result:* **(a) $4\ Fe(s) + 3\ O_2(g) \longrightarrow 2\ Fe_2O_3(s)$ (b) 7.98 g (c) 2.40 g**

Analyze: Given the products and reactants of a reaction and the mass of one reactant, balance the chemical equation, determine the mass of the product produced, and determine the mass of the other reactant.

Plan: Balance the equation from the given formulas of the reactants and product. Use the molar mass of the reactant to find the moles of that substance. Then use the mole ratio in the equation to determine the quantity (in moles) of the product produced and the quantity (in moles) of the other reactant required. Then use the molar mass of each to find the grams.

Execute:

(a) Balance O atoms, then Fe atoms: $4\ Fe(s) + 3\ O_2(g) \longrightarrow 2\ Fe_2O_3(s)$

(b) Molar Mass Fe_2O_3= 2(55.845 g/mol Fe) + 3(15.9994 g/mol O) = 159.688 g/mol Fe_2O_3

The balanced equation says: 4 mol Fe produces 2 mol Fe_2O_3.

$$5.58\text{ g Fe}\times\frac{1\text{ mol Fe}}{55.845\text{ g Fe}}\times\frac{2\text{ mol Fe}_2\text{O}_3}{4\text{ mol Fe}}\times\frac{159.688\text{ g Fe}_2\text{O}_3}{1\text{ mol Fe}_2\text{O}_3}=7.98\text{ g Fe}_2\text{O}_3$$

(c) The balanced equation says: 4 mol Fe reacts with 3 mol O_2.

$$5.58\text{ g Fe}\times\frac{1\text{ mol Fe}}{55.845\text{ g Fe}}\times\frac{3\text{ mol O}_2}{4\text{ mol Fe}}\times\frac{31.9988\text{ g O}_2}{1\text{ mol O}_2}=2.40\text{ g O}_2$$

☑ *Reasonable Result Check:* The sum of the reactant masses is: 5.58 g + 2.40 g = 7.98 g. This is the same as the total mass of the product compound.

65. *Result:* **(a) $CCl_2F_2 + 2\ Na_2C_2O_4 \longrightarrow C + 4\ CO_2 + 2\ NaCl + 2\ NaF$ (b) 170. g $Na_2C_2O_4$ (c) 112 g CO_2**

Analyze: Given the reactants and products of a reaction and the mass of a reactant, balance the chemical equation, determine the mass of another reactant, and determine the mass of one product produced.

Plan: Given the formulas of the reactants and products, balance the equation, selecting appropriate order for the systematic balancing of all the atoms. Then use the molar mass of the reactant to find the moles of that substance. Then use the mole ratios in the equation to determine the quantity (in moles) of the other reactant and product produced. Then use the molar mass of each to find the grams.

Execute:

(a) $\_\ CCl_2F_2 + \_\ Na_2C_2O_4 \longrightarrow \_\ C + \_\ CO_2 + \_\ NaCl + \_\ NaF$ Select order: Cl, F, Na, O, C

$\underline{1}\ CCl_2F_2 + \_\ Na_2C_2O_4 \longrightarrow \_\ C + \_\ CO_2 + \underline{2}\ NaCl + \_\ NaF$ 2 Cl

$\underline{1}\ CCl_2F_2 + \_\ Na_2C_2O_4 \longrightarrow \_\ C + \_\ CO_2 + \underline{2}\ NaCl + \underline{2}\ NaF$ 2 F

$\underline{1}\ CCl_2F_2 + \underline{2}\ Na_2C_2O_4 \longrightarrow \_\ C + \_\ CO_2 + \underline{2}\ NaCl + \underline{2}\ NaF$ 4 Na

$\underline{1}\ CCl_2F_2 + \underline{2}\ Na_2C_2O_4 \longrightarrow \_\ C + \underline{4}\ CO_2 + \underline{2}\ NaCl + \underline{2}\ NaF$ 8 O

$\underline{1}\ CCl_2F_2 + \underline{2}\ Na_2C_2O_4 \longrightarrow \underline{1}\ C + \underline{4}\ CO_2 + \underline{2}\ NaCl + \underline{2}\ NaF$

(b) Molar Mass CCl_2F_2 = 12.0107 g/mol C + 2(35.453 g/mol Cl) + 2(18.9984) = 120.914 g/mol CCl_2F_2

The balanced equation gives: 1 mol CCl_2F_2 requires 2 mol $Na_2C_2O_4$.

Molar Mass $Na_2C_2O_4$ = 2(22.9898 g/mol Na) + 2(12.0107 g/mol C)

+ 2(35.453 g/mol Cl) + 4(15.9994 g/mol O) = 133.9986 g/mol $Na_2C_2O_4$

$$76.8\text{ g CCl}_2\text{F}_2\times\frac{1\text{ mol CCl}_2\text{F}_2}{120.914\text{ g CCl}_2\text{F}_2}=0.635\text{ mol CCl}_2\text{F}_2$$

$$0.635\text{ mol CCl}_2\text{F}_2\times\frac{2\text{ mol Na}_2\text{C}_2\text{O}_4}{1\text{ mol CCl}_2\text{F}_2}\times\frac{133.9986\text{ g Na}_2\text{C}_2\text{O}_4}{1\text{ mol Na}_2\text{C}_2\text{O}_4}=170.\text{ g Na}_2\text{C}_2\text{O}_4$$

(c) Molar mass CO_2 = (12.0107 g/mol C) + 2(15.9994 g/mol O) = 44.0095 g/mol CO_2

The balanced equation gives: 1 mol CCl_2F_2 produces 4 mol CO_2.

$$0.635 \text{ mol } CCl_2F_2 \times \frac{4 \text{ mol } CO_2}{1 \text{ mol } CCl_2F_2} \times \frac{44.0095 \text{ g } CO_2}{1 \text{ mol } CO_2} = 112 \text{ g } CO_2$$

☑ *Reasonable Result Check:* (a) Check the number of atom of each type in the reactants and products: 5 C, 2 Cl, 2 F, 4 Na, and 8 O. The equation is properly balanced. (b) The masses of $Na_2C_2O_4$ and CO_2 are both larger than the mass of CCl_2F_2.

67. *Result:* **(a) 699 g (b) 526 g**

Analyze: Given a balanced chemical equation and the mass of a reactant in kilograms, determine the mass of one of the products produced and the mass of another reactant required.

Plan: Use metric conversion factors to convert kilograms to grams. Use the molar mass of the reactant to find the moles of that substance. Then use the mole ratio in the equation to determine the quantity (in moles) of product produced and the other reactant required. Then use molar masses to find the grams.

Execute: Molar Mass Fe_2O_3= 2(55.845 g/mol Fe) + 3(15.9994 g/mol O) = 159.688 g/mol Fe_2O_3

$$1.00 \text{ kg } Fe_2O_3 \times \frac{1000 \text{ g } Fe_2O_3}{1 \text{ kg } Fe_2O_3} \times \frac{1 \text{ mol } Fe_2O_3}{159.688 \text{ g } Fe_2O_3} = 6.26 \text{ mol } Fe_2O_3$$

(a) The balanced equation gives: 1 mol Fe_2O_3 produces 2 mol Fe.

$$6.26 \text{ mol } Fe_2O_3 \times \frac{2 \text{ mol Fe}}{1 \text{ mol } Fe_2O_3} \times \frac{55.845 \text{ g Fe}}{1 \text{ mol Fe}} = 699 \text{ g Fe}$$

(b) The balanced equation says: 1 mol Fe_2O_3 requires 3 mol CO.

Molar Mass CO = (12.0107 g/mol C) + (15.9994 g/mol O) = 28.0101 g/mol

$$6.26 \text{ mol } Fe_2O_3 \times \frac{3 \text{ mol CO}}{1 \text{ mol } Fe_2O_3} \times \frac{28.0101 \text{ g CO}}{1 \text{ mol CO}} = 526 \text{ g CO}$$

☑ *Reasonable Result Check:* The mass of iron produced is less than the mass of iron(III) oxide it was produced from. The mass of CO used is less than the mass of iron(III) oxide even with a larger stoichiometric coefficient because CO contains lighter-weight atoms.

Limiting Reactant (Section 3-7)

69. *Result:* **(a) Cl_2 is limiting. (b) 5.08 g Al_2Cl_6 (c) 1.67 g Al unreacted**

Analyze: Given a balanced chemical equation and the masses of both reactants, determine the limiting reactant, the mass of the product produced, and the mass remaining of the excess reactant when the reaction is complete.

Plan: Here, we use a slight adaptation of what the text calls "the mole method." We will calculate a directly comparable quantity, the moles of product. (a) Use the molar mass of the reactants to find the moles of the reactant substances. Then use the mole ratio in the equation to determine the moles of the product produced in each case. Identify the limiting reactant from the reactant that produces the least number of products. (b) Use the moles of product produced from the limiting reactant and the molar mass of the product to find the grams. (c) From the limiting reactant quantity, determine the moles of the other reactant needed for complete reaction. Convert that number to grams using the molar mass. Then subtract the quantity used from the initial mass given to get the mass of excess reactant.

Execute:

(a) The balanced equation says: 2 mol Al produces 1 mol Al_2Cl_6.

$$2.70 \text{ g Al} \times \frac{1 \text{ mol Al}}{26.9815 \text{ g Al}} \times \frac{1 \text{ mol } Al_2Cl_6}{2 \text{ mol Al}} = 0.0500 \text{ mol } Al_2Cl_6$$

The balanced equation says: 3 mol Cl_2 produces 1 mol Al_2Cl_6.

Molar Mass Cl_2 = 2(35.453 g Cl) = 70.906 g/mol

$$4.05 \text{ g } Cl_2 \times \frac{1 \text{ mol } Cl_2}{70.906 \text{ g } Cl_2} \times \frac{1 \text{ mol } Al_2Cl_6}{3 \text{ mol } Cl_2} = 0.0190 \text{ mol } Al_2Cl_6$$

The number of Al_2Cl_6 moles produced from Cl_2 is smaller (0.0190 mol < 0.0500 mol), so Cl_2 is the limiting reactant and Al is the excess reactant.

(b) Find the mass of 0.0190 mol Al_2Cl_6:

Molar Mass Al_2Cl_6 = 2(26.9815 g Al) + 6(35.453 g Cl) = 266.682 g/mol

$$0.0190 \text{ mol } Al_2Cl_6 \times \frac{266.682 \text{ g } Al_2Cl_6}{1 \text{ mol } Al_2Cl_6} = 5.08 \text{ g } Al_2Cl_6$$

(c) The balanced equation says: 3 mol Cl_2 react with 2 mol Al.

$$4.05 \text{ g } Cl_2 \times \frac{1 \text{ mol } Cl_2}{70.906 \text{ g } Cl_2} \times \frac{2 \text{ mol Al}}{3 \text{ mol } Cl_2} \times \frac{26.9815 \text{ g Al}}{1 \text{ mol Al}} = 1.03 \text{ g Al}$$

2.70 g Al initial – 1.03 g Al used up = 1.67 g Al remains unreacted

☑ *Reasonable Result Check:* There are fewer moles of Cl_2 present than Al, and the equation needs more Cl_2 than Al, so it makes sense that Cl_2 is the limiting reactant. In addition, the calculation in (c) proved that the initial mass of Al was larger than required to react with all of the Cl_2. The sum of the masses of the reactants that reacted (4.05 g + 1.03 g = 5.08 g) equals the mass of the product produced (5.08 g).

71. *Result:* **(a) CH_4 (b) 188 g (c) 739 g**

Analyze: Given a balanced chemical equation and the masses of both reactants, determine the limiting reactant, the mass of the product produced, and the mass remaining of the excess reactant when the reaction is complete.

Plan: For part (a): Use the molar mass of the reactants to find the moles of the reactant substances. Then use the mole ratio in the equation to determine the moles of the product produced in each case. Identify the limiting reactant from the reactant that produces the least number of products. For part (b): Use the moles of product produced from the limiting reactant and the molar mass of the product to find the grams. For part (c): From the quantity of limiting reactant, determine the moles of the other reactant needed for complete reaction. Convert that number to grams using the molar mass. Then subtract the quantity used from the initial mass given to get the mass of excess reactant.

Execute:

(a) The balanced equation says: 1 mol CH_4 produces 3 mol H_2.

Molar Mass CH_4 = (12.0107 g/mol C) + 4(1.0067 g/mol H) = 16.043 g/mol CH_4

$$500. \text{ g } CH_4 \times \frac{1 \text{ mol } CH_4}{16.0423 \text{ g } CH_4} \times \frac{3 \text{ mol } H_2}{1 \text{ mol } CH_4} = 93.5 \text{ mol } H_2$$

The balanced equation says: 1 mol H_2O produces 3 mol H_2.

Molar Mass H_2O =2(1.0067 g/mol H) + (15.9994 g/mol O) = 18.0152 g/mol H_2O

$$1300. \text{ g } H_2O \times \frac{1 \text{ mol } H_2O}{18.0152 \text{ g } H_2O} \times \frac{3 \text{ mol } H_2}{1 \text{ mol } H_2O} = 216.5 \text{ mol } H_2$$

The number of H_2 moles produced from CH_4 is smaller (93.5 mol < 216.5 mol), so CH_4 is the limiting reactant and H_2O is the excess reactant.

(b) Find the mass H_2: Molar Mass H_2 = 2(1.0067 g/mol H) = 2.0158 g/mol H_2

$$93.5 \text{ mol H}_2 \times \frac{2.0158 \text{ g H}_2}{1 \text{ mol H}_2} = 188 \text{ g H}_2$$

(c) $$500. \text{ g CH}_4 \times \frac{1 \text{ mol CH}_4}{16.0423 \text{ g CH}_4} \times \frac{1 \text{ mol H}_2\text{O}}{1 \text{ mol CH}_4} \times \frac{18.0152 \text{ g H}_2\text{O}}{1 \text{ mol H}_2\text{O}} = 561 \text{ g H}_2\text{O}$$

1300. g H_2O initial – 561 g H_2O used up = 739 g H_2O remains unreacted

☑ *Reasonable Result Check:* The small number of significant figures makes the uncertainty in this calculation somewhat high. However, the general expectations are still met. There is a larger mass of H_2O present, and the molar masses of the reactants are about the same size. Since the equation indicates that equal moles of H_2O and CH_4 react, it makes sense that CH_4 is the limiting reactant. In addition, the calculation in (c) proves that the initial mass of H_2O was larger than required to react with all the CH_4.

Percent Yield (Section 3-8)

73. *Result:* **699 g, 93.5%**

Analyze: Given the balanced chemical equation, the mass of the limiting reactant and the actual yield, determine the theoretical yield and the percent yield.

Plan: First, calculate the theoretical yield by determining the maximum mass of product that could have been made from the given quantity of reactant: Take the mass of the limiting reactant and convert it to moles. Then use mole ratio to find the moles of product. Then convert to grams using molar mass. Take the given actual yield and divide by the calculated theoretical yield and multiply by 100% to get percent yield.

Execute: The limiting reactant is Fe_2O_3. From its mass, find the maximum grams of Fe that could be made. The mole ratio comes from the balanced equation.

Molar Mass Fe_2O_3 = 2(55.845 g/mol Fe) + 3(15.9994 g/mol O) = 159.688 g/mol Fe_2O_3

$$1.00 \text{ kg Fe}_2\text{O}_3 \times \frac{1000 \text{ g Fe}_2\text{O}_3}{1 \text{ kg Fe}_2\text{O}_3} \times \frac{1 \text{ mol Fe}_2\text{O}_3}{159.688 \text{ g Fe}_2\text{O}_3} \times \frac{2 \text{ mol Fe}}{1 \text{ mol Fe}_2\text{O}_3} \times \frac{55.845 \text{ g Fe}}{1 \text{ mol Fe}} = 699 \text{ g Fe}$$

The given mass of Fe is the actual yield. Use these two masses to calculate the percent yield.

$$\frac{654 \text{ g Fe actual}}{699 \text{ g Fe theoretical}} \times 100\% = 93.5\% \text{ yield}$$

☑ *Reasonable Result Check:* Close to the maximum quantity of iron was produced, so it makes sense that the percent yield is over ninety percent.

74. *Result:* **56.0%**

Analyze: Given the theoretical yield and the actual yield, determine the percent yield.

Plan and Execute: Divide the actual yield by the theoretical yield and multiply by 100% to get percent yield. *(Note that the balanced equation is given, too, but you don't need to use it to answer this question.)*

$$\frac{36.7 \text{ g CaO actual}}{65.5 \text{ g CaO theoretical}} \times 100\% = 56.0\% \text{ yield}$$

☑ *Reasonable Result Check:* A little more than half the maximum quantity of quicklime was produced, so it makes sense that the percent yield is a little more than 50%.

76. *Result:* **5.3 g SCl_2**

Analyze: Given the balanced chemical equation, the desired mass of the product, and the percent yield, determine the amount of limiting reactant that must be used.

Plan: Interpret the percent yield as the relationship between the actual grams and theoretical grams. Use that relationship to find the theoretical yield mass of the product. Then use the mole ratio to find the moles of limiting reactant. Then, using molar mass, convert to actual grams of reactant needed.

Execute: The 51% percent yield tells us the following:

To make 51 grams of S_2Cl_2, we need to have enough limiting reactant to make 1.19 grams of S_2Cl_2.

$$1.19 \text{ g } S_2Cl_2 \text{ actual} \times \frac{100.\text{ g } S_2Cl_2}{51 \text{ g } S_2Cl_2 \text{ actual}} = 2.3 \text{ g } S_2Cl_2$$

Use this mass of product to determine what mass of limiting reactant to use.

Molar Mass S_2Cl_2 = 2(32.065 g/mol S) + 2(35.453 g/mol Cl) = 135.036 g/mol S_2Cl_2

Molar Mass SCl_2 = (32.065 g/mol S) + 2(35.453 g/mol Cl) = 102.971 g/mol SCl_2

$$2.3 \text{ g } S_2Cl_2 \times \frac{1 \text{ mol } S_2Cl_2}{135.036 \text{ g } S_2Cl_2} \times \frac{3 \text{ mol } SCl_2}{1 \text{ mol } S_2Cl_2} \times \frac{102.971 \text{ g } SCl_2}{1 \text{ mol } S_2Cl_2} = 5.3 \text{ g } SCl_2$$

☑ *Reasonable Result Check:* The yield suggests that we need to try to make about twice as much. The mole ratio is 3:1, so it makes sense that a larger mass of SCl_2 is needed than the mass of S_2Cl_2 formed.

Composition and Empirical Formulas (Section 3-9)

78. *Result:* $\mathbf{C_3H_6O_2}$

Analyze: Given the mass of a compound, the identity of the elements in the compound, and the identity and masses of all the products produced, determine the empirical formula.

Plan: Combustion uses oxygen. When the compound also contains oxygen, determine the amount of oxygen after the other elements. Use the molar mass of the products to find their moles and use the mole ratio in their formulas to determine the moles of the elements that are not oxygen in the compound. Find the masses of those elements, and subtract them from the total mass of the compound to get the mass of oxygen in the compound. Then use the molar mass to calculate the moles of oxygen in the compound. Then use a whole-number mole ratio to determine the empirical formula.

Execute: The compound contains C, H, and O: $C_iH_jO_k$. Its combustion produced H_2O and CO_2. Use molar mass and the mole ratio from the formula to determine the moles of C and H. Use the whole number ratio for the subscripts in the formula.

Molar Mass CO_2 = 12.0107 g/mol C + 2(15.9994 g/mol O) = 44.0095 g/mol CO_2

$$0.421 \text{ g } CO_2 \times \frac{1 \text{ mol } CO_2}{44.0095 \text{ g } CO_2} \times \frac{1 \text{ mol C}}{1 \text{ mol } CO_2} = 9.56 \times 10^{-3} \text{ mol C}$$

Molar Mass H_2O =2(1.0067 g/mol H) + (15.9994 g/mol O) = 18.0152 g/mol H_2O

$$0.172 \text{ g } H_2O \times \frac{1 \text{ mol } H_2O}{18.0152 \text{ g } H_2O} \times \frac{2 \text{ mol H}}{1 \text{ mol } H_2O} = 1.91 \times 10^{-2} \text{ mol H}$$

Calculate the masses of C and H.

$$9.56 \times 10^{-3} \text{ mol C} \times \frac{12.0107 \text{ g C}}{1 \text{ mol C}} = 0.115 \text{ g C} \qquad 0.0191 \text{ mol H} \times \frac{1.0079 \text{ g H}}{1 \text{ mol H}} = 0.0192 \text{ g H}$$

Calculate the masses of O by subtracting the masses of C and H from the given total compound mass.

$$0.236 \text{ g } C_iH_jO_k - 0.115 \text{ g C} - 0.0192 \text{ g H} = 0.102 \text{ g O}$$

Calculate the moles of O. $\quad 0.102 \text{ g O} \times \frac{1 \text{ mol O}}{15.9994 \text{ g O}} = 6.37 \times 10^{-3} \text{ mol O}$

Set up mole ratio and simplify by dividing by the smallest number of moles:

$$9.56 \times 10^{-3} \text{ mol C} : 1.91 \times 10^{-2} \text{ mol H} : 6.37 \times 10^{-3} \text{ mol O}$$

$$1.5 \text{ C} : 3 \text{ H} : 1 \text{ O}$$

Multiply by 2, to get a whole number ratio: 3 C : 6 H: 2 O

Use the whole number ratio for the subscripts in the formula.

The empirical formula is $C_3H_6O_2$.

☑ *Reasonable Result Check:* The mole ratio came out very close to whole number values.

80. *Result:* **(a) $C_9H_{11}NO_4$ (b) $C_9H_{11}NO_4$**

Analyze: Given the percent mass of elements in an organic compound and the compounds molar mass, determine the empirical formula and the molecular formula.

Plan: Choose a convenient sample mass of product, $C_xH_yN_zO_w$, such as 100.00 g. Find the mass of C and H in the sample, using the given mass percent, then subtract those masses from the total sample mass to get the mass of O. Use the molar mass of the elements to find their moles, then use a whole-number mole ratio to determine the empirical formula. For combustion, use the formula of the organic compound and O_2 as reactants and CO_2 and H_2O as products, then balance the equation.

Execute:

(a) 100.00 g of $C_xH_yN_zO_w$ contains 54.82 g C, 7.10 g N, and 32.46 g O.

$$100.00 \text{ g of } C_xH_yN_zO_w - 54.82 \text{ g C} - 7.10 \text{ g N} - 32.46 \text{ g O} = 5.62 \text{ g H}$$

$$54.82 \text{ g C} \times \frac{1 \text{ mol C}}{12.0107 \text{ g C}} = 4.564 \text{ mol C} \qquad 7.10 \text{ g N} \times \frac{1 \text{ mol N}}{14.0067 \text{ g N}} = 0.507 \text{ mol N}$$

$$32.46 \text{ g O} \times \frac{1 \text{ mol O}}{15.9994 \text{ g O}} = 2.029 \text{ mol O} \qquad 5.62 \text{ g H} \times \frac{1 \text{ mol H}}{1.0079 \text{ g H}} = 5.58 \text{ mol H}$$

Set up mole ratio and simplify by dividing by the smallest number of moles:

$$4.564 \text{ mol C} : 5.58 \text{ mol H} : 0.507 \text{ mol N} : 2.029 \text{ mol O}$$

$$9 \text{ C} : 11 \text{ H} : 1 \text{ N} : 4 \text{ O}$$

Use the whole number ratio for the subscripts in the formula. The empirical formula is $C_9H_{11}NO_4$.

(b) The molar mass of the empirical formula $C_9H_{11}NO_4$ = 9(12.0107 g/mol C) + 11(1.0067 g/mol H)

+ (14.0067 g/mol N) + 4(15.9994 g/mol O) = 197.1875 g/mol $C_9H_{11}NO_4$

$$\frac{197.19 \text{ g/mol compound}}{197.1875 \text{ g/mol emp. formula}} = 1 \text{ emp. formula/compound}$$

The molecular formula is $C_9H_{11}NO_4$.

☑ *Reasonable Result Check:* The mole ratio is quite close to whole number values. The empirical formula is very close to the molecular formula.

Solution Concentrations (Section 3-10)

82. *Result:* **(a) 0.254 M Na_2CO_3 (b) 0.508 M Na^+, 0.254 M CO_3^{2-}**

Analyze: Given the mass of the solute and the volume of the solution, find the molarity of the solute, and the concentrations of the ions.

Plan: Use the molar mass to determine the quantity (in moles) of solute, convert the volume to liters from milliliters, and divide the moles of solute by the volume in liters to get molarity. Use the mole ratio from the formula to find the concentrations of the ions.

Execute:

(a) Molar Mass Na_2CO_3 = 2(22.9898 g/mol Na) + 12.0107 g/mol C + 3(15.9994 g/mol O) = 105.9885 g/mol Na_2CO_3

Determine the quantity (in moles) of solute:

$$6.73\text{ g Na}_2\text{CO}_3 \times \frac{1\text{ mol Na}_2\text{CO}_3}{105.9885\text{ g Na}_2\text{CO}_3} = 6.35\times10^{-2}\text{ mol Na}_2\text{CO}_3$$

$$250.\text{ mL solution} \times \frac{1\text{ L}}{1000\text{ mL}} = 0.250\text{ L solution}$$

$$\text{Molarity} = \frac{6.35\times10^{-2}\text{ mol Na}_2\text{CO}_3}{0.250\text{ L solution}} = 0.254\text{ M Na}_2\text{CO}_3$$

Notice: The three sequential calculations shown above can be combined into one calculation:

$$\frac{6.73\text{ g Na}_2\text{CO}_3}{250.\text{ mL solution}} \times \frac{1\text{ mol Na}_2\text{CO}_3}{105.9885\text{ g Na}_2\text{CO}_3} \times \frac{1000\text{ mL}}{1\text{ L}} = 0.254\text{ M Na}_2\text{CO}_3$$

This combined version prevents having to write unnecessary intermediate answers and helps reduce round-off errors in significant figures. Both ways give the right answer, but it is helpful to consolidate your work, when you can.

(b) Na_2CO_3 has a mole ratio from the formula that looks like this:

1 mol Na_2CO_3 : 2 mol Na^+ ions: 1 mol CO_3^{2-} ions.

So, the 0.254 M Na_2CO_3 contains 2(0.254 M) Na^+ ion = 0.508 M Na^+ and 0.254 M CO_3^{2-}.

☑ *Reasonable Result Check:* The number 0.0635 is about one quarter of .250; this value looks right. The concentration of Na^+ is double the concentration of CO_3^{2-}.

84. *Result:* **5.08 × 10³ mL**

Analyze: Given the mass of the solute and the solution's molarity, find the volume of the solution.

Plan: Use the molar mass to determine the quantity (in moles) of solute. Then use the molarity as a conversion factor to determine volume of the solute in liters. Then convert the volume of solution to milliliters from liters.

Execute: Molar Mass NaOH = 22.9898 g/mol Na + 15.9994 g/mol O + 1.0067 g/mol H = 39.9971 g/mol NaOH

$$25.0\text{ g NaOH} \times \frac{1\text{ mol NaOH}}{39.9971\text{ g NaOH}} \times \frac{1\text{ L solution}}{0.123\text{ mol NaOH}} \times \frac{1000\text{ mL}}{1\text{ L}} = 5.08\times10^3\text{ mL solution}$$

☑ *Reasonable Result Check:* The units cancel appropriately. The relatively large number of milliliters seems appropriate, since this relatively dilute solution contains a relatively large mass of solute.

86. *Result:* **Method (b), because it is the only one with the correct concentration**

Analyze: Given the desired volume and the molarity of a dilute solution, determine which of several dilution methods produces this solution.

Plan: Looking at each choice, it's easy to see that each solution results in a total volume of 1.00 L, which is the desired volume. So, we should focus our attention on which of the choices provides a solution that contains the proper number of moles. Convert milliliters to liters. Then calculate the moles in each choice by multiplying the volume in liters by the concentration of the concentrated solution.

Execute: We want 1.00 L of 0.125 M H_2SO_4

$$1.00\text{ L dil} \times \frac{0.125\text{ mol H}_2\text{SO}_4}{1\text{ L dil}} = 0.125\text{ moles of H}_2\text{SO}_4$$

So, determine the quantity (in moles) H_2SO_4 in each choice and compare it to the 0.125 moles H_2SO_4 desired.

(a) $36.0 \text{ mL conc} \times \frac{1 \text{ L}}{1000 \text{ mL}} \times \frac{1.25 \text{ mol } H_2SO_4}{1 \text{ L conc}} = 0.00450 \text{ mol } H_2SO_4$ No

(b) $20.8 \text{ mL conc} \times \frac{1 \text{ L}}{1000 \text{ mL}} \times \frac{6.00 \text{ mol } H_2SO_4}{1 \text{ L conc}} = 0.125 \text{ mol } H_2SO_4$ Yes

(c) $50.0 \text{ mL conc} \times \frac{1 \text{ L}}{1000 \text{ mL}} \times \frac{3.00 \text{ mol } H_2SO_4}{1 \text{ L conc}} = 0.150 \text{ mol } H_2SO_4$ No

(d) $500. \text{ mL conc} \times \frac{1 \text{ L}}{1000 \text{ mL}} \times \frac{0.500 \text{ mol } H_2SO_4}{1 \text{ L conc}} = 0.250 \text{ mol } H_2SO_4$ No

Only choice (b) will make the desired solution.

Notice: You also could have used the *Molarity*(conc) × *V*(conc) = *Molarity*(dil) × *V*(dil) dilution equation to answer this question.

☑ *Reasonable Result Check:* Compare moles of solute in the concentrated and dilute solutions:

$$Molarity(\text{conc}) \times V(\text{conc}) = (6.00 \text{ M } H_2SO_4 \text{ conc}) \times (20.8 \text{ mL conc}) \times \frac{1 \text{ L}}{1000 \text{ mL}} = 0.125 \text{ mol } H_2SO_4$$

$$Molarity(\text{dil}) \times V(\text{dil}) = (0.125 \text{ M } H_2SO_4 \text{ dil}) \times (1.00 \text{ L dil}) = 0.125 \text{ mol } H_2SO_4$$

Within three significant figures, the moles in the concentrated solution (0.125 mol) are the same as the moles in the dilute solution (0.125 mol), as expected.

88. *Result:* **39.4 g $NiSO_4 \cdot 6H_2O$**

Analyze: Given the volume of the solution and its molarity, find the mass of solute in the solution.

Plan: Use the molarity as a conversion factor to determine the quantity (in moles) of solute. Then use the mole ratio from the formula to relate the solute to the solid hydrate, then use the molar mass of the hydrate to find grams of the hydrate.

Execute: Molar mass of $NiSO_4 \cdot 6\, H_2O$ = 58.6034 g/mol Ni + 32.065 g/mol S +

4(15.9994 g/mol O) + 6[2(1.0079 g/mol H) + 15.9994 g/mol O] = 262.757 g/mol $NiSO_4 \cdot 6\, H_2O$

$$0.500 \text{ L solution} \times \frac{0.300 \text{ mol } NiSO_4}{1 \text{ L solution}} \times \frac{1 \text{ mol } NiSO_4 \cdot 6H_2O}{1 \text{ mol } NiSO_4} \times \frac{262.847 \text{ g } NiSO_4 \cdot 6H_2O}{1 \text{ mol } NiSO_4 \cdot 6H_2O}$$

$$= 39.4 \text{ g } NiSO_4 \cdot 6H_2O$$

☑ *Reasonable Result Check:* The mass of solute is a reasonable size.

Stoichiometry in Aqueous Solutions (Sections 3-11)

90. *Result:* **121 mL HNO_3**

Analyze: Given the mass of one reactant, the balanced chemical equation for a reaction, and the molarity of a solution containing the other reactant, determine the volume of the second solution for a complete reaction.

Plan: The mole ratio in a balanced chemical equation dictates how the moles of reactants combine, so we will commonly look for ways to calculate moles. Here, the mass and molar mass can be used to find the moles of one reactant. Then we will use the mole ratio in the equation to find out moles of the other reactant needed. Then we will use the moles and molarity to find volume in liters and convert liters into milliliters.

Notice: It is NOT appropriate to use the dilution equation when working with reactions!

Execute: We learn from the balanced equation that 1 mol of $Ba(OH)_2$ reacts with 2 mol HNO_3.

Molar mass $Ba(OH)_2$ = 137.326 g/mol Ba + 2(15.9994 g/mol O) + 2(1.0079/mol g H)

= 171.3416 g/mol $Ba(OH)_2$

$$1.30 \text{ g } Ba(OH)_2 \times \frac{1 \text{ mol } Ba(OH)_2}{171.3416 \text{ g } Ba(OH)_2} \times \frac{2 \text{ mol } HNO_3}{1 \text{mol } Ba(OH)_2}$$

$$\times \frac{1 \text{ L } HNO_3 \text{ solution}}{0.125 \text{ mol } HNO_3} \times \frac{1000 \text{ mL}}{1 \text{ L}} = 121 \text{ mL } HNO_3 \text{ solution}$$

93. *Result:* **(a) Step (ii) is not correct. (Steps (iii) and (iv) show correct calculations but use the wrong numbers.) (b) 3.94 × 10^{-3} g citric acid**

Analyze: Given the volume and molarity of a solution containing one reactant, the balanced chemical equation, and a series of steps describing the calculation, determine which of the steps is not correct, and correctly determine the mass of the other reactant in a specific volume of its solution.

Plan: Check each step to see if it is right or wrong. If it is wrong, correct it.

Execute:

(a) **Step (i) is correct.** $6.42 \text{ mL} \times \frac{1 \text{ L}}{1000 \text{ mL}} \times \frac{9.580 \times 10^{-2} \text{ mol NaOH}}{1 \text{ L}} = 6.15 \times 10^{-4} \text{ mol NaOH}$

Step (ii) is not correct. The mole ratio in the equation gives 1 mol citric acid reacting with 3 mol of NaOH, not 3 mol citric acid reacting with 1 mol of NaOH.

$$6.15 \times 10^{-4} \text{ mol NaOH} \times \frac{1 \text{ mol citric acid}}{3 \text{ mol NaOH}} = 2.05 \times 10^{-4} \text{ mol citric acid}$$

Step (ii) has a 3 in front of the moles of citric acid and a 1 in front of the NaOH, resulting in multiplication by 3 instead of division by 3, so the moles of citric acid calculated there are wrong.

Step (iii) is not correct, because it uses the erroneous answer from Step (ii); however, the calculation it shows is correct:

$$2.05 \times 10^{-4} \text{ mol citric acid} \times \frac{192.12 \text{ g citric acid}}{1 \text{ mol citric acid}} = 0.0394 \text{ g citric acid}$$

Step (iv) is not correct, because it uses a different answer than the correct one from Step (iii) and the significant figures on the volume are incorrect; however, the calculation it shows is correct:

$$\frac{0.0394 \text{ g citric acid}}{10.0 \text{ mL}} = 3.94 \times 10^{-3} \text{ g citric acid in 1 mL of soft drink}$$

(b) The correct answer is 3.94 × 10^{-3} g citric acid in 1 mL of soft drink.

☑ *Reasonable Result Check:* The moles of citric acid that react must be less than the moles of NaOH that react.

95. *Result:* 16.1% $H_2C_2O_4$

Analyze: Given the volume and concentration of a solution containing one reactant, the balanced chemical equation for a reaction, and the mass of a mixture containing two salts—one of which is the salt of the second reactant, determine the weight percent of the reactive salt in the mixture.

Plan: The volume and molarity of the first reactant are used to calculate the moles. The mole ratio in the equation can be used to find the moles of the reactive compound. Then we will use the molar mass of the reactive compound to find the mass. Then we will compare that to the mass of the mixture to determine the weight percent. *Notice: It is NOT appropriate to use the dilution equation when working with reactions!*

Execute: We learn from the balanced equation that 1 mol of $H_2C_2O_4$ reacts with 2 mol NaOH.

Molar Mass $H_2C_2O_4$ = 2(1.0067 g H) + 2(12.0107 g C) + 4(15.9994 g O) = 90.0348 g/mol

$$29.58 \text{ mL NaOH solution} \times \frac{1 \text{ L}}{1000 \text{ mL}} \times \frac{0.550 \text{ mol NaOH}}{1 \text{ L NaOH solution}}$$

$$\times \frac{1 \text{ mol } H_2C_2O_4}{2 \text{ mol NaOH}} \times \frac{90.0348 \text{ g } H_2C_2O_4}{1 \text{ mol } H_2C_2O_4} = 0.732 \text{ g } H_2C_2O_4$$

Weight percent is calculated by dividing the mass of $H_2C_2O_4$ by the mixture mass and multiplying by 100%.

$$\frac{0.732 \text{ g } H_2C_2O_4}{4.554 \text{ g mixture}} \times 100\% = 16.1\% \ H_2C_2O_4$$

☑ *Reasonable Result Check:* The mass of $H_2C_2O_4$ is significantly smaller than the mass of the sample, so it makes sense that the percentage is so low.

97. *Result:* **104.0 g/mol**

Analyze: Given the mass of an acid sample, the balanced chemical equation for neutralization, and the volume and molarity of a solution containing the base reactant needed to neutralize the sample, determine the molar mass of the acid.

Plan: Use the volume and molarity to calculate the amount (in moles) of base, then use the mole ratio in the balanced chemical equation to relate moles of base to moles of acid, then divide the mass of acid in the sample by the moles of acid in the sample to determine the molar mass.

Execute: We learn from the balanced equation that 1 mol of H_2A reacts with 2 mol NaOH.

$$36.04 \text{ mL NaOH solution} \times \frac{1 \text{ L}}{1000 \text{ mL}} \times \frac{0.509 \text{ mol NaOH}}{1 \text{ L NaOH solution}} \times \frac{1 \text{ mol } H_2A}{2 \text{ mol NaOH}} = 9.172 \times 10^{-3} \text{ mol } H_2A$$

$$\text{Molar Mass} = \frac{0.954 \text{ g } H_2A}{9.172 \times 10^{-3} \text{ mol } H_2A} = 104.0 \text{ g/mol}$$

General Questions

98. *Result:* **(b) Cu**

Analyze: Given the mass of the only reactant, the formula of the reactant with one unknown element, a balanced equation showing its decomposition, and the mass and identity of one of the products, determine the identity of the unknown element from a given list.

Plan: Use the molar mass of the product to determine the quantity (in moles) of product, then use the mole ratio in the equation to determine the quantity (in moles) of reactant. Then divide the grams of reactant by the calculated moles of reactant to determine the molar mass of the reactant. Subtract the molar masses of the known elements in the compound to determine the molar mass of the unknown element. Compare the molar masses of the elements on the list and find the one that most closely matches it.

Execute: Use the mass of CO_2 to determine the quantity (in moles) of MCO_3:

$$\text{Molar Mass } CO_2 = 12.0107 \text{ g C} + 2(15.9994 \text{ g O}) = 44.0095 \text{ g/mol}$$

$$0.376 \text{ g } CO_2 \times \frac{1 \text{ mol } CO_2}{44.0095 \text{ g } CO_2} \times \frac{1 \text{ mol } MCO_3}{1 \text{ mol } CO_2} = 8.54 \times 10^{-3} \text{ mol } MCO_3$$

$$\text{Divide given mass by moles: } \frac{1.056 \text{ g } MCO_3}{8.54 \times 10^{-3} \text{ mol } MCO_3} = 124 \frac{\text{g}}{\text{mol}} = \text{molar mass of } MCO_3$$

Subtract the molar mass of C and three times the molar mass of O from this molar mass to find the molar mass of M:

$$\frac{124 \text{ g } MCO_3}{\text{mol } MCO_3} - \frac{12.0107 \text{ g C}}{\text{mol } MCO_3} - \frac{3 \times 15.9994 \text{ g O}}{\text{mol } MCO_3} = \frac{64 \text{ g M}}{\text{mol } MCO_3}$$

The element in the given list with the closest molar mass to 64 g/mol is (b) Cu.

☑ *Reasonable Result Check:* The molar mass of Cu (63.5 g/mol) is closest to 64. None of the others in the list (Ni at 58.7 g/mol, Zn at 65.4 g/mol, or Ba at 137.2 g/mol) are this close, though with an uncertainty of ±1 g/mol
zinc almost qualifies. If we carry more decimal places than strictly allowed by the rules of significant figures, the molar mass of MCO_3 is 123.65 g/mol and the molar mass of M is 63.6 g/mol. It might make us feel better selecting Cu, although the additional significant figures are unknown with the limited data provided.

100. *Result:* **SiH_4**

Analyze: Given the mass of a compound, the identity of the elements in the compound, and the identity and masses of all the products produced, determine the empirical formula.

Plan: Use the molar mass of the products to find their moles and use the mole ratio in their formulas to determine the moles of the elements in the compound. Then use a whole-number mole ratio to determine the empirical formula.

Execute: The compound is Si_xO_y. Its combustion produced SiO_2 and H_2O. Use molar mass and the mole ratio from the formula to determine the moles of Si and H.

Molar Mass SiO_2 = 28.0855 g/mol Si + 2(15.9994 g/mol O) = 60.0843 g/mol SiO_2

$$11.64 \text{ g SiO}_2 \times \frac{1 \text{ mol SiO}_2}{60.0843 \text{ g SiO}_2} \times \frac{1 \text{ mol Si}}{1 \text{ mol SiO}_2} = 0.1937 \text{ mol Si}$$

Molar Mass H_2O = 2(1.0067 g/mol H) + 15.9994 g/mol O = 18.0152 g/mol H_2O

$$6.980 \text{ g H}_2\text{O} \times \frac{1 \text{ mol H}_2\text{O}}{18.0152 \text{ g H}_2\text{O}} \times \frac{2 \text{ mol H}}{1 \text{ mol H}_2\text{O}} = 0.7749 \text{ mol H}$$

Set up mole ratio and simplify by dividing by the smallest number of moles: 0.1937 mol Si : 0.7749 mol H

1 Si : 4.000 H

Use the whole number ratio for the subscripts in the formula. The empirical formula is SiH_4.

☑ *Reasonable Result Check:* These are not necessary calculations, but we can calculate the masses of each element, Si and H.

$$0.1937 \text{ mol Si} \times \frac{28.0855 \text{ g Si}}{1 \text{ mol Si}} = 5.440 \text{ g Si} \qquad 0.7749 \text{ mol H} \times \frac{1.0079 \text{ g H}}{1 \text{ mol H}} = 0.7810 \text{ g H}$$

The sum of these masses (5.440 g + 0.7810 g = 6.221 g) add up to the mass of the original compound (6.22 g), to the given significant figures.

102. *Result:* **KOH, KOH**

Analyze: Given equal moles of all the reactants, determine the limiting reactant. Given equal masses of all the reactants, determine the limiting reactant.

Plan: When the reactants are present in equal quantities, the reactant with the largest stoichiometric coefficient is going to be the limiting reactant. In the second part, we use the molar mass of each reactant to find their relative quantity (in moles), then we use the mole ratio in the reaction to determine the moles of a product formed from each reactant. The reactant that produces the least amount of product is the limiting reactant.

Execute:

First Question: If 5 mol of each reactant is present, then the stoichiometric coefficients are: 4 for KOH, 2 for MnO_2, 1 for O_2 and 1 for Cl_2. The reactant with the largest stoichiometric coefficient is KOH. So, **KOH** is the limiting reactant.

Second Question: If 5 grams of each reactant is present, then determine the moles of KCl they each form:

Molar Mass KOH = 39.0983 g/mol K + 15.9994 g/mol O + 1.0067 g/mol H = 56.1056 g/mol KOH

$$5\text{ g KOH} \times \frac{1\text{ mol KOH}}{56.1056\text{ g KOH}} \times \frac{2\text{ mol KCl}}{4\text{ mol KOH}} = 0.04\text{ mol KCl}$$

Molar Mass MnO_2 = 54.9380 g/mol Mn + 2(15.9994 g/mol O) = 86.9368 g/mol MnO_2

$$5\text{ g MnO}_2 \times \frac{1\text{ mol MnO}_2}{86.9368\text{ g MnO}_2} \times \frac{2\text{ mol KCl}}{2\text{ mol MnO}_2} = 0.06\text{ mol KCl}$$

Molar Mass O_2 = 2(15.9994 g/mol O) = 31.9988 g/mol O_2

$$5\text{ g O}_2 \times \frac{1\text{ mol O}_2}{31.9988\text{ g O}_2} \times \frac{2\text{ mol KCl}}{1\text{ mol O}_2} = 0.3\text{ mol KCl}$$

Molar Mass Cl_2 = 2(35.453 g/mol Cl) = 70.906 g/mol Cl_2

$$5\text{ g Cl}_2 \times \frac{1\text{ mol Cl}_2}{70.906\text{ g Cl}_2} \times \frac{2\text{ mol KCl}}{1\text{ mol Cl}_2} = 0.1\text{ mol KCl}$$

Only 0.04 mol of HCl is produced from KOH, so the **KOH** is the limiting reactant here, too.

☑ *Reasonable Result Check:* The first question is easy. The balanced equation indicates that a greater quantity of KOH is needed that any other reactant, so when equal quantities of reactants are present, KOH runs out first. The differences in molar mass are insufficient to keep the large stoichiometric coefficient for KOH from making it the limiting reactant when equal masses of the reactants are present.

104. *Result:* **(a) K^+ is a spectator ion. (b) 5 H_2S(g) + 2 MnO_4^-(aq) + 6 H^+(aq) ⟶ 2 Mn^{2+}(aq) + 5 S(s) + 8 $H_2O(\ell)$ (c) elements are changing oxidation states (d) MnO_4^- is the oxidizing agent. H_2S is the reducing agent.**

Analyze, Plan, and Execute:

(a) The permanganate ion is a reactant in the equation. The soluble compound potassium permanganate is the source of permanganate. Its cation is a spectator ion.

(b) Using the rules provided in Section 3-5c, determine the oxidation state for S and Mn in the reactants and products to see what electrons are being transferred. Then follow the method described in the solution for Question 20.

In the reactants, the S atom is in the H_2S molecular compound with hydrogen.
Rule 6(d) says Ox. #H = +1.

2(Ox. #H) + (Ox. # S) = 0

2(+1) + (Ox. # S) = 0 Ox. # S = –2

In the products, Ox. # S = 0, since it is in elemental form.

In the reactants, the Mn atom is in the polyatomic MnO_4^- ion with oxygen. Rule 7(a) says Ox. #O = –2.
Ox. # Au + 2(Ox. # Au) = –1

(Ox. #Mn) + 4(Ox. #O) = –1

(Ox. #Mn) + 4(–2) = –1 Ox. # Mn = +7

In the products, the Mn atom is in the monatomic Mn^{2+}. Rule 2 says Ox. #Mn = +2

The difference between an element's oxidation states determines how many electrons it gains or loses.

Ox. # product atom – Ox. # reactant atom = electrons gained (+) or lost (–)

7 – 2 = +5. Mn is gaining 5 electrons.

–2 – 0 = –2. S is losing 2 electrons.

If we use 2 Mn and 5 S, then 10 electrons would pass from S atoms to Mn atoms.

Balance the equation starting with 2 Mn and 5 S, then complete the balancing of O and H.

__ $H_2S(g)$ + __ $MnO_4^-(aq)$ + __ $H^+(aq)$ ⟶ __ $Mn^{2+}(aq)$ + __ S(s) + __ $H_2O(\ell)$

Order: Mn, S, O, H

5 $H_2S(g)$ + 2 $MnO_4^-(aq)$ + __ $H^+(aq)$ ⟶ 2 $Mn^{2+}(aq)$ + 5 S(s) + __ $H_2O(\ell)$ 2 Mn, 5 S

5 $H_2S(g)$ + 2 $MnO_4^-(aq)$ + __ $H^+(aq)$ ⟶ 2 $Mn^{2+}(aq)$ + 5 S(s) + 8 $H_2O(\ell)$ 8 O

5 $H_2S(g)$ + 2 $MnO_4^-(aq)$ + 6 $H^+(aq)$ ⟶ 2 $Mn^{2+}(aq)$ + 5 S(s) + 8 $H_2O(\ell)$ 16 H

Check the balance: 16 H, 5 S, 2 Mn, 8 O, +4 charge (6 – 2 = 2×2) on each side.

(c) This is a redox equation because the element Mn is being reduced (gaining electrons) and the element S is being oxidized (losing electrons).

(d) **MnO_4^- is the oxidizing agent**, since the manganese in this ion is gaining electrons, assisting in the oxidation of the sulfur.

H_2S is the reducing agent, since the sulfur in this compound is losing electrons, assisting in the reduction of manganese.

107. *Result:* **(a) acid-base reaction; H_3PO_4 is the acid; NaOH is the base.**
(b) acid-base reaction; CO_2 and H_2O combine as the acid; NH_3 is the base.
(c) redox reaction; oxidizing agent is Ti; reducing agent is Mg. (d) gas-forming

Analyze: Given several reactions, determine if they are redox (oxidation-reduction) reactions and identify the oxidizing and reducing agents in those that are redox reactions.

Plan: To decide if a reaction is a redox reaction, we need to see if any of the elements change oxidation state. In redox reactions, the atoms change oxidation states during the reaction. If no change in oxidation state is observed, then the reaction is not a redox reaction. Using the rules provided in Section 3-5c, determine the oxidation states.

The oxidizing agent is the reactant that assists an oxidation by being reduced, so it will be the reactant whose atoms gain electrons, and end up with a lower (more negative or less positive) oxidation number. The reducing agent is the reactant that assists a reduction by being oxidized, so it will be the reactant whose atoms lose electrons, and end up with a higher (more positive or less negative) oxidation number.

Acids provide a source of H^+. Bases usually contain oxygen and nitrogen, such as OH^- or NH_3.

Execute:

Rule 7(a) says Ox. #O = –2. Rule 6(d) says Ox. #H = +1. Rule 5 indicates that we treat cations and anions independently.

(a) Look at the oxidation numbers for the reactants:

NaOH contains a monatomic cation, Na^+, and a diatomic anion, OH^-.

Rule 2 gives Ox. #Na = +1. Ox. #O = –2 and Ox. #H = +1.

H_3PO_4 is a molecular compound containing H, P and O. Using Rule 3:

$$3 \times (\text{Ox. \#H}) + (\text{Ox. \# P}) + 4(\text{Ox. \# O}) = 0$$

$$3(+1) + (\text{Ox. \# P}) + 4\,(-2) = 0$$

Ox. #P = +5

Look at the oxidation numbers for the products:

NaH_2PO_4 contains the monatomic cation, Na^+, and the polyatomic anion, $H_2PO_4^-$.

Rule 2 says Ox. #Na = +1.

$H_2PO_4^-$ is a polyatomic ion containing H, P and O. Using Rule 4 on $H_2PO_4^-$:

$$2(\text{Ox. \#H}) + (\text{Ox. \#P}) + 4(\text{Ox. \#O}) = -1$$

$$2(+1) \times (\text{Ox. \#P}) + 4(-2) = -1$$

Ox. #P = +5

H_2O is a molecular compound containing H and O. Ox. # O = –2 and Ox. # H = +1.

The oxidation numbers in this reaction don't change, so this is NOT a redox reaction.

The reaction is an **acid-base reaction**. The acid is $\mathbf{H_3PO_4}$. The base is **NaOH**.

(b) Look at the oxidation numbers for the reactants:

NH_3 is a molecular compound containing N and H. Using Rule 3:

$$(\text{Ox. \#N}) + 3(\text{Ox. \#H}) = 0$$

$$(\text{Ox. \# N}) + 3(+1) = 0 \qquad \text{Ox. \# N} = -3$$

CO_2 is a molecular compound containing C and O. Using Rule 3:

$$(\text{Ox. \#C}) + 2(\text{Ox. \#H}) = 0$$

$$(\text{Ox. \# C}) + 2(-2) = 0 \qquad \text{Ox. \# C} = +4$$

H_2O is a molecular compound contains oxygen and hydrogen. Ox. # O = –2 and Ox. # H = +1.

Look at the oxidation numbers for the products:

NH_4HCO_3 contains a polyatomic cation, NH_4^+, and a polyatomic anion, HCO_3^-.

Using Rule 4 on NH_4^+:

$$(\text{Ox. \#N}) + 4(\text{Ox. \#H}) = +1$$

$$(\text{Ox. \# N}) + 4(+1) = +1 \qquad \text{Ox. \# N} = -3$$

Using Rule 4 on HCO_3^-.

$$(\text{Ox. \#H}) + (\text{Ox. \#C}) + 3(\text{Ox. \#O}) = -1$$

$$(+1) + (\text{Ox. \# C}) + 3(-2) = -1 \qquad \text{Ox. \# C} = +4$$

The oxidation numbers don't change, so this is NOT a redox reaction.

The reaction is an **acid-base reaction**. The acid is a combination of $\mathbf{H_2O}$ **and** $\mathbf{CO_2}$. The base is $\mathbf{NH_3}$.

(c) Look at the oxidation numbers for the reactants:

$TiCl_4$ contains a monatomic cation, Ti^{4+}, and a monatomic anion, Cl^-. Rule 2 gives Ox. # Ti = +4 and Ox. # Cl = –1.

Rule 1 indicates the oxidation number of a substance in elemental form is zero, so Ox. # Mg = 0.

Look at the oxidation numbers for the products:

Rule 1 indicates that the Ox. # Ti = 0.

$MgCl_2$ contains a monatomic cation, Mg^{2+}, and a monatomic anion, Cl^-. Rule 2 gives us Ox. # Mg = +2 and Ox. # Cl = –1.

The oxidation numbers of Mg and Ti do change, so this is a **redox reaction**.

The oxidizing agent is **Ti** since its Ox. # goes from +4 to zero.

The reducing agent is **Mg** since its Ox. # goes from zero to +2.

(d) Look at the oxidation numbers for the reactants:

NaCl contains a monatomic cation, Na^+, and a monatomic anion, Cl^-.

Rule 2 gives us Ox. # Na = +1 and Ox. # Cl = –1.

$NaHSO_4$ contains a monatomic cation, Na^+, and a polyatomic anion, HSO_4^-.

Rule 2 gives us Ox. # Na = +1.

HSO_4^- is a polyatomic ion containing H, S, and O. Using Rule 4:

(Ox. #H) + (Ox. #S) + 4(Ox. #O) = –1

(+1) + (Ox. # S) + 4(–2) = –1 Ox. # S = +6

Look at the oxidation numbers for the products:

HCl is a molecular compound containing H and Cl. Using Rule 3:

(Ox. #H) + (Ox. #Cl) = 0

(+1) + (Ox. # Cl) = 0 Ox. # Cl = –1

Na_2SO_4 contains a monatomic cation, Na^+, and a polyatomic anion, SO_4^{2-}.

Rule 2 gives us Ox. # Na = +1.

Using Rule 4 on SO_4^{2-}:

(Ox. #S) + 4(Ox. #O) = –2

(Ox. # S) + 4(–2) = –2 Ox. # S = +6.

The oxidation numbers don't change, so this is NOT a redox reaction.

The reaction is a **gas-forming reaction**. The gas is HCl(g).

☑ *Reasonable Result Check:* The oxidation numbers are consistent with typical oxidation states of these elements. The reactants and products are common acids, ionic compounds and ions.

108. *Result:* **(a) $CaF_2(s) + H_2SO_4(aq) \longrightarrow 2\,HF(g) + CaSO_4(s)$; calcium fluoride, sulfuric acid, hydrogen fluoride, calcium sulfate (b) precipitation (c) carbon tetrachloride, antimony(V) pentachloride, hydrogen monochloride (d) CCl_3F**

Analyze, Plan and Execute:

(a) $CaF_2(s) + H_2SO_4(aq) \longrightarrow 2\,HF(g) + CaSO_4(s)$

The reactants are called calcium fluoride and sulfuric acid.

The products are called hydrogen fluoride and calcium sulfate.

(b) Determine if a reaction is an acid base reaction, an oxidation-reduction reaction, or a precipitation reaction.

To decide if a reaction is an acid-base reaction, we need to see if an acid is reacting with a base. To decide if a reaction is an oxidation-reduction reaction, we need to see if any of the elements change oxidation state. To decide if a reaction is a precipitation reaction, look for an insoluble ionic compound as a product.

Execute:

First, check to see if it is an acid-base reaction: The strong reactant acid, H_2SO_4, reacts to form a weak product acid, HF. We don't see a hydroxide compound in the reaction, but in some sense of the word, the ionic fluoride compound is serving as a base. So, there is an **acid-base reaction** happening here. We'll learn more about these kinds of acid-base reactions in Chapter 14.

Second, check to see if it is an oxidation-reduction reaction: Using the rules provided in Section 3-5c, determine the oxidation states. Rule 7(a) says Ox. #O = –2. Rule 6(d) says Ox. #H = +1. Rule 5 indicates that we treat cations and anions independently.

Look at the oxidation numbers for the reactants:

CaF_2 contains a monatomic cation, Ca^{2+}, and a monatomic anion, F^-.

Rule 2 gives us Ox. # Ca = +2 and Ox. # F = –1.

H_2SO_4 is a molecular compound containing H, S, and O. Using Rule 3:

$$2(\text{Ox. \# H}) + (\text{Ox. \# S}) + 4(\text{Ox. \# O}) = 0$$

$$2(+1) + (\text{Ox. \# S}) + 4(-2) = 0 \qquad \text{Ox. \# S} = +6$$

Look at the oxidation numbers for the products:

HF is a molecular compound containing H and F. Using Rule 3:

$$(\text{Ox. \#H}) + (\text{Ox. \#F}) = 0$$

$$(+1) + (\text{Ox. \#F}) = 0 \qquad \text{Ox. \# F} = -1$$

$CaSO_4$ contains a monatomic cation, Ca^{2+}, and a polyatomic anion, SO_4^{2-}.

Rule 2 gives Ox. # Ca = +2.

Using Rule 4 on SO_4^{2-}:

$$(\text{Ox. \#S}) + 4(\text{Ox. \#O}) = -2$$

$$(\text{Ox. \# S}) + 4(-2) = -2 \qquad \text{Ox. \# S} = +6$$

The oxidation numbers don't change, so this is NOT an oxidation-reduction reaction.

Third, check to see if it is a precipitation reaction: An insoluble solid, $CaSO_4$(s) (Table 3.1, Rule 4), is produced in a solution, so the reaction can also be considered a **precipitation reaction**.

In conclusion, most people studying Chapter 3 would probably decide this was a **precipitation reaction**.

☑ *Reasonable Result Check:* The oxidation states determined are typical for these atoms in these compounds. With constant oxidation states, it is clearly not an oxidation-reduction reaction. It is logical to call this either a precipitation reaction or an acid-base reaction, but NOT an oxidation-reduction reaction. Because these kinds of acid-base reactions are described more thoroughly in a later chapter, most students will probably choose the precipitation reaction answer.

(c) CCl_4 is **carbon tetrachloride**, $SbCl_5$ is **antimony pentachloride**, HCl is **hydrogen monochloride**.

(d) *Analyze:* Given the identity and the percent mass of each element in a compound, find the empirical formula of the compound.

Plan: This question might require a review of the "Empirical Formula" calculations introduced in Chapter 2, Section 2-12. Choose a convenient mass sample of the chlorofluorocarbon, such as 100.00 g. Using the given mass percents, determine the mass of C, Cl, and F in the sample. Using molar masses of the elements, determine the moles of each element in the sample. Find the whole number mole ratio of the elements C, Cl, and F, to determine the subscripts in the empirical formula.

Execute: A 100.00 gram sample will have 8.74 grams of C, 77.43 grams of Cl, and 13.83 grams of F.

Determine the quantity (in moles) of C, Cl, and F in the sample:

$$8.74\text{ g C} \times \frac{1\text{ mol C}}{12.0107\text{ g C}} = 0.728\text{ mol C} \qquad 77.43\text{ g Cl} \times \frac{1\text{ mol Cl}}{35.453\text{ g Cl}} = 2.184\text{ mol Cl}$$

$$13.83\text{ g F} \times \frac{1\text{ mol F}}{18.9984\text{ g F}} = 0.7279\text{ mol F}$$

Mole ratio: 0.728 mol C : 2.184 mol Cl : 0.7279 mol F

Divide each term by the smallest number of moles, 0.7279 mol, to get atom ratio:

Atom ratio: 1 C : 3 Cl : 1 F

The empirical formula is $\mathbf{CCl_3F}$.

☑ *Reasonable Result Check:* This empirical formula makes sense, because four halogens can be bonded to one carbon.

109. *Result:* **6.28% impurity**

Analyze: Given the mass of a tablet containing vitamin C, the balanced neutralization equation, and the volume and molarity of a base solution used for neutralization, determine what percentage of the tablet is impurity.

Plan: Use the volume, molarity, mole ratio in the chemical equation and molar mass of the compound to calculate the mass of the vitamin C in the sample. Subtract the mass of vitamin C from the mass of the tablet, to determine that mass of the impurity. Divide the mass of the impurity into the mass of the soil sample and multiply by 100% to get percentage mass of the impurity.

Execute: Molar Mass of $HC_6H_7O_6 = 6(12.0107\text{ g/mol C}) + 8(1.0079\text{ g/mol H})$

$$+ 6(15.9994\text{ g/mol O}) = 176.1238\text{ g/mol }HC_6H_7O_6$$

$$21.30\text{ mL NaOH} \times \frac{1\text{ L NaOH}}{1000\text{ mL NaOH}} \times \frac{0.1250\text{ mol NaOH}}{1\text{ L NaOH}} \times \frac{1\text{ mol }HC_6H_7O_6}{1\text{ mol HCl}}$$

$$\times \frac{176.1238\text{ g }HC_6H_7O_6}{1\text{ mol }HC_6H_7O_6} \times \frac{1000\text{ mg }HC_6H_7O_6}{1\text{ g }HC_6H_7O_6} = 468.6\text{ mg }HC_6H_7O_6$$

$$500.0\text{ mg tablet} - 468.6\text{ mg }HC_6H_7O_6 = 31.4\text{ mg impurity}$$

$$\frac{31.4\text{ g impurity}}{500.0\text{ g tablet}} \times 100\% = 6.28\%\text{ impurity}$$

☑ *Reasonable Result Check:* A vitamin C tablet should be mostly vitamin C, so it makes sense that the mass percent if the impurity is small.

111. *Result:* **99.7% CH_3OH, 0.3% C_2H_5OH**

Analyze: Given the mass of a sample of a liquid containing unknown amounts of two organic compounds and given the mass of one of the products of a chemical reaction, determine the composition of the liquid.

Plan: First, balance the combustion equation for ethyl alcohol and methyl alcohol. Set two variables for the mass of methyl alcohol and the mass of ethyl alcohol. Then establish two equations, one describing the total mass of the sample related to the two components and one describing the moles of carbon dioxide produced when burning the sample. Algebraically solve for X and Y, then use those masses and the total sample mass to find the percentage by mass of the two compounds.

Execute:

Unbalanced: $\_\ C_2H_5OH(\ell) + \_\ O_2(g) \longrightarrow \_\ CO_2(g) + \_\ H_2O(\ell)$ Order: C, H, then O

$\underline{1}\ C_2H_5OH(\ell) + \_\ O_2(g) \longrightarrow \underline{2}\ CO_2(g) + \_\ H_2O(\ell)$ 2 C's

$\underline{1}\ C_2H_5OH(\ell) + \_\ O_2(g) \longrightarrow \underline{2}\ CO_2(g) + \underline{3}\ H_2O(\ell)$ 6 H's

$\underline{1}\ C_2H_5OH(\ell) + \underline{6}\ O_2(g) \longrightarrow \underline{2}\ CO_2(g) + \underline{3}\ H_2O(\ell)$ 7 O's

Unbalanced: $\_\ CH_3OH(\ell) + \_\ O_2(g) \longrightarrow \_\ CO_2(g) + \_\ H_2O(\ell)$ Order: C, H, then O

$\underline{1}\ CH_3OH(\ell) + \_\ O_2(g) \longrightarrow \underline{1}\ CO_2(g) + \_\ H_2O(\ell)$ 1 C's

$\underline{1}\ CH_3OH(\ell) + \_\ O_2(g) \longrightarrow \underline{1}\ CO_2(g) + \underline{2}\ H_2O(\ell)$ 4 H's

$\underline{1}\ CH_3OH(\ell) + \frac{3}{2}\ O_2(g) \longrightarrow \underline{1}\ CO_2(g) + \underline{2}\ H_2O(\ell)$ 4 O's

Let X = grams of ethyl alcohol and Y = grams of methyl alcohol

Total mass of sample = 0.280 g = X + Y

Solve for Y in terms of X: $Y = 0.280 - X$

Molar Mass of CO_2 = 12.0107 g/mol C + 2(15.9994 g/mol O) = 44.0095 g/mol CO_2

Moles of carbon dioxide produced when burned = $0.385 \text{ g } CO_2 \times \frac{1 \text{ mol } CO_2}{44.0095 \text{ g } CO_2} = 0.00875 \text{ mol } CO_2$

Molar Mass of C_2H_5OH = 2(12.0107 g/mol C) + 6(1.0079 g/mol H)

+ 15.9994 g/mol O = 46.0682 g/mol C_2H_5OH

Molar Mass of CH_3OH = 12.0107 g C + 4(1.0079 g H) + 15.9994 g O = 32.0417 g/mol

$$0.00875 \text{ mol } CO_2 = X \text{ g } C_2H_5OH \times \frac{1 \text{ mol } C_2H_5OH}{46.0682 \text{ g } C_2H_5OH} \times \frac{2 \text{ mol } CO_2}{1 \text{ mol } C_2H_5OH}$$

$$+ Y \text{ g } CH_3OH \times \frac{1 \text{ mol } CH_3OH}{32.0417 \text{ g } CH_3OH} \times \frac{1 \text{ mol } CO_2}{1 \text{ mol } CH_3OH}$$

$$0.00875 = 0.0434139 X + 0.0312093 Y$$

Plug Y = 0.280 – X into this equation and solve it for X:

$$0.00875 = 0.0434139 X + 0.0312093(0.280 - X)$$

$$0.00875 = 0.0434139 X + 0.000874 - 0.0312093 X$$

$$0.00875 - 0.000874 = (0.0434139 - 0.0312093) X$$

$$0.00001 = 0.0122046 X$$

$$X = 0.0009 \text{ g ethyl alcohol}$$

$$Y = 0.280 - X = 0.280 - 0.0009 = 0.279 \text{ g methyl alcohol}$$

$$\frac{0.279 \text{ g } CH_3OH}{0.280 \text{ g liquid}} \times 100\% = 99.7\% \ CH_3OH \qquad \frac{0.0009 \text{ g } C_2H_5OH}{0.280 \text{ g liquid}} \times 100\% = 0.3\% \ C_2H_5OH$$

☑ *Reasonable Result Check:* Assuming the liquid is pure methyl alcohol gives the following mass of CO_2:

$$0.280 \text{ g } CH_3OH \times \frac{1 \text{ mol } CH_3OH}{32.0417 \text{ g } CH_3OH} \times \frac{1 \text{ mol } CO_2}{1 \text{ mol } CH_3OH} \times \frac{44.0095 \text{ g } CO_2}{1 \text{ mol } CO_2} = 0.385 \text{ g } CO_2$$

So it makes sense that the liquid is almost one hundred percent methanol.

Applying Concepts

116. *Results:* **See below**

Analyze, Plan, and Execute: For the first case: LiCl(aq) and $AgNO_3$(aq)

(a) The two separate solutions of soluble salts would be clear and colorless like water. Once combined, insoluble white AgCl (Table 5.1, Rule 3) will precipitate. We would probably see it eventually sink to the bottom of the beaker, leaving a clear, probably colorless liquid containing aqueous $LiNO_3$ above it.

(b) Notice, for proper proportions, these diagrams really need many more water molecules.

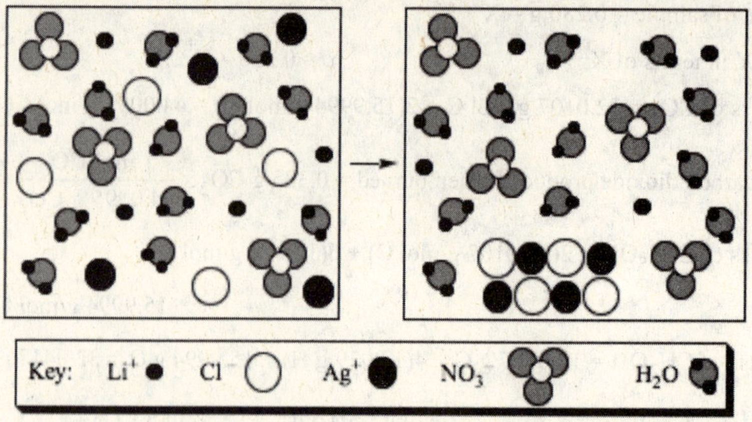

(c) $Li^+(aq) + Cl^-(aq) + Ag^+(aq) + NO_3^-(aq) \longrightarrow AgCl(s) + Li^+(aq) + NO_3^-(aq)$

For the second case: NaOH(aq) and HCl(aq)

(a) The separate acid and base solutions would be clear and colorless like water. Once combined, the solution would still be clear and colorless.

(b) Notice, for proper proportions, these diagrams really need many more water molecules.

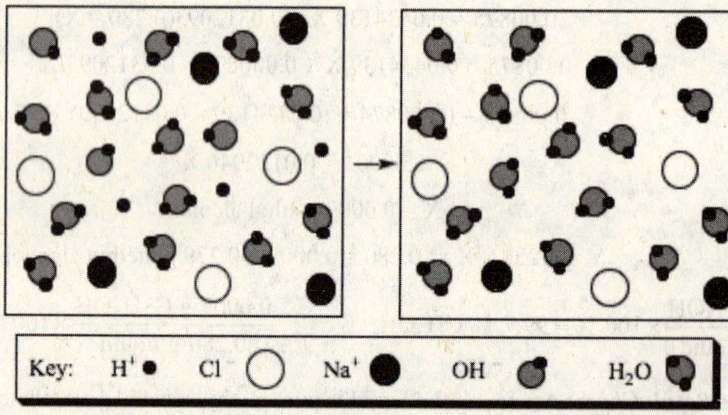

(c) $Na^+(aq) + OH^-(aq) + H^+(aq) + Cl^-(aq) \longrightarrow H_2O(\ell) + Na^+(aq) + Cl^-(aq)$

118. *Result/Explanation:* Two butane molecules react with 13 diatomic oxygen molecules to produce eight carbon dioxide molecules and ten water molecules.

Two mol of gaseous butane molecules react with 13 mol of gaseous diatomic oxygen molecules to produce eight mol of gaseous carbon dioxide molecules and ten mol of liquid water molecules.

120. *Result:* **Ag^+, Cu^{2+}, and NO_3^-**

Analyze: Given the moles of all the reactants, determine the limiting reactant and the ions present at the end of the reaction.

Plan and Execute: Use the mole ratio in the equation to determine the moles of a product formed from each reactant. The reactant that produces the least amount of product is the limiting reactant. The excess reactant and the products are present at the end of the reaction. From that information, determine what ions are in the solution after the reaction is complete.

$$1.5 \text{ mol Cu} \times \frac{1 \text{ mol Cu(NO}_3)_2}{1 \text{ mol Cu}} = 1.5 \text{ mol Cu(NO}_3)_2$$

$$4.0 \text{ mol AgNO}_3 \times \frac{1 \text{ mol Cu(NO}_3)_2}{2 \text{ mol AgNO}_3} = 2.0 \text{ mol Cu(NO}_3)_2$$

Only 1.5 mol $Cu(NO_3)_2$ can be formed, so the limiting reactant is Cu and the excess reactant is $AgNO_3$.

That means the solution contains: Ag^+ ions, Cu^{2+} ions, and NO_3^- ions after the reaction is over.

☑ *Reasonable Result Check:* The excess reactant is an ionic compound and one of the products is an ionic compound, so some of the ions present at the beginning are still present when the Cu runs out.

123. *Result:* **Equation (b)**

Analyze, Plan, and Execute: Four XY_3 molecules are made from two diatomic X_2 molecules and six diatomic Y_2 molecules. So, symbolically, the reaction is $2\ X_2 + 6\ Y_2 \longrightarrow 4\ XY_3$, and the stoichiometric equation representing that reaction is (b) $X_2 + 3\ Y_2 \longrightarrow 2\ XY_3$.

124. *Result:* $\mathbf{A_2 + 4\ BC_2 \longrightarrow 4\ C_2 + 2\ AB_2}$

Analyze, Plan, and Execute: Three A_2 molecules react with 12 BC_2 molecules to make 12 diatomic C_2 molecules and six AB_2 molecules. So, symbolically, the reaction is $3\ A_2 + 12\ BC_2 \longrightarrow 12\ C_2 + 6\ AB_2$ Dividing each coefficient by 3, gives the smallest integer coefficients: $A_2 + 4\ BC_2 \longrightarrow 4\ C_2 + 2\ AB_2$.

126. *Result:* **When the metal mass is less than 1.2 g, the metal is the limiting reactant. When the metal mass is greater than 1.2 g, the bromine is the limiting reactant.**

Analyze, Plan, and Execute: When masses smaller than 1.2 grams of the metal are added, the metal is the limiting reactant, so the mass of the compound produced is directly proportional to the mass of metal present (shown by the straight line with a positive slope). Adding a greater mass of metal produces proportionally more products whenever the mass of metal is less than 1.2 grams.

When the mass larger than 1.2 g of metal are added, the bromine is the limiting reactant, so any mass above 1.2 g of the metal will not affect how much compound is formed. Because the mass of bromine is held constant, the mass of compound formed is also a constant (shown by a horizontal line on the graph).

128. *Result:* **(a) Box 2 (b) Box 3 (c) Box 1**

Analyze: Identify the common ions present in the compounds.

Plan and Execute:

(a) The ions present in a solution of $MgCl_2$ are Mg^{2+} and 2 Cl^-, in a 1:2 ratio. **Box 2** has eight 1– charges and four 2+ charges, giving a 1:2 ratio.

(b) The ions present in a solution of K_2SO_4 are 2 K^+ and SO_4^{2-}, in a 2:1 ratio. **Box 3** has eight 1+ charges and four 2– charges, giving a 2:1 ratio.

(c) The ions present in a solution of NH_4Cl are NH_4^+ and Cl^-, in a 1:1 ratio. **Box 1** has five 1+ charges and five 1– charges, giving a 1:1 ratio.

131. *Result:* **(a) combine $\mathbf{H_2SO_4}$ and $\mathbf{Ba(OH)_2}$ (b) combine $\mathbf{Na_2SO_4}$ and $\mathbf{Ba(NO_3)_2}$ (c) combine $\mathbf{H_2SO_4(aq)}$ and $\mathbf{BaCO_3(s)}$**

Analyze: Prepare barium sulfate from a given list of chemicals by various means.

Plan and Execute:

(a) To make $BaSO_4$ from an acid-base reaction, use a base with the cation and an acid with the anion:

$$H_2SO_4(aq) + Ba(OH)_2(aq) \longrightarrow BaSO_4(s) + 2\ H_2O(\ell)$$

(b) To make $BaSO_4$ from a precipitation reaction, use a soluble salt containing the anion and a soluble salt containing the cation:

$$Na_2SO_4(aq) + Ba(NO_3)_2(aq) \longrightarrow BaSO_4(s) + 2\ NaNO_3(aq)$$

(c) To make $BaSO_4$ from a gas-forming reaction, use an acid with the anion and the carbonate salt of the cation:

$$H_2SO_4(aq) + BaCO_3(s) \longrightarrow BaSO_4(s) + H_2O(\ell) + CO_2(g)$$

☑ *Reasonable Result Check:* All of these reactants were provided in the list of chemicals. The neutralization reaction produces the solid and water. The precipitation reaction produces the solid and a soluble salt. The gas-forming reaction produces the solid and carbon dioxide gas.

133. *Result:* **Use HCl; precipitate forms if Pb^{2+} present, no precipitate if Ba^{2+} present.**

Analyze: Determine which of three reagents could be used to distinguish whether a solution contains lead(II) ions or barium ions.

Plan: Determine the anion of the added reagent. Determine the products of its reaction with Pb^{2+} and Ba^{2+}. If the reactions are visibly different, then the reagent could be used to distinguish them.

Execute: First reagent: HCl(aq) would provide the anion $Cl^-(aq)$ for a precipitation reaction.

$PbCl_2$ is insoluble, Table 3.1, Rule 3 $\quad 2\,Cl^-(aq) + Pb^{2+}(aq) \longrightarrow PbCl_2(s)$

$BaCl_2$ is soluble, Table 3.1, Rule 3 $\quad 2\,Cl^-(aq) + Ba^{2+}(aq) \longrightarrow$ N.R.

Insoluble lead(II) chloride precipitate would form if the unknown contained Pb^{2+}, but no precipitate would be seen if the unknown contained only Ba^{2+}. Thus, **HCl(aq) could be used** to distinguish these two ions.

Second reagent: $H_2SO_4(aq)$ would provide the $SO_4^{2-}(aq)$ anion to a precipitation reaction.

$PbSO_4$ is insoluble, Table 3.1, Rule 4 $\quad SO_4^{2-}(aq) + Pb^{2+}(aq) \longrightarrow PbSO_4(s)$

$BaSO_4$ is insoluble, Table 3.1, Rule 4 $\quad SO_4^{2-}(aq) + Ba^{2+}(aq) \longrightarrow BaSO_4(s)$

Both reactions produce insoluble sulfate precipitates, so the test would not differentiate the ions. $H_2SO_4(aq)$ **could NOT be used** to distinguish them.

Third reagent: $H_3PO_4(aq)$ would provide the $PO_4^{3-}(aq)$ anion to a precipitation reaction.

Table 3.1, Rule 8 $\quad 2\,PO_4^{3-}(aq) + 3\,Pb^{2+}(aq) \longrightarrow Pb_3(PO_4)_2(s)$

Table 3.1, Rule 8 $\quad 2\,PO_4^{3-}(aq) + 3\,Ba^{2+}(aq) \longrightarrow Ba_3(PO_4)_2(s)$

Both reactions produce insoluble phosphate precipitates, so the test would not differentiate the ions. $H_3PO_4(aq)$ could NOT be used to distinguish them.

☑ *Reasonable Result Check:* The selective solubility of lead(II), silver, and mercury(I) ions in chloride solutions makes chloride ion ideal for determining the presence or absence of these ions.

135. *Result:* **(d)**

Analyze, Plan, and Execute: Too much water was added, making the solution too dilute. So, (d) the concentration of the solution is less than 1 M because you added more solvent than necessary.

137. *Result:* **(a) and (d) are correct.**

Analyze, Plan, and Execute:

(a) The red color indicates that the solution is acidic, so there are more H^+ ions than OH^- ions in the mixture. This statement is **TRUE**.

(b) This statement is **FALSE**, for the same reason (a) was true.

(c) Only *equal quantities* of strong acid and strong base make a neutral solution. If either the acid or the base are weak or if the molarities are unequal, then the resulting solution will be basic or acidic. The test results indicate that this solution is acidic, so this statement is **FALSE**.

(d) Since the resulting solution was acidic, and equal volumes were added, that means the monoprotic acid's concentration must have been greater than the NaOH base's concentration. This statement is **TRUE**.

(e) While the concentration of H_2SO_4 might have been greater, it is not necessarily true that it MUST have been greater, since only half as many moles of the diprotic acid is required to neutralize the NaOH base; therefore, the acid concentration need only be more than half the base concentration. This statement is **FALSE**.

More Challenging Questions

140. *Result:* **(a) Groups C&D: $Ag^+ + Cl^- \longrightarrow AgCl\ (s)$; Groups A&B: $Ag^+ + Br^- \longrightarrow AgBr\ (s)$ (b) silver halide product is the same for A&B and C&D and A&B is different from C&D (c) Curve has upward slope while the Ag^+ is the limiting reactant. Bromide is heavier than chloride, so the curve levels out at different masses of product.**

Analyze, Plan, and Execute:

(a) Na^+ and NO_3^- are spectator ions.

Net Ionic for groups C and D	Net Ionic for groups A and B
$Ag^+ + Cl^- \longrightarrow AgCl\ (s)$	$Ag^+ + Br^- \longrightarrow AgBr\ (s)$

(b) The silver halide solid (AgCl) produced by groups C and D is the same. The silver halide solid (AgBr) produced by groups A and B is the same.

(c) Below the mass of 0.75 g, the graph line has a positive slope: more product mass forms with increasing masses of $AgNO_3$. In this region, Ag^+ is the limiting reactant. (See Question 126 and 127 for a more complete explanation.) These reactions both require the same mass of $AgNO_3$ to make their respective silver halide. Since bromide ion is heavier than chloride ion, the mass of the product will be different and the mass of product where the graph levels out will be different, because AgBr is heavier than AgCl. That means the products in groups A and B (AgCl) weigh less than the products in groups C and D (AgBr).

142. *Result:* **$H_2(g) + 3\ Fe_2O_3(s) \longrightarrow H_2O(\ell) + 2\ Fe_3O_4(s)$**

Analyze: Given the formulas of the reactants and one product and the percent mass of elements in the second product, determine the balanced chemical equation.

Plan: Choose a convenient sample mass of product, Fe_xO_y, such as 100.0 g. Find the mass of Fe and O in the sample, using the given mass percent. Use the molar mass of the elements to find their moles, then use a whole-number mole ratio to determine the empirical formula. Using the formulas of the reactants and products balance the equation.

Execute: 100.0 g of .Fe_xO_y contains 72.3 g Fe and 27.7 g O.

$$72.3\text{ g Fe} \times \frac{1\text{ mol Fe}}{55.845\text{ g Fe}} = 1.29\text{ mol Fe} \qquad 27.7\text{ g O} \times \frac{1\text{ mol O}}{15.9994\text{ g O}} = 1.73\text{ mol O}$$

Set up mole ratio and simplify by dividing by the smallest number of moles:

1.29 mol Fe : 1.73 mol O

1 Fe : 1.34 O

Multiply by 3 to get whole numbers: 3 Fe : 4 O

Use the whole number ratio for the subscripts in the formula. The empirical formula is Fe_3O_4.

$\_\ H_2(g) + \_\ Fe_2O_3(s) \longrightarrow \_\ H_2O(\ell) + \_\ Fe_3O_4(s)$ Select order: Fe, O, H

$\_\ H_2(g) + \underline{3}\ Fe_2O_3(s) \longrightarrow \_\ H_2O(\ell) + \underline{2}\ Fe_3O_4(s)$ 6 Fe

$\_\ H_2(g) + \underline{3}\ Fe_2O_3(s) \longrightarrow \underline{1}\ H_2O(\ell) + \underline{2}\ Fe_3O_4(s)$ 9 O

$\underline{1}\ H_2(g) + \underline{3}\ Fe_2O_3(s) \longrightarrow \underline{1}\ H_2O(\ell) + \underline{2}\ Fe_3O_4(s)$ 2 H

☑ *Reasonable Result Check:* Fe_3O_4 is a common oxide of iron. 6 Fe, 9 O and 2 H on each side.

144. *Result:* **44.9 u**

Analyze: Given the mass of a sample of a compound X_2S_3 that is then roasted to form a given mass of X_2O_3, determine the atomic mass of X.

Plan: Because all the X atoms from X_2S_3 end up in X_2O_3, then it must be true that each sample must contains the same number of moles of X. Because each compound has the same number of X atoms, it must be true that each sample must contain the same number of moles of compound. Using a variable, x, for the molar mass of X, write an expression that shows the calculation of moles for each compound and equate these two expressions. Solve for x.

Execute: Let x = molar mass of X and y = molar mass of Y

$$\text{mol of } X_2Y_3 = \text{mass of } X_2Y_3 \times \frac{1 \text{ mol } X_2Y_3}{(2x + 3y)}$$

$$10.00 \text{ g } X_2S_3 \times \frac{1 \text{ mol } X_2S_3}{\left[2x + 3(32.065) \text{ g S}\right]} = 7.410 \text{ g } X_2O_3 \times \frac{1 \text{ mol } X_2O_3}{\left[2x + 3(15.9994) \text{ g O}\right]}$$

$$10.00 \times \frac{1}{(2x + 96.195)} = 7.410 \times \frac{1}{(2x + 47.9982)}$$

$$10.00\,(2x + 47.9982) = 7.410\,(2x + 96.195)$$

$$20.00\,x + 479.982 = 14.82\,x + 712.805$$

$$20.00\,x - 14.82\,x = 712.805 - 479.982$$

$$5.18\,x = 232.813$$

$$x = 44.9 \text{ g/mol}$$

Atomic mass = 44.9 u

☑ *Reasonable Result Check:* The calculated atomic mass is very close to that of Sc, which does form 3+ ions and would combine with 2- ions like O^{2-} and S^{2-} to form Sc_2S_3 and Sc_2O_3.

145. *Result:* **0 g $AgNO_3$, 9.82 g Na_2CO_3, 6.79 g Ag_2CO_3, 4.19 g $NaNO_3$**

Analyze: Given the names of products and reactants of a chemical reaction, and the masses of the two reactants, determine the masses of all the products and reactants after the reaction is complete.

Plan: Determine the formulas of the reactants and products and set up and balance the chemical equation. Use the molar mass of the reactants to find the moles of the reactant substances. Then use the mole ratio in the equation to determine the moles of the product produced in each case. Identify the limiting reactant from the reactant that produces the least number of products. Use the moles of product produced from the limiting reactant, the molar masses of the products to find the grams of products. Use the moles of product produced, determine the moles of the other reactant needed for complete reaction. Convert that number to grams using the molar mass. Then subtract the quantity used from the initial mass given to get the mass of excess reactant.

Execute: Silver nitrate is $AgNO_3$. Sodium carbonate is Na_2CO_3. Silver carbonate is Ag_2CO_3. Sodium nitrate is $NaNO_3$.

$\_\_\ AgNO_3 + \_\_\ Na_2CO_3 \longrightarrow \_\_\ Ag_2CO_3 + \_\_\ NaNO_3$ Order: Ag, Na, N, C, O

$\underline{2}\ AgNO_3 + \_\_\ Na_2CO_3 \longrightarrow \underline{1}\ Ag_2CO_3 + \_\_\ NaNO_3$ 2 Ag

$\underline{2}\ AgNO_3 + \underline{1}\ Na_2CO_3 \longrightarrow \underline{1}\ Ag_2CO_3 + \underline{2}\ NaNO_3$ 2 Na, 2 N, 1 C, 9 O

The balanced equation says: 1 mol Na_2CO_3 produces 1 mol Ag_2CO_3.

Molar Mass Na_2CO_3 = 2(22.9898 g Na) + 12.0107 g C + 3(15.9994 g O) = 105.9885 g/mol

$$12.43\ \text{g}\ Na_2CO_3 \times \frac{1\ \text{mol}\ Na_2CO_3}{105.9885\ \text{g}\ Na_2CO_3} \times \frac{1\ \text{mol}\ Ag_2CO_3}{1\ \text{mol}\ Na_2CO_3} = 0.1173\ \text{mol}\ Ag_2CO_3$$

The balanced equation says: 2 mol $AgNO_3$ produces 1 mol Ag_2CO_3.

Molar Mass $AgNO_3$ = 107.8682 g Ag + 14.0067 g N + 3(15.9994 g O) = 169.8731 g/mol

$$8.37\ \text{g}\ AgNO_3 \times \frac{1\ \text{mol}\ AgNO_3}{169.8731\ \text{g}\ AgNO_3} \times \frac{1\ \text{mol}\ Ag_2CO_3}{2\ \text{mol}\ AgNO_3} = 0.0246\ \text{mol}\ Ag_2CO_3$$

The number of Ag_2CO_3 moles produced from $AgNO_3$ is smaller (0.0246 mol < 0.1173 mol), so $AgNO_3$ is the limiting reactant and Na_2CO_3 is the excess reactant. Therefore, at the end of the reaction, the mass of $AgNO_3$ present is zero grams.

Molar Mass Ag_2CO_3 = 2(107.8682 g Ag) + 12.0107 g C + 3(15.9994 g O) = 275.753 g/mol

Find the mass Ag_2CO_3: $0.0246\ \text{mol}\ Ag_2CO_3 \times \dfrac{275.7453\ \text{g}\ Ag_2CO_3}{1\ \text{mol}\ Ag_2CO_3} = 6.79\ \text{g}\ Ag_2CO_3$

Molar Mass $NaNO_3$ = 22.9898 g Na + 14.0067 g N + 3(15.9994 g O) = 84.9947 g/mol

Find the mass $NaNO_3$: $0.0246\ \text{mol}\ Ag_2CO_3 \times \dfrac{2\ \text{mol}\ NaNO_3}{1\ \text{mol}\ Ag_2CO_3} \times \dfrac{84.9947\ \text{g}\ NaNO_3}{1\ \text{mol}\ NaNO_3} = 4.19\ \text{g}\ NaNO_3$

Find the mass Na_2CO_3 used up, then subtract from initial for mass unreacted:

$$0.0246\ \text{mol}\ Ag_2CO_3 \times \frac{1\ \text{mol}\ Na_2CO_3}{1\ \text{mol}\ Ag_2CO_3} \times \frac{105.9885\ \text{g}\ Na_2CO_3}{1\ \text{mol}\ Na_2CO_3} = 2.61\ \text{g}\ Na_2CO_3\ \text{used up}$$

12.43 g Na_2CO_3 initial – 2.61 g Na_2CO_3 used up = 9.82 g Na_2CO_3 remains unreacted

☑ *Reasonable Result Check:* The total mass before the reaction = 12.43 g + 8.37 g = 20.80 g is equal to the total mass after the reaction = 6.79 g + 4.19 g + 9.82 g = 20.80 g, in accordance with the conservation of mass.

148. *Result:* **0.28 M NaCl**

Analyze: Known volumes of two NaCl solutions with different concentrations are combined then diluted. Determine the NaCl concentration in the new solution.

Plan: Calculate the moles of NaCl from each solution, then add them and divide by the new volume to calculate the new concentration.

Execute:

$$\frac{\left(60.0\ \text{mL NaCl} \times \dfrac{1\ \text{L NaCl}}{1000\ \text{mL NaCl}} \times \dfrac{2.00\ \text{mol NaCl}}{1\ \text{L NaCl}}\right) + \left(40.0\ \text{mL NaCl} \times \dfrac{1\ \text{L NaCl}}{1000\ \text{mL NaCl}} \times \dfrac{0.500\ \text{mol NaCl}}{1\ \text{L NaCl}}\right)}{500.0\ \text{mL} \times \dfrac{1\ \text{L}}{1000\ \text{mL}}}$$

$$= 0.28\ \text{M NaCl}$$

150. *Result:* **Combine the solutions pair-wise and look for bubbles from forming CO_2 and insoluble precipitates, black $Ag_2O(s)$, white AgCl, and yellow Ag_2CO_3**

Analyze: Determine the identity of four solutes using their solutions.

Plan: Systematically combine the solutions pair-wise and look for evidence of specific products resulting from reactions, such as precipitation and gas-forming.

Execute:

Systematically, start by mixing solutions of $AgNO_3$, Na_2CO_3, and NaOH with HCl:

- Mixing solutions of $AgNO_3$ and HCl will cause a precipitation reaction

  Complete Equation: $AgNO_3(aq) + HCl(aq) \longrightarrow AgCl(s) + HNO_3(aq)$

  Net Ionic Equation: $Ag^+(aq) + Cl^-(aq) \longrightarrow AgCl(s)$

  Figure 3.9 shows that AgCl is a white solid.

- Mixing solutions of Na_2CO_3 and HCl will cause a gas-forming reaction

  Complete Equation: $Na_2CO_3(aq) + 2\ HCl(aq) \longrightarrow 2\ NaCl(s) + H_2O(\ell) + CO_2(g)$

  Net Ionic Equation: $CO_3^{2-}(aq) + 2\ H^+(aq) \longrightarrow H_2O(\ell) + CO_2(g)$

  Figure 3.4 shows how acids react with carbonates to make bubbles of $CO_2(g)$.

- Mixing solutions of NaOH and HCl will cause a neutralization reaction

  Complete Equation: $NaOH(aq) + HCl(aq) \longrightarrow NaCl(s) + H_2O(\ell)$

  Net Ionic Equation: $OH^-(aq) + H^+(aq) \longrightarrow H_2O(\ell)$

  Without an acid-base indicator, this reaction has no color or phase change.

Systematically, mix solutions of Na_2CO_3 and NaOH with $AgNO_3$:

- Mixing solutions of Na_2CO_3 and $AgNO_3$ will cause a precipitation reaction

  Complete Equation: $Na_2CO_3(aq) + 2\ AgNO_3(aq) \longrightarrow Ag_2CO_3(s) + 2\ NaNO_3(aq)$

  Net Ionic Equation: $CO_3^{2-}(aq) + 2Ag^+(aq) \longrightarrow Ag_2CO_3(s)$

  The Ag_2CO_3 precipitate will be yellow. (http://en.wikipedia.org/wiki/Silver_carbonate)

- Mixing solutions of NaOH and $AgNO_3$ will cause a precipitation reaction:

  Complete Equation: $NaOH(aq) + AgNO_3(aq) \longrightarrow Ag(OH)(s) + NaNO_3(aq)$

  Net Ionic Equation: $OH^-(aq) + Ag^+(aq) \longrightarrow AgOH(s)$

  AgOH reacts to form stable solid $Ag_2O(s)$ is black. (http://en.wikipedia.org/wiki/AgOH#Preparation)

  $$2\ Ag(OH)(s) \longrightarrow H_2O(\ell) + Ag_2O(s)$$

  The solid $Ag_2O(s)$ precipitate is black.

Finally, mix solution NaOH with Na_2CO_3:

- Mixing solutions of Na_2CO_3 and NaOH will cause no reaction. The resulting solution remains colorless.

Summary:

	HCl	$AgNO_3$	Na_2CO_3	NaOH
HCl		ppt white AgCl	Bubbles of CO_2	neutralization reaction; clear, colorless solution
$AgNO_3$	ppt white AgCl		ppt yellow Ag_2CO_3	ppt black Ag_2O
Na_2CO_3	Bubbles of CO_2	ppt yellow Ag_2CO_3		No reaction; clear, colorless solution
NaOH	neutralization reaction; clear, colorless solution	ppt black Ag_2O	No reaction, clear colorless solution	

Experimental Procedure: Select a solution and make three separate samples. Add one of each of the remaining solutions to a separate sample and record visible phase changes.
Note: Each solute produces a unique set of results:

- HCl will make one mixture with a white precipitate, one with bubbles, and one colorless solution.
- $AgNO_3$ will make three different colored precipitates: yellow, white, and black.
- Na_2CO_3 will make one mixture with bubbles, one with a yellow precipitate, and one colorless solution.
- NaOH will make one mixture with a black precipitate and two colorless solutions.

153. *Result:* (a) **$AsO_4^{3-}(aq) + FeCl_3(aq) \longrightarrow 3\ Cl^-(aq) + FeAsO_4(s)$**
(b) $AsO_4^{3-}(aq) + Fe^{3+}(aq) \longrightarrow FeAsO_4(s)$ (c) 4.2 mL (d) 0.12 g

Analyze: Given a description of a chemical reaction, write a balanced chemical equation and the net ionic equation. Given the volume and concentration of a reactant in a sample and the concentration of another reactant solution, calculate the volume required for complete reaction and the mass of product formed.

Plan: Determine the formulas of the reactants and products, then balance the complete precipitation equation and identify ions in the solution to get the net ionic equation. Use the volume and molarity of the sample to calculate quantity (in moles) of one reactant. Use mole ratios, concentration, and molar mass to determine the volume of the other reactant and the mass of the product.

Execute: Soluble iron(III) chloride has the formula $FeCl_3$. Arsenate is analogous to phosphate PO_4^{3-}, so it has the formula AsO_4^{3-}. Iron(III) arsenate is $FeAsO_4$, which is identified as the insoluble product.

(a) Balance the chemical equation: **$AsO_4^{3-}(aq) + FeCl_3(aq) \longrightarrow 3\ Cl^-(aq) + FeAsO_4(s)$**

(b) Write the complete ionic equation: $AsO_4^{3-}(aq) + Fe^{3+}(aq) + 3\ Cl^-(aq) \longrightarrow 3\ Cl^-(aq) + FeAsO_4(s)$

Eliminate the spectator ion, Cl^-, to form the net ionic equation:

$$AsO_4^{3-}(aq) + Fe^{3+}(aq) \longrightarrow FeAsO_4(s)$$

(c) $$25.00\ \text{mL} \times \frac{1\ \text{L}}{1000\ \text{mL}} \times \frac{0.025\ \text{mol}\ AsO_4^{3-}}{1\ \text{L}} = 6.3 \times 10^{-4}\ \text{mol}\ AsO_4^{3-}$$

$$6.3 \times 10^{-4}\ \text{mol}\ AsO_4^{3-} \times \frac{1\ \text{mol}\ FeCl_3}{1\ \text{mol}\ AsO_4^{3-}} \times \frac{1\ \text{L}}{0.150\ \text{mol}\ FeCl_3} \times \frac{1000\ \text{mL}}{1\ \text{L}} = 4.2\ \text{mL}$$

(d) Molar Mass of $FeAsO_4$ = 55.845 g/mol Fe + 74.9216 g/mol As

+ 4(15.9994 g/mol O) = 194.759 g/mol $FeAsO_4$

$$6.3 \times 10^{-4}\ \text{mol}\ AsO_4^{3-} \times \frac{1\ \text{mol}\ FeAsO_4}{1\ \text{mol}\ AsO_4^{3-}} \times \frac{194.759\ \text{g}\ FeAsO_4}{1\ \text{mol}\ FeAsO_4} = 0.12\ \text{g}\ FeAsO_4$$

☑ *Reasonable Result Check:* The AsO_4^{3-} concentration in the lake water is ten times less concentrated as the iron solution, so it makes sense that a smaller volume of the $FeCl_3$ is needed and a small mass precipitated.

155. *Result:* **84.5%**

Analyze: Given a description of a chemical reaction, the mass of a tablet containing a base, and the volume and concentration of the acid used to neutralize it, determine the mass percent of the base in the window cleaner.

Plan: Balance the complete neutralization/gas-forming equation. Use the volume and molarity of the acid to calculate quantity (in moles) of the acid reactant. Use the mole ratio from the equation and molar mass to determine the mass of the base in the tablet. Calculate the mass percent using the mass of the base in the tablet and the original mass of the tablet.

Execute: The neutralization equation involves using the H^+ from the acid, HCl, to neutralize of OH^- and form $CO_2(g)$ from CO_3^{2-}.

$$NaAl(OH)_2CO_3(s) + 4\ HCl(aq) \longrightarrow NaCl(aq) + AlCl_3(aq) + 3\ H_2O(\ell) + CO_2(g)$$

Molar Mass $NaAl(OH)_2CO_3$ = 22.9898 g/mol Na + 26.9815 g/mol Al + 2(15.9994 g/mol O)

+ 2(1.0079 g/mol H) + 12.0107 g/mol C + 3(15.9994 g/mol O) = 143.9948 g/mol $NaAl(OH)_2CO_3$

Mass of $NaAl(OH)_2CO_3$ titrated:

$$27.60\ \text{mL} \times \left(\frac{1\ \text{l}}{1000\ \text{mL}}\right) \times \left(\frac{0.425\ \text{mol HCl}}{1\ \text{L}}\right) \times \left(\frac{1\ \text{mol NaAl(OH)}_2\text{CO}_3}{4\ \text{mol HCl}}\right) \times \left(\frac{143.9948\ \text{g NaAl(OH)}_2\text{CO}_3}{1\ \text{mol NaAl(OH)}_2\text{CO}_3}\right)$$

$$= 0.4223\ \text{g NaAl(OH)}_2\text{CO}_3$$

Mass percent of $NaAl(OH)_2CO_3$ in tablet:

$$\frac{0.4223\ \text{g NaAl(OH)}_2\text{CO}_3}{0.500\ \text{g tablet}} \times 100\% = 84.5\%$$

☑ *Reasonable Result Check:* It is satisfying to see that a majority of the tablet is active ingredient.

Chapter 4: Energy and Chemical Reactions

Solutions for Red-Numbered Questions for Review and Thought

Topical Questions

The Nature of Energy (Section 4-1)

9. *Result:* **(a) 399 Cal (b) 5.0 × 10⁶ J/day**

Analyze: Convert a quantity of kilojoules (provided by a piece of cake) into food Calories, and convert a quantity of food Calories into joules.

Plan: Use metric and energy conversion factors to achieve the conversions. A food Calorie is a kilocalorie.

Execute:

(a) $$1670\ \text{kJ} \times \frac{1000\ \text{J}}{1\ \text{kJ}} \times \frac{1\ \text{cal}}{4.184\ \text{J}} \times \frac{1\ \text{kcal}}{1000\ \text{cal}} \times \frac{1\ \text{Cal}}{1\ \text{kcal}} = 399\ \text{Cal}$$

(b) $$\frac{1200\ \text{Cal}}{1\ \text{day}} \times \frac{1\ \text{kcal}}{1\ \text{Cal}} \times \frac{1000\ \text{cal}}{1\ \text{kcal}} \times \frac{4.184\ \text{J}}{1\ \text{cal}} = \frac{5.0 \times 10^6\ \text{J}}{1\ \text{day}}$$

☑ *Reasonable Answer Check:* The food Calorie is about four times bigger than a kilojoule. So, the energy quantity in Calories should be about four times smaller than in kilojoules.

11. *Result:* **1.04 × 10⁴ J**

Strategy and Explanation: Given the calories required to melt one gram of lead, determine the number of joules needed to melt a larger mass of lead.

Plan: Start with the sample, the mass of the lead. Use metric and energy conversion factors to achieve the conversions.

Execute:
$$454\ \text{g} \times \frac{5.50\ \text{cal}}{1\ \text{g}} \times \frac{4.184\ \text{J}}{1\ \text{cal}} = 1.04 \times 10^4\ \text{J}$$

☑ *Reasonable Answer Check:* A larger mass of lead will require more energy, and the smaller unit of joules will also make the answer larger in size, so it makes sense that the result ends up large.

13. *Result:* **4 × 10⁶ J, $0.03**

Analyze: Determine the number of joules in a kilowatt-hour and the cost of a megajoule of electricity given the cost per kilowatt-hour.

Plan: Use metric and energy conversion factors to achieve the conversions.

Execute:
$$1\ \text{kW - hr} \times \frac{1000\ \text{W}}{1\ \text{kW}} \times \frac{1\ \frac{\text{J}}{\text{s}}}{1\ \text{W}} \times \frac{3600\ \text{s}}{1\ \text{hr}} = 3.60 \times 10^6\ \text{J} = \mathbf{4 \times 10^6\ J}$$

$$1\ \text{MJ} \times \frac{10^6\ \text{J}}{1\ \text{MJ}} \times \frac{1\ \text{kW - hr}}{3.60 \times 10^6\ \text{J}} \times \frac{\$0.09}{1\ \text{kW - hr}} = \$0.03$$

☑ *Reasonable Answer Check:* A kilowatt-hour is much bigger than a joule. So, the energy quantity in kilojoules should be much smaller than in kW-hr. A megajoule is similar in size but smaller than a kilowatt-hour, so the cost would be slightly lower.

Conservation of Energy (Section 4-2)

15. *Result/Explanation:*

(a) In the process of lighting the match, the kinetic energy of moving the match across the striking surface is converted into thermal energy due to friction. This thermal energy causes the match to "light." This "lighting" is the result of a combustion process, where the chemical energy stored in the reactants is converted into heat energy and light energy. The heat energy of the match is used to light the fuse, and the chemical energy of the fuse is converted into heat energy and light energy. The heat from the fuse ignites the chemical propellants in the rocket. The chemical energy here is converted into heat, light, and the kinetic energy and potential energy of the rocket as the rocket's speed and altitude increase. When the rocket explodes, more chemical energy is converted into light, heat, and kinetic energy.

(b) As the fuel is pumped from the underground storage tank, its potential energy is increased by the mechanical energy of the pump. By using the fuel to drive 25 miles, some of the chemical potential energy stored in the fuel is converted into kinetic energy that moves the car and into heat energy that warms the engine and passenger compartment, when necessary.

17. *Result/Explanation:* Potential energy of the water is converted to kinetic energy as it falls, then is converted to heat energy when it reaches the bottom of the water fall warming the water slightly.

Energy Transfer (Section 4-3)

19. *Analyze:* Describe and explain your choice of the system and the surroundings, describe transfer of energy and materials into and out of the system, and determine if the process is exothermic or endothermic.

Plan: The system is identified as precisely what we are studying. The surroundings are everything else. The important aspects of the surroundings are usually those things in contact with the system or in close proximity. The process is exothermic if the system loses energy. The process is endothermic if the system gains energy.

Execute:

(a) The System: NH_4Cl

This choice was made following the discussion in Section 4-3 when describing a reaction, "the system is usually defined as all the atoms that make up the reactants." Here the equation for the reaction being studied is: $NH_4Cl(s) \longrightarrow NH_4^+(aq) + Cl^-(aq)$

The Surroundings: Anything not NH_4Cl, including the water.

The choice to consider water as part of the surroundings instead of part of the system is based upon the fact that it is neither a reactant nor a product in the given dissolving equation. Although water undeniably has a strong interaction with the products of the reaction, there are still no H_2O molecules in this equation. Defining the system to include the water is also feasible. Data collected during dissolving experiments will not actually be able to functionally separate the water from the material dissolving.

(b) To study the release of energy during the phase change of this ionic compound, we must isolate it and see how it interacts with the surroundings.

(c) The system's interaction with the surroundings causes heat energy to be transferred into the surroundings and out of the system. There is no material transfer in this process, but there is a change in the specific interaction between the water and system.

(d) Since changes in the system cause energy to be gained by the system, the process is endothermic.

☑ *Reasonable Answer Check:* This definition of the system is restrictive enough to allow us to learn more about the relationship between the energy required to break the ionic bonds in the solid and the energy involved with the products' increased interaction with the surrounding solvent molecules.

21. *Result:* **(see diagram below), ΔE_{system} = 715.6 kJ**

Analyze: Make a heat flow diagram for the given heat energy and work energy changes in a system and use that to help determine the change in energy of the system ΔE_{system}.

Plan: Work energy (w) and heat energy (q) cross the boundary between the system and the surroundings. In chemistry, both q and w are both positive values, if energy flows from the surroundings into the system. Alternatively, q and w are both negative values, if energy flows out of the system into the surroundings. ΔE is calculated using Equation 4.1: $\Delta E = q + w$.

Execute:

Work is done on the surroundings, so energy flows out of the system and w is negative: $w = -127.6$ kJ

Heat energy is transferred into the system, so energy flows into the system and q is positive: $q = 843.2$ kJ

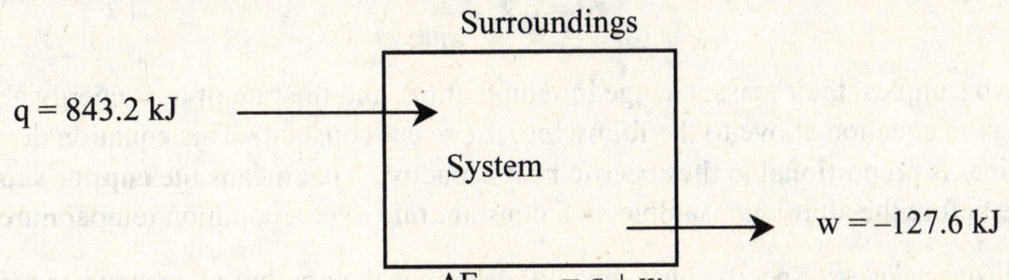

$$\Delta E_{system} = q + w$$

$$\Delta E_{system} = 843.2 \text{ kJ} + (-127.6 \text{ kJ}) = 715.6 \text{ kJ}$$

☑ *Reasonable Answer Check:* More heat energy is entering the system than work energy leaving the system, so it makes sense that the ΔE is positive.

Heat Capacity (Section 4-4)

23. *Result:* **Process (a)**

Analyze: Given the mass and temperature change for two samples, determine which process requires a greater transfer of energy.

Plan: Get the specific heat capacity of water from Table 4.1, then use Equation 4.2' to calculate the heat energy required in each scenario.

Execute: Table 4.1 gives the specific heat capacity of liquid water as 4.184 J $g^{-1}°C^{-1}$.

$$q = c \times m \times \Delta T.$$

(a) $\Delta T = T_f - T_i = 50\text{ °C} - 20\text{°C} = 30\text{ °C}$ *(must be rounded to tens place)*

$$q_{H_2O} = (4.184 \text{ J g}^{-1}\text{°C}^{-1}) \times (10.0 \text{ g}) \times (30 \text{ °C}) = 1 \times 10^3 \text{ J}$$

(b) $\Delta T = T_f - T_i = 37\text{ °C} - 25\text{°C} = 12\text{ °C}$

$$q_{Cu} = (0.385 \text{ J g}^{-1}\text{°C}^{-1}) \times (20.0 \text{ g}) \times (12 \text{ °C}) = 92 \text{ J}$$

The cooling described in (a) requires a greater transfer of energy than that described in (b).

☑ *Reasonable Answer Check:* The specific heat capacity of water is more that ten times the specific heat capacity of copper and the mass and the temperature change for the copper sample was also smaller. The water sample should thus require a much greater transfer of energy.

25. *Result:* **Copper. The specific heat capacity for Al is greater than for Cu, so more energy is required to heat the Al by 1°C and the Cu warms faster.**

Analyze: Given the mass and initial temperature of two samples made out of two different metals, also given the assumption that heat energy is transferred at the same rate when held in your hand, determine which metal will warm up to body temperature faster.

Plan: First, look up the specific heat capacities of the substances being heated in Table 4.1. The rate of absorption of heat energy can be related to heat capacity, mass, temperature increase, and elapsed time using Equation 4.2. The elapsed time can then be related to the specific heat capacity, the mass, the change in temperature, and the rate of absorption of heat energy. The fastest elapsed time will reach body temperature first.

Execute: Table 4.1 gives the specific heat capacities (c): c_{Al} = 0.902 J g^{-1}°C^{-1} and of c_{Cu} = is 0.385 J g^{-1}°C^{-1} from Table 4.1. Use Equation 4.2, then divide by Δt to write an equation relating the rate of heat energy absorption to the specific heat capacity, mass, temperature increase, and elapsed time:

$$\text{rate of heat energy absorption} = \frac{q}{\Delta t} = c \times m \times \frac{\Delta T}{\Delta t}$$

Solve this equation for the elapsed time, Δt:

$$\Delta t = c \times \frac{m \times \Delta T}{\text{rate}}$$

Comparing the two samples, their mass, change in temperature, and the rate of heat energy absorption are all the same, simplifying the equation above to the following: Δt = c × constant. This equation demonstrates that the elapsed heating time is proportional to the specific heat capacity. That means the **copper sample** will warm to body temperature before the aluminum sample, at a constant rate over a common temperature range.

☑ *Reasonable Answer Check:* Specific heat capacity describes the amount of energy needed to increase the temperature of a 1-gram sample by 1 °C. Since these samples have the same mass and temperature change, the object with the smaller specific heat capacity will be faster to heat.

27. *Result:* **136 J mol^{-1} K^{-1}**

Analyze: Given the specific heat capacity of a known substance, determine the molar heat capacity.

Plan: Calculate the molar mass of C_6H_6, and use that to calculate the molar heat capacity

Execute: The molar mass of C_6H_6 = 6 (12.0107 g/mol C) + 6 (1.0079 g/mol H) = 78.1116 g/mol C_6H_6

$$1.74 \frac{\text{J}}{\text{g K}} \times \frac{78.1116 \text{ g}}{1 \text{ mol}} = 136 \frac{\text{J}}{\text{mol K}}$$

☑ *Reasonable Answer Check:* The molar mass is larger than 1 gram, so it makes sense that the molar heat capacity is larger than the specific heat capacity.

29. *Result:* **0.25 J g^{-1} °C^{-1}**

Analyze: Given the masses and initial temperatures of a piece of metal and a quantity of water, and given the final temperature of the water after the metal is added, determine the specific heat capacity of the metal.

Plan: Thermal equilibrium is reached when the water and the metal reach the same temperature, so we know the final temperature of the metal is the same as the final temperature of the water. Use Equation 4.2' (q = c × m × ΔT) to calculate the heat gained by the water, which also represents the heat lost by the metal. Use Equation 4.2 to calculate the specific heat capacity of the metal.

Execute: Table 4.1 gives the specific heat capacity (c) of water as 4.184 J g^{-1}°C^{-1}.

$$\Delta T = T_f - T_i = 14.9\ °\text{C} - 10.0\ °\text{C} = 4.9\ °\text{C}$$

$$q_{water} = (4.184 \text{ J g}^{-1}°\text{C}^{-1}) \times (244 \text{ g}) \times (4.9\ °\text{C}) = 5.0 \times 10^3 \text{ J (two sig figs)}$$

$$q_{water} = -q_{Mo}$$

$$c_{Mo} = \frac{q_{Mo}}{m_{Mo} \times \Delta T} = \frac{-5.0 \times 10^3 \text{ J}}{237 \text{ g} \times \left(14.9\ °\text{C} - 100.0\ °\text{C}\right)} = 0.25 \frac{\text{J}}{\text{g °C}}$$

☑ *Reasonable Answer Check:* Because water has a much larger specific heat capacity than Mo, we expect a smaller temperature increase in the water than the decrease experienced by the molybdenum. Most metals have a heat capacity in this range.

31. *Result:* **330. °C**

Analyze: Given the mass of a piece of hot metal, the mass and temperature of a sample of cool water, and the final temperature after the two are combined and heat energy is transferred, determine the initial temperature of the metal.

Plan: First, look up the specific heat capacities of the metal and the water. Assuming that the temperature stopped dropping once the water and the metal reached the same temperature, set the final temperature of the metal to be the same as the final temperature of the water. Use Equation 4.2' to calculate the heat energy gained by the water from the metal. The heat energy gained by the water is the heat energy lost by the metal. Finally, rearrange Equation 4.2 to calculate the initial temperature of the metal.

Execute: Table 4.1 gives the specific heat capacities (c): $c_{Fe} = 0.451\ J\ g^{-1}°C^{-1}$ and $c_{water} = 4.184\ J\ g^{-1}°C^{-1}$.

$$\Delta T_{water} = T_f - T_i = 32.8\ °C - 20.0\ °C = 12.8\ °C$$

$$q_{water} = (4.184\ J\ g^{-1}°C^{-1}) \times (1.00\ kg) \times \left(\frac{1000\ g}{1\ kg}\right) \times (12.8\ °C) = 5.36 \times 10^4\ J$$

Heat energy is gained by the water, so q_{water} is positive. Heat energy is lost by the iron, so q_{iron} is negative. The quantity of heat energy lost by the hot iron is absorbed by the cold water: $q_{iron} = -q_{water} = -5.36 \times 10^4\ J$

$$\Delta T_{iron} = \frac{q_{iron}}{c \times m} = \frac{-5.36 \times 10^4\ J}{0.451\ J\ g^{-1}°C^{-1} \times 400.\ g} = -297\ °C$$

$$T_i = T_f - \Delta T_{iron} = 32.8\ °C - (-297\ °C) = 330.\ °C$$

☑ *Reasonable Answer Check:* Because water has a much larger specific heat capacity than iron, a smaller T increase in the water than T decrease in the iron.

33. *Result:* **Gold**

Analyze: Given the mass, initial and final temperatures, and energy needed to heat an unknown element, determine its most probable identity using Table 4.1.

Plan: Use Equation 4.2, then compare to elements in Table 4.1:

Execute:

$$c = \frac{q}{m \times \Delta T} = \frac{34.7\ J}{23.4\ g \times (28.9\ °C - 17.3\ °C)} = 0.128\ \frac{J}{g\ °C}$$

Looking at Table 4.1, the element whose specific heat capacity is closest to this is Au.

☑ *Reasonable Result Check:* The specific heat capacity of Au ($0.128\ J\ g^{-1}°C^{-1}$) matches the calculated value to three significant figures ($0.128\ J\ g^{-1}°C^{-1}$). The similarity in the two values gives us confidence in the answer.

Energy and Enthalpy (Section 4-5)

35. *Result:* **$\Delta T_{surroundings}$ = positive, ΔE_{system} = negative**

Analyze: Given the direction of transfer of thermal energy between a system and the surroundings with no work done, determine the algebraic sign of $\Delta T_{surroundings}$ and ΔE_{system}.

Plan: If thermal energy enters the surroundings as heat energy, then the temperature in the surroundings, $T_{surroundings}$, rises. If thermal energy enters the system as heat energy, then the temperature in the surroundings, $T_{surroundings}$, drops. If the system gets energy from the surroundings, then the internal energy of the system, E_{system}, rises. If the surroundings get energy from the system, then the internal energy of the system, E_{system}, lowers.

Execute: The thermal energy enters the surroundings as heat energy, so the temperature in the surroundings, $T_{surroundings}$, rises:

$$T_{f,surroundings} > T_{i,surroundings}$$

$$\Delta T_{surroundings} = T_{f,surroundings} - T_{i,surroundings} = \text{positive}$$

Here, the surroundings gets energy from the system, so the internal energy of the system, E_{system}, lowers:

$$E_{f,system} < E_{i,system}$$

$$\Delta E_{system} = E_{f,system} - E_{i,system} = \text{negative}$$

☑ *Reasonable Result Check:* Energy leaves the system, so a negative ΔE_{system} makes sense. The energy must show up in the surroundings, so the fact that these two algebraic signs are opposite signs also makes sense.

37. *Result:* **9.98 kJ**

Analyze: Given the moles of a sample of mercury and the heat of fusion of mercury at the freezing point, determine the quantity of energy transferred when freezing the sample at the freezing point.

Plan: Calculate the mass, then use the heat of fusion to find energy.

Execute: $$4.37 \text{ mol Hg} \times \frac{200.59 \text{ g Hg}}{1 \text{ mol Hg}} \times \frac{2.72 \text{ cal}}{1 \text{ g}} = 2.38 \times 10^3 \text{ cal}$$

$$2.38 \times 10^3 \text{ cal} \times \frac{4.184 \text{ J}}{1 \text{ cal}} \times \frac{1 \text{ kJ}}{1000 \text{ J}} = 9.98 \text{ kJ}$$

39. *Result:* **273 J**

Analyze: Given the volume and initial temperature of a liquid substance, the freezing point of the liquid, the final temperature of the solid substance, the density of the liquid, the specific heat capacity, and the enthalpy of fusion, determine the thermal energy (in joules) that must be released to the surroundings to complete the transition.

Plan: Use Equation 4.2', determine the heat energy that must be released when lowering the temperature to the freezing point. Then use the enthalpy of fusion ($\Delta_{fus}H$) to determine the thermal energy that must be released from the liquid to form the solid at the freezing point.

Execute: Define the system as the liquid mercury.

$$m_{mercury} = (1.00 \text{ mL}) \times \frac{1 \text{ cm}^3}{1 \text{ mL}} \times \frac{13.6 \text{ g Hg}}{1 \text{ cm}^3} = 13.6 \text{ g Hg}$$

To change the temperature of the system, use $q = c \times m \times \Delta T$

$$q_{T\text{-drop}} = c_{mercury} \times m_{mercury} \times \Delta T_{mercury} = (0.140 \text{ J g}^{-1}{}^{\circ}\text{C}^{-1}) \times (13.6 \text{ g Hg}) \times (-38.8 \,^{\circ}\text{C} - 23.0 \,^{\circ}\text{C}) = -117.7 \text{ J}$$

To change a phase from liquid to solid in the system, use $q_{freeze} = -q_{fusion} = -m \times \Delta_{fus}H$.

$$q_{freeze} = -m_{mercury} \times \Delta_{fus}H_{mercury} = -(13.6 \text{ g Hg}) \times (11.4 \text{ J/g}) = -155 \text{ J}$$

The sum gives the total heat energy required.

$$q_{total} = q_{T\text{-drop}} + q_{freeze} = (-117.7 \text{ J}) + (-155 \text{ J}) = -273 \text{ J}$$

The thermal energy that must be *transferred to the surroundings* (–) is **273 J**.

☑ *Reasonable Result Check:* To make the final product, the liquid's temperature needed to be lowered, and then the liquid needed to be frozen. Both of these changes require energy to be removed from the system.

41. *Result:* **(a) negative (b) positive**

Analyze and Plan: Work done when gas volumes change is sometimes referred to as PV work as shown in Section 4-5b, $w = -P\Delta V$ at constant pressure. When chemical reactions involving gases occur, there may be noticeable changes in volume because gas densities are typically much smaller than those of condensed states. Therefore, we will focus here on the consumption or production of gases in the reaction.

Execute:

(a) This reaction produces gaseous H_2S from condensed-phase reactants, so the volume of the system is greater than that of the reactants at fixed temperature and pressure. Work is done on the surroundings as the system's volume expands, so the sign of w is **negative**.

(b) This reaction consumes five moles of O_2 when it produces three moles of CO_2. All other reactants and products are condensed phases. This means the volume of the products is less than that of the reactants at fixed temperature and pressure. Work is done by the surroundings as a result of the volume decrease, so the sign of w is **positive**.

43. *Result:* **49.3 kJ**

Analyze: Given the mass of a sample of a compound, the compounds formula, and the enthalpy of vaporization for a compound at the boiling point, determine the energy required to vaporize the sample at the same temperature.

Plan: Use the molar mass of C_6H_6 and the enthalpy of vaporization as unit conversion factors.

Execute: Molar mass of C_6H_6 = 6 (12.0107 g/mol C) + 6 (1.0079 g/mol H) = 78.1116 g/mol C_6H_6

$$125 \text{ g} \times \frac{1 \text{ mol}}{78.1116 \text{ g}} \times \frac{30.8 \text{ kJ}}{1 \text{ mol}} = 49.3 \text{ kJ}$$

☑ *Reasonable Result Check:* This quantity of heat seems reasonable.

Reaction Enthalpies for Chemical Reactions (Section 4-6)

45. *Result:* **endothermic**

Analyze: Given a thermochemical expression, determine if the reaction is exothermic or endothermic.

Plan and Execute: Find the sign of $\Delta_r H$, and use that to determine if the reaction is exothermic or endothermic. From Section 4-5a: when $\Delta_r H$ is negative, the reaction is exothermic and when ΔH is positive, the reaction is endothermic.

$\Delta_r H$ = +20.5 kJ, so this reaction is **endothermic.**

☑ *Reasonable Result Check:* Endothermic reactions have positive $\Delta_r H$.

47. *Result:* $\mathbf{CaO(s) + H_2O(\ell) \longrightarrow Ca^{2+}(aq) + 2\ OH^-(aq)}$**; exothermic**

Analyze: Given the reactants of a chemical reaction and an observation of temperature change, write a balanced chemical equation for the reaction and indicate if it is exothermic or endothermic.

Plan and Execute: The reactant CaO contains a base that reacts with water molecules to make hydroxide ions.

$$\mathbf{CaO(s) + H_2O(\ell) \longrightarrow Ca^{2+}(aq) + 2OH^-(aq)}$$

The reaction mixture gets hot, meaning that heat energy is being transferred to the surroundings. That means the reaction is **exothermic**.

☑ *Reasonable Result Check:* Acid-base reactions often give off heat. The exothermic themochemical expression for this reaction is given in Question 64.

49. *Result:* **(a) – 2.1 × 10² kJ (b) –33 kJ**

Analyze: Given a thermochemical expression for a phase change, determine the quantity of energy transferred to the surroundings when two different samples undergo that phase change.

Plan: When necessary, convert the sample quantity into moles. Then use the thermochemical expression to create a unit conversion factor relating energy to moles of reactant.

Execute: Reversing the given balanced thermochemical expression to show the freezing reaction, causes a change of sign of the given $\Delta_r H°$:

$$H_2O(\ell) \longrightarrow H_2O\ (s) \qquad \Delta_{freezing}H° = -6.0 \text{ kJ}$$

The thermochemical expression tells us when 1 mol $H_2O(\ell)$ is frozen, that 6.0 kJ are transferred to the surroundings.

(a) $$34.2\ \text{mol}\ H_2O(s) \times \frac{-6.0\ \text{kJ}}{1\ \text{mol}\ H_2O(s)} = -2.1 \times 10^2\ \text{kJ}$$

(b) Molar mass of H_2O = 2 (1.0079 g/mol H) + 15.9994 g/mol O = 18.0152 g/mol H_2O

$$100.0\ \text{g}\ H_2O(\ell) \times \frac{1\ \text{mol}\ H_2O(s)}{1\ \text{mol}\ H_2O(\ell)} \times \frac{1\ \text{mol}\ H_2O(s)}{18.0152\ \text{g}\ H_2O(s)} \times \frac{-6.0\ \text{kJ}}{1\ \text{mol}\ H_2O(s)} = -33\ \text{kJ}$$

☑ *Reasonable Result Check:* The freezing reaction is exothermic, since heat energy must be removed from the reactants to make the products. Also, 34 mol of water weigh more than 100 grams, so it makes sense that the answer in (a) is larger than the answer in (b).

51. *Result:* **(a) 4 mol rxn (b) 0.115 mol rxn (c) 0.0788 mol rxn**

Analyze: Given a chemical equation, determine the amount of reaction (in moles) for specific conditions.

Plan: When necessary, convert the known quantity into moles. Then use mole ratios from the balanced equation.

Execute: The equation says that exactly 2 mol NO(g) and ½ mol $O_2(g)$ react to make exactly 1 mol $N_2O_3(g)$.

(a) $$2\ \text{mol}\ O_2(g) \times \frac{1\ \text{mol rxn}}{\frac{1}{2}\ \text{mol}\ O_2(g)} = 4\ \text{mol rxn}$$

(b) $$0.115\ \text{mol}\ N_2O_3 \times \frac{1\ \text{mol rxn}}{1\ \text{mol}\ N_2O_3} = 0.115\ \text{mol rxn}$$

(c) Molar Mass NO = 14.0067 g N + 15.9994 g O = 30.0061 g/mol

$$4.73\ \text{g NO} \times \left(\frac{1\ \text{mol NO}}{30.0061\ \text{g NO}}\right) \times \left(\frac{1\ \text{mol rxn}}{2\ \text{mol NO}}\right) = 0.0788\ \text{mol rxn}$$

☑ *Reasonable Result Check:* The larger samples have greater amount of reaction.

53. *Result:* **–3.3 × 10⁴ kJ**

Analyze: Given a thermochemical expression for a reaction and a specific volume of a liquid reactant and its density, determine the enthalpy change for the reaction.

Plan: Convert the volume to mass, then the mass into moles. Then use the thermochemical expression to create a unit conversion factor relating energy to moles of reactants or products.

Execute: The balanced thermochemical expression says: burning exactly 2 mol isooctane results in the evolution of 10,922 kJ of energy at constant pressure.

Molar mass of C_8H_{18} = 8(12.0107 g/mol C) + 18(1.0079 g/mol H) = 114.2278 g/mol C_8H_{18}

$$1.00\ \text{L}\ C_8H_{18} \times \frac{1000\ \text{mL}}{1\ \text{L}} \times \frac{0.69\ \text{g}}{1\ \text{mL}\ C_8H_{18}} \times \frac{1\ \text{mol}\ C_8H_{18}}{114.2278\ \text{g}\ C_8H_{18}} \times \frac{10922\ \text{kJ}}{2\ \text{mol}\ C_8H_{18}} = 3.3 \times 10^4\ \text{kJ evolved}$$

The reaction is exothermic, so the enthalpy change is -3.3×10^4 kJ.

☑ *Reasonable Result Check:* The combustion of a fuel should produce a large amount of energy.

55. *Result:* **-1.45×10^3 kJ/mol**

Analyze: Given a chemical equation for the combustion of a fuel, the mass of fuel burned, and the thermal energy evolved at constant pressure for the reaction, determine the molar enthalpy of combustion of the fuel.

Plan: Combustion enthalpy ($\Delta_c H°$) is identical to the thermal energy released at constant pressure per mol of substance. So, convert the mass into moles, then divide the thermal energy by the moles to get the molar enthalpy of combustion.

Execute: q = –3.62 kJ (It is negative since heat energy is evolved, rather than absorbed.)

Molar mass C_2H_5OH = 2(12.0107 g/mol C) + 6(1.0079 g/mol H) + 15.9994 g/mol O

= 46.0682 g/mol C_2H_5OH

$$n = 0.115 \text{ g } C_2H_5OH \times \frac{1 \text{ mol } C_2H_5OH}{46.0682 \text{ g } C_2H_5OH} = 0.00250 \text{ mol } C_2H_5OH$$

$$\Delta_c H° = \frac{q}{n} = \frac{-3.62 \text{ kJ}}{0.00250 \text{ mol}} = -1.45 \times 10^3 \text{ kJ/mol}$$

☑ *Reasonable Result Check:* Ethanol is used as a fuel, so it makes sense that a large amount of heat energy is evolved.

57. *Result:* **2.35 × 10³ kJ**

Analyze: Given a thermochemical expression for a reaction and a specific volume of a liquid product and its density, determine how much energy is transferred out of the system during its production.

Plan: Convert the volume to mass, then the mass into moles. Then use the thermochemical expression to create a unit conversion factor relating energy to moles of reactants or products.

Execute: The balanced thermochemical expression gives the production of exactly 1 mol acetic acid that results in the evolution of 135.3 kJ of energy at constant pressure.

Molar mass of CH_3COOH =

2(12.0107 g/mol C) + 4(1.0079 g/mol H) + 2(15.9994 g/mol O) = 60.0518 g/mol CH_3COOH

$$1.00 \text{ L } CH_3CO_2H \times \frac{1000 \text{ mL}}{1 \text{ L}} \times \frac{1.044 \text{ g } CH_3CO_2H}{1 \text{ mL } CH_3CO_2H} \times \frac{1 \text{ mol } CH_3CO_2H}{60.0518 \text{ g } CH_3CO_2H} \times \frac{-135.4 \text{ kJ}}{1 \text{ mol } CH_3CO_2H}$$

$$= -2.35 \times 10^3 \text{ kJ}$$

2.35 × 10³ kJ are transferred out (–) of the system.

☑ *Reasonable Result Check:* The production of more than 15 moles of acetic acid at about 400 kJ per mole is going to take more than 6000 kJ.

Where Does the Energy Come From? (Section 4.7)

59. *Result:* **HF**

Analyze, Plan, and Execute: Refer to the chart given above Question 59. The bond with the largest bond enthalpy is the strongest, so H–F (566 kJ/mol) is the strongest of the four hydrogen halide bonds. The others are weaker with smaller bond enthalpies: H–Cl (431 kJ/mol), H–Br (366 kJ/mol), and H–I (299 kJ/mol).

61. *Result:* **For reaction with fluorine: (a) 594 kJ (b) –1132 kJ (c) –538 kJ; For reaction with chlorine: (a) 678 kJ (b) –862 kJ (c) –184 kJ (d) Reaction of fluorine with hydrogen is more exothermic.**

Analyze: Given a table of bond enthalpies (above Question 59) and the description of two chemical reactions, determine for each reaction (a) the enthalpy change for breaking all the bonds in the reactants, (b) the enthalpy change for forming all the bonds in the products, (c) the enthalpy change for the reaction, and (d) determine which reaction is most exothermic.

Plan: Balance the equations and determine how many moles of bonds are broken and formed. Use each bond's bond enthalpy ($\Delta_b H$) as a conversion factor to determine energy per bond type, then add up all the energies.

Execute: The two balanced chemical equations look like this:

$$H_2 + F_2 \longrightarrow 2\ HF \qquad H_2 + Cl_2 \longrightarrow 2\ HCl$$

(a) In both reactions, one H–H bond and one halogen–halogen bond is broken.

$\Delta_r H_{reactants} = (\Delta_b H_{H\text{–}H}) + (\Delta_b H_{halogen\text{–}halogen})$

For fluorine, $\Delta H_{reactants} = (436\ kJ/mol) + (158\ kJ/mol) = 594\ kJ$

For chlorine, $\Delta H_{reactants} = (436\ kJ/mol) + (242\ kJ/mol) = 678\ kJ$

(b) In both reactions, two H–halogen bonds form. The enthalpy of forming a bond is the opposite sign of the enthalpy for breaking a bond, so we add a minus sign in the equation.

$$\Delta_r H_{products} = -2\ (\Delta_b H_{H\text{–}halogen})$$

For fluorine, $\Delta_r H_{products} = -2\ (566\ kJ/mol) = -1132\ kJ$

For chlorine, $\Delta_r H_{products} = -2\ (431\ kJ/mol) = -862\ kJ$

(c) To get the enthalpy change for the reaction, add the enthalpy change of the reactants to the enthalpy change of the products

$$\Delta_r H_{total} = \Delta_r H_{reactants} + \Delta_r H_{products}$$

For fluorine, $\Delta_r H_{total} = (594\ kJ) + (-1132\ kJ) = -538\ kJ$

For chlorine, $\Delta_r H_{total} = (678\ kJ) + (-862\ kJ) = -184\ kJ$

(d) The reaction of fluorine with hydrogen is more exothermic (–538 kJ is more negative) than the reaction of chlorine with hydrogen (–184 kJ is less negative).

☑ *Reasonable Result Check:* We expect the fluorine to be more reactive than chlorine, so more energy would be released.

63. *Result/Explanation:* The breaking of a C-C bond is **endothermic**. Separating two atoms that are bonded together requires a transfer of energy into the system, because work must be done against the force holding the pair of atoms together, as described in Section 4-7a.

Measuring Reaction Enthalpies: Calorimetry (Section 4-8)

64. *Result:* **23.9 °C**

Analyze: Given an exothermic chemical reaction, the mass of a reactant, and the mass and temperature of a sample of water, determine the final temperature after the two are combined causing the reaction to occur.

Plan: Use the mass and molar mass of the reactant to calculate the moles of reactant. Heat energy from an exothermic reaction raises the temperature of the water. Use the balanced thermochemical expression to set up a relationship between the quantity of reactant (in moles) to the quantity of heat transferred to water. Then, calculate the amount of heat energy transferred to the water during the reaction. Finally, relate ΔT to $T_f - T_i$.

Execute: Calculate the heat absorbed by the water:
Molar mass of CaO = (40.078 g/mol Ca) + (15.9994 g/mol O) = 56.077 g/mol CaO

$$q_{water} = 0.100\ g\ CaO \times \frac{1\ mol\ CaO}{56.077\ g\ CaO} \times \frac{81.9\ kJ\ transfered\ to\ water}{1\ mol\ CaO} \times \frac{1000\ J}{1\ kJ} = 146\ J$$

The specific heat capacity (c) of water is 4.184 J $g^{-1}°C^{-1}$ according to Table 4.1.

Rearrange Equation 4.2 to solve for ΔT_{water}

$$\Delta T_{water} = \frac{q_{water}}{m_{water} \times c_{water}} = \frac{146\ \text{J}}{125\ \text{g} \times \left(4.184\ \frac{\text{J}}{\text{g}\ ^\circ\text{C}}\right)} = 0.279\ ^\circ\text{C}$$

$$\Delta T_{water} = T_f - T_i$$

$$T_f = 23.6\ ^\circ\text{C} + (0.279\ ^\circ\text{C}) = 23.9\ ^\circ\text{C}$$

☑ *Reasonable Result Check:* The final temperature is greater than initial temperature because the reaction is exothermic.

66. *Result:* **(a) 1.4×10^4 J transferred (other answers are possible depending on the choice of system)**

(b) – 42 kJ/mol

Analyze: Given the mass of a soluble ionic solid and the volume and temperature of a sample of water, determine the heat energy transfer from the system to the surroundings and the reaction enthalpy ($\Delta_r H°$) by calculating the energy change per mol of ionic solid.

Plan: Define the system as the NaOH, with $\Delta_{dissolving}H = -q_{water}$ at constant pressure. Use the temperature of the surroundings (the water solution) and the specific heat capacity of water solution to calculate q_{water}. First, look up the specific heat capacity of water. Use Equation 4.2', find heat energy gained by the water. Heat energy from the reaction is used to raise the temperature of the water, so relate the heat energy gained by the water to that lost in the reaction, $\Delta_{dissolving}H$. Finally, divide the $\Delta_{dissolving}H$ by the mass of the ionic compound and convert grams to moles to obtain $\Delta_r H°$.

Execute:

(a) Table 4.1 gives the specific heat capacity (c) of water as 4.184 J g^{-1}°C^{-1}.

For a temperature change: $q = c \times m \times \Delta T$. The water represents the part of the surroundings affected by a change in the system.

$$q_{water} = c_{water} \times m_{water} \times \Delta T_{water}$$

$$\Delta T_{water} = T_{f,water} - T_{i,water} = 30.7\ ^\circ\text{C} - 22.6\ ^\circ\text{C} = 8.1\ ^\circ\text{C}$$

$$q_{water} = (4.184\ \text{J g}^{-1}{}^\circ\text{C}^{-1}) \times (400.0\ \text{mL}) \times \frac{1.00\ \text{g H}_2\text{O}}{1\ \text{mL H}_2\text{O}} \times (8.1\ ^\circ\text{C})$$

$$= 1.4 \times 10^4\ \text{J transferred from the system to the surroundings}$$

(b) Heat energy is lost by the reaction, so $q_{dissolving}$ is negative. Heat energy is gained by the water, so $q_{dissolving}$ is positive. The quantity of heat energy gained by the water is produced by the reaction.

$$q_{dissolving} = -q_{water} = -1.4 \times 10^4\ \text{J}$$

The reaction occurs at constant pressure and there is no work done, so $\Delta H = q_{dissolving}$ per mol of NaOH.

Molar mass NaOH = (22.9898 g/mol Na) + 15.9994 g/mol O + 1.0079 g/mol H = 39.9971 g/mol NaOH

$$\Delta_r H° = \frac{-1.4 \times 10^4\ \text{J}}{13.0\ \text{g NaOH}} \times \frac{39.9971\ \text{g NaOH}}{1\ \text{mol NaOH}} \times \frac{1\ \text{kJ}}{1000\ \text{J}} = -42\ \frac{\text{kJ}}{\text{mol NaOH}}$$

☑ *Reasonable Result Check:* The ionization of an ionic compound involves separating the cations and anions, then hydrating them. The enthalpy of this change could conceivably be positive or negative, but it is expect that it will be small compared to reactions where more significant bond rearrangement is happening. –42 kJ/mol is smaller (closer to zero) than the rest of the $\Delta_r H°$ values for other kinds of reactions recently studied.

Notice: A very common, more-empirical approach to solving this question is to identify the water and the salt as the system. With that definition for the system: $\Delta E_{system} = q_{dissolving} + q_{water}$. If that is the case, the answer to the question in (a) will be different. (a) If the system is insolated, $\Delta E = 0$, so there is no transfer of energy from the system to the surroundings. (b) $q_{dissolving} = -q_{water}$. Measuring the system's temperature change indicates a gain of thermal energy by the water and q_{water} is positive. Since the reaction occurs at the same time and no heat energy escapes the insolated system, it proves that the dissolving reaction produces energy, $q_{reaction}$ is negative, and the reaction is exothermic. The numerical result is the same as provided above.

68. *Result:* **$\Delta_r E$ per mole is -2.80×10^3 kJ**

Analyze: Given the mass of a reactant from a known reaction in a bomb calorimeter, the mass of water in the calorimeter, the initial and final temperatures, and the heat capacity of the bomb, determine the **ΔE** per mol of reactant.

Plan: $\Delta_r E = q + w = q + 0$ at constant volume

Use the specific heat capacity of water, the mass of water and the temperature changes, use Equation 4.2', to find heat energy gained by the water. Calculate the heat energy gained by the bomb using the heat capacity (C_{bomb}). Equate the heat energy gained, by the water and the bomb, to that lost in the reaction. Convert grams to moles of the reactant. Divide the heat by the moles of reactant to get **ΔE** per mole.

Execute: Table 4.1 gives the specific heat capacity (c) of water as 4.184 J $g^{-1}°C^{-1}$.

For a temperature change in water: $q = c \times m \times \Delta T$. For a change in temperature in the bomb with the heat capacity: $q = C_{bomb} \times \Delta T$. The bomb's heat capacity is 650 J/K = 650 J/°C, since ΔT in Kelvin is equal to ΔT in Celsius.

$$q_{total} = c_{water} \times m_{water} \times \Delta T + C_{bomb} \times \Delta T$$

$$\Delta T = T_f - T_i = 25.22\ °C - 21.70\ °C = 3.52\ °C$$

$$q_{total} = (4.184\ J\ g^{-1}°C^{-1}) \times (575\ g) \times (3.52\ °C) + (650\ J°C^{-1}) \times (3.52\ °C)$$

$$q_{total} = 1.08 \times 10^4\ J = -\,q_{reaction}$$

$$\Delta_r E = -\,1.08 \times 10^4\ J \times \frac{1\ kJ}{1000\ J} = -10.8\ kJ$$

Molar mass of $C_6H_{12}O_6$ = 6(12.0107 g/mol C) + 12(1.0079 g/mol H) + 6(15.9994 g/mol O)

= 180.1554 g/mol $C_6H_{12}O_6$

$$0.692\ g\ C_6H_{12}O_6 \times \frac{1\ mol\ C_6H_{12}O_6}{180.1554\ g\ C_6H_{12}O_6} = 3.84 \times 10^{-3}\ mol\ C_6H_{12}O_6$$

$$\Delta_r E\ \text{per mol} = \frac{-10.8\ kJ}{3.84 \times 10^{-3}\ mol\ C_6H_{12}O_6} = -2.80 \times 10^3\ \frac{kJ}{mol\ C_6H_{12}O_6}$$

☑ *Reasonable Result Check:* The change in internal energy is negative, since the reaction's energy loss causes the temperature in the water and the bomb to increase.

70. *Result/Explanation:* The experiment is described in Section 4-8a. The combustion (bomb) calorimeter apparatus is shown in Figure 4.15. A known amount of liquid $C_8H_{18}(\ell)$ is placed in the sample dish and the interior chamber is filled with oxygen gas. The interior chamber is immersed in water inside an insulated outer chamber. The sample is ignited by electric heating. The heat transfer from combustion of the sample raises the temperature of the bomb and the surrounding water. Measure the temperature change of the water and the bomb. Calculate q using the equations: $q = c_{water} \times m \times \Delta T$ and $q = C_{bomb} \times \Delta T$, then divide q by the amount (in moles) of $C_8H_{18}(\ell)$ to obtain the reaction enthalpy.

Hess's Law (Section 4-9)

72. *Result:* **$\Delta_r H° = -43.6$ kJ**

Analyze: Given two thermochemical expressions, determine the enthalpy change of a reaction.

Plan: Identify unique occurrences of the reactants and/or products in the given equations to help you determine which and how many of a given equation to use. If an equation's reactant shows in the products of the desired equation (or vice versa) then reverse the equation and change the sign of $\Delta_r H°$. If the substance has a different stoichiometric coefficient than the desired coefficient, determine an appropriate multiplier for the equation and multiply the $\Delta_r H°$ by the same multiplier.

Execute: Look at the first reactant: $C_2H_4(g)$. The only given equation that has $C_2H_4(g)$ is the first one, where $C_2H_4(g)$ is also a reactant, so we must include that equation, as written, with its given $\Delta_r H°$. Look at the second reactant: $H_2O(\ell)$. This chemical is found in both of the given reactions, so let's skip it. Look at the product: $C_2H_5OH(\ell)$. The only given equation that has $C_2H_5OH(\ell)$ is the second one, however, $C_2H_5OH(\ell)$ is a reactant, and we need it to be a product, so we must reverse that equation and change the sign of the given $\Delta_r H°$. We now have a plan to use both equations, so let's do it:

Add the equations, and eliminate reactants that also show up as products: all $O_2(g)$, all $CO_2(g)$, and two of the $H_2O(g)$ molecules. The remaining reactants and products should form the net equation for the reaction, and the sum of the individual $\Delta_r H°$ values will give us the $\Delta_r H°$ for that reaction:

$$C_2H_4(g) + \cancel{3\,O_2(g)} \longrightarrow \cancel{2\,CO_2(g)} + 2\,\cancel{H_2O(\ell)} \qquad \Delta_r H° = -1411.1\text{ kJ}$$

$$+ \quad \cancel{2\,CO_2(g)} + \cancel{3}\,H_2O(\ell) \longrightarrow C_2H_5OH(\ell) + \cancel{3\,O_2(g)} \qquad \Delta_r H° = -(-1367.5\text{ kJ})$$

$$C_2H_4(g) + H_2O(\ell) \longrightarrow C_2H_5OH(\ell) \qquad \Delta_r H° = -43.6\text{ kJ}$$

☑ *Reasonable Result Check:* The net equation adds up the equation we are looking for.

74. *Result:* **– 320. kJ/mol**

Analyze: Given a chemical reaction with unknown $\Delta_r H°$ and two thermochemical expressions, determine the enthalpy change of a named reaction for a given quantity of product.

Plan: Write the equation for the named reaction. Follow the plan described in the solution to Question 72.

Execute: Look at the first reactant: $P_4(s)$. The only given equation that has $P_4(s)$ is the first one which uses one mole of $P_4(s)$ as a reactant, so you must include that equation, as written, with its given $\Delta_r H°$. Look at the second reactant: Cl_2. This chemical is in both of the given reactions, so let's skip it. Look at the product: $PCl_3(\ell)$. The only given equation that has $PCl_3(\ell)$ is the second one, but one mole of $PCl_3(\ell)$ is a product in the equation and we need four of them. Therefore, we write the reverse of the equation with the sign of $\Delta_r H°$ changed, and multiply the equation and the $\Delta_r H°$ by four. We have now planned how we will use both the given reactions, so let's do it:

Add all the reactions as planned, eliminate reactants that also show up as products: four of the $Cl_2(g)$ and all the $PCl_5(s)$. Add all the remaining reactants and products to form the net equation for the desired reaction, and add all the individual $\Delta_r H°$ values to get the $\Delta_r H°$ for the reaction:

$$P_4(s) + \cancel{10}\,Cl_2(g) \longrightarrow 4\,\cancel{PCl_5(s)} \qquad \Delta_r H° = -1774.0\text{ kJ}$$

$$+ \quad 4\,\cancel{PCl_5(s)} \longrightarrow 4\,PbCl_3(\ell) + \cancel{4\,Cl_2(g)} \qquad + \;\Delta_r H° = -4\,(-123.8\text{ kJ})$$

$$P_4(s) + 6\,Cl_2(g) \longrightarrow 4\,PCl_3(\ell) \qquad \Delta_r H° = -1278.8\text{ kJ}$$

Formation reaction is the production of one mole of product from standard state elements.

$$\frac{1}{4}P_4(s) + \frac{3}{2}\,Cl_2(g) \longrightarrow PCl_3(\ell)$$

The given equation is four times the formation reaction. To get the enthalpy of formation, we'll divide the calculated $\Delta_r H$ by four.

$$\Delta_f H° = \frac{-1278.8\text{ kJ}}{4\text{ mol }PCl_3} = -319.7\text{ kJ}$$

$$1.00 \text{ mol } PCl_3 \times \frac{-319.7 \text{ kJ}}{1 \text{ mol } PCl_3} = -320. \text{ kJ}$$ *(Notice: we keep only 3 sig figs)*

☑ *Reasonable Result Check:* The net equation is the desired reaction. The enthalpy of formation of $PCl_3(\ell)$ is given in Appendix J as –319.7 kJ.

Standard Formation Enthalpies (Section 4-10)

76. *Result:* **(a)** $\mathbf{2\ Al\ (s) + \frac{3}{2}\ O_2\ (g) \longrightarrow Al_2O_3\ (s)\ \ \Delta_fH° = -1675.7\ kJ}$

(b) $\mathbf{Ti(s) + 2\ Cl_2\ (g) \longrightarrow TiCl_4(\ell)\ \ \Delta_fH° = -804.2\ kJ}$

(c) $\mathbf{N_2(g) + 2\ H_2\ (g) + \frac{3}{2}\ O_2\ (g) \longrightarrow NH_4NO_3(s)\ \ \Delta_fH° = -365.56\ kJ}$

(d) $\mathbf{C(s) + 2\ H_2\ (g) + \frac{1}{2}\ O_2\ (g) \longrightarrow CH_3OH(\ell)\ \ \Delta_fH° = -238.66\ k}$

Analyze, Plan, and Execute:

(a) The formation of $Al_2O_3(s)$ is written with the reactants as standard state elements, Al(s) and $O_2(g)$, and the product as one mole of standard state compound:

$$2\ Al\ (s) + \frac{3}{2}\ O_2\ (g) \longrightarrow Al_2O_3\ (s) \qquad \Delta_fH° = -1675.7\ kJ$$

(b) The formation of $TiCl_4(\ell)$ is written with the reactants as standard state elements, Ti(s) and $Cl_2(g)$, and the product as one mole of standard state compound:

$$Ti(s) + 2\ Cl_2\ (g) \longrightarrow TiCl_4(\ell) \qquad \Delta_fH° = -804.2\ kJ$$

(c) The formation of $NH_4NO_3(s)$ is written with the reactants as standard state elements, $N_2(g)$, $H_2(g)$, and $O_2(g)$, and the product as one mole of standard state compound:

$$N_2(g) + 2\ H_2\ (g) + \frac{3}{2}\ O_2\ (g) \longrightarrow NH_4NO_3(s) \qquad \Delta_fH° = -365.56\ kJ$$

(d) The formation of $CH_3OH(\ell)$ is written with the reactants as standard state elements, C(s), $H_2(g)$, and $O_2(g)$, and the product as one mole of standard state compound:

$$C(s) + 2\ H_2\ (g) + \frac{1}{2}\ O_2\ (g) \longrightarrow CH_3OH(\ell) \qquad \Delta_fH° = -238.66\ kJ$$

77. *Result:* **(a) –1675.7 kJ (b) –205.4 kJ**

Analyze, Plan, and Execute:

(a) The given equation is for the decomposition of two moles of $Al_2O_3(s)$. The formation reaction is the reverse of the decomposition. When the equation is reversed, the sign of $\Delta_rH°$ changes. The molar formation reaction produces one mole of $Al_2O_3(s)$, so divide the $\Delta_rH°$ by two.

$$2\ Al(s) + \frac{3}{2}\ O_2(g) \longrightarrow Al_2O_3(s) \qquad \Delta_fH° = \frac{1}{2}\ (-3351.4\ kJ) = -1675.7\ kJ$$

(b) Calculate the moles of $Al_2O_3(s)$ formed, then use the formation enthalpy to determine the $\Delta_fH°$ for this sample.

Molar mass of Al_2O_3 = 2(26.9815 g/mol Al) + 3(15.9994 g/mol O) = 101.9612 g/mol Al_2O_3

$$12.50\ g\ Al_2O_3 \times \frac{1\ mol\ Al_2O_3}{101.9612\ g\ Al_2O_3} \times \frac{3351.4\ kJ}{2\ mol\ Al_2O_3} = -205.4\ kJ$$

☑ *Reasonable Result Check:* The calculated formation enthalpy in (a) matches that given in Appendix J. The sample in (b) is smaller than a mole, so it makes sense that that enthalpy change is also smaller than the molar formation enthalpy.

79. *Result:* **–584 kJ**

Analyze: Given the mass of a reactant in a described reaction and the heat given off during the process, calculate the molar enthalpy of formation of the product.

Plan: Balance the chemical equation. Use the molar mass of lithium to determine the moles of lithium in the sample, then use the mole ratio from the balanced equation to determine the moles of Li_2O formed. Divide the heat by the moles of Li_2O to get the molar enthalpy of the reaction.

Execute:

$$2\ Li(s) + \frac{1}{2}\ O_2(g) \longrightarrow Li_2O(s)$$

$$3.47\ g\ Li \times \frac{1\ mol\ Li}{6.941\ g\ Li} \times \frac{1\ mol\ Li_2O}{2\ mol\ Li} = 0.250\ mol\ Li_2O$$

$$\Delta_f H° = \frac{-146\ kJ}{0.250\ mol\ Li_2O} = -584\ kJ/mol$$

☑ *Reasonable Result Check:* The calculated molar enthalpy of formation is not given in Appendix J, but a quick search on the internet gives the value – 20.01 kJ/g, which translates to -597.9 kJ/mol. This answer is in the same range, but it is not within the stated uncertainty.

81. *Result:* **(a) 178.32 kJ/mol (b) –595.2 kJ/mol (c) 44 kJ/mol**

Analyze: Given a balanced chemical equation for a reaction and a table of molar enthalpies of formation, determine the enthalpy change of the reaction.

Plan: Use the stoichiometric coefficient of the balanced equation to describe the moles of each of the reactants and products. Set up a specific version of Equation 4.10:

$$\Delta_r H° = \sum\{(\text{coefficient of product}) \times \Delta_f H°(\text{product})\} - \sum\{(\text{coefficient of reactant}) \times \Delta_f H°(\text{reactant})\}$$

Look up the $\Delta_f H°$ for each. Plug them into the equation and solve for $\Delta_r H°$.

Execute:

(a) The products are $CaO(s)$ and $CO_2(g)$. The reactant is $CaCO_3(s)$.

$$\Delta_r H° = \Delta_f H°\{CaO(s)\} + \Delta_f H°\{CO_2(g)\} - \Delta_f H°\{CaCO_3(s)\}$$

Look up the $\Delta_f H°$ values in Table 4.2 and Appendix J.

$$\Delta_r H° = (-635.09\ kJ/mol) + (-393.509\ kJ/mol) - (-1206.92\ kJ/mol) = 178.32\ kJ/mol$$

(b) The products are $HF(g)$ and $I_2(s)$. The reactants are $HI(g)$ and $F_2(g)$.

$$\Delta_r H° = 2\ \Delta_f H°\{HF(g)\} + \Delta_f H°\{I_2(s)\} - 2\ \Delta_f H°\{HI(g)\} - 2\ \Delta_f H°\{F_2(g)\}$$

Look up the $\Delta_f H°$ values in Table 4.2 and Appendix J.

$$\Delta_r H° = 2\ (-271.1\ kJ/mol) + (0\ kJ/mol) - 2\ (26.48\ kJ/mol) - 2\ (0\ kJ/mol) = -595.2\ kJ/mol$$

(c) The products are $HF(g)$ and $SO_3(g)$. The reactants are $SF_6(g)$ and $H_2O(\ell)$.

$$\Delta_r H° = 6\ \Delta_f H°\{HF(g)\} + \Delta_f H°\{SO_3(g)\} - \Delta_f H°\{SF_6(g)\} - 3\ \Delta_f H°\{H_2O(\ell)\}$$

Look up the $\Delta_f H°$ values in Table 4.2 and Appendix J.

$$\Delta_r H° = 6\ (-271.1\ kJ/mol) + (-395.72\ kJ/mol) - (-1209\ kJ/mol) - 3\ (-285.83\ kJ/mol)] = 44\ kJ/mol$$

83. *Result:* **103.6 kJ/mol**

Analyze: Given a balanced thermochemical expression for a reaction and a table of molar enthalpies of formation, determine the formation enthalpy of one of the reactants.

Plan: Use the stoichiometric coefficient of the balanced equation to describe the moles of each of the reactants and products. Set up a specific version of Equation 4.10 (shown in the solution to Question 81). Look up the

$\Delta_f H°$ for each and solve for $\Delta_f H°$.

Execute: The products are $CO_2(g)$ and $H_2O(\ell)$. The reactants are $C_8H_8(\ell)$ and $O_2(g)$.

$\Delta_r H° = 8\ \Delta_f H°\{CO_2(g)\} + 4\ \Delta_f H°\{H_2O(\ell)\} - \Delta_f H°\{C_8H_8(\ell)\} - 10\ \Delta_f H°\{O_2(g)\}$

Look up the $\Delta_f H°$ values in Table 4.2.

$-4395.0\text{ kJ/mol} = 8\ (-393.509\text{ kJ/mol}) + 4\ (-285.83\text{ kJ/mol}) - \Delta_f H°\{C_8H_8(\ell)\} - 10\ (0\text{ kJ/mol})$

Solve for $\Delta_f H°\{C_8H_8(\ell)\}$

$$\Delta_f H°\{C_8H_8(\ell)\} = +4395.0\text{ kJ} + 8\ (-393.509\text{ kJ}) + 4\ (-285.83\text{ kJ})]/\text{mol}$$

$$\Delta_f H°\{C_8H_8(\ell)\} = 103.6\text{ kJ/mol}$$

☑ *Reasonable Result Check:* Using the index of the textbook we find styrene described in Table 10.6. It includes a benzene ring, C_6H_6, attached to ethylene, $H_2C{=}CH_2$. Looking at Appendix J, $\Delta_f H°(C_6H_6) = 49.03$ kJ/mol and $\Delta_f H°(C_2H_4) = 52.26$ kJ/mol. It makes sense that the enthalpy of formation of styrene is positive and close to the sum of these two enthalpies, 101.29 kJ/mol.

85. *Result:* **41.2 kJ**

Analyze: Given the mass of a reactant, the description of a reaction, and formation enthalpies, determine how much thermal energy is transferred out of the system (at constant pressure).

Plan: Balance the equation and use it to describe the moles of each of the reactants and products. Set up a specific version of Equation 4.10 (shown in the solution to Question 81). Look up the $\Delta_f H°$ for each and solve for the reaction enthalpy ($\Delta_r H°$), which represents the thermal energy transferred out at constant pressure per mole. Convert mass to moles, and use the reaction enthalpy as a conversion factor to get thermal energy transferred out for the reactant sample.

Execute: The product is Fe_2O_3. The reactants are both elements: Fe and O_2.

$$2\ Fe(s) + \frac{3}{2}\ O_2(g) \longrightarrow Fe_2O_3(s)$$

$\Delta_r H° = \Delta_f H°\{Fe_2O_3(s)\} - 2\ \Delta_f H°\{Fe(s)\} - \frac{3}{2}\ \Delta_f H°\{O_2(g)\}$

Look up the $\Delta_f H°$ values in Table 4.2.

$$\Delta_r H° = (-824.2\text{ kJ/mol}) - 2\ (0\text{ kJ/mol}) - \frac{3}{2}\ (0\text{ kJ/mol}) = -824.2\text{ kJ/mol}$$

That means the thermochemical expression looks like this:

$$2\ Fe(s) + \frac{3}{2}\ O_2(g) \longrightarrow Fe_2O_3(s) \qquad \Delta_r H° = -824.2\text{ kJ/mol}$$

For 2 mol of Fe, 824.2 kJ needs to be transferred out. Now, we'll start the calculation for how much thermal energy is transferred out using the sample given:

$$5.58\text{ g Fe} \times \frac{1\text{ mol Fe}}{55.845\text{ g Fe}} \times \frac{824.2\text{ kJ}}{2\text{ mol Fe}} = 41.2\text{ kJ is transferred out}$$

☑ *Reasonable Result Check:* The sample is a tenth of a mole. The equation shows two mol of Fe are needed. So the resulting heat energy is half of a tenth of the original molar enthalpy.

Fuels (Section 4-11)

87. *Result:* **1.2×10^2 g CH_4**

Analyze: Determine the mass of fuel required to raise the temperature of the air in a house to a specified value, given the dimensions of rooms in the house, the molar heat capacity of air, the average molar mass of air and the density of the air.

Plan: Use the dimensions of the house, the molar mass, and the density of the air to determine the mass of air being heated. Then use Equation 4.2' to calculate the heat energy needed to raise the temperature of the air. Get the balanced equation describing the combustion of the fuel. Use Equation 4.10 (shown in the solution to Question 81), a table of formation enthalpies, and the mole ratio in the balanced equations to determine the molar enthalpy change for each combustion reaction. Finally, use that enthalpy and the molar mass as conversion factors to determine the mass of fuel needed.

Execute:

$$\text{Volume of air} = (275\ \text{m}^2) \times 2.50\ \text{m} \times \left(\frac{100\,\text{cm}}{1\ \text{m}}\right)^3 \times \frac{1\ \text{L}}{1000\ \text{cm}^3} = 6.88 \times 10^5\ \text{L}$$

$$\text{Mol of air} = 6.88 \times 10^5\ \text{L} \times \frac{1.22\ \text{g air}}{1\ \text{L air}} \times \frac{1\ \text{mol air}}{28.9\ \text{g air}} = 2.90 \times 10^4\ \text{mol air}$$

$$q_{air} = c_{air} \times n_{air} \times \Delta T_{air}$$

ΔT in Kelvin is the same as ΔT in °C.

$$q_{air} = (29.1\ \text{J mol}^{-1}\text{K}^{-1})\ (2.90 \times 10^4\ \text{mol})\ (22.0\ °\text{C} - 15.0\ °\text{C}) \times \frac{1\ \text{K change}}{1\ °\text{C change}} = 5.9 \times 10^6\ \text{J}$$

The thermochemical expression for the combustion of methane is given in Section 4-11:

$$CH_4(g) + 2\ O_2(g) \longrightarrow CO_2(g) + 2\ H_2O(g)$$

$$\Delta_rH° = \Delta_fH°\{CO_2(g)\} + 2\ \Delta_fH°\{H_2O(g)\} - \Delta_fH°\{CH_4(g)\} - 2\ \Delta_fH°\{O_2(g)\}$$

Look up the $\Delta_fH°$ values in Table 4.2.

$$\Delta_rH° = (-393.509\ \text{kJ/mol}) + 2\ (-241.818\ \text{kJ/mol}) - (-74.81\ \text{kJ/mol}) - 2\ (0\ \text{kJ/mol}) = -\ 802.34\ \text{kJ/mol}$$

Now, we'll figure out the mass needed for warming the air sample.

Molar mass of CH_4 = 12.0107 g/mol C + 4(1.0079 g/mol H) = 16.0423 g/mol CH_4

$$5.9 \times 10^6\ \text{J} \times \frac{1\ \text{kJ}}{1000\ \text{J}} \times \frac{1\ \text{mol CH}_4}{802.34\ \text{kJ}} \times \frac{16.0423\ \text{g CH}_4}{1\ \text{mol CH}_4} = 120\ \text{g CH}_4$$

☑ *Reasonable Result Check:* The quantity of air in the house is relatively large, so the amount of joules needed to raise its temperature by seven degrees is also relatively large. The enthalpy of the combustion reaction is exothermic and large, which makes sense because methane is a good fuel.

89. *Result:* **44.422 kJ/g octane > 19.927 kJ/g methanol**

Analyze: Compare the quantity of thermal energy evolved per gram when burning two different organic fuels.

Plan:

Balance the combustion equations. Then use Equation 4.10, a table of formation enthalpies, and the stoichiometry of the balanced equations to determine the combustion enthalpies. Then use the molar mass to determine enthalpy per gram of each fuel and compare them.

Execute:

The equations for the reactions describing the combustion of methanol and octane are similar to those described in Section 4-11. At standard temperature, both fuels are initially in the liquid state.

$$CH_3OH(\ell) + \tfrac{3}{2}\ O_2(g) \longrightarrow CO_2(g) + 2\ H_2O(g)$$

$$C_8H_{18}(\ell) + \tfrac{25}{2}\ O_2(g) \longrightarrow 8\ CO_2(g) + 9\ H_2O(g)$$

$$\Delta_cH°_{methanol} = \Delta_fH°\{CO_2(g)\} + 2\ \Delta_fH°\{H_2O(g)\}] - \Delta_fH°\{CH_3OH(g)\} - \tfrac{3}{2}\ \Delta_fH°\{O_2(g)\}$$

$$\Delta_cH°_{octane} = 8\ \Delta_fH°\{CO_2(g)\} + 9\ \Delta_fH°\{H_2O(g)\} - \Delta_fH°\{C_8H_{18}(\ell)\} - \tfrac{25}{2}\ \Delta_fH°\{O_2(g)\}$$

Look up the $\Delta_f H°$ values in Table 4.2 and Appendix J.

$$\Delta_c H°_{methanol} = (-393.509 \text{ kJ/mol}) + 2\,(-241.818 \text{ kJ/mol}) - (-238.66 \text{ kJ/mol}) - \frac{3}{2}\,(0 \text{ kJ/mol}) = -638.49 \text{ kJ/mol}$$

$$\Delta_c H°_{octane} = 8\,(-393.509 \text{ kJ/mol}) + 9\,(-241.818 \text{ kJ/mol}) - (-250.1 \text{ kJ/mol}) - \frac{25}{2}\,(0 \text{ kJ/mol}) = -5074.3 \text{ kJ/mol}$$

Now, calculate the enthalpy evolved per gram:

Molar mass of CH_3OH = (12.0107 g/mol C) + 4 (1.0079 g/mol H) + (15.9994 g/mol O)

= 32.0417 g/mol CH_3OH

Molar mass of C_8H_{18} = 8 (12.0107 g/mol C) + 18 (1.0079 g/mol H) = 114.2278 g/mol C_8H_{18}

$$\frac{638.49 \text{ kJ produced}}{1 \text{ mol } CH_3OH} \times \frac{1 \text{ mol } CH_3OH}{32.0417 \text{ g } CH_3OH} = 19.927 \frac{\text{kJ produced}}{\text{g } CH_3OH}$$

$$\frac{5074.3 \text{ kJ produced}}{1 \text{ mol } C_8H_{18}} \times \frac{1 \text{ mol } C_8H_{18}}{114.2278 \text{ g } C_8H_{18}} = 44.422 \frac{\text{kJ produced}}{\text{g } C_8H_{18}}$$

The enthalpy evolved per gram of octane (44.422 kJ/g) is larger than the enthalpy evolved per gram of methanol (19.927 kJ/g).

☑ *Reasonable Result Check:* Since the combustion process is an oxidation process, and the methanol is more oxidized than the hydrocarbon, it makes sense that its combustion will produce a smaller amount of energy.

91. *Result:* **2.2 hours walking**

Analyze: Determine how long you must walk to burn off the Calories of a quarter-pound hamburger.

Plan: Use the caloric value of hamburger (from Table 4.3) to determine the Calories in the hamburger. Use the basal metabolic rate (BMR) for a 70 kg person and the multiplier for exercising vigorously given in Section 4-11, determine the energy burned for that activity per hour. Then use that result to determine how many hours the person would need to play soccer.

Execute:

$$0.25 \text{ lb hamburger} \times \frac{454 \text{ g}}{1 \text{ lb}} \times \frac{3.60 \text{ Cal}}{1 \text{ g hamburger}} = 410 \text{ Cal}$$

Section 4-11 gives the following information: A 70-kg person has a BMR of 1750 Cal/day. Walking gives him a multiplier of 2.5 × BMR.

$$2.5 \times \frac{1750 \text{ Cal}}{1 \text{ day}} \times \frac{1 \text{ day}}{24 \text{ hr}} = 182 \frac{\text{Cal}}{\text{hr}}$$

$$410 \text{ Cal} \times \frac{1 \text{ hr}}{182 \text{ Cal}} = 2.2 \text{ hr}$$

☑ *Reasonable Result Check:* Every dieter knows that hamburgers (and most high Calorie junk food) require significant exercise to work off.

General Questions

93. *Result:* **4.82×10^4 J**

Analyze: Given the mass and initial temperature of a sample of tin, the melting point of tin, the specific heat capacity, and the heat of fusion, determine the quantity of energy (in Joules) required to complete a phase transition.

Plan and Execute: Use Equation 4.2' to determine the heat energy required to increase the temperature of the tin to the melting point. Then use the enthalpy of fusion ($\Delta_{fus}H$) to determine the thermal energy required to melt the solid at the melting point.

Use the equation, $q = c \times m \times \Delta T$, to determine heat transfer resulting from warming the solid.

Execute: ΔT in Kelvin is the same as ΔT in °C.

$$q_{warm} = c_{Sn} \times m_{Sn} \times \Delta T_{Sn} = (0.227 \text{ J g}^{-1}\text{K}^{-1}) \times \frac{1 \text{ K change}}{1\ °\text{C change}} (454 \text{ g}) (231.9\ °\text{C} - 25.0\ °\text{C}) = 2.13 \times 10^4 \text{ J}$$

To change solid to liquid in the system, use $q = m \times \Delta_{fus}H$.

$$q_{melt} = m_{Sn} \times \Delta_{fus}H_{Sn} = (454 \text{ g}) (59.2 \text{ J/g}) = 2.69 \times 10^4 \text{ J}$$

The sum gives the total heat energy required.

$$q_{total} = q_{warm} + q_{melt} = (2.13 \times 10^4 \text{ J}) + (2.69 \times 10^4 \text{ J}) = 4.82 \times 10^4 \text{ J}$$

☑ *Reasonable Result Check:* To reach the final state, the tin's temperature needed to be raised and then the solid needed to be melted. Both of these changes require energy to be added to the system.

95. *Result:* **75.4 g**

Analyze: Given the mass and initial temperature of a sample of ice, the mass and initial temperature of a sample of water, determine how much ice melted after the mixture reaches a common final temperature.

Plan: Using Equation 4.2' and Table 4.1, determine the heat energy required to lower the temperature of the water. Since the heat energy lost by the water comes from melting ice, relate those energies. Use the enthalpy of fusion ($\Delta_{fus}H$) from Section 4-5c to determine the amount of ice that melted. Subtract the calculated mass from the initial mass of ice.

To change the temperature of the water, use $q = c \times m \times \Delta T$:

$$q_{water} = c_{water} \times m_{water} \times \Delta T_{water}$$

Execute: Table 4.1 gives the specific heat capacity of water to be 4.184 J g^{-1}°C^{-1}.

$$q_{water} = (100.0 \text{ g}) (4.184 \text{ J g}^{-1}°\text{C}^{-1}) (0.00\ °\text{C} - 60.0\ °\text{C}) = -2.51 \times 10^4 \text{ J}$$

$$-q_{water} = q_{ice} = 2.51 \times 10^4 \text{ J}$$

To melt the ice, 333 J/g of heat energy are needed:

$$2.51 \times 10^4 \text{ J} \times \frac{1 \text{ g ice melted}}{333 \text{ J}} = 75.4 \text{ g melted}$$

☑ *Reasonable Result Check:* The ice did stop melting before it was all gone, because the thermal energy needed to reach the freezing point was reached in the solution and thermal equilibrium was established.

97. *Result:* **–96.4 kJ/mol**

Analyze: Given three thermochemical expressions, determine the molar enthalpy of another reaction.

Plan: Write the equation for the desired net reaction. Follow the plan described in the solution to Question 72.

Execute: Look at the first reactant: Ca^{2+}(aq). The only given equation that has Ca^{2+}(aq) is the third one, but we need Ca^{2+} to be a reactant, so the equation must be reversed. Therefore, we must write the reverse equation with the sign of $\Delta_rH°$ changed. Look at the second reactant: OH^-(aq). This reactant is also in equation 3, so let's skip it. Look at third reactant: CO_2(g). The only given equation that has CO_2(g) is the first one, but CO_2(g) is a product, so the equation must be reversed. Therefore, we must also write the reverse equation with the sign of $\Delta_rH°$ changed. Look at first product: $CaCO_3$(s). The only given equation that has $CaCO_3$(s) is the first one, and we already have that one so skip this product. Look at the second product: $H_2O(\ell)$. The only given equation that has $H_2O(\ell)$ is the second one, but $H_2O(\ell)$ is a reactant, so the equation must be reversed. Therefore, we must also write the reverse equation with the sign of $\Delta_rH°$ changed. We have now planned how to use all the equations, so let's do it:

Add all the equations as planned, eliminate reactants that also show up as products: $Ca(OH)_2$(s) and CaO(s). Add all the remaining reactants and products to form the net equation for the reaction, and add all the individual $\Delta_rH°$ values to get the $\Delta_rH°$ for the reaction:

$$Ca^{2+}(aq) + 2\ OH^-(aq) \longrightarrow \cancel{Ca(OH)_2}(s) \qquad \Delta_r H° = -(-16.7\ kJ)$$

$$\cancel{CaO(s)} + CO_2(g) \longrightarrow CaCO_3(s) \qquad \Delta_r H° = -(+178.3\ kJ)$$

$$+\quad \cancel{Ca(OH)_2(s)} \longrightarrow \cancel{CaO}(s) + H_2O(\ell) \qquad +\ \Delta_r H° = -(-65.2\ kJ)$$

$$Ca^{2+}(aq) + 2\ OH^-(aq) + CO_2(g) \longrightarrow CaCO_3(s) + H_2O(\ell) \qquad \Delta_r H° = -96.4\ kJ$$

☑ *Reasonable Result Check:* The appropriate sum of the given reactions does indeed recreate the desired reaction, and Hess's law allows us to calculate the $\Delta_r H°$ from that sum.

101. *Result:* **(a) –36.03 kJ/mol (b) 1.18 × 10⁴ kJ evolved**

Analyze: Given a balanced chemical equation for two reactions and a table of molar enthalpies of formation, determine the enthalpy change of the first reaction and the thermal energy evolved by the second.

Plan: Use the stoichiometric coefficient of the balanced equation to describe the moles of each of the reactants and products. Set up a version of Equation 4.10. Look up the $\Delta_f H°$ for each and solve for $\Delta_r H°$.

Execute:

(a) The products of the first reaction are $N_2O(g)$ and $H_2O(g)$. The reactant is $NH_4NO_3(s)$.

$$\Delta_r H° = \Delta_f H°\{N_2O(g)\} + 2\ \Delta_f H°\{H_2O(g)\} - \Delta_f H°\{NH_4NO_3(s)\}$$

Look up the $\Delta_f H°$ values in Table 4.2 and Appendix J.

$$\Delta_r H° = (82.05\ kJ/mol) + 2\ (-241.818\ kJ/mol) - (-365.56\ kJ/mol) = -36.03\ kJ/mol$$

(b) The products of the second reaction are N_2, $H_2O(g)$, and O_2. The reactant is $NH_4NO_3(s)$. N_2 and O_2 are elemental forms, with $\Delta_f H°$ exactly zero.

$$\Delta_r H° = 4\ \Delta_f H°(H_2O(g)) - 2\ \Delta_f H°(NH_4NO_3(s))$$

Look up the $\Delta_f H°$ values in Table 4.2 and Appendix J.

$$\Delta_r H° = 4\ (-241.818\ kJ/mol) - 2\ (-365.56\ kJ/mol) = -236.15\ kJ$$

The balanced chemical equation says that 2 mol NH_4NO_3 react with the evolution of 236.15 kJ of thermal energy.

Molar mass of NH_4NO_3 = 2 (14.0067 g/mol N) + 4 (1.0079 g/mol H) + 3 (15.9994 g/mol O)
= 80.0432 g/mol NH_4NO_3

$$8.00\ kg\ NH_4NO_3 \times \frac{1000\ g}{1\ kg} \times \frac{1\ mol\ NH_4NO_3}{80.0432\ g\ NH_4NO_3} \times \frac{236.15\ kJ}{2\ mol\ NH_4NO_3} = 1.18 \times 10^4\ kJ$$

☑ *Reasonable Result Check:* These explosive reactions produce large amounts of heat energy. The sign and size of these answers make sense.

Applying Concepts

103. *Result:* **endothermic, exothermic**

Analyze, Plan, and Execute:

Ice in a system melts in warm conditions and the molecules are moving around more quickly in the liquid state than in the solid state, so energy must be absorbed by the surroundings. Hence, melting ice is an **endothermic** process.

Freezing water in a system happens only in cold conditions. The molecules are moving more slowly in the solid state than in the liquid. That means energy must be lost by the water to freeze it. Hence, freezing water is an **exothermic** process.

105. *Result:* **(a) X(ℓ) (b) heat of fusion (c) positive**

Analyze, Plan, and Execute:

(a) When adding heat energy to a fixed-mass sample, the temperature change is directly proportional to heat capacity, according to the equation: $q = c \times m \times \Delta T$. If the heat capacity(c) is smaller, then a larger change in temperature will result when a specific amount of energy is used and the slope of the line shown on the heating graph (T vs. energy transfer) will be steeper. Thus, the line on the graph with the steepest slope has the smallest heat capacity and vice versa. Since the slope of the line representing the heating of the liquid is not as steep as the slopes of the lines representing the heating of the solid or the gas, **X(ℓ)** has the largest specific heat capacity.

(b) The horizontal lines on the graph represent phase transitions. The longer the horizontal line on the graph, the more energy will be required for that phase transition. Since the line representing the melting of the solid to liquid is shorter than the line representing the vaporization of the liquid to gas, the **heat of fusion is smaller** than the enthalpy of vaporization.

This is not surprising, since the vaporization of a liquid requires enough energy to overcome all the attractions between the molecules in the liquid state; whereas the melting process only requires some of the attractive forces be overcome. We always find that the fusion enthalpy is smaller than the vaporization enthalpy.

(c) The algebraic sign of the enthalpy of vaporization is always **positive**, since heat energy is required to overcome all of the attractions between the molecules in the liquid state.

107. *Result:* **Gold**

Analyze, Plan, and Execute: The metal with the smaller specific heat capacity will increase in temperature faster than a metal with a larger specific heat capacity. From Table 4.1, $c_{Cu} = 0.385$ J g^{-1}°C^{-1} and $c_{Au} = 0.128$ J g^{-1}°C^{-1}, so the **gold**, Au, will reach 100 °C first.

108. *Result:* **Substance A**

Analyze and Plan: Using a graph, determine which substance has the highest specific heat capacity.

Plan: Use Equation 4.2': $q = m \times c \times \Delta T$ to compare the slopes of the lines.

Execute: The equation says that energy transferred is proportional to ΔT for constant mass samples, using the specific heat capacity, c. So, the slower the temperature rises with the transfer of energy, the larger the specific heat capacity. On the graph, the shallowest line (the line with the least steep slope) has the highest specific heat capacity; here, that is **Substance A**.

110. *Result/Explanation:* Thermal energy content is **greater** in Beaker 1 than in Beaker 2, because a larger mass of water will contain larger quantity of thermal energy at a given temperature.

112. *Result/Explanation:* Enthalpy change is an extensive property. The given equation produces 2 mol SO_3. Formation enthalpy from Table 4.2 is for the production of 1 mol SO_3, so the enthalpy values must be different by a factor of two.

114. *Result:* **$\Delta_r E = 310$ J; w = 0 J**

Analyze, Plan, and Execute: If the balloon is fully inflated, the volume will not change. If $\Delta V = 0$, then $w = -P\Delta V = 0$ kJ. The heating of the contents of the balloon increases its thermal energy, so q = 310 J. Because $\Delta_r E = q + w$, $\Delta_r E = 310\text{ J} + 0\text{ J} = 310$ J.

More Challenging Questions

116. *Result:* **$\Delta_f H°(OF_2) = 25$ kJ/mol**

Analyze: Given a balanced thermochemical expression for a reaction and some molar enthalpies of formation, determine an unknown molar enthalpy of formation.

Plan: Find the moles of each reactant to determine the limiting reactant, then divide the heat transferred by the quantity of reaction to obtain the reaction enthalpy.

Use the stoichiometric coefficient of the balanced equation to describe the moles of each of the reactants and products. Set up a specific version of Equation 4.10, and solve it for the unknown $\Delta_f H°$. Look up the $\Delta_f H°$ for the rest of the reactants and products. Plug them into the equation and solve for $\Delta_f H°$.

Execute: Molar mass of OF_2 = 15.9994 g O + 2(18.9984 g F) = 53.9962 g/mol

Molar Mass of H_2O = 2(1.0079 g H) + 15.9994 g O = 18.0152 g/mol

$$4.32 \text{ g } OF_2 \times \left(\frac{1 \text{ mol } OF_2}{53.9962 \text{ g } OF_2}\right) \times \left(\frac{1 \text{ mol rxn}}{1 \text{ mol } OF_2}\right) = 0.0800 \text{ mol rxn}$$

$$3.37 \text{ g } H_2O \times \left(\frac{1 \text{ mol } H_2O}{18.0152 \text{ } H_2O}\right) \times \left(\frac{1 \text{ mol rxn}}{1 \text{ mol } H_2O}\right) = 0.187 \text{ mol rxn}$$

The limiting reactant is OF_2.

Use that quantity (in moles) and the heat energy evolved to calculate the exothermic (–) reaction enthalpy:

$$\Delta_r H° = \frac{-26.0 \text{ kJ}}{0.0800 \text{ mol rxn}} = -325 \text{ kJ / mol}$$

The products are HF(g) and O_2(g). The reactants are OF_2(g) and H_2O(g). O_2(g) is in elemental form.

$$\Delta_r H° = 2\ \Delta_f H°\{HF(g)\} + \Delta_f H°\{O_2(g)\} - \Delta_f H°\{OF_2(g)\} - \Delta_f H°\{H_2O(g)\}$$

Solve for $\Delta_f H°\{OF_2(g)\}$:

$$\Delta_f H°\{OF_2(g)\} = -\ \Delta_r H° + 2\ \Delta_f H°\{HF(g)\} + \Delta_f H°\{O_2(g)\} - \Delta_f H°\{H_2O(g)\}$$

Look up the $\Delta_f H°$ values for H_2O(g) and HF(g) in Table 4.2.

$\Delta_f H°\{OF_2(g)\}$ = – (–325 kJ/mol) + 2 (–271.1 kJ/mol) + (0 kJ/mol) – (–241.818 kJ/mol) = 25 kJ/mol

☑ *Reasonable Result Check:* The high exothermicity of the reaction can be almost completely accounted for with the destruction of the other reactant and formation of the products. So, it makes sense that the formation enthalpy of OF_2, the reactant, is small. This standard formation enthalpy is given in Question 84 to be 24.6 kJ/mol.

119. *Result:* **0.096 kJ; 4.0 × 10^{-3} g peanuts**

Analyze: Given the speed of an object, and the instruction to look up the mass of such an object on the Internet, determine its kinetic energy. Given the caloric value of a specific food, determine the minimum amount of energy needed to produce that kinetic energy.

Plan: Kinetic energy is discussed in Section 4-1. The equation relating the kinetic energy, mass and speed of an object is:

$$E_k = \frac{1}{2} m v^2$$

Use the unit relationship described in Section 4-1a, 1 J = 1 kg m^2/s^2

Execute: $$v = \frac{130 \text{ mi}}{1 \text{ h}} \times \frac{1 \text{ h}}{3600 \text{ s}} \times \frac{1 \text{ km}}{0.62137 \text{ mi}} \times \frac{1000 \text{ m}}{1 \text{ km}} = 58.115 \text{ m/s (two sig figs)}$$

At the time of the production of this manual in early 2014, the mass of a tennis ball was summarized from a variety of sources on this web site: http://hypertextbook.com/facts/2000/ShefiuAzeez.shtml Four of the five values given include the mass 56.7 g.

$$E_k = \frac{1}{2}(56.7 \text{ g}) \times \frac{1 \text{ kg}}{1000 \text{ g}} (58.115 \text{ m/s})^2 \times \frac{1 \text{ J}}{1 \text{ kg m}^2/\text{s}^2} \times \frac{1 \text{ kJ}}{1000 \text{ J}} = 0.096 \text{ kJ}$$

Table 4.3 in Section 4-11 give the caloric value of peanuts to be 23.91 kJ/g

$$0.096\ \text{kJ} \times \frac{1\ \text{g}}{23.91\ \text{kJ}} = \mathbf{4.0 \times 10^{-3}\ g\ peanuts}$$

☑ *Reasonable Result Check:* Hitting one tennis ball should not take very much energy, so it make sense that a very small amount of food is required to provide that caloric content.

120. *Result:* **(a) 26.6 °C, temperature leveled off from Exp. 3 on (b) ascorbic acid is limiting in Exp. 1, 2, and 3; sodium hydroxide is limiting in Exp. 3,4, and 5 (c) One; equal quantities of each reactant are present in Exp. 3 at the stoichiometric equivalence point**

Analyze and Execute:

(a) We predict the temperature of experiment to be 26.6 °C, because, above $C_6H_8O_6$ masses of 8.81g, the other reactant NaOH appears to be the limiting reactant.

(b) $C_6H_8O_6$ limits in Experiments 1, 2, and 3, which is why the final temperature increases proportionally to the mass of $C_6H_8O_6$ from experiment to experiment. NaOH limits in Experiments 3,4, and 5, as seen by the constant temperature increase in the available data for those three experiments, even though larger masses of $C_6H_8O_6$ are used.

(c) Experiment 3 seems to be the stoichiometric equivalence point, where both reactants run out at the same time. Before that experiment $C_6H_8O_6$ limits, and after that experiment, NaOH limits. Use the mass of $C_6H_8O_6$ and the volume and molarity of the NaOH to determine the moles present in that experiment.

$$100.\ \text{mL NaOH} \times \frac{1\ \text{L}}{1000\ \text{mL}} \times \frac{0.500\ \text{mol NaOH}}{1\ \text{L NaOH}} = 0.0500\ \text{mol NaOH}$$

Molar mass $C_6H_8O_6$ = 6 (12.0107 g/mol C) + 8 (1.0079 g/mol H) + 6 (15.9994 g/mol O)
= 176.1238 g/mol $C_6H_8O_6$

$$8.81\ \text{g}\ C_6H_8O_6 \times \frac{1\ \text{mol}\ C_6H_8O_6}{176.1238\ \text{g}\ C_6H_8O_6} = 0.0500\ \text{mol}\ C_6H_8O_6$$

Because equal quantities of each reactant are present at the stoichiometric equivalence point, ascorbic acid, $C_6H_8O_6$, must have just one hydrogen ion.

122. *Result:* **Results will vary depending on sources used: (a) Approx. 4.6×10^9 kJ (b) Approx. 1.0 metric kilotons TNT (c) 15 kilotons for Hiroshima bomb, 20 kilotons for Nagasaki; max ever = 57 megatons; approx. 6.7% Hiroshima bomb or 5.0% Nagasaki bomb; it was a good idea to evacuate (d) Same energy as approx. 3.1 seconds of a mature hurricane**

Analyze: Given five railroad tank cars full of liquid propane undergoing complete combustion at room temperature. (a) Estimate the necessary details (such as tank car volume and density of liquid propane near the boiling point), to determine the energy transferred if all the propane burned at once. (b) Given the enthalpy of decomposition of TNT, $C_7H_5N_3O_6$, determine how many metric tons of TNT would provide the same energy as in (a). (c) Find the energy transfer (in kilotons of TNT) resulting from the nuclear fission bombs dropped on Hiroshima and Nagasaki. Find the largest nuclear weapon thought to have been detonated to date. Compare these explosions to the answer in (b). Determine if it was wise to evacuate the town. (d) Find the energy of a hurricane and compare that energy to the answer in (a).

Many different results are possible for this question, depending on the various assumptions made.

(a) *Plan:* Determine the mass of propane in the tank cars using the volume and the density. Use Hess' Law to determine the enthalpy of combustion for propane, and use molar mass of propane and the thermochemical stoichiometric relationship to determine the total energy released when the propane reacts.

Execute:

In 2013, the following Wikipedia site: http://en.wikipedia.org/wiki/Tank_car#Tank_Containers gives dimensions for a railcar tank container, also known as an ISO tank, which is described as "a specialized type of container designed to carry bulk liquids, such as chemicals, liquid hydrogen, gases, and food-grade products. Both hazardous and non-hazardous products can be shipped in a tank container. A standard tank container is 20 feet (6.1 m) long, 8 feet (2.4 m) high and 8 feet (2.4 m) wide."

$$V = 6.1\text{ m} \times 2.4\text{ m} \times 2.4\text{ m} = 35\text{ m}^3$$

The density of liquid propane at its boiling point is cited as 582 kg/m^3 on the following web site: http://encyclopedia.airliquide.com/Encyclopedia.asp?GasID=53

$$5\text{ cars} \times 35\text{ m}^3 \times \frac{582\text{ kg } C_3H_8}{1\text{ m}^3\ C_3H_8} = 1.0 \times 10^5\text{ kg } C_3H_8$$

Balance the chemical equation for complete combustion of propane:

$$C_3H_8(g) + 5\ O_2(g) \longrightarrow 3\ CO_2(g) + 4\ H_2O(g)$$

$\Delta_rH° = 3\ \Delta_fH°\{CO_2(g)\} + 4\ \Delta_fH°\{H_2O(g)\} - \Delta_fH°\{C_3H_8(g)\} - 5\ \Delta_fH°\{O_2(g)\}$

Look up the $\Delta_fH°$ values in Table 4.2.

$\Delta_rH° = 3\ (-393.509\text{ kJ/mol}) + 4\ (-241.818\text{ kJ/mol}) - (-103.8\text{ kJ/mol}) - 5\ (0\text{ kJ/mol}) - 2044.0\text{ kJ}$

Now, calculate the energy evolved by the mass of propane in the 5 tank cars:

Molar mass of C_3H_8 = 3(12.0107 g/mol C) + 8(1.0079 g/mol H) = 44.0953 g/mol C_3H_8

$$1.0 \times 10^5\text{ kg } C_3H_8 \times \frac{1000\text{ g}}{1\text{ kg}} \times \frac{1\text{ mol } C_3H_8}{44.0953\text{ g } C_3H_8} \times \frac{2044.0\text{ kJ}}{1\text{ mol } C_3H_8} = 4.6 \times 10^9\text{ kJ}$$

(b) *Plan:* Use the answer in (a) and the energy of the TNT per mol to determine the moles of TNT, then use the molar mass and metric mass relationships to determine the metric kilotons (m.k.t.) of TNT responsible for generating such energy.

Execute:

Molar mass of TNT, $C_7H_5N_3O_6$ = 7(12.0107 g/mol C) + 5(1.0079 g/mol H)
+ 3(14.0067 g/mol N) + 6(15.9994 g/mol O) = 227.1309 g/mol TNT

$$4.7 \times 10^9\text{ kJ} \times \frac{1\text{ mol TNT}}{1066.1\text{ kJ}} \times \frac{227.1309\text{ g TNT}}{1\text{ mol TNT}} \times \frac{1\text{ Mg}}{1 \times 10^6\text{ g}} \times \frac{1\text{ metric ton}}{1\text{ Mg}} \times \frac{1\text{ m.k.t.}}{1000\text{ metric ton}} =$$

1.0 metric kilotons.

(c) *Plan:* Look up the energy output of the two bombs released above Japan and find how the energy reported in (b) compares by calculating percentages.

Execute: The following web site: http://hypertextbook.com/facts/2000/MuhammadKaleem.shtml cited a quote from the source *Encyclopedia Americana*. Danbury, CT: Grolier, 1995: 532. "By today's standards the two bombs dropped on a Japan were small -- equivalent to 15,000 tons of TNT in the case of the Hiroshima bomb and 20,000 tons in the case of the Nagasaki bomb."

Therefore, the Hiroshima bomb is 15 kilotons and the Nagasaki bomb 20 kilotons.

The website http://en.wikipedia.org/wiki/Tsar_Bomba identifies the biggest hydrogen bomb ever tested was the Tsar Bomba detonated at 11:32 on October 30, 1961, over the Mityushikha Bay nuclear testing range (Sukhoy Nos Zone C), north of the Arctic Circle over the Novaya Zemlya archipelago in the Arctic Sea." It released a total energy equivalent to 57 million tons of TNT, or 57 Megatons.

Use percentage to compare the energy from the propane in the railroad tank cars with these bombs:

$$\frac{1.0\text{ m.k.t.}}{15\text{ m.k.t.}} \times 100\,\% = 6.7\,\%\text{ the energy of the Hiroshima bomb}$$

$$\frac{1.0\text{ m.k.t.}}{20\text{ m.k.t.}} \times 100\,\% = 5.0\,\%\text{ the energy of the Nagasaki bomb}$$

It was a good idea to evacuate the town.

(d) *Plan:* Look up the energy associated with a hurricane and compare it with the energy calculated in (a).

Execute: Reported on the following web site: http://www.aoml.noaa.gov/hrd/tcfaq/D7.html , the energy evolved by the kinetic energy of the wind in a hurricane is about 1.3×10^{17} Joules/day, calculated using 40 m/s (90 mph) average wind speed in the inner core of the hurricane scale of radius 60 km (40 mi.).

$$4.6 \times 10^{9}\ \text{kJ} \times \frac{1000\ \text{J}}{1\ \text{kJ}} \times \frac{1\ \text{day}}{1.3 \times 10^{17}\ \text{J}} \times \frac{24\ \text{h}}{1\ \text{day}} \times \frac{3600\ \text{s}}{1\ \text{h}} = \mathbf{3.1\ seconds}$$

3.1 seconds of kinetic energy produced by a mature hurricane

✓ *Reasonable Result Check:* There are many different answers for this question depending on the assumptions made. The assumptions may be as varied and controversial as the results. Checking an answer against the one presented here may be fruitless.

Chapter 5: Electron Configurations and the Periodic Table

Solutions for Red-Numbered Questions for Review and Thought

Topical Questions

Electromagnetic Radiation (Sections 5.1)

11. *Result:* **(a) Visible green light (b) 5.71 × 10^{14} Hz**

Analyze: Given a wavelength, identify the region of the electromagnetic spectrum and determine the frequency range.

Plan: Check Figure 5.2 to identify the color. Follow the plan in Problem-Solving Example 5.1 Section 5-1 using $\nu\lambda = c$

(a) Figure 5.2 shows that the 525 nm is in the green region of the spectrum.

(b) $$\nu = \frac{2.998\times10^{8}\,\text{m/s}}{525\text{ nm}} \times \frac{1\text{ nm}}{1\times10^{-9}\text{ m}} = 5.71\times10^{14}\text{ s}^{-1} = 5.71\times10^{14}\text{ Hz}$$

☑ *Reasonable Result Check*: The wavelength is in the visible region, so it makes sense that the frequencies are also in the visible range.

13. *Result:* **(a) Radio waves (b) Microwaves**

Explanation: Use Figure 5.2

(a) **Radio waves** are lower frequency, thus have less energy, than infrared light.

(b) **Microwaves** have higher frequency than radio waves.

Planck's Quantum Theory (Sections 5-2)

15. *Result:* **4.4 × 10^{-19} J/photon**

Analyze and Plan: Adapt the plan in Problem-Solving Example 5.1 Section 5-1 using $\nu\lambda = c$. Use the equation relating energy to frequency given in Section 5-2: $E = h\nu$:

Execute: $$\nu = \frac{2.998\times10^{8}\,\text{m/s}}{450\text{ nm}} \times \frac{1\text{ nm}}{1\times10^{-9}\text{ m}} = 6.7\times10^{14}\text{s}^{-1} = 6.7\times10^{14}\text{ Hz}$$

$$E = (6.626\times10^{-34}\text{ J·s})(6.7\times10^{14}\text{ s}^{-1}) = 4.4\times10^{-19}\text{ J}$$

☑ *Reasonable Result Check*: The energy of this blue photon is similar to that calculated in Section 5-2 for another blue photon (with a wavelength of 434.1 nm).

17. *Result:* **1.11 × 10^{15} Hz, 7.36 × 10^{-19} J/photon**

Analyze: Given the wavelength of electromagnetic radiation, determine the frequency and energy of this radiation.

Plan: Use equations in Sections 5-1 and 5-2, and appropriate metric conversion factors. The energy calculated is the energy of one photon of this light.

Execute: $\lambda = 270.\text{ nm}$

$$\nu = \frac{c}{\lambda} = \frac{2.998\times10^{8}\,\text{m/s}}{270.\text{ nm}} \times \frac{1\text{nm}}{1\times10^{-9}\text{ m}} = 1.11\times10^{15}\text{ s}^{-1} = 1.11\times10^{15}\text{ Hz}$$

$$E_{photon} = h\nu = \left(6.626 \times 10^{-34}\ J \cdot s\right) \times \left(1.11 \times 10^{15}\ s^{-1}\right) = 7.36 \times 10^{-19}\ J \text{ for one photon}$$

☑ *Reasonable Result Check*: The wavelength calculated coincides with that of ultraviolet light, according to Figure 5.2. This ultraviolet photon's energy is also more than those of visible light calculated in Section 5-2.

20. *Result/Explanation:* Photons of light with long wavelength are low in energy. It is clear from the description that the **energy of these photons is insufficient** to cause electrons to be ejected. Increasing the intensity only increases the number of photons, not their energy.

22. *Result:* **No, the photon energy is too low**

Analyze and Plan: Given the wavelength of light and the minimum energy for ejecting electrons from a metal surface, determine if the light has sufficient energy to cause the photoelectric effect (i.e., eject electrons from that metal's surface). We use appropriate length conversions and the equation in Section 5-2 to find the energy of the photon. Compare the photon energy to the minimum energy. If the photon energy is larger, then the photon can eject electrons; if the energy is smaller, it cannot.

Execute: The wavelength is $\lambda = 600.$ nm.

$$E_{photon} = \frac{hc}{\lambda} = \frac{6.626 \times 10^{-34}\ J \cdot s \times 2.998 \times 10^{8}\ m/s}{600.\ nm} \times \frac{1\ nm}{1 \times 10^{-9}\ m} = 3.31 \times 10^{-19}\ J$$

$$3.31 \times 10^{-19}\ J < 3.69 \times 10^{-19}\ J$$

$$E_{photon} < E_{minimum}$$

Therefore the 600.-nm photon has insufficient energy to eject an electron.

☑ *Reasonable Result Check*: The minimum energy is similar to ones calculated for visible light photons in Section 5-2. These two energies are close, so we probably would not have been able to predict this result before doing the calculation.

The Bohr Model of the Hydrogen Atom (Section 5-3)

27. *Result/Explanation:* Energy is emitted from an atom when an electron moves from the **higher-energy** (excited) state to a **lower-energy** state (maybe the ground state, maybe another less-excited state). The energy of the emitted radiation corresponds to the **difference** between the two energy states.

28. *Result:* **(a) absorbed (b) emitted (c) absorbed (d) emitted**

Analyze: Given the values of n for the initial and final states of an electron in hydrogen, determine whether energy is absorbed or emitted.

Plan: The size of n indicates the relative energy of the state. If the final state has a larger n value than the initial state, then energy must be absorbed. If the final state has a smaller n value than the initial state, then energy must be emitted.

Execute:

(a) n = 1 to n = 3. Energy is absorbed.

(b) n = 5 to n = 1. Energy is emitted.

(c) n = 2 to n = 4. Energy is absorbed.

(d) n = 5 to n = 4. Energy is emitted.

31. *Result:* **(a), (b), and (d)**

Analyze, Plan and Execute: The difference between the energies of the two levels is the energy emitted. The smaller the energy, the longer the wavelength. When we look at Figure 5.9, we need to find which of the given levels is closer together than the energy levels represented by n = 1 and n = 4. The energies levels represented by (a) n = 2 and n = 4, (b) n = 1 and n = 3, and (d) n = 3 and n = 5, are closer together than n = 1 and n = 4. So, **(a), (b), and (d)** will require radiation with longer wavelength.

32. *Result:* **4.576×10^{-19} J absorbed and 434.0 nm**

Analyze: Given the values of n for the initial and final states of an electron in hydrogen, calculate the energy and wavelength of the photon associated with the transition.

Plan: Using equations given in Section 5-3, calculate the energy of the photon emitted during the transition. Use the equation given in Section 5-1 and appropriate metric conversions to calculate the wavelength in nanometers.

Execute: Here, $n_i = 2$ and $n_f = 5$:

$$\Delta E_{e-} = \left(2.179 \times 10^{-18}\ \text{J}\right)\left(\frac{1}{n_i^2} - \frac{1}{n_f^2}\right) = \left(2.179 \times 10^{-18}\ \text{J}\right)\left(\frac{1}{2^2} - \frac{1}{5^2}\right) = 4.576 \times 10^{-19}\ \text{J}$$

The energy of the photon absorbed is equal to the energy gained by the electron as it moves to a higher-energy state. $E_{photon} = \Delta E_{e-}$. The wavelength of the photon absorbed relates to the energy:

$$\lambda = \frac{hc}{E_{photon}} = \frac{\left(6.626 \times 10^{-34}\ \text{J} \cdot \text{s}\right)\left(2.998 \times 10^{8}\ \text{m/s}\right)}{4.576 \times 10^{-19}\ \text{J}} \times \frac{1\ \text{nm}}{1 \times 10^{-9}\ \text{m}} = 434.0\ \text{nm}$$

☑ *Reasonable Result Check*: The wavelength of radiation produced from the reverse of this transition is reported in Figure 5.8.

34. *Result:* **6.198×10^{-19} J, 320.5 nm, UV region**

Analyze: Given the values of n for the initial and final states of an electron in a helium cation, He^+, and an equation describing the relationship between the electron energy level, the atomic number, Z, and the quantum number, n, determine the transition energy of the electron and what region of the electromagnetic spectrum it is in.

Plan: Adapt the method shown in Section 5-3. Subtract the initial energy (n = 5) of the electron from the final energy (n = 3) of the electron in helium. Z = 2 for helium. Use Figure 5.1 to determine the region of the electromagnetic spectrum.

Execute:

$$E_3 = -\frac{2^2}{3^2}\left(2.179 \times 10^{-18}\ \text{J}\right) = -9.684 \times 10^{-19}\ \text{J}$$

$$E_5 = -\frac{2^2}{5^2}\left(2.179 \times 10^{-18}\ \text{J}\right) = -3.486 \times 10^{-19}\ \text{J}$$

$$\Delta E_{e-} = E_3 - E_5 = -3.486 \times 10^{-19}\ \text{J} - (-9.684 \times 10^{-19}\ \text{J}) = -6.198 \times 10^{-19}\ \text{J}$$

$$E_{photon} = -\Delta E_{e-} = \mathbf{6.198 \times 10^{-19}\ J}$$

$$\nu = \frac{E_{photon}}{h} = \frac{6.198 \times 10^{-19}\ \text{J}}{6.626 \times 10^{-34}\ \text{J} \cdot \text{s}} = 9.354 \times 10^{14}\ \text{s}^{-1}$$

$$\lambda = \frac{c}{\nu} = \frac{2.998 \times 10^{8}\ \text{m/s}}{9.354 \times 10^{14}\ \text{s}^{-1}} \times \frac{1\ \text{nm}}{10^{-9}\ \text{m}} = \mathbf{320.5\ nm}$$

This wavelength indicates that the emission produces electromagnetic radiation in the **UV region.**

☑ *Reasonable Result Check*: The electron levels in helium are closer together than those in hydrogen because the nucleus has a more-positive charge.

Quantum Mechanical Model of the Atom (Section 5-4)

37. *Result:* moon, bowling ball, baseball, neon atom, electron

Analyze: Order the items listed from smallest to largest de Broglie wavelength.

Plan: The deBroglie wavelength (λ) is larger when the mass (m) is small and/or the velocity (v) is small, as shown in this equation from Section 5-4: $\lambda = \frac{h}{mv}$.

Execute: The heaviest of these objects is the moon, followed by a bowling ball, then base ball, then neon atom. The lightest item is an electron. They have variable velocities, but the differences in their masses are far more dramatic.

39. *Result/Explanation:* The uncertainty principle states that it is impossible to simultaneously determine the exact position and the exact momentum of an electron. The dot-density diagram **shows where the electron is found in space and does not require the simultaneous knowledge of the electron's position and momentum**. This explains why the dot-density diagram in Figure 5.12 is better than the well-defined circular paths described in the Bohr model, which violate the uncertainty principle.

Quantum Numbers (Section 5-5)

41. *Result:* (a) (iii) and (iv) (b) (i) and (iii)

Analyze: Given sets of four quantum numbers for electrons, identify which two have the same spin and which two are in the same atomic orbital.

Plan: The spin is related to the value of m_s. Two electrons in the same orbital must have the same n, ℓ, and m_ℓ.

Execute:

(a) (iii) and (iv) both have $m_s = -\frac{1}{2}$. The other two electrons do not.

(b) (i) and (iii) both have n = 2, $\ell = 1$, $m_\ell = 1$. The other two have $\ell = 0$.

43. *Result:* **(a) 1,0,0,$\frac{1}{2}$; 1,0,0,$-\frac{1}{2}$;2,0,0,$\frac{1}{2}$; 2,0,0,$-\frac{1}{2}$;2,1,1,$\frac{1}{2}$ (other answers with different m_ℓ and m_s are also possible); (b) 3,0,0,$\frac{1}{2}$; 3,0,0,$-\frac{1}{2}$; (c) 3,2,2,$\frac{1}{2}$ (other answers with different m_ℓ and m_s are also possible)**

Analyze: Give the four quantum numbers for various electrons.

Plan: The set of three quantum numbers representing each orbital in an atom must be different (Pauli Exclusion Principle). ℓ values must be less than n, but not negative. m_ℓ values must be between $-\ell$ and $+\ell$. m_s values must be $+\frac{1}{2}$ or $-\frac{1}{2}$. Electrons with the same n, ℓ, and m_ℓ must have opposite signs for m_s. Electrons with the same n and ℓ, but different m_ℓ must have the same m_s value (Hund's rule) until all values of m_ℓ have been used once. Be systematic.

Execute:

(a) There are five electrons in boron:

First electron	n = 1	$\ell = 0$	$m_\ell = 0$	$m_s = +\frac{1}{2}$
Second electron	n = 1	$\ell = 0$	$m_\ell = 0$	$m_s = -\frac{1}{2}$
Third electron	n = 2	$\ell = 0$	$m_\ell = 0$	$m_s = +\frac{1}{2}$
Fourth electron	n = 2	$\ell = 0$	$m_\ell = 0$	$m_s = -\frac{1}{2}$
Fifth electron	n = 2	$\ell = 1$	$m_\ell = 1$	$m_s = +\frac{1}{2}$

(The fifth electron could have different m_ℓ and m_s values also, as long as they are within the appropriate ranges.)

(b) Electrons in the outermost shell are the valence electrons. In elements without partially filled d-shells, they will be the electrons with the highest n-value. Magnesium has two valence electrons:

First valence electron of Mg	n = 3	$\ell = 0$	$m_\ell = 0$	$m_s = +\frac{1}{2}$
Second valence electron of Mg	n = 3	$\ell = 0$	$m_\ell = 0$	$m_s = -\frac{1}{2}$

(c) A 3d electron in iron atom:

A 3d electron in Fe	n = 3	$\ell = 2$	$m_\ell = 2$	$m_s = +\frac{1}{2}$

(The values of m_ℓ and m_s can be different here, as long as they are within the appropriate ranges.)

45. *Result:* **(a) cannot occur, m_ℓ must be between $-\ell$ and $+\ell$ (b) can occur (c) cannot occur, m_s cannot be one, here (d) cannot occur, ℓ must be less than n (e) can occur**

Analyze and Plan: Determine which sets could not occur. Related to requirements for range and limits of quantum numbers.

Execute:

(a) $n = 2, \ell = 1, m_\ell = 2, m_s = +\frac{1}{2}$ This quantum number set can not occur because m_ℓ must be between $-\ell$ and $+\ell$ and 2 is not between -1 and $+1$.

(b) $n = 3, \ell = 2, m_\ell = 0, m_s = -\frac{1}{2}$ This quantum number set can occur.

(c) $n = 1, \ell = 0, m_\ell = 0, m_s = 1$ This quantum number set can not occur because m_s must be $+\frac{1}{2}$ or $-\frac{1}{2}$. Here $m_s = 1$.

(d) $n = 3, \ell = 3, m_\ell = 2, m_s = -\frac{1}{2}$ This quantum number set can not occur because ℓ must be less than n. Here, 3 is not among the possible values of 2, 1, or 0.

(e) $n = 2, \ell = 0, m_\ell = 0, m_s = +\frac{1}{2}$ This quantum number set can occur.

47. *Result/Explanation:*

(a) The up-spin electron in the 4s orbital has the following four quantum numbers:

n = 4	$\ell = 0$	$m_\ell = 0$	$m_s = +\frac{1}{2}$

(b) The down-spin electron in the first of three 3p orbitals has the following four quantum numbers:

n = 3	$\ell = 1$	$m_\ell = 1$	$m_s = -\frac{1}{2}$

Notice: Two other values of m_ℓ (0 and -1) would also be valid.

(c) The up-spin electron in the third of five 3d orbitals has the following four quantum numbers:

n = 3	$\ell = 2$	$m_\ell = 0$	$m_s = +\frac{1}{2}$

Notice: Other answers are also correct: Four other values of m_ℓ (+2, +1, -1, and -2) and one other value of m_s ($-\frac{1}{2}$) would also be valid.

49. *Result/Explanation:* **18** elements are in the fourth period of the periodic table. It is not possible for there to be another element in this period because **all possible orbital electron combinations are already used**.

Shapes of Atomic Orbitals (Section 5-6)

50. *Result: See sketches below*

Analyze: Sketch the shapes of three orbitals

Plan: All s orbitals are spherical and are drawn as a circle. All p orbitals are dumbbell-shaped and are drawn like a figure "8". The p-orbital alignment axis is identified in the subscript, as described in Section 5-6b. Four of the d orbitals each have four lobes, drawn like a four-leafed clover. The d-orbital alignment axes are identified in the subscript as described in Section 5-6c.

Execute: The $d_{x^2-y^2}$ orbital is aligned along the x- and y-axes. The p_y orbital is aligned along the y-axis.

(a) $4d_{x^2-y^2}$ (b) 2s (c) $3p_y$

52. *Result/Explanation:* There are n subshells in the electron shell with principle quantum number n; they have $\ell = n - 1, \ldots, 0$. That means **four** different subshells (designated $\ell = 3, 2, 1$, and 0) are found in the electron shell with principle quantum number 4.

Atom Electron Configurations (Section 5-7)

54. *Result/Explanation:* Electrons fill atomic orbitals starting with the 1s orbital and working upward in the subshell energy order shown in Figure 5.15. As show, the **4s orbital is filled before any of the 3d orbitals**, therefore titanium has the electron configuration [Ar] $3d^2$ $4s^2$. In Section 5-8b, we learn that **atoms lose the outermost shell (highest n) electrons first** to form a transition metal cation. Thus chromium atom ([Ar] $3d^5$ $4s^1$) will lose its 4s electron first, and the configuration for Cr^{2+} is [Ar] $4d^4$.

56. *Result/Explanation:*

$_{32}$Ge electron configuration: $\mathbf{1s^22s^22p^63s^23p^63d^{10}4s^24p^2}$

58. *Result/Explanation:* $_8$O, **oxygen**, is an element in Group 6A. (Several other possible elements that can correctly answer this question.) The valance electron configuration of this element is $2s^22p^4$. All elements in this group have **six valence electrons**, so their electron configurations are: $\mathbf{ns^2np^4}$

60. *Result:* **(a) Cs^+, Se^{2-}; (b) Cs^+, Se^{2-}**

Analysis, Plan, and Execute:

(a) Cations likely to form are ones that have lost only valence electrons. Any ion that has a more positive charge than its number of valence electrons is not likely to form. Anions likely to form are ones that add enough electrons to the atom to increase the number of valence electrons to eight. Any ion that has a more negative charge than the difference between eight and the number of valence electrons is not likely to form.

K has one valence electron; K^{2+} is **not** likely to form because a core electron must be removed.

Cs has one valence electron; Cs^+ **is** likely to form.

Al has three valence electrons; Al^{4+} is **not** likely to form because a core electron must be removed.

F has seven valence electrons. It needs 8 - 7 = 1 electron. F^{2-} is **not** likely to form because a new valence shell must be started to add the second electron.

Se has six valence electrons. It needs 8 – 6 = 2 electrons. Se^{2-} **is** likely to form.

(b) The only element above that loses all of its valence electrons and none of its core electrons is Cs. **Cs^+** has a [Xe] electron configuration. The only element above that gains just enough electrons to fill its valence shell without adding any more is Se. **Se^{2-}** has a [Kr] electron configuration.

61. *Result/Explanation:*

(a) The diagram in the book shows a total of 28 electrons and six 3d sublevels, but Fe atom should have only 26 electrons and the 3d sublevel should have only five orbitals. Also, in the ground state of Fe atom, the lower-energy **4s orbital must be full.**

	3d	4s
[Ar]	[↑↓][↑][↑][↑][↑]	[↑↓]

(b) The **orbital labels must be 3, not 2**. The electrons in 3p subshell of P should be in separate orbitals with parallel spin (Hund's Rule).

	3s	3p
[Ne]	[↑↓]	[↑][↑][↑]

(c) The diagram in the book shows total of 50 electrons and six 3d sublevels, but the cation, Sn^{2+} should have only 48 electrons and the 3d sublevel should have only five orbitals. Also, electrons should be removed from the higher-energy valence 5p orbitals of Sn to make the cation, Sn^{2+}, not from the lower-energy 4d orbitals. The **electrons lost must be from the 5p orbitals**, not the 4d orbitals.

	4d	5s
[Kr]	[↑↓][↑][↑][↑][↑]	[↑↓]

62. *Result/Explanation:* Find the number of electrons in the subshell, and add arrows according to Hund's rule.

	3d	4s
(a) V	[↑][↑][↑][ ][ ]	[↑↓]
(b) V^{2+}	[↑][↑][↑][ ][ ]	
(c) V^{4+}	[↑][ ][ ][ ][ ]	

64. *Result/Explanation:* When transition metals form cations, the electrons lost first are those from their valence (highest n) s orbitals.

$_{25}$Mn electron configuration: **$1s^22s^22p^63s^23p^63d^54s^2$**. Mn has **5 unpaired electrons**:

	3d	4s
[Ar]	[↑][↑][↑][↑][↑]	[↑↓]

$_{25}Mn^{2+}$ electron configuration: **$1s^22s^22p^63s^23p^63d^5$**. Mn^{2+} has **5 unpaired electrons:**

	3d
[Ar]	[↑][↑][↑][↑][↑]

$_{25}Mn^{3+}$ electron configuration: **$1s^22s^22p^63s^23p^63d^4$**. Mn^{3+} has **4 unpaired electrons:**

	3d
[Ar]	[↑][↑][↑][↑][ ]

67. *Result/Explanation:*

(a) $_{63}$Eu electron configuration: **[Xe]$4f^76s^2$** or **[Xe]$4f^65d^16s^2$**

(b) $_{70}$Yb electron configuration: **[Xe]$4f^{14}6s^2$** or **[Xe]$4f^{13}5d^16s^2$**

68. *Result/Explanation:* Put one dot for each valance electron:

(a) ·Sr· (b) :B̈r: (c) ·Ġa· (d) :Ṡb·

Ion Electron Configurations (Section 5-8)

70. *Result/Explanation:*

(a) $_{20}Ca^{2+}$ electron configuration: $\mathbf{1s^22s^22p^63s^23p^6}$ or **[Ar]**

(b) $_{19}K^{+}$ electron configuration: $\mathbf{1s^22s^22p^63s^23p^6}$ or **[Ar]**

(c) $_{8}O^{2-}$ electron configuration: $\mathbf{1s^22s^22p^6}$ or **[Ne]**

$_{20}Ca^{2+}$ and $_{19}K^{+}$ are isoelectronic. They have the same number of electrons.

72. *Result/Explanation:* Follow the order of subshell filling shown in Figure 5.21:

(a) $_{50}Sn$ electron configuration: $\mathbf{[Kr]4d^{10}5s^25p^2}$

(b) $_{50}Sn^{2+}$ electron configuration: $\mathbf{[Kr]4d^{10}5s^2}$

$_{50}Sn^{4+}$ electron configuration: $\mathbf{[Kr]4d^{10}}$

74. *Result/Explanation:*

(a) Diamagnetic elements have completely filled subshells. While the subshell is incompletely full, Hund's Rule requires that the electrons be as unpaired as possible. Therefore, only the element with completely filled subshells will be diamagnetic. The only period four transition element with completely filled s and d subshell is **Zn** ($[Ar]3d^{10}4s^2$).

(b) While the subshell is incompletely full, Hund's Rule requires that the electrons be as unpaired as possible. That means a transition element will have at most five unpaired d electrons. When transition elements have six outer electrons, they have one s electron and five d electrons to make a half filled d shell. The period four transition element with five unpaired d electrons and one unpaired s electron is **Cr**.

76. *Result/Explanation:* In both paramagnetic and ferromagnetic substances, atoms have unpaired spins and thus are attracted to magnets. Ferromagnetic substances retain their aligned spins after an external magnetic field has been removed, so they can function as permanent magnets. Paramagnetic substances lose their aligned spins after a time and therefore cannot be used as permanent magnets.

77. *Result:* **(a) TiO_2; it has no unpaired electrons (diamagnetic) (b) TiO; it has unpaired electrons (paramagnetic)**

Analyze and Plan: The oxide compounds contain different titanium ions: Ti^{2+} and Ti^{4+} Ions with no unpaired electrons are diamagnetic and those that have unpaired electrons are attracted to magnets. The one with more unpaired electrons will have the greater attraction to a magnetic field.

$_{22}Ti^{2+}$ electron configuration: $\mathbf{1s^22s^22p^63s^23p^63d^2}$. Ti^{2+} has **2 unpaired electrons.**

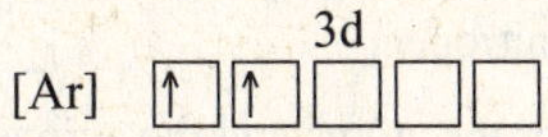

$_{22}Ti^{4+}$ electron configuration: $\mathbf{1s^22s^22p^63s^23p^6}$. Ti^{4+} has **no unpaired electrons**.

(a) TiO_2 is **diamagnetic**, because the Ti^{4+} ion has **no unpaired electrons**.

(b) **TiO** has more attraction to a magnetic field, because the Ti^{2+} ion has **unpaired electrons,** which makes it **paramagnetic**.

Periodic Trends (Sections 5-9 to 5-12)

79. *Result:* **(a) Cl has one more proton in its nucleus than S and adding one electron gives it a noble-gas configuration making its electron affinity greater than S, so S has a lower electron affinity. (b) Be has a full subshell, so it takes more energy to ionize it than B. (c) F electron is in a lower-energy 2p; whereas, Cl electron is in a higher-energy 3p orbital. (d) N has a half-filled subshell, so N takes more energy to ionize than O. (e) Br is smaller than I, so electrons added to Br end up closer to the nucleus than in I.**

Analyze and Plan: Write the electron configuration for the two ions, and consider what changes occur when the electrons are added (electron affinity) or removed (ionization).

Electron affinity is the ΔE for this equation: $A(g) + e^- \longrightarrow A^-(g)$

Ionization energy is the ΔE for this equation: $M(g) \longrightarrow M^+(g) + e^-$

Execute:

(a) Write the electron configuration for the atoms and the anions produced by adding an electron:

$_{17}Cl$ electron configuration: $[_{10}Ne]3s^23p^5$ $_{17}Cl^-$ electron configuration: $[_{18}Ar]$

$_{16}S$ electron configuration: $[_{10}Ne]3s^23p^4$ $_{16}S^-$ electron configuration: $[_{10}Ne]3s^23p^5$

$_{17}Cl$ has one more proton in its nucleus than $_{16}S$. Adding one electron to $_{17}Cl$ gives it a noble-gas configuration, making its electron affinity greater than S. Thus, sulfur has a lower (less negative) electron affinity than chlorine.

(b) Write the electron configuration for the atoms and the cations produced by removing an electron:

$_5B$ electron configuration: $[_2He]2s^22p^1$ $_5B^+$ electron configuration: $[_2He]2s^2$

$_4Be$ electron configuration: $[_2He]2s^2$ $_4Be^+$ electron configuration: $[_2He]2s^1$

The $_4Be$ atom has all subshells full, so removing an electron requires more energy, Thus, boron has a lower first ionization energy than beryllium.

(c) Write the electron configuration for the atoms and the cations produced by removing an electron:

$_{17}Cl$ electron configuration: $\mathbf{[_{10}Ne]3s^23p^5}$ $_{17}Cl^+$ electron configuration: $\mathbf{[_{10}Ne]3s^23p^4}$

$_9F$ electron configuration: $\mathbf{[_2He]2s^22p^5}$ $_9F^+$ electron configuration: $\mathbf{_2He]2s^22p^4}$

Removing an electron from a lower-energy subshell will require more energy. The electron removed from Cl comes from a 3p orbital and the one from F comes from a 2p orbital. Thus, chorine has a lower first ionization energy than fluorine.

(d) Write the electron configuration for the atoms and the cations produced by removing an electron:

$_8O$ electron configuration: $[_2He]2s^22p^4$ $_8O^+$ electron configuration: $[_2He]2s^22p^3$

$_7N$ electron configuration: $[_2He]2s^22p^3$ $_7N^+$ electron configuration: $[_2He]2s^22p^2$

The $_7N$ atom has a half-full p-subshell, so removing an electron requires more energy. Thus, oxygen has a lower first ionization energy than nitrogen.

(e) Write the electron configuration for the atoms and the anions produced by adding an electron:

$_{53}I$ electron configuration: $[_{36}Kr]5s^24d^{10}5p^5$ $_{53}I^-$ electron configuration: $[_{54}Xe]$

$_{35}Br$ electron configuration: $[_{18}Ar]4s^23d^{10}4p^5$ $_{35}Br^-$ electron configuration: $[_{36}Kr]$

Adding an electron to a lower-energy subshell will produce more energy. The electron added to Br goes into a 4p orbital and the one added to I goes into a 5p orbital. Thus, iodine has a lower (less negative) electron affinity than bromine.

81. *Result:* **B, C, N, O, F**

Analyze: Order elements in order of increasing effective nuclear charge.

Plan and Execute: In Section 5-9c, we learn that the effective nuclear charge increases across a period, so we will order these five second period elements from right to left as they appear on the period table: B, C, N, O, F

The closer an element is to a noble gas, the greater proportion of valence electrons there are. The inner electrons screen the valence electrons from the charge of the protons in the nucleus. Valance electrons are not effective at screening, so elements with more valance electrons will have a higher effective nuclear charge.

83. *Result:* **P < Ge < Ca < Sr < Rb**

Analyze and Plan: Without using the table of numerical sizes, list elements in order of increasing size. Without using the table that gives numerical sizes, we are asked to list elements in order of increasing size. *Plan:* Periodic Trends in atom size are related to the radius of the atoms:

- Comparing atoms across the period (row) of the periodic table, atom sizes increase from right to left.
- Comparing atoms down a group (column) of the periodic table, atom sizes increase from top to bottom.

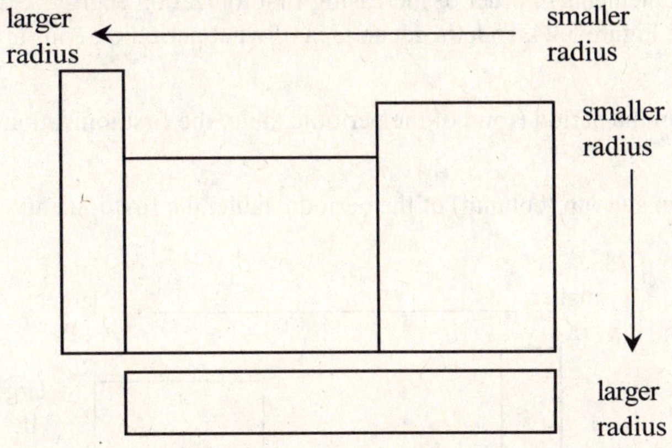

Execute:

Looking at a periodic table, the smallest of the ions given is P, since it is the closest to the top right corner. Next is Ge, diagonally down and to the left of P. Then comes Ca, in the same period as Ge, but further to the left. Then comes Sr, in Group 2A, below Ca. Then comes Rb, in the same period as Sr, but further to the left. So, the order is: P < Ge < Ca < Sr < Rb.

☑ *Reasonable Result Check:* Figures 5.25 confirms these predictions.

85. *Result:* **(a) Rb (b) O (c) Br (d) Ba^{2+} (e) Ca^{2+}**

Analyze and Plan: Without using the table that gives numerical sizes, determine which element of a pair of atoms and ions has larger radius. Use the periodic table for atom size trends (as described in the solution to Question 81) and the following periodic trends.

Periodic trends comparing atoms to ions:

- The cation of an element will always be smaller than the atom of the same element (since fewer electrons are attracted more closely to the nucleus).
- The anion of an element will always be larger than the atom of the same element (since more electrons will be less attracted to the nucleus).

Periodic trend comparing ions to ions:

- When comparing isoelectronic ions, the element with the larger atomic number is smaller (since it has more protons to attract the electrons).
- When comparing ions of the same element, the ion with the larger positive charge is smaller (since fewer electrons are attracted more closely to the nucleus).

Execute:

(a) Using the periodic table, Rb has a smaller radius than Cs atom, because they are both in the same group and Cs is below Rb.

(b) The O atom has a smaller radius than O^{2-} ion, since anions are larger than the neutral atom. (8 protons attract 8 electrons better than they can attract 10 electrons.)

(c) Using the periodic table, Br has a smaller radius than As atom, because they are both in the same period and Br is further to the right.

(d) The Ba^{2+} ion has a smaller radius than Ba atom, since cations are smaller than the neutral atom. (56 protons attract 54 electrons better than they can attract 56 electrons.)

(e) The Ca^{2+} ion has a smaller radius than Cl^- ion, since they are isoelectronic and the atomic number of Ca (20) is larger than the atomic number of Cl (17). (20 protons can attract 18 electrons better than 17 protons can.)

☑ *Reasonable Result Check:* Use Figures 5.25 - 5.27 to confirm these predictions.

87. *Result:* **Choice (c)**

Analyze and Plan: List elements in order of increasing first ionization energy. Sometimes we will not have a chart with numbers like Figure 5.24, so let's get an idea of what periodic trends are exhibited in first ionization energies:

- Comparing atoms in the period (row) of the periodic table, the first ionization energy increases from left to right.
- Comparing atoms in a group (column) of the periodic table, the first ionization energy increases from bottom to top.

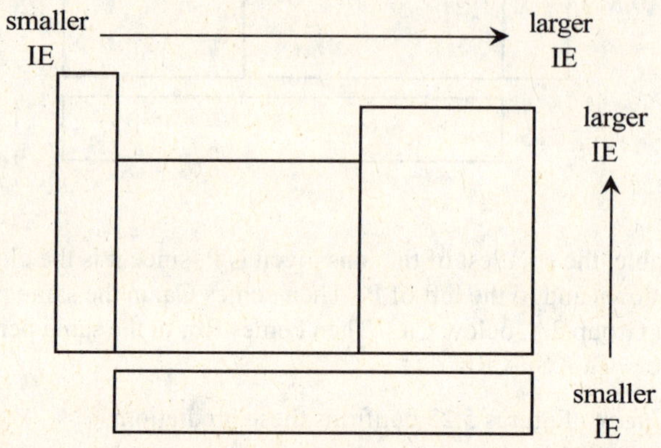

- Elements in group 2A, 2B, and 5A show exceptions to this general trend, having higher IE than elements to their immediate right.

Execute:

Looking at a periodic table, the element the furthest to the left is Li, so, we'll predict that it has the lowest first ionization energy. Then comes Si, to the right of the metal. Then comes C, above Si in the same group. Then comes Ne, to the right of C in the same period. So the correct order is: Li < Si < C < Ne. That is choice (c).

☑ *Reasonable Result Check*: Use Figure 5.30 to confirm these predictions.

89. *Result/Explanation:* Looking at Table 5.6, it is clear that the second ionization energies are all higher than the first ionization energies and once core electrons are being removed, the ionization energy rises dramatically. To have a large gap between the first and second ionization energies, the element **must have one valence electron**. The only element given that has one valence electron is **Na**, from Group 1A.

90. *Result:* **(a) Al (b) Al (c) Al < B < C**

Analyze and Plan: Given a list of elements, determine which is most metallic, determine which has largest radius, and place the list in order of increasing of ionization energy. Metallic character follows these general trends:

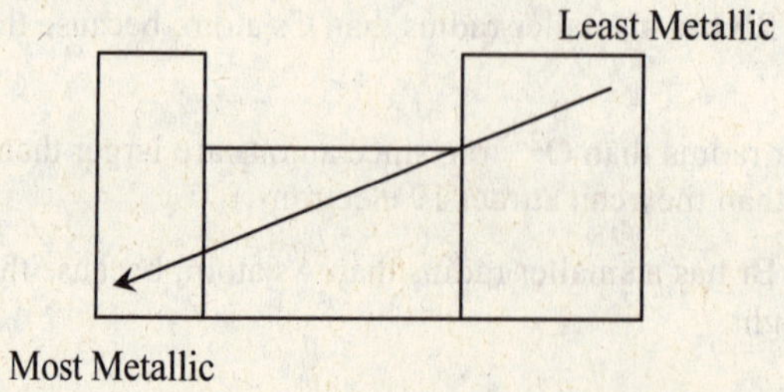

Use the periodic table for atom size trends and ionization energy trends (as described in the solutions to Questions 83 and 87).

Execute: Locate the elements: B, Al, C, and Si on a periodic table.

(a) We find Al closest to the bottom left corner. That means Al is most metallic.

(b) We find Al closest to the bottom left corner. That means Al is the largest of these four atoms.

(c) We find Al closest to the bottom left corner. That means Al has the smallest first ionization energy. B is next, because it is above Al in the same group. Last comes C, to the right of B in the same period. So the order is: Al < B < C

☑ *Reasonable Result Check:* Use Figures 5.25 - 5.27 to confirm these predictions.

92. *Result/Explanation:* Adding a free electron to the oxygen atom is exothermic and produces an anion. When the second electron is add, its negative charge is repelled by the anion's negative charge, according to Coulomb's Law. Additional energy is required to overcome the columbic charge repulsion, making the second electron affinity endothermic.

Energy, Ions, and Ionic Compounds (Section 5-13)

95. *Result:* **–862 kJ/mol**

Analyze and Plan: Given enthalpy of sublimation, ionization energy, bond energy, electron affinity, and formation enthalpies, determine the lattice energy. Adapt the method in Section 5-13, using Hess's Law.

Construct a diagram similar to Figure 5.33, using the enthalpy values given.

Execute: Combine the given chemical equations to make the desired chemical equation:

$$\cancel{Li(s)} + \tfrac{1}{2}\cancel{Cl_2(g)} \longrightarrow LiCl(s) \qquad \Delta_f H = -408.7 \text{ kJ/mol}$$

$$\cancel{Li(g)} \longrightarrow \cancel{Li(s)} \qquad -\Delta_{subl}H = -(161 \text{ kJ/mol})$$

$$\cancel{Cl(g)} \longrightarrow \tfrac{1}{2}\cancel{Cl_2(g)} \qquad \Delta_r H = -\tfrac{1}{2}BE = -\tfrac{1}{2}(242 \text{ kJ/mol})$$

$$\cancel{e^-} + Li^+(g) \longrightarrow \cancel{Li(g)} \qquad \Delta_r H = -IE_1 = -(520. \text{ kJ/mol})$$

$$+ \quad Cl^-(g) \longrightarrow \cancel{e^-} + \cancel{Cl(g)} \qquad \Delta_r H = -EA_1 = -(-349 \text{ kJ/mol})$$

$$Li^+(g) + Cl^-(g) \longrightarrow LiCl(s) \qquad \Delta_r H = \text{Lattice Energy} = -862 \text{ kJ/mol}$$

☑ *Reasonable Result Check*: Lithium ion is smaller than sodium ion and both chloride compounds are composed by combining a +1-cation and a –1-anion, so, as expected, this lattice energy is similar in size but slightly more negative than that seen for sodium chloride (which is given in the text (–786 kJ/mol).

97. *Result/Explanation:* Compounds with high lattice energies have high melting points.

Because all the compounds in the list, LiCl, NaBr, and KCl are composed of +1 cations and –1 anions, we'll look at the relative sizes of the ions. Compounds composed of small ions have larger lattice energies. Ion sizes increase down a period for ions in the same periodic group. Therefore, $Li^+ < Na^+ < K^+$ and $Cl^- < Br^-$. Because **LiCl** is composed of the smallest ions in each of these series, it probably has the greatest lattice energy and therefore the highest melting point.

General Questions

99. *Result:* **(a) He (b) Sc (c) Na**

Analyze, Plan, and Execute:

(a) For a ground state element to be in Group 8A but have no p electrons, it must have fewer than five electrons, because the fifth electron goes into a 2p orbital (i.e., $1s^2 2s^2 2p^1$). This Group 8 element is **He** ($1s^2$).

(b) For a ground state element to have a single electron in the 3d subshell, it must have this electron configuration: $1s^22s^22p^63s^23p^63d^1$. This element is **Sc**.

(c) For a ground state element's cation to have a +1 charge and an electron configuration of $1s^22s^22p^6$, the element's electron configuration would have one more electron: $1s^22s^22p^63s^1$. This element is **Na**.

102. *Result:* **(a) F < O < S using the relative positions of atoms on the periodic table (b) S has largest first ionization energy because it is closest to the top right corner of the periodic table.**

Analyze and Plan: Given two lists of elements, rank order one by increasing atomic radius and determine which one has the largest ionization energy. Use the periodic table, atom size trends, and ionization energy trends (as described in the solutions to Questions 83 and 87).

Execute:

(a) Locate the elements: O, S, and F on a periodic table. We find F closest to the top right corner. That means F is the smallest of these three atoms. Then comes O, to the left of the F in the same period. Then comes S, below O in the same group. So the order is: F < O < S

(b) Locate the elements: P, Si, S, and Se on a periodic table. We find S closest to the top right corner. That means S has the largest first ionization energy.

☑ *Reasonable Result Check:* Using Figures 5.23, 5.24, and 5.25, we can confirm these predictions.

104. *Result:* **(a) Sulfur (b) Radium (c) N (d) Ruthenium (e) Copper**

Analyze: Name the element corresponding to specified characteristics

Plan and Execute:

(a) The ground state element whose atoms have an electron configuration of $1s^22s^22p^63s^23p^4$ has 16 electrons and 16 protons. This is **sulfur**, S, with the atomic number of 16.

(b) The largest *known* element in the alkaline earth group (Group 2A) is the one at the bottom of the group on the periodic table. This is **radium**, Ra.

(c) The largest first ionization energy of elements in the group is the group member at the top of the group on the periodic table. In Group 5A, this is **nitrogen**, N.

(d) For a ground state element's cation to have a +2 charge and an electron configuration of $[Kr]4d^6$, the element's electron configuration would have two more electrons: $[Kr]4d^65s^2$. This element is **ruthenium**, Ru.

(e) The ground state element whose neutral atoms have an electron configuration of $[Ar]3d^{10}4s^1$ has 29 electrons and 29 protons. This is **copper**, Cu, with the atomic number of 29.

106. *Result:* **(a) Z (b) Z**

Analyze: Given some electron configurations, predict some properties.

Plan and Execute: Considering $X = [Ar]3d^84s^2 = Ni$ and $Z = [Ar]3d^{10}4s^24p^5 = Br$

(a) Using the periodic trends for sizes of first ionization energies (as described in the solution to Question 98), we find that Br has a larger first ionization energy than Ni, because they are both in period 4, and Br is found further to the right. That means **Z** is the answer.

(b) Using the periodic trends for atom sizes (as described in the solution to Question 93), we find that Br has a smaller radius than the Ni, because they are both in period 4, and Br is found further to the right. That means **Z** is the answer.

108. *Result:* **In^{4+}, Fe^{6+}, Sn^{5+}, very high successive ionization energies**

Analyze and Plan: Cations likely to form are ones that have lost only valence electrons. Any ion that has a more positive charge than its number of valence electrons is not likely to form. Anions likely to form are ones that add enough electrons to the atom to increase the number of valence electrons to eight. Any ion that has a more negative charge than the difference between eight and the number of valence electrons is not likely to form. Free ions with charges larger than ±3 are rare, because making the ion requires us to remove electrons from ions that are already very positive or add electrons to ions that are already very negative (i.e.,

Notice how large the high successive ionization energies get in Table 5.6).

Execute:

Cs has one valence electrons; Cs^{+} **is** likely to form.

In has three valence electrons; In^{4+} is **unlikely** to form because a core electron must be removed.

Fe has two valence electrons and six d electrons, so core electrons are being removed; however, Fe^{6+} is **unlikely** to form, because the positive charge is too large.

Te has six valence electrons. It needs 8 – 6 = 2 electrons. Te^{2-} **is** likely to form.

Sn has four valence electrons; Sn^{5+} is **unlikely** to form because a core electron must be removed and the positive charge is too large.

I has seven valence electrons. It needs 8 – 7 = 1 electron. I^{-} **is** likely to form.

110. *Result:* **(a) Boxes B, D, and I (b) Box E (c) Box D (d) Boxes A and G (e) Boxes H and C**

Analyze: Infrared light has wavelengths greater than 700 nm (Box C). Light with wavelengths between 400 nm and 700 nm are visible (e.g., Box G). Ultraviolet light has a wavelengths less than 300 nm (Box I).

Plan: Figure 5.8 in Section 5-3 on page 281 shows several transitions of electrons in the hydrogen atom from higher energy states to lower energy states resulting in emission of different wavelengths of light. From it we learn the following:

Execute:

Box A ($n_4 \rightarrow n_2$) represents a visible emission.

Box B ($n_3 \rightarrow n_1$) represents an ultraviolet emission.

Box D ($n_4 \rightarrow n_1$) represents an ultraviolet emission.

Box H ($n_6 \rightarrow n_3$) represents an infrared emission.

When an electron undergoes a transition from a lower energy state to a higher energy state, photons of various frequencies must be absorbed.

Box E ($n_2 \rightarrow n_7$) represents a visible absorption.

Box F ($n_3 \rightarrow n_5$) represents an infrared absorption.

(a) The ultraviolet emissions are Boxes B and D. Emissions less than 400 nm could also be ultraviolet (Box I), provided they are low enough energy to not be X-ray or gamma-rays emissions.

(b) There are two absorptions identified (Boxes E and F), the highest energy of the two is Box E, because visible light has higher energy than infrared light.

(c) There are four specific emissions identified (Boxes A, B, D, and H) and two of them (Boxes B and D) have ultraviolet wavelengths. The highest energy of that pair is Box D, because the energy level represented by n_4 is higher energy that that with n_3 and they are both making a transition to the energy level represented by n_1.

(d) The visible emission is Box A. Any emissions between 400 nm and 700 nm will also be visible (e.g., Box G), so if the wavelength described here is the result of an emission, it is visible.

(e) The infrared emission is Box H. Any emissions above 700 nm could also be infrared (e.g., Box C), provided they are high enough energy to not be called microwaves or radiowaves.

112. *Result:* **(a) replace "inversely" with "directly" (b) it is inversely proportional to the square of n (c) replace "as soon as" with "before" (d) replace "frequency" with "wavelength"**

Analyze, Plan, and Execute:

(a) This statement: "The energy of a photon is inversely related to its frequency." is **false**. Actually, the energy of a photon is **directly** related to its frequency according to the equation: $E = h\nu$.

(b) This statement: "The energy of the hydrogen electron is inversely proportional to its principle quantum number n." is **false**. Actually, the energy of the hydrogen electron is **inversely proportional to the square of** its principle quantum number, n.

$$E = -\frac{2.179 \times 10^{-18}\ J}{n^2}$$

(c) This statement: "Electrons start to fill the fourth energy level as soon as the third level is full." is **false**. Several different slight changes can make it into a true statement, for example: "Electrons start to fill the fourth energy level **before** the third level is full" This can be seen in the electron configuration for potassium and calcium: $1s^22s^22p^63s^23p^64s^1$ and $1s^22s^22p^63s^23p^64s^2$ where the 4s sublevel fills before the 3d sublevel starts to fill.

(d) This statement: "Light emitted by an n = 4 to n = 2 transition will have a longer frequency than that from an n = 5 to n = 2 transition." is **false**. Several different slight changes can make it into a true statement, for example: "Light emitted by an n = 4 to n = 2 transition will have a longer **wavelength** than that from an n = 5 to n = 2 transition." The energy difference is smaller in the first transition than the second, so the frequency of the emitted photon is smaller, and its wavelength is longer.

115. *Result:* **(a) $[Rn]5f^{14}6d^{10}7s^27p^68s^2$ (b) magnesium (or any other Group 2A elements) (c) XO, XCl_2**

Analyze and Plan: This question asks us to use the predictive power of the periodic table. Take the atom with atomic number 120 and write an electron configuration, predict a periodic table group member, and write the compounds of some predicted compounds.

Execute: The neutral element has 120 protons and 120 electrons.

(a) The ground state electron configuration has 120 electrons: **$[_{86}Rn]5f^{14}6d^{10}7s^27p^68s^2$**. Adding the Rn 86 electrons in the noble gas core to the 14 + 10 + 2 + 6 + 2 outer electrons, makes a total of 120 electrons.

(b) The valence electrons are the two electrons with n = 8. We will predict this element to be a member of Group 2A. There are several elements in this group, including **magnesium**.

(c) Use X for the symbol for unbinilium. Group 2A metals will combine with nonmetals forming ionic compounds. The metallic elements in that group can be predicted to form +2 cations They will combine with the anions of oxygen and chlorine (oxide, O^{2-}, and chloride, Cl^-) to form **XO** and **XCl_2**, respectively.

Applying Concepts

121. *Result:* **(a) Br^- product has a noble gas electron configuration, so it is very stable. (b) Mg atom has all full subshells. (c) P atom has a half-full p-subshell. (d) Cl electron comes from a 3p orbital and Br comes from a 4p orbital.**

Analyze and Plan: Write the electron configuration for the two ions, and consider what changes occur when the electrons are added (electron affinity) or removed (ionization).

Electron affinity is the ΔE for this equation:

$$A(g) + e^- \longrightarrow A^-(g)$$

Ionization energy is the ΔE for this equation:

$$A(g) \longrightarrow A^+(g) + e^-$$

Execute:

(a) Write the electron configuration for the atoms and the anions produced by adding an electron:

$_{34}Se$ electron configuration: $[_{18}Ar]4s^23d^{10}4p^4$ $_{34}Se^-$ electron configuration: $[_{18}Ar]4s^23d^{10}4p^5$

$_{35}Br$ electron configuration: $[_{18}Ar]4s^23d^{10}4p^5$ $_{35}Br^-$ electron configuration: $[_{36}Kr]$

$_{53}$I electron configuration: $[_{36}Kr]5s^2 4d^{10} 5p^5$ $\quad$ $_{53}I^-$ electron configuration: $[_{54}Xe]$

The **Br^- product has a noble gas electron configuration, so it is very stable.** Thus, selenium has a lower (less negative) electron affinity than bromine.

Adding an electron to a lower-energy subshell will produce more energy.

The electron added to Br goes into a 4p orbital and the one added to I goes into a 5p orbital.

Thus, iodine has a lower (less negative) electron affinity than bromine.

(b) Write the electron configuration for the atoms and the cations produced by removing an electron:

$_{13}$Al electron configuration: $[_{10}Ne]3s^2 3p^1$ $\quad$ $_{13}Al^+$ electron configuration: $[_{10}Ne]3s^2$

$_{12}$Mg electron configuration: $[_{10}Ne]3s^2$ $\quad$ $_{12}Mg^+$ electron configuration: $[_{10}Ne]3s^1$

The **$_{12}$Mg atom has all full subshells** so removing an electron requires more energy. Thus, aluminum has a lower first ionization energy than magnesium.

(c) Write the electron configuration for the atoms and the cations produced by removing an electron:

$_{16}$S electron configuration: $[_{10}Ne]3s^2 3p^4$ $\quad$ $_{16}S^+$ electron configuration: $[_{10}Ne]3s^2 3p^3$

$_{15}$P electron configuration: $[_{10}Ne]3s^2 3p^3$ $\quad$ $_{15}P^+$ electron configuration: $[_{10}Ne]3s^2 3p^2$

The **$_{15}$P atom has a half-full p-subshell** so removing an electron requires more energy. Thus, sulfur has a lower first ionization energy than phosphorus.

(d) Write the electron configuration for the atoms and the cations produced by removing an electron:

$_{35}$Br electron configuration: $[_{18}Ar]4s^2 3d^{10} 4p^5$ $\quad$ $_{35}Br^+$ electron configuration: $[_{18}Ar]4s^2 3d^{10} 4p^4$

$_{17}$Cl electron configuration: $[_{10}Ne]3s^2 3p^5$ $\quad$ $_{17}Cl^+$ electron configuration: $[_{10}Ne]3s^2 3p^4$

Removing an electron from a lower-energy subshell will require more energy. The electron removed from **Cl comes from a 3p orbital and the one removed from Br comes from a 4p orbital**. Thus, bromine has a lower first ionization energy than chlorine.

123. *Result/Explanation:* The element X has an electron configuration: $1s^2 2s^2 2p^6 3s^1$ with one valence electron ($3s^1$), so X is in Group 1A. Group 1A metals will combine with nonmetals forming ionic compounds. The Group 1A elements form cations with a 1+ charge. X^+ will combine with the anionic form of chlorine (chloride, Cl^-) to form **XCl**.

125. *Result:* **(a) ground state (b) could be ground state or excited state (c) excited state (d) impossible (e) excited state (f) excited state**

Analyze, Plan, and Execute:

(a) $1s^2 2s^1$ is a ground state electron configuration. The 1s sublevel has the lowest energy it is full. The 2s sublevel has the second lowest energy and the remaining electron is there. This is the lowest energy electron configuration, so it would be called the **ground state**.

(b) $1s^2 2s^2 2p^3$ could be **ground state or excited state**. If each of the p electrons is in a separate orbital with the same spin, that is the ground state. However, if any of the spins are reversed or if any of the electrons are paired, this is a different and higher energy excited state. We would need to see an orbital energy diagram of this state to determine an unambiguous answer:

Ground state 2p orbital energy diagram: 2p [↑][↑][↑]

Some excited state 2p orbital energy diagrams: 2p [↓][↑][↑] $\quad$ 2p [↑↓][↑][ ]

(c) $[Ne]3s^23p^34s^1$ is an **excited state**. The lower-energy 3p sublevel needs to be full ($3p^6$) before the 4s sublevel gets any electrons; therefore, this possible electron configuration is not the lowest energy state, and would be called an excited state. The ground state electron configuration would be $[Ne]3s^23p^4$.

(d) $[Ne]3s^23p^64s^33d^2$ is **impossible**. The 4s sublevel has only one orbital, so the maximum number of electrons in 4s is 2. This electron configuration has three electrons in that orbital.

(e) $[Ne]3s^23p^64f^4$ is an **excited state**. Several lower-energy sublevels need to be full before the 4f sublevel gets any electrons; therefore, this possible electron configuration is not the lowest energy state, and would be called an excited state. The ground state electron configuration would be $[Ne]3s^23p^63d^24s^2$.

(f) $1s^22s^22p^43s^2$ is an **excited state**. The lower-energy 2p sublevel needs to be full ($2p^6$) before the 3s sublevel gets any electrons; therefore, this possible electron configuration is not the lowest energy state, and would be called an excited state. The ground state electron configuration would be $1s^22s^22p^6$.

127. *Result:* **$[Rn]5f^{14}5g^{18}6d^{10}6f^{14}7s^27p^67d^{10}8s^28p^2$; bottom of Group 4A**

Analyze and Plan: This question asks us to use the predictive power of the periodic table. Appendix D can be of some assistance, also. Since the atomic number of our undiscovered element is 164, the number of protons and electrons in a neutral atom is 164. We need to extend Figure 5.19 by extending Figure 5.21, to predict what sublevels the new electrons fill.

Execute:

After 7p, the 8s orbitals fill. Then we fill the first g orbitals, 5g. Then we fill: 6f, 7d, 8p, 9s, and so on. The g sublevel has $\ell = 4$ and nine possible m_ℓ values, so there are nine orbitals and the g sublevel's capacity is 18. Constructing the electron configuration in the order in which the electrons fill the sublevels looks like this:

$$[Rn]7s^25f^{14}6d^{10}7p^68s^25g^{18}6f^{14}7d^{10}8p^2$$

(86 + 2 + 14 + 10 + 6 + 2 + 18 + 14 + 10 + 2 =) 164 electrons. We have added enough electrons now. To conform to those shown in Appendix D, we then rearrange the sublevels into shell order:

$$\mathbf{[Rn]5f^{14}5g^{18}6d^{10}6f^{14}7s^27p^67d^{10}8s^28p^2}$$

There are four valence electrons in this element ($8s^28p^2$), so we will predict it is at the bottom of Group 4A.

128. *Result:* **(a) increase, decrease (b) helium (c) 5 and 13 (d) no electrons left (e) It is a core electron. (f) $Mg^{2+}(g) \longrightarrow Mg^{3+}(g) + e^-$**

Analyze, Plan, and Execute:

(a) Based on the graphic data, ionization energies **increase** left to right and **decrease** top to bottom on the periodic table.

(b) The element with the largest first ionization energy is **helium** (atomic number = 2).

(c) The peaks of the fourth ionization energies will occur at the atomic number of **5** (boron) and at the atomic number of 13 (aluminum).

(d) He has only two electrons, so after two ionizations, there are **no electrons left**.

(e) First electron in lithium is a valence electron, but the **second electron is a core electron**, which takes a greater amount of energy to ionize.

(f) The atomic number of 12 belongs to magnesium, $_{12}Mg$. The third ionization is removing an electron from the +2 gas-phase ion.

$$\mathbf{Mg^{2+}(g) \longrightarrow Mg^{3+}(g) + e^-}$$

More Challenging Questions

130. *Result:* **(a) 3.35 (b) 3.35 (c) 10.85**

Analyze: Calculate the effective nuclear charge, Z*, on three electrons in a tin atom, $_{50}Sn$.

Plan: To determine σ in the effective nuclear charge equation $Z^* = Z - \sigma$, write the electron configuration for an atom. Group the subshells this way: (1s) (2s, 2p) (3s, 3p) (3d) (4s, 4p) (4d) (4f) (5s, 5p) . . . Choose the electron for which you want to calculate Z*. Sum these contributions to σ for all other electrons: Electrons in groups to the right of the selected electron contribute zero. Electrons in the same group as the selected electron contribute 0.35 per electron (except for (1s) where the contribution is 0.30). If the selected electron is in an (ns, np) group, electrons in the n − 1 shell contribute 0.85 and electrons in the shells with lower n contribute 1.0. If the selected electron is in an (nd) or (nf) group, all electrons in groups to the left count 1.0.

Execute: Write the electron configuration and group it as explained:

$$(1s^2)(2s^22p^6)(3s^23p^6)(3d^{10})(4s^24p^6)(4d^{10})(5s^25p^2)$$

(a) Look at the electron in a 5s orbital: There is no electron group to the right of the 5s group. Three electrons are in the same group as the selected electron; they contribute 0.35 per electron. Since the selected electron is in a (5s, 5p) group, electrons in the n = 4 shell contribute 0.85 and electrons in the 3, 2, and 1 shells contribute 1.0.

$$\sigma = 3 \times (0.35) + 18 \times (0.85) + 28 \times (1.00) = 46.65$$

$$Z^* = Z - \sigma = 50 - 46.65 = \mathbf{3.35}$$

(b) Look at the electron in a 5p orbital: The electron is in the same group as 5s, so the shielding and effective nuclear charge will be the same as 5s. σ = 46.65 Z* = **3.35**

(c) Look at the electron in a 4d orbital: There are 4 electrons in a group to the right of 4d; they contribute zero. Nine electrons are in the same group as the selected electron; they contribute 0.35 per electron. Since the selected electron is in a (4d) group, all 36 electrons in groups to the left of the (4d) group contribute 1.0.

$$\sigma = 9 \times (0.35) + 36 \times (1.00) = 39.15$$

$$Z^* = Z - \sigma = 50 - 39.15 = \mathbf{10.85}$$

☑ *Reasonable Result Check:* It makes sense that the n=4 shell electron experiences a greater effective nuclear charge than the n=5 shell electrons.

133. *Result:* **(a) 4.34×10^{-19} J (b) 6.54×10^{14} Hz**

Analyze: Given wavelengths of light absorbed by an electron ejected from a metal and the kinetic energy of the ejected electron, determine the binding energy and the minimum frequency required for the photoelectric effect.

Plan: Use the equation in Section 5-2 to find the energy of each photon. Subtract the kinetic energy of the electron from the photon energy to find the binding energy. Use E = hν to find the lowest frequency photon.

Execute:

(a) $\lambda = 4.00\times10^{-7}$ m $\quad E_{photon} = \dfrac{hc}{\lambda} = \dfrac{6.626\times10^{-34}\ \text{J}\cdot\text{s}\times2.998\times10^{8}\,\text{m/s}}{4.00\times10^{-7}\,\text{m}} = 4.97\times10^{-19}\ \text{J}$

Binding Energy = Photon Energy − Kinetic Energy $= 4.97\times10^{-19}\ \text{J} - 0.63\times10^{-19}\ \text{J} = 4.34\times10^{-19}\ \text{J}$

(b) $\nu_{photon} = \dfrac{E}{h} = \dfrac{4.34\times10^{-19}\,\text{J}}{6.626\times10^{-34}\ \text{J}\cdot\text{s}} = 6.55\times10^{14}\,\text{s}^{-1} = 6.55\times10^{14}\ \text{Hz}$

☑ *Reasonable Result Check*: The binding energy is lower than the photon energy, since the photon required to eject a zero kinetic energy electron needs to have less energy than one with 6.3×10^{-20} J.

135. *Result:* **2.18×10^{-18} J**

Analyze: Given an equation describing the relationship between the electron energy level, the atomic number, Z, and the quantum number, n, determine the ionization energy of the electron.

Plan: Subtract the ionization energy state (n = ∞) of the electron from ground state (n = 1) of the electron in hydrogen.

Execute: Z = 1 for hydrogen

$$E_{\infty} = -\frac{1^2}{(\infty)^2}\left(2.18 \times 10^{-18}\ \text{J}\right) = 0$$

$$E_1 = -\frac{1^2}{(1)^2}\left(2.18 \times 10^{-18}\ \text{J}\right) = 2.18 \times 10^{-18}\ \text{J}$$

$$E_{\text{ionization}} = E_{\infty} - E_1 = 0 - (-2.18 \times 10^{-18}\ \text{J}) = 2.18 \times 10^{-18}\ \text{J}$$

☑ *Reasonable Result Check*: The ionization of hydrogen is given in Figure 5.10 as the energy required to go from n = 1 to n= = ∞.

137. *Result/Explanation:* To remove an electron from an N atom requires disruption of a half filled sub-shell (p^3), which is relatively stable, so the ionization of oxygen will be less energy than the ionization of nitrogen.

139. *Result:* **(a) Ir (b) [Rn]$7s^2 5f^{14} 6d^7$**

Analyze, Plan, and Execute:

(a) Another transition element with the parallel outer electron configuration as Mt, is iridium, **Ir**. It is found directly above Mt on the periodic table. These two elements have the same number of electrons in the same types of orbitals. (Note, of course, that the core is different, and the values of n for electrons in Ir are one less than the n values in Mt.) The ground state electron configuration of Ir using the xenon core is [$_{54}$Xe]$6s^2 4f^{14} 5d^7$.

(b) The ground state electron configuration of Mt using the radon core is **[$_{86}$Rn]$7s^2 5f^{14} 6d^7$**

141. *Result/Explanation:* Copernicium is known and its existence is verified.

(a) Element 112 will reside in Group 2B, as one of the **transition metals**, directly below mercury.

(b) The ground state electron configuration of element 112 using the radon core is **[$_{86}$Rn]$7s^2 5f^{14} 6d^{10}$**.

Chapter 6: Covalent Bonding

Solutions for Red-Numbered Questions for Review and Thought

Topical Questions

Covalent Bonding (Section 6-1)

13. *Result/Explanation:* When two H atoms come close enough, the sharing of electrons lowers their energy since each H atom can achieve a noble gas electron configuration of He. This explains why they have lower energy at 74 pm than at 100 pm. If the atoms are pushed too close together, the repulsive interaction between the positive nuclei raises the energy. This explains why the energy at 25 pm is higher than at 74 pm.

Lewis Structures (Sections 6-2, 6-4)

15. *Result:* (a) 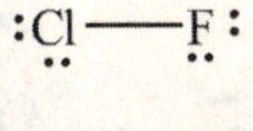 (b) 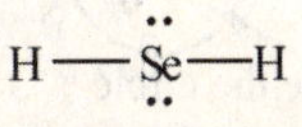 (c) 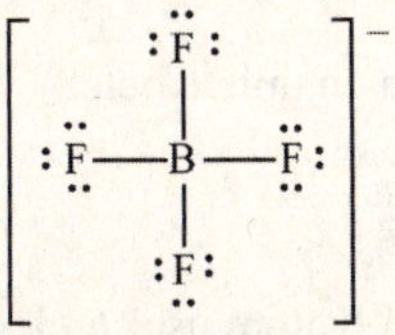 (d) 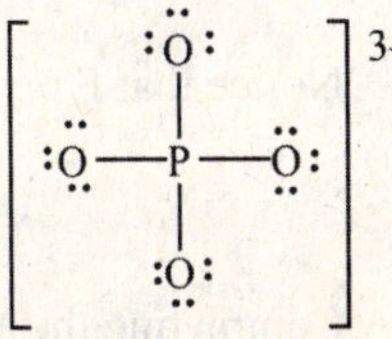

Analyze: Write Lewis structures for a list of ions and molecules.

Plan: Following a systematic plan will give you reliable results every time. Trial and error often works for small molecules, but as molecules get more complex, it's better to follow a procedure than to try to guess where electrons will end up. There are a number of methods. The one described here is the same as that described in the text and it works reliably.

[A] Count the total number of valence electrons. If there is a nonzero charge, adjust the electron count appropriately. Add electrons for negative charges and subtract electrons for positive charges.

[B] Use atomic symbols to draw a skeleton structure by joining the atoms with shared pairs of electrons (a single line for each shared pair). As a general rule, the inner, central atom is usually the first element in the formula, and it will often have the smallest electronegativity (see Section 6-7). There are some exceptions, for example: H will **never** be the central atom. The tendency of certain atoms to make certain numbers of bonds will also help inform you of how to arrange the skeleton structure (e.g., we often see elements in Group 4A make four bonds, Group 5A make 3 bonds, Group 6A make two bonds, and Group 7A and H make one bond.) Connect each atom in the skeleton structure with a single bond.

[C] Each single bond (–) is formed using two electrons. Subtract the electrons used for single bonds from the total number of electrons.

[D] Starting with the terminal "outer" atoms, place lone pairs of electrons around each atom (except H) to satisfy the octet rule.

[E] Subtract the electrons used for lone pairs from the remaining total.

[F] Place all remaining electrons on the central atom, even if it will give the central atom more than an octet. **Any atom that gets more than eight electrons must be from the third period or higher.**

[G] Once all the electrons are on the structure, check the octet of the inner, central atom(s).

(i) If each atom has eight or more electrons, then the structure is complete.

(ii) If the number of electrons around the central atom is less than eight, change single bonds to the central atom into multiple bonds. Do this by moving a lone pair from one of the outer atoms to make a new shared-pair, forming a multiple bond to the central/inner atom. Repeat this procedure until all atoms have an octet (or more). You can only make more than the minimum number of multiple bonds if you have an explainable reason (See Section 6-9 and 6-10).

[H] If the structure is an ion, put it in brackets and designate the net charge outside the bracket at the upper right corner.

[I] Check the octets of every atom in the structure (and make sure H atoms have only two electrons) then count the electrons and make sure the total is right.

Execute:

(a) [A] Cl and F atoms are both in Group 7A: $7\ e^- + 7\ e^- = 14\ e^-$

[B] Choose the Cl atom as the central atom with the F atom bonded to it, since it has the lower electronegativity.

Cl—F

[C] Connecting the two atoms uses two electrons, so subtract 2 electrons from the current electron count: $14\ e^- - 2\ e^- = 12\ e^-$

[D] F has two electrons already (in the single bond, so add six dots to the F atom: 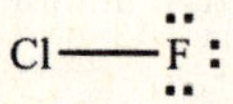

Notice that F now has a complete octet:

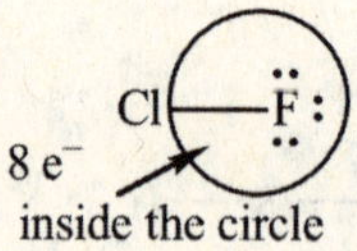

[E] Completing the octet of F atom uses 6 electrons, so subtract 6 electrons from the current electron count: : $12\ e^- - 6\ e^- = 6\ e^-$

[F] Put the last six electrons on Cl 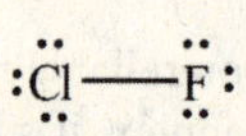

[G] Cl now has a complete octet.

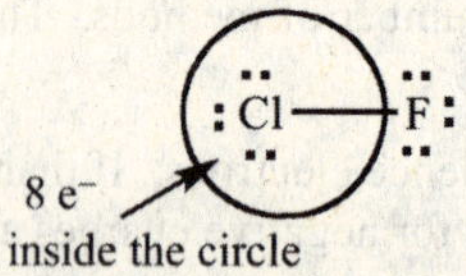

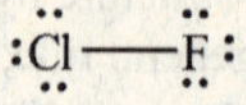

So, the structure is complete.

[H] The structure is not an ion.

[I] Cl has eight electrons (one pair in a single bond and six dots). F has eight electrons (one pair in a single bond and six dots). Looking at the structure the total number of electrons can be counted: $2\ e^-$ in a bond + $12\ e^-$ as dots = $14\ e^-$ total.

(b) [A] H atom is in Group 1A and Se atom is in Group 6A: $2 \times (1\ e^-) + 6\ e^- = 8\ e^-$

[B] Se atom is the central atom with H atoms around it.

H—Se—H

[C] Connecting the H and Se atoms uses four electrons: $8\ e^- - 4\ e^- = 4\ e^-$

[D] H atoms need no more electrons.

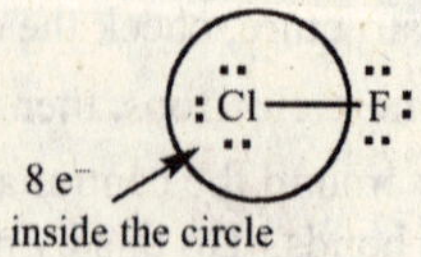

[E] No electrons used in this step. So we still have $4\ e^-$.

[F] Put the last four electrons on Se.

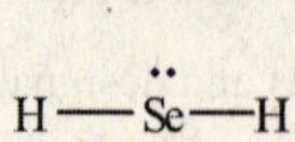

[G] Se has eight electrons, so the structure is complete.

[H] The structure is not an ion.

[I] Se has eight electrons (two pairs in single bonds and four dots). Each H has two electrons (in one single bond). The total number of electrons can be counted: 4 e^- in 2 bonds + 4 e^- as dots = 8 e^- total.

(c) [A] B atom is in Group 3A and F atoms are in Group 7A. The 1– charge **adds** one extra electron: 3 e^- + 4 × (7 e^-) + 1 = 32 e^-

[B] B atom is the central atom with the F atoms around it:

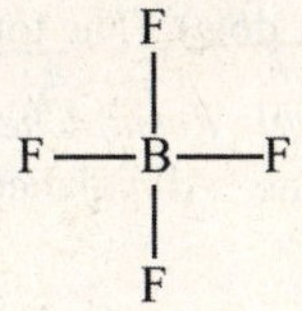

[C] Connecting four F atoms uses eight electrons: 32 e^- – 8 e^- = 24 e^-

[D] Complete the octets of all F atoms:

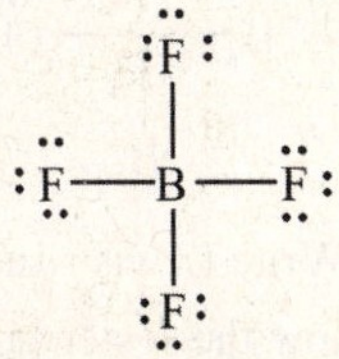

[E] Completing the octets of the four F atoms uses 24 electrons: 24 e^- – 24 e^- = 0 e^-

[F] All electrons are used.

[G] B has eight electrons, so the structure is complete.

[H] The structure is an ion, so put it in brackets and add the – charge.

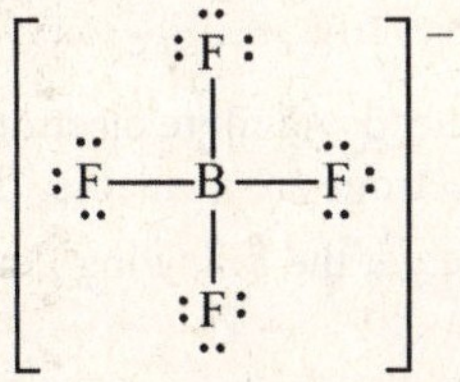

[I] B has eight electrons (four pairs in single bonds). F has eight electrons (one pair in a single bond and six dots). The total number of electrons can be counted: 8 e^- in four bonds + 24 e^- in dots = 32 e^- total.

(d) [A] P atom is in Group 5A and O atoms are in Group 6A. The 3– charge **adds** three extra electrons: 5 e^- + 4 × (6 e^-) + 3 = 32 e^-

[B] P atom is the central atom with the O atoms around it.

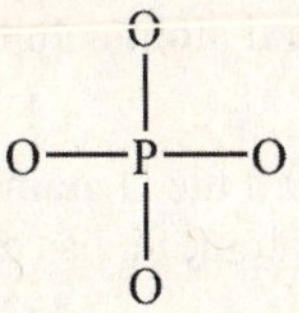

[C] Connecting four O atoms uses eight electrons: 32 e^- – 8 e^- = 24 e^-

[D] Complete the octets of all the O atoms:

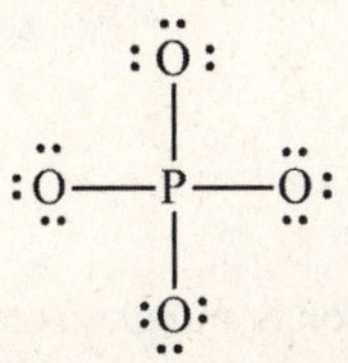

[E] Completing the octets of the four O atoms uses 24 electrons: 24 e^- – 24 e^- = 0 e^-

[F] All electrons are used.

[G] P has eight electrons, so the structure is complete.

[H] The structure is an ion, so put it in brackets and add the 3– charge.

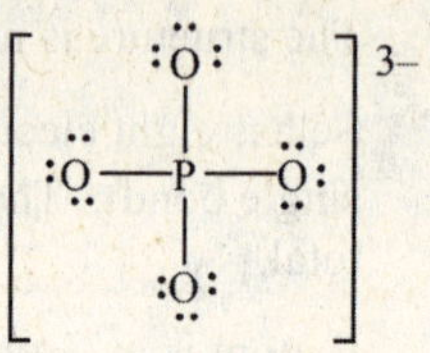

[I] P has eight electrons (four pairs in single bonds). Each O has eight electrons (one pair in a single bond and six dots). The total number of electrons can be counted: 8 e^- in four bonds + 24 e^- in dots = 32 e^- total.

☑ *Reasonable Result Check:* All the Period 2 and 3 elements have an octet of electrons. The H atoms have two electrons. All valance electrons are accounted for in the structures, either as members of shared pairs or of lone pairs.

17. *Result:* **(a)** 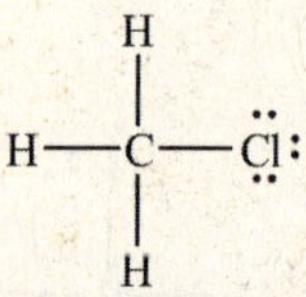 **(b)** 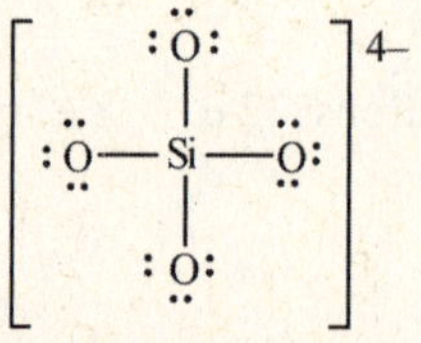 **(c)** 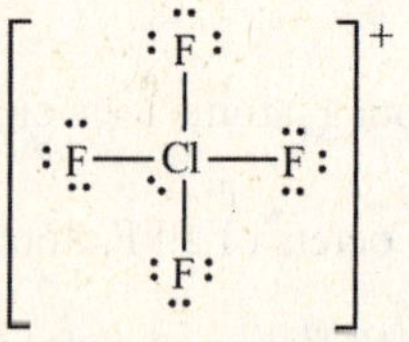 **(d)** 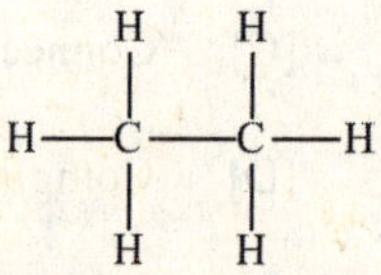

Analyze: Write Lewis structures for a list of ions and molecules.

Plan: Follow the systematic plan given in the solution to Question 15. Consult Chapter 2 Section 2-9 for details about structural formulas of organic compounds.

Execute:

(a) C is in Group 4A, H is in Group 1A, and Cl is in Group 7A: $4\ e^- + 3 \times (1\ e^-) + 7\ e^- = 14\ e^-$.

The C atom is the central atom with the other atoms around it. Connecting four atoms uses eight electrons: $14\ e^- - 8\ e^- = 6\ e^-$.

H needs no more electrons. Completing the octet of Cl uses six electrons: $6\ e^- - 6\ e^- = 0\ e^-$. No more electrons are available. The central atom, C, has eight electrons in shared pairs, so it needs no more.

We get the following Lewis structure:

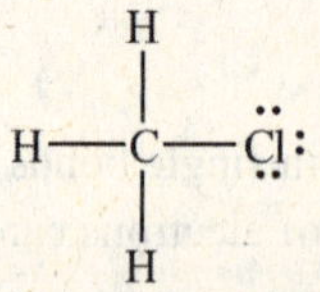

Check: Octet on Cl. Each H has two e^-. Total 8 e^- in bonds + 6 e^- as dots = 14 e^-.

(b) Si is in Group 4A and O is in Group 6A. The 4– charge **adds** four extra electrons: $4\ e^- + 4 \times (6\ e^-) + 4\ e^- = 32\ e^-$.

The Si atom is the central atom with the O atoms around it. Connecting four O atoms uses eight electrons: $32\ e^- - 8\ e^- = 24\ e^-$.

Complete the octets of all the O atoms using 24 electrons. $24\ e^- - 24\ e^- = 0\ e^-$. No more electrons are available. The central atom, Si, has eight electrons in shared pairs, so it needs no more.

The Lewis structure is an ion, so put it in brackets and add the 4– charge:

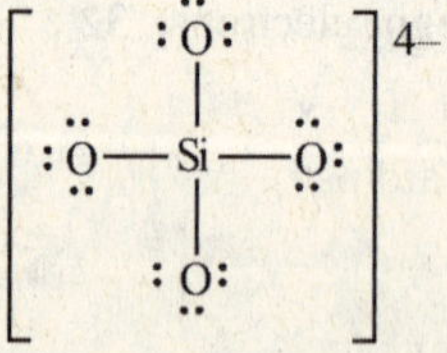

Check: Octets on S and O. Total 8 e^- in bonds + 24 e^- as dots = 32 e^-.

(c) The atoms Cl and F are both in Group 7A. The + charge **removes** one electron: $7\ e^- + 4 \times (7\ e^-) - 1\ e^- = 34\ e^-$.

The Cl atom is the central atom with the F atoms around it. Connecting four F atoms uses eight electrons: $34\ e^- - 8\ e^- = 26\ e^-$.

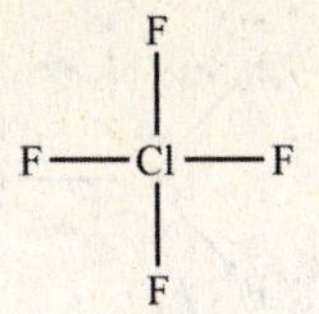

Complete the octets of all the F atoms, using 24 electrons: 26 e⁻ – 24 e⁻ = 2 e⁻. Put the last two electrons on the Cl. The central atom, Cl, has ten electrons already, so it needs no more.

The structure is an ion, so put it in brackets and add the + charge.

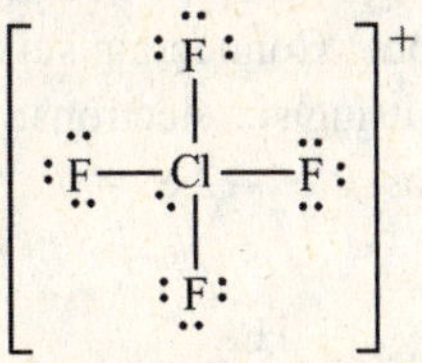

Check: Octets on F atoms. Ten e⁻ on Cl (third-period). Total 8 e⁻ in bonds + 24 e⁻ as dots = 32 e⁻.

(d) C in Group 4A and H in Group 1A: 2 × (4 e⁻) + 6 × (1 e⁻) = 14 e⁻.
The structural formula of ethane molecule has three H atoms bonded to each C, and the two carbons bonded to each other. Connecting eight atoms uses 14 electrons:
14 e⁻ – 14 e⁻ = 0 e⁻ No more electrons are available. Each C has eight electrons in shared pairs, so they need no more.

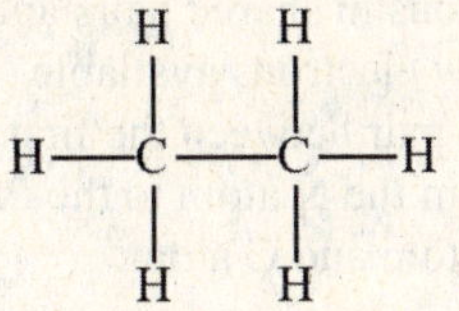

Check: Octet on Cl. Each H has two e⁻. Total 14 e⁻ in 7 bonds = 14 e⁻.

☑ *Reasonable Result Check:* All the Period 2 elements have an octet of electrons. The H atoms have two electrons. All the Period 3 elements have eight or more electrons. All valance electrons are accounted for in the structures, either as members of shared pairs or of lone pairs.

19. *Result:* **(a)**

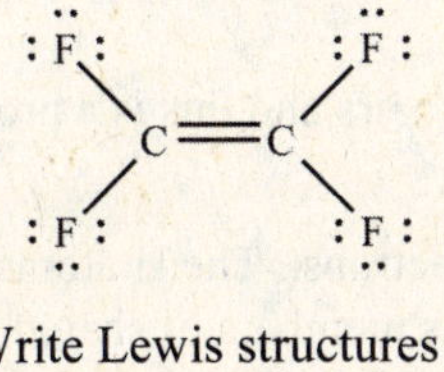

(b)

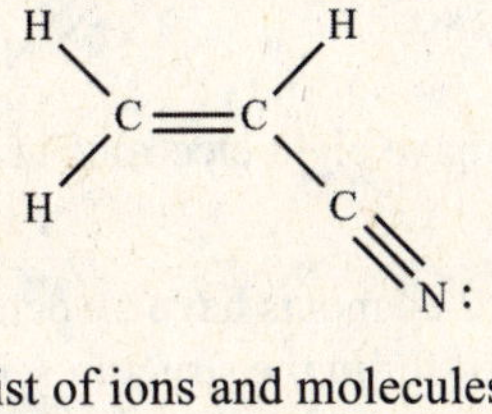

Analyze: Write Lewis structures for a list of ions and molecules.

Plan: Follow the systematic plan given in the solution to Question 15. Consult Chapter 2 Section 2-9 for details about structural formulas of organic compounds.

Execute:

(a) 2 × (4 e⁻) + 4 × (7 e⁻) = 36 e⁻. This organic molecule looks like ethene with all the H atoms changed to F atoms. The two carbon atoms are bonded to each other, and each has two F atoms bonded to it. Connecting six atoms with single bonds uses 10 electrons: 36 e⁻ – 10 e⁻ = 26 e⁻. Complete the octets of the F atoms, using 24 electrons: 26 e⁻ – 24 e⁻ = 2 e⁻. Put the last pair on one of the C atom, using the last of the electrons: 2 e⁻ – 2 e⁻ = 0 e⁻. At this point, the structure looks like this:

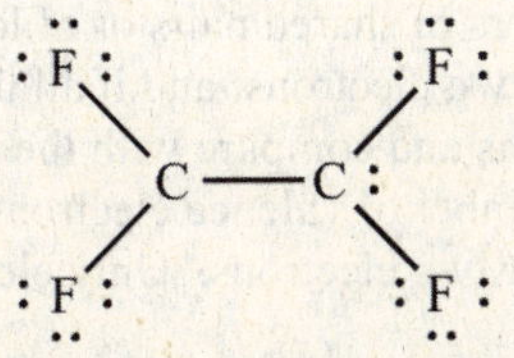

(incomplete)

The first C atom has only six electrons in shared pairs, so it needs two more, but there are no more electrons available for lone pairs. Therefore, we must move one lone pairs of electrons from the right C atom to make a new shared-pair, forming a double bond between the two C atoms:

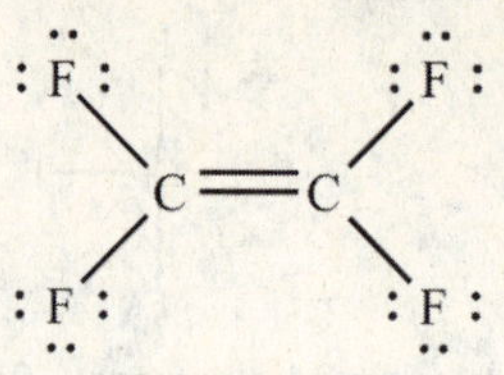

The first C atom now has eight electrons in shared pairs and this is the proper Lewis structure.

(b) $3 \times (4) + 3 \times (1) + 5 = 20\ e^-$. The structural formula given here tells us that this organic compound has two H atoms and a C atom bonded to the first C atom, an H atom and a C atom bonded to the second C atom, and an N atom bonded to the third C atom. Connecting seven atoms uses 12 electrons: $20\ e^- - 12\ e^- = 8\ e^-$. Complete the octet of the N atom using six electrons: $8\ e^- - 6\ e^- = 2\ e^-$. Put the last pair on one of the C atom, using the last of the electrons: $2\ e^- - 2\ e^- = 0\ e^-$. At this point, we have the following structure:

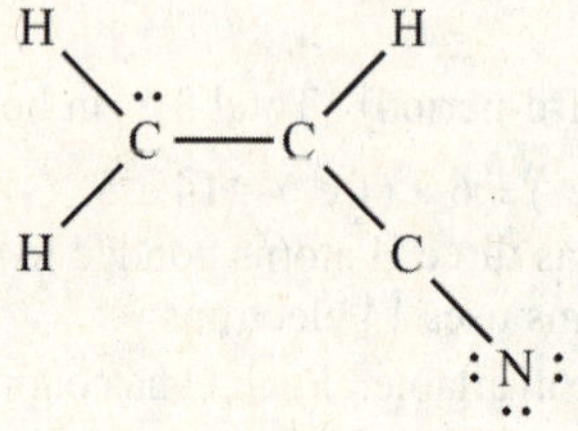

(incomplete)

The second C atom has only six electrons in shared pairs and the third C atom has only four electrons in shared pairs, but there are no more new electrons available. Therefore, we must move one lone pair from the first C atom to make a new shared-pair between the first and second C atoms, making a double bond. We must also move two lone pairs from the N atom to the third C atom to make two new shared pairs, forming a triple bond between the N atom and C atom:

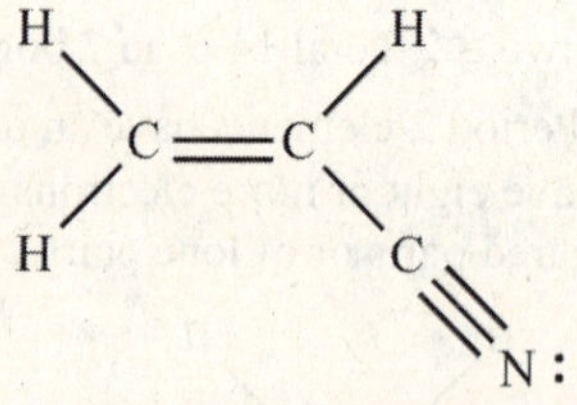

The second and third C atoms now both have eight electrons in shared pairs and this is a proper Lewis structure.

☑ *Reasonable Result Check:* All the Period 2 elements have an octet of electrons. The H atoms have two electrons. All valance electrons are accounted for in the structures, either as members of shared pairs or of lone pairs.

21. *Result:* **All incorrect. (a) The structure has two few electrons and violates the octet rule on both F atoms. (b) The structure has too few electrons. (c) The structure has an in appropriate arrangement for the atoms, is missing one H atom, and violates the rule of two for hydrogen. (d) The structure has too many electrons; carbon atom has nine electrons, so the single electron should be deleted; oxygen has ten electrons so one of the lone pairs should be removed. (e) The structure has too few electrons and violates the octet rule.**

Analyze: Determine if given Lewis structures are correct and explain what is wrong with the incorrect ones.

Plan: A Lewis structure is correct if it has the correct type and number of atoms, if all valance electrons are accounted for in the structures, either as members of shared pairs or of lone pairs; if all the Period 2 elements have an octet of electrons; if the H atoms have two electrons; and if all the Period 3 and higher elements have eight or more electrons. So, first count the atoms and compare with the formula. Then count the electrons in the structure and compare that with the total number of valence electrons. If the count is correct, then check the octets of all the elements and check H atoms for two electrons. Only elements in period 3 and higher are allowed to exceed eight electrons.

Execute:

(a) OF_2 **Check electron count**: $6 + 2 \times (7) = 20\ e^-$. The given structure has 8 electrons (two lone pairs and two single bonds), so the total electron count is wrong. This structure is incorrect.

(b) O_2 **Check electron count**: $2 \times (6) = 12\ e^-$. The given structure has 10 electrons (two lone pairs and a triple bond), so the total electron count is wrong. This structure is incorrect.

(c) CH_3Cl **Check type and number of atoms:** The structure has only two H atoms, not three.

Check electron count: $4 + 3 \times (1) + 7 = 14\ e^-$. The given structure has 16 electrons (four lone pairs, and four single bonds), so the total electron count is wrong.

Check octets, etc.: The structure also has one hydrogen atom with too many electrons (4, not 2).

This structure is incorrect.

(d) CCl_2O **Check electron count**: $4 + 2 \times (7) + 6 = 24\ e^-$. The given structure also has 27 electrons (nine lone pairs, one unpaired electron, two single bonds, and one double bond), so the total electron count is wrong.

Check octets, etc.: Both Cl atoms each have eight electrons, but the C atom has 9 electrons and the O atom has 10 electrons, so the octet rule is not satisfied.

(e) NO_2^- **Check electron count**: $5 + 2 \times (6) + 1 = 18\ e^-$. The given structure has 16 electrons (five lone pairs, one single bond, and one double bond), so the total electron count is wrong. This structure is incorrect. It is also missing the brackets and the net charge.

☑ *Reasonable Result Check:* We used what we know about Lewis structures to determine the incorrect structures. Let's draw correct Lewis structures for those that were incorrect, (a), (b), (d) and (e).

The correct structure for (a) OF_2 needs 12 more electrons; each F atom needs three more lone pairs:

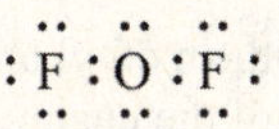

The correct structure for (b) O_2 needs 2 more electrons, two more lone pairs, and one less shared pair:

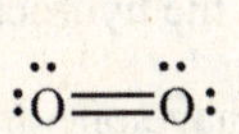

The correct structure for (d) has two fewer electrons and the carbon bonded to the chlorine without an H atom between them. Remember, H atom only shares two electrons:

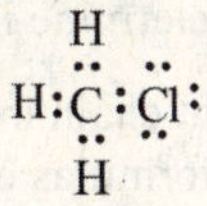

The correct structure for (e) has two more electrons on N (and brackets with a charge):

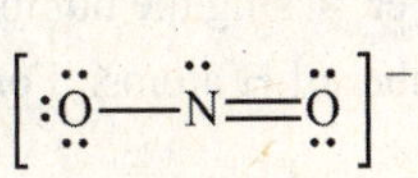

The corrected structures are different from the incorrect ones, so these answers look right.

Bonding in Hydrocarbons (Sections 6-3, 6-5)

23. *Result:* **(see structures below)**

Analyze: Write structural formulas for all the branched-chain compounds with a given formula.

Plan: Be systematic. Start with a long chain that has only one methyl branch. Move the methyl around (but don't put it on the end carbon and don't put it on a carbon past the first half of the chain). Then make two methyl branches, and move them around similarly. Continue this process until the main chain is too short for methyl branches. Then make an ethyl branch and move it around (but don't put it on the end carbon or the carbon next to the end carbon and don't put it on a carbon past the first half of the chain). Follow a similar pattern with ethyls as with methyls.

Execute: The straight chain isomer of C_6H_{14} has six carbons, so start with a five-carbon chain.

One methyl branch can go on the five-carbon chain in two different ways. The methyl branch can go on the second carbon or on the third carbon.

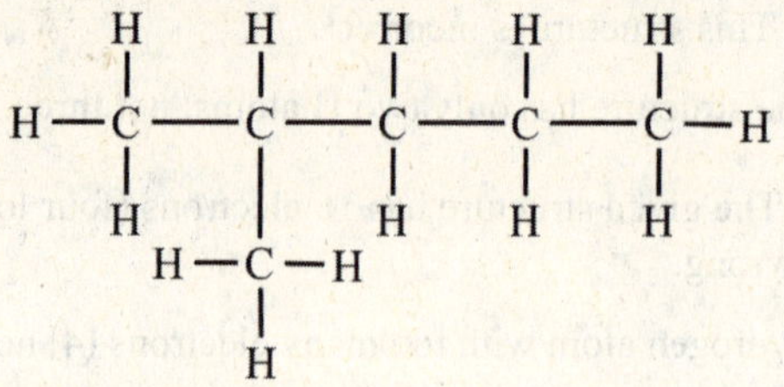
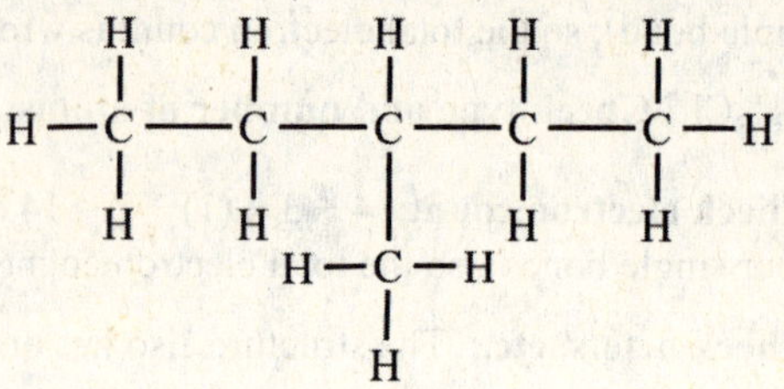

Two methyl branches can go on the four-carbon chain in two ways. They can both go on the second carbon or one can go on the second carbon and one can go on the third carbon.

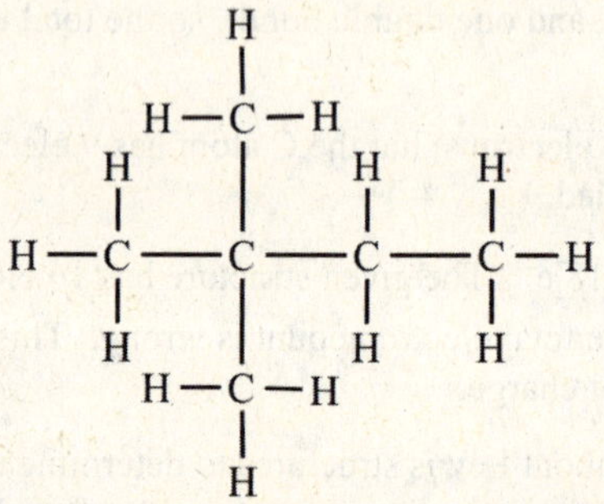
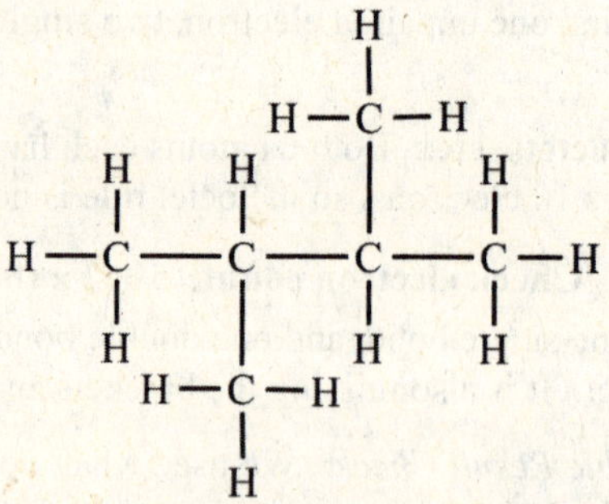

Three methyl branches cannot all three be attached to the one middle carbon in a three-carbon chain, so we're done with methyl branches.

We could try to make a four-carbon chain with an ethyl branch. However, ethyl branches can't be attached to chain-carbons within two carbons of the end of the chain. (If you do that, the molecules "longest" chain will actually include your branch!) A four-carbon chain isn't long enough and therefore we can't use ethyl branches. So we have found all the branched isomers of the hydrocarbon with the formula C_6H_{14}.

☑ *Reasonable Result Check:* The molecules have one or more branches off the three-carbon chain. They also all have six C atoms and 14 H atoms.

25. *Result:* **(a) alkyne (b) alkane (c) alkene**

Analyze: Given formulas of hydrocarbons determine if they are alkane, alkene, or alkyne.

Plan: If we assume that each of these straight chain molecules has at most one multiple bond, we can use the formula pattern to determine the class. The formulas of alkanes are C_nH_{2n+2}, for n = 1 and higher. Each multiple bond removes two electrons. The formulas of alkenes are C_nH_{2n}, for n = 2 and higher. The formulas of alkynes are C_nH_{2n-2}, for n = 2 and higher. Using the number of C atoms, set n. Then determine 2n+2, 2n, or 2n–2. Compare these numbers to the number of H atoms. Form a conclusion based on that comparison.

Execute:

(a) C_5H_8 n = 5, with this n, 2n+2 = 12, 2n = 10, 2n–2 = 8. This hydrocarbon is an alkyne. Notice, that it could also be an alkene, if there are two double bonds in the molecule.

(b) $C_{24}H_{50}$ n = 24, with this n, 2n+2 = 50, 2n = 48, 2n–2 = 46. This hydrocarbon is an alkane.

(c) C_7H_{14} n = 7, with this n, 2n+2 = 16, 2n = 14, 2n–2 = 12. This hydrocarbon is an alkene.

☑ *Reasonable Result Check:* One and only one of the three calculations using n gave a matching number.

27. *Result:* **See structures below**

Analyze: Given the condensed structural formulas for some organic compounds, determine if *cis-* and *trans-* isomers exist. For those that do, write the structural formula for the isomers and label them. For those that don't, explain why.

Plan: If there is a double bond, identify the fragments attached to each carbon of the double bond. If both carbons have two different fragments, then *cis* and *trans*-isomers can exist. If either of the carbons has identical fragments, then there is no *cis*- and *trans*-isomerism.

Execute:

(a) Br_2CH_2 does not have a double bond.

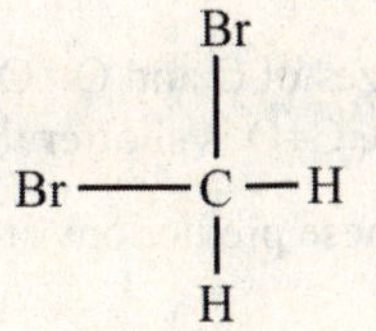

(b) $CH_3CH_2CH{=}CHCH_2CH_3$

Switch the positions of the —H and the —CH_2CH_3 fragments on the right carbon to make the isomers:

cis (ethyls on the same side) *trans* (ethyls on the opposite sides)

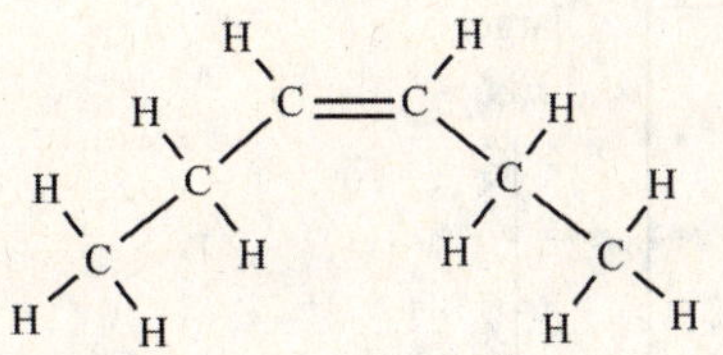

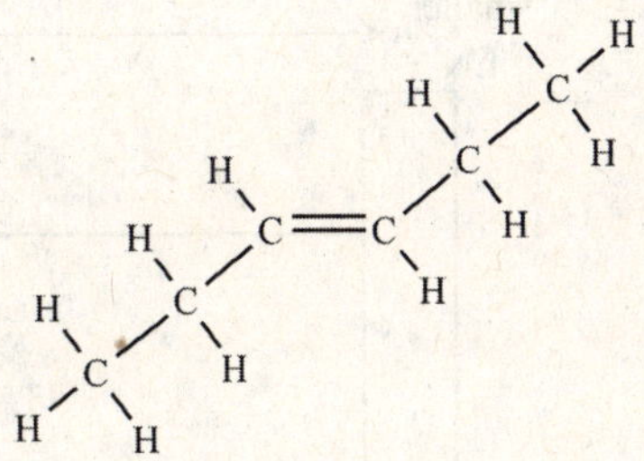

(c) $CH_3CH{=}CHCH_3$

Switch the positions of the —H and the —CH_3 fragments on the right carbon to make the isomers:

cis (methyls on the same side) *trans* (methyls on the opposite sides)

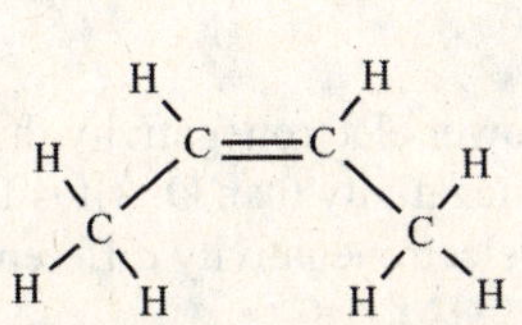

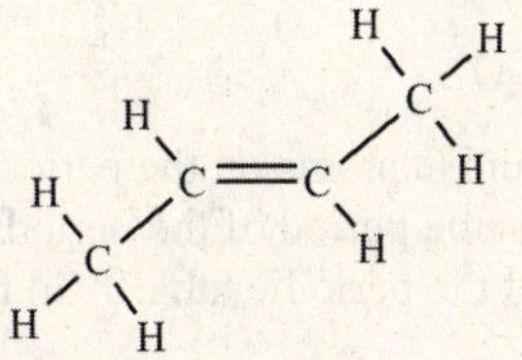

(d) The two fragments (—H) on the first carbon are the same in $CH_2{=}CHCH_2CH_3$

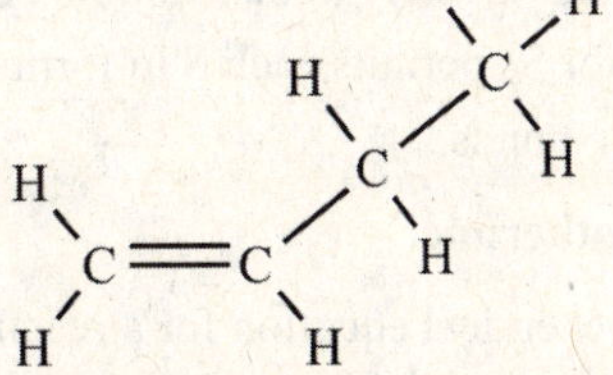

Bond Length and Bond Energy (Section 6-6)

29. *Result:* **(a) B–Cl (b) C–O (c) P–O (d) C=O**

Analyze: Given a series of pairs of bonds, predict which of the bonds will be shorter.

Plan: Use periodic trends in atomic radii to identify the smaller atoms. The bond with the smaller atoms, will have a shorter bond. In cases where bonds between the same atoms are compared, triple bonds are shorter than double bonds, which are shorter than single bonds.

Execute:

(a) Both bonds have Cl, so compare the sizes of B and Ga. B is smaller than Ga. (It is higher in the same group of the periodic table.) So, B–Cl is shorter than Ga–Cl.

(b) Both bonds have O, so compare the sizes of C and Sn. C is smaller than Sn. (It is higher in the same group of the periodic table.) So, C–O is shorter than Sn–O.

(c) Both bonds have P, so compare the sizes of O and S. O is smaller than S. (It is higher in the same group of the periodic table.) So, P–O is shorter than P–S.

(d) Both bonds have C, so compare the sizes of C and O. O is smaller than C. (It is further to the right in the same period of the periodic table.) So, C=O is shorter than C=C.

☑ *Reasonable Result Check:* Several of these predictions are confirmed in Table 6.1.

31. *Result:* **(a)**

Analyze: Given some bonds and only a periodic table, predict which of the bonds will be strongest.

Ionic bonds are stronger than polar covalent bonds. Polar covalent bonds are stronger than purely covalent bonds. Use the periodic trend for electronegativity (EN):

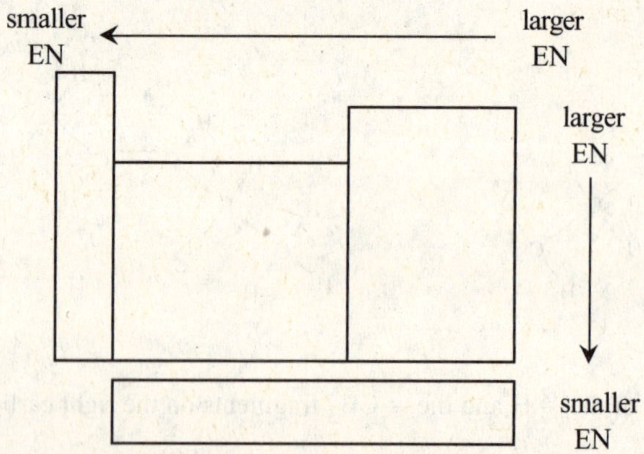

The larger the difference in electronegativity, the stronger the bond will be.

(a) Si–F, (b) P–S, (c) P–O

Looking up Si, F, P, S, and O atoms on the periodic table, we find that Si has lower electronegativity than P. (It is further to the left in the same period of the periodic table.) F has higher electronegativity than O. (It is further to the left in the same period of the periodic table.) So the **Si–F** bond has the largest electronegativity difference; hence, it is the strongest bond.

☑ *Reasonable Result Check:* These predictions are confirmed in Figure 6.8.

32. *Result/Explanation:* The lone S_2 bond is likely to be a double bond, since elements in Group 6A need two electrons each. The ring structure for S_8 permits each S to form single bonds to adjacent S atoms. Double bonds are always shorter than single bonds.

34. *Result:* **–92 kJ; the reaction is exothermic.**

Analyfze: Given a description of a chemical equation for a reaction and a table of bond enthalpies (Table 6.2), estimate the standard reaction enthalpy and determine whether the reaction is exothermic or endothermic.

Plan: First, balance the chemical equation. Then use Hess's Law to estimate $\Delta_r H°$ by breaking all the bonds in the reactants (with bond enthalpies) and then we will form all the bonds—the opposite of breaking—in the products (by removing their bond enthalpy from the system). That is the logic behind Equation 6.1.

$$\Delta_r H° = \sum\left[(\text{number of bond}) \times D(\text{bond broken})\right] - \sum\left[(\text{number of bond}) \times D(\text{bond formed})\right]$$

Get the balanced equation and use it to describe the relative number (in moles) of each of the reactants and products. Count the bonds of each type that break and form.

Execute: Set up a specific version of the above equation. Use the D for each bond in Table 6.2 and solve for $\Delta_r H°$.

The reactants are nitrogen, N_2, and hydrogen, H_2. The product is ammonia, NH_3. The balanced equation is

$$N_2 \quad + \quad 3\,H_2 \quad \longrightarrow \quad 2\,NH_3$$

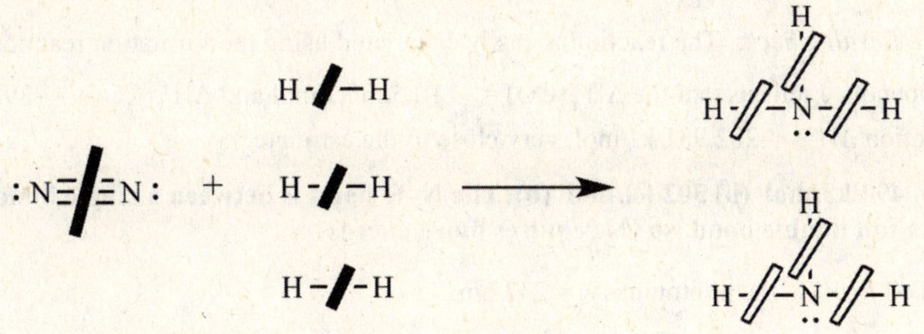

Hence, we break 1 N≡N bond and 3 H–H bonds, then form 6 N–H bonds:

$$\Delta_r H° = (1\ \text{N}{\equiv}\text{N}) \times D_{N\equiv N} + (3\ \text{H–H}) \times D_{H-H} - [(6\ \text{N–H}) \times D_{N-H}]$$

Look up the D values in Table 6.2.

$$\Delta_r H° = (1)(946\ \text{kJ/mol}) + (3)(436\ \text{kJ/mol}) - (6)(391\ \text{kJ/mol}) = -92\ \text{kJ}$$

This reaction is exothermic.

☑ *Reasonable Result Check:* The reaction is twice the formation reaction for NH_3. Appendix J tells us that NH_3 has $\Delta_f H°= -46.11$ kJ/mol. Twice this value produces a $\Delta_r H° = -92.22$ kJ/mol. This is very close to the estimate calculated here.

35. *Result:* **–284 kJ; exothermic**

Analyze: Given a description of a chemical equation for a reaction and a table of bond enthalpies (Table 6.2), estimate the standard reaction enthalpy and determine whether the reaction is exothermic or endothermic.

Plan: Use the plan described in the solution for Question 34.

Execute: Set up a specific version of the above equation. Use the D for each bond in Table 6.2 to solve for $\Delta_r H°$.

The reactants are carbon monoxide, CO, and molecular oxygen, O_2. The product is carbon dioxide, CO_2. The balanced equation is

$$CO \quad + \quad \frac{1}{2}\,O_2 \quad \longrightarrow \quad CO_2$$

At the molecular scale, we must double this reaction:

$$2\,CO \quad + \quad O_2 \quad \longrightarrow \quad 2\,CO_2$$

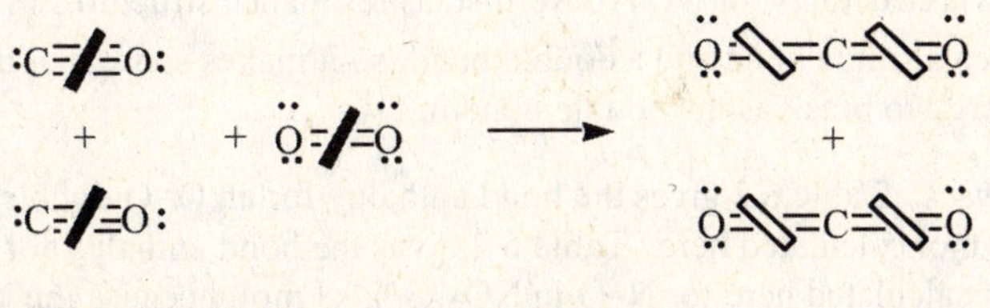

We break 2 C≡O bond and one O=O bonds to form 4 C=O bonds.

$$2(\Delta_r H°) = (2\ \text{C}{\equiv}\text{O}) \times D_{C\equiv O} + (1\ \text{O=O}) \times D_{O=O} - (4\ \text{C=O}) \times D_{C=O}$$

We can divide the equation by 2 at the macroscopic-level, and break one mol of C≡O bond and half a mol of O=O bonds. We form two mol C=O bonds:

$$\Delta_r H° = (1\ \text{C}{\equiv}\text{O}) \times D_{C\equiv O} + (\tfrac{1}{2}\ \text{O=O}) \times D_{O=O} - (2\ \text{C=O}) \times D_{C=O}$$

Look up the D values in Table 6.2.

$\Delta_r H° = (1)(1073 \text{ kJ/mol}) + (\frac{1}{2})(498 \text{ kJ/mol}) - (2)(803 \text{ kJ/mol}) = -284 \text{ kJ}$

This reaction is exothermic.

☑ *Reasonable Result Check:* The reaction is easily calculated using the formation reactions and $\Delta H_f°$ for CO and CO_2. Appendix J tells us that the $\Delta H_f°(CO) = -110.525$ kJ/mol and $\Delta H_f°(CO_2) = -393.509$ kJ/mol. That gives this reaction $\Delta H° = -282.981$ kJ/mol, very close to the estimate.

36. *Result:* **(a) (i) 494 kJ/mol (ii) 302 kJ/mol (b) The N–O bond is between a single bond and a double bond and O=O is a full double bond, so O_2 requires more energy.**

(a) (i) To break O_2 the photon has $\lambda = 242$ nm:

$$E_{photon} = \frac{hc}{\lambda} = \frac{6.626 \times 10^{-34} \text{ J} \cdot \text{s} \times 2.998 \times 10^{8} \text{ m/s}}{242 \text{ nm}} \times \frac{1 \text{ nm}}{1 \times 10^{-9} \text{ m}} = 8.21 \times 10^{-19} \text{ J}$$

Each photon breaks one O=O molecule:

$$\left(\frac{8.21 \times 10^{-19} \text{ J}}{1 \text{ photon}}\right) \times \left(\frac{1 \text{ photon}}{1 \text{ H}_2 \text{ molecule}}\right) \times \left(\frac{6.022 \times 10^{23} \text{ H}_2 \text{ molecules}}{1 \text{ mol H}_2}\right) \times \left(\frac{1 \text{ kJ}}{1000 \text{ J}}\right) = 494 \text{ kJ/mol}$$

(i) To break an N–O bond in NO_2 the photon has $\lambda = 395$ nm:

$$E_{photon} = \frac{hc}{\lambda} = \frac{6.626 \times 10^{-34} \text{ J} \cdot \text{s} \times 2.998 \times 10^{8} \text{ m/s}}{395 \text{ nm}} \times \frac{1 \text{ nm}}{1 \times 10^{-9} \text{ m}} = 5.03 \times 10^{-19} \text{ J}$$

Each photon breaks one N–O molecule in NO_2:

$$\left(\frac{5.03 \times 10^{-19} \text{ J}}{1 \text{ photon}}\right) \times \left(\frac{1 \text{ photon}}{1 \text{ H}_2 \text{ molecule}}\right) \times \left(\frac{6.022 \times 10^{23} \text{ H}_2 \text{ molecules}}{1 \text{ mol H}_2}\right) \times \left(\frac{1 \text{ kJ}}{1000 \text{ J}}\right) = 302 \text{ kJ/mol}$$

(b) Follow the guidelines in Section 6-2a and the solution to Question 15:

The Lewis structure for O_2 has 12 electrons.

The Lewis structure for NO_2 ion has 17 electrons.

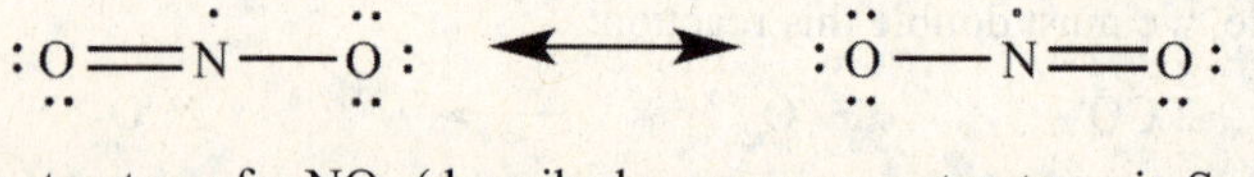

Two equivalent Lewis structures for NO_2 (described as resonance structures in Section 6-9) show that the N–O bond is between a single bond and a double bond, so it makes sense that the N–O bond does not require as much energy to break as the double bond in O_2.

☑ *Reasonable Result Check:* Table 6.2 gives the bond enthalpy for an O=O double bond to be 498 kJ/mol, which is close to 494 kJ/mol calculated here. Table 6.2 gives the bond enthalpy for an N–O single bond to be 201 kJ. This is less than calculated here for N–O in NO_2, 302 kJ/mol, because the NO_2 Lewis structure has a N=O bond, also. As described in (b), it also makes sense that the enthalpy for breaking an N–O bond is lower than the enthalpy for breaking an O=O bond.

Electronegativity and Bond Polarity (Section 6-7)

37. *Result:* **(a)** $\underset{\delta^+}{C}$—$\underset{\delta^-}{O}$ **(b)** $\underset{\delta^+}{B}$—$\underset{\delta^-}{O}$ **(c)** $\underset{\delta^+}{P}$—$\underset{\delta^-}{N}$ **(d)** $\underset{\delta^+}{B}$—$\underset{\delta^-}{Cl}$

Analyze and Plan: Bonds are more polar when the electronegativity difference is larger. Look up electronegativity values for the atoms in these pairs.

Execute: $EN_C = 2.4$, $EN_O = 3.4$, $EN_N = 3.0$, $EN_B = 1.9$, $EN_P = 2.0$, $EN_S = 2.3$, $EN_H = 2.1$, $EN_{Cl} = 2.7$

The partial negative (δ^-) end of a bond is the atom with the larger electronegativity.

The partial positive (δ^+) end of a bond is the atom with the smaller electronegativity.

(a) $\Delta EN_{C–O} = EN_O - EN_C = 3.4 - 2.4 = 1.0$ more polar

$\Delta EN_{C–N} = EN_N - EN_C = 3.0 - 2.4 = 0.6$

$\underset{\delta^+}{C} — \underset{\delta^-}{O}$

(b) $\Delta EN_{B–O} = EN_O - EN_B = 3.4 - 2.0 = 1.4$ more polar

$\Delta EN_{P–S} = EN_S - EN_P = 2.3 - 2.0 = 0.3$

$\underset{\delta^+}{B} — \underset{\delta^-}{O}$

(c) $\Delta EN_{P–H} = EN_H - EN_P = 2.1 - 2.0 = 0.1$

$\Delta EN_{P–N} = EN_N - EN_P = 3.0 - 2.0 = 1.0$ more polar

$\underset{\delta^+}{P} — \underset{\delta^-}{N}$

(d) $\Delta EN_{B–H} = EN_H - EN_B = 2.1 - 1.9 = 0.2$

$\Delta EN_{B–Cl} = EN_{Cl} - EN_B = 2.7 - 1.9 = 0.8$

$\underset{\delta^+}{B} — \underset{\delta^-}{Cl}$

38. *Result:* **(a) B–O (b) O–Se (c) O–Se (d) B–F (e) N–F**

Analyze and Plan: Bonds are more polar when the electronegativity difference is larger. Look up electronegativity values for the atoms in these pairs.

Analyze: $EN_B = 2.0$, $EN_{Cl} = 2.7$, $EN_O = 3.4$, $EN_F = 4.0$, $EN_{Se} = 2.4$, $EN_S = 2.3$, $EN_N = 3.0$, $EN_H = 2.1$

(a) $\Delta EN_{B–Cl} = EN_{Cl} - EN_B = 2.7 - 2.0 = 0.7$

$\underset{\delta^+}{B} — \underset{\delta^-}{Cl}$

$\Delta EN_{B–O} = EN_O - EN_B = 3.4 - 2.0 = 1.4$ B–O is more polar than B–Cl.

$\underset{\delta^+}{B} — \underset{\delta^-}{O}$

(b) $\Delta EN_{O–F} = EN_F - EN_O = 4.0 - 3.4 = 0.4$

$\underset{\delta^+}{O} — \underset{\delta^-}{F}$

$\Delta EN_{O–Se} = EN_O - EN_{Se} = 3.4 - 2.4 = 1.0$ O–Se is more polar than O–F.

$\underset{\delta^+}{Se} — \underset{\delta^-}{O}$

(c) $\Delta EN_{S–Cl} = EN_{Cl} - EN_S = 2.7 - 2.3 = 0.4$

$\underset{\delta^+}{S} — \underset{\delta^-}{Cl}$

$\Delta EN_{B\text{–}F} = EN_F - EN_B = 4.0 - 2.0 = 2.0$ B–F is more polar than S–Cl.

B —— F
δ^+ δ^-

(d) $\Delta EN_{N\text{–}H} = EN_N - EN_H = 3.0 - 2.1 = 0.9$

H —— N
δ^+ δ^-

$\Delta EN_{N\text{–}F} = EN_F - EN_N = 4.0 - 3.0 = 1.0$ N–F is more polar than N–H.

N —— F
δ^+ δ^-

39. *Result:* **(a) all bonds are somewhat polar (b) C=O bond is the most polar; O atom is partial negative**

Analyze, Plan, and Execute:

(a) Bonds are polar if the atoms' electronegativities are different.

$$EN_H = 2.1,\ EN_N = 3.0,\ EN_C = 2.4,\ EN_O = 3.4$$

$$\Delta EN_{N\text{–}H} = EN_N - EN_H = 3.0 - 2.1 = 0.9$$

$$\Delta EN_{C\text{–}N} = EN_N - EN_C = 3.0 - 2.4 = 0.6$$

$$\Delta EN_{C\text{–}O} = EN_O - EN_C = 3.4 - 2.4 = 1.0$$

All the electronegativity differences are greater than zero, so the bonds are **all somewhat polar**. **None** of the bonds **are nonpolar**.

(b) Bonds are more polar when the electronegativity difference is larger.

The largest electronegativity difference is for the **C=O bond**, so it is the most polar.

C —— O
δ^+ δ^-

The O atom is the partial negative end of this bond, because it has the larger electronegativity.

40. *Result:* **(a) C–H is somewhat polar, C=O is polar; C–C is nonpolar (b) C=O bond is the most polar; O atom is partial negative**

Analyze, Plan, and Execute:

(a) Bonds are polar if the atoms' electronegativities are different.

$$EN_H = 2.1,\ EN_C = 2.4,\ EN_O = 3.4$$

$$\Delta EN_{C\text{–}H} = EN_C - EN_H = 2.4 - 2.1 = 0.3$$

$$\Delta EN_{C\text{–}C} = EN_C - EN_C = 2.4 - 2.4 = 0.0$$

$$\Delta EN_{C\text{–}O} = EN_O - EN_B = 3.4 - 2.0 = 1.4$$

The **C–H** bond is very **slightly polar**. The **C=O** bond is quite **polar**. The **C–C bond is nonpolar**, since both C atoms have the same electronegativity.

(b) Bonds are more polar when the electronegativity difference is larger.

The largest electronegativity difference is for the **C=O bond**, so it is the most polar.

C —— O
δ^+ δ^-

The O atom is the partial negative end of this bond, because it has the larger electronegativity.

Formal Charge (Section 6-8)

41. *Answer:* (a) 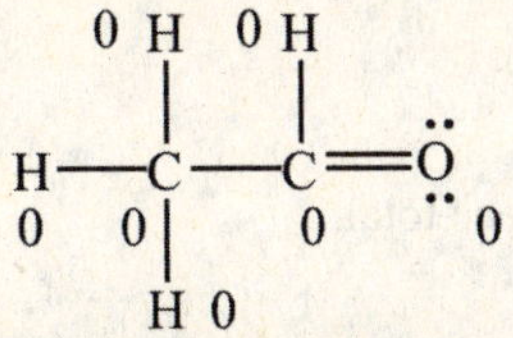 (b) 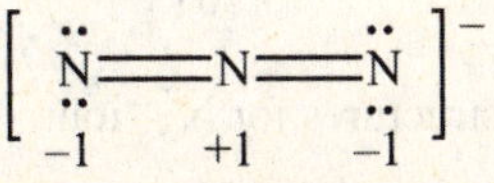 (c) 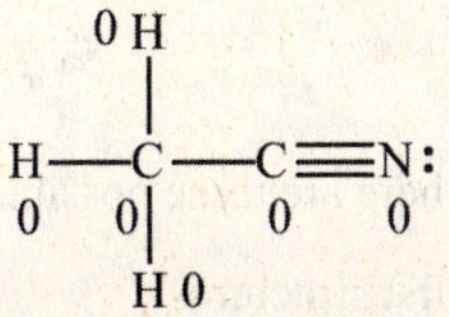

Strategy and Explanation: Given the formulas of molecules or ions, write the correct Lewis structure and assign formal charges to each atom.

Write the Lewis structures. Then determine the number of lone pair electrons and bonding electrons around each atom. Use the method described in Section 6-8 to determine the formal charges on each atom.

Formal charge = (number of valence electrons in an atom)

$$- [(\text{number of lone pair electrons}) + (\frac{1}{2}\text{ number of bonding electrons})]$$

To get the *(number of lone pair electrons)* just count all the dots. To get *(number of bonding electrons)* just count two times the number lines representing covalent bonds.

Formal Charge (Section 6-8)

41. *Answer:* (a) 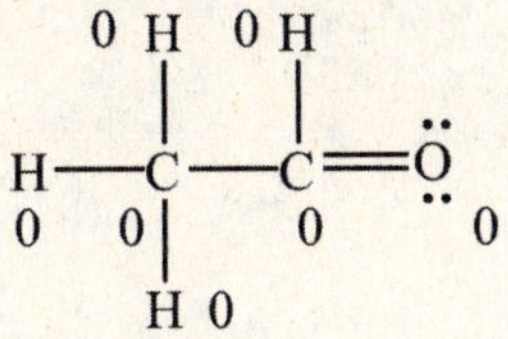 (b) 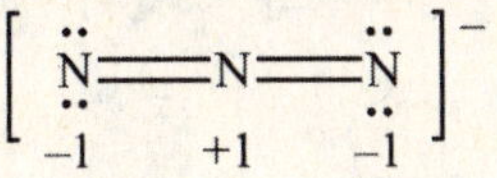 (c) 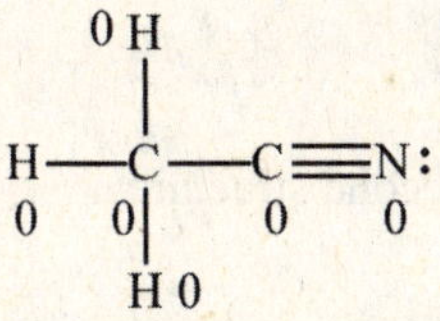

Strategy and Explanation: Given the formulas of molecules or ions, write the correct Lewis structure and assign formal charges to each atom.

Write the Lewis structures. Then determine the number of lone pair electrons and bonding electrons around each atom. Use the method described in Section 6-8 to determine the formal charges on each atom.

Formal charge = (number of valence electrons in an atom)

$$- [(\text{number of lone pair electrons}) + (\frac{1}{2}\text{ number of bonding electrons})]$$

To get the *(number of lone pair electrons)* just count all the dots. To get *(number of bonding electrons)* just count two times the number lines representing covalent bonds.

(a) The Lewis structure for CH_3CHO molecule: 18 electrons total.

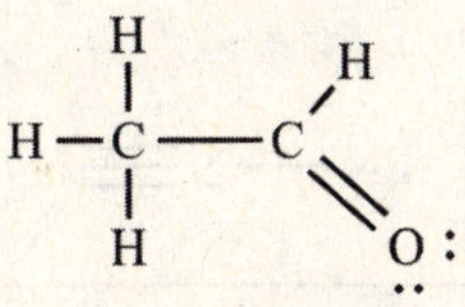

Set up the chart:

	C–	C=	H	O
Valence electrons	4	4	1	6
Lone pair electrons	0	0	0	4
Bonding electrons	8	8	2	4
Formal charge	4 – (0 + 4) = 0	4 – (0 + 4) = 0	1 – (0 + 1) = 0	6 – (4 + 2) = 0

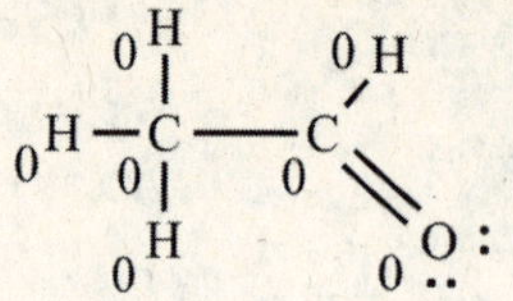

(b) There are three possible Lewis structures for N_3^- ion: 16 electrons total.

First structure:

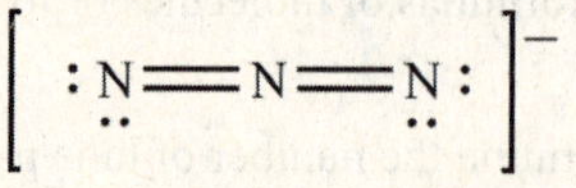

Set up the chart:

	N=	=N=
Valence electrons	5	5
Lone pair electrons	4	0
Bonding electrons	4	8
Formal charge	5 – (4 + 2) = –1	5 – (0 + 4) = +1

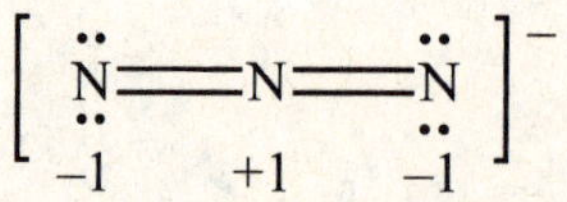

Second structure:

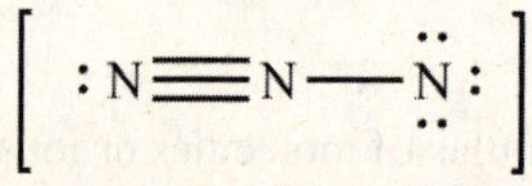

Set up the chart:

	Left-most N	Middle N	Right-most N
Valence electrons	5	5	5
Lone pair electrons	2	0	6
Bonding electrons	6	8	2
Formal charge	5 – (2 + 3) = 0	5 – (0 + 4) = +1	5 – (6 + 1) = –2

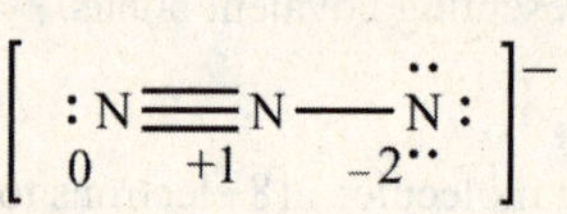

Third structure:

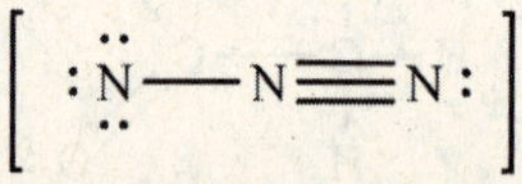

Set up the chart:

	Left-most N	Middle N	Right-most N
Valence electrons	5	5	5
Lone pair electrons	6	0	2
Bonding electrons	2	8	6
Formal charge	5 – (6 + 1) = –2	5 – (0 + 4) = +1	5 – (2 + 3) = 0

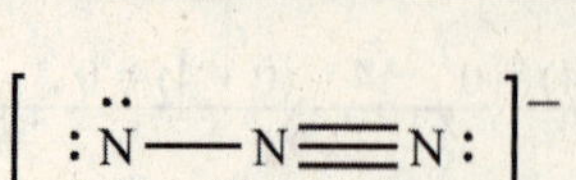

The first of these three structures is the best, since the formal charges on atoms in that structure are smaller (–1, +1) than in the other two structures (–2, +1).

(c) Lewis structure for CH_3CN molecule: 16 electrons total

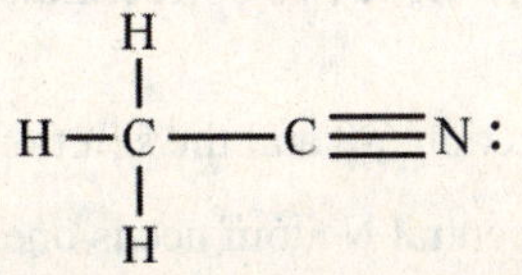

Set up the chart:

	C	H	N
Valence electrons	4	1	5
Lone pair electrons	0	0	2
Bonding electrons	8	2	6
Formal charge	4 – (0 + 4) = 0	1 – (0 + 1) = 0	5 – (2 + 3) = 0

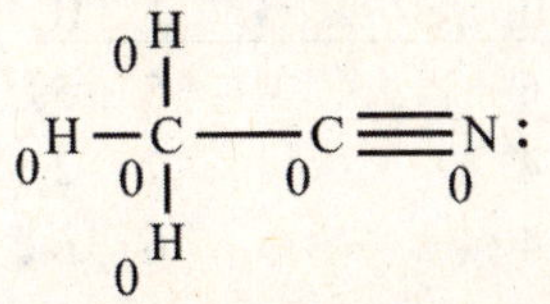

✓ *Reasonable Result Check:* The sum of the formal charges is zero for the neutral molecules and the ionic charge for the charged ion. The atoms with more bonds have more positive formal charges than those with fewer bonds and more lone pairs.

43. *Result:* 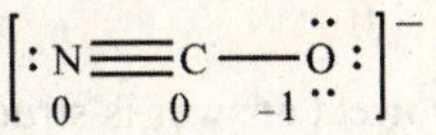

Analyze: Given the formula of an ion, write the correct Lewis structure and assign formal charges to each atom.

Plan: Use the method described Section 6-8 and the solution to Question 41 to get the formal charges.

Execute: With 16 electrons total, the central N atom needs two multiple bonds:

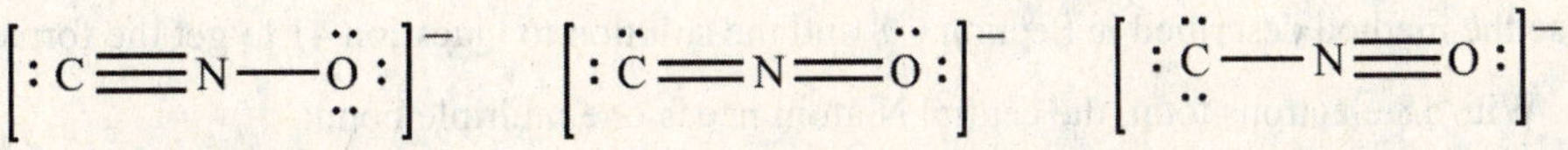

I II III

Set up a chart:

	C			N			O		
	I	II	III	I	II	III	I	II	III
Valence electrons	4	4	4	5	5	5	6	6	6
Lone pair electrons	2	4	6	0	0	0	6	4	2
Bonding electrons	6	4	2	8	8	8	2	4	6
Formal charge	–1	–2	–3	+1	+1	+1	–1	0	+1

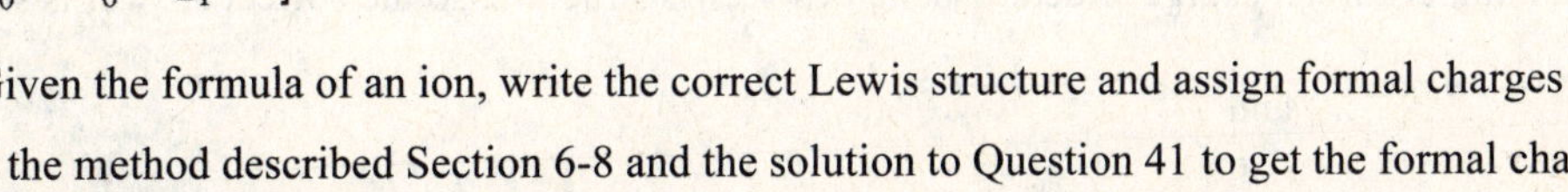

I II III

The lowest-formal-charge structure is the best Lewis structure, so the right answer is structure (I). All the atoms are from period 2, so we cannot lower the formal charges any more.

45. *Result:* 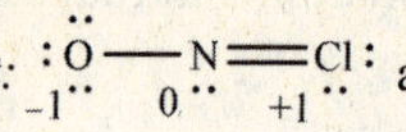 and 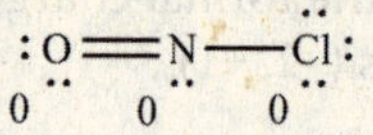

Analyze: Given the atoms in a molecule, write two correct Lewis structures and assign formal charges to each atom.

Plan: Use the method described in Section 6-8 and the solution to Question 41 to get the formal charges.

Execute: With 18 electrons total, the central N atom needs one multiple bond:

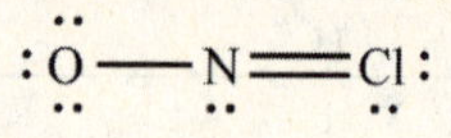

I

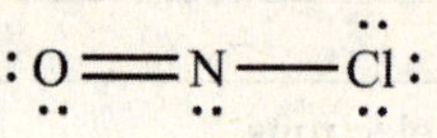

II

Set up a chart:

	O I	O II	N I	N II	Cl I	Cl II
Valence electrons	6	6	5	5	7	7
Lone pair electrons	6	4	2	2	4	6
Bonding electrons	2	4	6	6	4	2
Formal charge	−1	0	0	0	+1	0

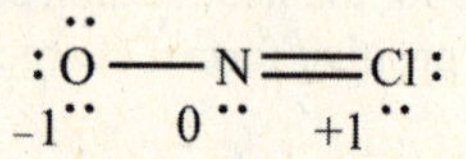

I

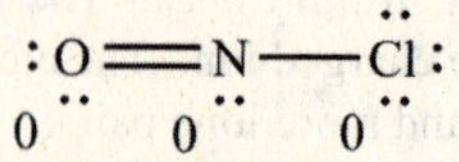

II

The lowest number formal charge structure is the best Lewis structure, so the correct answer is structure (II).

46. *Result:* 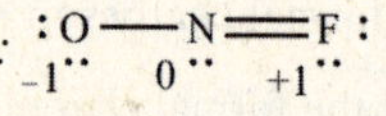 and 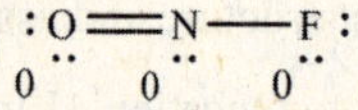

Analyze: Given the atoms in a molecule, write two correct Lewis structures and assign formal charges to each atom.

Plan: Use the method described in Section 6-8 and the solution to Question 41 to get the formal charges.

Execute: With 18 electrons total, the central N atom needs one multiple bond:

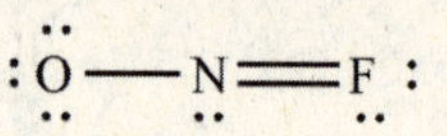

I

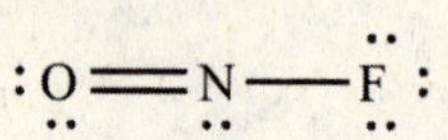

II

Set up a chart:

	O I	O II	N I	N II	F I	F II
Valence electrons	6	6	5	5	7	7
Lone pair electrons	6	4	2	2	4	6
Bonding electrons	2	4	6	6	4	2
Formal charge	−1	0	0	0	+1	0

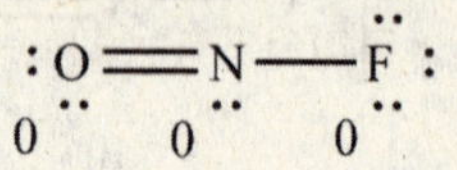

I

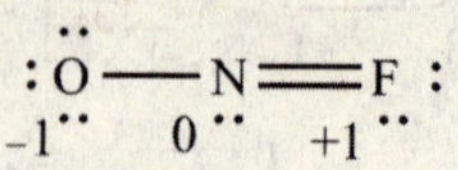

II

The zero-formal-charge structure is the best Lewis structure, so the correct answer is structure (I).

Resonance (Section 6-9)

49. *Result:* **See structures below**

Analyze: Given the formulas of molecules or ions, write all the resonance structures.

Plan: Write the Lewis structure. Each resonance structure differs only by where the electrons for a multiple bond come from. When two or more atoms with lone pairs are bonded to an atom that needs more electrons, any one of them can supply the electron pair for a multiple bond. To write all resonance structures, systematically and sequentially supply the central atom with needed electrons from each of the possible sources. Separate these different structures with a double-headed arrow to show that they are resonance structures.

Execute:

(a) There are three plausible resonance structures for nitric acid, HNO_3, with 24 valence electrons. They are formed using one lone pair from a different one of the outer O atoms to make the second bond in the double bond to complete the octet of the N atom.

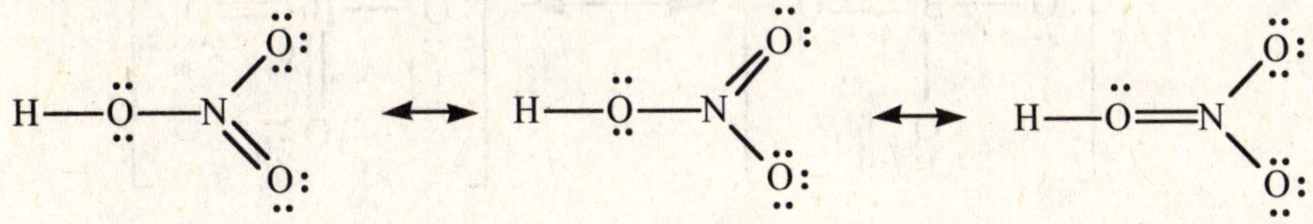

These structures are not equally plausible.

(b) There are three plausible resonance structures for nitrate ion, NO_3^-, with 24 valence electrons. They are formed using one lone pair from a different one of the outer O atoms to make the second bond in the double bond to complete the octet of the N atom.

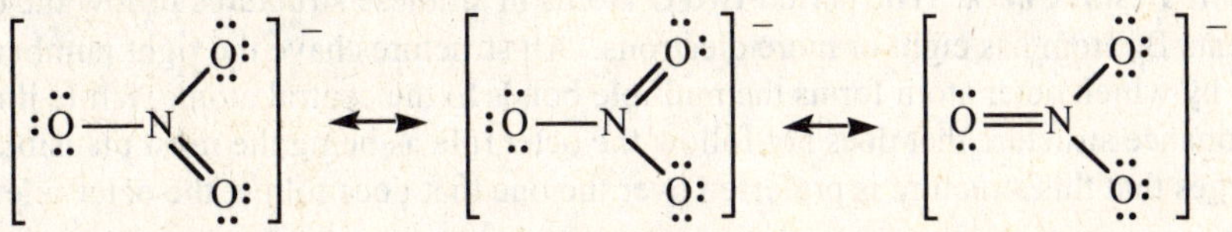

☑ *Reasonable Result Check:* The structures all follow the octet rule and have the right number of valence electrons. They differ by which outer atom is double bonded to the N atom.

51. *Result:* **see structures below; the fourth one is the most plausible.**

Analyze: Write resonance structures using all single bonds, one, two and three double bonds, and use formal charges to predict the most plausible one.

Plan: Write the Lewis structure. To write all resonance structures, systematically and sequentially supply the central atom with needed electrons from each of the possible sources. Keep in mind that period-two elements must have an octet, but period-three elements can have 8 or more electrons. Separate these different structures with a double-headed arrow to show that they are resonance structures. To determine the relative plausibility of the structures, determine the formal charges (as described in the solution to Question 41) then use the rules described in Section 6-9:

- Smaller formal charges are more favorable than larger ones.
- Negative formal charges should reside on the more electronegative atoms.
- Positive formal charges should reside on the least electronegative atoms.
- Like charges should not be on adjacent atoms.

Execute: BrO_4^- has 32 electrons. Formal charges for single bonded O atoms are always –1. Formal charges for double bonded O are always 0. Each time electrons are moved from lone pairs into bonding pairs the positive formal charge on Br goes down:

The Lewis structure that follows the octet rule for all the atoms is the first one:

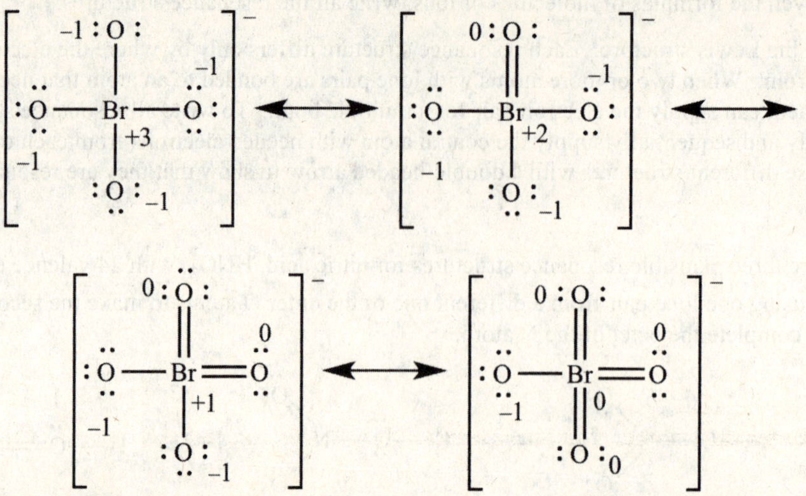

Several other equivalent resonance structures (not shown) can be written by altering the specific location(s) of the double bonds in the second and third resonance structures.

Since smaller formal charges are more favorable than larger ones, we predict that this **fourth resonance structure**, with three double bonds and the largest number of atoms with zero formal charge, is the most plausible.

☑ *Reasonable Result Check:* The period-two O atoms in all these structures follow the octet rule. The period-three elements Br atom has eight or more electrons. All structures have the right number of valence electrons. They differ by which outer atom forms the multiple bonds to the central atom. While it might feel strange to select a resonance structure that does not follow the octet rule as being the most plausible, it is clear from the formal charges that this structure is preferred over the one that does follow the octet rule.

53. *Result:* **See structures below; structures with two S-O single bonds are the most important**

Analyze: Given the formula of an ion, write the Lewis structures of the possible resonance forms and predict which structure makes the most important contribution to the resonance hybrid.

Plan: Follow the systematic plan given in the solution to Question 51.

Execute: One Lewis structures for $S_2O_3^{2-}$ follows the octet rule. The molecule has 32 valence electrons.

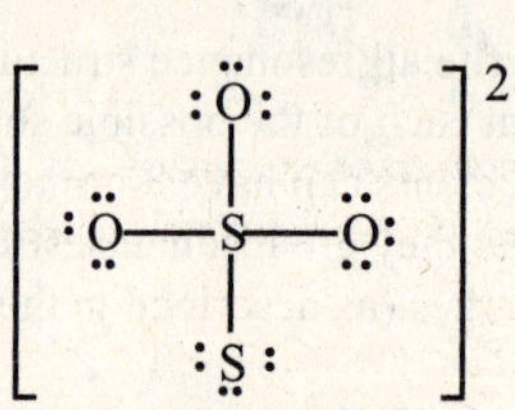

Several other plausible Lewis structures for $S_2O_3^{2-}$ do not follow the octet rule on the S atom from Period 3, using two lone pairs from the outer atoms to make the more multiple bonds to the S atom:

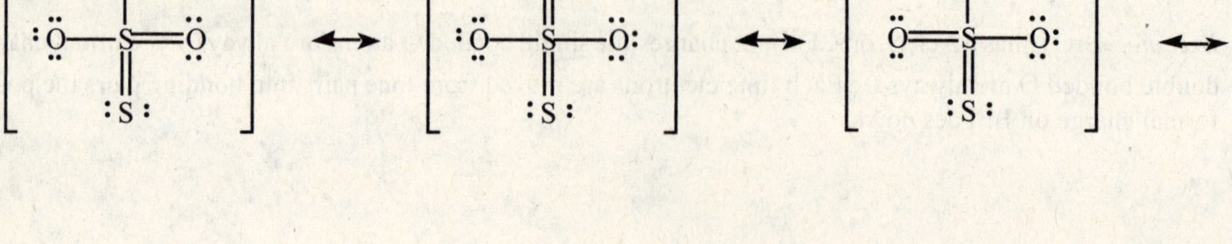

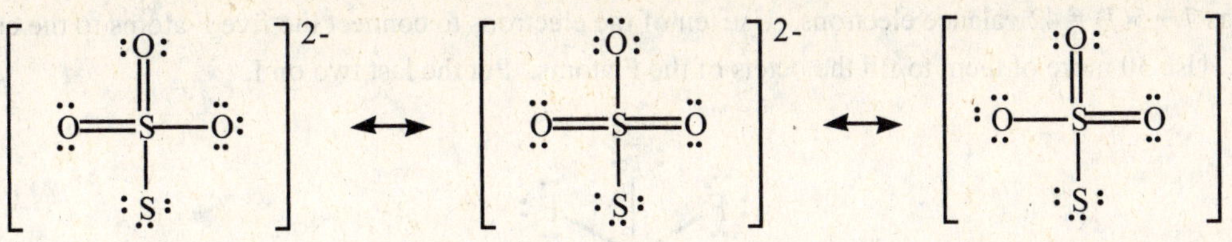

The electronegativity of S (2.3) is lower than that of O (3.4), so it is preferable to have the –1 formal charge, therefore the single bonds, on the O atom. That means the **first row in the series above are equally plausible** and the second row of structures is not as good, but better than the structure that follows the octet rule.

54. *Result:* **CO_3^{2-} has longer C–O bonds; it has more resonance structures with C–O single bonds**

Analyze: Given two chemical formulas, predict which has the shorter carbon-oxygen bond.

Plan: First, write resonance structures for the two molecules as in the solution to Question 51. In cases where bonds between the same two atoms are compared, triple bonds are shorter and stronger than double bonds, which are shorter and stronger than single bonds. If more than one resonance structure is possible, average their contribution.

Execute: Consider formate ion first. Two equivalent plausible Lewis structures exist for HCO_2^- with 18 valence electrons:

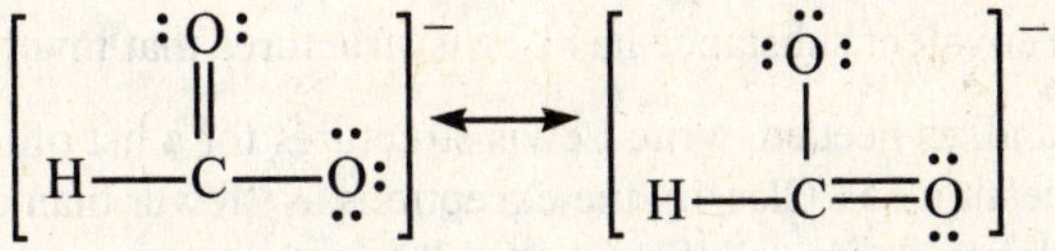

In the formate ion, one resonance structure has a single carbon-oxygen bond and one has a double carbon-oxygen bond. We predict that the carbon-oxygen bond will be halfway between a single and double bond.

There are three plausible equivalent Lewis structures for carbonate ion, CO_3^{2-}, with 24 valence electrons:

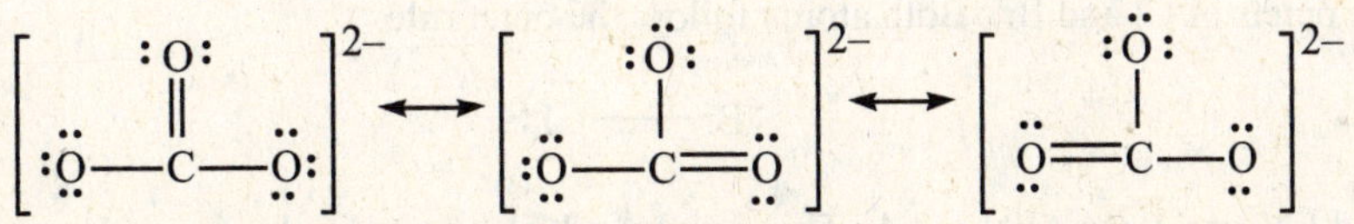

In the carbonate, one resonance structure has a double carbon-oxygen bond and two have single carbon-oxygen bond. We predict that the carbon-oxygen bond will be closer to a single bond than a double bond.

That means the longer bond is the bond in CO_3^{2-}.

☑ *Reasonable Result Check:* The Lewis structures obey the octet rule and have the right number of valance electrons. These predictions are upheld in general with values given in Table 6.1.

Exceptions to the Octet Rule (Section 6-10)

58. *Result:* **See structures below**

Analyze and Plan: Given formulas, write Lewis Structures by following the systematic plan given in the solutions to Question 15.

Execute:

(a) BrF_5 has 7 + 5(7) = 42 valance electrons. Use ten of the electrons to connect the six F atoms to the central Br atom. Use 30 more of them to fill the octets of the F atoms. Put the last two on Br.

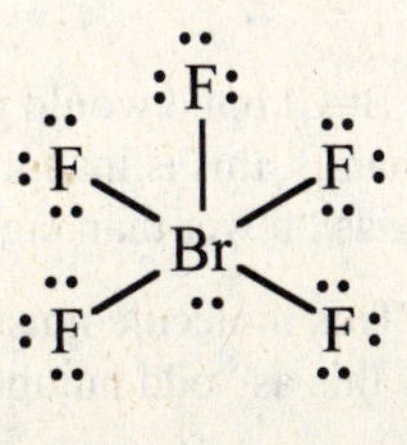

(b) IF_5 has 7 + 5(7) = 42 valance electrons. Use ten of the electrons to connect the five F atoms to the central I atom. Use 30 more of them to fill the octets of the F atoms. Put the last two on I.

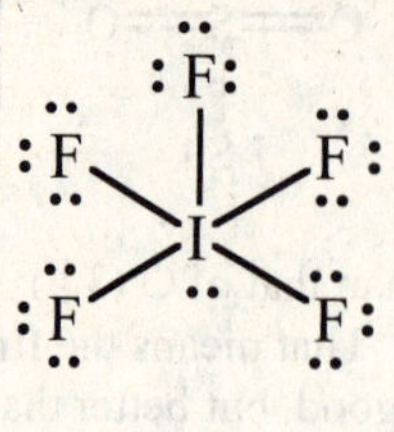

(c) IBr_2^- has 7 + 2(7) + 1 = 22 valance electrons. Use four of the electrons to connect the Br atoms to the central I atom. Use 12 more of them to fill the octets of the Br atoms. Put the last six on I.

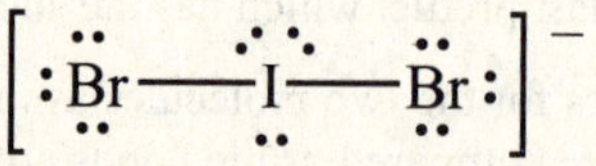

☑ *Reasonable Result Check:* All the period-two elements have an octet of electrons. All the period-three elements have eight or more electrons. All valance electrons are accounted for in the structures, either as members of shared pairs or of lone pairs.

60. *Result:* **(c) has fewer than eight valence electrons; (d) has an odd number of valence electrons**

Analyze: Determine which covalent substance has Lewis structures that involve exceptions to the octet rule.

Plan: Count the electrons and, as needed, write Lewis structures for a list of ions and molecules using the plan given in the solutions to Question 15. Classify the exceptions as "fewer than eight valence electrons" (Section 6-10a), "odd number of valence electrons" (Section 6-10b), or "more than eight valence electrons" (Section 6-10c)

Execute:

(a) ClBr has 7 + 7 = 14 valance electrons. Use two to connect the three F atoms to the central Br atom. Use the rest fill the octets of Cl and Br. Both atoms follow the octet rule:

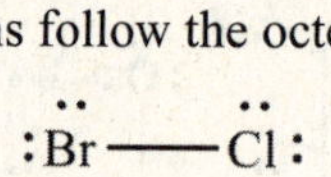

(b) EN_{Sn} = 1.6 and EN_F = 4.0, so ΔEN = 2.4. This compound is expected to be ionic. However, the question asserts that these are all covalent compounds, so we will treat it as such.

SnF_4 has 4 + 4(7) = 32 valance electrons. The rest are used for lone pairs on the F atoms. All atoms follow the octet rule:

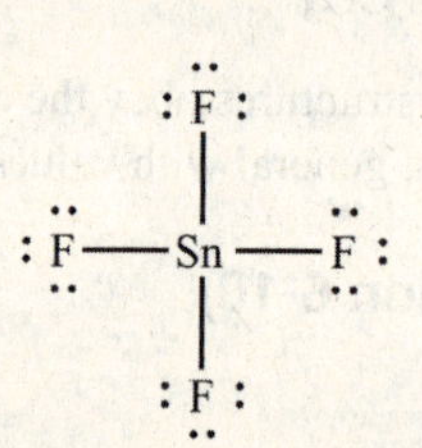

(c) BCl_3 has 3 + 3(7) = 24 valance electrons. Use six of them to connect the three Cl atoms to the central B atom. Use the rest to fill the octets of the Cl atoms. This structure has zero formal charge on every atom:

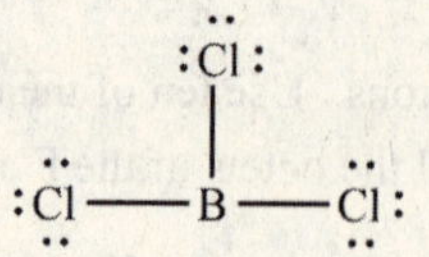

Moving a lone pair from Cl to form a B=Cl bond would place a positive formal charge on Cl. Because Cl has a much higher electronegativity than B, this is inappropriate. The structure shown is preferred and boron is electron deficient, classifying as "fewer than eight valence electrons"

(c) NO has 5 + 6 = 11 valance electrons. This molecule must involve exceptions to the octet rule since it has an odd number of electrons. We classify this as "odd number of valence electrons."

☑ *Reasonable Result Check:* The exceptions fit one of the three classifications described in Section 6-10.

Aromatic Compounds (Section 8.11)

63. *Result/Explanation:* The statement "All carbon-to-carbon bond lengths are the same." argues **against** the presence of C=C bonds in benzene, since **the C=C bonds would be shorter than the C–C bonds.**

65. *Result:* **see structures below; *ortho*-dibromobenzene, *meta*-dibromobenzene, and *para*-dibromobenzene**

Analyze, Plan, and Execute: Lewis Structures for the dibromobenzene compounds and the naming definitions for the constitutional isomers are patterned after the xylene example in Section 6-11a. Each of these Lewis structures has a second resonance structure with the alternating double bonds shifted.

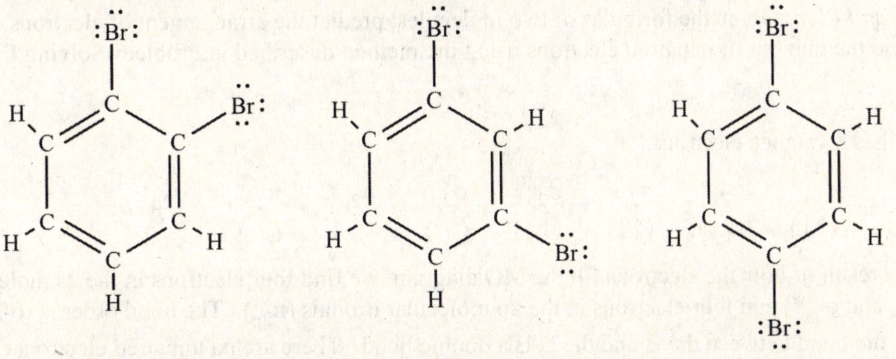

ortho-dibromobenzene *meta*-dibromobenzene *para*-dibromobenzene

Molecular Orbital Theory (Section 6-12)

67. *Result:* **(a) See MO diagram below; three bonds, no unpaired electrons, (b) See MO diagram below; half bond; one unpaired electron (c) See MO diagram below; half a bond, one unpaired electron (d) See MO diagram below; 2.5 bonds; one unpaired electrons**

Analyze and Plan: Given the formulas of several ions, predict the MO diagram, the number of bonds, and the number of unpaired electrons using the method described in Problem-Solving Example 6.12.

Execute:

(a) The peroxide ion, O_2^{2-}, has 14 electrons.

	σ_{2s}	σ_{2s}^*	π_{2p}	π_{2p}	σ_{2p}	π_{2p}^*	π_{2p}^*	σ_{2p}^*
O_2^{2-}	(↑↓)	(↑↓)	(↑↓)	(↑↓)	(↑↓)	(↑↓)	(↑↓)	( )

As a result of how the electrons fill the MO diagram, the bond order is (8-6)/2 = 1, meaning that the bond between the O and the O is a single bond. There are no unpaired electrons.

(b) B_2^+ has 5 electrons.

	σ_{2s}	σ_{2s}^*	π_{2p}	π_{2p}	σ_{2p}	π_{2p}^*	π_{2p}^*	σ_{2p}^*
B_2^+	(↑↓)	(↑↓)	(↑)	( )	( )	( )	( )	( )

As a result of how the electrons fill the MO diagram, the bond order is (3-2)/2 = 0.5, meaning that the bond between B and B is a half bond. There is one unpaired electron.

(c) Li_2^+ has 1 electron.

	σ_{2s}	σ_{2s}^*	π_{2p}	π_{2p}	σ_{2p}	π_{2p}^*	π_{2p}^*	σ_{2p}^*
Li_2^+	(↑)	( )	( )	( )	( )	( )	( )	( )

As a result of how the electrons fill the MO diagram, the bond order is (1-0)/2 = 0.5, meaning that the bond between Li and Li is a half bond. There is one unpaired electron.

(d) O_2^+ has 11 electrons.

	σ_{2s}	σ_{2s}^*	π_{2p}	π_{2p}	σ_{2p}	π_{2p}^*	π_{2p}^*	σ_{2p}^*
O_2^+	(↑↓)	(↑↓)	(↑↓)	(↑↓)	(↑↓)	(↑)	( )	( )

As a result of how the electrons fill the MO diagram, the bond order is (8-3)/2 = 2.5, meaning that the bond between the N and the O is half way between a double bond and triple bond. There is one unpaired electron.

69. *Result:* **(a) 4 electrons in molecular orbitals made from 2s orbitals, 6 electrons in molecular orbitals made from 2p orbitals; double bond, no unpaired electron (b) 4 electrons in molecular orbitals made from 2s orbitals, 6 electrons in molecular orbitals made from 2p orbitals; three bonds; no unpaired electrons**

Analyze and Plan: Given the formulas of two molecules, predict the arrangement of electrons in MOs, the bond order, and the number of unpaired electrons using the method described in Problem-Solving Example 6.12.

Execute:

(a) BN has 8 valance electrons.

	σ_{2s}	σ_{2s}^*	π_{2p}	π_{2p}	σ_{2p}	π_{2p}^*	π_{2p}^*	σ_{2p}^*
BN	(↑↓)	(↑↓)	(↑↓)	(↑↓)	( )	( )	( )	( )

As a result of how the electrons fill the MO diagram, we find four electrons in the 2s molecular orbitals (σ_{2s} and σ_{2s}*) and four electrons in the 2p molecular orbitals (π_{2p}). The bond order is (6-2)/2 =2, meaning that the bond between the B and the N is a double bond. There are no unpaired electrons.

(b) CN^- has 10 valance electrons.

	σ_{2s}	σ_{2s}^*	π_{2p}	π_{2p}	σ_{2p}	π_{2p}^*	π_{2p}^*	σ_{2p}^*
CN^-	(↑↓)	(↑↓)	(↑↓)	(↑↓)	(↑↓)	( )	( )	( )

As a result of how the electrons fill the MO diagram, we find four electrons in the 2s molecular orbitals (σ_{2s} and σ_{2s}*) and six electrons in the 2p molecular orbitals (π_{2p} and σ_{2p}). The bond order is (8-2)/2 = 3, meaning that the bond between the C and the N is a triple bond. There are no unpaired electrons.

General Questions

71. *Result:* **Si–F is most polar; Si is farthest from F on the periodic table, so it is larger and has a lower electronegativity than O or C.**

Analyze: Given several bonds, determine which is likely to be most polar.

Plan: Use the periodic table and trends in electronegativities to determine which bond is the most polar. The electronegativity of fluorine is the highest, so look at the other three atoms. The most polar bond has atoms with electronegativities that are the most different.

Execute: Here, all the choices have F atom in common. Both O and C are in the second period and Si is in the third period, so it is larger and has a lower electronegativity. Therefore, (c) Si–F is more polar than these other choices (a) C–F, (b) S–F, (d) O–F.

73. *Result/Explanation:* **Yes**, this is a good generalization, because **elements close together in the periodic table have similar electronegativities and the bonds would be formed by shared electrons** (polar covalent). If the elements are **far apart on the periodic table, their electronegativities will likely be very different and the bond more likely to be ionic.**

A few exceptions exist, of course. For example, H atom is typically displayed on the periodic table fairly far away from most of the nonmetals on the periodic table, yet it is also a nonmetal and forms polar covalent bonds.

74. *Result:* **(a) C=C (b) C=C (c) C≡N; N is the partial negative end.**

Analyze: Given the Lewis structure for a molecule, identify the shortest and strongest bond between carbons and identify the partial negative end of the most polar bond.

Plan: Follow the plans in the solutions to Questions 29, 31, and 37

Execute: The Lewis structure given in the question is the only resonance structure with zero formal charges, so it is the dominant form for the molecule.

(a) There are two carbon-carbon bonds, a C=C and a C–C. Double bonds are shorter than single bonds, so C=C is the shorter carbon-carbon bond.

(b) Double bonds are stronger than single bonds, so C=C is the stronger carbon-carbon bond.

(c) The most polar bond is the bond with atoms that have the largest electronegativity difference. The difference between the electronegativities of C (EN = 2.5) and H (EN = 2.1) is 0.4. The difference between the electronegativities of C and N (EN = 3.0) is 0.5. So, the carbon-nitrogen bond is slightly more polar. Nitrogen has the higher electronegativity, so it is the partial negative end of the bond.

$$\overset{\delta^+}{C}\equiv\overset{\delta^-}{N}$$

76. *Result:* 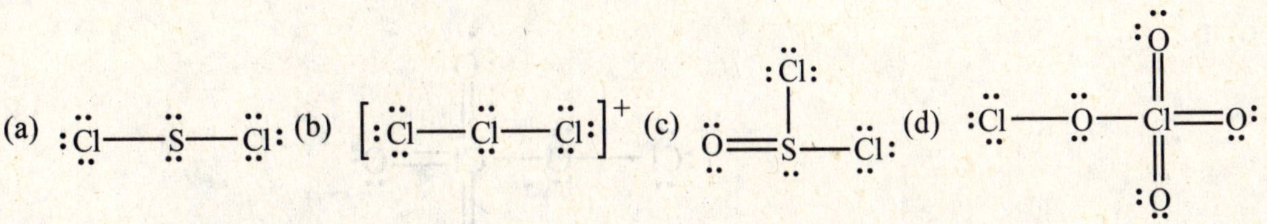

Analyze and Plan: Given several formulas of ions and molecules, write the Lewis structure as outlined in the solution to Question 15.

Execute:

(a) SCl_2 has 20 valence electrons. Use four of them to attach the two Cl atoms to the central S atom. Use the 12 of them for the lone pairs on the Cl atoms. Use the remaining four of the electrons for the lone pairs on the S atom.

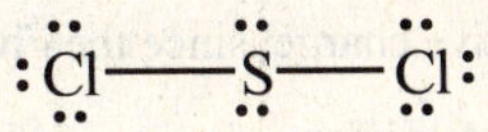

(b) The Cl_3^+ ion has 20 valence electrons, just as in (a). One of the Cl atoms is the central atom, while the other two Cl atoms are single bonded to the central Cl atom:

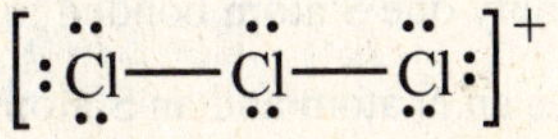

(c) $SOCl_2$ has 26 valence electrons. Use six of them to attach the O atoms to the central Cl atom. Use the remaining 18 electrons to complete the octets of the O atoms. Use the remaining 2 electrons to complete the octet on S.

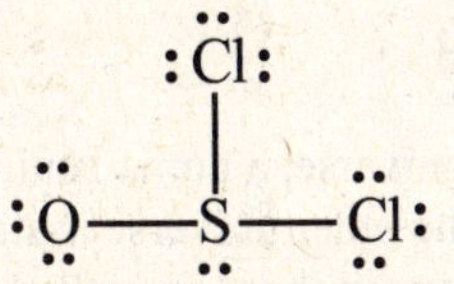

The S atom has a +1 formal charge in this Lewis structure, so we can make a lower formal charge resonance structure for this molecule by expanding the octet of the period-three elements S atom and making one double bond to the S atom, similarly to what we did in (c) for Cl.

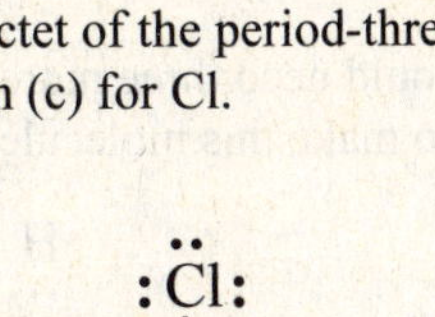

(d) $ClOClO_3$ has 38 valence electrons. We are told there is a Cl–O–Cl bond, so we will put the last three O atoms on the second Cl. Use ten electrons to attach the atoms together using single bonds. Use the remaining 28 of them to complete the octets.

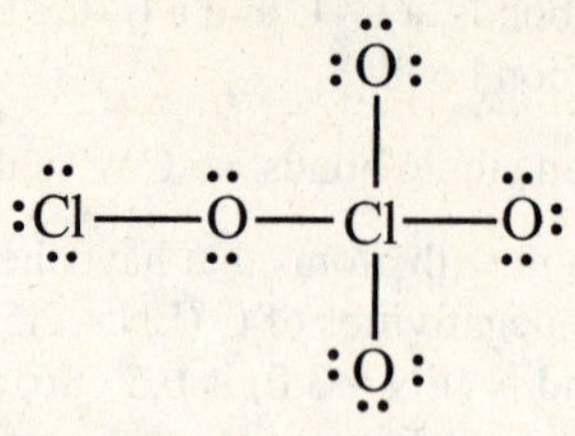

The second Cl atom has a +3 formal charge in this Lewis structure, so we can make a lower formal charge resonance structure for this molecule by expanding the octet of the period-three elements Cl atom, as we did for Br atom in Question 51.

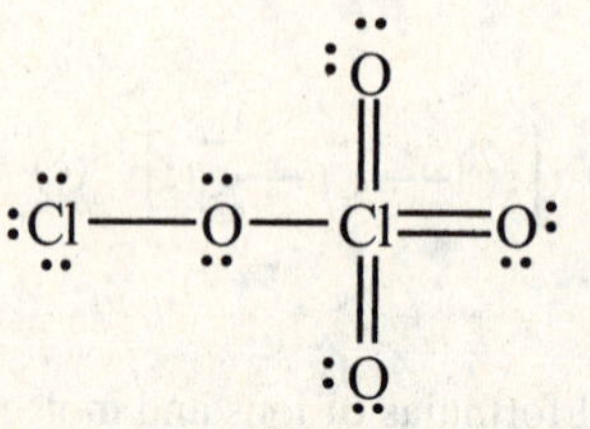

Now, all the formal charges are zero.

Applying Concepts

80. *Result:* **One electron was not subtracted for the positive charge**

Explanation: The number of valence electrons in SF_5^+ is $6 + 5 \times (7) - 1 = 40$ e^-. It looks like the student forgot to subtract one electron for the positive charge, since the given structure has $6 + 5 \times (7) = 41$ e^-.

83. *Result:* **their bonding arrangements differ**

Explanation: Resonance structures must only be different by where the electrons are. The bonding arrangement of the atoms (what atom is bonded to what atom) must be the same from structure to structure.

In the first structure, the C atom has only one S atom bonded to it.

In the second structure the C atom has an N atom and an S atom bonded to it.

These **bonding arrangements differ**; therefore these structures represent different molecules, not resonance structures of one molecule.

84. *Result:* **(a)** 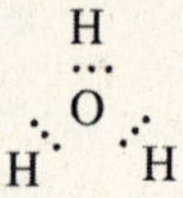 **(b)** 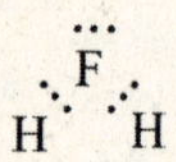

Analyze and Plan: In this fictional universe, a nonet is nine electrons. We will assume that the number of electrons in the atom called "O" in the other universe is still six, the number of electrons in the atom called "H" is still one, and the number of electrons in the atom called "F" is still seven.

Execute:

(a) The 6 electrons in the O atom would need three more electrons to make a total of nine. That means it would combine with three H atoms to make this molecule:

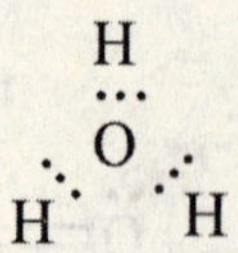

(b) The 7 electrons in the F atom would need two more electrons to make a total of nine. That means it would combine with two H atoms to make this molecule:

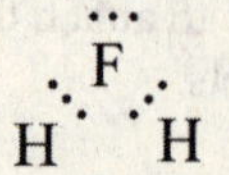

85. *Result:* **Cl: 2.7; As: 2.1; Br: 2.6?; S: 2.4?; and Se: 2.3? The last three are uncertain.**

Analyze and Plan: Assign several electronegativities to the atoms in a list, using the periodic trend for electronegativity (EN):

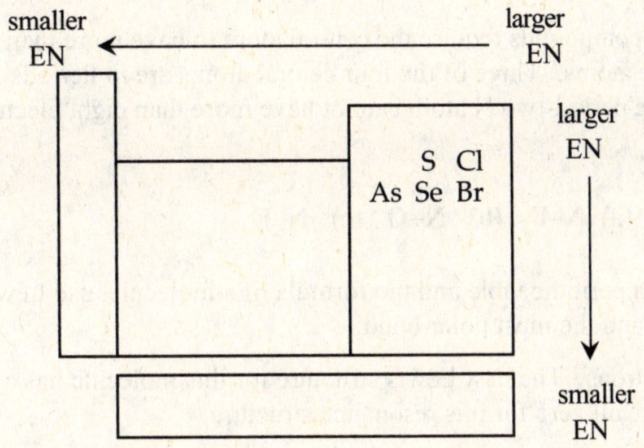

Execute:

We certainly expect the EN of Cl to be the largest, so we'll assign it the value of 2.7. We expect that the EN of As to be the lowest, so we'll assign it value of 2.1.

That leaves the EN values of 2.3, and 2.4, and 2.6 for Se, S, and Br. S and Br are both adjacent to Cl, so we'd expect them to be close to the same and probably higher than Se, so we'll assign the 2.6 and 2.3 to Br and S, leaving 2.3 for Se. Se is adjacent to both Br and S, so these three are really uncertain.

87. *Result:* **(a) Box 7 (b) Box 1 (c) Boxes 4 and 9 (d) Box 10 (e) Box 12 (f) Box 2**

Analyze: Identify items in a grid to answer several question.

Plan: First, figure out which boxes have answers that should be checked, then figure out which of them is correct. Use methods shown in Section 6-6b and Table 6.2 for bond enthalpies. Use periodic trends to predict bond lengths and Table 6.1 to confirm them. Use the methods described in the solution to Question 15 for Lewis Structures. Use methods shown in Section 6-7 for polarity.

Execute:

(a) Three boxes have Si-containing bonds, Box 2, 7 and 8. H is a much smaller atom than F or Si, so from periodic trends, we predict that Si-H has the shortest bond. Table 6.1 can assist in confirming that Si-H (145 pm) is shorter than Si-F (181 pm) or Si-Si (234 pm). Si-H is in **Box 7**.

(b) The two choices for this answer are Box 1 and Box 5. The atoms get slightly smaller as you look across the period, so the prediction is that the Cl-Br bond will be the smallest and the Si-Br bond will be the longest. Table 6.1 can assist in confirming that Cl-Br (213 pm) is the shortest and Si-Br (231 pm) is the longest. The correct order is given in **Box 1**.

(c) Three boxes have structures of $FClO_2$: Boxes 4, 6, and 9. The structures each have 26 electrons. The structure in Box 6 has the wrong structural arrangement; O atom should not be the central atom. The structure in Box 4 follows the octet rule and the structure in Box 9 shows the expanded octet on the central Cl atom. Expanding the octet is a legitimate way to lower formal charges for elements in Periods 3 or higher, so the structure in Box 4 is the most important, but both of these structures qualify as resonance hybrids for $FClO_2$. **Boxes 4 and 9** are correct.

(d) The two choices for this answer are Box 3 and Box 10. Polarity is related to the difference in electronegativities. The electronegativity from Figure 6.8 are: EN(Cl) = 2.7, EN(S) = 2.3, EN(Se) = 2.4, and EN(O) = 3.4. So, the most polar will be Cl-O (ΔEN = 0.7), then S-Cl(ΔEN = 0.4), then Se-Cl(ΔEN = 0.3). The correct order is given in **Box 10**.

(e) The two choices for this answer are Box 11 and Box 12. Bond enthalpy of multiple bonds tends to be quite high, so the best prediction would be Box 12 where the double bond is listed as the highest. Table 8.2 can assist in confirming that C=N (616 kJ/mol) is stronger than C-F (486 kJ/mol) or C-P (264 kJ/mol). This order is listed in **Box 12**.

(f) Three boxes have Si-containing bonds, Box 2, 7 and 8. The bond enthalpies are given in Table 6.2. It shows that Si-Si (226 kJ/mol) is weaker than Si-F (582 kJ/mol) or Si-H (323 kJ/mol) or . Si-Si is in **Box 8**.

89. *Result:* **NF_5; N cannot expand its octet**

Explanation: All of these compounds require the central atom to have more than eight electrons around it in order to have bonds to five atoms. Three of the four central atoms are in Periods 3 or higher, making it possible to expand their octets. The period-two N atom cannot have more than eight electrons, so NF_5 is least likely to exist.

90. *Result:* :F̤̈—N̈=Ö: **(a) N–F (b) N=O (c) N–F**

Analyze and Plan: Given a periodic table and the formula of a molecule, use Lewis structures to predict the longer and stronger bonds and the most polar bond.

Execute: FNO has 18 electrons. The best Lewis structure for this molecule has a double bond on the oxygen, since the formal charges are all zero for this resonance structure.

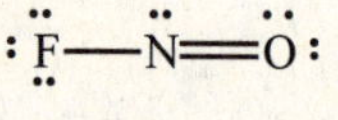

(a) Single bonds are longer than double bonds, so N–F is the longer bond.

(b) Double bonds are stronger than single bonds, so N=O is the stronger bond.

(c) Both bonds include N, and the periodic trend for electronegativity indicates that EN_F is greater than EN_O, so the most polar bond is N–F.

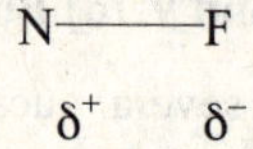

92. *Result:* **(a) C—O (b) C≡N (c) C—O**

Analyze and Plan: Given the Lewis structure of a molecule, predict the weaker and stronger carbon-containing bonds, using the plans in the solution to Questions 29 and 31 and the most polar bond using the plan in the solution to Question 37.

Execute:

(a) Bond strength increases from single to double to triple, so the weakest bond will be a single bond. Table 8.2 gives the various bond enthalpies for single bonds. The C–C bond enthalpy is 356 kJ/mol. The C–O bond enthalpy is 336 kJ/mol. Therefore, the C–O bond is the weakest carbon-containing bond.

(b) The triple C≡N bond is the strongest carbon-containing bond.

(c) The electronegativity of O atom is greater than that of any other element in this structure, so the C–O bond is the most polar.

94. *Result:* **See structure below**

Analyze: Given the name and formula of a compound, write the Lewis structure.

Plan: Examine the components of the compound's name to help figure out the atom arrangement, then use the method described in the solution to Question 15 for writing the Lewis structure.

Execute: The name is thiopropionaldehyde-S-oxide:

"Prop" in the name of an organic compound means there will be a three-carbon chain: C—C—C

The aldehyde functional group is R-CHO:

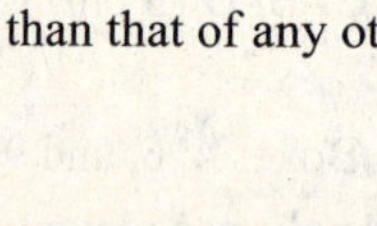

The prefix "thio" means one O atom has been replaced with an S atom:

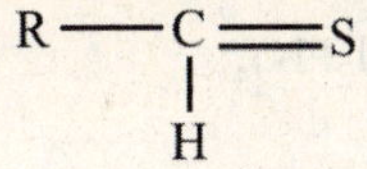

The "S-oxide" part of the name suggests that there will be an O atom bonded to the S atom.

The valence electron count is 3(4) + 6(1) + 6 + 6 = 30. All atoms except S atom must satisfy the octet rule.

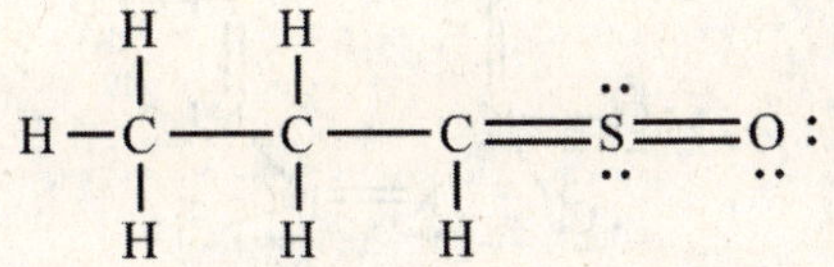

☑ *Reasonable Result Check:* At the time of production (Summer 2013), this molecule is accurately shown on the following web site: http://pubchem.ncbi.nlm.nih.gov/summary/summary.cgi?cid=441491

95. *Result:* **see structure below**

Analyze: Given the name of a compound, determine the formula and write the Lewis structure.

Execute: Tetraphosphorus trisulfide is P_4S_3, with 38 valance electrons.

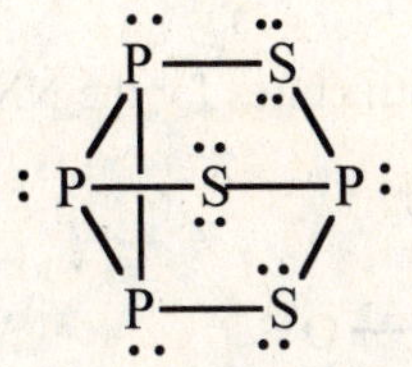

☑ *Reasonable Result Check:* At the time of production (Summer 2013), this molecule is accurately shown on the following web site: http://en.wikipedia.org/wiki/Phosphorus_sesquisulfide

More Challenging Questions

100. *Result:* **(a) see structure below (b) see structure below**

Analyze: Given the formula of two compounds, write their Lewis structures.

Plan: Use the method described in the solution to Question 15 for writing the Lewis structure.

Many different and creative guesses could result from attempting to arrange the atoms for these molecules. There is limited information provided, so several arrangements could be equally good.

Cl has a high electronegativity and it has zero formal charge with one bond, so we'll plan to make the Cl atoms terminal.

The N atoms must follow the octet rule. It is also not likely that the N atoms will bond to other N atoms or to Cl atoms, since their electronegativities are relatively high, so try a ring of alternating N and P atoms.

Execute:

(a) There are 72 valence electrons in $(Cl_2PN)_3$.

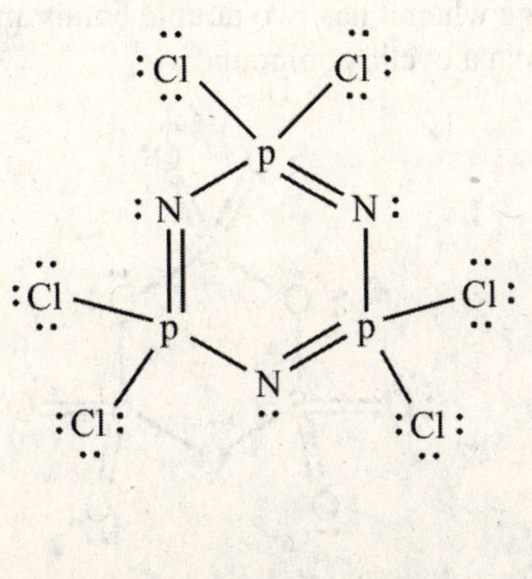

(b) There are 96 valence electrons in $(Cl_2PN)_4$.

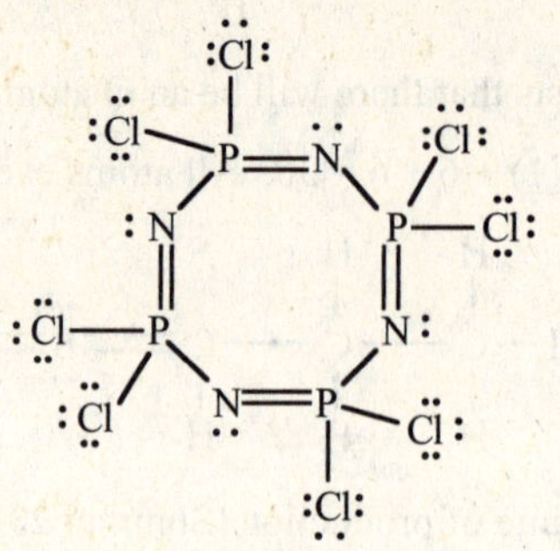

☑ *Reasonable Result Check:* At the time of production (Summer 2013), $(Cl_2PN)_3$ is accurately shown on the following web site: http://en.wikipedia.org/wiki/Hexachlorocyclotriphosphazene

101. *Result:* **The NON structures have higher formal charges than the NNO structures.**

Analyze: Given the formula of a compound and two variations in the arrangement of the atoms, identify why one arrangement is not as good as the other.

Plan and Execute: Draw the best Lewis structures for the NNO and NON structures of N_2O, and calculate the formal charges:

One other resonance structure exists for each of the molecules in question. The NON structures all have higher formal charge than the NNO structures, in addition to the positive formal charge being on the more electronegative atom. Both of these reasons explain why the structure of N_2O is not NON, with the N atoms each bonded to the O atom.

☑ *Reasonable Result Check:* These Lewis structures all have the correct number of electrons and all atoms follow the octet rule. The formal charges help explain the preferred form.

104. *Result:* **(a) See structures below (b) See structures below**

Analyze and Plan: Given the formulas of two compounds, write Lewis structure for them. We will use some of the same strategies applied in the solution to Question 100 to determine the atom arrangement.

Execute:

(a) $(SO_3)_3$ has 72 electrons.

We must make sure that the O atoms satisfy the octet rule. O has a zero formal charge when it has two bonds and two lone pairs. As a terminal atom, it will be double bonded. As an inner atom it must have 2 single bonds and two lone pairs.

Sulfur has a zero formal charge when it has two double bonds and two single bonds, so we can try alternating the O and S atoms in a cyclic compound:

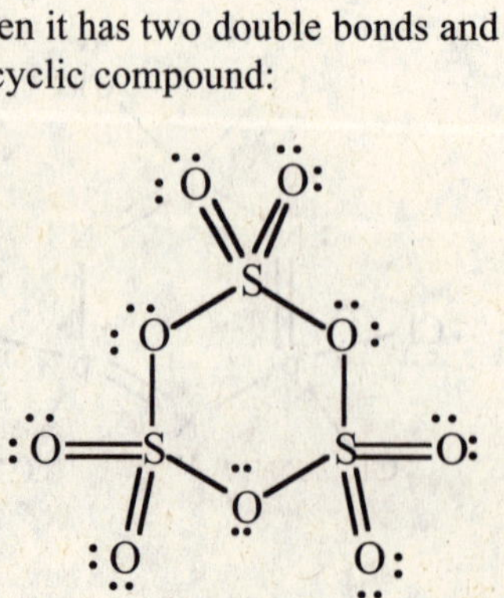

(b) $FXeN(SO_2F)_2$ has 70 electrons.

The arrangement of atoms in the molecular formula suggests that two $–SO_2F$ fragments are bonded to N.

Fluorine has a zero formal charge only if it has 1 bond pair and 3 lone pairs, so F will always be a terminal atom.

Oxygen has a zero formal change only if it has 2 bond pairs and 2 lone pairs. Terminal Oxygen will have 1 double bond and two lone pairs.

Nitrogen has a zero formal charge when it has 3 bonds and 1 lone pair, so it can bond to Xe, and both S atoms.

Sulfur has a zero formal charge when it has two double bonds and two single bonds.

Xenon, from Group 8, can have a large number of electrons.

Putting all these features together, we find that one plausible Lewis structure is:

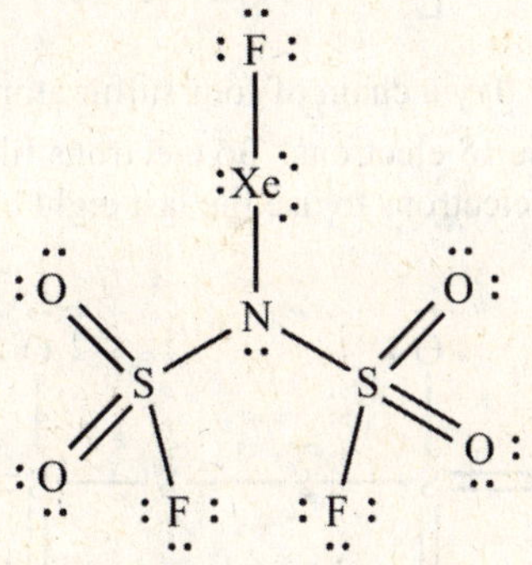

Xenon, from Group 8, and all the other atoms have a zero formal charge in this Lewis structure.

☑ *Reasonable Result Check:* These Lewis structures all have the correct number of electrons. The O, N, and F atoms all follow the octet rule.

105. *Result:* H—C≡C—C≡N:

Analyze: Given the formula of a compound, write a plausible Lewis structure.

Plan and Execute: The C and N atoms in this molecule must all follow the octet rule. Make a straight chain, and add as many multiple bonds to fulfill the octet rule for both period two atoms.

H—C≡C—C≡N:

All the atoms in this structure have zero formal change.

109. *Analyze and Plan:* Given the formulas of two compounds, write Lewis structure for them. We will use some of the same strategies applied in the solution to Questions 53 and 100 to figure out the atom arrangement.

Execute: To write correct Lewis structures for $S_2O_4^{2-}$, $S_2O_6^{2-}$ and $S_4O_6^{2-}$ we will seek structural elements that resemble the Lewis structures for SO_3^{2-} and SO_4^{2-}.

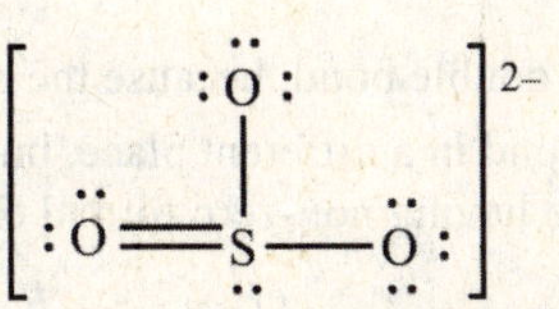

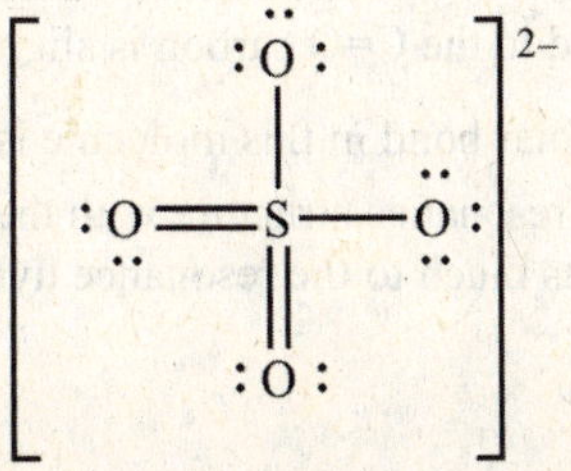

The $S_2O_4^{2-}$ ion has 38 electrons. It must have a bond between the two sulfur atoms and two O atoms bonded to each S atom. The first bonds use 10 electrons. 24 electrons fill the octets of the O atoms and each S atom needs two more electrons, which uses up the last four electrons. Forming double bonds between O and S reduces the formal charges:

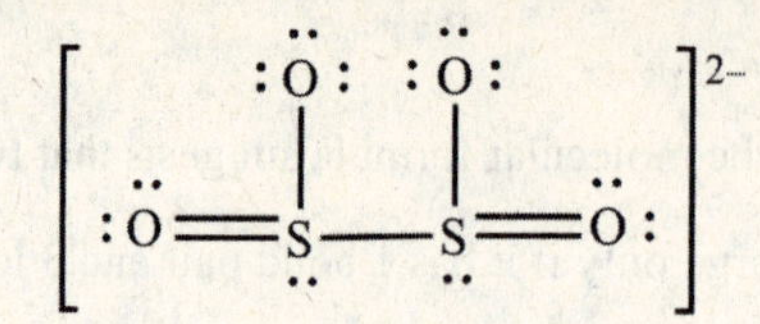

The $S_2O_6^{2-}$ ion has 50 electrons. It must have a bond between the two sulfur atoms and three O atoms bonded to each S atom. The first bonds use 14 electrons. 36 electrons fill the octets of the O atoms. Forming double bonds between O and S reduces the formal charges:

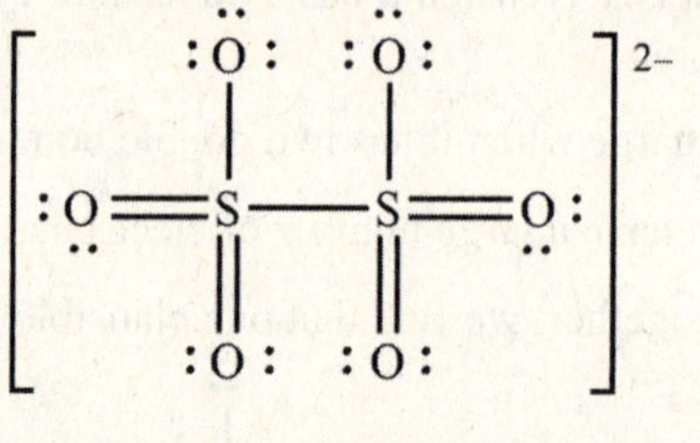

The $S_4O_6^{2-}$ ion has 62 electrons. Try a chain of four sulfur atoms and three O atoms on the far left and far right S atoms. The first bonds use 18 electrons. 36 electrons fill the octets of the O atoms. Each of the two middle S atoms needs four more electrons to use the last eight electrons. Forming double bonds between O and S reduces the formal charges:

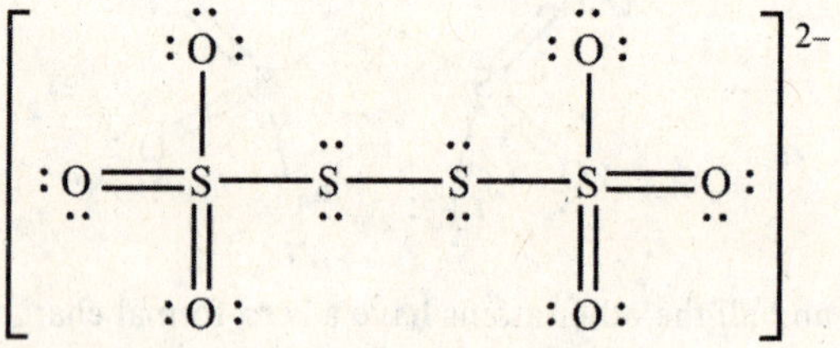

111. *Result:* **(a) see structure below (b) C–C (c) C=O (d) C=O**

Analyze: Given the name and formula for a molecule, write the Lewis structure and identify the longest, most polar, and strongest bonds.

Plan: Use the information in Section 6-3 to help interpret the structural formula of this organic molecule. Use the plans in the solutions to Questions 29, 31, 33, and 37, for judging bond length, strength, and polarity.

(a) This four-carbon chain has the amine functional group, $-NH_2$, and a carboxylic acid functional group, –COOH. Gamma is the third letter of the Greek alphabet, so we'll put the amine on the last carbon in the chain, three carbons away from the carboxylic acid.

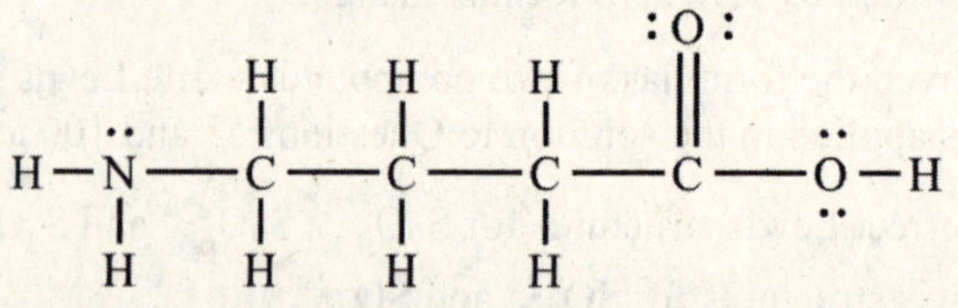

(b) The longest bonds in this molecule are the C–C single bonds in the alkane section of the chain. The C–C bonded to the C=O carbon is slightly shorter.

(c) The most polar bond in this molecule is the C=O double bond, because the ΔEN_{C-O} is the largest. There is a second resonance structure with the double bond in a different place, but this structure does not contribute as much to the resonance hybrid due to having non-zero formal charges:

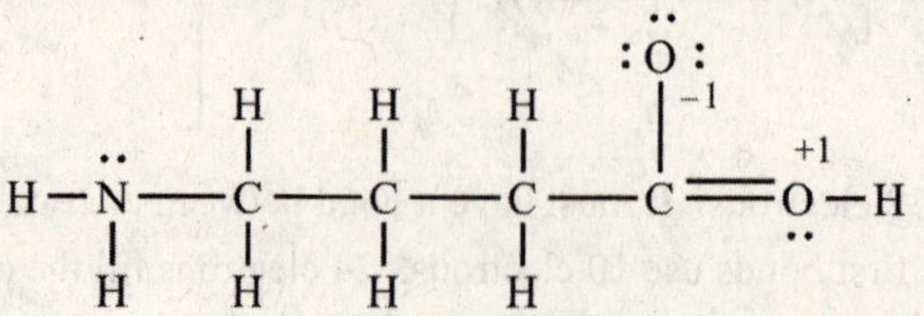

(d) The strongest bond in this molecule is the C=O double bond, because double bonds are stronger than single bonds.

Chapter 7: Molecular Structures

Solutions for Red-Numbered Questions for Review and Thought

Topical Questions

Molecular Shape (Section 7-2)

13. *Result:* **See structures below; (a) linear (b) triangular planar (c) octahedral (d) triangular pyramidal**

Analysis: Write Lewis structures for a list of chemical formulas and identify their shape.

Plan: Follow the systematic plan for Lewis structures given in the solution to Question 6.15, then determine the number of bonded atoms and lone pairs on the central atom, determine the designated type (AX_nE_m) and use Table 7.1. *Notice: It is not necessary to take the time to expand the octet of a central atom solely for the purposes of lowering its formal charge, because the shape of the molecule will not change. Hence, most Lewis structures here are written to follow the octet rule, unless the atom must have more than eight electrons for another reason.*

Execute:

(a) BeH_2 (4 e^-) H—Be—H The type is AX_2E_0, so it is **linear**.

(b) CH_2Cl_2 (20 e^-) The type is AX_4E_0, so it is **tetrahedral**.

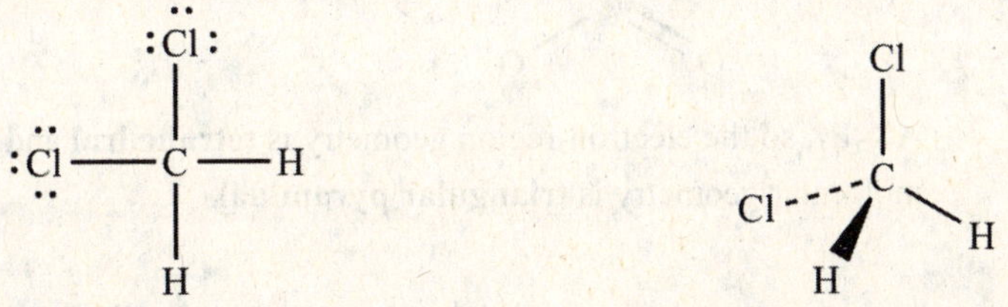

(c) BH_3 (6 e^-) The type is AX_3E_0, so it is triangular planar.

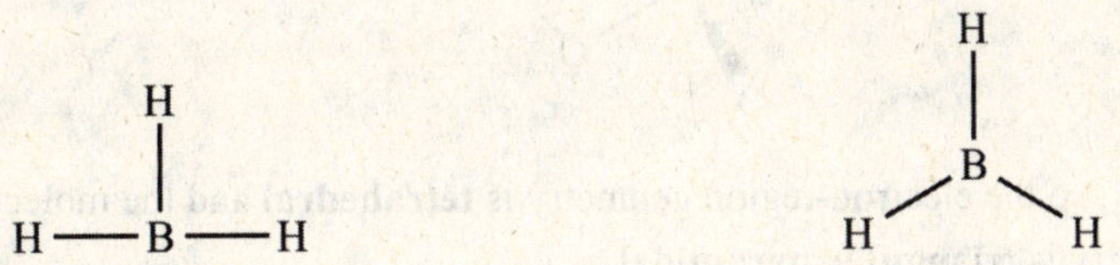

(d) $SeCl_6$ (48 e^-) The type is AX_6E_0, so it is octahedral.

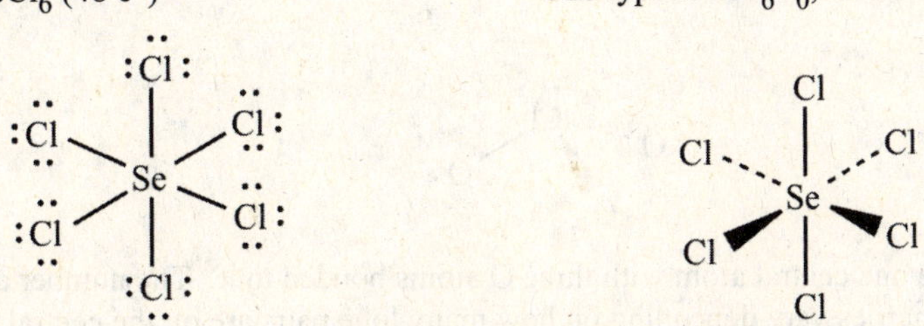

(e) PF_3 (26 e^-) The type is AX_3E_1, so it is triangular pyramidal.

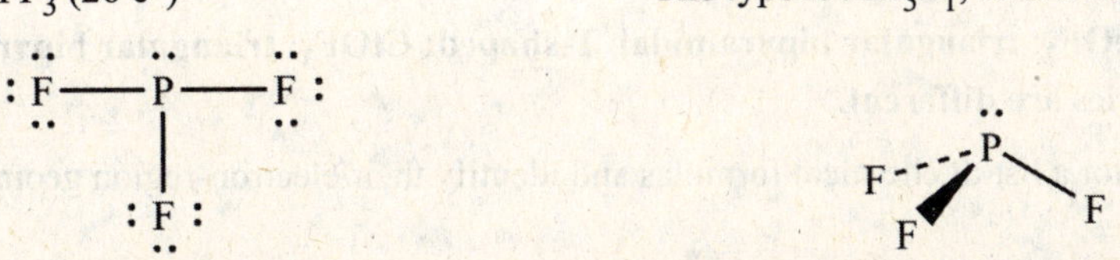

15. *Result:* **See structures below; (a) triangular planar, triangular planar (b) triangular planar, triangular planar (c) tetrahedral, triangular pyramidal (d) tetrahedral, triangular pyramidal**

Analysis: Write Lewis structures for a list of chemical formulas and identify their electron-region geometry and the molecular geometry.

Plan: Follow the systematic plan for Lewis structures given in the solution to Question 6.15, then determine the number of bonded atoms and lone pairs on the central atom, determine the designated type (AX_nE_m) and use Tables 7.1, 7.2, or 7.3.

Execute:

(a) BO_3^{3-} (24 e^-) The type is AX_3E_0, so both the electron-region geometry and the molecular geometry are **triangular planar**.

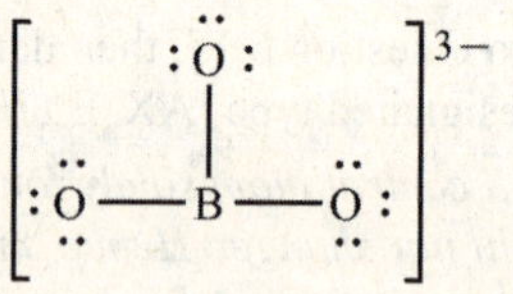

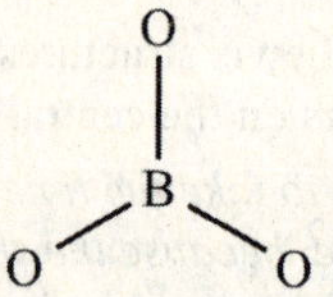

(b) CO_3^{2-} (24 e^-) AX_3E_0, so both the electron-region geometry and the molecular geometry are **triangular planar**.

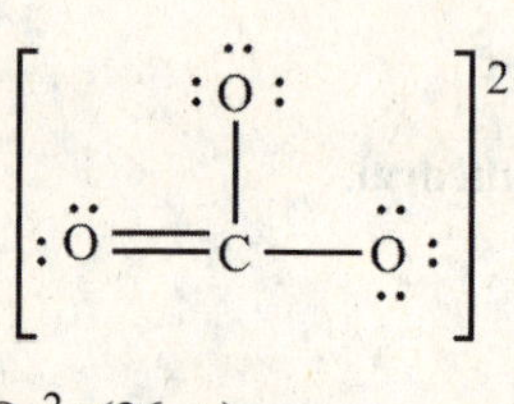

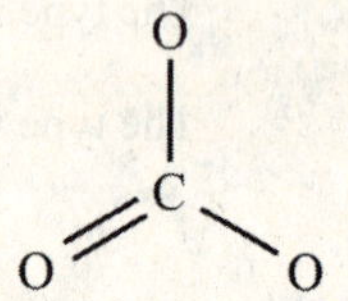

(c) SO_3^{2-} (26 e^-) AX_3E_1, so the electron-region geometry is **tetrahedral** and the molecular geometry is **triangular pyramidal.**

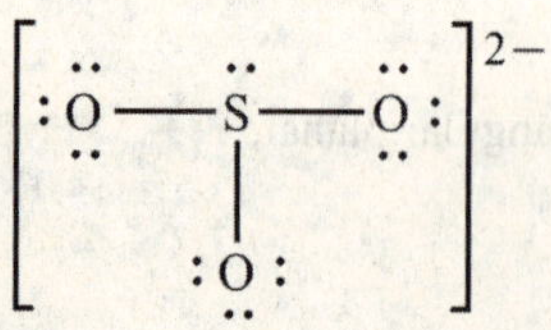

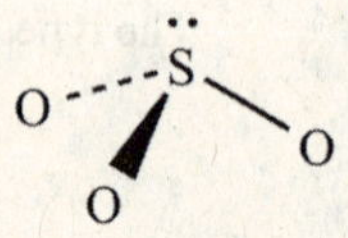

(d) ClO_3^- (26 e^-) AX_3E_1, so the electron-region geometry is **tetrahedral** and the molecular geometry is **triangular pyramidal.**

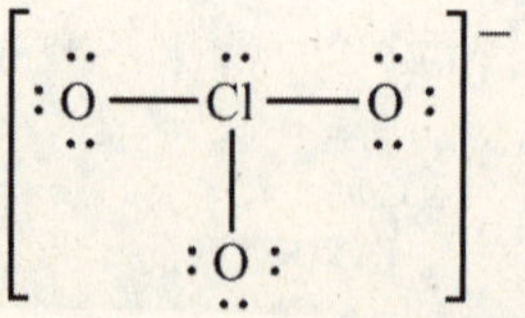

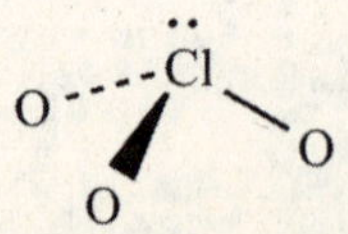

All of these ions and molecules have one central atom with three O atoms bonded to it. The number and type of bonded atoms is constant. The geometries vary depending on how many lone pairs are on the central atom. The structures with the same number of valance electrons all have the same geometry.

17. *Result:* **See structures below; $XeOF_2$: triangular bipyramidal, T-shaped; $ClOF_3$: triangular bipyramidal, seesaw; their molecular geometries are different.**

Analysis: Write Lewis structures for a list of chemical formulas and identify their electron-region geometry and the molecular geometry.

Plan: Adapt the method given in the solution to Question 15.

Execute:

$XeOF_2$ (28 e^-) The type is AX_3E_2, so the electron-region geometry is **triangular bipyramidal** and the molecular geometry is **T-shaped**.

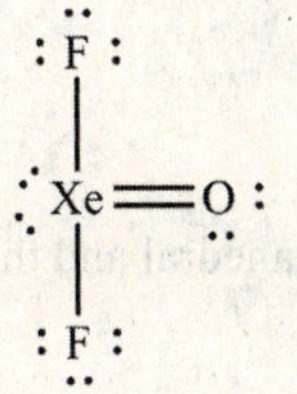

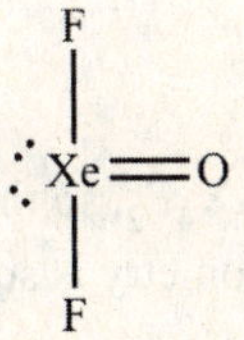

$ClOF_3$ (34 e^-) The type is AX_4E_1, so the electron-region geometry is **triangular bipyramidal** and the molecular geometry is **seesaw**.

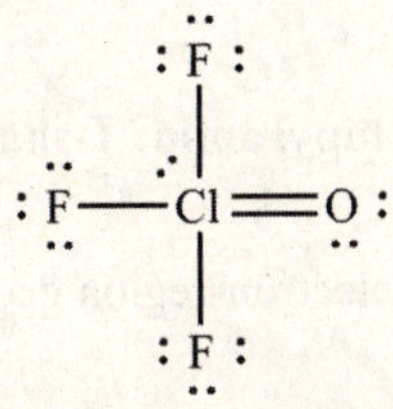

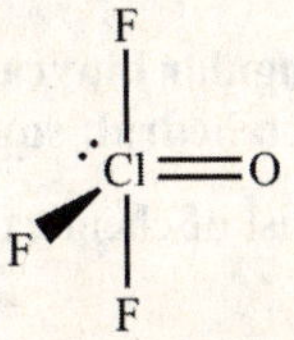

The two molecules have the same electron-region geometry, but **their molecular geometries are different.**

19. *Result:* **See structures below; (a) octahedral, octahedral (b) triangular bipyramid; seesaw (c) triangular bipyramidal; triangular bipyramidal (d) octahedral; square planar**

Analysis: Write Lewis structures for a list of chemical formulas and identify their electron-region geometry and the molecular geometry.

Plan: Adapt the method given in the solution to Question 15.

Execute:

(a) SiF_6^{2-} (48 e^-) The type is AX_6E_0, so both the electron-region geometry and the molecular geometry are octahedral.

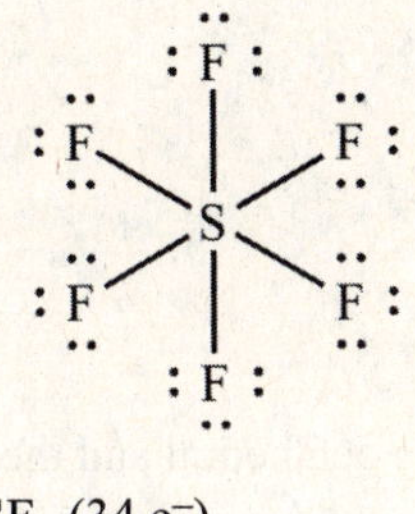

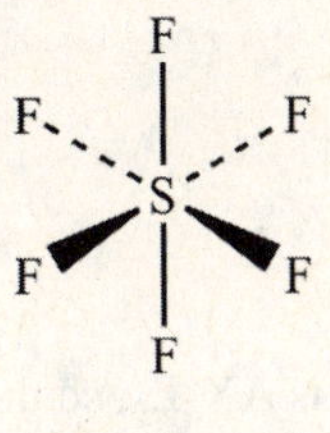

(b) SF_4 (34 e^-) The type is AX_4E_1, so the electron-region geometry is triangular bipyramid and the molecular geometry is seesaw.

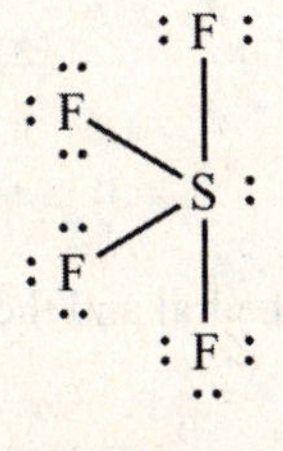

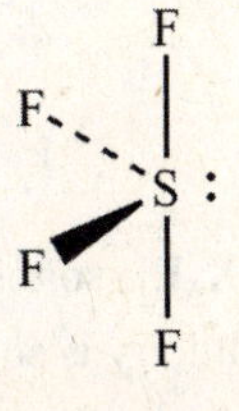

(c) PF_5 (40 e^-) The type is AX_5E_0, so the electron-region geometry is triangular bipyramidal and the molecular geometry is triangular bipyramidal.

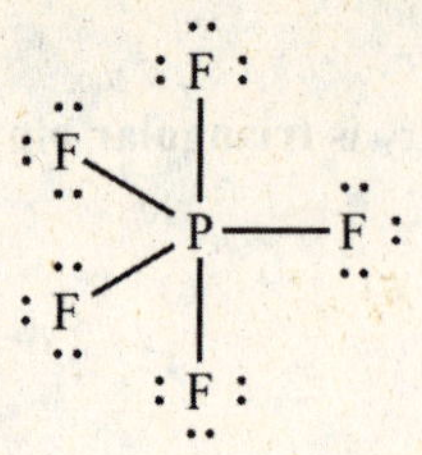

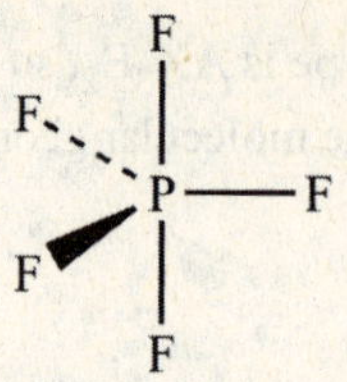

(d) XeF_4 (36 e^-) The type is AX_4E_2, so the electron-region geometry is octahedral and the molecular geometry is square planar.

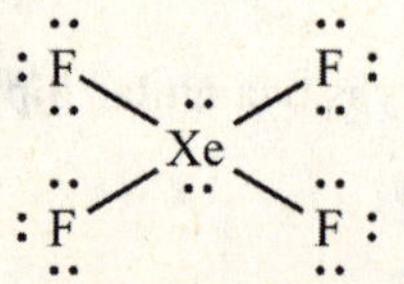

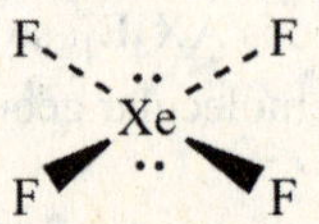

20. *Result:* **See structures below; (a) triangular bipyramid; linear (b) triangular bipyramid; T-shaped (c) octahedral; square planar (d) octahedral; square pyramidal**

Analysis: Write Lewis structures for a list of chemical formulas and identify their electron-region geometry and the molecular geometry.

Plan: Adapt the method given in the solution to Question 15.

Execute:

(a) ClF_2^- (22 e^-) The type is AX_2E_3, so the electron-region geometry is triangular bipyramid and the molecular geometry is linear.

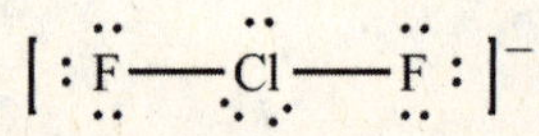

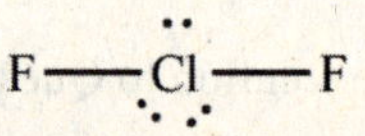

(b) ClF_3 (28 e^-) The type is AX_3E_2, so the electron-region geometry is triangular bipyramid and the molecular geometry is T-shaped.

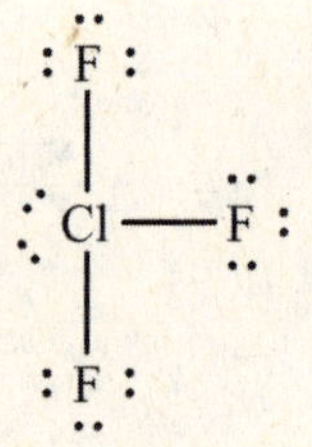

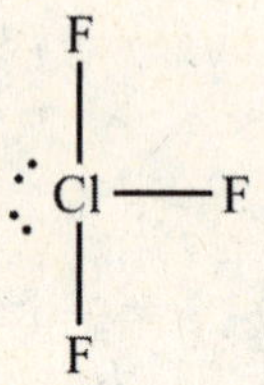

(c) ClF_4^- (36 e^-) The type is AX_4E_2, so the electron-region geometry is octahedral and the molecular geometry is square planar.

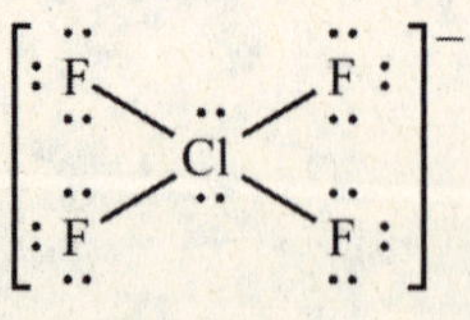

(d) ClF_5 (42 e^-) The type is AX_5E_1, so the electron-region geometry is octahedral and the molecular geometry is square pyramidal.

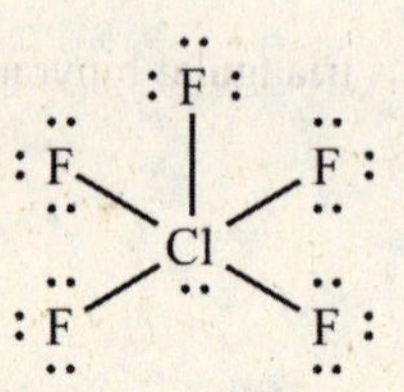

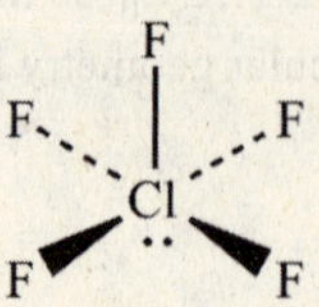

22. *Result:* **(a) O–S–O angle is 120° (b) F–B–F angle is 120° (d) H–C–H angle is 120°; C–C–N angle is 180° (c) N–O–H angle is 109.5°; O–N–O is 120°**

Analysis: Given chemical formulas and some structural information, predict some approximate bond angles.

Plan: We will adapt the method given in the solution to Question 15 to get the electron-region geometry of the second atom in the bond description (e.g., A in X–A–Y). We use this to predict the approximate bond angle.

Execute:

(a) SO_2 (18 e^-)

The type is AX_2E_1, so the electron-region geometry is triangular planar and the approximate O–S–O angle is 120°.

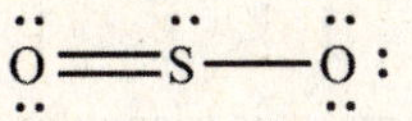

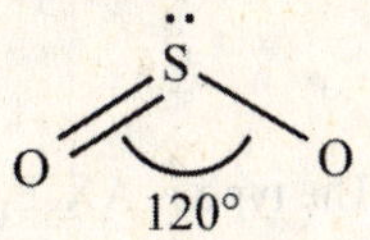

(b) BF_3 (24 e^-)

The type is AX_3E_0, so the electron-region geometry is triangular planar and the approximate F–B–F angle is 120°.

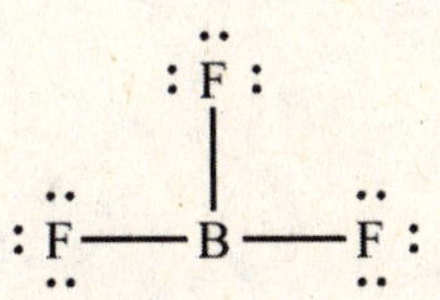

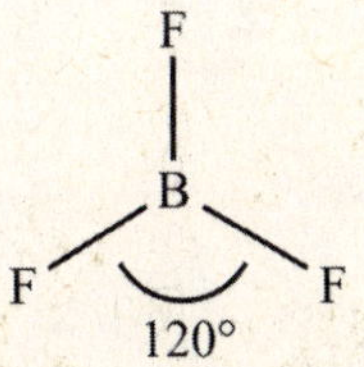

(c) CH_2CHCN (20 e^-)

Look at the first C atom: The type is AX_3E_0, so the electron-region geometry is triangular planar and the approximate H–C–H angle is 120°. Look at the third C atom: The type is AX_2E_0, so the electron-region geometry is linear and the approximate C–C–N angle is 180°.

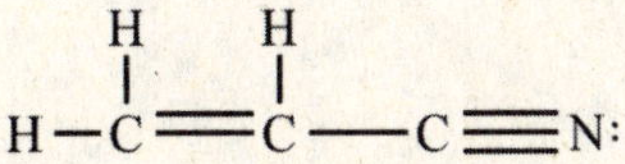

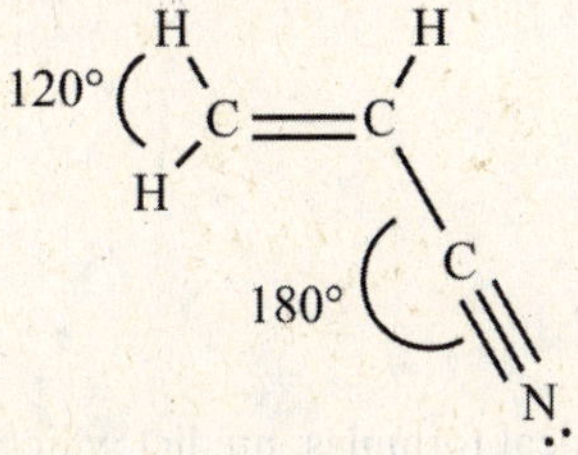

(d) HNO_3 (24 e^-)

Look at the first O atom: The type is AX_2E_2, so the electron-region geometry is tetrahedral and the approximate N–O–H angle is 109.5°. Look at the N atom: The type is AX_3E_0, so the electron-region geometry is triangular planar and the approximate O–N–O angle is 120°.

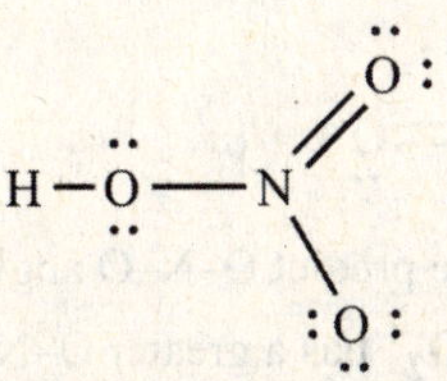

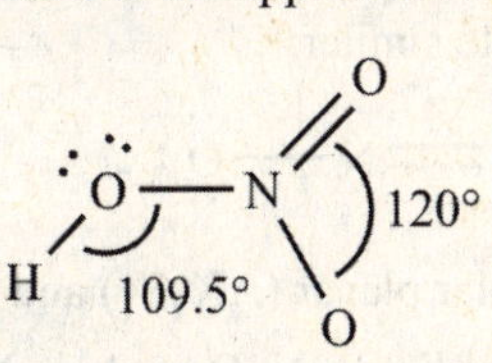

24. *Result:* **(a) four at 90°, one at 120°, and one at 180° (b) 90° and 120° (c) eight at 90° and two at 180°**

Analysis: Given chemical formulas and some structural information, predict some approximate bond angles.

Plan: Adapt the method given in the solution to Question 22.

Execute:

(a) SeF_4 (34 e^-)

The type is AX_4E_1, so the electron-region geometry is triangular bipyramid. The $F_{equitorial}$–Se–$F_{equitorial}$ angle is 120°, the $F_{equitorial}$–Se–F_{axial} angles are 90° and F_{axial}–Se–F_{axial} angle is 180°.

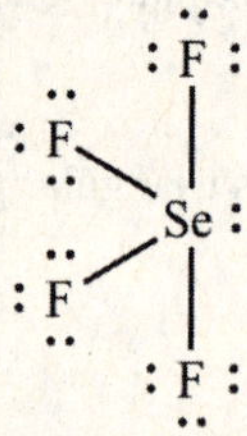

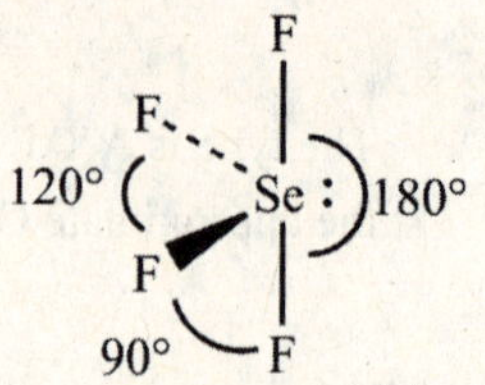

(b) SOF_4 (40 e^-)

The type is AX_5E_0, so the electron-region geometry is triangular bipyramidal, equatorial-F–S–O angles are 120° and the axial-F–S–O angles are 90°.

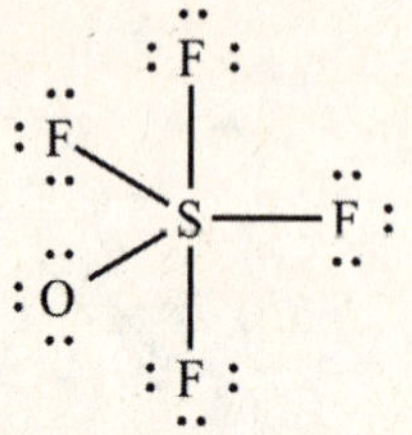

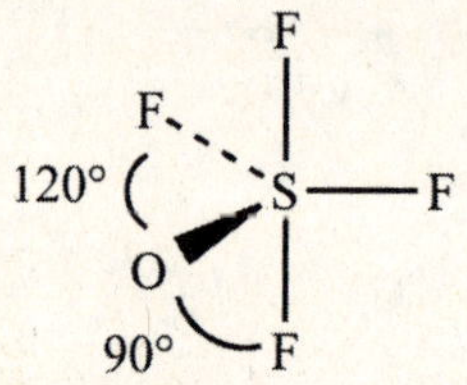

(c) BrF_5 (42 e^-)

The type is AX_5E_1, so the electron-region geometry is octahedral and all the F–Br–F angles are 90°. The angles across the bottom of the structure will be 180°.

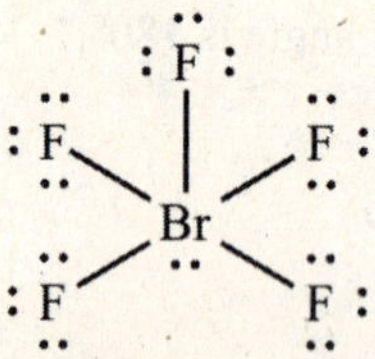

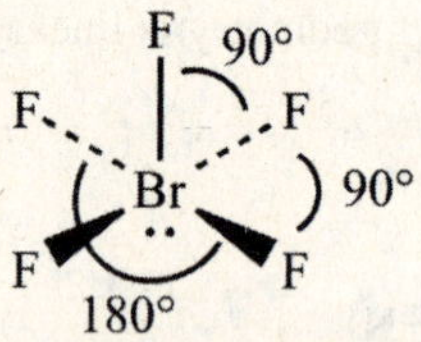

26. *Result:* $\mathbf{NO_2^+}$

Analysis: Given two chemical formulas, predict which has a greater O–N–O bond angle.

Plan: We adapt the method given in the solution to Question 22.

Execute:

NO_2 molecule and NO_2^+ ion differ by only one electron. NO_2 has 17 electrons and NO_2^+ has 16 electrons. Their Lewis structures are quite similar:

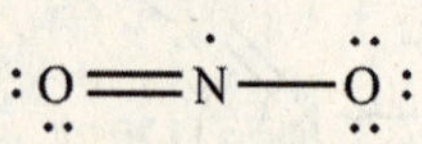

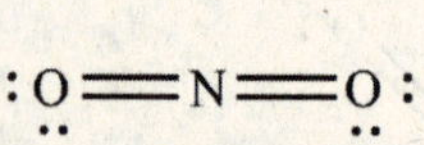

Using VESPR, NO_2 is triangular planar (AX_2E_1) and NO_2^+ is linear (AX_2). The predict O–N–O angle in NO_2 is approximately 120°. The predict O–N–O angle in NO_2^+ ion is 180°. So, $\mathbf{NO_2^+}$ has a greater O–N–O angle.

Chirality in Organic Compounds (Section 7-2)

28. *Result:* **See circles on structures below**

Analysis: Given structural formulas, identify the chiral centers.

Plan: Chiral centers are C atoms with four bonds, where each fragment bonded is different. Cirle the chiral centers.

Execute:

(a) The second and third C atoms have four different fragments: HOOC–, HO–, –H, and –CH(OH)COOH.

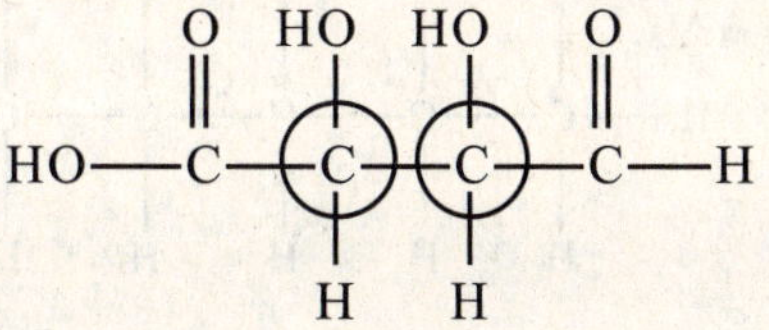

(b) This molecule has no chiral centers. The two inner C atoms do not have four bonds. The first C atom has three H atoms bonded.

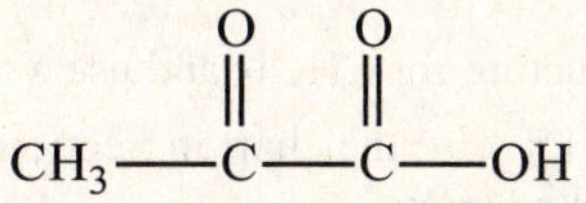

(c) The third C atom has four different fragments: CH_3CH_2–, –H, –COOH, and $–NH_2$. The first and second C atoms have three and two H atoms bonded, respectively, so they are not chiral centers.

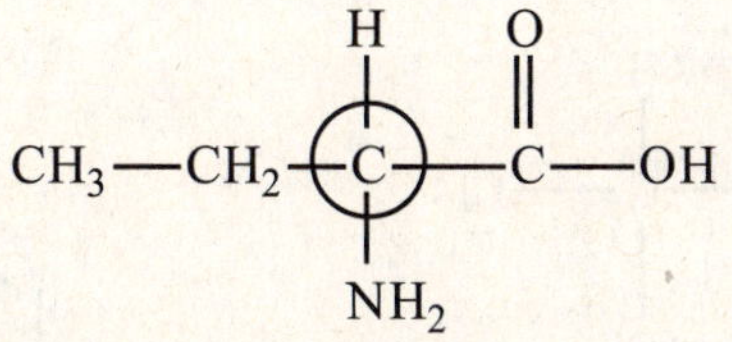

30. *Result:* **See circle on structure below**

Analysis: Given structural formulas, identify the chiral centers.

Plan: We will adapt the plan from the solution to Question 28.

Execute:

(a) This molecule has no chiral centers. The end C atoms have three H atoms bonded, and the second C atom has three CH_3 fragments.

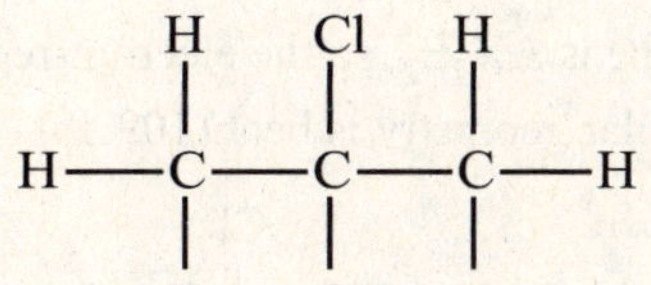

(b) This molecule has no chiral centers. The end C atoms have two or three H atoms bonded, and the second C atom has two CH_3 fragments.

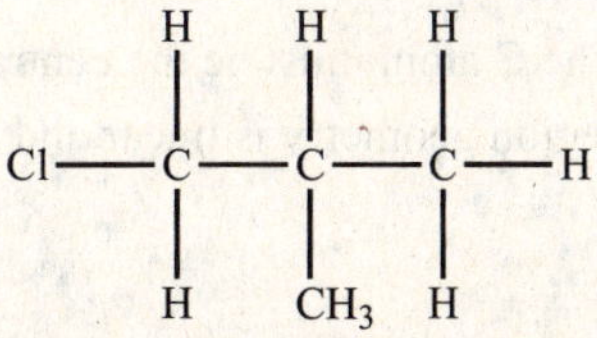

(c) The second C atom has four different fragments: CH_3–, Br–, $–CH_2CH_2CH_3$, and –H. The other C atoms have more than one H atom bonded, so they are not chiral centers.

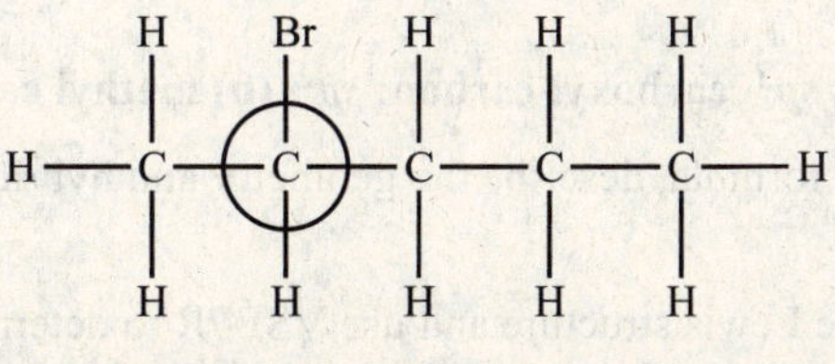

(d) This molecule has no chiral centers. Most of the C atoms have two or three H atoms, and the second C atom has two $–CH_2CH_3$ fragments.

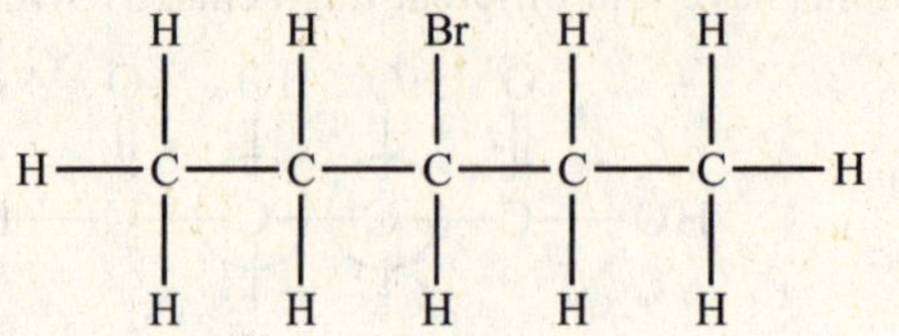

Hybridization (Section 7-3, 7-4)

32. *Result:* **tetrahedral, sp^3-hybridization**

Analyze and Plan: Write a Lewis structure for $CHCl_3$ and use VSEPR to determine the molecular geometry as described in the solution to Question 15. Use the chart in Section 7-3c to determine the hybridization of the central atom using the electron-region geometry.

Execute: The molecule is AX_4E_0 type, so its electron-region geometry and its molecular geometry are **tetrahedral.**

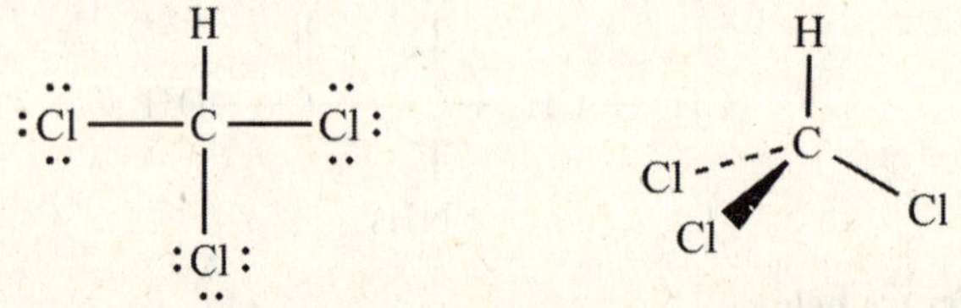

To make four equal bonds with an electron-region geometry of tetrahedral, the C atom must be ***sp^3*-hybridized.**

34. *Result:* **The central S atom in SCl_2 has *sp^3* hybridization; the central C atom in OCS is *sp*-hybridized. SCl_2 has two single bonds. OCS has two double bonds.**

Analyze: Describe the geometry and hybridization of several atoms in a structural formula.

Plan and Execute: Follow the procedure described in the solution to Question 32.

SCl_2 (20 e^-) The type is AX_2E_2, so the electron-region geometry is tetrahedral and the molecular geometry is bent (109.5°).

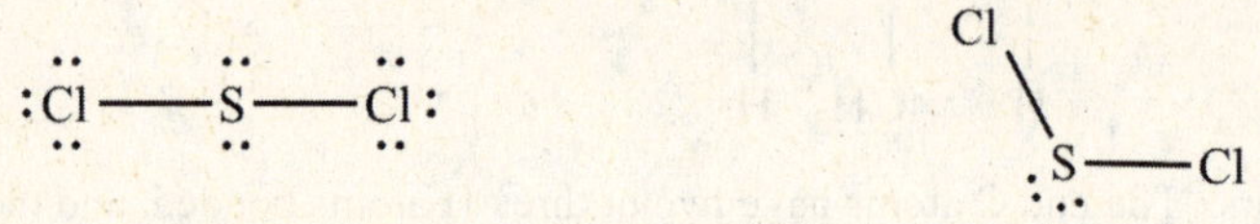

The S atom must be sp^3-hybridized.

OCS (16 e^-) The C atom must be the central atom. The type is AX_2E_0, so the electron-region geometry is linear and the O–C–S angle is 180°.

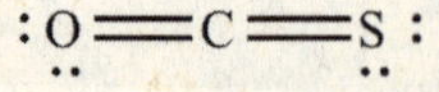

The C atom must be *sp*-hybridized.
The bonding in these molecules is polar covalent. The two bonds in SCl_2 are **single bonds.** The two bonds in OCS are **double bonds.**

35. *Result:* **(a) methyl carbon: *sp^3* carboxyl carbon: *sp^2* (b) methyl carbon: 109.5°, carboxyl carbon: 120°**

Analyze: Given a structural formula, describe the geometry and hybridization of several atoms and determine the approximate bond angle.

Plan and Execute: Draw the Lewis structure and use VSEPR to determine the electron-region geometry of each of the C atoms as described in the solution to Question 15. The bond angles and hybridization are associated with the electron-region geometry using Figures 7.1 and 7.2 and the chart in Section 7-3c.

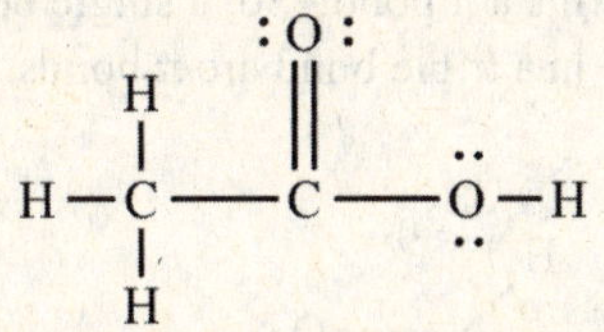

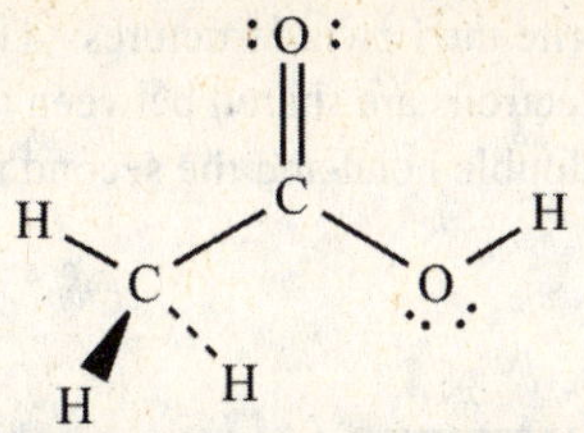

(a) Look at the left C atom: The type is AX_4E_0, so the electron-region geometry is tetrahedral and this C atom's hybridization is ***sp³***. Look at the right C atom: The type is AX_3E_0, so the electron-region geometry is triangular planar and this C atom's hybridization is ***sp²***.

(b) The *sp³*-hybridized C atom has tetrahedral H–C–H and H–C–C bond angles of approximately **109.5°**. The *sp²*-hybridized C atom has triangular planar C–C–O and O–C–O bond angles of approximately **120°**.

37. *Result:* **N: *sp³*, 109.5°; first two C atoms (from left to right): *sp³*, 109.5°; third C: *sp²*, 120°; top O: *sp²*; right O: *sp³*, 109.5°. The shortest carbon-to-oxygen bond is C=O.**

Analyze: Given a structural formula, describe the geometry and hybridization of several atoms, determine the approximate bond angles, and identify the shortest CO bond.

Plan and Execute: Adapt the plan described in the solution to Question 35. Consult Section 6-6c for assistance in determining the shortest carbon-to-oxygen bond.

For convenient reference, some atoms are numbered below from left to right: C_1, C_2, C_3, O_1, and O_2.

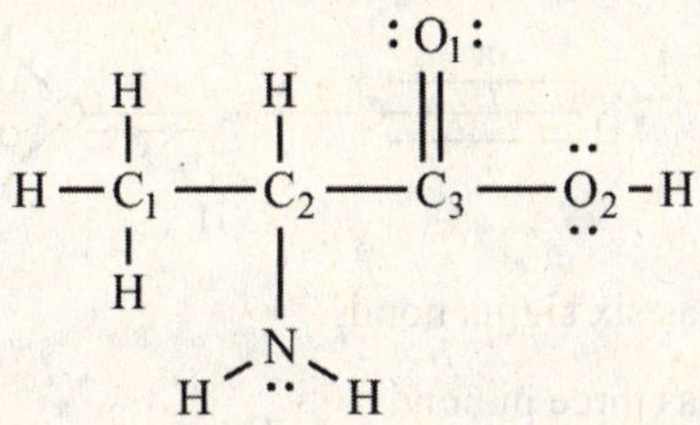

Look at C_1: The type is AX_4E_0, so the electron-region geometry is tetrahedral, its hybridization is ***sp³***, and has bond angles are approximately **109.5°**.

Look at C_2: The type is AX_4E_0, so the electron-region geometry is tetrahedral, its hybridization is ***sp³***, and the bond angles are approximately **109.5°**.

Look at the N atom: The type is AX_3E_1, so the electron-region geometry is tetrahedral, its hybridization is ***sp³***, and the bond angles are approximately **109.5°**.

Look at C_3: The type is AX_3E_0, so the electron-region geometry is triangular planar, its hybridization is ***sp²***, and the bond angles are approximately **120°**.

Look at the N atom: The type is AX_3E_1, so the electron-region geometry is tetrahedral, its hybridization is ***sp³***, and the bond angles are approximately **109.5°**.

Look at O_1: The type is AX_1E_2, so the electron-region geometry is triangular planar, its hybridization is ***sp²***. This oxygen has only one bond to carbon, so there are no associated bond angles.

Look at O_2: The type is AX_2E_2, so the electron-region geometry is tetrahedral, its hybridization is ***sp³***, and the bond angle is approximately **109.5°**.

The shortest carbon-to-oxygen bond is the double bond, C_3=O_1. This molecule does have a resonance structure with the double bond on O_2, but that structure has non-zero formal charges, so it contributes little to the resonance hybrid.

39. *Result:* **See structures with σ and π bonds designated below**

Analyze: Given structural formulas, identify the sigma and pi bonds.

Plan: First, we write the Lewis Structures. The first bond between two atoms must be a σ bond. When more than one pair of electrons are shared between two atoms, it forms a π bond. So, a single bond is a σ bond; the second bond in a double bond and the second and third bonds in a triple bond are π bonds.

Execute:

(a) 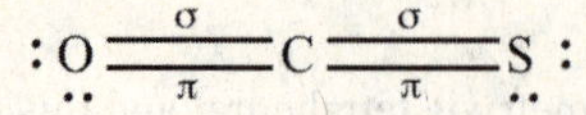

(b) 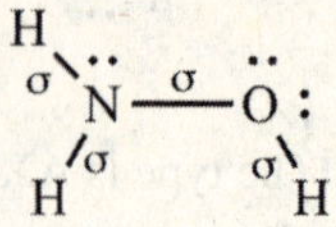

(c) 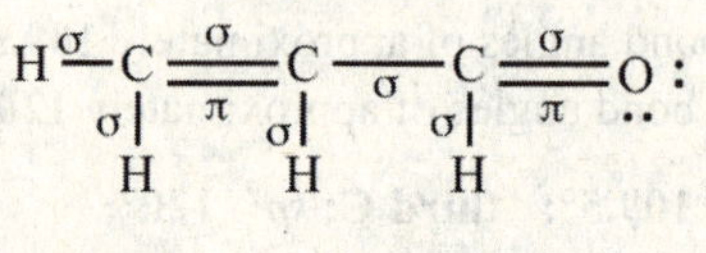

(d) 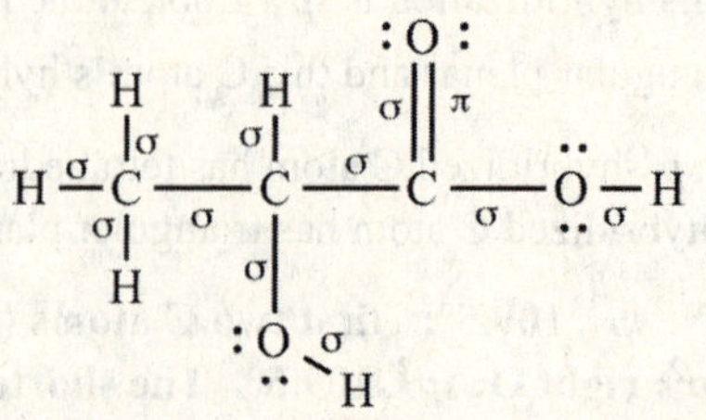

41. *Result:* **see structure below for location of sigma and pi bonds; (a) 6 sigma bonds (b) 3 pi bonds (c) *sp*-hybridization (d) *sp*-hybridization (e) both have sp^2-hybridization**

Analyze: Given a Lewis structure, identify the sigma and pi bonds, and the hybridization of several atoms.

Plan: Adapt the method described in the solution to Question 37 and Question 39.

Execute:

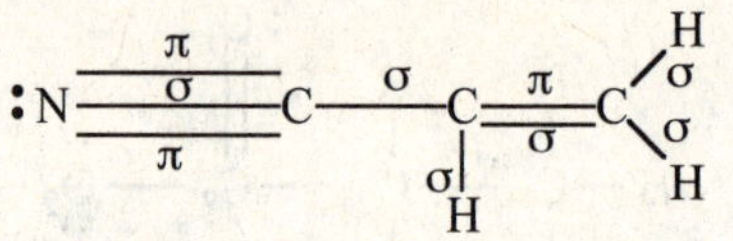

(a) As seen above, the structure has six sigma bonds.

(b) As seen above, the structure has three pi bonds.

(c) Look at the C atom bonded to the N atom: It is of type AX_2E_0, so the electron-region geometry is linear and these C atoms have *sp*-hybridization.

(d) Look at the N atom: It is of type AX_1E_1, so the electron-region geometry is linear and it has *sp*-hybridization.

(e) Both H-bearing C atoms are type AX_3E_0, so the electron-region geometry is triangular planar and both of them have a hybridization of sp^2.

Molecular Polarity (Section 7-5)

43. *Result:* **Molecules (b) and (c) are polar; HBF_2 has the F atoms on partial negative end and the H atom on the partial positive end; CH_3Cl has the Cl atom on partial negative end and the H atoms on the partial positive end.**

Analyze: Identify which molecules are polar and identify the direction of polarity for those that are polar.

Plan: Adapt the plan given in Problem-Solving Example 7.7: Write the correct Lewis structure for each molecule and use it to assess whether the bonds are arranged symmetrically or unsymmetrically. Use the periodic table or a table of electronegativities (Figure 6.8) to predict the polarity of the bonds. If the bonds are polar and arranged unsymmetrically around the central atom, the molecule is polar. The dipole points from the more electronegative atoms to the less electronegative atoms.

Execute:

First write the Lewis structures and determine the molecular shapes for the molecules:

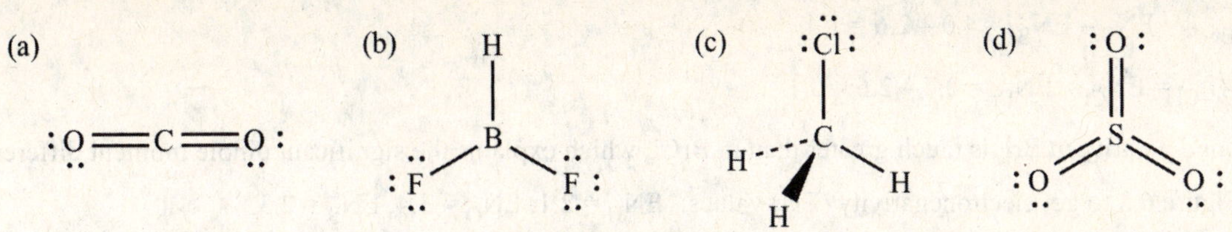

While the bonds in CO_2 and SO_3 are polar, the molecules are composed of identical atoms symmetrically arranged around the central atom, thus (a) and (d) are nonpolar:

HBF_2 and CH_3Cl have atoms unsymmetrically arranged around the central atom, so we must predict the polarity of the poles. The bond polarities related to the difference in electronegativity. Use Figure 6.8 to get electronegativity (EN) values: $EN_B = 1.9$, $EN_H = 2.1$, $EN_F = 4.0$, $EN_C = 2.4$, $EN_{Cl} = 2.7$

$$\Delta EN_{B\text{–}H} = EN_H - EN_B = 2.1 - 1.9 = 0.2$$

$$\Delta EN_{B\text{–}F} = EN_F - EN_B = 4.0 - 1.9 = 2.1$$

$$\Delta EN_{C\text{–}H} = EN_C - EN_H = 2.4 - 2.1 = 0.3$$

$$\Delta EN_{C\text{–}Cl} = EN_{Cl} - EN_C = 2.7 - 2.4 = 0.3$$

All the bonds are polar, so **(b) and (c) are polar.**

The bond pole arrows' lengths are related to their ΔEN and the arrows points toward the atom with the larger EN. So, the B–F arrows are much longer than the HB arrow, and a net dipole points toward the F atoms' side of the molecule, making the **F atoms' side of the molecule the partial negative end** and the **H atom's side of the molecule the partial positive end.**

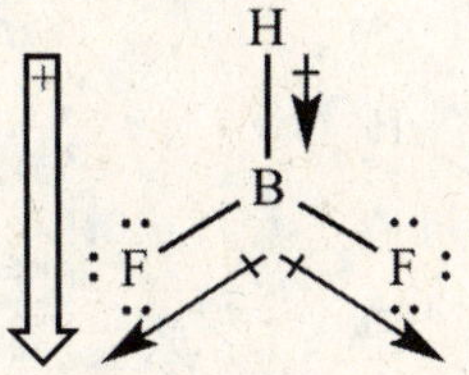

The C–Cl arrow points toward Cl, and the C–H points toward the C, so all the arrows point toward the Cl. That means a net dipole points toward the Cl atom's side of the molecule, making the **Cl atom's side of the molecule the partial negative end** and the **H atoms' side of the molecule the partial positive end.**

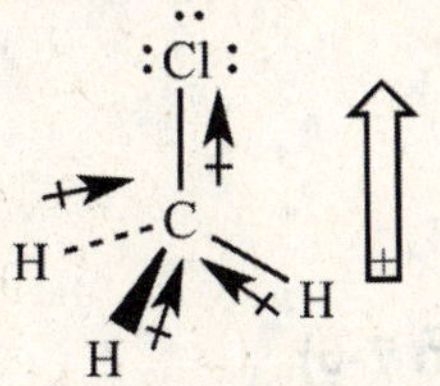

45. *Result:* **(a) The Br–F bond has a larger electronegativity difference. (b) The H–O bond has a larger electronegativity difference.**

Analyze: Given chemical formula and dipole moment, explain the differences in the dipole moments.

Plan: Adapt the plan described in the solution to Question 43.

Execute: The bond polarities are related to the difference in electronegativity.

Use Figure 6.8 to get electronegativity (EN) values: $EN_{Br} = 2.6$, $EN_F = 4.0$, $EN_{Cl} = 2.7$

$\Delta EN_{Br-F} = EN_F - EN_{Br} = 4.0 - 2.6 = 1.4$

$\Delta EN_{Br-Cl} = EN_{Cl} - EN_{Br} = 2.7 - 2.6 = 0.1$

The bond polarity in BrF is much greater than in BrCl, which explains the significant dipole moment difference.

Use Figure 6.8 to get electronegativity (EN) values: $EN_H = 2.1$, $EN_O = 3.4$, $EN_S = 2.3$

$\Delta EN_{O-H} = EN_O - EN_H = 3.4 - 2.1 = 1.3$

$\Delta EN_{S-H} = EN_S - EN_H = 2.3 - 2.1 = 0.2$

The bond polarity of the O–H bonds is much greater than that of the S–H bonds, which explains the dipole moment difference.

46. *Result:* **Both molecules have dipole moments. (a) NH_2OH has the O atom on the partial negative end and the H atoms on the partial positive end. (b) S_2F_2 has the F atoms on the partial negative end and the S atoms on the partial positive end.**

Analyze: Identify which molecules are polar and identify the direction of the dipole moment.

Plan: Adapt the plan described in the solution to Question 43.

Execute:

(a) The Lewis structure/geometry has polar O–H, N–H, and N–O bonds.

$\Delta EN_{O-H} = EN_O - EN_H = 3.4 - 2.1 = 1.3$

$\Delta EN_{N-H} = EN_N - EN_H = 3.0 - 2.1 = 0.9$

$\Delta EN_{N-O} = EN_O - EN_N = 3.4 - 3.0 = 0.4$

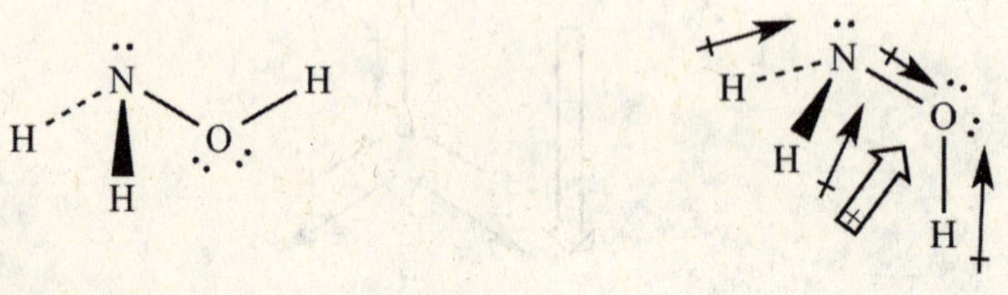

(b) The Lewis structure/geometry has polar S–Cl bonds.

$\Delta EN_{Cl-S} = EN_{Cl} - EN_S = 2.7 - 2.3 = 0.4$

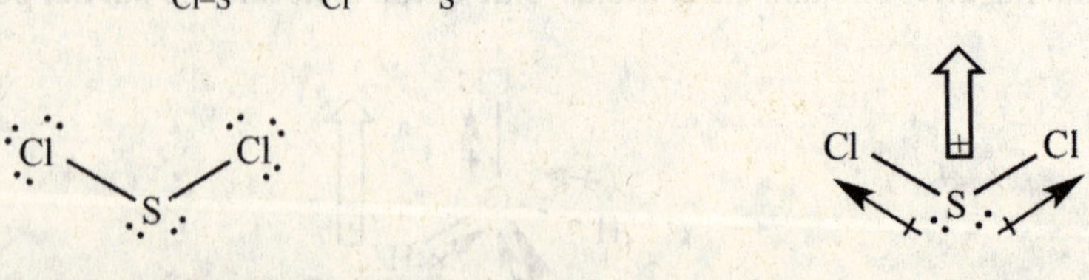

Noncovalent Interactions (Section 7-6)

48. *Result/Explanation:* Use information provided in Section 7-6 to construct this table:

Noncovalent Interaction	Strength	Example
ion-ion	Greatest strength	Na^+ interaction with Cl^-
ion-dipole	High strength	Na^+ ions in H_2O
dipole-dipole	medium strength	H_2O interaction with H_2O
dipole-induced dipole	Low strength	H_2O interaction with Br_2
induced dipole-induced dipole	Lowest strength	Br_2 interaction with Br_2

49. *Result:* **Water and ethanol interact using the same kinds of intermolecular forces (i.e., both experience H-bonding); whereas, water and cyclohexane interact with significantly different intermolecular forces.**

Analyze: Given molecular formulas and a description of variable miscibility, explain using molecular structure and noncovalent interactions the miscibility.

Plan and Execute: The electronegativity of O and C are fairly different (3.4 and 2.4, respectively), so the C–O bonds in $(CH_3)_2O$ are quite polar. Thus, H_2O and $(CH_3)_2O$ will be able to interact using hydrogen bonding (shown below) and these two substances would remain mixed together when combined (i.e., they are miscible).

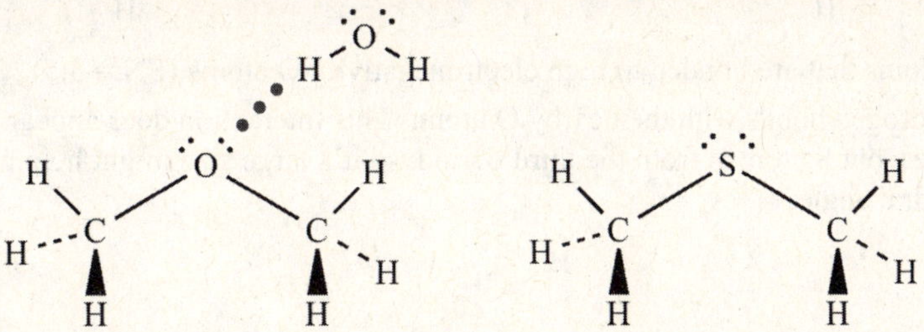

The bonds in $(CH_3)_2S$ are not very polar, because the electronegativities of S and C are similar (2.3 and 2.4, respectively). None of the atoms in $(CH_3)_2S$ can form hydrogen bonds with water. Water molecules are highly polar. To interact with $(CH_3)_2S$, water would have to sacrifice hydrogen bonding with other water molecules. Without comparably strong interactions, these two substances will not stay mixed together when combined. Thus it makes sense that $(CH_3)_2S$ is only slightly water soluble.

51. *Result:* **Wax is non-polar. Water droplets will form so that the polar water molecules can avoid interaction with the wax. Dirty, unwaxed cars have surface soils and salts that interact well with water.**

Analyze, Plan, and Execute: Wax is made up of **nonpolar** molecules that interact almost exclusively with London forces. The water molecules are highly **polar** and would have to give up hydrogen bonds between other water molecules to interact with wax using the much weaker London forces. These two substances would not interact very well. The water "beads up" in an attempt to have the smallest possible necessary surface interaction with the wax. On a dirty, unwaxed car, the soils and salts that are often found on cars contain ions and polar compounds. They interact much more readily with water, so the water would not "bead up" on the dirty surface.

53. *Result:* **(c) and (d)**

Analyze: Given condensed structural formulas, determine which can form intramolecular hydrogen bonds.

Plan: A hydrogen bond forms between (1) a very electronegative atom and (2) a H atom that is bonded to a very electronegative atom (EN ≥ 3.0). Intramolecular hydrogen bonding occurs only within a molecule that has both of these features nearby each other positioned so they can interact without significantly straining the geometric structure.

Execute:

(a) CH_2Br_2 does not contain any atoms with EN ≥ 3.0, so this molecule cannot form any hydrogen bonds.

(b) $CH_3OCH_2CH_3$ has only one O atom and all its H atoms are bonded to C atoms, so this molecule cannot form intramolecular hydrogen bonds.

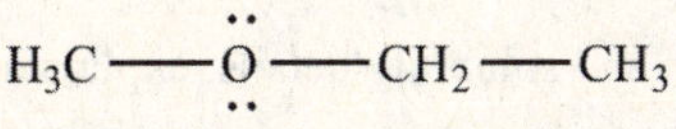

(c) H_2NCH_2COOH has three different H atoms are bonded to high electronegativity atoms (EN_O = 3.5 and EN_N = 3.0).

One of the H atoms bonded to the N atom can form an intramolecular hydrogen bond with the nearby O atom. One of those H atoms bonded to O can form an intramolecular hydrogen bond with the nearby N atom in the structure. Two of the possible structures are shown here:

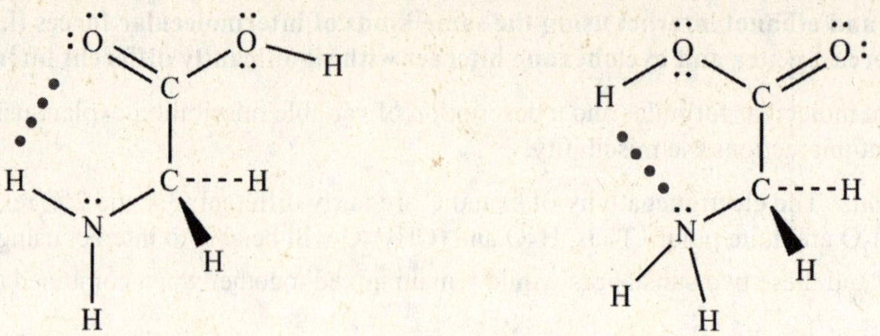

(d) H_2SO_3 has H atoms that are bonded to high electronegativity O atoms (EN = 3.5). Those H atoms might be able to form hydrogen bonds with the nearby O atom. This interaction does appear to put some strain on the 109.5° angles, but S atom is from the third period, so it's large size might help permit the O atom to keep its tetrahedral angle.

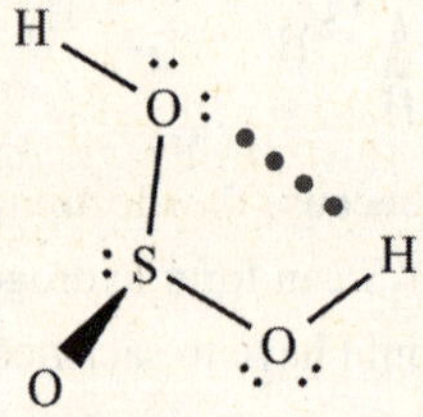

(e) CH_3CH_2OH has only one O atom, so this molecule cannot form intramolecular hydrogen bonds.

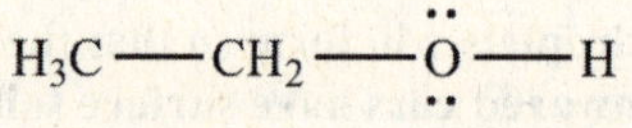

55. *Result:* **(b) < (c) < (d) < (a); Hydrogen bonding interactions are stronger than dipole-dipole interactions and London forces are the weakest. London forces increase as the number of electrons increases.**

Analyze: Given condensed structural formulas, explain the order of their increasing boiling point.

Plan: Boiling points are higher when the interactive forces are stronger. Determine the structural elements that relate to polarity and the types of noncovalent intermolecular forces each molecule can employ.

Execute:

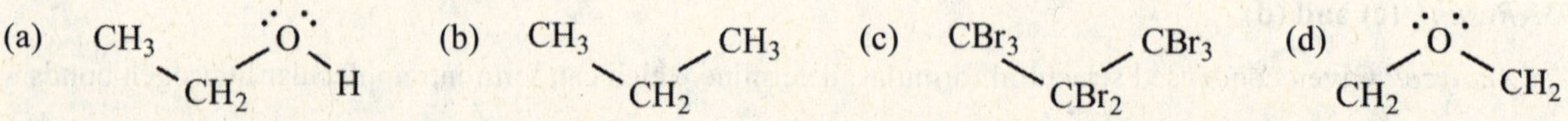

CH_3CH_2OH is only molecule that can interact with hydrogen bonding, the strongest intermolecular force, due to the O–H bond. Of the remaining three molecules, CH_3OCH_3 is the most polar, since the C–O bonds are very polar. It interacts with dipole-dipole forces. The remaining two molecules are nonpolar and they interact via London forces. London forces are stronger when the molecule has more electrons, so the $CBr_3CBr_2CBr_3$ has stronger London forces than $CH_3CH_2CH_3$. The molecule with the lowest boiling point is $CH_3CH_2CH_3$, followed by CH_3OCH_3, and the highest boiling point is CH_3CH_2OH, so the order of increasing boiling point is: **(b) < (c) < (d) < (a)**

57. *Result:* **Vitamin C is capable of forming hydrogen bonds with water, thus vitamin is miscible in H_2O.**

Analyze: Given a structural formula and solubility details, explain at the molecular level the solubility.

Plan: Boiling points are higher when the interactive forces are stronger. Determine the structural elements that relate to polarity and the types of noncovalent intermolecular forces each molecule can employ.

Execute: Four H atoms are bonded to high electronegativity O atoms (EN = 3.5). Those H atoms (circled in the structure below) can form hydrogen bonds with H_2O molecules.

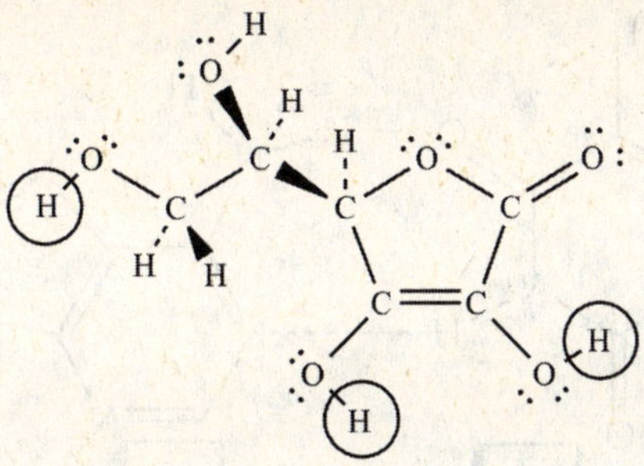

This molecule is quite polar and would not interact well with fats, because fats interact primarily using London forces.

60. *Result:* **(a) London forces (b) London forces (c) Covalent bonds (d) Dipole-dipole forces (e) Intermolecular hydrogen bonding forces**

Analyze, Plan, and Execute:

(a) Hydrocarbons have no capability of forming hydrogen bonds and the bonds are close to non-polar. The molecules interact primarily with London forces. The London forces between the molecules must be overcome to sublime $C_{10}H_8$.

(b) As described in (a), propane molecules, a hydrocarbon, would need to overcome London forces to melt.

(c) Decomposing molecules of nitrogen and oxygen require breaking covalent bonds, so the forces that must be overcome are the covalent intramolecular forces.

(d) To evaporate polar PCl_3 requires that the molecules overcome dipole-dipole forces.

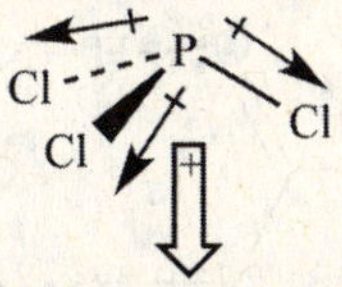

(e) The double strands of DNA are shown in Figure 7.27. The interaction between the two strands are hydrogen bonds. These hydrogen bonding forces must be overcome to "unzip" the DNA double helix.

Biomolecules (Section 7-7)

61. ***Answer/Explanation:*** The structure of a section of DNA with T-C-G (See Figure 1, next page.)

Molecular Structure Determination by Spectroscopy

63. *Result:* **Compared to UV: (a) IR has lower energy, (b) IR has longer wavelength, (c) IR has lower frequency.**

Explanation:

(a) IR radiation is lower in energy than UV radiation. (IR is below visible spectrum and UV is above it).

(b) IR radiation have longer wavelength than UV radiation. (Lower energy waves have longer wavelength than higher energy waves.).

(c) IR radiation has lower frequency than UV radiation. (Lower energy waves have smaller frequency than higher energy waves).

65. *Result:* **(a) C–C stretch (b) O–H stretch**

Explanation: The Tools of Chemistry: Infrared Spectroscopy in Section 7-2d describes infrared spectroscopy. Wavelength, λ, is the variable on the horizontal axis of the spectrum shown. Wavelength is inversely proportional to energy. So, longer wavelength light has lower energy and shorter wavelength light has higher energy.

(a) The lowest energy motion described in the spectrum is C–C stretch.

(b) The highest energy motion described in the spectrum is O–H stretch.

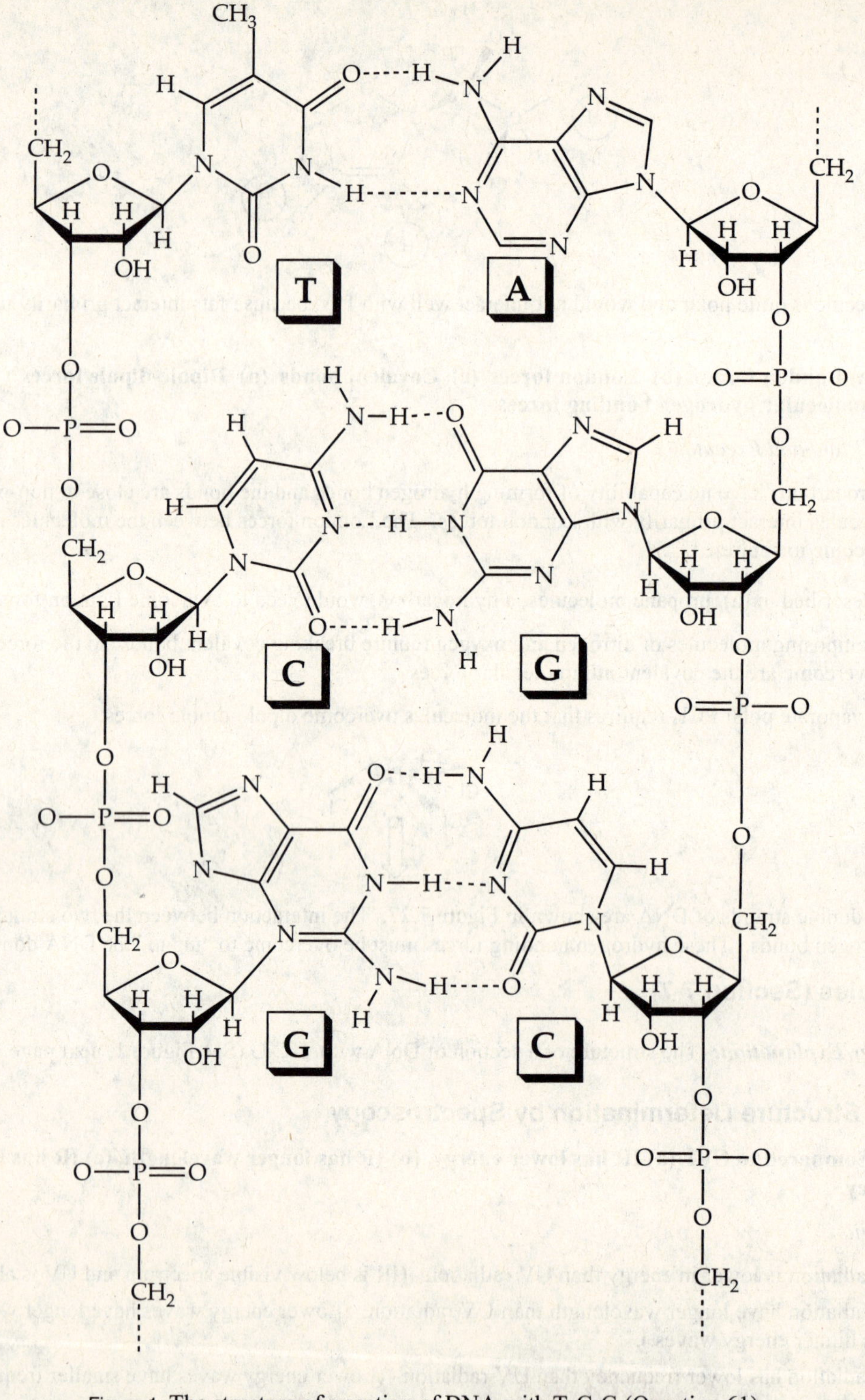

Figure 1: The structure of a section of DNA with T-C-G (Question 61)

General Questions

67. *Result:* **(a) Angle 1: 180° Angle 2: 109.5° Angle 3: 109.5° (b) C=O (c) C=O (d) C=C**

Analyze: Given a Lewis Structure, determine a few indicated bond angles and identify the most polar bond, and the shortest CO and CC bonds.

Plan and Execute: Follow the plans from the solution to Questions 32 - 35 and 43.

(a) Angle 1: Look at the C atom bonded to the N atom: It is of type AX_2E_0, so the electron-region geometry is linear and this C atom has *sp*-hybridization. The *sp*-hybridized C atoms have linear bond angles of approximately 180°. So, the CCN angle is 180°.

Angles 2 and 3: Look at the C atom bonded to only one O atom: The type is AX_4E_0, so the electron-region geometry is tetrahedral this C atom has sp^3-hybridization. The sp^3-hybridized C atom has tetrahedral bond angles of approximately 109.5°. So, the HCH angle (Angle 2) and the OCH angle (Angle 3) are both 109.5°.

(b) The most polar bond is the C=O bond.

(c) Multiple bonds are shorter than single bonds, so the C=O bond is the shortest CO bond:

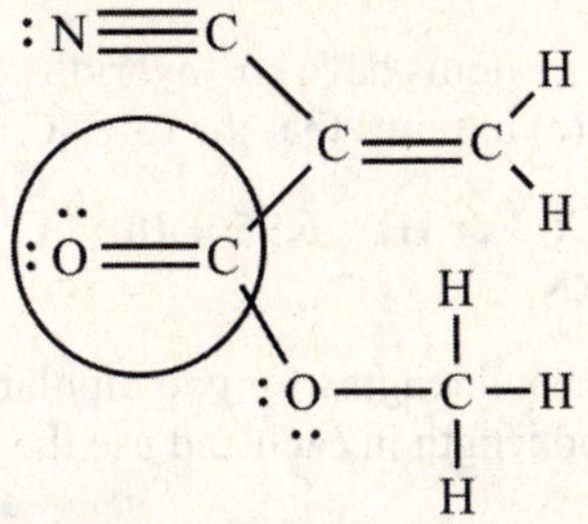

(d) Multiple bonds are shorter than single bonds, so the C=C bond is the shortest CC bond:

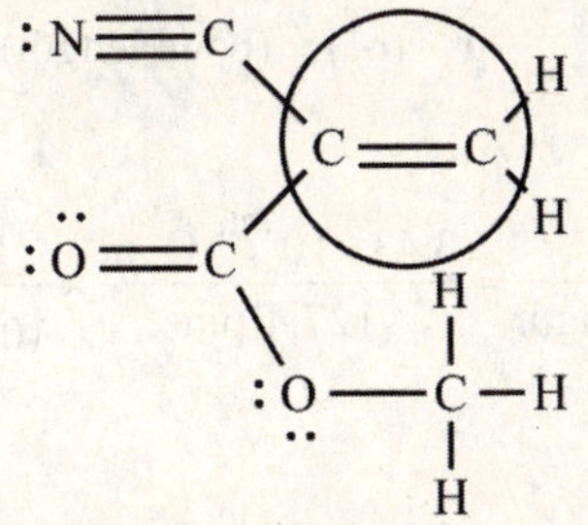

69. *Result:* **(a) See structure below (b) Angle 1: 180° (c) Angle 2: 180°**

Analyze: Given a chemical formula, write the Lewis structure and determine the CCO and CCC bond angles.

Plan and Execute: Follow the plans from the solutions to Question 6.15 and Questions 32 - 35..

(a) 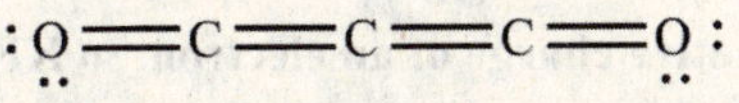

(b) Angle 1: Look at a C atom bonded to the O atom: It is of type AX_2E_0, so the electron-region geometry is linear with *sp*-hybridization and CCO bond angle of approximately 180°.

(c) Angle 2: Look at the C atom bonded only to the C atoms: It is of type AX_2E_0, so the electron-region geometry is linear with *sp*-hybridization and CCO bond angle of approximately 180°.

72. *Result:* **It will absorb visible light because it has an extended pi-system.**

Explanation: The Tools of Chemistry: Ultraviolet-Visible Spectroscopy in Section 7-5 describes ultraviolet-visible spectroscopy. Compounds with an extended pi system formed by conjugation between a series of alternating double bonds absorb visible light. This molecule also has an extended pi system, so we would predict it absorbs visible light.

73. *Result:* **(a) see structure below (b) each C has sp^2-hybridization. (c) Inconsistent, since angles are 90° not 120°**

Analyze: Given a chemical formula, write the Lewis structure and determine the hybridization of indicated atoms, then discuss the inconsistency between the molecular shape and the predicted bond angles.

Plan and Execute: Follow the plans from the solutions to Question 6.15 and Questions 32 - 35.

(a) $C_4O_4^{2-}$ (42 e^-) has four equivalent resonance structures. One of these resonance structures looks like this:

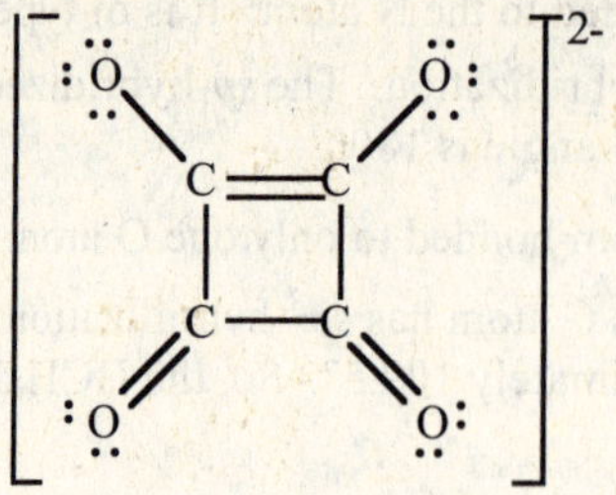

(b) Look at each C atom: The type for each C atom is AX_3E_0, so the electron-region geometry is triangular planar and each C atom has sp^2-hybridization.

(c) There is an inconsistency. If the C atoms have sp^2-hybridization, the C–C–C bond angle would be 120°; however, the square structure in (a) requires that the C–C–C bond angles must be 90°.

75. *Result:* **(a) For HCl: 2.69 × 10^{-20} C; For HF: 6.95 × 10^{-20} C (b) partial negative charge on F is larger than that on Cl, so F has a higher EN.**

Analyze: Given dipole moments and bond lengths for two dipolar molecules, calculate the quantity of charge in coulombs that is separated by the bond length in each and use the result to show that one atom is more electronegative than another.

Plan and Execute: Dipole moment is determined by multiplying the partial charge by the bond length.

$$\mu = (\delta^+) \times (\text{bond length})$$

For HCl:

$$\delta^+ = \frac{\mu}{\text{bond length}} = \frac{3.43 \times 10^{-30}\ \text{C} \cdot \text{m}}{127.4\ \text{pm}} \times \frac{1\ \text{pm}}{10^{-12}\ \text{m}} = 2.69 \times 10^{-20}\ \text{C}$$

For HF:

$$\delta^+ = \frac{\mu}{\text{bond length}} = \frac{6.37 \times 10^{-30}\ \text{C} \cdot \text{m}}{91.68\ \text{pm}} \times \frac{1\ \text{pm}}{10^{-12}\ \text{m}} = 6.95 \times 10^{-20}\ \text{C}$$

The partial charges in the HF bond are larger than those of the HCl bond, indicating that the fluorine is more electronegative than chlorine.

76. *Result:* **1.32 × 10^{-19} C; this partial charge is 81.6% of a full negative charge of an electron, so KF is not really completely ionic.**

Analyze: Given dipole moments and bond lengths for two dipolar molecules, calculate the partial charge on each atom, then compare results with the charge of an electron and determine if a compound is completely ionic.

Plan and Execute: The definition of dipole moment was given in Question 75:

$$\mu = (\delta^+) \times (\text{bond length})$$

For KF: $$\delta^+ = \frac{\mu}{\text{bond length}} = \frac{28.7 \times 10^{-30}\ \text{C} \cdot \text{m}}{217.2\ \text{pm}} \times \frac{1\ \text{pm}}{10^{-12}\ \text{m}} = 1.32 \times 10^{-19}\ \text{C}$$

$$\delta^- = -1.32 \times 10^{-19}\ \text{C}$$

If the bond were completely ionic, then the partial negative charge on fluorine would be the same as the charge on the electron (–1.62x10^{-19} C). So, KF is not completely ionic.

$$\text{Percent of full negative charge} = \frac{-1.32 \times 10^{-19}\ \text{C}}{-1.62 \times 10^{-19}\ \text{C}} \times 100\% = 81.6\% = 81.6\%\ \text{of a full negative charge.}$$

78. *Result:* **If the two atoms are the same element, then the molecule will be nonpolar. If the two atoms have different EN's, the molecule will be polar.**

Explanation:

A diatomic molecule can have a dipole moment of zero if the atoms have the same electronegativity, such as when they are atoms of the same element. O_2 is a good example of a nonpolar diatomic molecule.

If the two atoms have different electronegativities, then the molecule will have a nonzero dipole moment and the molecule will be polar. CO is a good example of a polar diatomic molecule.

79. *Result:* **(a) Box A, B, D, and H (b) Boxes D, G, and H (c) Boxes A and G (d) Boxes E and I (e) Boxes B, C, D, E,F, H, and I (f) Box A (g) Box F (h) Box H (i) Box A (j) Box F and G**

Analyze and Plan: Given several choices of chemical formulas, find the answer for each prompt related to electron-region geometry, shape, hybridization, and physical properties. To answer several of these prompts, we must write the Lewis Structures and determine the electron-region geometries and molecular geometries for each structure. Follow the plans from the solutions to Question 6.15 and Questions 32 - 35.

Execute:

A (10 e^-) 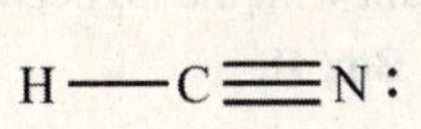	B (26 e^-) 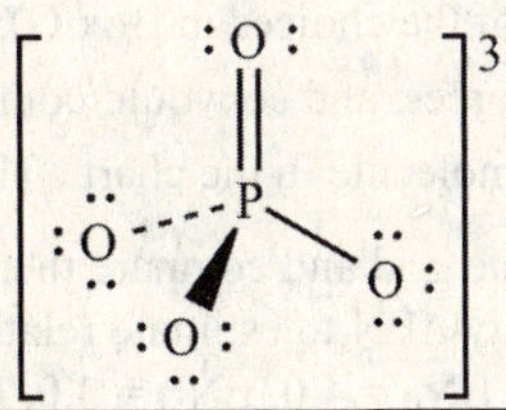	C (8 e^- and 26e^-) 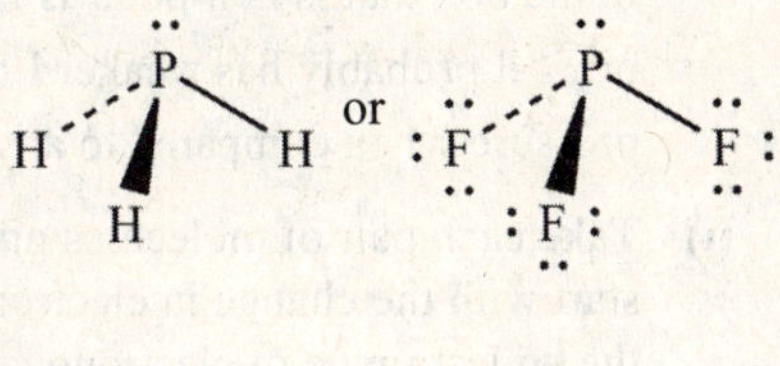
D (8 e^-) 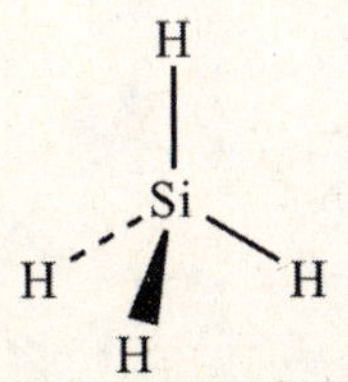	E (16 e^-) 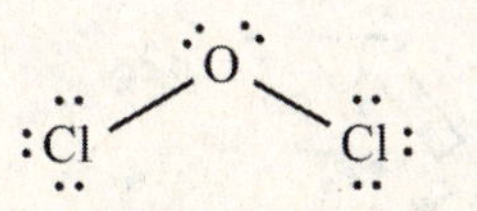	F (14 e^-) 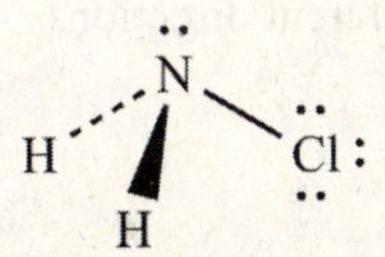
G (8 e^- and 14 e^-) 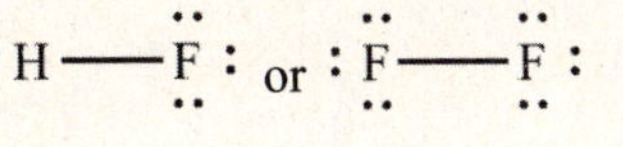	H (8 e^-) 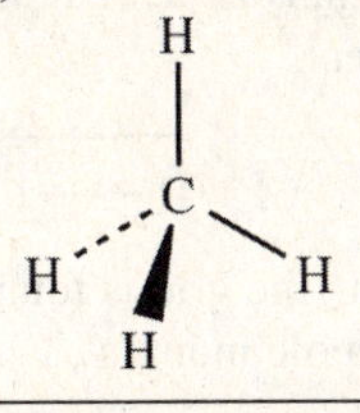	I (20 e^-) 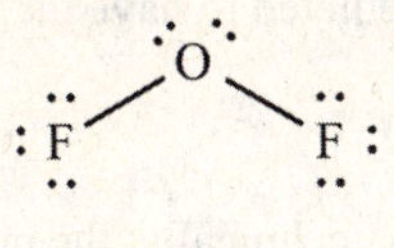

(a) The structures that have electron-region geometry the same as the molecular geometry are the structures with no lone pairs on the central atom. These are seen in **Box A** with a linear geometry (AX_2E_0) and **Boxes B, D and H** with a tetrahedral geometry (AX_4E_0). In Box G, both of the two structures qualify (with a linear geometry), so it is appropriate to include **Box G** in the answer, as well.

(b) Non-polar molecular structures have the same atoms bonded to the central atom and no lone pairs on the central atom, so that all the bond poles balance and result in a zero net dipole moment. These are represented in **Boxes D and H** with the bonding type of AX_4E_0. One of the two structures in Box G also qualifies (F_2). Since that box indicates that <u>one or the other</u> structure could answer the prompt, it is logical to include **Box G** in the answer, as well.

(c) The structures that have linear geometry have bonding type of AX_2E_0 or are diatomic molecules. These are seen in **Box A** with bonding type of AX_2E_0 and **Box G** with two diatomic molecules.

(d) The structures that have angular (bent) geometry have bonding type of AX_2E_2 or AX_2E_1. These are seen in **Box E and I**, both with bonding type of AX_2E_2.

(e) The central atom is sp^3-hybridized in structures with the bonding type of AX_4E_0, AX_3E_1, or AX_2E_2. **Boxes B, D and H** have the central atom with the AX_4E_0 bonding type. **Boxes C and F** have the central atom with the AX_3E_1 bonding type. **Boxes E and I** have the central atom with the AX_2E_2 bonding type.

(f) The central atom has *sp*-hybridization in structures with the bonding type of AX_2E_0. **Box A** has the central C atom with an AX_2E_0 bonding type.

(g) Take each pair of molecules given in the grid and compare them. The liquid with the lowest boiling point will be composed of molecules with the strongest intermolecular forces. Box F has a molecule that would interact using both dipole-dipole intermolecular forces and hydrogen bonding intermolecular forces. The only other molecule in the grid that can undergo hydrogen bonding forces is HF (one of the choices in Box G). Since HF is smaller and has fewer atoms, NH_2Cl would compose the liquid with the lower boiling point when compared to any other molecule in the chart. The answer is **Box F**.

(h) Take each pair of molecules given in the grid and compare them. The liquid with the highest vapor pressure will be composed of molecules with the weakest intermolecular forces. Boxes D and H have molecules that would interact using only London intermolecular forces. The only other nonpolar molecule in the box that is non-polar is F_2 (one of the choices in Box G). Since CH_4 has fewer electrons than SiH_4 or F_2 it probably has weaker London forces, and so would compose the liquid with the higher vapor pressure when compared to any other molecule in the chart. The answer is **Box H**.

(i) Take each pair of molecules given in the grid and compare them. To determine relative dipole moments, start with the change in electronegativity ΔEN to estimate relative bond polarity. Boxes A and F contain the widest range of electronegativities. $EN_F = 4.0$, $EN_N = 3.0$, $EN_{Cl} = 2.7$, $EN_C = 2.4$, $EN_H = 2.1$. While the N-H bonds are more polar (in Box F), the pair will partially cancel each other out, pulling electrons in different directions.

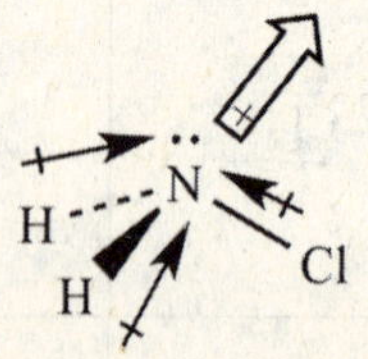

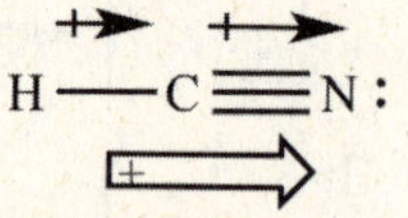

There is one other highly polar molecule in Box G: H-F has the largest ΔEN, and would therefore be predicted to have the higher dipole moment.

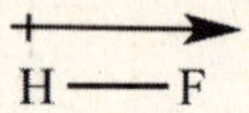

So, technically, the most polar molecule in the grid is found in Box G. However, there is a non-polar molecule also in Box G that has a zero dipole moment, F_2, and the molecules in both Box A and F have greater dipole moments than a molecule in Box G. Therefore, aside from HF, the most polar molecule in the grid is HCN in **Box A**.

(j) Several of the structures are polar, which is the prerequisite for dipole-dipole intermolecular forces; however only two molecules, HF (one of the two molecules in Box G) and NH_2Cl (in Box F) have an H atom bonded to a high-electronegativity atom; therefore, both **Boxes F and G** qualify to answer this prompt.

Applying Concepts

81. *Result:* **CH_3SH has weaker IM forces.**

Analyze: Given two chemical formulas and two boiling points, explain the differences.

Plan: Write the Lewis Structures and determine the electron-region geometries and molecular geometries for each structure. Follow the method described in the solutions to Question 6.15 and Questions 32 - 35.Determine intermolecular forces for each molecule and relate to the properties as described in Section 7-6.

Execute: The two molecules have the same geometry, since O and S are in the same periodic group:

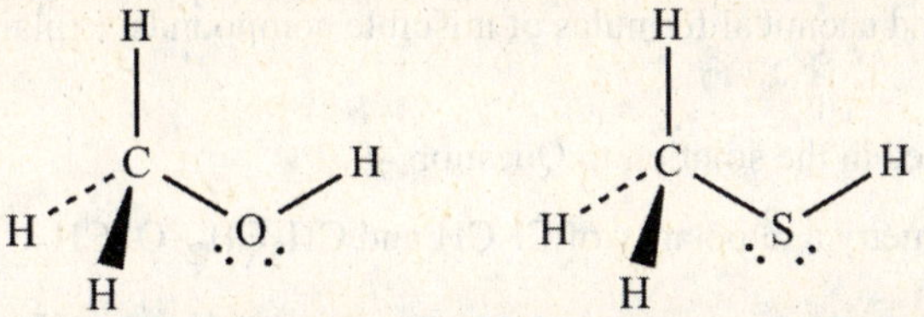

The lower electronegativity of S (2.3) makes the S–H bond essentially nonpolar, unlike the O–H bond. In the liquid state, a molecule of CH_3OH interacts with other CH_3OH molecules using hydrogen bonds, dipole-dipole, and London forces. Liquid CH_3SH molecules only interact with London forces, which are much weaker. The boiling point of CH_3SH is lower because it has **weaker intermolecular forces.**

83. *Result:* **(a) See structure below, 120° (b) 180° (c) sp², sp³ (d) See structures below; sp², sp², sp³; sp³; sp², sp²**

Analyze and Plan: Given a name and chemical formula for a molecules, write a Lewis structure, estimate a bond angle, identify the hybridization for one atom, draw resonance structures for a related compound, and identify the hybridization of some atoms in it. Adapt the plan used in the solution to Questions 32-35.

(a) Determine the Lewis structure knowing that C–N–N–N is part of the structure (based on angles requested in this and later parts). The structure has 40 e⁻. Two resonance structures exist. This is one of them:

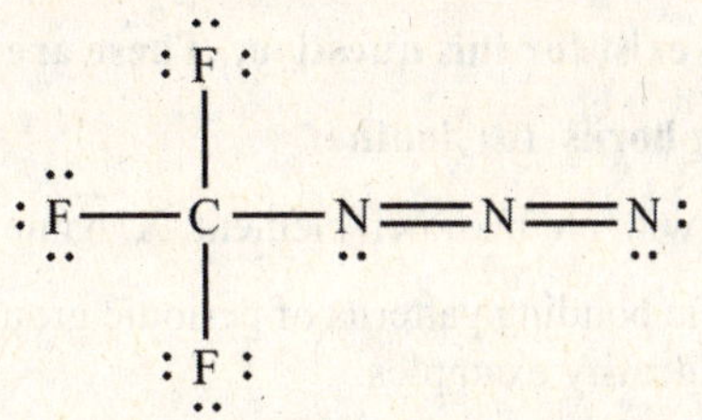

The central N in the N-N-C angle has AX_2E_1 type bonding, so the bond angle is **120°.**

(b) The central N in the N-N-N angle has AX_2E_0 type bonding, so the bond angle is **180°.**

(c) The N atoms next to the C atom has AX_2E_1 type bonding, so that means ***sp²***-hybridization is needed. The C atom has AX_4E_0 type bonding, so it needs ***sp³***-hybridization.

(d) The new H atoms will react with the electrons of a π-bond. The product has two predominant resonance structures with the lowest formal charges, shown here:

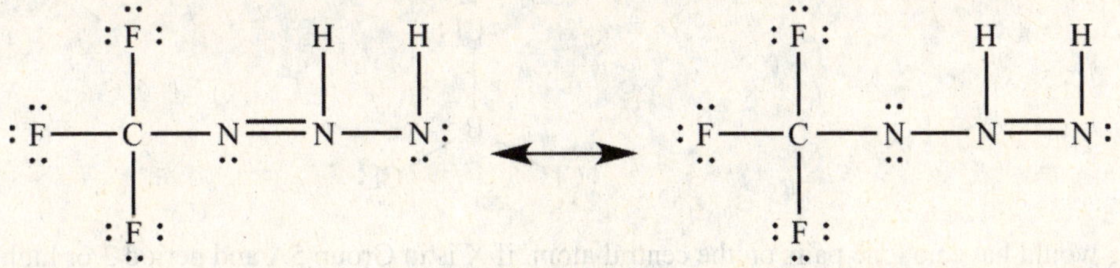

There are other possible answers.

From left to right: In the first structure, the first N atom has AX_2E_1 type bonding, so that means the N atom has *sp²*-hybridization, the next N atom has AX_3E_0 type bonding, so that means the N atom has *sp²*-hybridization, and the end N atom has AX_2E_2 type bonding, so that means the N atom has *sp³*-hybridization. In the second structure, the first N atom has AX_2E_2 type bonding, so that means the N atom has *sp³*-hybridization, the next N atom has AX_3E_0 type bonding, so that means the N atom has *sp²*-hybridization, and the end N atom has AX_2E_1 type bonding, so that means the N atom has *sp²*-hybridization.

88. *Result:* **The molecules are polar, so they interact using dipole-dipole forces.**

Analyze: Given names and chemical formulas of miscible compounds, explain, on a molecular bases, why they are miscible.

Plan: Adapt the plans used in the solution to Question 49.

Execute: Determine geometry and polarity of Cl_3CH and $CH_3CH_2–O–CH_2CH_3$:

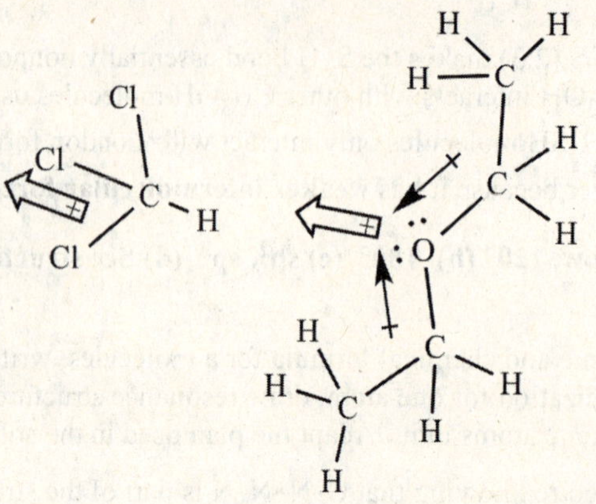

The molecules are both polar. So they can all interact with each other with dipole-dipole forces.

90. *Result:* **Several correct answers exist for this question. These are examples:**

(a) nitrogen (b) boron (c) phosphorus (d) iodine

Analyze: Given specific formulas with an unknown element, X, name a Group 1A-8A element that could be X.

Plan and Execute: Use the specific bonding patterns of periodic groups and the Lewis structures from other Questions in Chapters 6 and 7 to identify examples.

The answers given here are not the only correct answers. They are only examples.

(a) XH_3 would have a central atom with one lone pair of electrons if X is in Group 5A, such as: N, **nitrogen**.

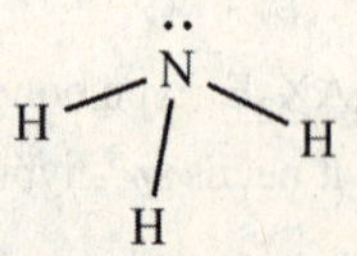

(b) XCl_3 would have no lone pairs on the central atom, if X is in Group 3A, such as: B, **boron**.

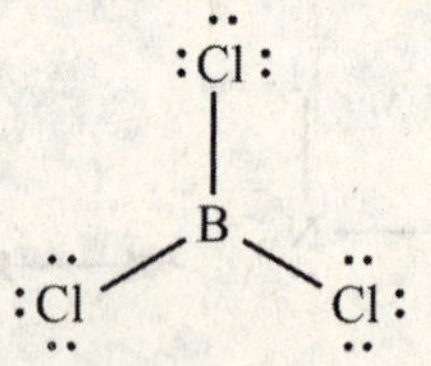

(c) XF_5 would have no lone pairs on the central atom, if X is in Group 5A and period 3 or higher, such as: P, **phosphorus**.

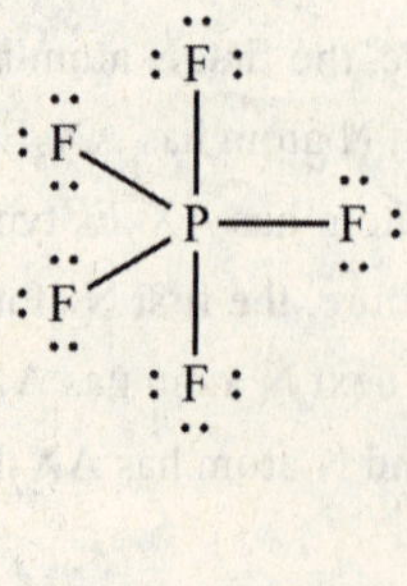

(d) XCl_3 would have two lone pairs on the central atom, if X is in Group 7A and period 3 or higher, such as: I, **iodine**.

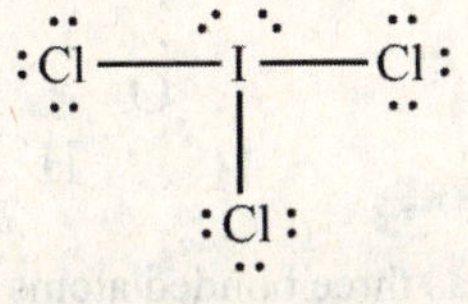

92. *Result:* **Five; see diagram below with dotted lines (• • •) showing intermolecular interactions.**

Analyze: Given a Lewis structure for a molecule, show how water interacts with it using hydrogen bonds.

Plan: Identify all the possible interactions between one of the lone pairs on an O atom and the H atom in an O–H bond and draw a dotted line between them O–H • • • O.

Execute: Five water molecules could hydrogen bond to an acetic acid molecule: one to each of the four lone pairs on O atoms in acetic acid and one to the H atom bonded to an O atom in acetic acid.

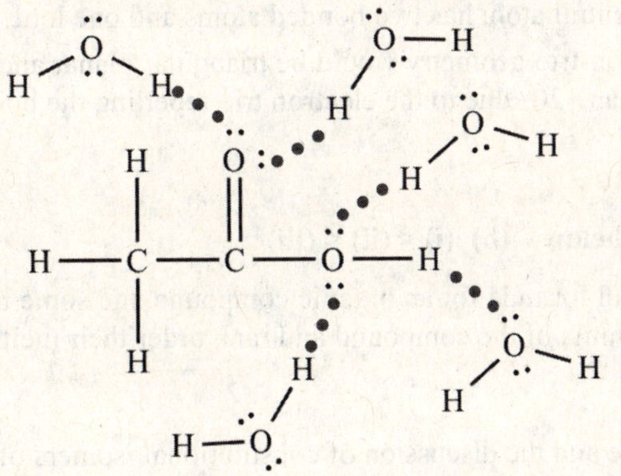

94. *Result:* **Diagram (d)**

Analyze: Given a list of examples, determine which correctly describes hydrogen bonding.

Plan: Hydrogen bonding is a noncovalent interaction between one of the lone pairs on an N, O, or F atom and the H atom in an O–H bond, a N–H bond, or a F–H bond. Compare each example to these requirements. If any of them are not met, the example is incorrect.

Execute:

(a) This is incorrect, because the H atoms are not bonded to highly electronegative atoms and H atoms would not be attracted to each other, either.

(b) This is incorrect. The H atoms are both bonded to O atom (EN = 3.4), but the two partial-positive H atoms will not be attracted to each other!

(c) This is incorrect. The H atom is bonded to an F atom. This is a covalent bond, not an example of the noncovalent interactive force called hydrogen bonding.

(d) This is **correct**. The H atom is bonded to an O atom (EN = 3.4) and is hydrogen bonding to another O atom.

The only answer that is correct is **(d)**.

96. *Result:* 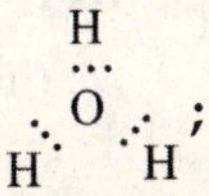 **; electron-trio geometry and molecular geometry both are triangular planar;** 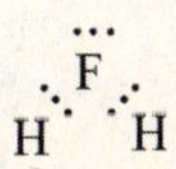 **;**

electron-trio geometry is triangular planar; molecular geometry is angular (120°).

Analyze, Plan, and Execute: In this fictional universe, a nonet is nine electrons. We will assume that the number of electrons in the atom called "O" in the other universe is still six, the number of electrons in the atom called "H" is still one, and the number of electrons in the atom called "F" is still seven. The equivalent to our VSEPR model is the VSTPR model (Valence Shell Electron Trio Repulsion Theory), where electron trios repel each other. The 6 electrons in the O atom would need three more electrons to make a total of nine. That means it would combine with three H atoms to make this molecule:

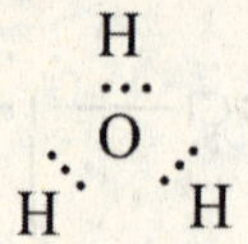

In this H_3O molecule, the central atom has three bonded atoms and no lone trios. That would make it AX_3E_0 type. As a result, the electron-trio geometry and the molecular geometry would both be triangular planar.

The 7 electrons in the F atom would need two more electrons to make a total of nine. That means it would combine with two H atoms to make this molecule:

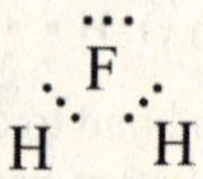

In this H_3F molecule, the central atom has two bonded atoms and one lone trio. That would make it AX_2E_1 type. As a result, the electron-trio geometry would be triangular planar and the molecular geometry would be angular at somewhat less than 120° due to the electron trio repelling the bonding trios to a narrower angle.

More Challenging Questions

98. *Result:* **(a) See diagrams below (b) (i) < (ii) < (iii)**

Analyze: Given the chemical formula for an organic compound and some information about its structure, determine the structural formula of the compound and rank order their melting points.

Plan:

Using the compound's name and the discussion of constitutional isomers of aromatic compounds in Section 6-11a, to assist in designing the Lewis structure and locating the *ortho-*, *meta-*, and *para-* positions of the –OH groups in isomers.

Adapt the plans from the solutions of Questions 49 and 53 to help decide the order for the melting points.

Execute:

(a) The Lewis structures for the three isomers of $C_6H_6O_2$ (each with one benzene ring and two –OH groups attached) look like this:

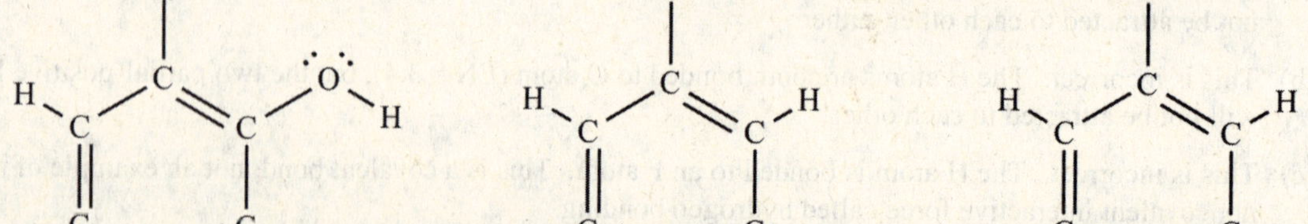

(i) (ii) (iii)

(b) All of these molecules have the same kind and strength of hydrogen bonding; however, because of the proximity of the two –OH groups in structure (i), these molecules are able to experience intramolecular hydrogen bonding:

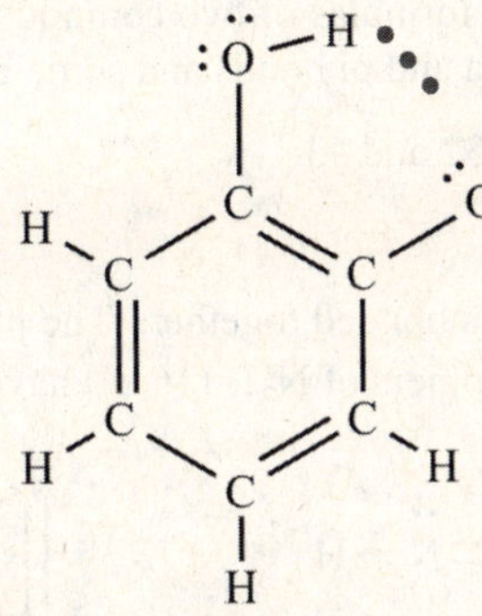

Less disruption of hydrogen bonding will be experienced when the (i) molecules undergo a transition to the liquid state; hence, (i) will melt at a lower temperature than the other two.

The asymmetry of structure (ii) might suggest that the intermolecular forces could be somewhat weaker in the solid, compared to those experienced among the higher-symmetry molecules in structure (iii).

Therefore, the predicted order for the melting points of these three solids would be (i) < (ii) < (iii).

100. *Result:* **(a) see structure below (b) First C: triangular planar, H–C–C and H–C–H bond angles are 120°; Second C: linear, C–C–O bond angle is 180° (c) First C: sp^2-hybridized; Second C: *sp*-hybridized; O: sp^2-hybridized (d) polar, because the polar C=O bond contributes to a nonzero dipole moment.**

Analyze: Given the chemical formula and some information about the molecular structure, write the Lewis structure, identify geometries, bond angles, hybridization of several atoms and explain polarity.

Plan: Adapt the solutions to Questions 32 and 43.

Execute:

(a) Ketene, C_2H_2O (16 e^-) has no –OH bond, so it must have the following Lewis structure.

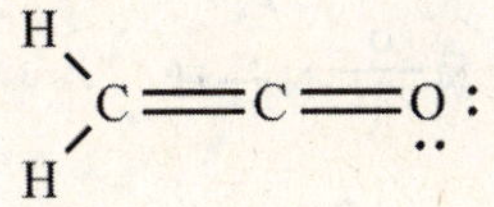

(b) 1st C atom: Its type is AX_3E_0, so the electron-region geometry is triangular planar with H–C–C and H–C–H bond angles of 120°.

2nd C atom: Its type is AX_2E_0, so the electron-region geometry is linear with a C–C–O bond angle of 180°.

(c) Look at the 1st C atom: Its type is AX_3E_0, so the C atom must be sp^2-hybridized.

Look at the 2nd C atom: Its type is AX_2E_0, so the C atom must be *sp*-hybridized.

Look at the O atom: Its type is type AX_1E_2, so the O atom must be sp^2-hybridized.

(d) The C=O bond in this asymmetric molecule is a very polar bond which makes the molecule polar.

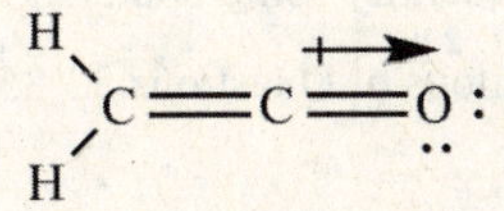

102. *Result:* **(a) see diagrams below (b) *sp*; *sp*; sp^2 (c) sp^2, sp^2, sp^3 (d) hydrogen azide: 3 σ bonds, cyclotriazene: 4 σ bonds (e) hydrogen azide: 2 π bonds; cyclotriazene: 1 π bond (f) hydrogen azide: 180°; cyclotriazene: 60°**

Analyze: Given the names and chemical formulas of two compounds, write their Lewis structures, identify the hybridization of several atoms, the sigma and pi bonds and some bond angles.

Plan: Adapt the solutions to Questions 32 and 43.

Execute:

(a) The name "azide" suggests three N's bonded together. The prefixes "cyclo" and "tri" suggests a three-nitrogen ring. Therefore, the two isomers of N_3H (16 e^-) have the following Lewis structures:

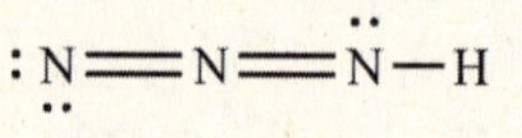

hydrogen azide

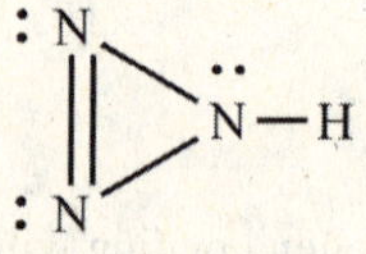

cyclotriazene

(b) Look at the 1st N atom (the left-most one) in hydrogen azide: It is of type AX_1E_1, so this N must be *sp*-hybridized.

Look at the 2nd N atom in hydrogen azide: It is of type AX_2E_0, so this N is *sp*-hybridized.

Look at the 3rd N atom in hydrogen azide: It is of type AX_2E_2, so we'd think that the N is sp^3-hybridized. However, to accommodate the other resonance structure (shown in (c) with two double bonds between the N atoms), one p-orbital must remain unhybridized, so the N is sp^2-hybridized

(c) Look at the first two N atom (the two on right-side of the triangle) in cyclotriazene: Each N is of type AX_2E_1, so each N atom must be sp^2-hybridized.

Look at the 3rd N atom in cyclotriazene: It is of type AX_3E_1, so this N is sp^3-hybridized.

For answering questions (d) and (e), recall that every single bond is a sigma-bond; every double bond is one sigma bond and one pi bond. So the sigma and pi bonds in this molecule are shown here:

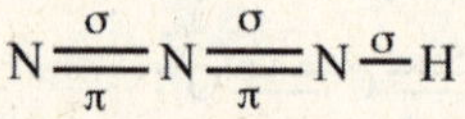

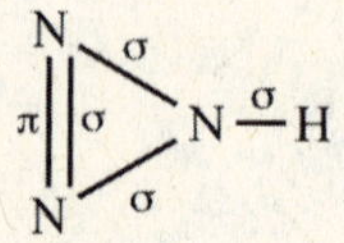

(d) In hydrogen azide, there are three sigma bonds. In cyclotriazene, there are four sigma bonds.

(e) In hydrogen azide, there are two pi bonds. In cyclotriazene, there is one pi bond.

(f) In hydrogen azide, the second N dictates the N–N–N bond angle. In (b), we ascertained that this N atom is sp-hybridized, so the bond angle must be 180°.

In cyclotriazene, the VSEPR method of this chapter indicate that the N=N–N bond angles should be 120° and the N–N–N angle should be 109.5°; however, because the three N atoms form a triangle, geometric constraints dictate that these bond angles must be approximately 60°.

104. *Result:* **(a) Angle 1: 120°, Angle 2: 120°, Angle 3: 109.5° (b) sp^3 (c) single-bonded O: sp^3, double-bonded O: sp^2**

Analyze: Given the Lewis structure, identify some bond angles and the hybridization of several atoms.

Plan: Adapt the plans from the solutions of Questions 32-35.

Execute:

(a) For the N–C–O Angle 1: Look at the C atom: Its type is AX_3E_0, so it must have triangular planar electron geometry, and the bond angle is approximately 120°.

For the C–C–C Angle 2: Look at the second of the three C atoms: Its type is type AX_3E_0, so it must have triangular planar electron geometry, and the bond angle is approximately 120°.

For the O–C–O Angle 3: Look at the C atom: Its type is type AX_4E_0, so it must have tetrahedral electron geometry, and the bond angle is approximately 109.5°.

(b) The N atom is of type AX_3E_1, so it must be sp^3-hybridized.

(c) Look at each of the two O atoms with two single bonds: They both have type AX_2E_2, so they must each be sp^3-hybridized.

Look at the double bonded O atom: Its type is AX_1E_2, so it must be sp^2-hybridized.

106. *Result:* **(a) see structures below (b) 120°, 60°**

Analyze: Given the names and chemical formulas of two compounds, write their Lewis structures and identifysome bond angles.

Plan: Adapt the plan from the solutions to Questions 32-35.

Execute:

(a) The root- "az" means three N atoms bonded together. The suffix "ene" means double bond, so use this to help write two Lewis structures for N_3H_3 compound with 18 electrons:

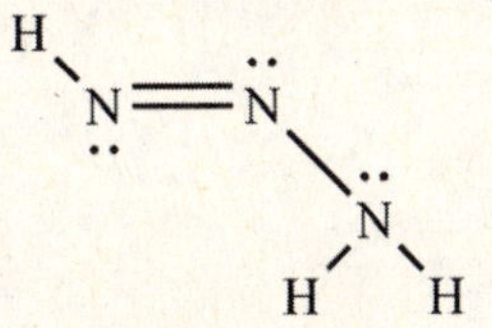

Triazene

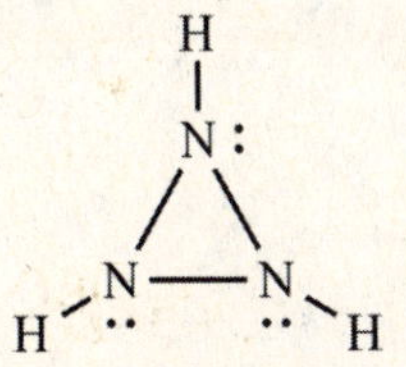

Triaziridine

(b) The central N atom in the N-N-N bond of triazene is of type AX_2E_1, so its bond angle is approximately **120°**. The N atoms in triaziridine form a three-membered ring, so they are geometrically constrained to has a **60°** angle, and not the VSEPR predicted bond angles.

108. *Result:* **(a) sp^3 (b) sp^2 (c) sp^2 (d) Eight σ-bonds and two π-bonds (e) See structure below (f) See structure below.**

Analyze: Given the Lewis structure for a compound, identify hybridization of several atoms, and sigma and pi bonds, and write one other resonance structure.

Plan: Adapt the solutions to Questions 32 and 43.

Execute:

(a) Looking at the structure, the single-bonded N atoms have type AX_2E_2, so the N atoms must be sp^3-hybridized.

(b) Looking at the structure, the double-bonded S atoms have type AX_2E_1, so the S atoms must be sp^2-hybridized.

(c) Looking at the structure, the double-bonded N atoms have type AX_2E_1, so the N atoms must be sp^2-hybridized.

(d) Each single bond is one σ-bond. Each double bond is one π-bond and one σ-bond.

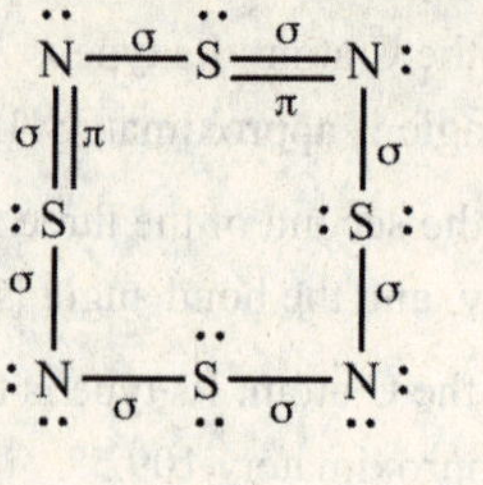

The molecule has eight σ-bonds and two π-bonds.

(e) See structure drawn in (d).

(f) As described in Section 7-2c, Make a new double bond with one electron pair on a single-bonded N atom with a double bonded S atom and move one pair of the original double bond electrons to a lone pair on the next N atom in the chain, as shown here on the left side of the structure.

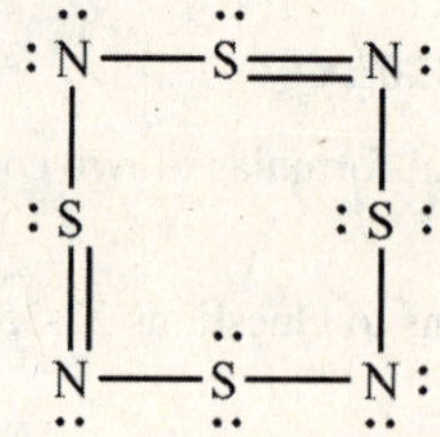

Chapter 8: Gases and the Atmosphere
Solutions for Red-Numbered Questions for Review and Thought

Topical Questions

Gas Pressure (Section 8-1)

10. *Result:* **(a) 0.947 atm (b) 950. mm Hg (c) 542 torr (d) 98.7 kPa (e) 6.91 atm**

Analyze: Convert a series of pressure quantities into other pressure units.

Plan: Use Table 8.1 to design conversion factors to achieve the conversions.

Execute:

(a) $$720.\ \text{mmHg} \times \frac{1\ \text{atm}}{760.\ \text{mmHg}} = 0.947\ \text{atm}$$

(b) $$1.25\ \text{atm} \times \frac{760\ \text{mmHg}}{1\ \text{atm}} = 950.\ \text{mmHg}$$

(c) $$542\ \text{mmHg} \times \frac{760\ \text{torr}}{760\ \text{mmHg}} = 542\ \text{torr}$$

(d) $$740.\ \text{mmHg} \times \frac{1\ \text{atm}}{760.\ \text{mmHg}} \times \frac{101.325\ \text{kPa}}{1\ \text{atm}} = 98.7\ \text{kPa}$$

(e) $$700.\ \text{kPa} \times \frac{1\ \text{atm}}{101.325\ \text{kPa}} = 6.91\ \text{atm}$$

☑ *Reasonable Result Check:* The unit of atm represents a lot more pressure than the units of kPa, torr, and mm Hg, so it makes sense that the numbers of atmospheres are always much smaller than the pressure expressed in these other units. The unit of kPa represents more pressure than the units of torr and mm Hg, so it makes sense that the numbers of kPa are always smaller than the pressure expressed in the other units of torr and mm Hg. Torr and mm Hg are the same size, so their quantities should be identical.

12. *Result:* **14 m**

Analyze: Given the density of mercury, the density of an oil used to construct a barometer, and the atmospheric pressure, determine the height in meters of the oil column in the oil barometer.

Plan: Use Table 8.1 to convert the pressure to mm Hg. Pressure on the liquid pushes the liquid up inside the barometer until the column's mass exerts the same force per unit area as the air pressure:

$$P_{\text{liquid}} = \text{Force/Area} = \text{mg}^2\text{/Area}$$

Since g is a constant, the force per unit area of the liquid counteracting the air pressure is proportional to just the mass per unit area. Relate the mass per unit area of mercury to the mass per unit area of the oil for 1 atm pressure. Relate the mass of each liquid to its respective density and volume. Relate the volume of each liquid to the dimensions of its respective barometers, including the height. Relate the height of mercury in the mercury barometer to the height of the oil in the oil barometer.

Execute: Let m = mass of the liquid, A = cylindrical area of the barometer's column, d = density of the liquid, V = volume of the liquid, and h = the height of the liquid in the barometer. Now, equate the mass per unit area of each barometer and derives an equation relating their heights:

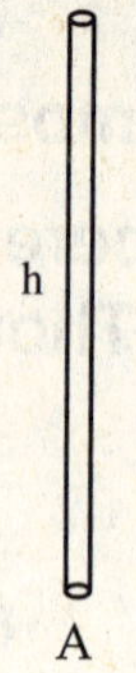

$m_{Hg}/A_{Hg} = m_{oil}/A_{oil}$

$d_{Hg}V_{Hg}/A_{Hg} = d_{oil}V_{oil}/A_{oil}$

$V = (\text{Area}) \times (\text{height}) = Ah$

$d_{Hg}A_{Hg}h_{Hg}/A_{Hg} = d_{oil}A_{oil}h_{oil}/A_{oil}$

$d_{Hg}h_{Hg} = d_{oil}h_{oil}$

According to Table 8.1, at 1.0 atm the height of a column of mercury in a mercury barometer is 760 mm.

$$h_{oil} = \frac{d_{Hg}h_{Hg}}{d_{oil}} = \frac{\left(13.55\ \text{g/cm}^3\right)\left(760\ \text{mm}\right)}{\left(0.75\ \text{g/cm}^3\right)} \times \frac{1\ \text{m}}{1000\ \text{mm}} = 13.7\ \text{m} = 14\ \text{m}$$

☑ *Reasonable Result Check:* A larger mass of oil will be needed to counterbalance the atmospheric pressure because it is less dense than mercury, so a higher column of oil makes sense.

13. *Result:* **Atmospheric pressure can only push water up to about 33.8 feet.**

Analyze and Plan: With a perfect vacuum at the top of the well this system would resemble a water barometer. The densities of mercury and water are given in Table 1.1. Using the equation derived in the solution to Question 12, then convert from mm to feet.

Execute:

$$h_{water} = \frac{d_{Hg}h_{Hg}}{d_{water}} = \frac{\left(13.55\ \text{g/cm}^3\right)\left(760.00\ \text{mm}\right)}{\left(0.998\ \text{g/cm}^3\right)} \times \frac{1\ \text{m}}{1000\ \text{mm}} \times \frac{3.281\ \text{feet}}{1\ \text{m}} = 33.8\ \text{feet}$$

That means atmospheric pressure can only push water up to about 33.8 feet. The well cannot be deeper than that, not even with the highest quality vacuum pump. So, it would not help much at all to have such a vacuum pump.

Kinetic-Molecular Theory (Section 8-2)

15. *Result:* **See I-V below; postulate III and IV become false at high P or low T; postulate I becomes false at high P; postulate II is most nearly correct.**

Analyze: List the postulates of kinetic-molecular theory and identify which is incorrect at high pressures, which is incorrect at low temperatures, and which is most nearly correct.

Plan and Execute: The five basic postulates of kinetic-molecular theory are given in Section 8-2:

I. A gas is composed of molecules whose size is much smaller than the distances between them.

II. Gas molecules move randomly at various speeds and in every possible direction.

III. Except when gas molecules collide, forces of attraction and repulsion between them are negligible; gas pressure results from the force exerted on a container's walls by molecules colliding with the walls.

IV. When collisions between molecules occur, the collisions are elastic.

V. The average kinetic energy of gas molecules is proportional to the absolute temperature.

A discussion of nonideal behavior is provided in Section 8.8. Of these five basic postulates, the ones that become false at very high pressures or very low temperatures are postulates III and IV. Slow molecules crowded together are much more likely to interact with each other. Collisions may not be elastic under these circumstances, because colliding molecules might stick together due to large enough attractive forces between them or repel due to repulsive forces. Postulate I may become false at very high pressures, because the molecules are crowded together reducing the distance between them.

The assumption that seems most likely to always be most nearly correct is assumption II. As long as a substance is a gas, it retains the capacity to disperse to uniformly fill any container it is introduced into, though the rate of dispersal may vary.

The Behavior of Ideal Gases: Gas Laws (Section 8-3)

17. *Result:* **25.5 L**

Analyze: Given the mass, volume, temperature and pressure of a sample of gas, determine the volume of a different mass of the gas at the same temperature and pressure).

Plan: Use Avogadro's law to relate volume to the number of moles.

$$\frac{V_1}{n_1} = \frac{V_2}{n_2} \qquad \text{(P and T constant)}$$

For any given substance, the number of moles is directly proportional to the mass. (because n = m/M, where m is the mass and M is the molar mass). Therefore, we can derive a similar relationship between volume and mass:

$$\frac{V_1}{m_1} = \frac{V_2}{m_2} \quad \text{(for different samples made of the same substance, and P and T constant)}$$

Execute: Solve for V_2

$$V_2 = \frac{V_1 \times m_2}{m_1} = (10.0\ \text{L}) \times \frac{(129.3\ \text{g})}{(50.75\ \text{g})} = 25.5\ \text{L}$$

Notice: This question could also be answered by calculating the amount (in moles) for each sample, using the molar mass, then applying Avogadro's Law.

☑ *Reasonable Result Check:* It makes sense that the sample with the larger gas mass has a larger volume.

19. *Result:* **172 mm Hg**

Analyze: Given initial volume, temperature, and pressure of a gas and final volume and temperature, determine the final pressure of the gas.

Plan and Execute: Use Boyle's law: $P_1V_1 = P_2V_2$ (unchanging T and n).

$$P_2 = \frac{P_1V_1}{V_2} = \frac{(735\ \text{mm Hg}) \times (3.50\ \text{L})}{(15.0\ \text{L})} = 172\ \text{mm Hg}$$

☑ *Reasonable Result Check:* The larger volume should have a smaller pressure.

21. *Result:* **26.5 mL**

Analyze: Given the initial volume and temperature of a sample (at atmospheric pressure) and the final temperature of the sample, determine the new volume (presumably at the same atmospheric pressure).

Plan: Convert the temperatures to Kelvin. Use Charles' law to relate volume to absolute temperature:

$$\frac{V_1}{T_1} = \frac{V_2}{T_2} \qquad \text{(P and n constant)}$$

Execute:

$$T_1 = 20.\ °\text{C} + 273.15 = 293\ \text{K}$$

$$T_2 = 37\ °\text{C} + 273.15 = 310.\ \text{K}$$

$$V_2 = V_1 \times \frac{T_2}{T_1} = (25.0\ \text{mL}) \times \frac{(310.\ \text{K})}{(293\ \text{K})} = 26.5\ \text{mL}$$

☑ *Reasonable Result Check:* Gas at higher temperature should have a larger volume.

23. *Result:* **4.00 atm**

Analyze: Given the original pressure and temperature of gas sample, the assumption that volume is unchanged, and the new temperature, determine the new pressure exerted by the gas.

Plan: Convert the temperature to Kelvin. Use the combined gas law described in Section 8-3d to relate pressure to absolute temperature.

$$\frac{P_1V_1}{T_1} = \frac{P_2V_2}{T_2} \qquad \text{(n constant)}$$

At constant volume the equation is: $$\frac{P_1}{T_1} = \frac{P_2}{T_2} \qquad \text{(V and n constant)}$$

Execute: $$T_1 = 15\ °C + 273.15 = 288\ K$$

$$T_2 = 35\ °C + 273.15 = 308\ K$$

$$P_2 = P_1 \times \frac{T_2}{T_1} = (3.74\ atm) \times \frac{(308\ K)}{(288\ K)} = 4.00\ atm$$

☑ *Reasonable Result Check:* A gas with a higher temperature should exert a higher pressure.

25. *Result:* **501 mL**

Analyze: Given the original volume, temperature and pressure of gas sample, determine the new volume once the pressure and temperature are raised.

Plan: Convert the temperature to Kelvin. Use the combined gas law described in Section 8-3d to relate pressure to absolute temperature.

$$\frac{P_1V_1}{T_1} = \frac{P_2V_2}{T_2} \qquad \text{(n constant)}$$

Execute: $$T_1 = 22\ °C + 273.15 = 295\ K$$

$$T_2 = 42\ °C + 273.15 = 315\ K$$

$$V_2 = V_1 \times \frac{T_2}{T_1} \times \frac{P_1}{P_2} = (754\ mL) \times \frac{(315\ K)}{(295\ K)} \times \frac{(165\ mmHg)}{(265\ mmHg)} = 501\ mL$$

☑ *Reasonable Result Check:* Increasing the temperature increases the volume. Increasing pressure decreases the volume. Clearly these two changes counteract each other, but the pressure change affects the volume more than the temperature change does.

27. *Result:* **0.51 atm**

Analyze: Given the mass, chemical formula, temperature, and volume of a gas sample, determine the pressure of the sample.

Plan: Use the ideal gas law: $$PV = nRT \qquad R = 0.08206\ \frac{L \cdot atm}{mol \cdot K}$$

Execute: The units of R indicate that we must get the quantity (in moles), temperature in Kelvin, and volume in liters.

$$1.55\ g\ Xe \times \frac{1\ mol\ Xe}{131.29\ g\ Xe} = 0.0118\ mol\ Xe$$

$$T = 20.\ °C + 273.15 = 293\ K$$

$$V = 560\ mL \times \frac{1\ L}{1000\ mL} = 0.56\ L \quad \textit{(2 sig figs)}$$

$$P = \frac{nRT}{V} = \frac{(0.0118 \text{ mol}) \times \left(0.08206 \frac{\text{L} \cdot \text{atm}}{\text{mol} \cdot \text{K}}\right) \times (293 \text{ K})}{(0.56 \text{ L})} = 0.51 \text{ atm}$$

Notice: Other equations derived in Section 8-4 may also be used to obtain this result.

☑ *Reasonable Result Check:* The small fractional quantity (in moles) is expected with the small mass. The units in the pressure calculation cancel properly to give atm.

29. *Result:* **Largest number in sample (d); smallest number in sample (c)**

Analyze: Given a set of gas samples, determine which has the largest number of molecules and which has the smallest number of molecules.

Plan: Some of the samples are at STP, so use the molar volume of a gas at STP to determine the number of molecules (in units of moles). One mol of **any** gas occupies 22.414 L.

$$1.0 \text{ L} \times \frac{1 \text{ mol gas}}{22.414 \text{ L}} = 0.045 \text{ mol gas}$$

In the other cases, use the ideal gas law as described in the solution to Question 27. Once all the moles are calculated, identify which sample has the most moles and which sample has the least moles.

Execute:

(a) 0.045 mol H_2 at STP

(b) 0.045 mol N_2 at STP

(c) T = 27 °C + 273.15 = 300. K

$$P = 760. \text{ mmHg} \times \frac{1 \text{ atm}}{760. \text{ mmHg}} = 1.00 \text{ atm}$$

$$n = \frac{PV}{RT} = \frac{(1.0 \text{ atm}) \times (1.00 \text{ L})}{\left(0.08206 \frac{\text{L} \cdot \text{atm}}{\text{mol} \cdot \text{K}}\right) \times (300. \text{ K})} = 0.0406 \text{ mol}$$

(d) T = 0. °C + 273.15 = 273 K

$$P = 800. \text{ mmHg} \times \frac{1 \text{ atm}}{760. \text{ mmHg}} = 1.05 \text{ atm}$$

$$n = \frac{PV}{RT} = \frac{(1.05 \text{ atm}) \times (1.0 \text{ L})}{\left(0.08206 \frac{\text{L} \cdot \text{atm}}{\text{mol} \cdot \text{K}}\right) \times (273 \text{ K})} = 0.047 \text{ mol}$$

Of these samples, sample (d) has the most molecules and sample (c) has the smallest number of molecules.

☑ *Reasonable Result Check:* To keep a 1.0-L gas sample at standard temperature and still have larger than standard pressure suggests that there must be more molecules hitting the walls than a sample at STP. To have a 1.0-L gas sample at higher than standard temperature and still stay at standard pressure suggests that there must be fewer molecules hitting the walls harder, than a sample at STP.

31. *Result:* **2.6 × 10³ mL**

Analyze: Given the mass of a gas, the identity of the gas, its temperature, and its pressure, determine the volume occupied by the gas.

Plan: Use the mass to get the quantity (in moles), then use the plan from the solution to Question 30.

Execute:

$$4.4 \text{ g } CO_2 \times \frac{1 \text{ mol } CO_2}{44.0095 \text{ g } CO_2} = 0.10 \text{ mol } CO_2$$

$$T = 27\text{ °C} + 273.15 = 300.\text{ K}$$

$$730.\text{ mmHg} \times \frac{1\text{ atm}}{760.\text{ mmHg}} = 0.961\text{ atm}$$

$$V = \frac{nRT}{P} = \frac{(0.10\text{ mol}) \times \left(0.08206 \frac{\text{L} \cdot \text{atm}}{\text{mol} \cdot \text{K}}\right) \times (300.\text{ K})}{(0.961\text{ atm})} = 2.6\text{ L}$$

$$2.6\text{ L} \times \frac{1000\text{ mL}}{1\text{ L}} = 2.6 \times 10^3\text{ mL}$$

☑ *Reasonable Result Check:* The molar volume of a gas under these conditions is going to be similar to that at STP (22.4 L/mol), because T is slightly higher than standard T and P is slightly lower than standard P. When using one tenth of a mole of CO_2, one would expect approximately one tenth of the molar volume, roughly 2.2 liters.

Gas Density, Molar Mass, Ideal Gas Law (Section 8-4)

33. *Result:* **130. g/mol**

Analyze: Given the density of a gas at STP, determine the molar mass of the gas.

Plan: Density is related to mass and volume: $d = \frac{m}{V}$. Because we know the molar volume of a gas at STP (22.414 liters per mol), we can use this equation to get the molar mass (M = grams per mole).

$$d \times (\text{molar volume}) = M$$

Execute:

$$5.79 \frac{\text{g}}{\text{L}} \times 22.414 \frac{\text{L}}{\text{mol}} = 130. \frac{\text{g}}{\text{mol}}$$

☑ *Reasonable Result Check:* This molar mass is an appropriate size for a small gaseous molecule.

35. *Result:* **3.7×10^{-4} g/L**

Analyze: Given the average molar mass of a sample of air and its pressure and temperature, determine the density of the sample.

Plan: Use one of the equations derived in Section 8-4 to relate density to gas properties:

$$d = \frac{m}{V} = \frac{PM}{RT}$$

Execute:

$$T = -23\text{ °C} + 273.15 = 250.\text{ K}$$

$$0.20\text{ mmHg} \times \frac{1\text{ atm}}{760\text{ mmHg}} = 2.6 \times 10^{-4}\text{ atm}$$

$$d = \frac{MP}{RT} = \frac{\left(29.0 \frac{\text{g}}{\text{mol}}\right) \times (2.6 \times 10^{-4}\text{ atm})}{\left(0.08206 \frac{\text{L} \cdot \text{atm}}{\text{mol} \cdot \text{K}}\right) \times (250.\text{ K})} = 3.7 \times 10^{-4} \frac{\text{g}}{\text{L}}$$

☑ *Reasonable Result Check:* The low gas density makes sense at the very low pressure.

37. *Result:* **P_{He} is 7.000 times greater than P_{N_2}**

Analyze: Given the formulas of the compounds composing two gas samples with equal density, volume and temperature, determine the relationship between the pressures of the two gas samples.

Plan: Use one of the equations derived in Section 8-4 to relate density to gas properties:

$$d = \frac{m}{V} = \frac{PM}{RT}$$

$$PM = dRT$$

Therefore, at constant d and T: $P_2M_2 = M_1P_1$ or $\frac{P_2}{P_1} = \frac{M_1}{M_2}$

Execute: $$\frac{P_{He}}{P_{N_2}} = \frac{M_{N_2}}{M_{He}} = \frac{(28.0134\ g/mol)}{(4.0026\ g/mol)} = 7.000$$

The pressure of helium is **seven times greater** than the pressure of nitrogen.

☑ *Reasonable Result Check:* A much larger number of atoms would need to be in the light-weight helium sample for it to have the same density as the heavier nitrogen sample. Because pressure is proportional to number of particles, it makes sense that the pressure in the helium sample is much larger than in the nitrogen sample.

Quantities of Gases in Chemical Reactions (Section 8-5)

39. *Result:* **8.9 L H_2**

Analyze: Given the balanced chemical equation for a chemical reaction, the mass of one reactant, and excess other reactant, determine the volume of the product produced at a specified pressure and temperature.

Plan: Convert from mass to quantity (in moles). Use the 2:3 mole ratio from the balanced equation to determine moles of product. Use the ideal gas law to determine the pressure of the product.

Execute: $$6.5\ g\ Al \times \frac{1\ mol\ Al}{26.9815\ g\ Al} \times \frac{3\ mol\ H_2}{2\ mol\ Al} = 0.36\ mol\ H_2$$

$$T = 22\ °C + 273.15 = 295\ K \qquad P = 742\ mmHg \times \frac{1\ atm}{760\ mmHg} = 0.976\ atm$$

$$P = \frac{n_{H_2}RT}{V} = \frac{(0.36\ mol\ H_2) \times \left(0.08206 \frac{L \cdot atm}{mol \cdot K}\right) \times (295\ K)}{(0.976\ atm)} = 8.9\ L\ H_2$$

☑ *Reasonable Result Check:* All units cancel properly in the calculation of liters. Approximately 1/3 mole of gas occupies a volume approximately 1/3 the molar volume of a gas at STP. This answer make sense, since the sample's temperature is larger than the STP temperature.

41. *Result:* **10.4 L O_2, 10.4 L H_2O**

Analyze: Given a balanced chemical equation and the volume of one reactant, determine the volume of the other reactant and the volume of one of the products, assuming all volumes are measured at the same temperature and pressure.

Plan: As described in Section 8-5, Avogadro's law allows us to interpret a balanced equation with gas reactants in terms of gas volumes, as long as their temperatures and pressures are the same. All gases here are at the same T and P, so volumes of gases involved in the reaction are proportional to the coefficients in the balanced chemical equation. Use the ratio of coefficients for the volume ratio.

Execute: The balanced equation tells us that one volume of $SiH_4(g)$ reacts with two volumes of $O_2(g)$ to make two volumes of $H_2O(g)$, with all the volumes are measured at the same temperature and pressure.

$$5.2\ L\ H_2(g) \times \frac{2\ L O_2(g)}{1\ L SiH_4(g)} = 10.4\ L O_2(g)$$

$$5.2\ L\ H_2(g) \times \frac{2\ L H_2O(g)}{1\ L SiH_4(g)} = 10.4\ L H_2O(g)$$

Notice: Multiplying a number by an exact whole number, n, is like adding that number to itself n times. 2 × (5.2) = 5.2 L + 5.2 L = 10.4 L, that is why the addition rule was used for assessing the sig. fig., here.

☑ *Reasonable Result Check:* Twice as many O_2 molecules are needed compared to the number of SiH_4 molecules, forming twice as many H_2O molecules. So, it makes sense that both the volume of O_2 and the volume of H_2O are twice the volume of SiH_4.

43. *Result:* **0.38 atm**

Analyze: Given the mass of a metallic element and the identity of the gas-phase molecule it is converted to, determine the pressure of the molecular gas at a specified volume and temperature.

Plan: Convert from grams to moles. Use the stoichiometric relationship of the molecular formula to determine moles of molecules. Use the ideal gas law to determine the pressure of the gas.

Execute:

$$1.0\times10^3 \text{ g U}\times\frac{1 \text{ mol U}}{238.03 \text{ g U}}\times\frac{1 \text{ mol UF}_6}{1 \text{ mol U}}=4.2 \text{ mol UF}_6$$

$$T = 62.\ ^\circ\text{C} + 273.15 = 335 \text{ K}$$

$$P=\frac{n_{UF_6}RT}{V}=\frac{(4.2 \text{ mol UF}_6)\times\left(0.08206\frac{\text{L}\cdot\text{atm}}{\text{mol}\cdot\text{K}}\right)\times(335 \text{ K})}{(3.0\times10^2 \text{ L})}=0.38 \text{ atm}$$

☑ *Reasonable Result Check:* The relative quantities of U and UF_6 seem sensible. All units cancel properly in the calculation of atmospheres. This is a relatively low pressure, but the container is fairly large.

45. *Result:* **88 g/mol**

Analyze: Given the balanced chemical equation for a chemical reaction, the mass of the solid ionic reactant, the formula of the ionic compound and the periodic group that its metal is found in, the pressure of a gas-phase product with a specified volume and temperature, determine the molar mass of the reactant.

Plan: The formula of the ionic compound, $M_x(CO_3)_y$, can be used to determine the values of x and y, since M is from Group 2A and the anion has a known charge. The ideal gas law can be used to find the moles of CO_2 produced. The mole ratio is then used to find the quantity (in moles) of solid that reacted. Divide the mass by the quantity (in moles) for the molar mass.

Group 2A metals have 2+ charges and carbonate ion, CO_3^{2-}, has a 2– charge:

$$x\,M^{2+} + y\,CO_3^{2-} \longrightarrow M_x(CO_3)_y$$

From this equation, we can see that x = y = 1, and the heating of the metal oxide reaction can be simplified to the following equation:

$$MCO_3(s) \longrightarrow MO(s) + CO_2(g)$$

$$T = 25\ ^\circ\text{C} + 273.15 = 298 \text{ K}$$

$$69.8 \text{ mm Hg}\times\frac{1 \text{ atm}}{760 \text{ mm Hg}}=0.0918 \text{ atm} \qquad 285 \text{ mL}\times\frac{1 \text{ L}}{1000 \text{ mL}}=0.285 \text{ L}$$

$$n_{CO_2}=\frac{PV}{RT}=\frac{(0.0918 \text{ atm})\times(0.285 \text{ L})}{\left(0.08206\frac{\text{L}\cdot\text{atm}}{\text{mol}\cdot\text{K}}\right)\times(298 \text{ K})}=1.07\times10^{-3} \text{ mol CO}_2$$

$$1.07\times10^{-3} \text{ mol CO}_2\times\frac{1 \text{ mol MCO}_3}{1 \text{ mol CO}_2}=1.07\times10^{-3} \text{ mol MCO}_3$$

The given mass (0.158 g) contains 1.07×10^{-3} mol MCO_3

$$\text{Molar Mass} = \frac{0.158 \text{ g } MCO_3}{1.07 \times 10^{-3} \text{ mol } MCO_3} = 148 \frac{\text{g}}{\text{mol}}$$

☑ *Reasonable Result Check:* If this metal M is really a Group 2A element, it's molar mass should be similar to one of them. We can subtract the molar mass of one C atom and three O atoms from the molar mass of the solid and get the molar mass of M:

$$148 \text{ g/mol} - 12 \text{ g/mol} - 3 \times (16 \text{ g/mol}) = 88 \text{ g/mol M}$$

This is close to that of Sr (molar mass = 87.62 g/mol), a group 2A metal.

47. *Result:* **(a) $2\ C_8H_{18}(\ell) + 25\ O_2(g) \longrightarrow 16\ CO_2(g) + 18\ H_2O(g)$ (b) 1.4×10^3 L CO_2**

(a) The balanced combustion equation is:

$$2\ C_8H_{18}(\ell) + 25\ O_2(g) \longrightarrow 16\ CO_2(g) + 18\ H_2O(g)$$

(b) *Analyze:* Given the length in miles of a trip, the fuel efficiency of a car, the density of the liquid fuel, and the temperature and pressure, determine the volume of a gas-phase product produced during the trip.

Plan: Use the miles, the fuel efficiency, volume conversions, density, and mole ratios to find the moles of the product. Use the ideal gas law to determine the volume of the product.

Execute:

$$10. \text{ miles} \times \frac{1 \text{ gal gasoline}}{32 \text{ miles}} \times \frac{3.785 \text{ L gasoline}}{1 \text{ gal gasoline}} \times \frac{1000 \text{ mL}}{1 \text{ L}} \times \frac{1 \text{ cm}^3}{1 \text{ mL}} = 1.2 \times 10^3 \text{ cm}^3$$

$$1.2 \times 10^3 \text{ cm}^3 \text{ gasoline} \times \frac{0.703 \text{ g gasoline}}{1 \text{ cm}^3 \text{ gasoline}} \times \frac{1 \text{ mol gasoline}}{114.22 \text{ g gasoline}} \times \frac{16 \text{ mol } CO_2}{2 \text{ mol gasoline}} = 58 \text{ mol } CO_2$$

$$T = 25\ °C + 273.15 = 298 \text{ K}$$

$$V = \frac{n_{CO_2}RT}{P} = \frac{(58 \text{ mol } CO_2) \times \left(0.08206 \frac{\text{L} \cdot \text{atm}}{\text{mol} \cdot \text{K}}\right) \times (298 \text{ K})}{(1.0 \text{ atm})} = 1.4 \times 10^3 \text{ L}$$

☑ *Reasonable Result Check:* This is the same question as Chapter 10 Question 97 and produces the same result. This is a large volume of CO_2! However, it is not an unreasonable quantity considering how many gallons of gasoline are used and how much CO_2 is generated from each octane molecule. Comparing the results to the results from Question 48, less CO_2 is generated by methanol as a fuel than octane.

Gas Mixtures and Partial Pressures (Section 8-6)

50. *Result:* **4.51 atm total**

Analyze: Given the masses of gases in a mixture at a specified volume and temperature, determine the total pressure.

Plan: Convert mass to quantity (in moles). Use the ideal gas law to determine the partial pressure of each gas. Use Dalton's law of partial pressures to determine the total pressure. Dalton's law and its applications are described in Section 8-6. Dalton's law of partial pressures states that the total pressure (P_{tot}) exerted by a mixture of gases is the sum of their partial pressures (P_1, P_2, P_3, etc.), if the volume (V) and temperature (T) are constant.

$$P_{tot} = P_1 + P_2 + P_3 + \ldots \qquad \text{(V and T constant)}$$

Execute:

$$1.50 \text{ g } H_2 \times \frac{1 \text{ mol } H_2}{2.0158 \text{ g } H_2} = 0.744 \text{ mol } H_2$$

$$T = 25\ °C + 273.15 = 298 \text{ K}$$

$$P_{H_2} = \frac{n_{H_2}RT}{V} = \frac{(0.744\text{ mol } H_2) \times \left(0.08206 \frac{\text{L} \cdot \text{atm}}{\text{mol} \cdot \text{K}}\right) \times (298\text{ K})}{(5.00\text{ L})} = 3.64\text{ atm } H_2$$

$$5.00\text{ g } N_2 \times \frac{1\text{ mol } N_2}{28.0134\text{ g } N_2} = 0.178\text{ mol } N_2$$

$$P_{N_2} = \frac{n_{N_2}RT}{V} = \frac{(0.178\text{ mol } H_2) \times \left(0.08206 \frac{\text{L} \cdot \text{atm}}{\text{mol} \cdot \text{K}}\right) \times (298\text{ K})}{(5.00\text{ L})} = 0.873\text{ atm } N_2$$

Using Dalton's law: $P_{tot} = P_{H_2} + P_{N_2} = 3.64\text{ atm } H_2 + 0.873\text{ atm } N_2 = 4.51\text{ atm total}$

☑ *Reasonable Result Check:* The relative pressure of H_2 and N_2 makes sense from the relative number of moles. All un-needed units cancel properly in the calculation of atmosphere.

52. *Result:* **(a) 154 mm Hg (b) X_{N_2}=0.777, X_{O_2} = 0.208, X_{Ar} = 0.0093, X_{CO_2} = 0.0003, X_{H_2O} = 0.0053 (c) 77.7% N_2, 20.8% O_2, 0.93% Ar, 0.03% CO_2, 0.54% H_2O; slight difference, since this sample is wet**

Analyze: Given the partial pressure of several gases in a sample of the atmosphere, and the total pressure of the sample, determine the partial pressure of O_2, the mole fraction of each gas, and the percent by volume. Compare the percentages to Table 8.6.

Plan: Dalton's law and its applications are described in Section 8-6. Dalton's law of partial pressures states that the total pressure (P) exerted by a mixture of gases is the sum of their partial pressures (P_1, P_2, P_3, etc.), if the volume (V) and temperature (T) are constant.

$$P_{tot} = P_1 + P_2 + P_3 + \ldots \qquad \text{(V and T constant)}$$

$P_i = X_i P_{tot}$, so we use the total pressure and the partial pressure of a component to determine its mole fraction:

$$X_i = \frac{P_i}{P_{tot}}$$

According to Avogadro's law, moles and gas volumes are proportional, so the mole fraction is equal to the volume fraction. To get percent, multiply the volume fraction by 100%.

Execute:

(a)

$$P_{tot} = P_{N_2} + P_{O_2} + P_{Ar} + P_{CO_2} + P_{H_2O}$$

$$P_{O_2} = P_{tot} - P_{N_2} - P_{Ar} - P_{CO_2} - P_{H_2O}$$

$$P_{O_2} = (740.\text{ mm Hg}) - (575\text{ mm Hg}) - (6.9\text{ mm Hg}) - (0.2\text{ mm Hg}) - (4.0\text{ mm Hg}) = 154\text{ mm Hg}$$

(b)

$$X_{N_2} = \frac{P_{N_2}}{P_{tot}} = \frac{575\text{ mmHg}}{740.\text{ mmHg}} = 0.777 \qquad X_{O_2} = \frac{P_{O_2}}{P_{tot}} = \frac{154\text{ mmHg}}{740.\text{ mmHg}} = 0.208$$

$$X_{Ar} = \frac{P_{Ar}}{P_{tot}} = \frac{6.9\text{ mmHg}}{740.\text{ mmHg}} = 0.0093 \qquad X_{CO_2} = \frac{P_{CO_2}}{P_{tot}} = \frac{0.2\text{ mmHg}}{740.\text{ mmHg}} = 0.0003$$

$$X_{H_2O} = \frac{P_{H_2O}}{P_{tot}} = \frac{4.0\text{ mmHg}}{740.\text{ mmHg}} = 0.0054$$

(c)

$$\%\ N_2 = X_{N_2} \times 100\% = 0.777 \times 100\% = 77.7\%\ N_2$$

$$\%\ O_2 = X_{O_2} \times 100\% = 0.208 \times 100\% = 20.8\%\ O_2$$

$$\%\ Ar = X_{Ar} \times 100\% = 0.0093 \times 100\% = 0.93\%\ Ar$$

$$\% \ CO_2 = X_{CO_2} \times 100\% = 0.0003 \times 100\% = 0.03\% \ CO_2$$

$$\% \ H_2O = X_{H_2O} \times 100\% = 0.0054 \times 100\% = 0.54\% \ H_2O$$

The Table 8.6 figures are slightly different. This sample is wet, whereas the percentages given in the table are for dry air.

☑ *Reasonable Result Check:* The percentages are very close to those provided in Table 8.6, and the variations are explainable due to water vapor. The sum of the mole fractions is 1, and the sum of the percentages is 100%.

54. *Result:* **(a) 1.98 atm (b) P_{O_2} = 0.438 atm, P_{N_2} = 0.182 atm, P_{Ar} = 1.36 atm**

Analyze: Given three containers of gases, with known volume and pressure, determine the total pressure and partial pressures of the gases when the three containers are opened to each other at constant temperature.

Plan: Use Boyle's law to determine how the pressure of each gas changes with the increase in total volume and then use Dalton's law to determine the total pressures after mixing. Use the definition of partial pressure to show that the pressures calculated are the partial pressure.

Execute: The gas in one chamber is allowed to diffuse into all three chambers. Assuming that the connecting tube has negligible volume, its final volume increases to the total of the three chambers' volumes:

$$V_{tot} = 3.00 \text{ L} + 2.00 \text{ L} + 5.00 \text{ L} = 10.00 \text{ L}$$

Boyle's law: $$P_{f,gas} = \frac{P_{i,gas} V_{i,gas}}{V_{f,gas}}$$

For O_2, $$P_{f,O_2} = \frac{P_i V_i}{V_f} = \frac{(1.46 \text{ atm}) \times (3.00 \text{ L})}{(10.00 \text{ L})} = 0.438 \text{ atm}$$

For N_2, $$P_{f,N_2} = \frac{(0.908 \text{ atm}) \times (2.00 \text{ L})}{(10.00 \text{ L})} = 0.182 \text{ atm}$$

For Ar, $$P_{f,Ar} = \frac{(2.71 \text{ atm}) \times (5.00 \text{ L})}{(10.00 \text{ L})} = 1.36 \text{ atm}$$

(a) $P_{f,tot} = P_{f,O_2} + P_{f,N_2} + P_{f,Ar} = 0.438 \text{ atm} + 0.182 \text{ atm} + 1.36 \text{ atm} = 1.98 \text{ atm}$

(b) Partial pressure is the pressure each gas would cause if it were alone in the container.
$P_{f,O_2} = P_{O_2} = 0.438 \text{ atm}, \ P_{f,N_2} = P_{N_2} = 0.182 \text{ atm}, \ P_{f,Ar} = P_{Ar} = 1.36 \text{ atm}$

☑ *Reasonable Result Check:* All the individual final pressures are less than the initial pressures, which makes sense because the volume is larger. The total pressure is a weighted average of the three pressures, and is influenced most by the gas present in the largest quantity (Ar).

56. *Result:* **Membrane irritation: 1 × 10^{-4} atm; fatal narcosis: 0.02 atm**

Analyze: Given two concentrations of a gas as pressure ratios, determine the partial pressure of the gas at a given temperature and pressure.

Plan and Execute: Use the ppm pressure ratio values to get the partial pressure:

$$P_i = (\text{pressure ratio}) \times P_{tot}$$

$$P_{tot} = 1 \text{ atm at STP}$$

Membrane irritation: $$P_{benzene} = \frac{100 \text{ atm}}{1{,}000{,}000 \text{ atm}} \times 1 \text{ atm} = 1 \times 10^{-4} \text{ atm}$$

Fatal narcosis: $$P_{benzene} = \frac{20{,}000 \text{ atm}}{1{,}000{,}000 \text{ atm}} \times 1 \text{ atm} = 0.02 \text{ atm}$$

☑ *Reasonable Result Check:* The larger pressure ratio has a larger partial pressure.

57. *Result:* $\mathbf{X_{H_2O} = 0.0041}$; $\mathbf{P_{H_2O} = 3.1}$ **mm Hg; the mean partial pressure includes humid and dry air, summer and winter, worldwide**

Analyze: Given the mean fraction by mass of water in the atmosphere and the molar mass of "air," determine the mean mole fraction of water in the atmosphere.

Plan: The mean fraction by weight describes the average number of grams of water per gram of air. Convert grams to moles in both cases, to get mole fraction. Use $P_i = X_iP_{tot}$, to get its partial pressure.

Execute: The molar mass of air is given at 29.2 g/mol. The molar mass of water is 18.0152 g/mol.

$$X_{H_2O} = \frac{0.0025 \text{ g } H_2O}{1 \text{ g air}} \times \frac{29.2 \text{ g air}}{1 \text{ mol air}} \times \frac{1 \text{ mol } H_2O}{18.0152 \text{ g } H_2O} = 0.0041 \frac{\text{mol } H_2O}{\text{mol air}}$$

Assume that the $P_{tot} = 760$ mm Hg, standard pressure.

$$P_{H_2O} = X_{H_2O}P_{tot} = (0.0041) \times (760 \text{ mm Hg}) = 3.1 \text{ mm Hg}$$

This number represents the mean partial pressure. This average pressure includes all of earth's air, humid and dry, summer and winter, at all altitude, so it is unsurprising that the mean water vapor pressure is much smaller than that of a rainy summer day at Earth's surface (25 mm Hg).

☑ *Reasonable Result Check:* The mole fraction is small, meaning the partial pressure is small compared to the total pressure.

58. *Result:* **74.7%**

Analyze: Given the balanced equation for a reaction, the mass of a reactant, the volume, temperature and pressure of a product collected over water, and the vapor pressure of water at that same temperature, determine the percent yield of the reaction.

Plan: Calculate the theoretical yield from the mass of reactant. Use Dalton's law to calculate the pressure of the product gas. Use the ideal gas law to determine the moles of product. Convert the moles to grams to get actual yield. Divide the actual yield by the theoretical yield and multiply by 100% to get percent yield.

Execute: 1 mole CaC_2 produces 1 mol C_2H_2.

$$2.62 \text{ g } CaH_2 \times \frac{1 \text{ mol } CaH_2}{64.10 \text{ g } CaH_2} \times \frac{1 \text{ mol } C_2H_2}{1 \text{ mol } CaH_2} \times \frac{26.0372 \text{ g } C_2H_2}{1 \text{ mol } C_2H_2} = 1.06 \text{ g } C_2H_2$$

$$P_{tot} = P_{C_2H_2} + P_{H_2O}$$

$$P_{C_2H_2} = P_{tot} - P_{H_2O} = (735.2 \text{ mm Hg}) - (23.8 \text{ mm Hg}) = 711.4 \text{ mm Hg}$$

$$711.4 \text{ mmHg} \times \frac{1 \text{ atm}}{760 \text{ mmHg}} = 0.9361 \text{ atm}$$

$$T = 25.0 \text{ °C} + 273.15 = 298.2 \text{ K}$$

$$795 \text{ mL} \times \frac{1 \text{ L}}{1000 \text{ mL}} = 0.795 \text{ L}$$

$$n_{C_2H_2} = \frac{PV}{RT} = \frac{(0.9361 \text{ atm}) \times (0.795 \text{ L})}{\left(0.08206 \frac{\text{L} \cdot \text{atm}}{\text{mol} \cdot \text{K}}\right) \times (298.2 \text{ K})} = 0.0304 \text{ mol } C_2H_2$$

$$0.0304 \text{ mol } C_2H_2 \times \frac{26.0372 \text{ g } C_2H_2}{1 \text{ mol } C_2H_2} = 0.792 \text{ g } C_2H_2$$

$$\frac{0.792 \text{ g } C_2H_2 \text{ actual}}{1.06 \text{ g } C_2H_2 \text{ theoretical}} \times 100\ \% = 74.7\ \%$$

☑ *Reasonable Result Check:* The percent yield is a realistic size for collection of a gas.

Kinetic Molecular Theory: Velocities of Gas Molecules (Section 8-7)

61. *Result:* **$CH_2Cl_2 < Kr < N_2 < CH_4$**

Analyze: Given the formulas of various atoms and molecules and their common temperature, put their gases in order of increasing average molecular speed.

Plan: Kinetic energy is proportional to temperature. With all of the samples at the same temperature, their average kinetic energies are the same. Kinetic energy is related to mass and velocity. $E_{kin} = \frac{1}{2}mv^2$. Velocity is related to kinetic energy and mass: $v^2 = \frac{2E}{m}$. Therefore, molecules with smaller mass have the faster molecular speed. To rank the molecules with increasing speed, rank them from the largest molar mass to the smallest.

Execute: Estimate the molar masses: Kr molar mass 83.8 g/mol, CH_4 molar mass = 16.0 g/mol, N_2 molar mass is 28.0 g/mol, and CH_2Cl_2 molar mass = 84.9 g/mol.

$$\text{slowest speed: } CH_2Cl_2 < Kr < N_2 < CH_4 \text{ fastest speed}$$

☑ *Reasonable Result Check:* It makes sense that molecules with lower mass go faster.

63. *Result:* **Ne; it has the lowest mass, so fastest average velocity and reaches the end first.**

Analyze: Given the formulas of various gases introduced into one end of a tube, determine which gas will reach the end of the tube first.

Plan: Velocity is related to kinetic energy and mass: $v^2 = \frac{2E}{m}$. Therefore, molecules with smaller mass have the faster molecular speed. The molecule with the fastest speed will travel a fixed distance the quickest. To determine which molecule arrives first, rank them from the largest molar mass to the smallest.

Execute: Estimate the molar masses: Ar molar mass = 40 g/mol, Ne molar mass = 20 g/mol, Kr molar mass = 84 g/mol, and Xe molar mass = 131 g/mol. **Ne will arrive first because it has the lowest mass, so fastest average velocity and reaches the end first.**

☑ *Reasonable Result Check:* Molecules with lower mass go faster, so they will arrive sooner.

The Behavior of Real (Non-Ideal) Gases (Section 8-8)

66. *Result:* **18 mL $H_2O(\ell)$; 22.4 L $H_2O(g)$; No we cannot achieve a pressure of 1 atm at this temperature, because the vapor pressure of water at 0 °C to be 4.6 mm Hg; at any higher pressures, water liquefies.**

Analyze: Using the density and molar mass of water, calculate the volume occupied by one mole of liquid water. Determine the volume occupied by one mole of gaseous water at STP. Explain if it is possible to achieve standard state conditions for water vapor.

Plan and Execute: Use standard conversion factors:

$$1 \text{ mol } H_2O \times \frac{18.02 \text{ g } H_2O}{1 \text{ mol } H_2O} \times \frac{1 \text{ mL } H_2O}{1.0 \text{ g } H_2O} = 18 \text{ mL } H_2O \text{ liquid}$$

1 mol H_2O at STP occupies 22.4 L.

Table 8.4 gives the vapor pressure of water at 0 °C to be 4.6 mm Hg. At any higher pressures, water would liquefy. We cannot achieve 1 atm pressure of water vapor at such a low temperature, so we cannot achieve the standard state condition for water vapor.

67. *Result/Explanation:* The non-ideal behavior of real gases is discussed in Section 8.8. As external pressure increases, the gas volume decreases, the molecules are squeezed closer together, and the attractions among the molecules get stronger. Figure 8.13 shows that a gas molecule strikes the walls of the container with less force due to the attractive forces between it and its neighbors. Less force exerted by the molecules leads to lower the pressure (P), and makes the mathematical product PV smaller than the mathematical product nRT.

70. *Result:* **N_2**

Analyze: Use the van der Waals constants to predict the ideal gas behavior of two gases.

Plan: Table 8.5 gives the values for the van der Waals constants, *a* and *b*. The *a* constant is related to the pressure correction. The smaller the value of *a*, the closer to ideal the gas is.

Execute: $a_{N_2} = 1.39$ $\quad a_{CO_2} = 3.59$

N_2 is more like an ideal gas at high pressures.

☑ *Reasonable Result Check:* It makes sense that the smaller diatomic molecule with a nonpolar bond is more like an ideal gas than the larger triatomic molecule with polar bonds.

The Atmosphere (Section 8-9)

73. *Result/Explanation:* Nitrogen is stable and moderates the reactiveness of oxygen by diluting it. Oxygen sustains animal life as a reactant in the conversion of food to energy. Oxygen is produced by plants in the process of photosynthesis.

75. *Result:* (a) 1.2×10^4 L (b) 2 times more air in room than is breathed (c) 5.9×10^{25} molecules

Analysis: Given the number of breaths per minute and the volume of a breath, calculate the volume of air inhaled in a given amount of time. Compare the volume to the volume of air in a residence hall room given the cubic feet dimensions and a conversion factor between liters and cubic feet. Finally, calculate the number of oxygen molecules inhaled.

Plan: Use time, the breathing rate, and the volume per breath to calculate the total volume breathed. Use the dimensions of the room to calculate the volume of the room, then compare them. Use the percent oxygen in air from Table 8.6, the molar mass, and Avogadro's number to determine the number of molecules.

Execute:

(a)
$$24\text{ h} \times \frac{60\text{ min}}{1\text{ hr}} \times \frac{16\text{ breaths}}{1\text{ min}} \times \frac{0.50\text{ L}}{1\text{ breath}} = 1.2 \times 10^4\text{ L air}$$

(b)
$$864\text{ ft}^3 \times \frac{28\text{ L}}{1\text{ ft}^3} = 2.4 \times 10^4\text{ L}$$

1.2×10^4 L of air is half the air volume in the room (2.4×10^4 L). There is two times more air in the room than is breathed in a day.

(c) Using Table 8.6:

$$1.2 \times 10^4\text{ L air} \times \frac{20.948\text{ L O}_2}{100\text{ L air}} = 2.5 \times 10^3\text{ L O}_2$$

Use the ideal gas law to calculate moles of O_2.

$$T = 18.0\text{ °C} + 273.15 = 291\text{ K} \qquad P = 0.970\text{ atm.}$$

$$n = \frac{PV}{RT} = \frac{(0.970\text{ atm})(2.4 \times 10^4\text{ L O}_2)}{\left(0.08206\frac{\text{L}\cdot\text{atm}}{\text{mol}\cdot\text{K}}\right)(291\text{ K})} = 102\text{ mol O}_2 = 1.0 \times 10^2\text{mol O}_2$$

$$102\text{ mol O}_2 \times \frac{6.022 \times 10^{23}\text{ O}_2\text{ molecules}}{1\text{ mol O}_2} = 6.1 \times 10^{25}\text{ O}_2\text{ molecules}$$

Stratospheric Ozone Depletion (Section 8-10)

78. *Result:* **(a) $\cdot CF_3 + \cdot Cl$ (b) $ClO\cdot$ (c) $\cdot Cl + O_2$**

Analyze and Plan: Similar equations can be found in Section 8-10b. A photon of light separates chlorine atom from the radical, CF_3. When two radicals combine, a new bond forms between them.

Execute:

(a) $CF_3Cl \xrightarrow{h\nu} \cdot CF_3 + \cdot Cl$

(b) $\cdot Cl + \cdot O\cdot \longrightarrow ClO\cdot$

(c) $ClO\cdot + \cdot O\cdot \longrightarrow \cdot Cl + O_2$

80. *Result/Explanation:* CH_3F has no C–Cl bonds, which in CH_3Cl are readily broken when exposed to UV light. Looking at the bond enthalpies, C–Cl (327 kJ/mol) is much weaker than C–F (486 kJ/mol). In fact, the bond enthalpy of C–F is very close to the bond enthalpy of O=O (498 kJ/mol)!

82. *Result/Explanation:* CFCs are not toxic. Refrigerants used before CFCs were very dangerous. One example is NH_3, a strong-smelling, reactive chemical. In any web browser, type the keywords "CFCs" or "refrigerants" and you will get more than a million hits.

Greenhouse Gases and Global Warming (Section 8-11)

83. *Result/Explanation:* Greenhouse effect is the trapping of heat radiation by atmospheric gases. Global warming is the increase of the average global temperature. Global warming is caused by an increase in the amount of greenhouse gases in the atmosphere.

85. *Result/Explanation:* CO_2 gets into the atmosphere by animal respiration, by burning fossil fuels and other plant materials, and by the decomposition of organic matter. CO_2 gets removed from the atmosphere by plants during photosynthesis, when it is dissolved in rainwater, and when it is incorporated into carbonate and bicarbonate compounds in the oceans. Currently, atmospheric CO_2 production exceeds CO_2 removal.

Chemistry of Air Quality and Air Pollution (Section 8-12)

87. *Result:* **Primary pollutants (e.g., particle pollutants, including aerosols and particulates; sulfur dioxide; nitrogen oxides; hydrocarbons) secondary pollutants (e.g., ozone). (see Section 8-12)**

Explanation:

Primary pollutants are substances that are introduced into the air directly from their source.

- Particle pollutants: pollutants made out of particles:
  - aerosols: particles incorporated into water droplets.
  - particulates: larger solid particles
- Sulfur dioxide: pollutant produced when sulfur or sulfur compounds are burned in air.
- Nitrogen oxides: pollutant produced when nitrogen and oxygen react at high temperatures.
- Hydrocarbons: pollutants produced from many organic sources; their identity is small hydrocarbons from CH_4 to ones with six or seven carbons.

Secondary pollutants are substances produced from reactions of primary pollutants.

- Ozone: ozone in the troposphere is produced from a reaction of O_2 with O_2 in the presence of intense energy (e.g., spark, lightening, etc.)
- Sulfur trioxide: SO_3 is produced from the reaction of SO_2.
- PAN (peroxyacetylnitrate): produced by the reaction of various free radicals in urban air.

Read Section 8-12 for the specific ways these pollutants are harmful.

89. *Result:* **1.6×10^9 metric tons; 2.3×10^6 hr**

Analyze: Given the above information and the percentage of sulfur in the coal, determine the mass (in metric tons) of coal burned and determine the number of hours this quantity of coal will burn.

Plan: Determine the moles of sulfur in the product, then determine the mass of coal that contains this amount of sulfur. Use the metric tons of coal per hour to determine the hours one plant would need to burn this quantity.

Execute:

$$65\times10^{6}\text{ metric tons SO}_2\times\frac{1000\text{ kg SO}_2}{1\text{ metric tons SO}_2}\times\frac{1000\text{ g SO}_2}{1\text{ kg SO}_2}=6.5\times10^{13}\text{ g SO}_2$$

$$6.5\times10^{13}\text{ g SO}_2\times\frac{1\text{ mol SO}_2}{64.07\text{ g SO}_2}\times\frac{1\text{ mol S}}{1\text{ mol SO}_2}\times\frac{32.065\text{ g S}}{1\text{ mol S}}=3.3\times10^{13}\text{ g S}$$

$$3.3\times10^{13}\text{ g S}\times\frac{100\text{ g coal}}{2.0\text{ g S}}\times\frac{1\text{ kg coal}}{1000\text{ g coal}}\times\frac{1\text{ metric tons coal}}{1000\text{ kg coal}}=1.6\times10^{9}\text{ metric tons coal}$$

$$1.6\times10^{9}\text{ metric tons coal}\times\frac{1\text{ hr}}{700.\text{ metric tons coal}}=2.3\times10^{6}\text{ hr}$$

☑ *Reasonable Result Check:* This is about 30 decades; that seems like a long time. It is clear that there are many power plants adding SO_2 to the atmosphere for this amount to be produced in one year.

91. *Result/Explanation:* The atmospheric reaction that favors the formation of nitrogen monoxide, NO, is given in this chemical equation:

$$N_2 + O_2 \xrightarrow{\text{heat}} 2\,NO$$

The formation of NO in a combustion chamber is similar to the formation of NH_3 in a reactor designed to manufacture ammonia, because in both cases a reaction takes elemental nitrogen and makes a compound of nitrogen.

93. *Result:* **(a) 1.33×10^{-4} mm Hg (b) $X(SO_2) = 1.75\times10^{-7}$ (c) 500. μg SO_2**

Analyze: Given the concentration of an air pollutant, determine the partial pressure, the mole fraction, and the mass in micrograms contained in a given volume of air at STP. (We are given the time of exposure, too, but that information is not needed to answer the questions asked.)

Plan: Convert the concentration of SO_2 from ppm to mole fraction, using Avogadro's law to relate volume and moles. Use $P_i = X_iP_{tot}$ and the standard air pressure to find the partial pressure of SO_2. Use the molar volume of a gas to determine moles of SO_2 in the given volume at STP, then convert to micrograms.

Execute:

(a)
$$\frac{0.175\text{ L SO}_2}{1{,}000{,}000\text{ L air}}=1.75\times10^{-7}\ \frac{\text{L SO}_2}{\text{L air}}$$

A gas-volume ratio is equal to a mole ratio, according to Avogadro's law, so

$$\mathbf{X(SO_2)} = 1.75\times10^{-7}$$

(b) Assume that the P_{tot} = 760 mm Hg, standard pressure.

$$P_i = X_iP_{tot} = (1.75\times10^{-7})\times(760\text{ mm Hg}) = 1.33\times10^{-4}\text{ mm Hg}$$

(c)
$$1\text{ m}^3\text{ air}\times\left(\frac{100\text{ cm}}{1\text{ m}}\right)^3\times\frac{1\text{ mL}}{1\text{ cm}^3}\times\frac{1\text{ L}}{1000\text{ mL}}\times\frac{0.175\text{ L SO}_2}{1{,}000{,}000\text{ L air}}=1.75\times10^{-4}\text{ L SO}_2$$

$$1.75\times10^{-4}\text{ L SO}_2\times\frac{1\text{ mol SO}_2}{22.414\text{ L SO}_2}\times\frac{64.0638\text{ g SO}_2}{1\text{ mol SO}_2}\times\frac{1\ \mu\text{g SO}_2}{10^{-6}\text{ g SO}_2}=500.\ \mu\text{g SO}_2$$

☑ *Reasonable Result Check:* The concentration is expressed in small units (ppm), so a small partial pressure, the small mole fraction, and the small mass all make sense.

General Questions

94. *Result:* **(a) Before: $P_{H_2} = 3.7$ atm; $P_{Cl_2} = 4.9$ atm (b) Before: $P_{tot} = 8.6$ atm (c) After: $P_{tot} = 8.6$ atm (d) Cl_2; 0.5 mol remain (e) $P_{HCl} = 7.4$ atm; $p_{Cl_2} = 1.2$ atm (f) P = 8.9 atm**

Analyze: Given the description of a reaction, the masses of two gaseous reactants, and the volume and temperature of the mixture, determine the partial pressure of the reactants, the total pressure due to the gases in the flask before and after the reaction, the excess reactant, the number of moles of it left over, the partial pressures of the gases in the flask, and the total pressure after the temperature is increased to a given value.

Plan: Balance the equation. Find moles of each reactant and use the ideal gas law to determine the pressures. Use Dalton's law for total pressures. Find limiting reactant and excess reactant as described in Chapter 4. Use the combined gas law to determine the pressure at a different temperature.

Execute: $H_2(g) + Cl_2(g) \longrightarrow 2HCl(g)$

(a) $3.0\text{ g }H_2 \times \dfrac{1\text{ mol }H_2}{2.0158\text{ g }H_2} = 1.5\text{ mol }H_2$ $\qquad 140.\text{ g }Cl_2 \times \dfrac{1\text{ mol }Cl_2}{70.906\text{ g }Cl_2} = 1.97\text{ mol }Cl_2$

$$T = 28\ °C + 273.15 = 301\ K$$

$$P_{H_2} = \frac{nRT}{V} = \frac{(1.5\text{ mol }H_2)\times\left(0.08206\dfrac{L\cdot atm}{K\cdot mol}\right)\times(301\ K)}{10.\ L} = 3.7\text{ atm}$$

$$P_{Cl_2} = \frac{nRT}{V} = \frac{(1.97\text{ mol }Cl_2)\times\left(0.08206\dfrac{L\cdot atm}{K\cdot mol}\right)\times(301\ K)}{10.\ L} = 4.9\text{ atm}$$

(b) $P_{tot,\ before} = P_{H_2} + P_{Cl_2} = (3.7\text{ atm }H_2) + (4.9\text{ atm }Cl_2) = 8.6\text{ atm total}$

(c) The number of moles of gas reactants is equal to the number of moles of gaseous products, so the total pressure will not change. $P_{tot,\ after} = 8.6$ atm total

(d) The balanced chemical equation shows equal molar quantities of each reactant reacting, so the H_2 is the limiting reactant and Cl_2 is the excess reactant, and 1.5 mol of each reactant react. Subtracting the moles of Cl_2 that react from the moles of Cl_2 describes how many moles of Cl_2 remain.

$$2.0\text{ mol }Cl_2\text{ initial} - 1.5\text{ mol }Cl_2\text{ react} = 0.5\text{ mol }Cl_2\text{ remain}$$

(e) When 1.5 mol of each reactant react, 3.0 mol of HCl are formed.

$$P_{HCl} = \frac{nRT}{V} = \frac{(3.0\text{ mol HCl})\times\left(0.08206\dfrac{L\cdot atm}{K\cdot mol}\right)\times(301\ K)}{10.\ L} = 7.4\text{ atm}$$

$$P_{Cl_2} = PP_{tot,\ after} - P_{HCl} = (8.6\text{ atm total}) - (7.4\text{ atm HCl}) = 1.2\text{ atm }Cl_2$$

(f) $T_1 = 301$ K, $T_2 = 40.\ °C + 273.15 = 313$ K

$$P_2 = P_1 \times \frac{V_1}{V_2} \times \frac{T_2}{T_1} = P_1 \times \frac{T_2}{T_1} \text{ at constant volume}$$

$$P_2 = P_1 \times \frac{T_2}{T_1} = (8.6\text{ atm}) \times \frac{(313\text{K})}{(301\ K)} = 8.9\text{ atm}$$

☑ *Reasonable Result Check:* The partial pressure of Cl_2 calculated in (e) can also be found using the ideal gas law:

$$P_{Cl_2} = \frac{nRT}{V} = \frac{(0.5 \text{ mol } Cl_2) \times \left(0.08206 \frac{L \cdot atm}{K \cdot mol}\right) \times (301 \text{ K})}{10. \text{ L}} = 1. \text{ atm}$$

We could construct mole fractions of the products and multiply them by the total pressure to determine the partial pressures after the reaction was completed:

$$X_{HCl} = \frac{3.0 \text{ mol HCl}}{3.0 \text{ mol HCl} + 0.50 \text{ mol } Cl_2} = 0.86$$

$$P_{HCl} = X_{HCl}P_{tot} = (0.86) \times (8.6 \text{ atm}) = 7.4 \text{ atm}$$

$$X_{Cl_2} = \frac{0.50 \text{ mol } Cl_2}{3.0 \text{ mol HCl} + 0.5 \text{ mol } Cl_2} = 0.14$$

$$P_{Cl_2} = X_{Cl_2}P_{tot} = (0.14) \times (8.6 \text{ atm}) = 1.2 \text{ atm}$$

$$P_{tot,\ after} = P_{HCl} + P_{Cl_2} = 7.4 \text{ atm} + 1.2 \text{ atm} = 8.6 \text{ atm total}$$

These answers are the same as calculated above. The results are self-consistent and reasonable size (i.e., smaller number of moles produce smaller partial pressures and smaller mole fractions).

97. *Result:* **3 times bigger**

Analyze: Given the fraction of oxygen in air and the balanced equation for the reaction of methane with oxygen, determine how much bigger the cross section of the tube for air must be compared to that for methane.

Plan: The amount of gas is proportional to the volume of gas. As shown here, the volume of gas is proportional to the square of tube diameter, assuming the same rate of gas flow from each tube:

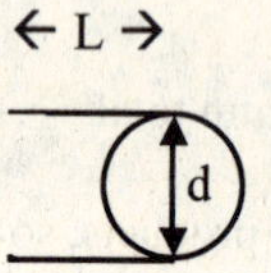

$$V = L \times A = L \times \pi r^2 = L \times \pi\left(\frac{d}{2}\right)^2$$

$$\frac{V_{air}}{V_{CH_4}} = \left(\frac{d_{air}}{d_{CH_4}}\right)^2$$

First, determine the volume of air per mL of methane, then relate the volume ratio to the diameter ratio.

Execute: Air is one-fifth oxygen by volume. The balanced equation shows that twice the volume of oxygen is needed compared to the volume of methane.

$$\frac{5 \text{ mL air}}{1 \text{ mL } O_2} \times \frac{2 \text{ mL } O_2}{1 \text{ mL } CH_4} = \frac{10 \text{ mL air}}{1 \text{ mL } CH_4}$$

$$\frac{d_{air}}{d_{CH_4}} = \sqrt{\frac{10 \text{ mL air}}{1 \text{ mL } CH_4}} = 3.16 \approx 3$$

Therefore, the air tube needs to be **3 times bigger** than the methane tube.

☑ *Reasonable Result Check*: More oxygen is required, so it makes sense that the air tube must be larger than the methane tube.

99. *Result:* **Use F = mg = PA, to derive m = PA/g and appropriate metric conversions**

Analyze: Given the surface area of the earth and the definition of a metric ton, show a calculation that verifies the mass of earth's atmosphere.

Plan and Execute: Use $F = mg = PA$, to derive $m = PA/g$ and then use appropriate metric conversions.

At sea level the pressure is 1 atmosphere, so:

$$P = 1\text{ atm} \times \left(\frac{101.325\text{ kPa}}{1\text{ atm}}\right) \times \left(\frac{1000\text{ Pa}}{1\text{ kPa}}\right) \times \left(\frac{1\frac{\text{kg}}{\text{m}\cdot\text{s}^2}}{1\text{ Pa}}\right) \times \left(\frac{1\text{ metric ton}}{10^3\text{ kg}}\right) = 101.325\frac{\text{metric tons}}{\text{m}\cdot\text{s}^2}$$

$$A = 5.1 \times 10^8\text{ km}^2 \times \left(\frac{1000\text{ m}}{1\text{ km}}\right)^2 = 5.1 \times 10^{14}\text{ m}^2$$

$$m = \frac{PA}{g} = \frac{\left(101.325\frac{\text{metric tons}}{\text{m}\cdot\text{s}^2}\right)\left(5.1 \times 10^{14}\text{ m}^2\right)}{\left(9.8\frac{\text{m}}{\text{s}^2}\right)} = 5.3 \times 10^{15}\text{ metric tons}$$

☑ *Reasonable Result Check*: This calculation results in the given mass.

Applying Concepts

102. *Result:* **3.2 × 10⁸ molecules/cm³**

Analyze: Given pressure and temperature of air in a flask, calculate the number of molecules of air per cm^3.

Plan: Use the ideal gas law.

Execute: $T = 25°C + 273.15 = 298\text{ K}$ $V = 1\text{ cm}^3$ (exact)

$$n = \frac{PV}{RT} = \frac{\left(1.0 \times 10^{-8}\text{torr} \times \frac{1\text{ atm}}{760\text{ torr}}\right)\left(1\text{ cm}^3 \times \frac{1\text{ L}}{1000\text{ cm}^3}\right)}{\left(0.08206\frac{\text{L}\cdot\text{atm}}{\text{mol}\cdot\text{K}}\right)(298\text{ K})} = 5.4 \times 10^{-16}\text{ mol/cm}^3$$

5.4×10^{-16} mol/cm³ × 6.022×10^{23} molecules/mol = **3.2×10^8 molecules/cm³**

105. *Result:* **Statements (a), (b), (c), and (d) are true.**

Analyze: We are considering a real gas, $N_2(g)$, whose conditions are such that it obeys the ideal gas law exactly. Explain which statements are true.

Plan: Use the kinetic-molecular theory to assist in answering this question.

Execute:

(a) N_2 is a larger than Ne and therefore N_2 is not as much like an ideal gas. If these conditions allow N_2 to behave as an ideal gas, they would certainly be sufficient for Ne to also behave like an ideal gas. Hence, this statement: "A sample of Ne(g) under the same conditions must obey the ideal gas law exactly." is true.

(b) All collisions are elastic, however, energy can be transferred during an elastic collision from one molecule to another. A faster molecule hitting a slower molecule might make the slow molecule go faster if the first one ends up slower. In addition, each time the molecule hits the wall, there is an instant when its speed is zero as it bounces off the wall and goes flying away in the opposite direction. Hence, the statement: "The speed at which one particular N_2 molecule is moving changes from time to time." is true.

(c) The average speed of the N_2 molecules will be faster than the average speed of the O_2 molecules, but ideal gas particles move at varying speeds, so some molecules in each sample will be moving very slowly and others very quickly. Hence, the statement: "Some of the N_2 molecules are moving more slowly than some of the molecules in a sample of $O_2(g)$ under the same conditions." is true.

(d) The average speed of the N_2 molecules will be slower than the average speed of the Ne molecules; hence, the statement: "Some of the N_2 molecules are moving more slowly than some of the molecules in a sample of Ne(g) under the same conditions." is true.

(e) All collisions are elastic, so collisions must conserve energy. There is no way that both molecules could be going faster, since that implies that energy has increased. Hence, the statement: "When two N_2 molecules collide, it is possible that both may be moving faster after the collision than they were before." is false.

107. *Result:* **See graph below**

Explanation: The molar masses of C_2H_6 and F_2 are 30 g/mol and 38 g/mol. That means the average speed of F_2 is somewhat less than that of C_2H_6. The total pressure of the gases is 720. mm Hg, and the partial pressure of F_2 is 540. mm Hg. Dalton's law tells us that the sum of the partial pressures is the total pressure, and so the partial pressure of C_2H_6 is 180. mm Hg, or one third that of the F_2 molecules. The partial pressure is directly proportional to the mole fraction of the molecules, in the container, so there should be one third as many C_2H_6 molecules as F_2 molecules. So, the graph of number of molecules verses molecular speed will have the F_2 curve peaking at a slightly smaller speed value and the C_2H_6 curve will have one third of the vertical rise.

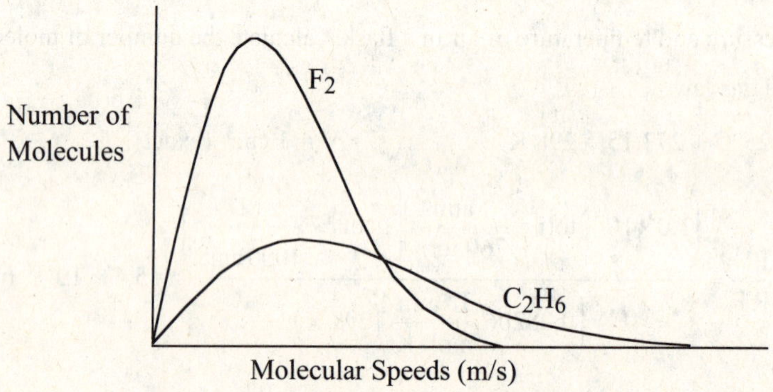

110. *Result:* **(a) Pressure: Box (iii) = Box (iv) > Box (i) = Box (ii) (b) Density: Box (iii) > Box (i) = Box (iv) > Box (ii) (c) Average kinetic energy: Box (i) = Box (ii) = Box (iii) = Box (iv) (d) Average molecular speed:** Box (ii) = Box (iv) > Box (i) = Box (iii)

Explanation:

(a) The pressure of a gas is related to how often the particles hit the walls of the container. Boxes (i) and (ii) have equal volume and equal number of particles, so the pressure in those boxes are equal. Boxes (iii) and (iv) have equal volume and equal number of particles, so the pressure in those boxes are equal. Boxes (iii) and (iv) are half the volume of Boxes (i) and (ii) with the same number of atoms, so their pressures will be greater. Therefore, the samples rank in this order: Box (iii) = Box (iv) > Box (i) = Box (ii)

(b) The density of a gas is the relationship between the mass of the gas particles and the volume of the container. Comparing molar masses: 20g/4g = 5, the mass of the neon atoms is five times larger than the mass of the helium atoms. The small boxes are half the volume of the large boxes.

Box (iii) has the heavy-weight atoms in the smaller volume, so the gas density in that box would be the largest. Box (iv) is half the size of Box (i) with a mass of atoms fives times smaller, so the density in Box (i) is greater than the density in Box (iv). The same number of light-weight atoms in the larger box, Box (ii), will result in the lowest density. Therefore, the samples rank in this order: Box (iii) > Box (i) > Box (iv) > Box (ii)

(c) $E = \frac{1}{2} mv^2$ and all the samples have the same temperature, so the average kinetic energy in each sample is equal, so the samples rank in this order: Box (i) = Box (ii) = Box (iii) = Box (iv)

(d) The speed of the atoms is related to the mass and the temperature: $T \propto mv^2$. All the samples have the same temperature and the average speed of the heavier atoms is slower than the average speed of the lighter-weight atoms. And the samples rank in this order: Box (ii) = Box (iv) > Box (i) = Box (ii)

112. *Result:* **See drawings below**

Analyze: Given a figure showing a syringe containing a gas at specific volume, temperature, and pressure, redraw figures of the syringe after the conditions are altered.

Plan: The speed of the molecules is related to the temperature. In particular, $T \propto mv^2$. So, when the temperature changes by a certain factor, the speed of the molecules (v) changes by the square root of that same factor. We'll indicate that fact by making the tails of the arrows proportionally shorter. The pressure describes how close together the molecules are. When the pressure is changed by a certain factor, the molecules will be represented as that much closer together. The volume will change the space occupied by the gas, and that will be indicated in the movement of the syringe plunger.

Execute: As a reference, the initial state looks like this:

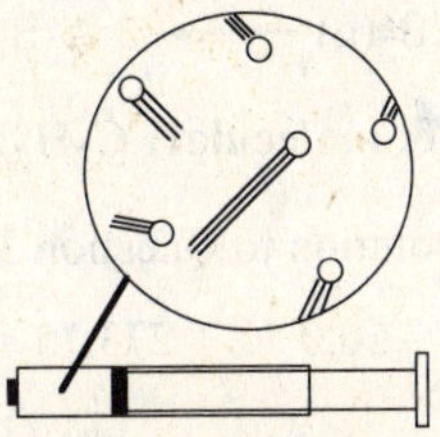

(a) When the temperature is decreased to one half of its original value, Charles' law tells us that the volume decreases by half. The molecules are just as far apart as they were (since the pressure is the same), but the plunger level goes down by half, and the tails on the molecules decrease in length by about 0.7 times.

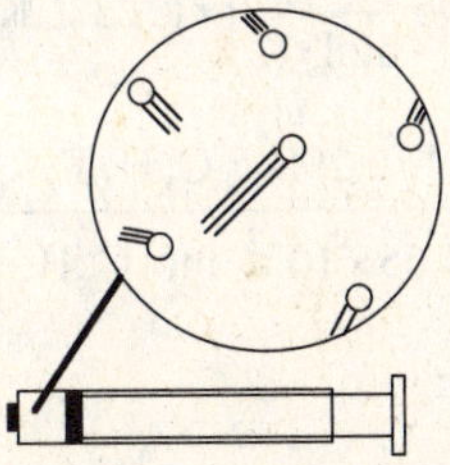

(b) When the pressure decreased to one half of its original value, Boyle's law tells us that the volume increases by half. The tails on the molecules are the same length, but the molecules are twice as far apart as they were (so we put half as many in this view), and the plunger level moves up by half.

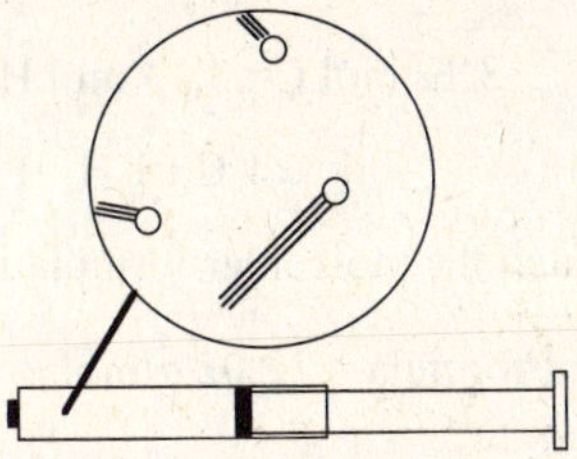

(c) When the temperature is tripled and the pressure is doubled, the combined gas law tells us that the volume will have a net increase by 3/2 (three times larger due to the temperature change and half as large due to the pressure change). The molecules are two times closer together than they were (so we'll add twice as many), the plunger level goes up by three halves, and the tails on the molecules increase in length by about 1.7 times.

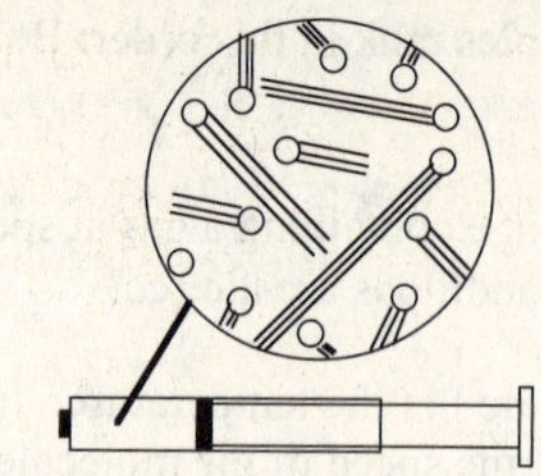

114. *Result:* **Box (b); for every two molecules of gas reactants there must be one molecule of gas products**

Explanation: The initial volume is 1.8 L and the final volume is 0.9 L. This 2:1 ratio in the gas volumes means for every two molecules of gas reactants there must be one molecule of gas products. The reactant count is six, so the box that has three product molecules fits these observations. The correct box is Box (b):

$$6\ AB_2(g) \longrightarrow 3\ A_2B_4(g)$$

116. *Result:* **(a) 64.1 g/mol (b) empirical: CHF, molecular: $C_2H_2F_2$ (c) see structures below**

Analyze: Use the methods described in the solution to Question 38, Chapter 2, and Chapter 7.

(a)

$$T = 50.0\ °C + 273.15 = 323.2\ K$$

$$750.\ mmHg \times \frac{1\ atm}{760.\ mmHg} = 0.987\ atm \qquad 125\ mL \times \frac{1\ L}{1000\ mL} = 0.125\ L$$

$$n_{C_xH_yF_z} = \frac{PV}{RT} = \frac{(0.987\ atm) \times (0.125\ L)}{\left(0.08206 \frac{L \cdot atm}{mol \cdot K}\right) \times (323.2\ K)} = 4.65 \times 10^{-3}\ mol\ C_xH_yF_z$$

$$\text{Molar Mass} = \frac{0.298\ g\ C_xH_yF_z}{4.65 \times 10^{-3}\ mol\ C_xH_yF_z} = 64.1\ g/mol\ C_xH_yF_z$$

(b)

$$37.5\ g\ C \times \frac{1\ mol\ C}{12.0107\ g\ C} = 3.12\ mol\ C \qquad 3.15\ g\ H \times \frac{1\ mol\ H}{1.0079\ g\ H} = 3.13\ mol\ H$$

$$59.3\ g\ F \times \frac{1\ mol\ F}{18.9984\ g\ F} = 3.12\ mol\ F$$

Set up a mole ratio and simplify:

$$3.12\ mol\ C : 3.13\ mol\ H : 3.12\ mol\ F$$

$$1\ C : 1\ H : 1\ F$$

The empirical formula is CHF and the molecular formula is $(CHF)_n$. So, n = x = y = z

The molar mass of the empirical formula = 32.02 g/mol

$$n = \frac{63\ g/mol}{32.02\ g/mol} = 2.0 \qquad \text{The molecular formula is } C_2H_2F_2.$$

(c) $C_2H_2F_2$ can have *cis*- and *trans*-isomers if the F atoms are on different C atoms. (described in Chapter 7)

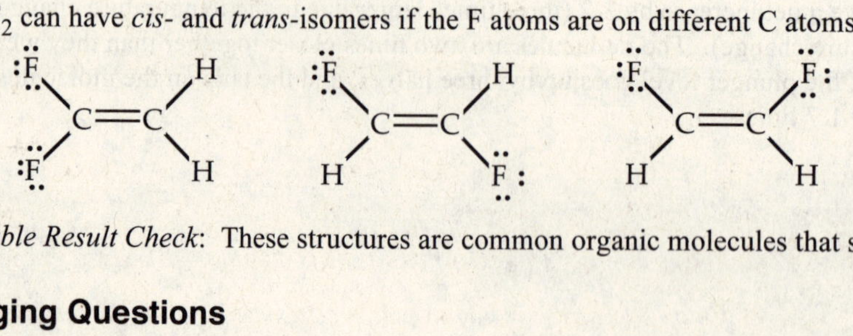

☑ *Reasonable Result Check:* These structures are common organic molecules that satisfy the octet rule.

More Challenging Questions

123. *Result:* **(a) 7.93 × 10⁻⁴ L (b) 0.16**

Analyze: Given the radius of an atom and the equation relating the radius to volume, calculate the volume of one mole of atoms. Given temperature and pressure, calculate the fraction of space occupied by the atoms.

Plan: Use the equation to determine the volume of one atom and Avogadro's number to determine the volume of a mole. Then use the ideal gas law to determine the total volume of the gas and find the fraction.

Execute:

(a) $$V = \frac{4\pi r^3}{3} = \frac{4(3.14159)\left(68\text{ pm} \times \frac{10^{-12}\text{ m}}{1\text{ pm}}\right)^3}{3} = 1.3 \times 10^{-30}\text{ m}^3$$

$$\frac{1.3 \times 10^{-30}\text{ m}^3}{\text{atom}} \times \frac{6.022 \times 10^{23}\text{ atoms}}{1\text{ mol}} \times \frac{1000\text{ L}}{1\text{ m}^3} = 7.9 \times 10^{-4}\ \frac{\text{L}}{\text{mol}}\text{ of Ne}$$

(b) $T = 20.°\text{C} + 273.15\ 293\text{ K}$

$$V = \frac{nRT}{P} = \frac{(1.0\text{ mol}) \times \left(0.08206 \frac{\text{L} \cdot \text{atm}}{\text{mol} \cdot \text{K}}\right)(293\text{ K})}{50.\text{ atm}} = 0.48\text{ L total}$$

$$\text{fraction} = \frac{7.9 \times 10^{-4}\text{ L He}}{0.48\text{ L total}} \times 100\% = 0.16$$

The Ne occupies only 0.16 times the total volume.

☑ *Reasonable Result Check:* The gas molecules occupy a very small fraction of the total space, as predicted by the first postulate of kinetic-molecular theory.

126. *Result:* $\frac{m_{Ne}}{m_{Ar}} = 1.0$

Analyze: Given two known gaseous substances in separate balloons at the same temperature and pressure with one volume designated as double the other, determine the mass ratio of the two substances.

Plan and Execute: Use the ideal gas law and the relationship between mass and moles to determine the mass ratio in terms of volume and molar mass: M=m/n

$$PV = nRT = \left(\frac{m}{M}\right)RT$$

Rearranging to solve for m gives: $$m = \frac{MPV}{RT}$$

At equal T and P: $$\frac{m_2}{m_1} = \frac{M_2 V_2}{M_1 V_1}$$

$$\frac{m_{Ne}}{m_{Ar}} = \frac{M_{Ne} V_{Ne}}{M_{Ar} V_{Ar}}$$

The volume of the neon balloon is twice that of the argon balloon. $V_{Ne} = 2\ V_{Ar}$ so $\frac{V_{Ne}}{V_{Ar}} = 2$

$$\frac{m_{Ne}}{m_{Ar}} = \frac{20.1797\text{ g/mol}}{39.948\text{ g/mol}}(2) = 1.0 = \text{mass ratio}$$

The mass of the neon in the orange balloon is approximately equal the mass of the argon in the blue balloon.

☑ *Reasonable Result Check:* Even though the neon balloon is twice the volume, the heavier argon atoms result in a more massive argon sample.

127. *Result:* **458 torr**

Analyze: Given the initial pressure of a gas, the relative amount of it that undergoes reaction, and the stoichiometric relationship between the reactants and products, determine the new pressure in the container.

Plan: Determine the pressure of the gas reactant that reacts. Use the equation stoichiometry interpreted in units of pressure for the gases to determine pressure of the gas products. Then add this number to the pressure of the unreacted reactant gas to calculate the final pressure.

Execute: Half of the reactant reacts, so the pressure of the reactant that undergoes reaction is:

$$0.5 \times (550.\ \text{torr}) = 275\ \text{torr}.$$

$$275\ \text{torr reactant} \times \frac{2\ \text{torr product}}{3\ \text{torr reactant}} = 183\ \text{torr product}$$

The unreacted reactant has a pressure of 550. torr – 275 torr = 275 torr

The total pressure after this reaction is complete = 275 torr + 183 torr = 458 torr

☑ *Reasonable Result Check:* The reaction reduces the number of gases, so it makes sense that the pressure after some of the reactant has reacted is less.

128. *Result:* **(a) More significant, because of more collisions (b) More significant, because of more collisions (c) Less significant, because the molecules will move faster**

Analyze: Determine the effects of changing temperature and pressures on intermolecular interactions.

Plan and Execute:

(a) When the gas is compressed to a smaller volume, the molecules will be closer together and they will collide with each other more often. The temperature is fixed, so the molecules will hit each other at the same average speed; however, with more collisions the interactions between the molecules will be more significant.

(b) When more molecules of the same gas are added to the container, the molecules will be closer together and they will collide with each other more often. The temperature is fixed, so the molecules will hit each other at the same average speed; however, with more collisions the interactions between the molecules will be more significant.

(c) When the temperature is increased, the average kinetic energy of the molecules is increased and the molecules will move faster. That means they will collide with each other more often. Because the speed has increased the time these molecules spend in proximity to each other will decrease, so, compared to the situation in (a) and (b), the effect of intermolecular interactions will be less significant.

Chapter 9: Liquids, Solids, and Materials

Solutions for Red-Numbered Questions for Review and Thought

Topical Questions

Liquids, Solids, and Intermolecular Forces (Section 9-1)

11. *Result:* **(b) < (a) < (d) < (c); (a) London Forces, (b) London forces, (c) H-bonding, dipole-dipole, and London forces, (d) Dipole-dipole and London forces**

    Strategy: Given four chemical formulas, rank them in order of increasing noncovalent intermolecular attractions.

    Plan: Look for compounds that contain O atom, since these will be polar. Look for compounds that contain O–H groupings, since these can experience hydrogen bonding interactions.

    Execute:

    (a) The compound $CH_3CH_2CH_2CH_3$ is nonpolar and experiences **London forces**.

    (b) The compound CH_3CH_3 is nonpolar and experiences **London forces**.

    (c) The compound $CH_3CH_2CH_2OH$ is polar with an O–H bond, so it experiences **hydrogen bonding, dipole-dipole, and London forces**.

    (d) The compound $CH_3CH_2CH_2OCH_3$ is polar, so experiences **dipole-dipole and London forces**.

    Because the compound in (b) is a smaller molecule than the other three, its London forces are weaker.

    Because the compounds in (b), (c), and (d) are of similar overall size, their London forces have similar strength.

    Of those three, the compound that experiences hydrogen-bonding forces has the greatest noncovalent intermolecular attractions. That is $CH_3CH_2CH_2OH$.

    The compound that can experience dipole-dipole forces as well as London forces has the second largest noncovalent intermolecular attractions. That is $CH_3CH_2CH_2OCH_3$.

    The molecule that experiences only London dispersion forces has the smallest noncovalent intermolecular attractions. That is $CH_3CH_2CH_2CH_3$.

    Therefore, the order of increasing noncovalent intermolecular attractions is **(b) < (a) < (d) < (c).**

Vaporization and Condensation (Section 9-2)

13. *Result/Explanation:* Some molecules have more kinetic energy than the potential energy of the intermolecular attractive forces holding the liquid molecules together. If such a molecule is at the surface of the liquid and moving in the right direction, it will leave the liquid phase and enter the gaseous phase. This describes vaporization at the molecular level.

    Molecules in the gas phase move randomly at various speeds and in every possible direction. That means some of them will eventually impact the surface of the liquid. This can drive the molecule back into the liquid phase where its high kinetic energy is absorbed by several of the molecules near the surface where it hit. The loss of kinetic energy and attractive intermolecular forces causes the molecule to reincorporate into the liquid state. This describes condensation at the molecular level.

15. *Result:* **181 kJ**

    Analyze: Given the vaporization enthalpy of a compound, determine the heat energy required to vaporize a given mass of a compound.

    Plan: Convert the mass to moles, and use the vaporization enthalpy to get total heat energy.

Execute:

$$1.00\text{ kg CCl}_3\text{F} \times \frac{1000\text{ g}}{1\text{ kg}} \times \frac{1\text{ mol CCl}_3\text{F}}{137.368\text{ g CCl}_3\text{F}} \times \frac{24.8\text{ kJ}}{1\text{ mol CCl}_3\text{F}} = 181\text{ kJ}$$

☑ *Reasonable Result Check:* Units cancel appropriately, and it makes sense that a quantity of about seven moles needs about seven times the enthalpy.

17. *Result:* **73.4 kJ**

Analyze: Given the vaporization enthalpy of a compound, determine the heat energy required to vaporize a given mass of a compound.

Plan: Convert the mass to moles, and use the vaporization enthalpy to get total heat energy.

Execute:

$$190.\text{ g C}_4\text{H}_{10} \times \frac{1\text{ mol C}_4\text{H}_{10}}{58.1218\text{ g C}_4\text{H}_{10}} \times \frac{24.3\text{ kJ}}{1\text{ mol C}_4\text{H}_{10}} = 73.4\text{ kJ}$$

☑ *Reasonable Result Check:* Units cancel appropriately, and it makes sense that a quantity of about three moles needs about three times the enthalpy.

Vapor Pressure (Section 9-3)

19. *Result/Explanation:* The molecules in the liquid state must gain sufficient energy to overcome the attractive noncovalent intermolecular forces holding the liquid molecules together in order to enter the gas phase; hence, vaporization is endothermic.

21. *Result/Explanation:* NH_3 has a relatively large boiling point because the molecules interact using relatively strong hydrogen bonding intermolecular forces. The increase in the boiling points of the series PH_3, AsH_3, and SbH_3 is related to the increasing London forces experienced due to the larger number of electrons in the molecule and thus a greater ability to be polarized. ($P < As < Sb$)

23. *Result/Explanation:* Methanol molecules are capable of hydrogen bonding, whereas formaldehyde molecules use dipole-dipole forces to interact. Molecules experiencing **stronger intermolecular forces** (such as methanol) will **have higher boiling points and lower vapor pressures** compared to molecules experiencing weaker intermolecular forces (such as formaldehyde), because a greater temperature is required to provide the energy required to disrupt those forces and permit the molecules to enter the gas phase.

24. *Result:* **0.21 atm, approximately 57 °C**

Analyze: Given the altitude on a mountain above sea level, a simple relationship between altitude and pressure changes, and Figure 9.4, determine the atmospheric pressure on the mountain and the boiling point of water at that altitude.

Plan: Use the altitude/pressure relationship to get the atmospheric pressure on the mountain, then look up the boiling temperature on the graph.

$$22834\text{ ft} \times \frac{3.5\text{ mbar decrease}}{100\text{ ft}} \times \frac{1\text{ bar}}{1000\text{ mbar}} \times \frac{10^5\text{ Pa}}{1\text{ bar}} \times \frac{1\text{ kPa}}{1000\text{ Pa}} \times \frac{1\text{ atm}}{101.325\text{ kPa}} = 0.79\text{ atm decrease}$$

$$1.00\text{ atm} - 0.79\text{ atm} = \mathbf{0.21\text{ atm}}$$

$$0.21\text{ atm} \times \frac{760\text{ mm Hg}}{1\text{ atm}} = 160\text{ mm Hg}$$

Looking at Figure 9.4, this corresponds to a boiling temperature between 40 °C and 60 °C, about **57 °C**.

☑ *Reasonable Result Check:* It makes sense that a lower pressure causes a liquid to have a lower boiling point.

26. *Result:* **Substance D; it has lowest vapor pressure**

Explanation: Vapor pressure increases as more molecules are able to escape the liquid state, so the substance with the **greatest intermolecular attractive forces** at a given temperature is the substance with the **lowest**

vapor pressure. Looking at the graph at 25°C, curve D has the lowest vapor pressure, so **substance D** has the greatest intermolecular attractive forces at 25°C (or any other given temperature) than A, B, or C.

27. *Result:* **5 × 10² mm Hg**

Analyze: Given the vaporization enthalpy and the boiling point at a given pressure, determine the vapor pressure at a given temperature.

Plan: Use the two-set Clausius-Claypeyron equation (from Section 9-3b): $\ln\left(\frac{P_2}{P_1}\right)=\frac{-\Delta_{vap}H}{R}\left(\frac{1}{T_2}-\frac{1}{T_1}\right)$

R is given in Table 8.3.

Execute: $T_1 = 110°C + 273.15 = 380.15\text{ K} \cong 380\text{ K}$ *(round to 10s place)*

$$P_1 = 1\text{ atm} = 760.0\text{ mm Hg}$$

$$T_2 = 97°C + 273.15 = 370.15\text{ K} \cong 370.\text{ K}$$

$$\ln\left(\frac{P_2}{760.0\text{ mmHg}}\right)=\frac{-(38.7\text{ kJ/mol})}{\left(8.314\frac{\text{J}}{\text{K}\cdot\text{mol}}\right)\left(\frac{1\text{ kJ}}{1000\text{ J}}\right)}\left(\frac{1}{370.\text{ K}}-\frac{1}{380\text{ K}}\right)$$

$$\ln\left(\frac{P_2}{760.0\text{ mmHg}}\right)=\frac{-(38.7)}{(0.008314)}(0.00270-0.0026)=\frac{-(38.7)}{(0.008314)}(0.0001)$$

$$\ln\left(\frac{P_2}{760.0\text{ mmHg}}\right)=-0.4$$

$$P_2=(760.0\text{ mmHg})e^{-0.4}=\mathbf{5\times10^2\text{ mm Hg}}$$

☑ *Reasonable Result Check:* Since the temperature was lower than the boiling temperature, the vapor pressure should be less than the atmospheric pressure.

29. *Result:* **39.4 kJ/mol**

Analyze: Given the boiling point and the vapor pressure at a given temperature, find the vaporization enthalpy.

Plan: Use the two-set Clausius-Claypeyron equation (given above in the solution to Question 27)

R is given in Table 8.3. $T_1 = 50.0°C + 273.15 = 323.2\text{ K}$ $T_2 = 78.3°C + 273.15 = 351.5\text{ K}$

$$\ln\left(\frac{760.0\text{ mmHg}}{233\text{ mmHg}}\right)=\frac{-\Delta_{vap}H}{\left(8.314\frac{\text{J}}{\text{K}\cdot\text{mol}}\right)\left(\frac{1\text{ kJ}}{1000\text{ J}}\right)}\left(\frac{1}{351.5\text{ K}}-\frac{1}{323.5\text{ K}}\right)$$

$$\ln(3.26)=\frac{-\Delta_{vap}H}{\left(0.008314\frac{\text{kJ}}{\text{mol}}\right)}(0.002845-0.003095)=\frac{-\Delta_{vap}H}{\left(0.008314\frac{\text{kJ}}{\text{mol}}\right)}(-0.000249)$$

$$1.18=-\Delta_{vap}H\left(-0.0300\frac{\text{mol}}{\text{kJ}}\right)$$

$$39.4\frac{\text{kJ}}{\text{mol}}=\Delta_{vap}H$$

☑ *Reasonable Result Check:* The vaporization enthalpy must be positive, since energy is required to take a liquid into the gas state. The size is similar to the vaporization enthalpies given in earlier questions.

Solids and Changes of Phase (Section 9-4)

31. *Result/Explanation:* A low fusion enthalpy tells us that not much energy is required to melt a solid. That is the case when the intermolecular interactions between the particles in the solid are weak, such as in **molecular solids** composed of small nonpolar molecules or atoms **held together with weak noncovalent intermolecular forces.**

33. *Result:* **I_2; stronger London dispersion forces**

Explanation: A higher fusion enthalpy occurs when the intermolecular forces are stronger. Because I_2 is larger than N_2, we expect the **London forces to be stronger in I_2** than in N_2, giving I_2 a higher fusion enthalpy.

35. *Result:* **27 kJ**

Analyze: Given the fusion and vaporization enthalpies of a compound, determine the heat energy required to raise a given number of moles of solid to the melting point, melt it, raise the liquid to the boiling point and boil it.

Plan and Execute: Follow the plan described for the heating curve in Section 9-4c. Add together the heat energy required for each stage to get the total heat energy.

$$0.50 \text{ mol } H_2O \times \frac{18.0152 \text{ g } H_2O}{1 \text{ mol } H_2O} = 9.0 \text{ g } H_2O$$

Warm the ice to the melting point (0 °C), melt the ice, warm the water to the boiling point (100. °C), and vaporize the water. Total heat energy is the sum of the heat energies to required for each transition:

$$q_{tot} = q_{heat\ ice} + q_{melt} + q_{heat\ water} + q_{boil}$$

$$q_{tot} = (c_{ice} \times m \times \Delta T) + (n\Delta_{fus}H) + (c_{liquid} \times m \times \Delta T) + (n\Delta_{vap}H)$$

$$q_{tot} = \left(\frac{2.06 \text{ J}}{\text{g °C}}\right) \times \left(\frac{1 \text{ kJ}}{1000 \text{ J}}\right) \times (9.0 \text{ g}) \times [0 \text{ °C} - (-5 \text{ °C})] \ + \ (0.50 \text{ mol}) \times \left(\frac{6.020 \text{ kJ}}{\text{mol}}\right)$$

$$+ \left(\frac{4.184 \text{ J}}{\text{g °C}}\right) \times \left(\frac{1 \text{ kJ}}{1000 \text{ J}}\right) \times (9.0 \text{ g}) \times [100. \text{ °C} - (0 \text{ °C})] \ + \ (0.50 \text{ mol}) \times \left(\frac{40.07 \text{ kJ}}{\text{mol}}\right)$$

$$q_{tot} = 0.09 \text{ kJ} + 3.0 \text{ kJ} + 3.8 \text{ kJ} + 20. \text{ kJ} = 27 \text{ kJ}$$

☑ *Reasonable Result Check:* The relative size of the four terms seems sensible, comparing the sizes of the heat capacities and the phase transition enthalpies.

37. *Result:* **51.9 g CCl_2F_2**

Analyze: Given the vaporization enthalpy of a compound, the fusion enthalpy for ice, and the specific heat capacity of water, determine what mass of the liquid compound must evaporate to lower the temperature of a sample of water to the freezing point and freeze it.

Plan and Execute: Adapt the plan described for the heating curve in Section 9-4c. Add together the heat energy required for each stage to get the total heat energy. Use the vaporization enthalpy of the compound to determine the mass.

$$2.00 \text{ mol } H_2O \times \frac{18.0152 \text{ g } H_2O}{1 \text{ mol } H_2O} = 36.0 \text{ g } H_2O$$

Total heat energy for freezing the water is the sum of the heat energy to cool the ice to the freezing point (0 °C) and the heat energy required to freeze it.

$$q_{tot\ for\ water} = q_{cool\ ice} + q_{freeze}$$

$$q_{tot\ for\ water} = (c_{water} \times m \times \Delta T) + (n\Delta_{cryst}H)$$

$$\Delta_{cryst}H = -\Delta_{fus}H_{fus} = -6.02 \text{ kJ/mol}$$

$$q_{\text{tot for water}} = \left(\frac{4.184\ \text{J}}{\text{g}\ °\text{C}}\right) \times \left(\frac{1\ \text{kJ}}{1000\ \text{J}}\right) \times (36.0\ \text{g}) \times [0\ °\text{C} - (-20.\ °\text{C})] \ + \ (2.00\ \text{mol}) \times \left(\frac{-6.02\ \text{kJ}}{\text{mol}}\right)$$

$$q_{\text{tot for water}} = -3.0\ \text{kJ} + (-12.0\ \text{kJ}) = -15.0\ \text{kJ}$$

15.0 kJ must be removed to freeze the water, so determine how many grams of CCl_2F_2 must evaporate to use up the 15.0 kJ of thermal energy.

$$15.0\ \text{kJ} \times \frac{1000\ \text{J}}{1\ \text{kJ}} \times \frac{1\ \text{g}\ CCl_2F_2}{289\ \text{J}} = 51.9\ \text{g}\ CCl_2F_2 \text{ must evaporate}$$

☑ *Reasonable Result Check:* The relative size of the two terms seems sensible, comparing the sizes of the heat capacity and the heat energy of the phase transition. The calculated mass of CCl_2F_2 would definitely fit inside a typical freezer compressor.

38. *Result/Explanation:* A higher melting point is a result of stronger interparticle forces. Coulomb's law describes the attraction between charged particles; the ion-ion coulombic interaction is stronger with higher ionic charges such as in MgO (with Mg^{2+} and O^{2-}). Interionic attractions in the solid MgO are much stronger than those in solid NaF (with Na^+ and F^-), so MgO has a higher melting point.

40. *Result:* **Highest melting point is (b) CaO. Lowest melting point is (c) CO.**

Explanation: **The highest melting point is a result of strongest interparticle forces**. Ionic interactions are the strongest individual forces in solids. Looking at the two ionic compounds LiBr and CaO, **CaO has higher ionic charges** (Ca^{2+} and O^{2-} verses Li^+ and Br^-). Coulomb's law describes the attraction between charged particles; the ion-ion coulombic interaction is stronger with higher ionic charges. Interionic attractions in the solid CaO are stronger than those in solid LiBr, so (b) CaO has a higher melting point.

The lowest melting point is a result of the weakest intermolecular forces. We need to compare the intermolecular forces in the two molecules, CO and CH_3OH. The strongest forces that CO molecules can experience are dipole-dipole forces. The strongest forces that CH_3OH molecules experience are hydrogen bonding. **Dipole-dipole intermolecular forces are weaker than hydrogen bonding interactions, so (c) CO has the lowest melting point.**

42. *Result/Explanation:* The freezer compartment of a frost-free refrigerator keeps the air so cold and dry that any **ice stored inside the freezer compartment undergoes sublimation**, the direct conversion of solid to gaseous form, at normal pressures. The ice in **the hailstones would eventually disappear** unless kept in a tightly sealed container.

43. *Result:* **(a) A = solid, B = liquid, C = gas (b) Point 1: solid and gas; point 2: solid, liquid and gas; point 3: liquid and gas; point 5: solid and liquid**

Explanation:

(a) The material has to be in the solid state at very low temperatures and very high pressures, so the area of the phase diagram identified by A represents the solid phase. The material has to be in the gas phase at very high temperatures and very low pressures, so the area of the phase diagram identified by C represents the gas phase. At other pressures and temperatures, the material can be in the liquid state, that leaves the area of the phase diagram identified by B represents the liquid phase.

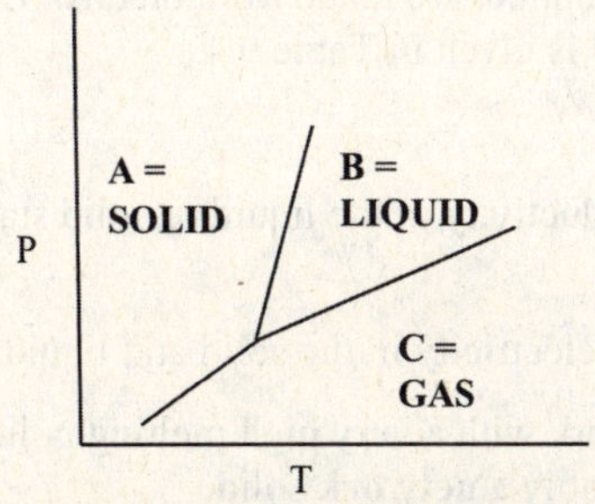

(b) The phases in equilibrium at points along the lines of the phase diagram are those that border the point. At point 1, the phases present will be solid and gas. At point 2 (the triple point), all three phases will be present, solid liquid and gas. At point 3, the phases present would be liquid and gas. At point 5, the phases present will be solid and liquid.

45. *Result:* **Ideal gas has density of 0.13 g/mL; much smaller than the real density of CO_2 (0.47 g/mL).**

Analyze: Using the phase diagram for CO_2 to get the temperature and pressure of the critical point. Calculate the density of an ideal gas with conditions the same as the critical point, then calculate the actual density of a real CO_2 gas from the given relationship between quantity (in moles) and V.

Plan: Use methods described in Section 8-4 to find the density of an ideal gas at a given temperature and pressure. Convert that number to g/cm^3. Use the moles and volume of the real gas to find the density.

Execute: Figure 9.14 shows the phase diagram for CO_2 with a triple point at 73 atm and 31 °C.

31 °C + 273.15 = 304 K

$$d = \frac{PM}{RT} = \frac{(73\text{ atm}) \times \left(44.0095 \frac{\text{g}}{\text{mol}}\right)}{\left(0.08206 \frac{\text{L}\cdot\text{atm}}{\text{mol}\cdot\text{K}}\right) \times (304\text{ K})} = 130\ \frac{\text{g}}{\text{L}}$$

$$\frac{130\text{ g}}{1\text{ L}} \times \frac{1\text{ L}}{1000\text{ mL}} \times \frac{1\text{ mL}}{1\text{ cm}^3} = 0.13\ \frac{\text{g}}{\text{cm}^3} = \text{density of an ideal gas}$$

$$\frac{1\text{ mol } CO_2}{94\text{ cm}^3} \times \frac{44.0095\text{ g } CO_2}{1\text{ mol } CO_2} = 0.47\ \frac{\text{g}}{\text{cm}^3} = \text{density of } CO_2$$

☑ *Reasonable Result Check:* The triple point conditions are not ideal gas conditions. Especially the extreme pressure makes it easy to believe that CO_2 under these conditions will not be ideal, so the much larger real density compared to the ideal gas density makes sense.

46. *Result:* **(a) ionic solid (b) molecular solid (c) amorphous solid (vitreous silica) or network solid (quartz) (d) network solid**

Explanation: Using characteristics described in the beginning of Section 9-4:

(a) KF is composed of common ions, K^+ and F^-, so it is an **ionic solid.**

(b) I_2 is a nonpolar molecule, so it is a **molecular solid.**

(c) SiO_2 comes in two forms. Vitreous silica is an **amorphous solid** (Section 9-11c), while quartz is a **network solid** (Section 9-7).

(d) BN is a nonoxide ceramic described in Section 9-11b. It is a **network solid.**

48. *Result:* **(a) molecular or network solid (b) metallic solid (c) network solid (d) molecular solid; see explanations below.**

Analyze and Plan: Pure substances have fixed melting temperatures, whereas mixtures and amorphous solids have ill-defined melting points. Network and ionic solids have much higher melting points, because of stronger interparticle interactions. Solids that conduct are made from metals. Liquids that conduct could be metals or ions. A summary of various properties is given in Table 9.4.

Execute:

(a) A brittle yellow solid with no conductivity in the liquid or solid state is probably a **molecular or network** solid.

(b) A soft, silvery solid that conducts electricity in the solid and liquid states is a **metallic** solid.

(c) A hard, colorless crystalline network with a very high melting point, with neither the solid nor the liquid able to conduct electricity, is probably a **network** solid.

(d) A soft slippery solid that has a definite melting point but does not conduct electricity is probably a **molecular** solid.

50. *Result:* **(a) molecular solid (b) ionic solid (c) metallic solid or network solid (d) amorphous solid**

Analyze and Plan: Use the characterizations described in the plan for Question 48:

Execute:

(a) A solid that melts below 100 °C and is insoluble in water is probably a nonpolar **molecular** solid.

(b) An **ionic** solid will conduct electricity only when melted.

(c) A solid that is insoluble in water and conducts electricity is probably a metallic solid, though it might be a **network** solid like graphite.

(d) A noncrystalline solid that has a wide melting point range is an **amorphous** solid.

Water: Its Important and Unusual Properties (Section 9-5)

52. *Result/Explanation:* The crystal structure in ice maximizes the hydrogen-bonding capacity and leaves considerable open spaces completed to the liquid. See discussion in Section 9-5a for more details.

54. *Result/Explanation:* Surface tension is based on the ability of a liquid to interact with other particles in the liquid. At higher temperatures, the molecules move around more. The increased random motion disrupts the intermolecular interactions responsible for surface tension.

55. *Result/Explanation:* Water molecules interact using relatively strong hydrogen-bonding intermolecular forces, so greater kinetic energy is needed for molecules to escape.

Crystalline Solids (Section 9-6)

58. *Result/Explanation:* See Figure 9.22 and its description.

59. *Result:* **220 pm**

Analyze: Given the length of the edge of a face-centered cubic (fcc) unit cell of xenon, determine the radius of the xenon atoms.

In Section 9-6b, we see that the fcc face diagonal distance is the hypotenuse of an equilateral triangle: Diagonal distance $= \sqrt{2} \times (\text{edge})$. We can see from the fcc figure drawn below that the diagonal distance represents four times the radius of the atom.

Use these two relationships to find the radius of the atom.

$$\text{Diagonal distance} = \sqrt{2} \times (\text{edge length}) = \sqrt{2} \times (620 \text{ pm}) = 880 \text{ pm}$$

$$\text{Radius} = \frac{\text{Diagonal distance}}{4} = \frac{880 \text{ pm}}{4} = 220 \text{ pm}$$

☑ *Reasonable Result Check:* The geometric relationships are logical. We expect the radius to be less than the edge length. The van der Waals radius for Xenon is 216 pm.

61. *Result/Explanation:* Examining Figure 9.24:

Four unit cells (including the pictured cell) share **each of the Na^+ ions** in the front face of the cell.

Eight unit cells (including the pictured cell) share the **corner Cl^- ion** in the front face of the cell.

Two unit cells (including the pictured cell) share the **face-centered Cl^- ion** in the front face of the cell.

63. *Result:* **No; the ratio of the ions in the unit cell must reflect the empirical formula of the compound.**

Explanation: The ratio of ions in the unit cell must reflect the empirical formula of the compound. Here, the compound $CaCl_2$ has a 1:2 ratio between Ca^{2+} and Cl^-. The compound NaCl has a 1:1 ratio between the Na^+ and Cl^- ions, so it is not possible to have the same structure.

65. *Result:* **0.533 g/cm³**

Analyze: Given the lengths of the edge of the unit cells for solid lithium metal, the type of structure, and the temperature, determine the density of the metal at this temperature.

Plan: Find the edge length in units of centimeters so that the density will come out in g/cm³. Determine the volume of the unit cell, and the mass of the atoms in that cell. Then, divide mass by volume to get density.

Execute: $$351\text{ pm} \times \frac{10^{-12}\text{m}}{1\text{ pm}} \times \frac{1\text{ cm}}{10^{-2}} = 3.51 \times 10^{-8}\text{ cm}$$

$$V = (\text{edge}) = (3.81 \times 10^{-8}\text{ cm})^3 = 4.32 \times 10^{-23}\text{ cm}^3$$

In Section 9-6a we learn that a body-centered cubic unit cell has 2 atoms in it.

$$2\text{ atoms Li} \times \frac{1\text{ mole Li}}{6.022 \times 10^{23}\text{ atoms Li}} \times \frac{6.941\text{ g Li}}{1\text{ mole Li}} = 2.305 \times 10^{-23}\text{ g Li}$$

$$\text{Density} = \frac{2.305 \times 10^{-23}\text{ g}}{4.32 \times 10^{-23}\text{ cm}^3} = 0.533\text{ g/cm}^3$$

☑ *Reasonable Result Check:* Lithium has a small molar mass. Table 1.1 shows various metals (not Li) and the metals with small molar masses have smaller densities.

Network Solids (Section 9-7)

67. *Result/Explanation:* Carbon atoms in diamond are sp^3 hybridized and are tetrahedrally bonded to four other carbon atoms. Carbon atoms in pure graphite are sp^2 hybridized and bonded with a trianglar planar shape to other carbon atoms. These bonds are partially double bonded so they are shorter than the single bonds in diamond. However, the planar sheets of sp^2 hybridized carbon atoms in graphite are only weakly attracted by intermolecular forces to adjacent layers, so these **interplanar distances in graphite are much longer than the C–C single bonds in the diamond**. The net result is that graphite is less dense than diamond.

69. *Result/Explanation:* Some examples of network solids are:

Graphite: A planar network solid. It is insoluble in water and nonpolar solvents.

Diamond: A three-dimensional network solid. It is insoluble in all solvents.

All of these solids are huge molecules held together by covalent bonds. Covalent bonds are much stronger than the intermolecular forces that can be formed with the solvent. **The energy released on mixing is insufficient to overcome the energy of these intramolecular bonds**. As a result, it is not surprising to find that **network solids are insoluble**.

Tools of Chemistry: X-Ray Crystallography

71. *Result:* **ν = 5.30 × 10¹⁷ s⁻¹; (a) 3.51 × 10⁻¹⁶ J for one photon (b) 2.11 × 10⁸ J; x-ray**

Analyze: Using the length of the edge of a unit cell as the wavelength of light, determine the frequency of light, the energy per photon, and the energy per mole.

Plan: Use methods described in Chapter 5 (Sections 5.1 and 5.2).

In Problem-Solving Example 9.9, the length of a unit cell of NaCl is found to be 566 pm. Use that as the wavelength for the electromagnetic radiation, and calculate the frequency.

Execute: $$\nu = \frac{c}{\lambda} = \frac{2.998 \times 10^8\ \frac{\text{m}}{\text{s}}}{566\text{ pm} \times \frac{10^{-12}\text{ m}}{1\text{ pm}}} = 5.30 \times 10^{17}\text{ s}^{-1}$$

(a) $$E = h\nu = (6.626 \times 10^{-34}\text{ J·s}) \times (5.30 \times 10^{17}\text{ s}^{-1}) = 3.51 \times 10^{-16}\text{ J for one photon}$$

(b) $$1 \text{ mol photon} \times \frac{3.51 \times 10^{-16} \text{ J}}{1 \text{ photon}} \times \frac{6.022 \times 10^{23} \text{ photons}}{1 \text{ mol photon}} = 2.11 \times 10^{8} \text{ J}$$

This photon is in the X-ray region of the electromagnetic spectrum.

☑ *Reasonable Result Check:* It makes sense that the electromagnetic radiation is an X-ray, since X-ray crystallography uses X-rays to examine atom arrangement in crystals.

73. *Result:* **361 pm**

Analyze: Given the second-order (n = 2) Bragg reflection angle and the wavelength of the X-ray beam, determine the spacing between the planes of the atoms in a metallic crystal.

Plan: Use methods described in the "Tools of Chemistry" about X-ray Crystallography in Section 9-7a:

$$n\lambda = 2d \sin\theta$$

Execute: The metal is copper. It is second order (n = 2). The angle of scattering (θ) is 27.35°. The wavelength (λ) is 166 pm.

$$d = \frac{n\lambda}{2\sin\theta} = \frac{(2)(166 \text{ pm})}{2\sin(27.35^\circ)} = 361 \text{ pm}$$

☑ *Reasonable Result Check:* The atomic radius of copper is 126 pm (see Chapter 5, Figure 5.26), and this distance is somewhat more than twice that number.

Material Science (Section 9-9)

75. *Result/Explanation:* The four major classes of materials are: **metals** (such as **iron**), **ceramics** (such as **brick**), **polymer** (such as **wool**), **composites** (such as **fibre-reinforced polymer**). Other examples are possible.

Metals, Semiconductors, and Insulators (Section 9-9)

78. *Result/Explanation:* **In a conductor, the valence band is only partially filled**, whereas, **in an insulator, the valence band is completely full**, the conduction band is empty, and there is a wide energy gap between the two. **In a semiconductor, the gap between the valence band and the conduction band is very small** so that electrons are easily excited into the conduction band.

80. *Result/Explanation:* Substance **(c) Ag** has the greatest electrical conductivity because **it is a metal**. Substance **(d) P_4** has the smallest electrical conductivity because **it is a nonmetal**. (The other two are metalloids.)

82. *Result/Explanation:* A superconductor is a substance that is able to conduct electricity with no resistance. Two examples are found in Section 9-9b: $LaBa_2Cu_3O_x$ and $Hg_{0.8}Tl_{0.2}Ba_2Ca_2Cu_3O_{8.23}$. The transition temperatures are 35 K for the lanthanum-barium-copper oxide compound and 138 K for the mercury compound.

Silicon and the Chip (Section 9-10)

84. Doping is described in Section 9-10. It is the intentional addition of small amounts of specific impurities into very pure silicon. **Group III elements are used because they have one less electron per atom than the group IV silicon. Group V elements are used because they have one more electron per atom.**

Cement, Ceramics, and Glass (Section 9-11)

87. *Result/Explanation:* Amorphous solids are compared to crystalline solids in Section 9-4 and glasses are discussed in more detail in Section 9-11c. The amorphous solids known as glasses are different from crystalline SiO_2, because they lack symmetry or long-range order, whereas ionic solids such as crystalline SiO_2 are extremely symmetrical. SiO_2 must be heated to melting temperatures, then cooled very slowly, to make a glass.

89. *Result/Explanation:* Ceramics are described in Section 9-11.

(a) Two examples of silicate ceramics are **aluminosilicates**, such as kaolinite, **and calcium silicate**, Ca_2SiO_4

(b) Two examples of oxide ceramics: **Al_2O_3 and MgO**.

(c) Two examples of nonoxide ceramics: **Si_3N_2 and SiC**.

Other correct answers are possible.

General Questions

90. *Result/Explanation:* With the lid on, hot water vapor is not able to escape the pan, so **the temperature and internal pressure increase faster, reaching a higher vapor pressure more quickly causing more vigorous boiling**.

93. *Result/Explanation:* The given pressure, 1.75 atm, is higher than 1 atm. The given temperature, –112°C, is above the triple point temperature, –121°C, and below the boiling point, –108°C. Therefore, xenon is probably in the **liquid** phase.

95. *Result:* **(a) 560 mm Hg (b) benzene (c) 73°C (d) methyl ethyl ether is 7°C; carbon disulfide is 47°C; benzene is 81°C (these are all approximate, due to the size of the graph provided)**

Explanation: Results to (a), (c) and (d) are obtained from the graph, by locating the given information on the appropriate axis and following that horizontal or vertical to the point on the appropriate curve.

(a) Finding 0°C on the horizontal axis and tracing a vertical line to the methyl ethyl ether curve shows that the vapor pressure of methyl ethyl ether at 0°C is approximately **560 mm Hg**.

(b) A substance will remain in the liquid state at higher temperatures if its molecules experience higher intermolecular forces; hence, its vapor pressure will be lower. Therefore, of the three liquids compared here, molecules in **benzene** experience the greatest intermolecular forces.

(c) Finding 600 mm Hg on the vertical axis and tracing a horizontal line to the benzene curve gives a temperature of approximately **73°C**.

(d) Normal boiling point is the boiling temperature at 1 atm (760 mm Hg). Finding 760 mm Hg on the vertical axis and tracing horizontal lines to each curve gives the following normal boiling points: for methyl ethyl ether, approximately **7°C**; for carbon disulfide, approximately **47°C**, for benzene, approximately **81°C**.

97. *Result:* **2.18 kJ**

Analyze: Given the normal boiling point, vaporization enthalpy, and the specific heat capacities of the gas and liquid states of a compound, determine the heat energy evolved when a given mass of the substance is cooled from an initial to a final temperature.

Plan: Use the enthalpies to get total heat energy transferred during the phase transition. Use Equation 4.2' for the temperature changes. Add together the heat energy for each stage to get the total quantity of heat energy.

Heat energy is evolved when cooling the *gas* from $T_i = 40°C$ to the boiling point ($T_f = -30°C$). Heat energy is evolved when condensing the gas to a liquid. Heat energy evolved when cooling the *liquid* from $T_i = -30°C$ to the final temperature ($T_f = -40°C$). The total heat energy evolved is the sum of these three steps.

A change in temperature in Kelvin degrees is the same as the change in temperature in Celsius degrees. So, use $c_{gas} = 0.61\ J\ g^{-1}\ °C^{-1}$ and $c_{liquid} = 0.97\ J\ g^{-1}\ °C^{-1}$ *(Let's assume T's are ± 1 °C.)*

$$q_{tot} = (c_{gas} \times m \times \Delta T) + (m \times \Delta H_{vap}) + (c_{liquid} \times m \times \Delta T)$$

Execute:
$$q_{tot} = \left(\frac{0.61\ J}{g \cdot °C}\right) \times (10.0\ g) \times (-30.\ °C - 40.\ °C) + (10.0\ g) \times \left(\frac{165\ J}{g}\right)$$
$$+ \left(\frac{0.97\ J}{g \cdot °C}\right) \times (10.0\ g) \times [-40.\ °C - (-30.\ °C)]$$

$$q_{tot} = [-430\ J + (-1650\ J) + (-97\ J)] \times \left(\frac{1\ kJ}{1000\ J}\right) = -2.18\ kJ$$

A negative q_{tot} means that heat energy is evolved. The amount of **heat energy evolved is 2.18 kJ**.

☑ *Reasonable Result Check:* The relative size of the three terms seems sensible, comparing the sizes of the heat capacities and the vaporization enthalpy.

99. *Result:* **Dipole-dipole forces and London forces**

Explanation: The molecular geometry of the SO_2 molecule is determined by determining the Lewis structure and using VSEPR theory:

SO_2 (18 e^-) The type is AX_2E_1, so the electron-pair geometry is triangular planar and the molecular geometry is angular (120°).

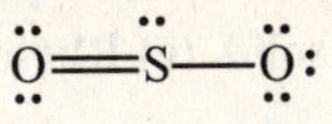

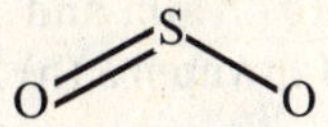

The asymmetric shape of the SO_2 molecule means that it is polar. Therefore, the strongest intermolecular forces between molecules in solid and liquid SO_2 are dipole-dipole forces. All molecules experience London forces, so London forces are part of the attractions in the liquid and solid states of SO_2, also.

Applying Concepts

103. *Result:* **(a) Ionic; solubility in water and high density suggests large heavy ions. (b) Molecular; solubility in benzene and low density suggests it is a nonpolar molecular compound.**

Analyze: Given physical characteristics of two compounds, determine if they are ionic or molecular compounds.

Plan: Compare the properties with characteristics described in Section 9-4 and Table 9.4.

Execute:

(a) Orpiment has a high density and solubility in water and basic solution consistent with a compound containing large ions. Thus, we predict it is an **ionic compound**.

(b) Zeaxanthin has a low density and solubility in benzene consistent with a nonpolar organic compound. Thus, we predict it is a **molecular compound**.

107. *Result/Explanation:* Vapor-phase water condenses on contact with cool skin. The condensation of steam into water is exothermic. That energy is also absorbed by the skin causing more burning, along with the burn resulting from the heat energy given off as the temperature drops (common in each case).

110. *Result/Explanation:* The butane in the lighter is under great enough pressure that the vapor pressure of butane at room temperature is less than the pressure inside the lighter. Hence, it exists as a liquid.

112. *Result:* **Diagram 1 and region C; diagram 2 and line E, diagram 3 and region B, diagram 4 and line F, diagram 5 and line G, diagram 6 and point H, diagram 7 and region A**

Explanation: Refer to Section 9-4d, if you need to be reminded about the different parts of a phase diagram.

Nanoscale **diagram 1** looks like the substance atoms are all in the gas phase. That is **region C** on the phase diagram.

Nanoscale **diagram 2** looks like some of the substance atoms are in the solid phase (atoms piled up in a regular array) and some of the substance atoms are in the gas phase. That is a point on the line described as **line E** on the phase diagram.

Nanoscale **diagram 3** looks like the substance atoms are all in the liquid phase. That is **region B** on the phase diagram.

Nanoscale **diagram 4** looks like some of the substance atoms are in the solid phase and some of the substance atoms are in the liquid phase. That is a point on the line described as **line F** on the phase diagram.

Nanoscale **diagram 5** looks like some of the substance atoms are in the liquid phase and some of the substance atoms are in the gas phase. That is a point on the line described as **line G** on the phase diagram.

Nanoscale **diagram 6** looks like some of the substance atoms are in the liquid phase, some of the substance atoms are in the liquid phase, and some of the substance atoms are in the gas phase. That is the point described as **point H** on the phase diagram.

Nanoscale **diagram 7** looks like the substance atoms are all in the solid phase. That is **region A** on the phase diagram.

114. *Result/Explanation:* Each has the same fraction of filled space. The fraction of spaces filled by closest packed equal-sized spheres is the same, no matter what the size of the spheres.

More Challenging Questions

116. *Result:* **(a) Two (b) Three (c) At 1 atm and 80 °C it is a rhombic solid. At 1 atm and 125 °C it is a liquid. (d) At low pressures, it is a gas, between 5 × 10^{-4} atm and 10^3 atm it is a liquid, and above 10^3 atm it is a rhombic solid. (e) P < 10^{-4} atm and T < 120 °C (f) Monoclinic solid (g) Rhombic to Monoclinic to monoclinic to liquid to vapor. (h) NMP = 92°C**

Analyze: Given a phase diagram for sulfur, interpret the diagram to answer various questions about the physical state of sulfur under various conditions.

Plan and Execute: Locate points on the phase diagram to answer the related questions.

(a) There are **two** solid phases, **rhombic and monoclinic**.

(b) There are **three** triple points, labeled B, C, and E.

(c) At 1 atm pressure and 80°C, sulfur is rhombic. At 1 atm and 125°C sulfur is liquid. Shown here:

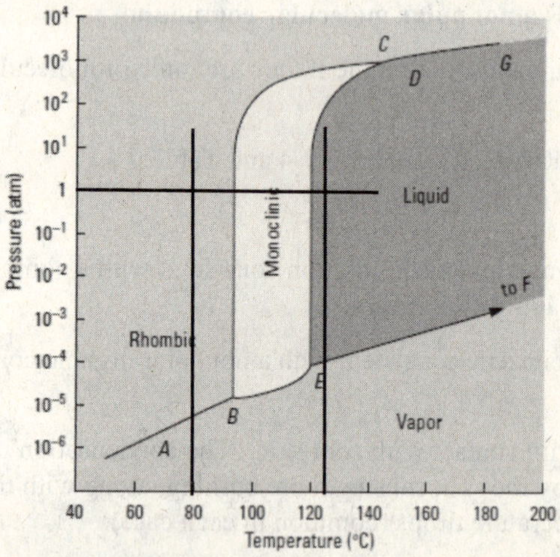

(d) At 151°C and pressures lower than 5 × 10^{-4} atm, sulfur is a gas, between 5 × 10^{-4} atm and 10^3 atm, it is a liquid, and above 10^3 atm it is a rhombic solid, as shown below:

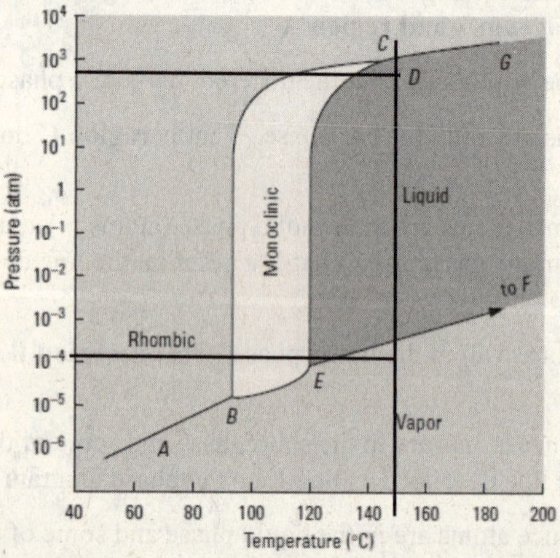

(e) Sulfur will sublime at a temperature and pressure where there is a solid/gas line on the phase diagram, when P < 10^{-4} atm and T < 120 °C.

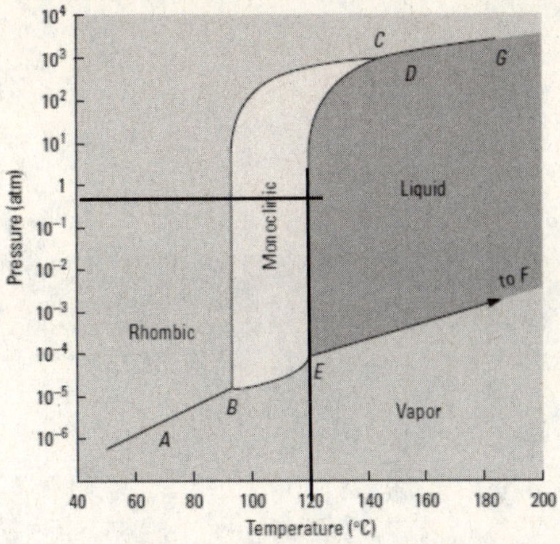

(f) The most stable phase at 1 atm and 100°C is monoclinic solid:

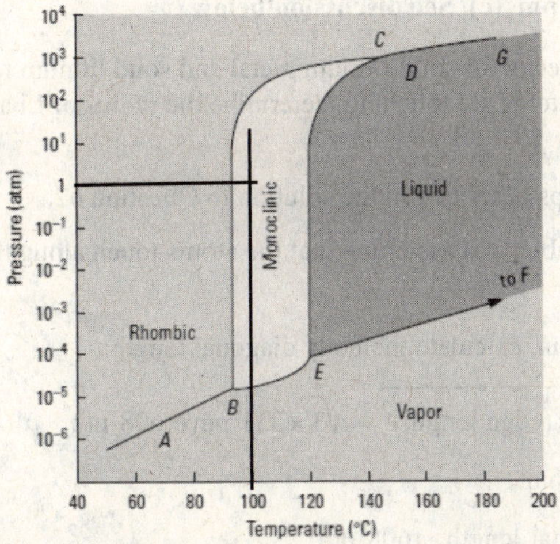

(g) In sequence from 50°C to 200°C, the solid starts as a rhombic solid, then changes to a monoclinic solid, then to a liquid, then to a vapor, as shown below:

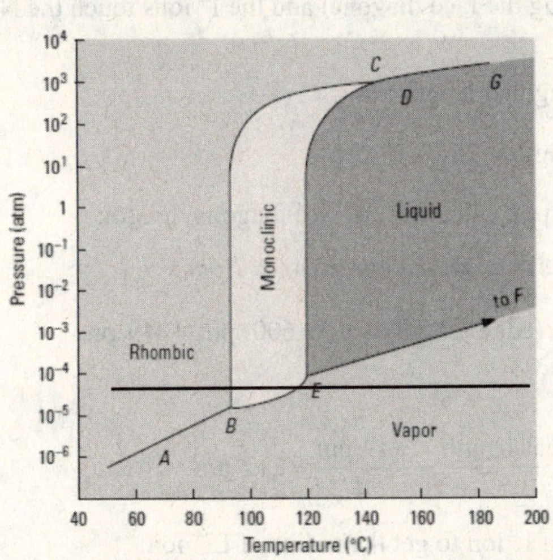

(h) The Normal Melting Point, the point where the solid melts to liquid at 1 atm pressure, is 120°C.

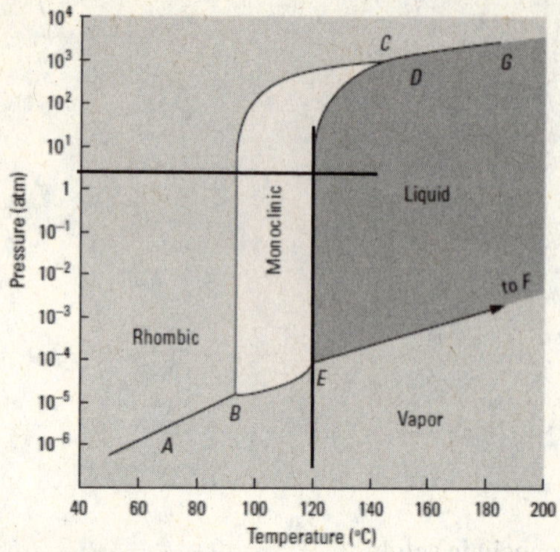

120. *Result:* **(a) 152 pm (b) r_{I^-} = 212 pm, r_{Li^+} = 88.0 pm (c) See discussion below**

Analyze: Given the lengths of the sides of the unit cells for solid lithium metal and solid lithium iodide, the type of structure, and some assumptions of which atoms are touching, determine the radius of Li atom and the radii of Li^+ and I^-.

Plan and Execute: Use the geometrical relationships described in the solution to Question 62.

(a) Looking at Figure 9.22 at the body-centered cubic, and assuming that the atoms touch along the body diagonal, we get: Body diagonal length = $4r_{Li}$

Given the edge length of the Li unit cell, 351 pm, calculate the body diagonal length:

$$\text{Body diagonal length} = \sqrt{3 \times (\text{edge length})^2} = \sqrt{3} \times 351 \text{ pm} = 608 \text{ pm}$$

Use body diagonal length to get radius of Li atom:

$$r_{Li} = \frac{\text{Body diagonal length}}{4} = \frac{608 \text{ pm}}{4} = 152 \text{ pm}$$

(b) Assuming that the I^- ions touch each other along the face diagonal and the I^- ions touch the Na^+ ion along the edge:

$$\text{Face diagonal length} = 4\, r_{I^-}$$

$$\text{Edge length} = 2\, r_{Li^+} + 2\, r_{I^-}$$

Given the edge length of the LiI unit cell, 600. pm, calculate the face diagonal length:

(For sig. figs, assume that this length is as precise as the length in (a), ± 1 pm.)

$$\text{Face diagonal length} = \sqrt{2} \times (\text{edge length}) = \sqrt{2} \times 600.\text{ pm} = 849 \text{ pm}$$

Use face diagonal length to get radius of I^- ion:

$$r_{I^-} = \frac{\text{Body diagonal length}}{4} = \frac{849 \text{ pm}}{4} = 212 \text{ pm}$$

Use the edge length of the LiI unit cell radius of I^- ion to get the radius of Li^+ ion:

$$2\, r_{Li^+} = \text{Edge length} - 2\, r_{I^-} = 600.\text{ pm} - 2 \times 212 \text{ pm} = 176 \text{ pm}$$

$$r_{Li^+} = 88.0 \text{ pm}$$

(c) It is reasonable that the Li atom is larger than the Li^+ cation. Figure 5.27 gives slightly larger value for the radius of the Li atom (157 pm), for Li^+ ion (90 pm) and for I^- (206 pm). The assumption that I^-

anions touch each other is reasonable geometrically, but their negative charges will make them repulsive. The Li^+ ions may not be large enough to span the entire gap between the I^- ions along the edge. Thus the assumptions that the I^- and I^- ions touch and the Li^+ and I^- ions touch are both likely to cause some errors.

☑ *Reasonable Result Check:* The geometric relationships are logical. The relative sizes of the atom and ion radii, the body diagonal length, and the edge length are all reasonable. The sum of the Figure 5.27 radii to back-calculate the edge length (2 × 90 pm + 2 × 206 pm) gives 592 pm, a little less than the actual measured length, suggesting that the ions do not touch. Similarly, the face diagonal length (4 × 206 pm) gives 824 pm, a little less than the actual measured length.

Chapter 10: Fuels, Organic Chemicals, and Polymers

Solutions for Red-Numbered Questions for Review and Thought

Topical Questions

Petroleum and Fuels (Section 10-1)

18. *Result:* **(a) 20 - 200 °C (b) No, it is 55 (c) No, the octane rating is too low and it would cause much more pre-ignition in the engine.**

Explanation:

(a) The gasoline fraction, with the hydrocarbons that will provide fuel for most people's cars, has a temperature range of 20 - 200 °C.

(b) The octane rating of the straight-run fraction is 55, so no it is not greater than 87.

(c) No, it is not advisable to use the straight-run fraction as a motor fuel. The octane rating is far lower than regular gasoline we buy at the pump (87 - 92), which means it would cause far more pre-ignition knocking than we expect from the gasoline. It would need to be reformulated to make it an acceptable motor fuel.

21. *Result/Explanation:* Removing the aromatics and alkenes from oxygenated gasolines allows the gas to burn more completely and as a result the carbon monoxide emissions are reduced. Also by removing some of the aromatic hydrocarbons, the gas has a lower volatility and fewer hydrocarbons get into the atmosphere from spills and normal fuel tank filling operations. This decreases urban air pollution.

U.S. Energy Sources and Consumption (Section 10-2)

23. *Result/Explanation:* Described in Section 10-2, the major shift from 2007 to 2011 has been the increase in the percent of energy derived from natural gas compared with that from coal, as shown in Figure 10.4.

25. *Result/Explanation:* Hydraulic fracturing ("fracking") increased the production of natural gas from 1% in 2000 to more than 20% in 2010, as described in Section 10-2a.

Tools of Chemistry: Gas Chromatography

28. *Result/Explanation:* Polar molecules would not be attracted to the non-polar liquid stationary phase, so they will exit the chamber earlier. Hence, they will show up earlier on the chromatograph.

Alcohols (Section 10-4)

30. *Result:* **See structures below (There are other correct answers; these are examples.)**

Analyze: Give an example if a primary alcohol, a secondary alcohol and a tertiary alcohol; draw Lewis structures for each example.

Plan: A primary alcohol has the –OH group attached to a C atom that is only bonded to one other C atom. A secondary alcohol has the –OH group attached to a C atom that is bonded to two other C atoms. A tertiary alcohol has the –OH group attached to a C atom that is bonded to three other C atoms.

Execute:

(a) An example of a primary alcohol is 1-propanol:

$$CH_3{-}CH_2{-}CH_2{-}OH$$

(b) An example of a secondary alcohol is 2-propanol:

$$\begin{array}{c} OH \\ | \\ CH_3{-}CH{-}CH_3 \end{array}$$

(c) An example of a tertiary alcohol is 2-methyl-2-propanol:

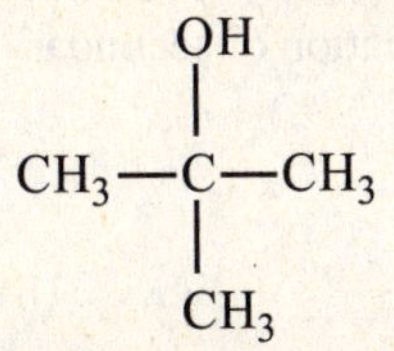

There are other correct answers. These are just examples.

32. *Result:* **(a) tertiary (b) primary (c) secondary (d) secondary (e) tertiary (f) secondary**

Analyze and Plan: Given several chemical formulas for alcohols, classify them as primary, secondary, or tertiary alcohols. Use the plan in the solution to Question 30.

Execute:

(a) The given structure has the –OH group attached to a C atom that is bonded to three other C atoms, so it is an example of a **tertiary** alcohol.

(b) The given structure has the –OH group attached to a C atom that is bonded to only one other C atom, so it is an example of a **primary** alcohol.

(c) The given structure has the –OH group attached to a C atom that is bonded to two other C atoms, so it is an example of a **secondary** alcohol.

(d) The given structure has the –OH group attached to a C atom that is bonded to two other C atoms, so it is an example of a **secondary** alcohol.

(e) The given structure has the –OH group attached to a C atom that is bonded to three other C atoms, so it is an example of a **tertiary** alcohol.

(f) The given structure has the –OH group attached to a C atom that is bonded to two other C atoms, so it is an example of a **secondary** alcohol.

34. *Result:* **See structures below**

Analyze: Given structures for alcohols, draw structures for their first and second oxidation products.

Plan: Oxidation of a primary alcohol gives an aldehyde. To accomplish this, change the $-CH_2OH$ to $-CH{=}O$. Oxidation of an aldehyde gives a carboxylic acid. On the C=O carbon in the aldehyde, change the C—H to a C—OH.

Execute:

(a)

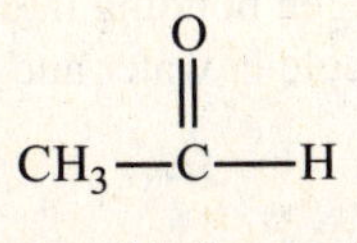

first oxidation product

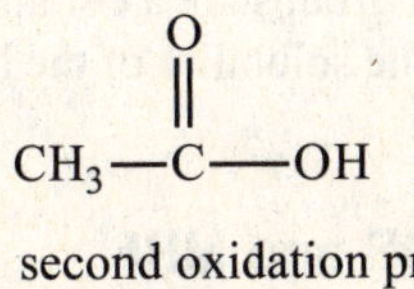

second oxidation product

(b)

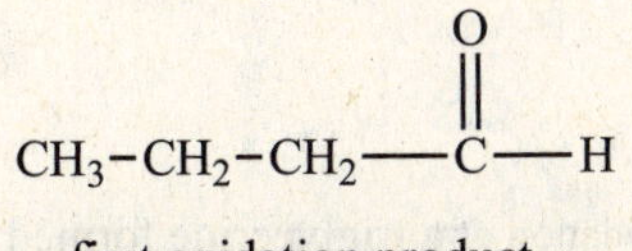

first oxidation product

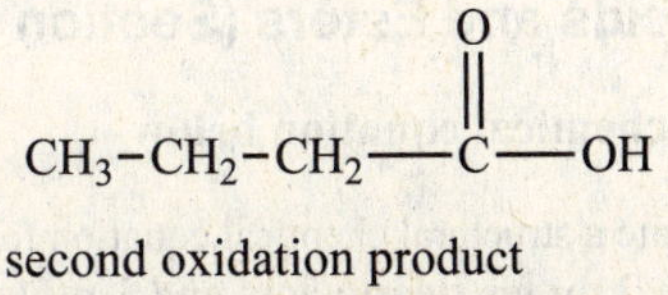

second oxidation product

35. *Result:* **See structures below**

Analyze and Plan: Given structures for alcohols, draw structures of the oxidation products. Adapt the plan given in the solution to Question 34. Oxidation of secondary alcohols results in ketones. Take the CH—OH and replace it with C=O.

Execute:

(a) Ketone product:

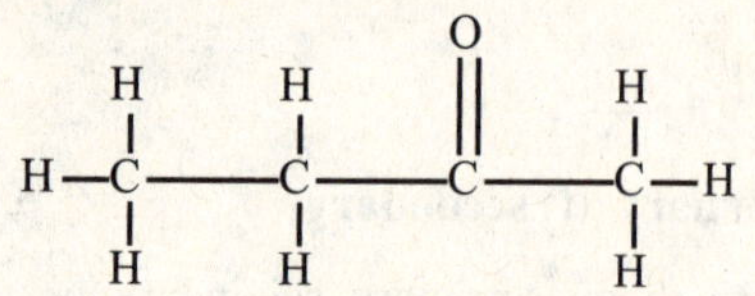

(b) Ketone product:

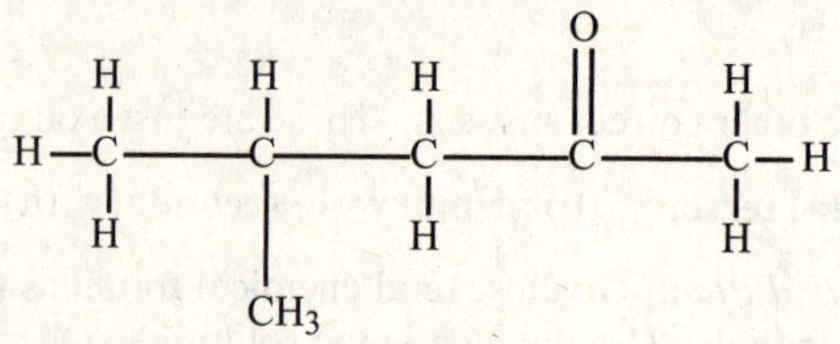

36. *Result:* **See structures below**

Analyze: Given aldehyde, ketone, and carboxylic acid, determine the alcohol that can be oxidized to make these compounds.

Plan: Adapt the plan given in Question 34. Oxidation of a secondary alcohol gives a ketone. Oxidation of a primary alcohol gives an aldehyde. Oxidation of an aldehyde gives a carboxylic acid. So, if we have an aldehyde, replace –CH=O with $–CH_2OH$. If we have a ketone, replace C=O with –CHOH. If we have a carboxylic acid, replace the COOH with $–CH_2OH$.

(a) The aldehyde given is produced from the oxidation of this primary alcohol:

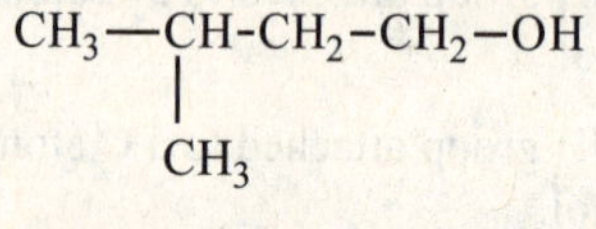

(b) The ketone given is produced from the oxidation of this secondary alcohol:

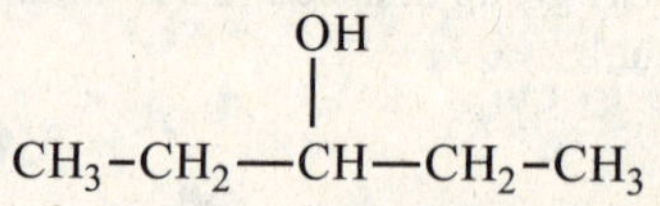

(c) This carboxylic acid is produced from the oxidation of a primary alcohol and subsequent oxidation of the resulting aldehyde. The original alcohol is this one:

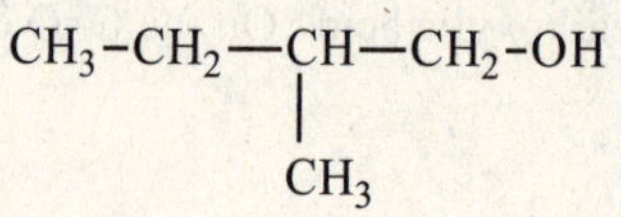

38. *Result/Explanation:* Wood alcohol (methanol) is made by heating hardwood such as beech, hickory, maple, or birch. Grain alcohol (ethanol) is made from the fermentation of plant materials, such as grains. (Section 10-4a)

40. *Result/Explanation:* –OH groups are a common site of hydrogen bonding intermolecular forces. Their presence would increase the solubility of the biological molecule in water and create specific interactions with other biological molecules.

Tools of Chemistry: NMR and MRI

41. *Result/Explanation:* The "Tools of Chemistry" given in Section 10-4d describes NMR as using **radio** frequencies.

Carboxylic Acids and Esters (Section 10-5)

43. *Result:* **See chemical equation below**

Analyze: Write a structural chemical equation for the formation of a triglyceride formed by the reaction of 1 mol glycerol with 2 moles stearic acid and 1 mole oleic acid.

Plan: Use the plan described in Problem-Solving Example 10.4 in Section 10-5c.

The structural formulas of glycerol, $CH_2OHCHOHCH_2OH$, stearic acid, $CH_3(CH_2)_{16}COOH$, and oleic acid, $CH_3(CH_2)_4CH{=}CHCH_2CH{=}CH(CH_2)_7COOH$, are given in Problem-Solving Example 10.4. An ester linkage forms between each –OH group on glycerol with one –COOH group on the acids.

Execute:

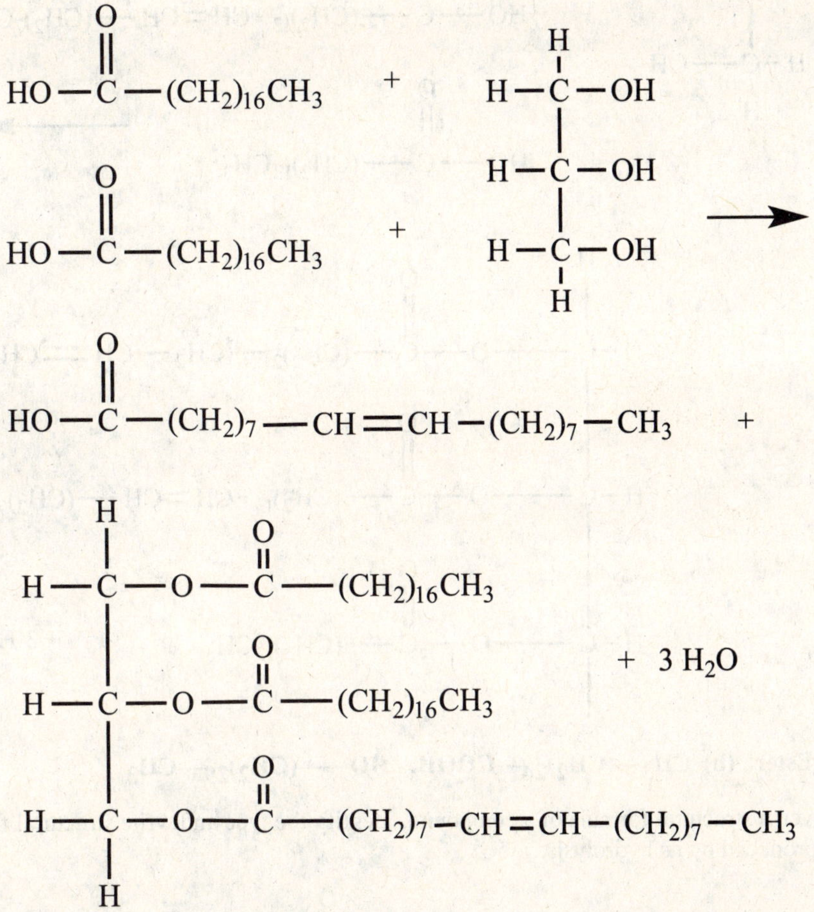

45. *Result:* **(a) (i) polyunsaturated (ii) monounsaturated (iii) saturated (b) See structure below**

Analyze: Classify each amino acid give as mono-, di-, or polyunsaturated and write a balanced equation for the formation of a triglyceride that incorporates all three.

Plan: A mono-unsaturated amino acid has only one double C=C bond. A di-unsaturated amino acid has two double C=C bonds. A polyunsaturated amino acid has more than two double C=C bonds. A saturated amino acid has no double C-C bonds. Use the plan from Question 43 for writing the balanced equation.

Execute:

(a) (i) This amino acid has four repeating units of (CH2—CH=CH), so it has four double C=C bonds. That means it is **polyunsaturated.**

(ii) This amino acid has one double C=C bond. That means it is **mono-unsaturated.**

(iii) This amino acid has no double C=C bonds. That means it is **saturated.**

(b) The three amino acids form ester bonds with the glycerol:

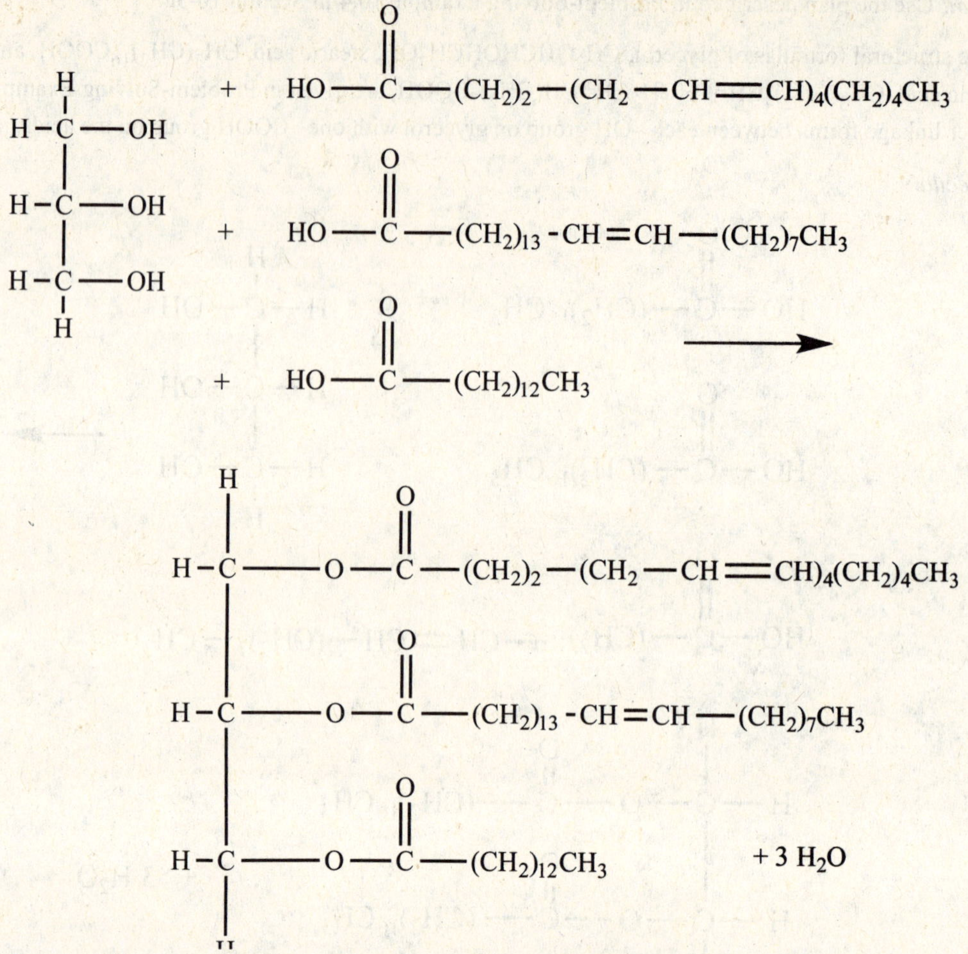

$$
\begin{array}{l}
\begin{matrix}
& H & \\
& | & \\
H-C&-OH & \\
& | & \\
H-C&-OH & \\
& | & \\
H-C&-OH & \\
& | & \\
& H &
\end{matrix}
\quad
\begin{matrix}
+ & HO-\overset{\overset{\displaystyle O}{\|}}{C}-(CH_2)_2-(CH_2-CH=CH)_4(CH_2)_4CH_3 \\
\\
+ & HO-\overset{\overset{\displaystyle O}{\|}}{C}-(CH_2)_{13}-CH=CH-(CH_2)_7CH_3 \\
\\
+ & HO-\overset{\overset{\displaystyle O}{\|}}{C}-(CH_2)_{12}CH_3
\end{matrix}
\quad \longrightarrow
\end{array}
$$

$$
\begin{matrix}
& H & & & \\
& | & & O & \\
H-C&-O-&\overset{\|}{C}&-(CH_2)_2-(CH_2-CH=CH)_4(CH_2)_4CH_3 \\
& | & & & \\
& & & O & \\
H-C&-O-&\overset{\|}{C}&-(CH_2)_{13}-CH=CH-(CH_2)_7CH_3 \\
& | & & & \\
& & & O & \\
H-C&-O-&\overset{\|}{C}&-(CH_2)_{12}CH_3 \qquad +\,3\,H_2O \\
& | & & & \\
& H & & &
\end{matrix}
$$

47. *Result:* **(a) Ester   (b) $CH_3-(CH_2)_{24}-COOH$, $HO-\!\!-(CH_2)_{29}-CH_3$**

Analyze: Given a structural formula of a compound, identify the type and write structural formulas for the compounds produced by its hydrolysis.

Plan: The compound containing the functional group: $-\overset{\overset{\displaystyle O}{\|}}{C}-O-R$ is identified in Section 10-5b. Hydrolysis is described in Section 10-5d, as the reaction of the ester and water to make a carboxylic acid and an alcohol, ROH.

Execute:

(a) The functional group indicates that the compound is an ester.

(b) The process of hydrolysis breaks the molecule into two molecules at the R—O—C bond, making an alcohol, R—OH, and a carboxylic acid, —COOH.

Here, the acid is: $CH_3-(CH_2)_{24}-COOH$

The alcohol is: $HO-\!\!-(CH_2)_{29}-CH_3$.

50. *Result:* **(a) $CH_3COOCH_2CH_3$ (b) $CH_3CH_2COOCH_2CH_2CH_3$ (c) $CH_3CH_2COOCH_3$**

Analyze: Given formulas of alcohols and acids, write the structural formulas of the esters that can form.

Plan: The carboxylic acid group loses OH, and the alcohol group loses H (in the formation of water). Connect the remaining fragments to get the product of the condensation reaction produces the resulting ester.

Execute:

(a) $CH_3CO\underline{OH} + \underline{H}OCH_2CH_3 \longrightarrow CH_3COOCH_2CH_3 + H_2O$

(b) $CH_3CH_2CO\underline{OH} + \underline{H}OCH_2CH_2CH_3 \longrightarrow CH_3CH_2COOCH_2CH_2CH_3 + H_2O$

(c) $CH_3CH_2CO\underline{OH} + \underline{H}OCH_3 \longrightarrow CH_3CH_2COOCH_3 + H_2O$

52. *Result:* **(a) CH_3CH_2COOH, CH_3OH (b) $HCOOH$, CH_3CH_2OH (c) CH_3COOH, CH_3CH_2OH**

Analyze: Given the formulas of esters, write the formulas for the alcohol and acid that react to form them.

Plan: Break the ester product of the condensation reaction between the two O atoms. Add OH to the C=O to make the carboxylic acid, and add H to the other fragment to make the alcohol group.

(a) $CH_3CH_2CO\underline{OH} + \underline{H}OCH_3 \longrightarrow CH_3CH_2COOCH_3 + H_2O$

(b) $HCO\underline{OH} + \underline{H}OCH_2CH_3 \longrightarrow HCOOCH_2CH_3 + H_2O$

(c) $CH_3CO\underline{OH} + \underline{H}OCH_2CH_3 \longrightarrow CH_3COOCH_2CH_3 + H_2O$

Synthetic Organic Polymers (Section 10-6)

55. *Result/Explanation:* Examples of thermoplastics are milk jugs (polyethylene), cheap sunglasses and toys (polystyrene) and CD audio discs (polycarbonates). Thermoplastics soften and flow when heated, permitting them to be formed to specific shape that sets when cooled.

57. *Result:* **See structures below**

Analyze: Given names of monomers, draw the structure of the repeating unit of its polymer.

Plan: Change the double bond into a single bond between two "linking" carbons then create linkage lines through the parenthesis to show that each of those carbon atoms connect to neighboring units. Put the symbol "n" on the outside of the parentheses to indicate that the units repeat many times.

Execute:

(a) 1-butene has a double bond between the first and second C atoms in the four-atom chain, $CH_2{=}CHCH_2CH_3$, where the addition polymerization occurs. The saturated two-carbon end of the butene will put an ethyl branch on every other carbon in the long polymer chain

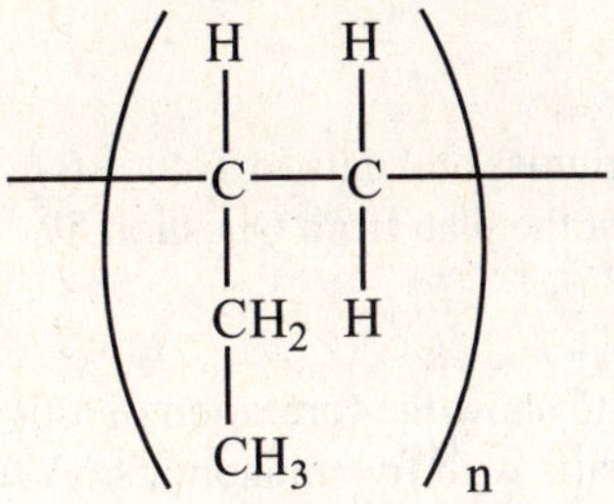

(b) 1,1-dichloroethylene has a double bond between the two C atoms where the addition polymerization occurs The two Cl atoms on the first C atom in the monomer mean that there will be two Cl atoms on every other carbon in the long polymer chain:

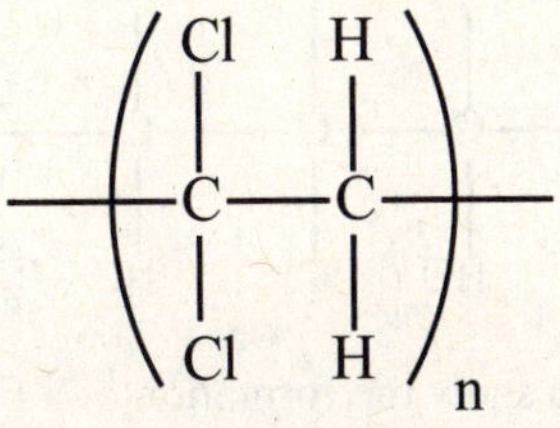

(c) Vinyl acetate, $CH_3COOCH{=}CH_2$, has a double bond between the two vinyl C atoms where the addition polymerization occurs. The acetate group in the monomer means that every other carbon in the long polymer chain will have an acetate group attached to it:

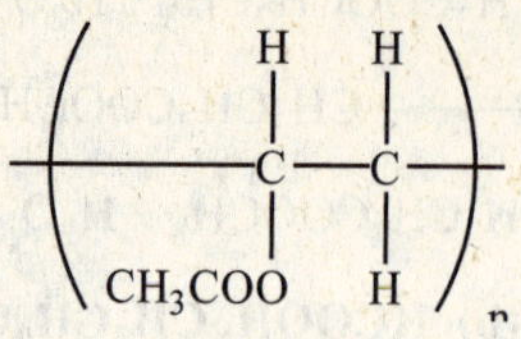

59. *Result:* **See equation and structure below**

Analyze and Plan: The monomer, methyl methacrylate, CH_2=C(CH_3)$COOCH_3$, has a double bond between the first two C atoms where the addition polymerization occurs. The monomer has a methyl branch and the –$COOCH_3$ group on the first carbon, so every other carbon in the polymer chain will have a methyl and a –$COOCH_3$ group attached to it.

Execute:

(a) The equation for the reaction to form a four-unit chain of the polymer looks like this.

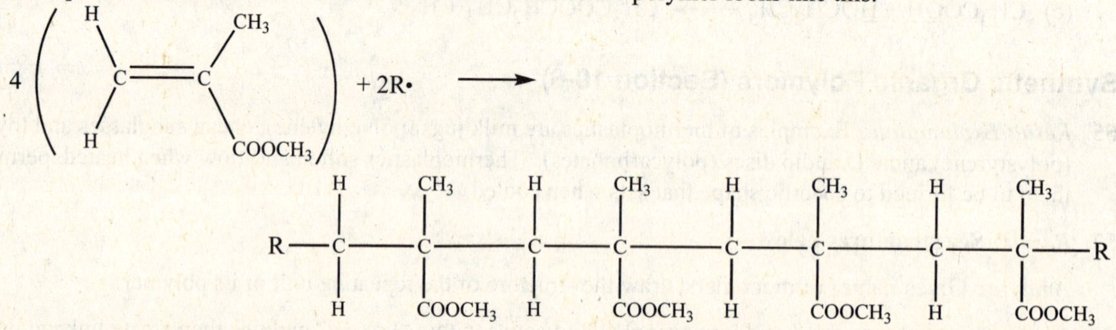

(b) The repeating unit looks like this:

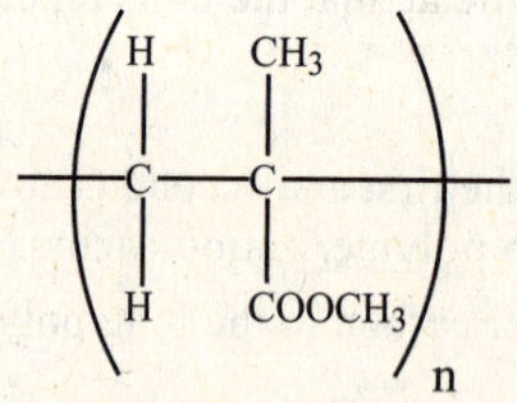

61. *Result:* **See equations below**

Analyze and Plan: Given structural formulas and data in Table 10.6, write the chemical equation for the formation of a five-unit polymer. Adapt the plan from Question 59.

Execute:

(a) The chemical equation that occurs to show the formation of a five-unit polymer chain with the monomer vinylidene chloride (H_2C=CCl_2) starts with five monomers. A free-radical initiator (•OR) is used to start the reaction and to terminate it after the five monomers are added.

5 H_2C=CCl_2 + 2 •OR ⟶

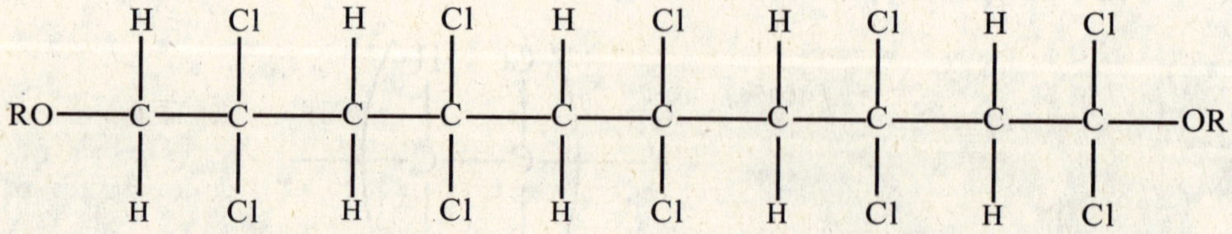

(b) The chemical equation that occurs to show the formation of a five-unit polymer chain with the monomer tetrafluoroethylene starts with five monomers. A free-radical initiator (•OR) is used to start the reaction and to terminate it after the five monomers are added.

$5\ F_2C{=}CF_2 + 2\ {\bullet}OR \longrightarrow$

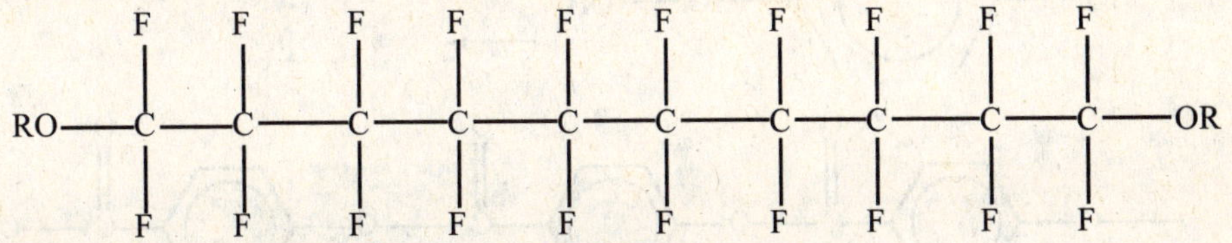

62. *Result:* **See structures below**

Analyze: Given structural formulas for polymer segments, write the structural unit for the monomer.

Plan: Find the repeating unit, then put a double bond between those two C atoms.

Execute:

(a) The repeating unit looks like this.

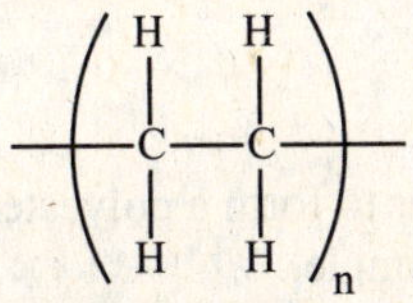

That means $CH_2{=}CH_2$ is the monomer.

(b) The repeating unit looks like this.

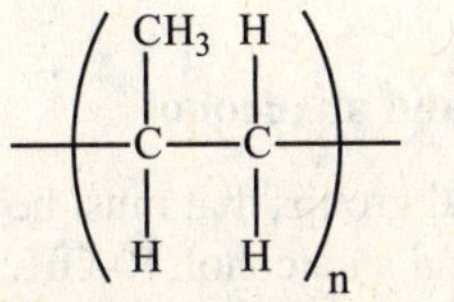

That means $CH_3CH{=}CH_2$ is the monomer.

(c) The repeating unit looks like this.

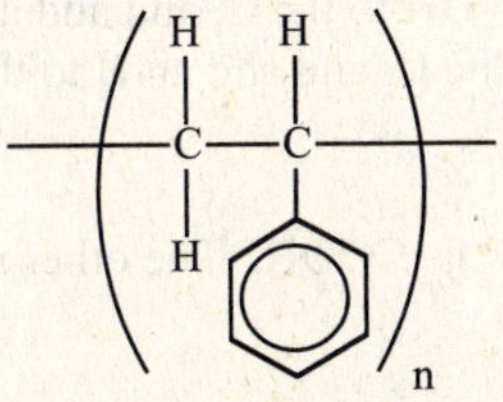

That means the monomer is:

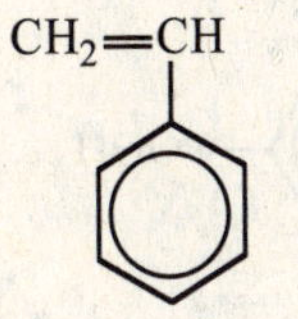

(d) The repeating unit looks like this.

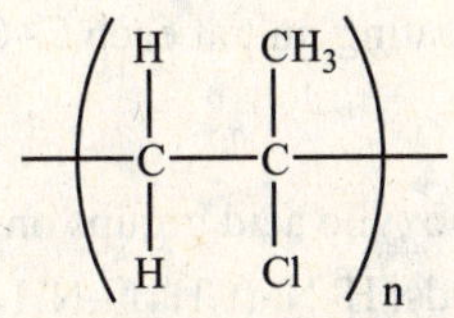

That means $CH_2{=}CClCH_3$ is the monomer.

(e) The repeating unit looks like this.

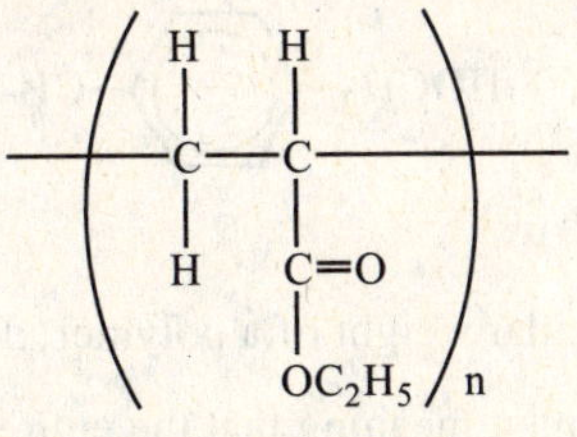

That means we need $CH_2{=}CHCOOC_2H_5$ for the monomer.

64. *Result:* **See structures below**

Analyze: Given a monomer that can form a condensation polymer, write a structural chemical equation for the formation of a four-unit chain of the polymer and indicate the repeating unit.

Plan: The polymer chain will form from making an ester bond between the OH group and the COOH group. The carboxylic acid group loses -OH, and the alcohol group loses -H (in the formation of water).

Execute:

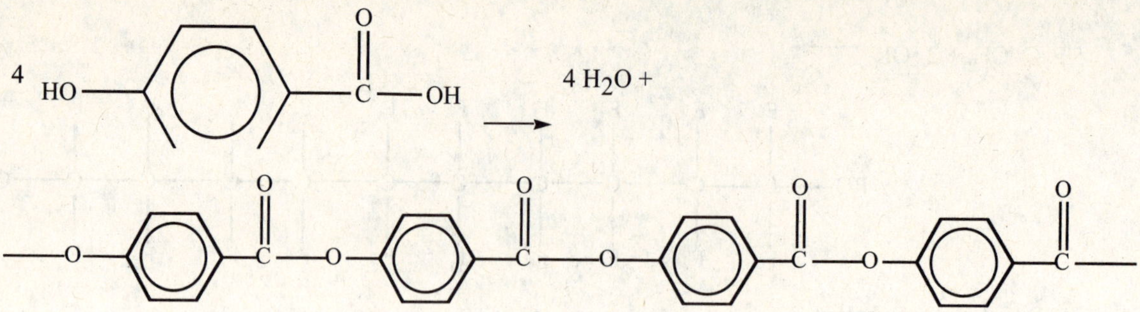

The repeating unit is:

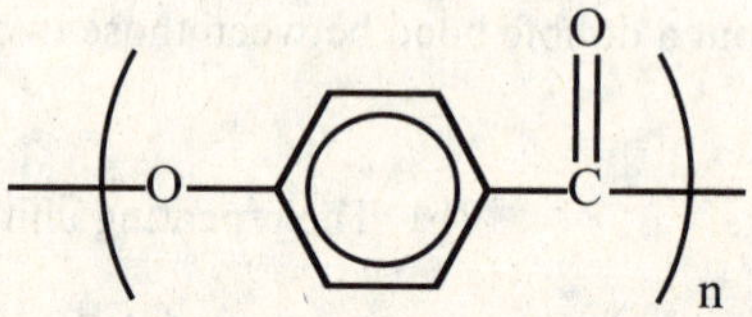

66. *Result:* **a carboxylic acid and an alcohol**

Explanation: The functional groups that must be present in a single monomer to form a polyester are a carboxylic acid, –COOH and an alcohol, R–OH. See Question 64 for an example.

69. *Result:* **(a) HOOC–$(CH_2)_8$–COOH, H_2N–$(CH_2)_6$–NH_2 (b) see structures below**

Analyze: Draw the structures of monomer that could form given condensation polymers.

Plan: In polyamides, break the repeating unit at each C–N bond, add an –OH to the C, and add an H to the N. In polyesters, break the repeating unit at each C–O bond, add an –OH to the C, and add an H to the O.

Execute:

(a) One monomer has carboxylic acid groups on both ends, HOOC–$(CH_2)_8$–COOH. The other monomer has NH_2 groups on both ends, H_2N–$(CH_2)_6$–NH_2.

(b) One monomer has carboxylic acid groups on both ends:

The other monomer has OH groups on both ends

71. *Result:* **1500 monomers**

Analyze: Given the approximate molecular weight of a polymer, determine how many monomers in it.

Plan: Polyethylene is an addition polymer, meaning that the entire monomer molecule is incorporated into the polymer chain. By calculating the molar mass of the monomer and dividing it into the molar mass (also known as "molecular weight" in units of grams per mole) of the polymer, we can estimate the number or monomers in the polymer chain.

Execute: The molar mass of the monomer, $H_2C{=}CH_2$, is 28.0530 g/mol.

$$\frac{42000\ \text{g/mol polymer}}{28.0530\ \text{g/mol monomer}} = 1500\ \frac{\text{monomers}}{\text{polymer}}$$

74. *Result:* **See structures below**

Analyze: Given structures of polymer units, determine the structural formulas for the repeating units.

Plan: Find the monomer used and then link the monomer appropriately.

Execute:

(a) Natural rubber is described in Section 10-6 as poly-*cis*-2-methyl-1,3-butadiene. To show the *cis*-features of the polymer, two monomer units need to be identified in the repeating unit.

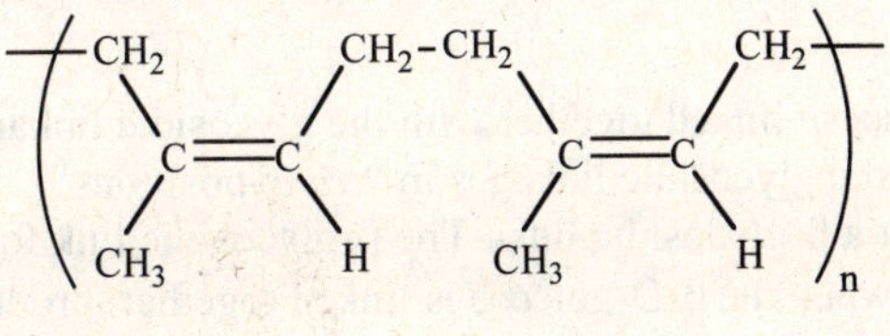

(b) The repeating unit for neoprene is:

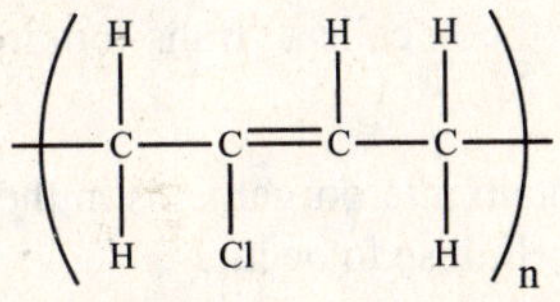

(c) Polybutadiene would be similar to (a) above, except that the alkene has no substitutions.

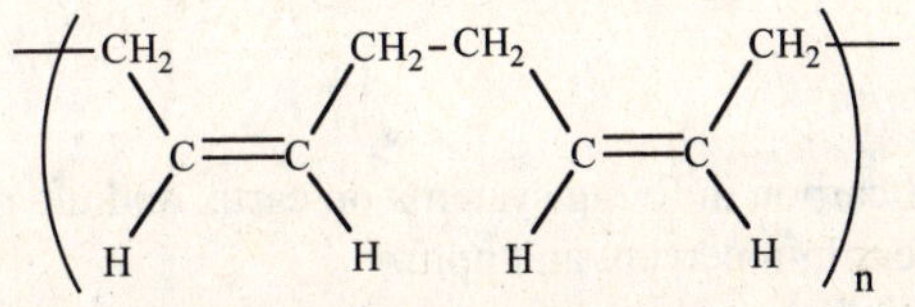

75. *Result/Explanation:* Major end uses for recycled PET include fiberfill for ski jackets and sleeping bags, carpet fibers, and tennis balls. HDPE is converted into fiber used for sportswear, insulating wrap for new buildings, and very durable shipping containers.

Polysaccharides and Proteins (Section 10-7)

76. *Result/Explanation:* One major difference between proteins and most other polyamides is that the protein polymer's monomers are not all alike. Different side chains on the fatty acids change the properties of the protein.

78. *Result/Explanation:* Cellulose and starch have monomers units (glucose sugars) that are all alike.

80. *Result:* **amine, amide, alcohol, carboxylic acid**

Analyze: Given the molecular structure of a tripeptide, identify all the functional goups.

Plan: Looking at the molecule from left to right we see several functional groups. NH_2 is an amine functional group. RCONHR' is an amide functional group. ROH is an alcohol functional group. RCOOH is a carboxylic acid. (A benzene ring with an –OH group on it is phenol and RSH is thiol, but these answers are not likely to be answers given by students studying this textbook.)

81. *Result:* **See structure below**

Analyze: Given the name of a polypeptide, draw the structural formula.

Plan: The monomers alanine, glycine, and phenylalanine are given in Table 10.8. We will link them using peptide linkages.

Execute:

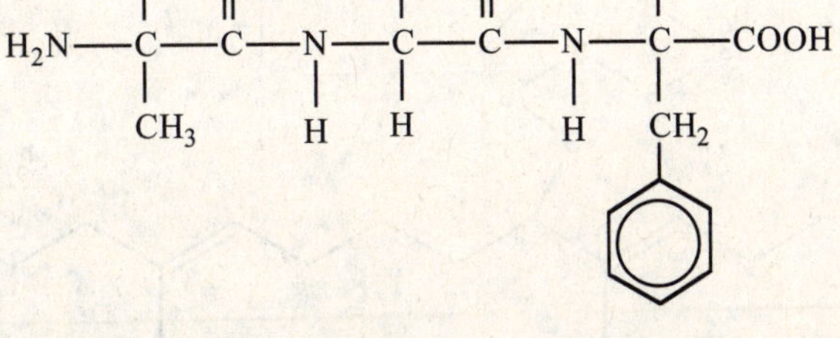

83. *Result/Explanation:*

(a) A monosaccharide is a molecule composed of one simple sugar molecule, while disaccharides are molecules composed of two simple sugar molecules.

(b) Disaccharides have only two simple sugar molecules; whereas, polysaccharides have many.

85. *Result/Explanation:* Starch and cellulose are polysaccharides that yield only D-glucose upon complete hydrolysis.

87. *Result/Explanation:*

(a) Glycogen contains glucose linked together with the glycosidic linkages in "*cis*-positions" and cellulose contains glucose with the glycosidic linkages in "*trans*-positions." The difference is generally described as an α-glycosidic link or a β-glycosidic link. The β-glycosidic link found in amylose, amylopectin, and glycogen is produced when the β-D-glucose is linked together, producing what the textbook calls a "cis-looking" stereoisomer. The β-glycosidic link found in cellulose is produced when the β-D-glucose is linked together, producing what the textbook calls a "trans-looking" stereoisomer. Humans lack the enzyme to hydrolyze the β-glycosidic link.

(b) Cows (and other ruminants) contain microorganisms in their digestive tract that produce the appropriate enzyme, allowing digestion of cellulose to occur.

General Questions

88. *Result:* **1.0×10^{-3} %, 10. ppm**

Analyze: Given the mass of carbon in living systems on earth, and the mass of carbon on earth, determine the concentration of carbon on earth in percent and ppm.

Plan: Parts per million (ppm) is described in margin note in Sections 8-11. To accomplish these conversions, use the usual relationship describing percent and ppm.

Execute: Mass fraction of carbon in living systems on earth =

$$\frac{\text{mass of C in living systems}}{\text{mass of C total}} = \frac{7.5\times10^{17}\text{ g C}}{7.5\times10^{22}\text{ g C total}} = 1.0\times10^{-5}$$

Percent of carbon in living systems on earth = (fraction) × 100 % = $(1.0 \times 10^{-5}) \times 100$ % = 1.0×10^{-3} %

Parts per million of carbon in living systems on earth

$$= (\text{fraction}) \times 10^6 \text{ ppm} = (1.0 \times 10^{-5}) \times 10^6 \text{ ppm} = 10.\text{ ppm}$$

☑ *Reasonable Result Check:* These proportional quantities appear properly represented. When the quantity varies by a factor of about 10^{-5}, then the recorded concentration in ppm will be a useful size.

91. *Result/Explanation:* $CH_3CH_2CH_2CH_2CH_2CH_2CH_2CH_2CH_2CH_2OH$ is a larger molecule than CH_3CH_2OH. The polar end of the alcohol group will interact well with the water; however, the nonpolar end of the molecule will not. The longer nonpolar end of the decanol will not be miscible in water, lowering the solubility compared to the smaller ethanol molecule.

93. *Result/Explanation:* Vulcanized rubber has short chains of sulfur atoms that bond together (crosslink) the polymer chains of natural rubber.

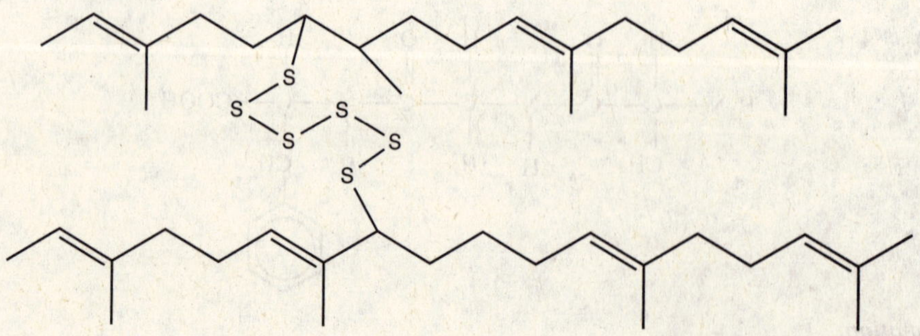

94. *Result:* **See structure below; tertiary**

Explanation: 3-ethyl-5-methyl-3-hexanol has a six-carbon main chain with an ethyl branch on the third C atom, a methyl branch attached to the fifth C atom, and an OH attached to the third C atom:

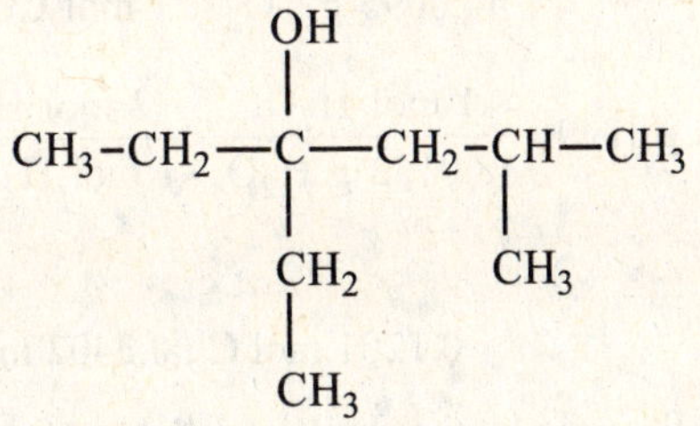

This is a tertiary alcohol, since the atom with the OH bonded to it is also bonded to three other C atoms.

97. *Result:* **(a)** $2\ C_8H_{18}(\ell) + 25\ O_2(g) \longrightarrow 16\ CO_2(g) + 18\ H_2O(g)$ **(b)** 1.4×10^3 **L** CO_2

Analyze: Given a fuel's chemical formula and density, fuel efficiency of a car, temperature, pressure, molar volume, and the length of a trip, write a balanced equation for its burning and calculate the volume of CO_2 generated on the trip.

(a) *Explanation*: Burning is combustion, the reaction with oxygen:

$$2\ C_8H_{18}(\ell) + 25\ O_2(g) \longrightarrow 16\ CO_2(g) + 18\ H_2O(g)$$

(b) Plan: Use the miles, the fuel efficiency, volume conversions, the density, and the stoichiometry to find the moles of the product. Use the ideal gas law (from Chapter 8) to determine the volume of the product.

Execute:

$$10.\ \text{miles} \times \frac{1\ \text{gal gasoline}}{32\ \text{miles}} \times \frac{3.785\ \text{L gasoline}}{1\ \text{gal gasoline}} \times \frac{1000\ \text{mL}}{1\ \text{L}} \times \frac{1\ \text{cm}^3}{1\ \text{mL}} = 1.2 \times 10^3\ \text{cm}^3$$

$$1.2 \times 10^3\ \text{cm}^3\ \text{gasoline} \times \frac{0.703\ \text{g gasoline}}{1\ \text{cm}^3\ \text{gasoline}} \times \frac{1\ \text{mol gasoline}}{114.22\ \text{g gasoline}} \times \frac{16\ \text{mol}\ CO_2}{2\ \text{mol gasoline}} = 58\ \text{mol}\ CO_2$$

$$58\ \text{mol}\ CO_2 \times \left(\frac{24.5\ \text{L}}{1\ \text{mol}}\right) = 1.4 \times 10^3\ \text{L}$$

☑ *Reasonable Result Check:* This is the same as Chapter 8 Question 47 and produces the same result. This is a large volume of CO_2. However, it is not an unreasonable quantity considering how many gallons of gasoline are used and how much CO_2 is generated from each octane molecule. Comparing the results to the answer in Question 98, less CO_2 is generated by methanol as a fuel than octane.

99. *Result:* **See structures below**

Explanation: Glycogen has a highly branched polymer of glucose so molecules of glycogen can pack together like spheres to form granules. Cellulose is a linear polymer with many –OH groups along the sides so the molecules can easily pack side-by-side held together by hydrogen bonds to form layers.

101. *Result:* **(a)** CH_2 **(b) Alkene (c) See structure below**

Analyze: Given the mass of a hydrocarbon that undergoes complete combustion to form given masses of carbon dioxide and water, determine the empirical formula, determine if the compound is an alkane or an alkene and write a plausible Lewis structure for it.

Plan: Calculate the moles of the two products, CO_2 and H_2O produced. Then use the stoichiometry of the product formulas to determine the moles of C and H. Then set up a mole ratio and find the simplest whole number relationship between the atoms. Use the ratio to determine the empirical formula.

Execute:

(a) The combustion produces 5.287 g CO_2 and 2.164 g H_2O. Determine the moles of C and O.

$$5.287\text{ g }CO_2 \times \frac{1\text{ mol }CO_2}{44.0095\text{ g }CO_2} \times \frac{1\text{ mol C}}{1\text{ mol }CO_2} = 0.1201\text{ mol C}$$

$$2.164\text{ g }H_2O \times \frac{1\text{ mol }H_2O}{18.0152\text{ g }H_2O} \times \frac{2\text{ mol H}}{1\text{ mol }H_2O} = 0.2402\text{ mol H}$$

Set up a mole ratio:

0.1201 mol C : 0.2402 mol O

Divide all the numbers in the ratio by the smallest number of moles, 0.1201 mol: 1 C : 2 H

Empirical formula: CH_2

(b) This compound will have a molecular formula: $(CH_2)_n$ which can also be written C_nH_{2n}. Alkanes have formulas fitting the pattern of C_nH_{2n+2}; whereas, alkenes have formulas fitting the pattern of C_nH_{2n}. (Refer to Chapter 3 and 8 in the textbook). So, the compound burned here is an alkene.

(c) The molecular formula is $(CH_2)_n$, however there is not enough information given in the Question to determine what the molecule is. That means there are a few different answers to this part of the Question. The smallest possible alkene has n = 2, $H_2C{=}CH_2$. Its Lewis Structure is:

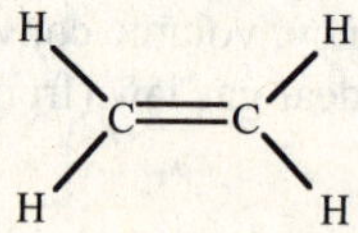

☑ *Reasonable Result Check:* The mole ratio is very close to an integer that corresponds to a known alkene. The given moles of carbon and hydrogen can be converted to mass and added together to produce the initial mass of the sample,

$$0.1201\text{ mol C} \times \frac{12.0107\text{ g C}}{1\text{ mol C}} = 1.442\text{ g C}$$

$$0.2402\text{ mol H} \times \frac{1.0079\text{ g H}}{1\text{ mol H}} = 0.2423\text{ g H}$$

$$1.442\text{ g C} + 0.2423\text{ g H} = 1.6485\text{ g compound}$$

So, we can be assured of 100% yield. With regard to the answer in (c), several other answers are just as plausible as the one that is published, including propylene, also given in Section 6-5. Student should be made aware that there is not one unique answer to this Question.

Applying Concepts

103. *Result:* **See structures below**

Analyze: Draw a structural formula for each of three possible repeating units of poly (vinyl chloride).

Plan: The monomer for PVC, vinyl chloride, $H_2C{=}CHCl$ can combine in three different ways. The first forms a chain with chlorine atoms on alternating carbon atoms, formed when the monomers come together always leading with the same side of the molecule.

Execute:

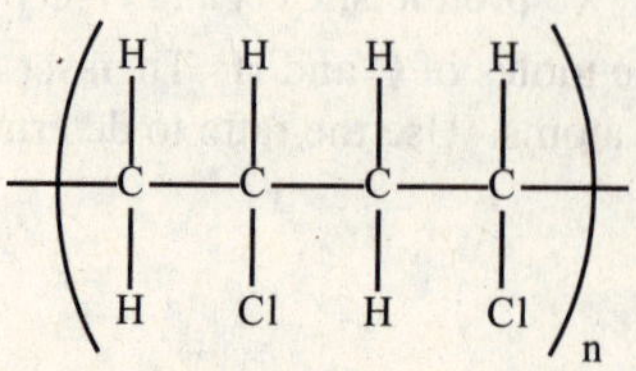

The second forms a chain with chlorine atoms on adjacent carbon atoms, formed when the monomers alternate which side of the monomer attaches to the polymer chain:

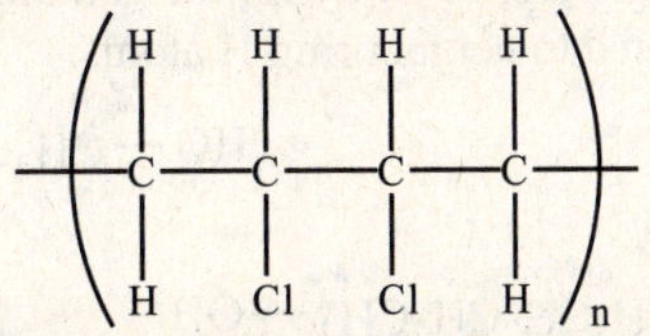

The third forms a chain with random orientation of the monomers and therefore variable orientation of the Cl atoms in the polymer chain:

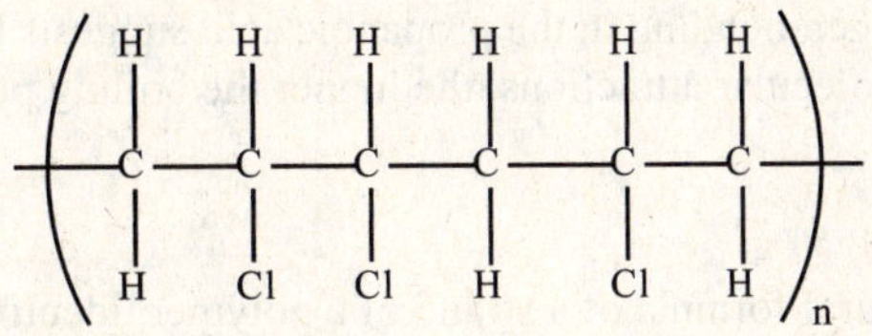

105. *Result:* **See four-unit polymer segment below**

Analyze and Plan: Given two compounds write a structural chemical equation for a condensation copolymer. Adapt the plan given in the solution to Question 64.

Execute: Lactic acid is given in Section 7-2e, $CH_3CH(OH)COOH$. It forms a condensation copolymer with glycolic acid, $HOCH_2COOH$, by the elimination of water:

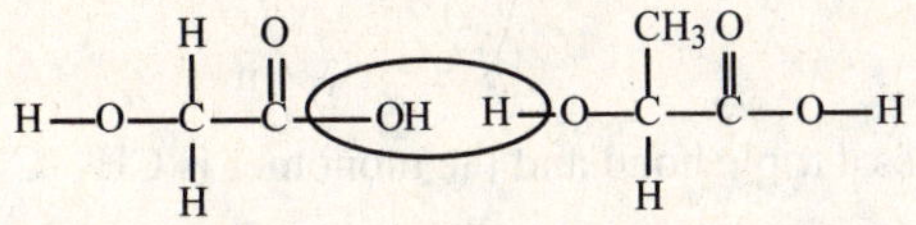

A four-monomer-unit chain of the polymer looks like this:

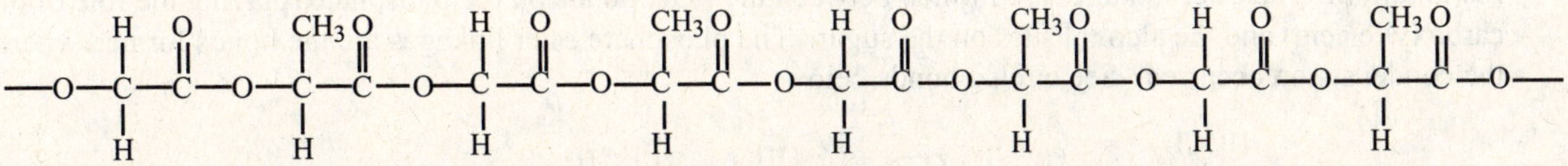

107. *Result:* **See structures below; propanoic acid boils at higher T, due to more H-bonding.**

Analyze and Plan: The hydrogen bonding between molecules occurs when a hydrogen-bonding H atom (one that is bonded to a highly electronegative element) is attracted to another highly electronegative element in a neighboring molecule.

Execute: In propanoic acid, there is one hydrogen-bonding H atom and two highly electronegative O atoms in the molecule. That means there will be four different kinds of hydrogen-bonding interactions between the given molecule (in the middle below) and its neighbors: two ways for the H atom to be attracted to neighboring O atoms and two ways for O atoms to be attracted to neighboring H atoms:

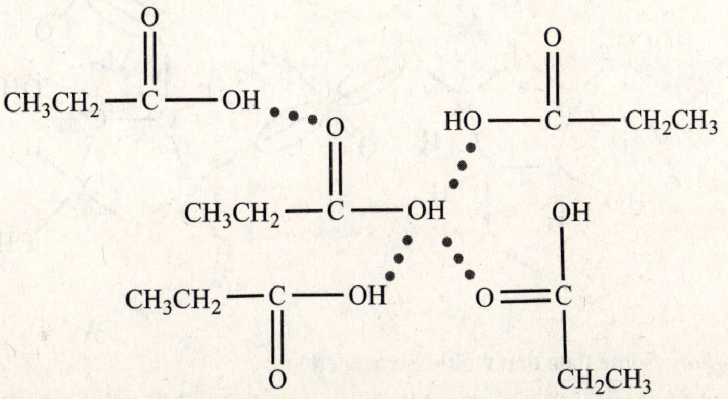

In 1-butanol, there is one hydrogen-bonding H atom and one highly electronegative O atom in the molecule. That means there will be two different kinds of hydrogen-bonding interactions between the given molecule (in the middle below) and its neighbors; one way for the H atom to be attracted to neighboring O atoms and one way for O atoms to be attracted to neighboring H atoms:

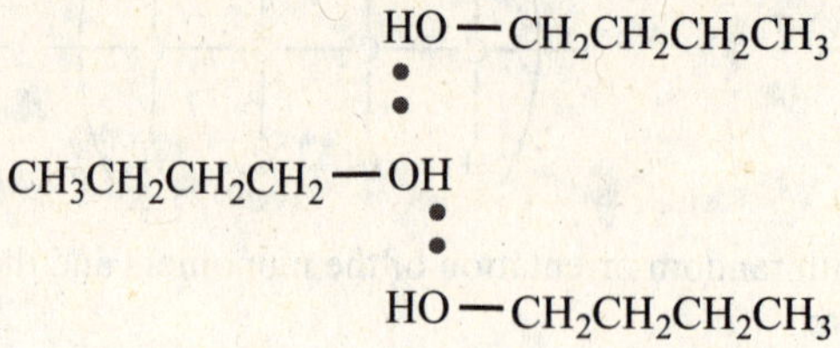

The more extensive hydrogen bonding in the propanoic acid suggests that it will have a higher boiling point, since the larger the intermolecular attractions, the higher the boiling point.

109. *Result:* **CH_3–C≡C–H**

Analyze: Given the structural formula of a strand of a polymer, identify the monomer.

Plan: Find the repeating unit, then take the connector bonds out and put an extra bond between the two C atoms. The repeating unit looks like this.

Execute: The repeating unit is:

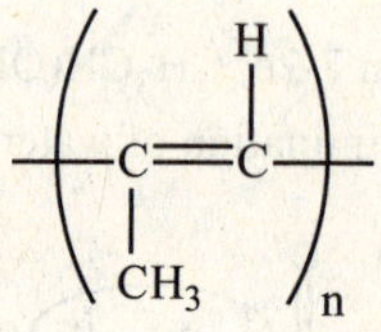

Thus, the double bond becomes a triple bond and the monomer is CH_3–C≡C–H.

111. *Result:* **See annotated structure, below**

Explanation: The ester linkages are formed between the POH bonds on the phosphate (playing the role of the carboxylic acid) and the alcohol sites on the sugar. The phosphate ester linkages are the bonds formed where the condensation occurred. See circled bonds below:

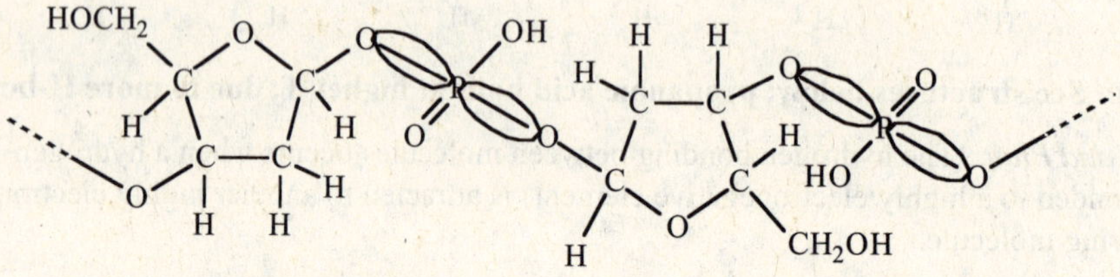

In the backbone of DNA, the phosphate ester linkages are made to the third and fifth –OH groups:

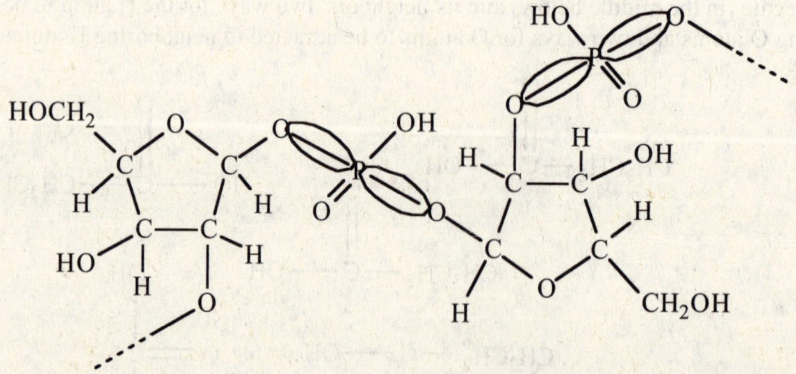

113. *Result/Explanation:* Some data that would be needed are:

- Sources and amounts of CO_2 generated over time to determine additional CO_2.

- Photosynthesis rate of depletion per tree per year.
- Average number of trees per acre.
- The number of acres of land in Australia that could support trees.
- The allowable tree density.

Other information would also be necessary. Students should try to think of other things they would need to know to determine the validity of the assertion.

115. *Result:* **(a) Box C (b) Box G (c) Box G (d) Box F (e) Box D (f) Box B (g) Boxes A and H (h) Box C (i) Box I (j) Box E**

Explanation: Given several prompts and set of choices in a grid of boxes, identify which box(es) answer the prompt.

(a) Triglycerides are described in Section 10-5c as an ester formed between three fatty acids and glycerol **Box C** shows a triglyceride.

(b) Oxidation of alcohols is discussed in Section 10-4. The final oxidation of a primary alcohol is a carboxylic acid. **Box G** has a molecule that contains the functional group carboxylic acid, –COOH.

(c) Condensation polymerization is discussed in Section 10-6b. Molecules with two carboxylic acid groups can serve as monomers for condensation polymerization. The molecule in **Box G** fits this description.

(d) Addition polymerization is discussed in Section 10-6a. The addition polymer typically is a long C–C chain molecule. A fragment of such a molecule is shown in **Box F**.

(e) Oxidation of alcohols is discussed in Section 10-4. The initial oxidation of a primary alcohol is an aldehyde. **Box D** has a molecule that contains the aldehyde functional group, –CHO.

(f) Amide bonds are discussed in Section 10-6b. **Box B** contains a molecule with an amide bond, –CONH.

(g) Addition polymerization is discussed in Section 10-6a. The monomer is a molecule with a C=C bond. Two such molecules are shown in **Boxes A and H**.

(h) Ester linkage is the C–O bond formed in the condensation of a carboxylic acid with an alcohol, as shown in the formation of a triglyceride in Section 10-5c. Therefore **Box C** has three ester linkages.

(i) Oxidation of alcohols is discussed in Section 10-4. The final oxidation of a secondary alcohol is a ketone. **Box I** has a molecule that contains the ketone functional group, –CO–.

(j) Alcohols are classified according to the number of carbon atoms directly bonded to the –C–OH carbon. If three carbon atoms are bonded directly, the compound is a tertiary alcohol. (See Section 10-4b). **Box E** shows a tertiary alcohol.

More Challenging Questions

117. *Result:* **(a) C_3H_8O (b) $C_3H_8O(\ell) + \frac{9}{2}\,O_2(g) \rightarrow 3\,CO_2(g) + 4\,H_2O(g)$ (c) Empirical formula, since it only identifies the smallest ratio of atoms (d) See structures below (e) See structures below (f) – 315 kJ/mol**

Analyze: Given the mass of a liquid containing only C, H, and O and the masses of combustion products, calculate the formula of the liquid, write the balanced equation for combustion, explain whether the formula is empirical or molecular, write three possible Lewis structures and identify the functional groups. Given the results of a reaction using this compound, determine which compound written in (d) is correct. Given the combustion enthalpy for a specified mass of the compound, calculate the standard formation enthalpy of the compound.

Plan and Execute:

(a) Calculate moles of each atom in the molecule, then set up and simplify a mole ratio, for the empirical formula. Follow the plan shown in the solution to Question 101.

The combustion produces 0.0979 g CO_2 and 0.0535 g H_2O. Determine the moles of C and O.

$$0.0979\text{ g } CO_2 \times \frac{1\text{ mol } CO_2}{44.01\text{ g } CO_2} \times \frac{1\text{ mol C}}{1\text{ mol } CO_2} = 2.22 \times 10^{-3}\text{ mol C}$$

$$0.0535 \text{ g } H_2O \times \frac{1 \text{ mol } H_2O}{18.02 \text{ g } H_2O} \times \frac{2 \text{ mol H}}{1 \text{ mol } H_2O} = 5.94 \times 10^{-3} \text{ mol H}$$

Calculate grams of C and O

$$0.0979 \text{ g } CO_2 \times \frac{1 \text{ mol } CO_2}{44.01 \text{ g } CO_2} \times \frac{1 \text{ mol C}}{1 \text{ mol } CO_2} = 2.22 \times 10^{-3} \text{ mol C}$$

$$2.22 \times 10^{-3} \text{ mol C} \times \frac{12.01 \text{ C}}{1 \text{ mol C}} = 0.0267 \text{ g C} \qquad 5.94 \times 10^{-3} \text{ mol H} \times \frac{1.008 \text{ g H}}{1 \text{ mol H}} = 0.00599 \text{ g H}$$

Subtract the mass of C and H from the sample mass to get the mass of O:

0.0446 g compound – 0.0267 g C – 0.00599 g H = 0.0119 g O

Calculate moles of O:

$$0.0199 \text{ g O} \times \frac{1 \text{ mol O}}{16.00 \text{ g O}} = 7.44 \times 10^{-4} \text{ mol O}$$

Set up a mole ratio:

0.00222 mol C : 0.00594 mol H : 0.000744 mol O

Divide all the numbers in the ratio by the smallest number of moles, 0.000744 mol: 3 C : 8 H : 1 O

Empirical formula: C_3H_8O

(b) Balance the combustion equation. Oxygen gas is the other reactant.

$$\mathbf{C_3H_8O(\ell) + \tfrac{9}{2}\, O_2(g) \rightarrow 3\, CO_2(g) + 4\, H_2O(g)}$$

(c) This is an **empirical formula**. More information would be needed to determine the molecular formula.

(d) Assuming the empirical equation is a molecular equation, draw three Lewis Structures by writing the C atoms in a chain, distributing the H atoms around the C atoms, then moving the O around to make different isomers. Start with the O on one end, as an OH, then move the OH to the center C atom, then move the O to between two C atoms in the chain:

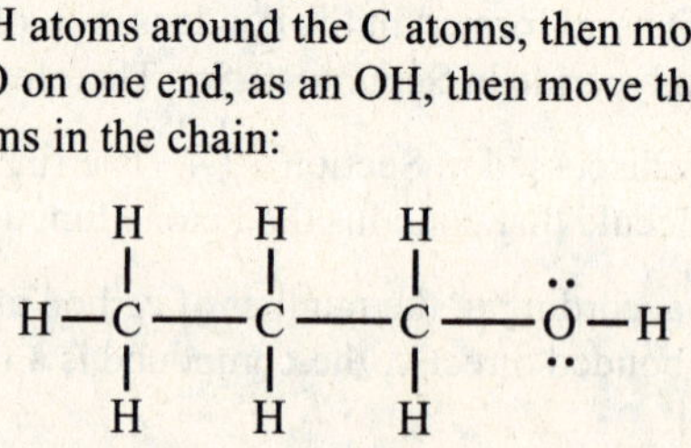

alcohol

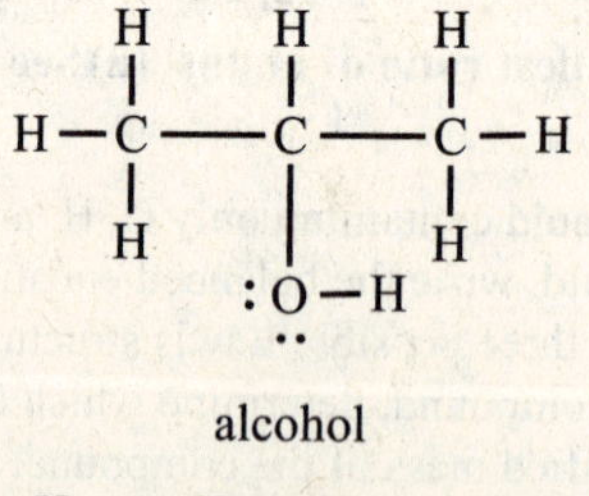

alcohol

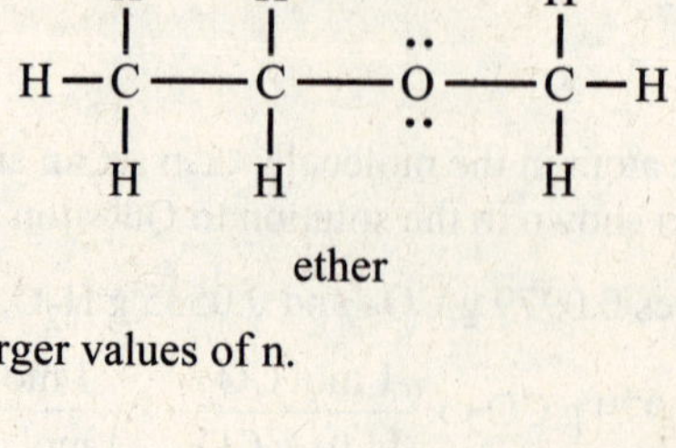

ether

Other answers are possible with larger values of n.

(e) In Section 10-4d, acetone is identified as CH_3COCH_3. Reaction of an alcohol with acidic potassium dichromate ion converts CH-OH into C=O. The second structures in (c) above would be oxidized to acetone, as shown in Conceptual Exercise 10.7 in Section 10-4 where it shows acetone produced by the oxidation of the secondary alcohol:

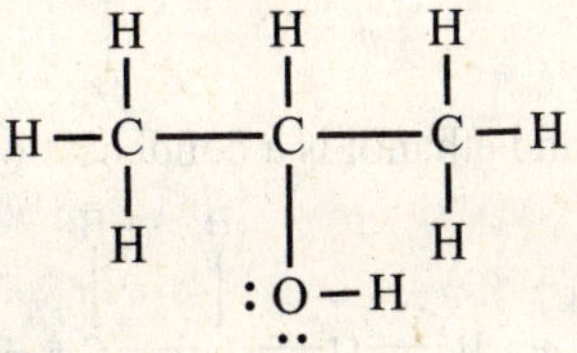

(f) Follow the Plan from the solution to Question 4.83. Calculate moles of C_3H_8O and use that to calculate the molar combustion enthalpy. Then use Hess's law on the combustion reaction and Appendix J to determine the $\Delta_f H$ of the compound.

$$10.00\text{ g } C_3H_8O \times \frac{1\text{ mol } C_3H_8O}{60.09\text{ g } C_3H_8O} = 0.1664\text{ mol } C_3H_8O$$

$$\Delta_c H = \frac{333.8\text{ kJ}}{0.1664\text{ mol } C_3H_8O} - 2006\text{ kJ/mol}$$

$$C_3H_8O(\ell) + \frac{9}{2}O_2(g) \rightarrow 3\,CO_2(g) + 4\,H_2O(g)$$

$$\Delta_c H° = 3\,\Delta_f H°\{CO_2(g)\} + 4\,\Delta_f H°\{H_2O(\ell)\} - \Delta_f H°\{C_3H_8O(\ell)\} - \frac{9}{2}\,\Delta_f H°\{O_2(g)\}$$

Look up the $\Delta_f H°$ values in Table 4.2.

$$-2006\text{ kJ/mol} = 3 \times (-393.509\text{ kJ/mol}) + 4 \times (-285.83\text{ kJ/mol}) - \Delta_f H°\{C_8H_8(\ell)\} + \frac{9}{2} \times (0\text{ kJ/mol})]$$

Solve for $\Delta_f H°\{C_3H_8O(\ell)\}$

$$\Delta_f H°\{C_3H_8O(\ell)\} = +2006\text{ kJ/mol} + 3 \times (-393.509\text{ kJ/mol}) + 4 \times (-285.83\text{ kJ/mol})$$

$$\Delta_f H°\{C_3H_8O(\ell)\} = -315\text{ kJ/mol } C_3H_8O$$

123. *Result:* **(a) X is ethanol, Y is acetic acid, Z is acetaldehyde; see Lewis structures below (b) See equations below (c) See explanation below.**

Analyze: Given three chemical names and some information about reactions characterizing them, determine which substances are reacting. Write Lewis structures and equations using structural formulas for the reactions observed. Explain information given regarding the oxidation of two of these substances.

(a) An ester is formed by the reaction of a carboxylic acid and an alcohol. So, X and Y must be acetic acid and ethanol. Because Y is also seen to produce an acid aqueous solution, Y must be acetic acid. That leaves X to be ethanol and Z to be acetaldehyde.

X = Ethanol is CH_3CH_2OH. It's Lewis structure looks like this:

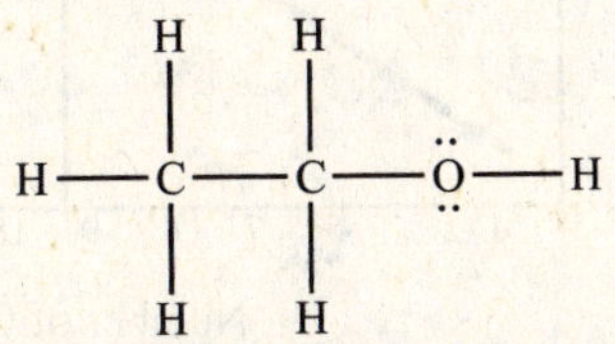

Y = Acetic acid is CH_3COOH. It's Lewis structure looks like this:

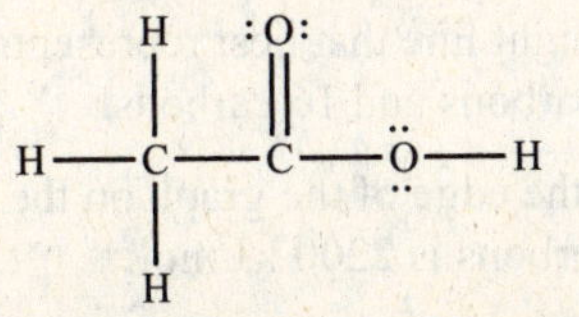

Z = Acetaldehyde is CH_3COH. It's Lewis structure looks like this:

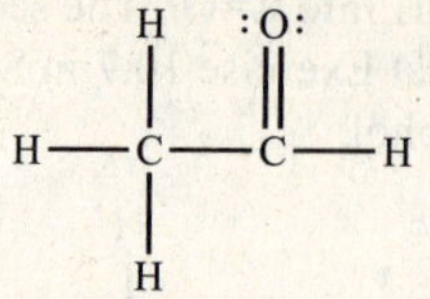

(b) The reaction of acetic acid and ethanol is a condensation reaction. This is the specific equation:

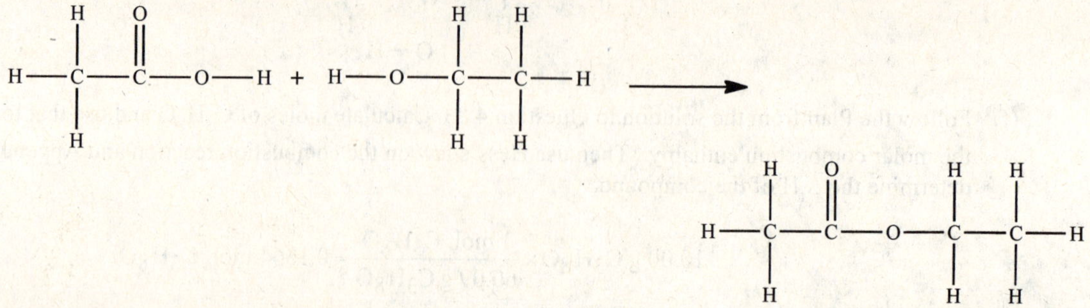

The ionization of acetic acid looks like this:

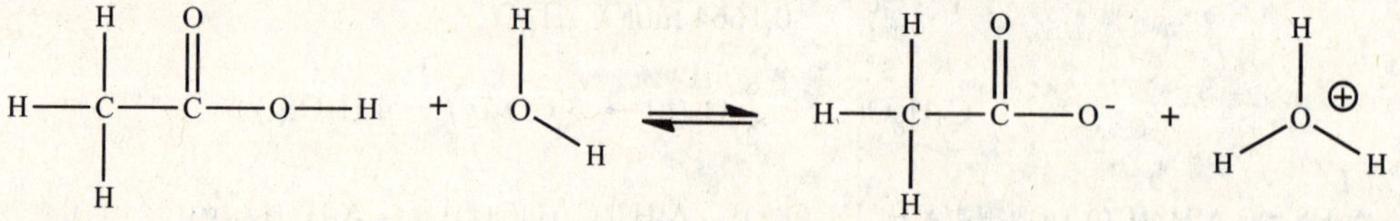

(c) Ethanol is a primary alcohol that is initially oxidized to acetaldehyde; further oxidation of acetaldehyde produces acetic acid. When acetaldehyde is the initial reactant, it is directly oxidized to acetic acid.

124. *Result:* **(a) See graph below; for C_3H_8: 2200 kJ/mol; for C_9H_{20}: 6100 kJ/mol; for $C_{16}H_{34}$: 10700 kJ/mol (b) The bond enthalpies increase because breaking C–H bonds require more energy than breaking C–C bonds.**

Explanation:

(a) Graph the enthalpy of combustion per mole of carbon atoms against the number of carbon atoms:

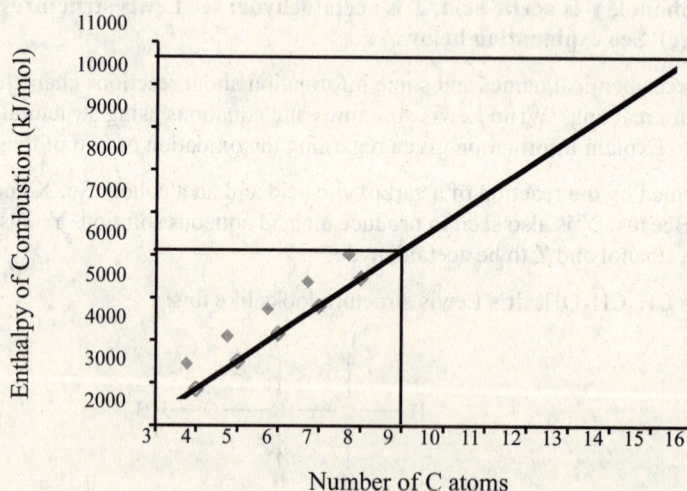

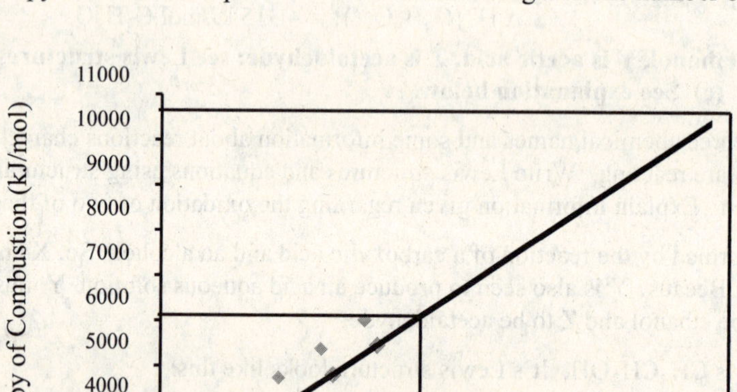

After extrapolating the straight line that best represents the linear relationship, we can determine an estimate for 3-carbons, 9-carbons and 16 carbons.

For three carbons (look at the edge of the graph on the left), an estimated enthalpy of vaporization per mole of carbon for three carbons is 2200 kJ/mol.

For nine carbons (see horizontal and vertical lines provided), an estimated enthalpy of vaporization per mole of carbon for three carbons is 6100 kJ/mol.

For 16 carbons (look at the edge of the graph on the right), an estimated enthalpy of vaporization per mole of carbon for three carbons is 10700 kJ/mol.

(a) Alkanes have the formula C_nH_{2n+2}, so larger alkane molecules have larger proportion of C–H bonds. Looking at the bond enthalpies from Table 6.2, $D_{C–C}$ = 356 kJ/mol and $D_{C–H}$ = 416 kJ/mol, the C–H bonds require somewhat more energy to break, causing a positive slope in the graph.

126. *Result:* **(a) See four-unit polymer segment below (b) Condensation polymerization**

Analyze:

(a) The monomers combine by losing a water molecule as shown here:

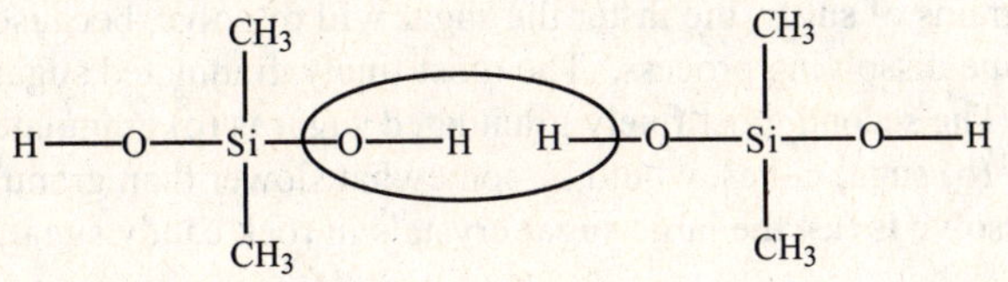

The four-monomer unit looks like this:

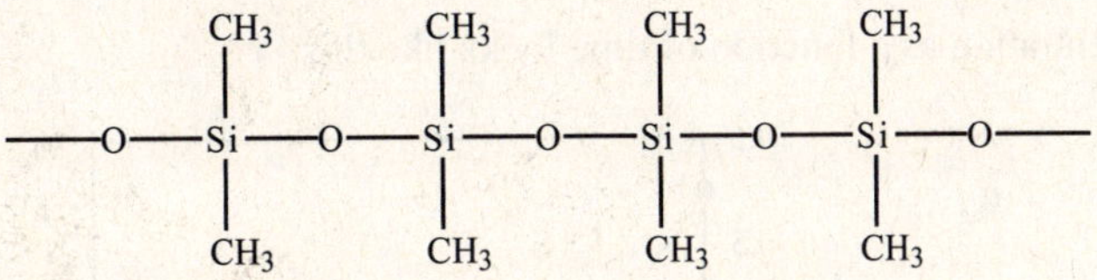

(b) This type of polymerization is called condensation polymerization.

chemical potential energy can be stored in them than the sugars that make up carbohydrates.

Chapter 11: Chemical Kinetics: Rates of Reactions

Solutions for Red-Numbered Questions for Review and Thought

Topical Questions

Reaction Rate (Section 11-1)

9. *Result:* **(a) dissolves fastest, (d) dissolves slowest because particle sizes affect surface area.**

   Explanation: As described in Section 11-1, heterogeneous reactions take place at the interface between the two phases. The smaller the grains of sugar, the faster the sugar will dissolve, because the greater surface contact with the water speeds up the dissolving process. The most finely granulated sugar is powdered sugar, so **(d) will dissolve the fastest**. The second most finely granulated sugar is (c) granulated sugar. Granulated sugar compressed into cubes, or (b) sugar cubes, would be somewhat slower than granulated sugar to dissolve. The **slowest of the four to dissolve is (a)**, the large sugar crystals in rock candy sugar.

11. *Result:* **(a) see graph below; declining nonlinear shape, similar to that in Figure 11.3 (b) 0.5 mol $L^{-1}min^{-1}$, 0.2 mol $L^{-1}min^{-1}$, concentration of reactant is decreasing (c) 0.2 mol $L^{-1}min^{-1}$**

   (a) The graph of concentration as a function of time looks like this:

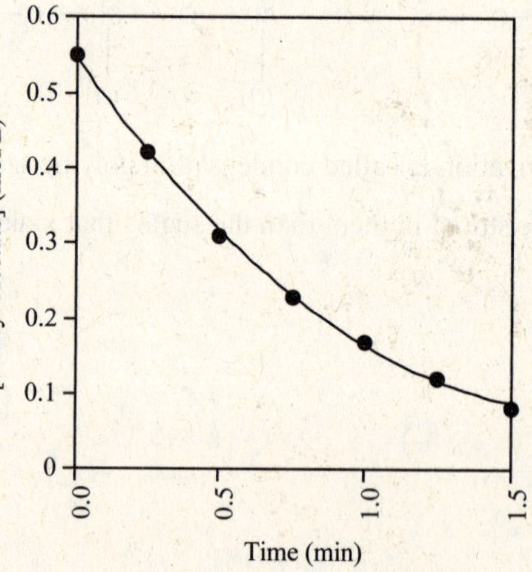

   The curve is qualitatively described as a nonlinear shape that declines monotonically. A more precise investigation of the shape suggests that the curve is best described with a second-order polynomial function. Figure 11.3 has a very similar qualitative shape; it is also a declining nonlinear curve.

   (b) *Analyze:* Given data of concentrations at various times during the course of a chemical reaction, find the rate of change for each time interval.

   Plan: The time intervals requested are not from the data given, so use the graph to interpolate the concentration at these times (as shown below). Then calculate the average rate for each time interval, use the method described in Section 11-1:

$$\text{Rate} = \frac{-\Delta[\text{reactant}]}{\Delta t} = \frac{-\left([\text{reactant}]_{\text{final}} - [\text{reactant}]_{\text{initial}}\right)}{t_{\text{final}} - t_{\text{initial}}} = \frac{[\text{reactant}]_{\text{initial}} - [\text{reactant}]_{\text{final}}}{t_{\text{final}} - t_{\text{initial}}}$$

   Interpolation of the four time points requested is done by drawing vertical lines at those times and finding the horizontal line that crosses at the curve:

   Execute:

   The crossing points give us the following values:

Time (min)	[Phenyl Acetate] (mol/L)
0.20	0.45
0.40	0.36
1.20	0.12
1.40	0.09

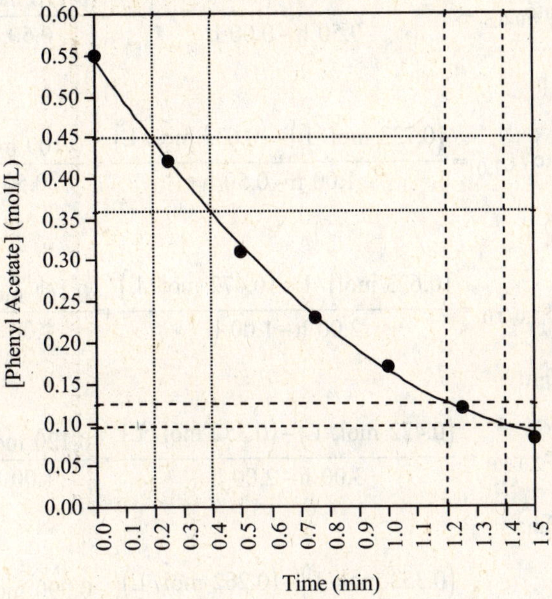

The rate of change of the reactant phenyl acetate concentration from 0.20 to 0.40 min:

$$\text{Rate}_{0.2\text{-}0.4} = \frac{(0.45\text{ mol/L}) - (0.36\text{ mol/L})}{0.40\text{ min} - 0.20\text{ min}} = \frac{0.09\text{ mol/L}}{0.20\text{ min}} = 0.5\ \frac{\text{mol}}{\text{L}\cdot\text{min}}$$

The rate of change of the reactant phenyl acetate concentration from 1.2 to 1.4 min:

$$\text{Rate}_{1.2\text{-}1.4} = \frac{(0.12\text{ mol/L}) - (0.09\text{ mol/L})}{1.40\text{ min} - 1.20\text{ min}} = \frac{0.03\text{ mol/L}}{0.20\text{ min}} = 0.2\ \frac{\text{mol}}{\text{L}\cdot\text{min}}$$

The rate of change decreases because, as time passes, the concentration of reactant is decreasing.

(c) The rate of change of the reactant phenyl acetate is the same as the rate of change of the product phenol, since the mole ratio between them is 1:1. So find the rate of change of the reactant phenyl acetate concentration as done in (b) and equate them.

The rate of change of the reactant phenyl acetate concentration from 1.00 to 1.25 min:

$$\text{Rate}_{1.00\text{-}1.25} = \frac{(0.17\text{ mol/L}) - (0.12\text{ mol/L})}{1.25\text{ min} - 1.00\text{ min}} = \frac{0.05\text{ mol/L}}{0.25\text{ min}} = 0.2\ \frac{\text{mol}}{\text{L}\cdot\text{min}}$$

$$\text{Rate}_{1.00\text{-}1.25} = -\frac{\Delta[\text{phenyl acetate}]}{\Delta t} = \frac{\Delta[\text{phenol}]}{\Delta t} = 0.2\ \frac{\text{mol}}{\text{L}\cdot\text{min}}$$

☑ *Reasonable Result Check:* According to the discussion of rate dependence on concentration in Section 11-2, it makes sense that the rate decreases as the reactant concentration decreases. Only the limitations in the significant figures prevents us from seeing how $\text{Rate}_{1.00\text{-}1.25}$ and $\text{Rate}_{1.2\text{-}1.4}$ differ.

13. *Result:* **(a) 0.23 mol L^{-1}h^{-1} (b) 0.20 mol L^{-1}h^{-1} (c) 0.161 mol L^{-1}h^{-1} (d) 0.12 mol L^{-1}h^{-1} (e) 0.090 mol L^{-1}h^{-1} (f) 0.066 mol L^{-1}h^{-1}**

Analyze: Given data of concentrations at various times during the course of a chemical reaction, find the rate of change for each time interval.

Plan: To calculate the average rate for each time interval, use the answers from Questions 11 and 12:

(a) From 0 to 0.50 h:

$$\text{Rate}_{0\text{-}0.5} = \frac{(0.849\ \text{mol/L}) - (0.733\ \text{mol/L})}{0.50\ \text{h} - 0.00\ \text{h}} = \frac{0.116\ \text{mol/L}}{0.50\ \text{h}} = 0.23\ \frac{\text{mol}}{\text{L}\cdot\text{h}}$$

(b) From 0.5 to 1.00 h:

$$\text{Rate}_{0.5\text{-}1.0} = \frac{(0.733\ \text{mol/L}) - (0.633\ \text{mol/L})}{1.00\ \text{h} - 0.50\ \text{h}} = \frac{0.100\ \text{mol/L}}{0.50\ \text{h}} = 0.20\ \frac{\text{mol}}{\text{L}\cdot\text{h}}$$

(c) From 1.00 to 2.00 h:

$$\text{Rate}_{1.0\text{-}2.0} = \frac{(0.633\ \text{mol})/\text{L} - (0.472\ \text{mol/L})}{2.00\ \text{h} - 1.00\ \text{h}} = \frac{0.161\ \text{mol/L}}{1.00\ \text{h}} = 0.161\ \frac{\text{mol}}{\text{L}\cdot\text{h}}$$

(d) From 2.00 to 3.00 h:

$$\text{Rate}_{2.0\text{-}3.0} = \frac{(0.472\ \text{mol/L}) - (0.352\ \text{mol/L})}{3.00\ \text{h} - 2.00\ \text{h}} = \frac{0.120\ \text{mol/L}}{1.00\ \text{h}} = 0.120\ \frac{\text{mol}}{\text{L}\cdot\text{h}}$$

(e) From 3.00 to 4.00 h:

$$\text{Rate}_{3.0\text{-}4.0} = \frac{(0.352\ \text{mol/L}) - (0.262\ \text{mol/L})}{4.00\ \text{h} - 3.00\ \text{h}} = \frac{0.090\ \text{mol/L}}{1.00\ \text{h}} = 0.090\ \frac{\text{mol}}{\text{L}\cdot\text{h}}$$

(f) From 4.00 to 5.00 h:

$$\text{Rate}_{3.0\text{-}4.0} = \frac{(0.262\ \text{mol/L}) - (0.196\ \text{mol/L})}{5.00\ \text{h} - 4.00\ \text{h}} = \frac{0.066\ \text{mol/L}}{1.00\ \text{h}} = 0.066\ \frac{\text{mol}}{\text{L}\cdot\text{h}}$$

☑ *Reasonable Result Check:* According to the discussion of rate dependence on concentration in Section 11-2, it makes sense that the rate decreases as the reactant concentration decreases.

14. *Result:* **(a) The average rate is directly proportional to average [N_2O_5] for each pair. (b) k = 0.29 hr^{-1} (c) 0.12 mol L^{-1}h^{-1}**

Analyze: Given data of concentrations at various times during the course of a chemical reaction and the average rate calculated over different time intervals, show that the reaction obeys a specific given rate law, evaluate the rate constant by averaging. Determine the instantaneous rate of reaction at a specific time.

Plan: The average rates over each time interval were calculated in the solution to Question 11. Find the average concentration over that interval. If the rate is proportional to the concentration, then a graph of concentration vs. rate should be linear.

Execute:

(a) From 0 to 0.50 h, the rate was 0.23 mol L^{-1}h^{-1}.

$$\text{Average concentration}_{0\text{-}0.5} = \frac{(0.849\ \text{mol/L}) + (0.733\ \text{mol/L})}{2} = 0.791\ \frac{\text{mol}}{\text{L}}$$

From 0.5 to 1.00 h, the rate was 0.20 mol L^{-1}h^{-1}.

$$\text{Average concentration}_{0.5\text{-}1.0} = \frac{(0.733\ \text{mol/L}) + (0.633\ \text{mol/L})}{2} = 0.683\ \frac{\text{mol}}{\text{L}}$$

From 1.00 to 2.00 h, the rate was 0.161 mol $L^{-1}h^{-1}$

$$\text{Average concentration}_{1.0\text{-}2.0} = \frac{(0.633\text{ mol/L}) + (0.472\text{ mol/L})}{2} = 0.553\ \frac{\text{mol}}{\text{L}}$$

From 2.00 to 3.00 h, the rate was 0.120 mol $L^{-1}h^{-1}$

$$\text{Average concentration}_{2.0\text{-}3.0} = \frac{(0.472\text{ mol/L}) + (0.352\text{ mol/L})}{2} = 0.412\ \frac{\text{mol}}{\text{L}}$$

From 3.00 to 4.00 h, the rate was 0.090 mol $L^{-1}h^{-1}$.

$$\text{Average concentration}_{3.0\text{-}4.0} = \frac{(0.352\text{ mol/L}) + (0.262\text{ mol/L})}{2} = 0.307\ \frac{\text{mol}}{\text{L}}$$

From 4.00 to 5.00 h, the rate was 0.066 mol $L^{-1}h^{-1}$.

$$\text{Average concentration}_{4.0\text{-}5.0} = \frac{(0.262\text{ mol/L}) + (0.196\text{ mol/L})}{2} = 0.229\ \frac{\text{mol}}{\text{L}}$$

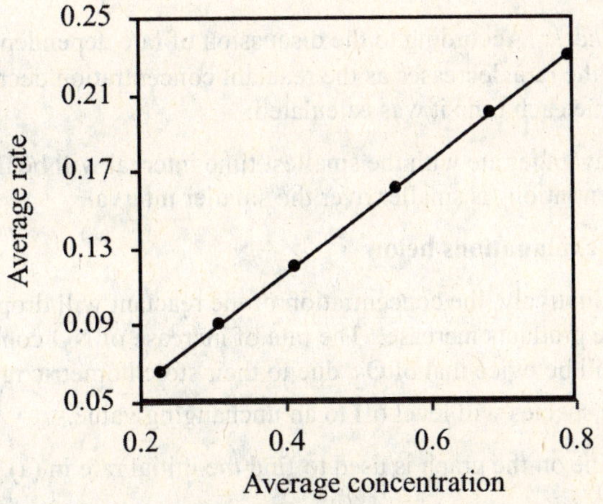

The linear relationship between rate and concentration satisfies the equation described in the question:

$$\text{Rate} = k\,[N_2O_5]$$

(b) From 0 to 0.50 h, the rate is 0.23 mol $L^{-1}h^{-1}$ and the average concentration is 0.791 mol/L.

$$k = \frac{\text{Rate}}{[N_2O_5]} = \frac{0.23\frac{\text{mol}}{\text{L}\cdot\text{h}}}{0.791\frac{\text{mol}}{\text{L}}} = 0.29\text{h}^{-1}$$

Similarly, for each of the other time increments:

Average rate, mol $L^{-1}h^{-1}$	Average concentration, mol/L	k (h^{-1})
0.23	0.791	0.29
0.20	0.683	0.29
0.161	0.553	0.291
0.120	0.412	0.291
0.090	0.307	0.29
0.066	0.229	0.29

The average value of k is 0.29 h^{-1}.

(c) We could use the graph to find the concentration at t = 2.50 hr, but the precision of the graph at this scale is limited. Section 11-3 shows us a way to calculate the concentration: The rate law gives us the order. From the order, we can select an integrated rate law from Section 11-3 to determine the concentration of the reactant after a certain elapsed time. Use that concentration to determine the rate of the reaction using the rate law.

The reaction is first-order (the power to which the concentration is raised in the rate law is one). The integrated first-order rate law for this reaction is:

$$\ln[N_2O_5]_t = -kt + \ln[N_2O_5]_0$$

Here, $[N_2O_5]_0 = 0.849$ mol/L, k = 0.29 h^{-1}, and t = 2.50 h.

$$\ln[N_2O_5]_t = -(0.29\ h^{-1}) \times (2.50\ h) + \ln(0.849)$$

$$\ln[N_2O_5]_t = -0.889$$

$$[N_2O_5]_t = 0.411\ mol/L$$

The rate law says: Rate $= k[N_2O_5]_t = (0.29\ h^{-1}) \times (0.411\ mol/L) = 0.12\ \frac{mol}{L \cdot h}$

☑ *Reasonable Result Check:* According to the discussion of rate dependence on concentration in Section 11-2, it makes sense that the rate decreases as the reactant concentration decreases. It also makes sense that the value of k is the same each time it was calculated.

It makes sense that the average rate with the smallest time interval will be the closest to the instantaneous rate, since the range of concentrations is smaller over the smaller interval.

15. *Result:* **See graph and explanations below**

Analyze and Plan: Qualitatively, the concentration of the reactant will drop nonlinearly (first-order reaction) as the concentrations of the products increase. The rate of increase of NO concentration will mirror the rate of decrease of NO_2, and will be twice that of O_2, due to their stoichiometric ratios being 2 : 2 : 1. At late times, the concentrations of all species will level off to an unchanging value.

Execute: (The dashed line on the graph is used to find the initial rate in (a).)

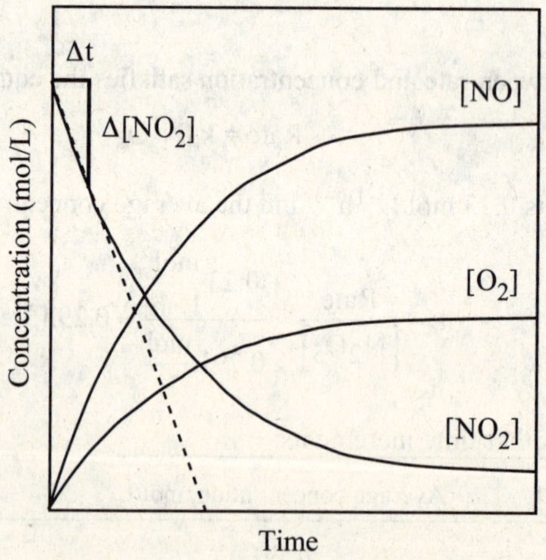

(a) As shown on the graph, the initial rate can be determined by taking a small time interval (Δt) near the initial time where the curve is still well approximated by a straight line (represented by the dashed line) and finding the slope of that line.

$$\text{Rate} = \frac{-\Delta[NO_2]}{\Delta t} = -\text{ slope of the straight line}$$

(b) As shown on the graph at very long times, the final rate will be zero because the concentration stops changing and the numerator of the rate expression is zero. That will happen for one of two reasons, either all the reactants will have been used up or the reaction will reach an equilibrium state.

17. *Result:* **(a)** $\frac{-\Delta[N_2O_4]}{\Delta t}$ **(b)** $\frac{\Delta[N_2]}{\Delta t}$

Analyze and Plan: Express the rate of the reaction in terms of changing concentrations of a reactant and a product. Rate must be a positive quantity. Δt is positive since time is increasing. Because the concentrations of reactants decrease with time, the rate of change of reactant concentration is negative; therefore, the numerator of the reactant rate expression will include a minus sign to make the rate a positive number. Because the concentrations of products increase with time, the rate of change of reactant concentration is positive as must be the rate.

Execute:

(a) The rate of depletion of the reactant N_2O_4 is $\frac{-\Delta[N_2O_4]}{\Delta t}$

(a) The rate of formation of the product N_2 is $\frac{\Delta[N_2]}{\Delta t}$

Effect of Concentration on Reaction Rates (Section 11-2)

19. *Result:* **(a) nine times faster (b) one quarter as fast**

Analyze: Given the rate law of a reaction, determine what happens to the rate when the concentration is changed.

Plan: Knowing the rate law, the rate increase can be determined by creating a ratio to provide a simple relationship between the rate changes and the concentration changes.

The rate law is: rate = $k[A]^2$. Make a ratio of the rate law of the reaction under two different conditions: the initial condition (designated by a subscript "1", $\text{rate}_1 = k[A]_1^2$) compared to the final condition (designated by subscript "2", $\text{rate}_2 = k[A]_2^2$)

$$\text{rate change factor} = \frac{\text{rate}_2}{\text{rate}_1} = \frac{k[A]_2^2}{k[A]_1^2} = \left(\frac{[A]_2}{[A]_1}\right)^2$$

Execute:

(a) When concentration is tripled, the concentration ratio is $\frac{[A]_2}{[A]_1} = 3$, so the rate change factor = $3^2 = 9$. The rate increases by a factor of nine.

(b) When concentration is halved, the concentration ratio is $\frac{[A]_2}{[A]_1} = \frac{1}{2}$, so the rate change factor = $\left(\frac{1}{2}\right)^2 = \frac{1}{4}$.

The rate will be quartered.

☑ *Reasonable Result Check:* It makes sense that the rate is faster when the concentration is larger and vice versa, also that the rate change is larger than the concentration change, because the reaction is second-order.

21. *Result:* **(a) rate = $k[NO_2]^2$ (b) rate is quartered (c) rate is unchanged**

Analyze and Plan: Adapt the plan from the solution to Question 19.

Execute:

(a) The reaction is second-order in NO_2 and zeroth-order in CO, so the rate law has powers of 2 and zero on the concentrations of NO_2 and CO, respectively.

$$\text{rate} = k[NO_2]^2[CO]^0$$

Because any number raised to the zero power results in 1, that says the rate is unaffected by the concentration of CO.

$$\text{rate} = k[NO_2]^2$$

(b) A ratio gives the rate change factor:

$$\text{rate change factor} = \frac{\text{rate}_2}{\text{rate}_1} = \frac{k[NO_2]_2^2}{k[NO_2]_1^2} = \left(\frac{[NO_2]_2}{[NO_2]_1}\right)^2$$

When the concentration of NO_2 is halved, the concentration ratio is $\frac{[NO_2]_2}{[NO_2]_1} = \frac{1}{2}$,

so the rate change factor $= \left(\frac{1}{2}\right)^2 = \frac{1}{4}$. The rate will be quartered.

(c) Because the rate law does not include concentration dependence for CO, the rate will be unchanged when the CO concentration is doubled.

☑ *Reasonable Result Check:* Lowering the $[NO_2]$ lowers the rate, but changing the CO concentration does not change the rate, as expected.

23. *Result:* **(a) (i) 9.0×10^{-4} M/h, (ii) 1.8×10^{-3} M/h, (iii) 3.6×10^{-3} M/h; (b)-(d) see explanations below**

Analyze: Given the rate law of a reaction, the rate constant, and several concentrations, determine the instantaneous rates at each of those concentrations. Relate changes in concentration to changes in rate. Related changes to rate law. Relate changes of a reactant to changes of a product.

(a) *Plan:* Plug the concentration and the rate constant into the rate law:

$$\text{rate} = k[Pt(NH_3)_2Cl_2]$$

Execute:

(i) $\text{rate}_{0.010} = (0.090\ h^{-1})(0.010\ M) = 9.0 \times 10^{-4}\ M/h$

(ii) $\text{rate}_{0.020} = (0.090\ h^{-1})(0.020\ M) = 1.8 \times 10^{-3}\ M/h$

(iii) $\text{rate}_{0.040} = (0.090\ h^{-1})(0.040\ M) = 3.6 \times 10^{-3}\ M/h$

(b) If initial concentration of $Pt(NH_3)_2Cl_2$ is high, the initial rate of disappearance of $Pt(NH_3)_2Cl_2$ will be high. If initial concentration of $Pt(NH_3)_2Cl_2$ is low, the initial rate of disappearance of $Pt(NH_3)_2Cl_2$ will be small. The rate of disappearance of $Pt(NH_3)_2Cl_2$ is directly proportional to the concentration of $Pt(NH_3)_2Cl_2$.

(c) The rate law shows direct proportionality between rate and the concentration of $Pt(NH_3)_2Cl_2$.

(d) The balanced equation shows that when one $Pt(NH_3)_2Cl_2$ reacts, one Cl^- forms. When the initial concentration of $Pt(NH_3)_2Cl_2$ is high, the rate of appearance of Cl^- will be high. When initial concentration is low, rate of appearance Cl^- will be small. The rate of appearance of Cl^- is directly proportional to the concentration of $Pt(NH_3)_2Cl_2$.

☑ *Reasonable Result Check:* The rate is faster when the concentration is higher. The rate of formation of a product depends on the concentration of the reactant.

25. *Result:* **(a) rate = k [I][II] (b) k = 1.04 L $mol^{-1}s^{-1}$ (c) First order in I and first order in II (d) 5.9×10^{-9} mol L^{-1} s^{-1}**

Analyze: Given the initial concentrations and initial rates of a reaction for several different experimental conditions for the same chemical reaction, determine the rate law and the rate constant for the reaction, the orders of the reactants, and the instantaneous rate under specific conditions.

Plan: In Section 11-2, the method of finding the rate law from initial rates is described for getting the orders. However, comparing pairs of experiments where only one of the concentrations is different and relating that to the changes in the rate does not give consistent results. So, here we will derive a linear equation relating the concentrations and the rates and graph the results. Once the orders are determined, plug the data into the rate law to determine the value of k.

Because I and II are reactants and are both varied in the different experiments, we will seek a rate law that looks like this: rate = $k[I]^i[II]^j$ where k, i, and j are currently unknown.

(a) Comparing experiments that have [I] constant, the rate law can be simplified to:

$$\text{rate} = k'[II]^j \qquad \text{where } k' = k[I]^i$$

If we take the log of both sides of the equation, we can derive a linear relationship:

$$\log(\text{rate}) = \log(k') + \log([II]^j)$$

$$\log(\text{rate}) = \log(k') + j\log[II]$$

$$\log(\text{rate}) = j\log[II] + \log(k')$$

Comparing these equations to the equation of a line: $y = m x + b$, we see that if we plot log(rate) against log[II], the slope of the line will be the reaction order, j.

Similarly comparing experiments with [II] constant, the rate law can be simplified to:

$$\text{rate} = k''[II]^I$$

$$\text{where } k'' = k[II]^j$$

This gives a similar linear relationship: $\log(\text{rate}) = i\log[I] + \log(k'')$

Comparing this equation to the equation of a line: $y = m x + b$, we see that if we plot log(rate) against log[I], the slope of the line will be the reaction order, i.

Execute: Four data sets have constant [I] and four data sets have constant [II].

log[II] with constant [I]	**log rate** with constant [I]	**log[I]** with constant [II]	**log rate** with constant [II]
-4.745	-8.509	-4.783	-8.824
-4.453	-8.201	-4.606	-8.569
-4.149	-7.951	-4.304	-8.345
-3.975	-7.752	-3.827	-7.752

log rate

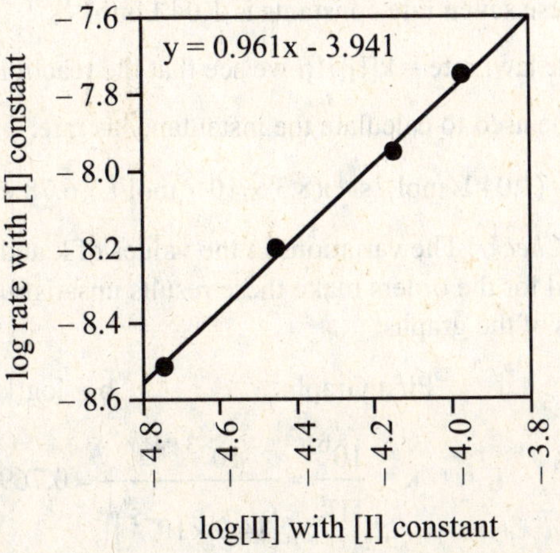

Therefore, the reaction order of reactant II (j) is one.

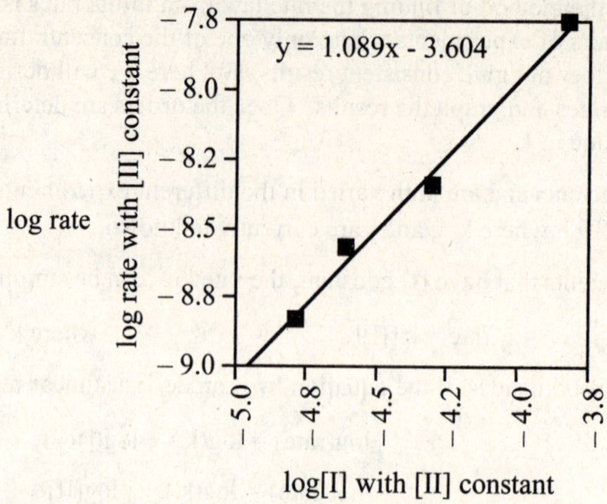

Therefore, the reaction order of reactant I (i) is also one. So, rate = k[I][II]

(b) Solve the rate law for k: $k = \dfrac{\text{rate}}{[\text{I}][\text{II}]}$

Plug in each experiment's data. Here is an example of the first experiment's calculation:

$$k_1 = \frac{1.50 \times 10^{-9}\ \text{mol L}^{-1}\text{s}^{-1}}{(1.65 \times 10^{-5}\ \text{mol/L})(10.6 \times 10^{-5}\ \text{mol/L})} = 0.853\ \text{L mol}^{-1}\text{s}^{-1}$$

$[\text{I}] \times 10^5$ (mol/L)	$[\text{I}] \times 10^5$ (mol/L)	Initial rate $\times 10^9$ (mol $L^{-1}s^{-1}$)	Rate constant (L $mol^{-1}s^{-1}$)
1.65	10.6	1.50	0.853
14.9	10.6	17.7	1.12
14.9	7.10	11.2	1.06
14.9	3.52	6.30	1.20
14.9	1.76	3.10	1.18
4.97	10.6	4.52	0.853
2.48	10.6	2.70	1.03

The average of these seven rate constants is 1.04 L $mol^{-1}s^{-1}$

(c) Examining the rate law, rate = k[I][II], we see that the reaction is **first-order in I** and **first-order in II**.

(d) The rate law can be used to calculate the instantaneous rate:

rate = k[I][II] = (1.04 L $mol^{-1}s^{-1}$)(8.3 × 10^{-5} mol/L)(6.78 × 10^{-5} mol/L) = **5.9 × 10^{-9} mol L^{-1} s^{-1}**

☑ *Reasonable Result Check:* The variations in the values of k and the deviation of the slopes from the exact integer values expected for the orders make these results unsatisfying. In addition, we can try to get the value of k from the y-intercepts of the graphs:

First Graph: $b = \log(k') = \log(k[\text{I}]^i)$

$$k = \frac{10^b}{[\text{I}]^i} = \frac{10^{-3.941}}{\left(14.9 \times 10^{-5}\right)^1} = 0.769\ \text{L mol}^{-1}\text{s}^{-1}$$

Second Graph: $b = \log(k'') = \log(k[\text{II}]^j)$

$$k = \frac{10^b}{[II]^j} = \frac{10^{-3.604}}{\left(10.6 \times 10^{-5}\right)^1} = 2.35 \text{ L mol}^{-1}\text{s}^{-1}$$

These are not very close to each other either. These wide variations might suggest a systematic error in the collection of the data under various conditions. The calculated instantaneous rate in (d) is a similar size to that described in the data table.

Rate Law and Order of Reaction (Section 11-3)

28. *Result:* **(a) First-order in A, third-order in B, and fourth-order overall (b) First-order in A, first-order in B, and second-order overall (c) First-order in A, zero-order in B, and first-order overall (d) Third-order in A, first-order in B, and fourth-order overall**

Analyze: Given some rate laws, determine the orders.

Plan: Reaction order is the power to which the concentration of a component is raised. If a reactant does not appear in the rate law at all, then its order is zero. The overall order is the sum of all the individual orders.

Execute:

(a) In the rate law: Rate = $k[A][B]^3$ the order with respect to A is **one** and the order with respect to B is **three**. The overall order is (1 + 3 = 4) **four**.

(b) In the rate law: Rate = $k[A][B]$ the order with respect to A is **one** and the order with respect to B is **one**. The overall order is (1 + 1 = 2) **two**.

(c) In the rate law: Rate = $k[A]$ the order with respect to A is **one** and the order with respect to B is **zero**. The overall order is (1 + 0 = 1) **one**.

(d) In the rate law: Rate = $k[A]^3[B]$ the order with respect to A is **three** and the order with respect to B is **one**. The overall order is (3+ 1 = 4) **four**.

☑ *Reasonable Result Check:* The orders match the powers. The overall order is the sum of individual orders.

30. *Result:* **Equation (a) cannot be correct**

Explanation: If a reaction is second-order in B, the rate law must include the concentration of B squared. That means "(a) Rate = k[A][B]" cannot be right, since it shows [B] raised to the first power.

32. *Result:* **(a)** $\frac{\Delta[NH_3]}{\Delta t} = k$ **(b) 2.5 × 10⁻⁴ mol L⁻¹min⁻¹**

Analyze: Given the order of a reaction, write the rate expression, calculate the rate at a given concentration, and determine at what concentration of the reactant would the rate be equal to the rate constant.

(a) The reactant is zero order, so $\frac{\Delta[NH_3]}{\Delta t} = k$

(b) Because the reaction is zero order, the rate is independent of the concentration.

$$\text{rate} = k = 2.5 \times 10^{-4} \text{ mol L}^{-1}\text{min}^{-1}$$

☑ *Reasonable Result Check:* Because the reaction is zero order, the rate is equal to the rate constant at any concentration above zero.

33. *Result:* **(a) rate = k [phenyl acetate] (b) First-order in phenyl acetate (c) k = 1.259 s⁻¹ (d) 0.13 mol L⁻¹s⁻¹**

Analyze: Given concentration data as a function of time, find the rate law, the reaction order, the rate constant, and the rate at a specific concentration.

Plan: In Section 11-3, the method of finding the rate law from linear graphs is described. Construct the three graphs described in Table 11.2, representing zero, first and second-order functions. Determine which one of these three graphs is the linear graph. The linear graph describes the order of the reaction. Once the order is determined, the slope of the straight line is used to determine the value of k.

(a) The reactants are phenyl acetate and water, and the question tells us to assume that the concentration of water does not change. Data are available for the concentration changes with time for phenyl acetate, so the rate law looks like this:

$$\text{rate} = k[\text{phenyl acetate}]^i \quad \text{where k and i are currently unknown.}$$

Make three graphs. The zero-order graph is [phenyl acetate] vs. time, the first-order graph is ln[phenyl acetate] vs. time, and the second-order graph is 1/[phenyl acetate] vs. time, then identify which one produces a straight line.

Execute:

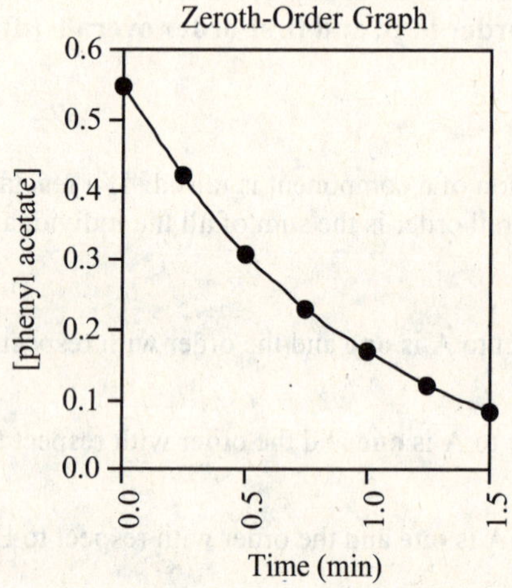

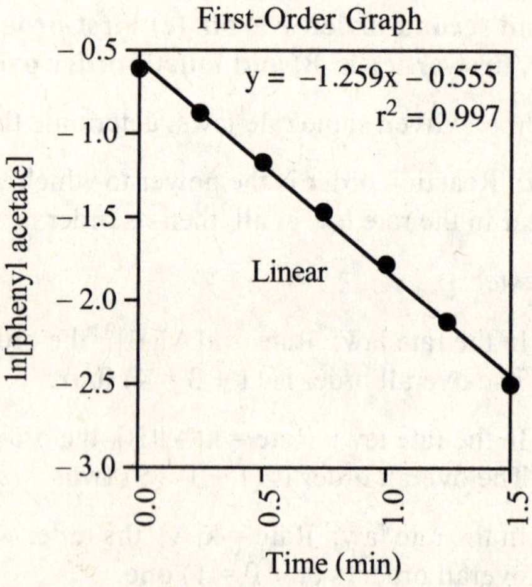

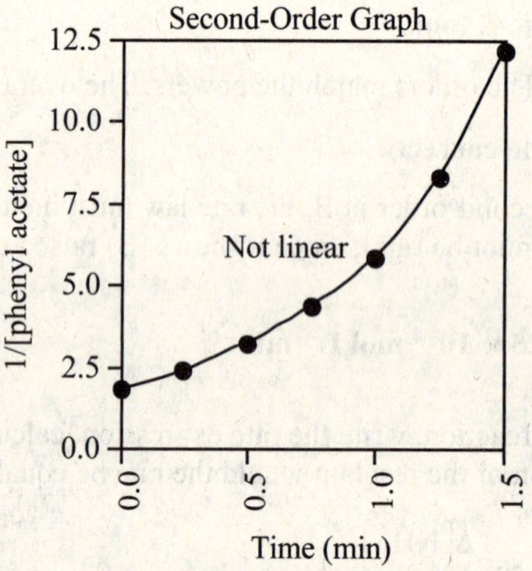

Of the three graphs, the only one that is linear is the ln[phenyl acetate] vs. time graph. That means the reaction is **first-order**. The rate law looks like this:

$$\text{rate} = k[\text{phenyl acetate}]^1$$

(b) The order of the reaction with respect to phenyl acetate is **one.**

(c) The slope of the straight-line graph is equal to −k. So, here: $k = 1.259 \text{ min}^{-1}$

(d) $\text{rate} = k[\text{phenyl acetate}]^1 = (1.259 \text{ min}^{-1})(0.10 \text{ mol/L})^1 = 0.13 \text{ mol L}^{-1}\text{min}^{-1}$

☑ *Reasonable Result Check:* It is satisfying when one graph is clearly more obviously linear than the others.

35. *Result:* **(a) rate = $k[A]^2$ (b) 1.6×10^{-5} L mol^{-1}min^{-1} (c) The order for A is two.**

Analyze and Plan: Given reaction data over time, determine the rate law, the rate constant and the order using the plan described in Question 33.

Execute:

(a) The reactants are two A molecules and the product is one B molecule. The percent reaction data can be converted into percent unreacted data, which is a description of the concentration of the A molecules. So the rate law looks like this: rate = $k[A]^i$ where k and i are currently unknown.

Graph [Reactant] vs. time, ln[Reactant] vs. time, and 1/[Reactant] vs. time, to see which one produces a straight line. *(Notice, the time axes here have been converted to days.)*

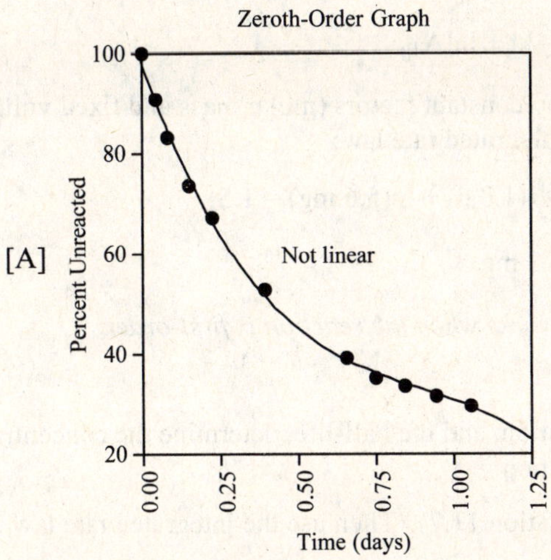

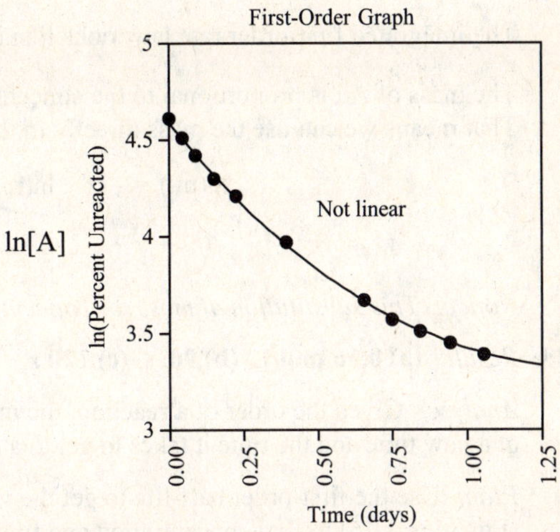

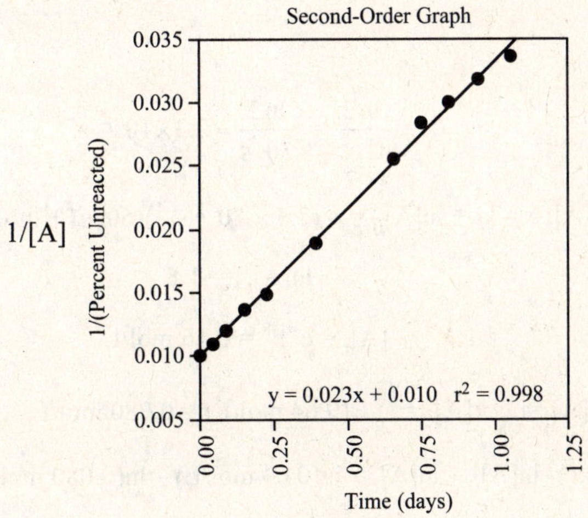

Of the three graphs, the linear one is the 1/[A] vs. time graph. That means the reaction is second-order. So the rate law is:

$$\text{rate} = k[A]^2$$

(b) The slope of the straight-line graph is equal to k. So, here:

$$k = 0.023 \text{ L mol}^{-1}\text{day}^{-1} \times \frac{1 \text{ day}}{24 \text{ hr}} \times \frac{1 \text{ hr}}{60 \text{ min}} =$$

$$k = \mathbf{1.6 \times 10^{-5} \text{ L mol}^{-1}\text{min}^{-1}}$$

(c) The order with respect to A is **two**.

☑ *Reasonable Result Check:* It is satisfying to obtain one graph that is much more linear than the other two. It makes sense that a dimerization reaction would be second-order.

37. *Result:* **4.3 mg**

Analyze: Given the order of a reaction, the initial mass, and the half-life, determine the amount remaining at a new time.

Plan: Use the first-order half-life to get the value of k (Equation 11.7). Then use the integrated rate law (Equation 11.5) and the relationship between mass and concentration to find the mass at the new time.

Execute: Calculate the rate constant:

$$k = \frac{\ln 2}{t_{1/2}} = \frac{\ln 2}{2.7\ \text{d}} = 0.26\ \text{d}^{-1}$$

The integrated first-order rate law looks like this: $\ln[A]_t = -kt + \ln[A]_0$

The mass of Au is proportional to the concentration of Au by constant factors (molar mass and fixed volume). That means we can use the mass directly in the first-order integrated rate law:

$$\ln(m_t) = -kt + \ln(m_0) = -(0.26\ \text{d}^{-1})(1.0\ \text{d}) + \ln(5.6\ \text{mg}) = 1.5$$

$$m_t = e^{1.5} = 4.3\ \text{mg}$$

Notice: This substitution of mass for concentration ONLY works when the reaction is first-order.

39. *Result:* **(a) 0.16 mol/L (b) 90. s (c) 120 s**

Analyze: Given the order of a reaction, the initial concentration, and the half-life, determine the concentration at a new time and the time it takes to get to a new concentration.

Plan: Use the first-order half-life to get the value of k (Equation 11.7). Then use the integrated rate law (Equation 11.5) to find concentration and time.

Execute:

Calculate the rate constant:

$$k = \frac{\ln 2}{t_{1/2}} = \frac{\ln 2}{30.\ \text{s}} = 2.3 \times 10^{-2}\ \text{s}^{-1}$$

(a)

$$\ln[A]_t = -kt + \ln[A]_0 = -(2.3 \times 10^{-2}\ \text{s}^{-1})(60.\ \text{s}) + \ln(0.64\ \text{mol/L})$$

$$\ln[A]_t = -1.8$$

$$[A]_t = e^{-1.8} = 0.16\ \text{mol/L}$$

(b)

$$[A]_t = \frac{1}{8} \times [A]_0 = \frac{1}{8} \times (0.64\ \text{mol/L}) = 0.080\ \text{mol/L}$$

$$kt = \ln[A]_0 - \ln[A]_t = \ln(0.64\ \text{mol/L}) - \ln(0.080\ \text{mol/L}) = 2.1$$

$$t = \frac{2.1}{k} = \frac{2.1}{2.3 \times 10^{-2}\ \text{s}^{-1}} = 90.\ \text{s}$$

(c)

$$[A]_t = 0.040\ \text{mol/L}$$

$$kt = \ln[A]_0 - \ln[A]_t = \ln(0.64\ \text{mol/L}) - \ln(0.040\ \text{mol/L}) = 2.8$$

$$t = \frac{2.8}{k} = \frac{2.8}{2.3 \times 10^{-2}\ \text{s}^{-1}} = 120\ \text{s}$$

☑ *Reasonable Result Check:* After some time has passed, the concentration of a reactant should be smaller than the initial concentration.

41. *Result:* **(a) 0.02 M (b) 14.0 yr (c) 5.55 yr**

Analyze and Plan: Given the order of a reaction, the rate constant, and the initial concentration, determine the concentration at a new time, the time it takes to get to a new concentration, and the half-life. Adapt the plan described in Question 39.

Execute:

(a) $$\ln[A]_t = -kt + \ln[A]_0$$

$$= -(3.42 \times 10^{-4}\ d^{-1})(2\ \text{month}\frac{30\ d}{1\ \text{month}}) + \ln(0.0200\ M)$$

$$\ln[A]_t = -3.9$$

$$[A]_t = e^{-3.9} = 0.02\ M$$

(b) $$kt = \ln[A]_0 - \ln[A]_t = \ln(0.0200\ M) - \ln(0.00350\ M) = 1.74$$

$$t = \frac{1.72}{k} = \frac{1.72}{3.42 \times 10^{-4}\ d^{-1}} = 5.10 \times 10^3\ d$$

$$5.10 \times 10^3\ d \times \frac{1\ yr}{365.25\ d} = 14.0\ yr$$

(c) $$t_{1/2} = \frac{\ln 2}{k} = \frac{\ln 2}{3.42 \times 10^{-4}\ d^{-1}} = 2.03 \times 10^3\ d$$

$$2.03 \times 10^3\ d \times \frac{1\ yr}{365.25\ d} = 5.55\ yr$$

☑ *Reasonable Result Check:* In part (a), the concentration did not change measurably, because two months is not much time when the half-life is 5.55 years. Looking at part (b), when we divide the original sample by two 3 times we get a sample about the size of the final sample, so the time (14 yr) should be about 3 half-lives, 3×(5.55 yr) = 16.7 yr.

A Nanoscale View: Elementary Reactions (Section 11-4)

42. *Result:* **Reaction (d) is unimolecular and elementary; reaction (b) is bimolecular and elementary; reactions (a) and (c) are not elementary.**

Analyze and Plan: Given several reactions, determine if they are unimolecular, bimolecular, or not elementary. A reaction with exactly one reactant is unimolecular and elementary. A reaction with exactly two reactants, with either two of the same reactants or one of each of two different reactants, is bimolecular and elementary. A reaction that has more than two reactants, in any combination, is not elementary. We will assume that these reactions fit one of these three descriptions.

Execute:

(a) The reaction has three reactants (one CH_4 and two O_2 molecules), so it is **not elementary**.

(b) The reaction has two reactants (one O_3 molecule and one O atom), so it is **bimolecular and elementary**.

(c) The reaction has three reactants (one Mg atom and two H_2O molecules), so it is **not elementary**.

(d) The reaction has one reactant (one O_3 molecule), so it is **unimolecular and elementary**.

☑ *Reasonable Result Check:* Elementary reactions can only have one or two reactants.

44. *Result:* **Reaction NO with O_3; NO is asymmetrical and Cl is symmetrical.**

Explanation: Given two reactions, explain which has a more important steric factor. The reaction of NO with O_3 will have a more important steric factor than the reaction of Cl with O_3, because NO is an asymmetrical molecule and Cl is a symmetrical atom. All collisions with a symmetrical atom could be effective, if they have enough energy. Collisions with the "wrong end" of the asymmetric molecule might be ineffective just because of which atoms came in contact during the collision.

Temperature and Reaction Rates: The Arrhenius Equation (Section 11-5)

46. *Result:* $\dfrac{\mathbf{rate_1}}{\mathbf{rate_2}} = \mathbf{1.8}$

Analyze: Given activation energy, concentrations, and A for a reaction, determine the ratio of the rates at two different Ts.

Plan: This is an elementary reaction so the rate law is: rate = $k[O_3][O]$

With constant concentrations, a ratio of the rates will be equal to a ratio of the rate constants. Convert the temperatures to Kelvin, then use the equation derived in Problem-Solving Example 11.9 to calculate the rate constant ratio.

$E_a = E_a(\text{forward}) = 19$ kJ/mol (from Problem-Solving Example 11.8).

$T_1 = 50.\ °C + 273.15 = 323\ K,\qquad T_2 = 25\ °C + 273.15 = 298\ K$

$$\frac{\text{rate}_1}{\text{rate}_2} = \frac{k_1}{k_2} = e^{\frac{E_a}{R}\left(\frac{1}{T_2}-\frac{1}{T_1}\right)} = e^{\frac{(19\ \text{kJ/mol})}{(0.008314\ \text{kJ/mol}\cdot\text{K})}\left(\frac{1}{(298\ \text{K})}-\frac{1}{(323\ \text{K})}\right)} = e^{0.6} = 1.8$$

Notice, it would be possible to calculate the individual rate at each temperature, using the rate law and the Arrhenius equation, then take a ratio of those two rates to answer the question. The answer would be the same as this one.

☑ *Reasonable Result Check:* The rate increases at a higher temperature.

48. *Result:* **10.7 times faster**

Analyze and Plan: As directed in Question 47, use the equation derived there (and also in Problem-Solving Example 11.9) to answer this question. Adapt the plan from the solution to Question 46.

$E_a = 76$ kJ/mol, $\quad T_1 = 50.\ °C + 273.15 = 323\ K,\qquad T_2 = 25\ °C + 273.15 = 298\ K$

$$\frac{\text{rate}_1}{\text{rate}_2} = \frac{k_1}{k_2} = e^{\frac{E_a}{R}\left(\frac{1}{T_2}-\frac{1}{T_1}\right)} = e^{\frac{(76\ \text{kJ/mol})}{(0.008314\ \text{kJ/mol}\cdot\text{K})}\left(\frac{1}{(298\ \text{K})}-\frac{1}{(323\ \text{K})}\right)}$$

$$\frac{\text{rate}_1}{\text{rate}_2} = e^{2.4} = 10.7$$ *With strict sig figs, this should be $1 \times 10^1 = 10$*

☑ *Reasonable Result Check:* The rate is greater at the higher temperature.

50. *Result:* $\mathbf{3 \times 10^2}$ **kJ/mol**

Analyze and Plan: As directed in Question 47, use the equation derived there (and also in Problem-Solving Example 11.9) to answer this question. Adapt the plan from the solution to Question 46.

Rearrange the equation derived in Question 47 by taking the natural logarithm of each side (See Appendix A, page A.12).

$$\ln\left(\frac{k_1}{k_2}\right) = \frac{E_a}{R}\left(\frac{1}{T_2}-\frac{1}{T_1}\right)$$

$T_1 = 600.\ K \qquad T_2 = 610.\ K \qquad k_2 = 3k_1$

$$\ln\left(\frac{k_1}{3k_1}\right) = \frac{E_a}{(0.008314\ \text{kJ/mol}\cdot\text{K})}\left(\frac{1}{610.\ \text{K}}-\frac{1}{600.\ \text{K}}\right)$$

Solve for E_a

$$E_a = 3 \times 10^2\ \text{kJ/mol}$$

51. (a) E_a = 115 kJ/mol, A = 1.98×10^{13} s^{-1} (b) k = 1.5×10^{-3} s^{-1}

Analyze and Plan: Given the values of the rate constant at two temperatures, determine the activation energy, the frequency factor and the rate constant at a new temperature. As directed in Question 47, use the equation derived there (and also in Problem-Solving Example 11.9) to answer this question. Adapt the plan from the solution to Question 46 and 50.

$$\ln\left(\frac{k_1}{k_2}\right) = \frac{E_a}{R}\left(\frac{1}{T_2} - \frac{1}{T_1}\right)$$

Execute: $T_1 = 56.2\ °C + 273.15 = 329.4\ K$, $T_2 = 78.2\ °C + 273.15 = 351.4\ K$

$$\ln\left(\frac{1.04 \times 10^{-5}}{1.45 \times 10^{-4}}\right) = \frac{E_a}{(0.008314\ kJ/mol \cdot K)}\left(\frac{1}{351.4\ K} - \frac{1}{329.4\ K}\right)$$

$$-2.63 = E_a(-2.29 \times 10^{-2}\ mol/kJ)$$

$$E_a = 115\ kJ/mol$$

$$k = A e^{-E_a/RT_1}$$

$$A = k e^{E_a/RT}$$

Using T_1

$$A = (1.04 \times 10^{-5}) e^{\left[\frac{115\ kJ\ mol^{-1}}{(0.008314\ kJ\ mol^{-1}K^{-1})(329.4\ K)}\right]} = (1.04 \times 10^{-5})e^{42.1} = 1.99 \times 10^{13}\ s^{-1}$$

Using T_2

$$A = (1.45 \times 10^{-4}) e^{\left[\frac{115\ kJ\ mol^{-1}}{(0.008314\ kJ\ mol^{-1}K^{-1})(351.4\ K)}\right]} = (1.45 \times 10^{-4})e^{39.5} = 1.97 \times 10^{13}\ s^{-1}$$

These are very close to the same. The average is A = 1.98×10^{13} s^{-1}

(b) Use the Arrhenius equation to estimate the rate constant.

$$T = 100.0\ °C + 273.15\ K = 373.2\ K$$

$$k = A e^{-E_a/RT} = (1.98 \times 10^{13}\ s^{-1}) e^{\left[\frac{-115\ kJ\ mol^{-1}}{(0.008314\ kJ\ mol^{-1}K^{-1})(373.2\ K)}\right]} = 1.5 \times 10^{-3}\ s^{-1}$$

☑ *Reasonable Result Check:* A is essentially constant, as it should be, and the rate constant is larger at a higher temperature.

53. *Result:* **(a) 22.2 kJ/mol, 6.66×10^7 L^{2}mol^{-2}s^{-1} (b) 8.39×10^4 L^{2}mol^{-2}s^{-1}**

Analyze: Given the values of the rate constant and several temperatures, determine the activation energy, the frequency factor and the rate constant at a new temperature.

Plan: As described in Section 11-5, we can make a linear graph by taking the natural logarithm of the rate constant and the reciprocal of the absolute temperature. The slope of the graph is related to the activation energy and the y-intercept is related to the frequency factor. The Arrhenius equation can then be used to find the rate constant at a new temperature.

Execute:

(a) According to the table heading in the question, the k values have each been multiplied by 10^{-5} so, for example, the first k is 1.12×10^5 L^{2}mol^{-2}s^{-1}.

$\frac{1}{T}$ (K^{-1})	ln(k)
0.002393	11.626
0.002080	12.468
0.001923	12.889
0.001579	13.752
0.001500	13.955
0.001408	14.292
0.001355	14.433

Now make a graph of 1/T versus ln k:

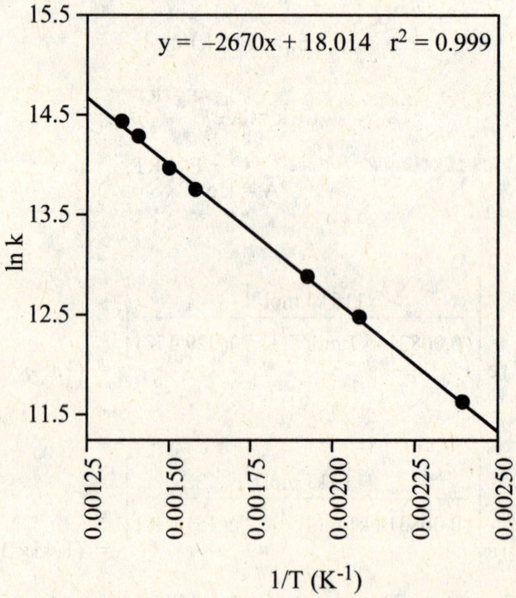

The slope of the graph (m) is –2670.

$$m = -\frac{E_a}{R}$$

$$E_a = -mR$$

$$E_a = -(-2670)(0.008314) = 22.2 \text{ kJ/mol}$$

The y-intercept of the graph (b) is –18.014.

$$b = \ln A$$

$$A = e^b$$

$$A = e^{18.014} = 6.66 \times 10^7 \text{ L}^2\text{mol}^{-2}\text{s}^{-1}$$

(b) Use the Arrhenius equation to estimate the rate constant at 400.0 K.

$$k = A\,e^{-E_a/RT}$$

$$k = (6.66 \times 10^7 \text{ L}^2\text{mol}^{-2}\text{s}^{-1})\,e^{\left[\frac{-22.2 \text{ kJ mol}^{-1}}{(0.008314 \text{ kJ mol}^{-1}\text{K}^{-1})(400.0 \text{ K})}\right]} = 8.39 \times 10^4 \text{ L}^2\text{mol}^{-2}\text{s}^{-1}$$

55. *Result:* **(a) 3×10^{-20} (b) 4×10^{-16} (c) 4×10^{-10} (d) 1.9×10^{-6}**

Analyze: Given the activation energy and several temperatures, determine the fraction of molecules whose energies would be energetic enough to react.

Plan: In Equation 11.8 of Section 11-5, the exponential term, $e^{-E_a/RT}$, is described as the fraction of sufficiently energetic molecules.

Execute: Convert temperatures to Kelvin: °C + 273.15 = Kelvin. An example of the calculation for item (a) is:

$$100.\ °C + 273.15 = 373\ K$$

$$\text{Fraction} = e^{-E_a/RT} = e^{\left[\frac{-139.7\ kJ\ mol^{-1}}{(0.008314\ kJ\ mol^{-1}K^{-1})(373\ K)}\right]} = e^{-45.0} = 3 \times 10^{-20}$$

	T (K)	Fraction of sufficiently energetic molecules
(a)	373	$e^{-45.0} = 3 \times 10^{-20}$
(b)	473	$e^{-35.5} = 4 \times 10^{-16}$
(c)	773	$e^{-21.7} = 4 \times 10^{-10}$
(d)	1273	$e^{-13.20} = 1.9 \times 10^{-6}$

☑ *Reasonable Result Check:* The fraction of molecules with enough energy to react increases with increasing temperature.

57. *Result:* **(a) 1×10^{-5} mol $L^{-1}s^{-1}$ (b) 25 mol $L^{-1}s^{-1}$**

Analyze: Given the activation energy, the concentration of the reactant, the frequency factor, and two temperatures, determine the rates of the reaction at those two temperatures.

Plan: The units of the frequency factor are also the units of the rate constant. Use that information to determine the order of the reaction. Write the rate law for the reaction in terms of k and concentration. Use the Arrhenius equation to find the value of k, then use the rate law to calculate rate.

$$k = A e^{-E_a/RT} \qquad \text{Rate} = k[CH_3CH_2I]^1$$

Execute: The reaction is first-order since the units of A, and therefore the units of k, are $time^{-1}$. (Table 11.2)

An example of the calculation for item (a):

Convert temperatures to Kelvin. 400. °C + 273.15 = 673 K

$$k = (1.2 \times 10^{14}\ s^{-1}) \times e^{\left[\frac{-221\ kJ\ mol^{-1}}{(0.008314\ kJ\ mol^{-1}K^{-1})(673\ K)}\right]} = (1.2 \times 10^{14}\ s^{-1}) \times e^{-39.5} = 8 \times 10^{-4}\ s^{-1}$$

$$\text{Rate} = (8 \times 10^{-4}\ s^{-1}) \times (0.012\ mol/L)^1 = 1 \times 10^{-5}\ mol\ L^{-1}s^{-1}$$

	T (K)	k (s^{-1})	Rate (mol $L^{-1}s^{-1}$)
(a)	673	8×10^{-4}	1×10^{-5}
(b)	1073	3×10^{1}	25

☑ *Reasonable Result Check:* It makes sense that the rate of the reaction increases with increasing temperature.

Rate Laws for Elementary Reactions (Section 11-6)

59. *Result* **(a) rate = k[Cl][ICl] (b) rate = $k[CH_3N{=}NCH_3]$ (c) rate = $k[N_2O_4]^2$**

Analyze: Given elementary equations, determine their rate laws.

Plan: Section 11-6 shows that it is possible to write the rate laws for elementary reactions using the stoichiometry. Unimolecular equations are first-order. Bimolecular equations are second-order, when the reactants are the same, and first order in each reactant when the reactants are different.

Execute:

(a) This is a bimolecular equation with reactants that are different: rate = k[Cl][ICl]

(b) This is a unimolecular equation: rate = $k[CH_3N{=}NCH_3]$

(c) This is a unimolecular equation: rate = $k[N_2O_4]^2$

61. *Result:* **See graphs below**

Analyze: Draw these diagrams using the information described in Problem-Solving Example 11.8 and Section 11-4b. Δ_rH and Δ_rE are identical for these diagrams.

(a) $E_{a,reverse} = E_{a,forward} - \Delta_rH = (75\ kJ\ mol^{-1}) - (-145\ kJ\ mol^{-1}) = 220.\ kJ\ mol^{-1}$

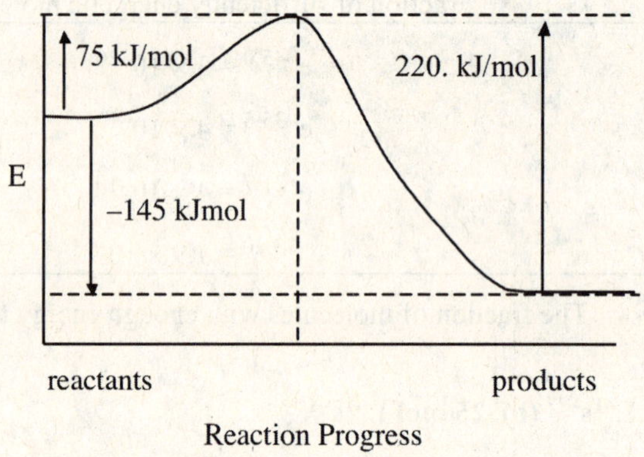

(b) $E_{a,reverse} = E_{a,forward} - \Delta_rH = (65\ kJ\ mol^{-1}) - (-70.\ kJ\ mol^{-1}) = 135\ kJ\ mol^{-1}$

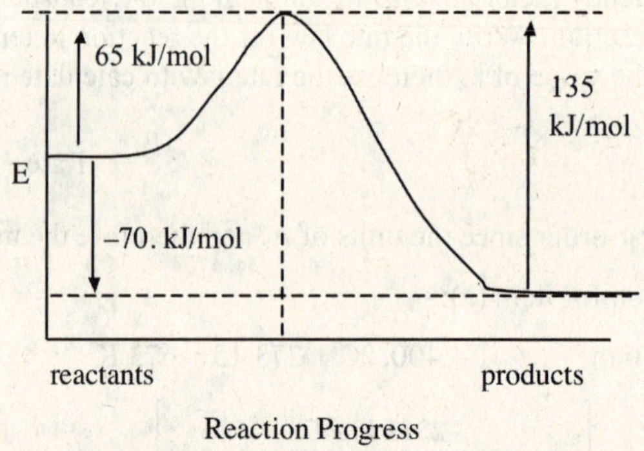

(c) $E_{a,reverse} = E_{a,forward} - \Delta_rH = (85\ kJ\ mol^{-1}) - (+70.\ kJ\ mol^{-1}) = 15\ kJ\ mol^{-1}$

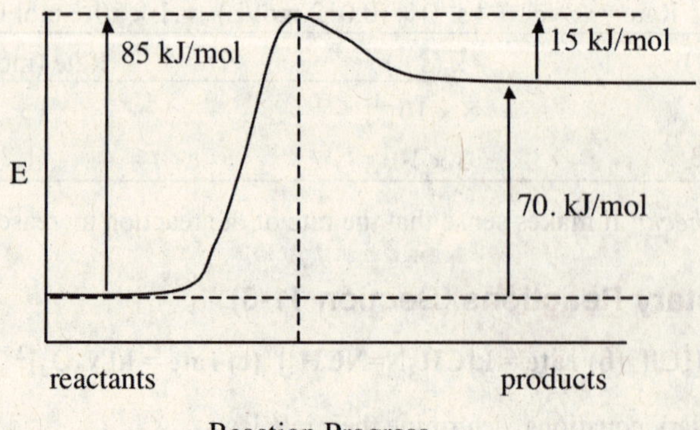

63. *Result:* **(a) Reaction (b), (b) Reaction (c)**

Analyze: Assuming everything about the forward reactions in this question is identical except the activation energy, the reaction with the largest $E_{a,forward}$ will be the slowest, and that with the smallest $E_{a,forward}$ will be the fastest.

(a) Of the three, the smallest $E_{a,forward}$ is 65 kJ mol^{-1}. That is reaction (b).

(b) Of the three, the largest $E_{a,forward}$ is 85 kJ mol^{-1}. That is reaction (c).

65. *Result:* **(a) Reaction (c), (b) Reaction (a)**

Analyze: Assuming everything about the reverse reactions in this question is identical except the activation energy, the reaction with the largest $E_{a,reverse}$ will be the slowest, and that with the smallest $E_{a,reverse}$ will be the fastest.

(a) Of the three, the smallest $E_{a,reverse}$ is 15 kJ mol^{-1}. That is the reverse of reaction (c).

(b) Of the three, the largest $E_{a,reverse}$ is 220 kJ mol^{-1}. That is the reverse of reaction (a).

67. *Answer:* **(a) rate = k[NO][NO_3] (b) rate = k[O][O_3] (c) rate = k[$(CH_3)_3CBr$] (d) rate = k[HI]^2**

Strategy and Explanation: Given the chemical equations of elementary reactions write their rate laws.

As described in Section 13.6, the stoichiometric coefficient of a reactant in an elementary equation is the reaction order for that reactant.

(a)	$NO + NO_3 \longrightarrow$ products	rate = $k[NO]^1[NO_3]^1$
(b)	$O + O_3 \longrightarrow$ products	rate = $k[O]^1[O_3]^1$
(c)	$(CH_3)_3CBr \longrightarrow$ products	rate = $[(CH_3)_3CBr]^1$
(d)	$2\ HI \longrightarrow$ products	rate = $k[HI]^2$

Reaction Mechanisms (Section 11-7)

69. *Result:* **(a) See equations below (b) The first step is rate-determining.**

Analyze: Given an equation, rate law and mechanism, show that the sum of the sequence gives the overall reaction and determine which step is rate-determining.

Plan and Execute:

(a) As we did in Chapter 4 with Hess's law questions, cancel anything that shows up on both sides of the equation and add up the rest.

$$NO_2 + F_2 \longrightarrow FNO_2 + \cancel{F}$$
$$+\ NO_2 + \cancel{F} \longrightarrow FNO_2$$
$$\overline{2\ NO_2 + F_2 \longrightarrow 2\ FNO_2}$$

(b) The rate law looks like this: rate = $k[NO_2][F_2]$

The reaction orders for NO and F_2 match the stoichiometric coefficients of NO and F_2 in the first reaction in the mechanism, so it must be the rate-determining step.

☑ *Reasonable Result Check:* It makes sense that the first reaction would be the rate-limiting step, since a fluorine free radical is going to be extremely reactive.

71. *Result:* **(a) Rate = k'[NO]^2[Cl_2] (b) See mechanism below (c) See mechanism below**

Analyze: Given an equation and mechanism, determine the rate law for the mechanism and suggest another mechanism that agrees with the same rate law and one that does not.

(a) *Plan and Execute:* This mechanism has a fast initial step. Following the description given in Section 11-7, write the rate law from the rate-determining step that includes an intermediate as a reactant. Set up an equation showing the steady state for the creation and destruction of the intermediate. Solve that equation for the intermediate concentration in terms of the concentrations of reactants (and possibly products). Use that equation to eliminate the intermediate concentration from the rate law.

The slow step is step 2, so the rate of the reaction is equal to the rate of this rate-determining step. Because this reaction is elementary, its rate law is related to the stoichiometry of its reactants:

$$\text{rate of reaction} = \text{rate}_{\text{reaction 2}} = k_2[NOCl_2][NO]$$

$NOCl_2$ is an intermediate, so we need to eliminate it from the proposed mechanism. Look for all the ways this intermediate is created and destroyed during the mechanism. It is only created in the forward reaction of step 1. It is destroyed in the reverse reaction of step 1 and also in the reaction of step 2. Because all three of these reactions are elementary, use their reactants' stoichiometric coefficients to write rate laws for each of these three reactions:

$$\text{rate}_{\text{forward reaction 1}} = k_1[NO][Cl_2]$$

$$\text{rate}_{\text{reverse reaction 1}} = k_{-1}[NOCl_2]^1$$

$$\text{rate}_{\text{reaction 2}} = k_2[NOCl_2][NO]$$

The rate of $NOCl_2$ creation is equal to the rate of its destruction once the steady state condition is reached:

$$\text{rate}_{\text{forward reaction 1}} = \text{rate}_{\text{reverse reaction 1}} + \text{rate}_{\text{reaction 2}}$$

$$k_1[NO][Cl_2] = k_{-1}[NOCl_2]^1 + k_2[NOCl_2][NO]$$

Because the rate of step 2 is presumed to be much smaller than the rate of step 1, the second term is presumed to be negligibly small compared to the first term:

$$\text{rate}_{\text{forward reaction 1}} \cong \text{rate}_{\text{reverse reaction 1}}$$

$$k_1[NO][Cl_2] = k_{-1}[NOCl_2]^1$$

$$[NOCl_2] = \frac{k_1}{k_{-1}}[NO][Cl_2]$$

Substitute the equality just derived into the rate law: $\text{rate} = k_2[NOCl_2][NO]$

$$\text{rate} = k_2\left(\frac{k_1}{k_{-1}}[NO][Cl_2]\right)[NO]$$

We will define the new group of rate constants by one variable, $k' = k_2\frac{k_1}{k_{-1}}$.

$$\text{rate} = k'[NO][Cl_2][NO]$$

$$\text{rate} = k'[NO]^2[Cl_2]$$

(b) *Plan and Execute:* Suggesting mechanisms involves some creativity. We need to make sure that the rate law is satisfied, but we often must use our imaginations to determine what may get formed during these reactions. Here is an example of a reaction that also satisfies the rate law: rate = $k'[NO]^2[Cl_2]$.

$$2\,NO \rightleftharpoons N_2O_2 \qquad \text{Fast}$$

$$N_2O_2 + Cl_2 \longrightarrow 2\,NOCl_2 \qquad \text{Slow}$$

To think this one up, we just switched the roles of one NO molecule and the Cl_2 molecule, putting both NO molecules in the fast reaction and making an intermediate, then having the Cl_2 react with that intermediate.

To confirm that it qualifies, let's derive the rate law for this new mechanism, as described in (a):

$$\text{rate of reaction} = \text{rate}_{\text{reaction 2}} = k_2[N_2O_2][Cl_2]$$

Setting up the steady state for the N_2O_2 intermediate gives this equation:

$$k_1[NO]^2 = k_{-1}[N_2O_2]^1 + k_2[N_2O_2][Cl_2]$$

Again the rate of step 2 is presumed to be much smaller than the rate of step 1:

$$k_1[NO]^2 = k_{-1}[N_2O_2]^1$$

$$[N_2O_2] = \frac{k_1}{k_{-1}}[NO]^2$$

Substitute the equality just derived into the rate law: $\text{rate} = k_2[N_2O_2][Cl_2]$

$$\text{rate} = k_2\left(\frac{k_1}{k_{-1}}[NO]^2\right)[Cl_2]$$

We again define the new group of rate constants by one variable, $k' = k_2\frac{k_1}{k_{-1}}$.

$$\text{rate} = k'[NO]^2[Cl_2]$$

(c) *Plan and Execute:* Many mechanisms will not satisfy this rate law: rate = $k'[NO]^2[Cl_2]$. Here is an example of a reaction that does not satisfy the rate law:

$$NO + Cl_2 \rightleftharpoons NOCl + Cl \quad \text{Slow}$$

$$NO + Cl \longrightarrow NOCl \quad \text{Fast}$$

This reaction has a rate law that looks like this: rate = $[NO][Cl_2]$

It is not the same as the observed rate law and cannot be the mechanism for this reaction.

☑ *Reasonable Result Check:* (a) The first step involves the reaction of NO and Cl_2 molecules in the formation of the steady state of the intermediate, so it makes sense that the concentration of Cl_2 is involved in the reaction's rate along with the concentration of NO. (b) The mechanism does recreate the rate law. (c) The mechanism produces a rate law different from that observed.

72. *Result:* **Only mechanism (a) is compatible with the observed rate law.**

Analyze and Plan: Given a reaction and its rate law, evaluate three possible mechanisms to see if they are compatible. Follow the plan described in Question 71.

Execute: The observed rate law is given:

$$\text{rate} = k[(CH_3)_3CBr]$$

(a) The rate law for this mechanism comes from the rate of the slow first step:

rate = $k[(CH_3)_3CBr]$, so it is compatible with the observed rate law.

(b) The single step mechanism suggests an elementary bimolecular reaction with a rate law that looks like this: rate = $k[(CH_3)_3CBr][OH^-]$ It is incompatible with the observed rate law.

(c) The rate law for this mechanism comes from the rate of the slow second step:

$$\text{rate} = k_2[(CH_3)_2(CH_2)CBr^-]$$

The steady state conditions for $(CH_3)_2(CH_2)CBr^-$ must be derived to replace the $[(CH_3)_2(CH_2)CBr^-]$ in this expression with an expression using the reactant concentrations. The steady state equation looks like this:

$$k_1[(CH_3)_3CBr][OH^-] = k_{-1}[(CH_3)_2(CH_2)CBr^-][H_2O] + k_2[(CH_3)_2(CH_2)CBr^-]$$

The second term on the right side ($k_2[(CH_3)_2(CH_2)CBr^-]$) is negligible, since the rate of the second step is considered to be small compared to the first step.

$$k_1[(CH_3)_3CBr][OH^-] = k_{-1}[(CH_3)_2(CH_2)CBr^-][H_2O]$$

Solve for $[(CH_3)_2(CH_2)CBr^-]$ in terms of in terms of reactant concentrations,

$$[(CH_3)_2(CH_2)CBr^-] = \frac{k_1}{k_{-1}}\frac{[(CH_3)_3CBr][OH^-]}{[H_2O]}$$

Substituting this equality back into the rate law gives: $\text{rate} = k_2[(CH_3)_2(CH_2)CBr^-]$

$$\text{rate} = k_2\frac{k_1}{k_{-1}}\frac{[(CH_3)_3CBr][OH^-]}{[H_2O]} = k'\frac{[(CH_3)_3CBr][OH^-]}{[H_2O]}$$

This is incompatible with the observed rate law.

So, only mechanism (a) is compatible.

Catalysts and Reaction Rate (Section 11-8)

74. *Result:* **(a) True. (b) False. "A catalyst must never be consumed in a reaction." (c) False. "A catalyst need not be the same phase as the reactants." (d) False. "A catalyst can change the course of a reaction, but the same products are always formed."**

Analyze and Plan: Given several statements, identify whether they are true or not and reword any that are not true to make it a true statement. Use the information in Section 11-8.

Execute:

(a) True: The concentration of a homogeneous catalyst may appear in the rate law.

(b) False: A catalyst may be consumed in a reaction, but must always be recreated. A related correct statement is: "A catalyst must never be consumed in a reaction."

(c) False: A homogeneous catalyst is in the same phase as the reactants, but a heterogeneous catalyst is always in a different phase. A related correct statement is: "A catalyst need not be the same phase as the reactants."

(d) False: A catalyst can change the mechanism of a reaction and allow different intermediates to be produced, but it cannot change the products formed. A chemical that changes the products is not called a catalyst. A related correct statement is: "A catalyst can change the course of a reaction, but the same products are always formed."

Notice that if a multiple set of reactions occurs simultaneously with the same reactants, a catalyst that speeds only one of these reactions would help produce one product in favor of some others. In such case, the statement might be considered true, but it is likely not something students studying this chapter will think of.

76. *Result:* **(a) $CH_3COOCH_3 + H_2O \rightleftharpoons CH_3COOH + CH_3OH$**
(b) Rate = $k'[CH_3COOCH_3][H^+][H_2O]$ (c) catalyst: H^+ (d) intermediates: $H_3C(OH)OCH_3^+$, $H_3C(H_2O)(OH)OCH_3^+$, $H_3C(OH)_2OHCH_3^+$, and H_2O

Analyze: Given a mechanism, write the overall equation, rate law, determine if there is a catalyst, and identify intermediates.

(a) *Plan and Execute:* As we did in Chapter 4 with Hess's law questions, cancel anything that ends up on both sides of the equation and add up the rest.

$$CH_3COOCH_3 + \cancel{H^+} \rightleftharpoons \cancel{CH_3C(OH)OCH_3^+}$$
$$+\ \cancel{CH_3C(OH)OCH_3^+} + \cancel{H_2O} \rightleftharpoons \cancel{CH_3C(H_2O)(OH)OCH_3^+}$$
$$+\ \cancel{CH_3C(H_2O)(OH)OCH_3^+} \rightleftharpoons \cancel{CH_3C(OH)_2OHCH_3^+}$$
$$+\ \cancel{CH_3C(OH)_2OHCH_3^+} \rightleftharpoons CH_3COOH + CH_3OH + \cancel{H^+}$$
$$\overline{CH_3COOCH_3 + H_2O \rightleftharpoons CH_3COOH + CH_3OH}$$

(b) *Plan and Execute:* Follow the same plan described in the solution to Question 71.

The slow step is step 2, so the rate of the reaction is equal to the rate of this rate-determining step. Because this reaction is elementary, its rate law is related to the stoichiometry of its reactants:

$$\text{rate of reaction} = \text{rate}_{\text{reaction 2}} = k_2[CH_3C(OH)OCH_3^+][H_2O]$$

$CH_3C(OH)OCH_3^+$ is an intermediate, so we need to eliminate it from the proposed mechanism. Look for all the ways this intermediate is created and destroyed during the mechanism. It is only created in the forward reaction of step 1. It is destroyed in the reverse reaction of step 1 and also in the forward reaction of step 2. Because all three of these reactions are elementary, use their reactants' stoichiometric coefficients to write rate laws for each of these three reactions:

$$\text{rate}_{\text{forward reaction 1}} = k_1[CH_3COOCH_3][H^+]$$

$$\text{rate}_{\text{reverse reaction 1}} = k_{-1}[CH_3C(OH)OCH_3^+]$$

$$\text{rate}_{\text{reaction 2}} = k_2[CH_3C(OH)OCH_3^+][H_2O]$$

The rate of $CH_3C(OH)OCH_3^+$ creation is equal to the rate of its destruction once the steady state condition is reached:

$$\text{rate}_{\text{forward reaction 1}} = \text{rate}_{\text{ reverse reaction 1}} + \text{rate}_{\text{reaction 2}}$$

$$k_1[CH_3COOCH_3][H^+] = k_{-1}[CH_3C(OH)OCH_3^+] + k_2[CH_3C(OH)OCH_3^+][H_2O]$$

Because the rate of step 2 is presumed to be much smaller than the rate of step 1, the second term is presumed to be negligibly small compared to the first term:

$$k_1[CH_3COOCH_3][H^+] \cong k_{-1}[CH_3COHOCH_3^+]$$

$$[CH_3C(OH)OCH_3^+] = \frac{k_1}{k_{-1}}[CH_3C(OH)OCH_3^+][H^+]$$

Substitute the equality just derived into the rate law:

$$\text{rate} = k_2[CH_3C(OH)OCH_3^+][H_2O]$$

$$\text{rate} = k_2\left(\frac{k_1}{k_{-1}}[CH_3C(OH)OCH_3^+][H^+]\right)[H_2O]$$

We will define the new group of constants by one variable, $k' = k_2\dfrac{k_1}{k_{-1}}$.

$$\text{rate} = k'[CH_3COOCH_3][H^+][H_2O]$$

(c) *Explanation:* A catalyst shows up in a mechanism as a reactant in an early step and then again as a product in a later step. There is a catalyst in this reaction. It is H^+, introduced in step 1 and reproduced in step 4.

(d) *Explanation:* An intermediate appears in a mechanism as a product in an early step and then again as a reactant in a later step. Three intermediates are formed in this reaction: $H_3C(OH)OCH_3^+$, $H_3C(H_2O)(OH)OCH_3^+$, and $H_3C(OH)_2OHCH_3^+$. A molecule of H_2O is also created then destroyed in this reaction, but another H_2O molecule is consumed, so water may be technically considered to be an intermediate, as well as a reactant.

(e) The potential energy diagram has four hills. The second one is the highest, representing the rate-determining step.

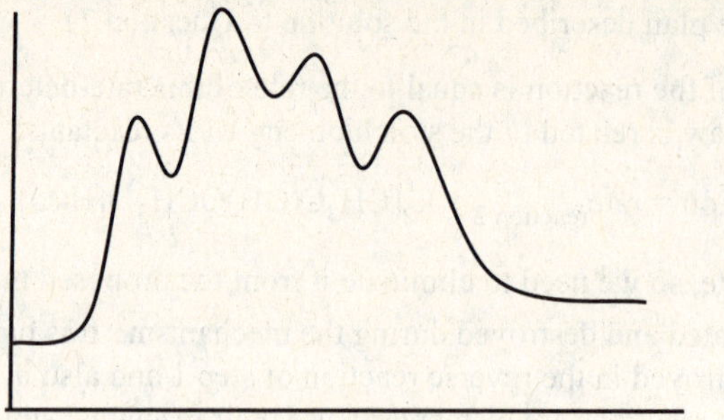

78. *Result/Explanation:* $H^+(aq)$ is a homogeneous catalyst in this aqueous reaction, which explains why it appears in the rate law but not the overall equation.

Enzymes: Biological Catalysts (Section 11-9)

79. *Result:* **38**

Analyze: Given the E_a before and after the addition of a catalyst, determine by what factor the rate constant would increase (assuming the frequency factor, A, remains unchanged).

Plan and Execute: The Arrhenius equation given in Section 11-5 is $k = Ae^{-E_a/RT}$ Set up a ratio:

$$\frac{k_2}{k_1} = \frac{Ae^{-E_{a2}/RT}}{Ae^{-E_{a1}/RT}} = e^{-E_{a2}/RT + E_{a1}/RT} = e^{(E_{a1} - E_{a2})/RT}$$

$$\frac{k_2}{k_1} = e^{\left[\frac{215\ \text{kJ mol}^{-1} - 206\ \text{kJ mol}^{-1}}{(0.008314\text{J/mol K})(25+273.15)\text{K}}\right]} = 38$$

80. *Result:* **The rate is 30. times faster.**

Analyze: We will use information described in Section 11-2, and methods similar to those in Question 26.

Plan and Execute: The reaction rate is proportional to the concentration in a first-order reaction, rate = k[E]. That means the rate ratio is equal to the concentration ratio:

$$\frac{\text{rate}_2}{\text{rate}_1} = \frac{[E]_2}{[E]_1} = \frac{4.5 \times 10^{-6}\ \text{M}}{1.5 \times 10^{-7}\ \text{M}} = 30.$$

The reaction rate is 30. times faster.

82. *Result/Explanation:* The succinate dehydrogenase enzyme catalyzes the reaction of substrate succinate ion, $^-OOCCH_2CH_2COO^-$. **The active site** (where the substrate binds to the enzyme in preparation for conversion into products) **is shaped to accommodate the four O atoms, two with negative charges, on the ends of the molecule in the formation of the enzyme-substrate complex.** The malonate ion, $^-OOCCH_2COO^-$, and the oxalate ion, $^-OOCCOO^-$, also have four O atoms, two with negative charges, at the ends of their structures. Evidently, **these other ions enter and occupy the active site of an enzyme preventing the substrate from binding, inhibiting the reaction, and reducing the rate** of the succinate dehydrogenation reaction.

84. *Result:* **(a) Rate = $k\frac{[X][HA]}{[A^-]}$ (b) First-order with respect to HA (c) Doubling the [HA] doubles the rate**

Analyze: Given a mechanism, derive the rate law, determine the orders, and determine how changing the concentration of a reactant affects the rate.

Plan: We will use information described in Section 11-2, and methods similar to those in Question 26.

(a) *Execute:* This mechanism has fast first and second step. Following the description given in Section 11-7, write the rate expression from the rate-determining step that includes an intermediate as a reactant. Set up equations showing the steady state for the creation and destruction of the intermediates. Combine those equations and solve for the intermediate concentrations in terms of the concentrations of reactants (and possibly products). Use those equations to eliminate the intermediate concentrations from the rate expression.

The slow step is step 3, so the rate of the reaction is equal to the rate of this rate-determining step. Because this reaction is elementary, its rate expression is related to the stoichiometry of its reactants:

$$\text{rate of reaction} = \text{rate}_{\text{reaction 3}} = k_3[XH^+]$$

XH^+ is an intermediate, so we need to eliminate it from the proposed mechanism. Look for all the ways this intermediate is created and destroyed during the mechanism. It is only created in the forward reaction of step 2. It is destroyed in the reverse reaction of step 1 and also in the reaction of step 3. Because all three of these reactions are elementary, use their reactants' stoichiometric coefficients to write rate expressions for each of these three reactions:

$$\text{rate}_{\text{forward reaction 2}} = k_2[X][H^+]$$

$$\text{rate}_{\text{reverse reaction 2}} = k_{-2}[XH^+]$$

$$\text{rate}_{\text{reaction 3}} = k_3[XH^+]$$

The rate of XH^+ creation is equal to the rate of its destruction once the steady state condition is reached:

$$\text{rate}_{\text{forward reaction 1}} = \text{rate}_{\text{reverse reaction 1}} + \text{rate}_{\text{reaction 2}}$$

$$k_2[X][H^+] = k_{-2}[XH^+] + k_3[XH^+]$$

Because the rate of step 3 is presumed to be much smaller than the rate of step 2, the second term is presumed to be negligibly small compared to the first term:

$$k_2[X][H^+] = k_{-2}[XH^+]$$

$$[XH^+] = \frac{k_2}{k_{-2}}[X][H^+]$$

Substitute the equality just derived for $[XH^+]$ into the rate expression: $\text{rate} = k_3[XH^+]$

$$\text{rate} = k_3\left(\frac{k_2}{k_{-2}}[X][H^+]\right)$$

$$\text{rate} = \frac{k_3 k_2}{k_{-2}}[X][H^+]$$

H^+ is an intermediate, so we need to eliminate it from the proposed mechanism. Look for all the ways this intermediate is created and destroyed during the mechanism. It is only created in the forward reaction of step 2. It is destroyed in the reverse reaction of step 1 and in the forward reaction of step 2. Because all three of these reactions are elementary, use their reactants' stoichiometric coefficients to write rate expressions for both reactions:

$$\text{rate}_{\text{forward reaction 1}} = k_1[HA]$$

$$\text{rate}_{\text{reverse reaction 1}} = k_{-1}[H^+][A^-]$$

$$\text{rate}_{\text{forward reaction 2}} = k_2[X][H^+]$$

The rate of H^+ creation is equal to the rate of its destruction once the steady state condition is reached:

$$\text{rate}_{\text{forward reaction 1}} = \text{rate}_{\text{reverse reaction 1}} + \text{rate}_{\text{forward reaction 2}}$$

$$k_1[HA] = k_{-1}[H^+][A^-] + k_2[X][H^+]$$

Because the rate of step 3 is presumed to be much smaller than the rate of step 2, the second term is presumed to be negligibly small compared to the first term:

$$k_1[HA] = k_{-1}[H^+][A^-]$$

$$[H^+] = \frac{k_1}{k_{-1}}\frac{[HA]}{[A^-]}$$

Substitute the equality just derived for $[H^+]$ in the rate expression derived above:

$$\text{rate} = \frac{k_3k_2}{k_{-2}}[X][H^+]$$

$$\text{rate} = \frac{k_3k_2}{k_{-2}}[X]\left(\frac{k_1}{k_{-1}}\frac{[HA]}{[A^-]}\right)$$

We will define the new group of rate constants by one variable, $k' = \dfrac{k_3k_2k_1}{k_{-2}k_{-1}}$.

$$\text{rate} = k'\frac{[X][HA]}{[A^-]}$$

(b) *Result/Explanation:* The reaction is first order with respect to HA.

(c) *Result/Explanation:* Because the reaction is first order with respect to HA, doubling the concentration of HA doubles the rate of the reaction.

Catalysts in Industry (Section 11-10)

86. *Result/Explanation:* Heterogeneous catalysts are in a different physical phase than the reactants and presumably the products, making them easier to separate by simple techniques (e.g., decantation, filtration, etc.). Homogeneous catalysts are in the same physical phase as the reactants and presumably the products. Simple physical separation techniques are ineffective in separating the homogenous catalyst from the other chemicals.

87. *Result/Explanation:*

(a) The grid-like arrangement of the ceramic support is most likely to **maximize the surface area** and increase contact with the heterogeneous catalyst.

(b) Catalysis happens only at the surface of the metal. So, using expensive platinum in the form of strips or rods would be inefficient and not cost effective, since **too little of the metal would be on the surface**.

General Questions

89. *Result:* **See diagram and energy relationships below**

Explanation: An exothermic reaction has the products at a lower-energy state than the reactants.

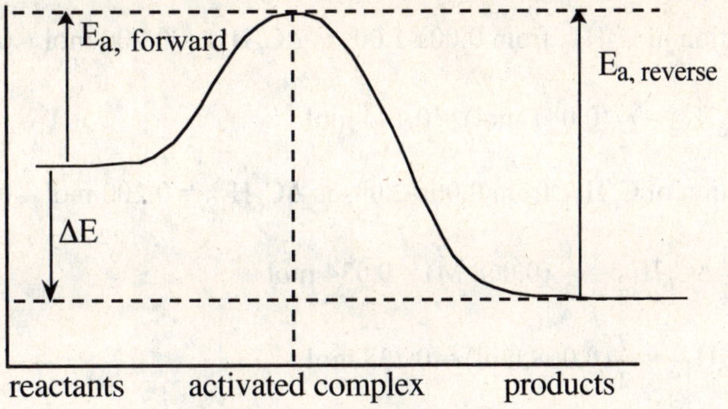

$$\Delta_r E = E_{a,forward} - E_{a,reverse}$$

90. *Result:* **See diagram and energy relationships below**

Explanation: An endothermic reaction has the products at a higher-energy state than the reactants.

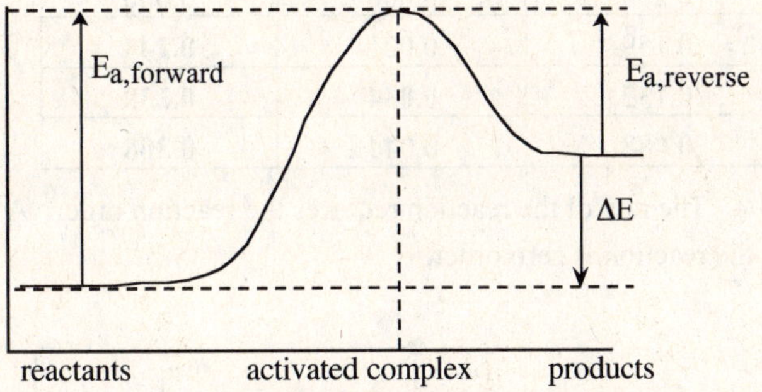

$$\Delta_r E = E_{a,forward} - E_{a,reverse}$$

91. *Result:* **(a)**

Time(s)	$[C_6H_{12}]$ ($\frac{mol}{V}$)	$[C_{12}H_{10}]$ ($\frac{mol}{V}$)	$[H_2]$ ($\frac{mol}{V}$)
0.0	0.200	0.000	0.000
1.00	0.159	0.021	**0.144**
2.00	0.132	**0.034**	**0.238**
3.00	**0.088**	0.044	**0.308**

(b) 0.036 mol V^{-1} s^{-1}

Analyze: Given a chemical equation and some concentration data collected at different times, determine the missing concentration information and calculate the rate of the reaction at a specified time.

(a) *Plan:* Use mole ratios in the balanced equation to relate reactants to products and use the time-dependence of the concentration changes to determine the reaction rate.

Execute: The balanced equation gives the ratio: 2 mol C_6H_{12} reactant: 1 mol $C_{12}H_{10}$ product: 7 mol H_2 product.

Time(s)	$[C_6H_{12}]$ ($\frac{mol}{V}$)	$[C_{12}H_{10}]$ ($\frac{mol}{V}$)	$[H_2]$ ($\frac{mol}{V}$)
0.0	0.200	0.000	0.000
1.00	0.159	0.021	(i)
2.00	0.132	(ii)	(ii)
3.00	(iii)	0.044	(iii)

Pick a convenient volume, V, and relate the stoichiometry of the reactants and products.

(i) The depletion of C_6H_{12} from 0.00s-1.00s is ΔC_6H_{12} = 0.200 mol – 0.159 mol = 0.041 mol C_6H_{12}

$$-\Delta H_2 = \frac{7}{2}\Delta C_6H_{12} = \frac{7}{2}(0.041 \text{ mol}) = 0.144 \text{ mol}$$

(ii) The depletion of C_6H_{12} from 0.00s-2.00s is ΔC_6H_{12} = 0.200 mol – 0.132 mol = 0.068 mol C_6H_{12}

$$-\Delta C_{12}H_{10} = \frac{1}{2}\Delta C_6H_{12} = \frac{1}{2}(0.068 \text{ M}) = 0.034 \text{ mol}$$

$$-\Delta H_2 = \frac{7}{2}\Delta C_6H_{12} = \frac{7}{2}(0.068 \text{ mol}) = 0.238 \text{ mol}$$

(iii) Formation of $C_{12}H_{10}$ from 0.00s-3.00s is – $\Delta C_{12}H_{10}$ = 0.044 mol – 0.000 mol = 0.044 M $C_{12}H_{10}$

$$\Delta C_{12}H_{10} = 2(-\Delta C_6H_{12}) = 2(0.044 \text{ mol}) = 0.088 \text{ mol}$$

$$-\Delta H_2 = 7(-\Delta C_6H_{12}) = 7(0.044 \text{ mol}) = 0.308 \text{ mol}$$

Time(s)	$[C_6H_{12}]\ (\frac{mol}{V})$	$[C_{12}H_{10}]\ (\frac{mol}{V})$	$[H_2]\ (\frac{mol}{V})$
0.0	0.200	0.000	0.000
1.00	0.159	0.021	**0.144**
2.00	0.132	**0.034**	**0.238**
3.00	**0.088**	0.044	**0.308**

(b) *Plan and Execute:* The rate of the reaction requires the reaction order. A plot of $[C_6H_{12}]$ vs. time gives a linear graph, so the reaction is zero order.

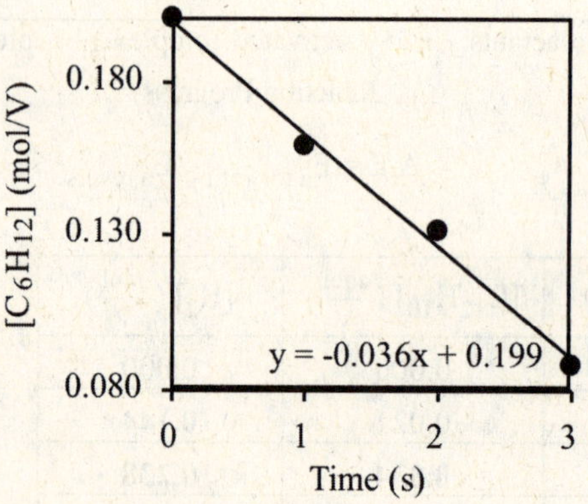

The slope of the graph is 0.036 mol V^{-1} s^{-1}, which also represents the value of k.

Since this is a zero-order reactions, $\frac{-\Delta[C_6H_{12}]}{\Delta t} = k = 0.036 \text{ mol } V^{-1} s^{-1}$

The zero-order reaction is independent of concentration, so the rate at 1.5 s is 0.036 mol M^{-1} s^{-1}

92. *Result:* **(a) False "The reaction might occur in a single step." (b) True (c) False. "Raising the temperature will cause the value of k to increase." (d) False. "The activation energy is independent of temperature." (e) False. "If the concentration of both reactants are doubled, the rate will quadruple." (f) True.**

Analyze: Given several statements related to a chemical equation and its rate law, determine which statements are correct and explain why incorrect statements are incorrect.

Plan: Use the information from Section 11-4 and 11-5 to help answer these questions.

Execute:

(a) False. What is true is: "The reaction *might* occur in a single step." It could be an elementary reaction, but it might not. For example, if one of the reactants binds with a catalytic substance in a fast reaction forming an intermediate, then the second reactant reacts with the intermediate in a slow reaction, the rate law given would be consistent with this two-step mechanism.

(b) True. This is a second-order reaction overall.

(c) False. What is true is: "Raising the temperature will cause the value of k to increase."

The Arrhenius equation ($k = Ae^{-E_a/RT}$) quantifies the increase in rate that occurs as a result of greater molecular motion. Raising the temperature increases the size of the denominator in the negative exponential. A smaller negative exponent makes the multiplier of the frequency factor (A) bigger. Another way to think about it is to use kinetic molecular theory: If the temperature increases, molecules move around more rapidly, permitting more collisions with enough energy to surmount the requisite activation energy.

(d) False. What is true is: "The activation energy is independent of temperature."

(e) False. What is true is: "If the concentration of both reactants are doubled, the rate will quadruple." Doubling the concentration of either reactant would double the rate, so doubling them both will make the rate four times faster.

(f) True. Adding a catalyst in the reaction will cause an increase in both the forward and reverse rates. The rate of the reaction initially will increase.

94. *Result:* **(a) NO is second order, O_2 is first order. (b) rate = $k[NO]^2[O_2]$ (c) 25 $L^2mol^{-2}s^{-1}$ (d) 7.8×10^{-4} mol $L^{-1}s^{-1}$ (e) rate for NO: 2.0×10^{-4} mol $L^{-1}s^{-1}$; rate for NO_2: 2.0×10^{-4} mol $L^{-1}s^{-1}$**

Analyze: Given a reaction and a table of concentrations and rates., determine the order with respect to each reactant, calculate the rate constant and the rate at given reactant concentrations.

Plan: Follow the method described in the solution to Question 25.

Execute: The reactants are NO and O_2. Data are available for the changes in each of these reactants' concentrations, so the rate law looks like this:

$$\text{rate} = k[NO]^i[O_2]^j \quad \text{where k, i and j are currently unknown.}$$

(a) Looking at Experiments 1 and 2, the initial concentration of NO doubles, the initial concentration of O_2 stays constant, and the initial rate changes by a factor of four ($1.0 \times 10^{-4}/2.5 \times 10^{-5}$). The rate change is the square of the concentration change, which suggests that the rate is proportional to the square of the concentration of NO, and the order with respect to NO is two.

Looking at Experiments 1 and 3, the initial concentration of O_2 doubles, the initial concentration of NO stays constant, and the initial rate changes by a factor of two. The rate change is the same as the concentration change, which suggests that the rate is proportional to the concentration of O_2, and the order with respect to O_2 is one.

(b) The rate law (also called the rate equation) now looks like this: $\text{rate} = k[NO]^2[O_2]^1$

(c) Solve the rate law for k:

$$k = \frac{\text{rate}}{[NO]^2[O_2]}$$

Plug in each experiment's data. Here is an example of the calculation for Experiment 1:

$$k_1 = \frac{2.5\times10^{-5}\ \text{mol L}^{-1}\text{s}^{-1}}{(0.010\ \text{mol/L})^2(0.010\ \text{mol/L})} = 25\ \text{L}^2\ \text{mol}^{-2}\text{s}^{-1}$$

[NO] (mol/L)	$[O_2]$ (mol/L)	Initial rate (mol $L^{-1}s^{-1}$)	Rate constant ($L^2mol^{-2}s^{-1}$)
0.010	0.010	2.5×10^{-5}	25
0.020	0.010	1.0×10^{-4}	25
0.010	0.020	5.0×10^{-5}	25

The average of these three rate constants is 25 $L^2mol^{-2}s^{-1}$.

(d) rate = $k[NO]^2[O_2]^1 = (25\ L^2mol^{-2}s^{-1})(0.025\ mol/L)^2(0.050\ mol/L)^1 = 7.8 \times 10^{-4} mol\ L^{-1}s^{-1}$

(e) The standard reaction rate is defined using the stoichiometric coefficients (as shown in Equation 11.3). Here, the stoichiometric relationship is 2 NO : 1 O_2 : 2 NO_2

$$\frac{1}{2}\left(\frac{\Delta[NO]}{\Delta t}\right) = -\frac{1}{1}\left(\frac{\Delta[O_2]}{\Delta t}\right) = \frac{1}{2}\left(\frac{\Delta[NO_2]}{\Delta t}\right)$$

$$-\frac{\Delta[NO]}{\Delta t} = 2\left(-\frac{\Delta[O_2]}{\Delta t}\right) = 2\left(1.0 \times 10^{-4}\ mol\,L^{-1}s^{-1}\right) = 2.0 \times 10^{-4}\ mol\,L^{-1}s^{-1}$$

$$\frac{\Delta[NO_2]}{\Delta t} = 2\left(-\frac{\Delta[O_2]}{\Delta t}\right) = 2\left(1.0 \times 10^{-4}\ mol\,L^{-1}s^{-1}\right) = 2.0 \times 10^{-4}\ mol\,L^{-1}s^{-1}$$

96. *Result:* **(a) First-order in $HCrO_4^-$, first-order in H_2O_2, and first-order in H^+ (b) See equations below (c) Second step, because the rate law derived from the mechanism is the same as the observed rate law**

Analyze and Plan: Given a chemical equation, the rate law, and a suggested mechanism, give the order with respect to each reactant, show that the steps of the mechanism agree with the overall equation, and explain which step is the rate-limiting step. Follow methods similar to those used in the solutions to Questions 57-58.

Execute:

(a) The observed rate law can be used to find the reaction orders of the reactants:

$$\text{rate} = k[HCrO_4^-][H_2O_2][H^+]$$

Since all the concentrations are raised to the same power, all three of them are first-order. That is, the reaction is first-order in $HCrO_4^-$, first-order in H_2O_2, and first-order in H^+.

(b) Cancel intermediates, H_2CrO_4 and $H_2CrO(O_2)_2$, and add the three reactions.

$$HCrO_4^- + H^+ \rightleftharpoons \cancel{H_2CrO_4}$$
$$\cancel{H_2CrO_4} + H_2O_2 \longrightarrow \cancel{H_2CrO(O_2)_2} + H_2O$$
$$+\ \cancel{H_2CrO(O_2)_2} + H_2O_2 \longrightarrow CrO(O_2)_2 + 2\,H_2O$$
$$HCrO_4^- + 2\,H_2O_2 + H^+ \longrightarrow CrO(O_2)_2 + 3\,H_2O$$

(c) It is clear, for two reasons, that the first step is not the rate limiting step. First, the double arrow used between the reactants and products indicates that the reaction is fast enough to reach a steady state. Second, the rate law would be: rate = $k[HCrO_4^-][H^+]$, if the first step was slow. That rate law is incompatible with the observed rate law.

If we derive the rate law for the mechanism assuming that the second step were slow, we have a step that looks like this: rate = $k_2[H_2CrO_4][H_2O_2]$

We set up a steady state condition for the intermediate, H_2CrO_4.

$$k_1[HCrO_4^-][H^+] = k_{-1}[H_2CrO_4] + k_2[H_2CrO_4][H_2O_2]$$

Assume that the second term is negligibly small since the rate of step 2 is small:

$$k_1[HCrO_4^-][H^+] = k_{-1}[H_2CrO_4]$$

Solve for the concentration of the intermediate: $[H_2CrO_4] = \frac{k_1}{k_{-1}}[HCrO_4^-][H^+]$

Plug into the rate law: $\text{rate} = k_2\left(\frac{k_1}{k_{-1}}[HCrO_4^-][H^+]\right)[H_2O_2]$

The observed rate constant is defined as $k = k_2\frac{k_1}{k_{-1}}$, giving **the derived rate law the same functional form as the observed rate law.**

$$\text{rate} = k[HCrO_4^-][H_2O_2][H^+]$$

Therefore, the second step is the rate-limiting step.

99. *Result:* **(a) 2.8×10^3 s (b) 1.4×10^4 s (c) 2.0×10^4 s**

Analyze and Plan: Given the order and rate constant for a reaction, calculate the half-life and how long it takes for the concentration to change in two different ways. Use methods and equations similar to ones described in the solution to Question 39.

Execute:

(a) $$t_{1/2} = \frac{\ln 2}{k} = \frac{\ln 2}{2.5 \times 10^{-4}\ s^{-1}} = 2.8 \times 10^3\ s$$

(b) $$[A]_t = \frac{1}{32} \times [A]_0$$

Notice, the initial concentration is not known, but the ratio is all we need:

$$kt = \ln[A]_0 - \ln[A]_t = \ln\left(\frac{[A]_0}{[A]_t}\right) = \ln\left(\frac{[A]_0}{\frac{1}{32}[A]_0}\right) = \ln(32) = 3.47$$

$$t = \frac{3.47}{k} = \frac{3.47}{2.5 \times 10^{-4}\ s^{-1}} = 1.4 \times 10^4\ s$$

Alternatively, we could also use the results derived in Question 112 to answer this Question:

$$[A]_t = [A]_0\left(\frac{1}{2}\right)^x \qquad x = \frac{t}{t_{1/2}}$$

Since $\left(\frac{1}{2}\right)^x = \frac{1}{32}$, when x = 5; therefore, t = 5 × (2.8×10^3 s) = 1.4×10^4 s.

(c) $$kt = \ln[A]_0 - \ln[A]_t = \ln(3.4 \times 10^{-3}\ mol/L) - \ln(2.3 \times 10^{-5}\ mol/L) = 5.00$$

$$t = \frac{5.00}{k} = \frac{5.00}{2.5 \times 10^{-4}\ s^{-1}} = 2.0 \times 10^4\ s$$

101. *Result:* **(a) 270 kJ (b) 2×10^3 s**

Analyze: Given two different temperatures and the rate constants at those temperatures, determine the activation energy of a reaction.

Plan and Execute:

(a) Convert the temperatures to Kelvin, then use the equation derived in Problem-Solving Example 11.9 to calculate the activation energy:

$$\ln\left(\frac{k_1}{k_2}\right) = \frac{E_a}{R}\left(\frac{1}{T_2} - \frac{1}{T_1}\right)$$

470.0 °C + 273.15 = 743.2 K 510.0 °C + 273.15 = 783.2 K

$$\ln\left(\frac{1.10 \times 10^{-4}\ s^{-1}}{1.02 \times 10^{-3}\ s^{-1}}\right) = \frac{E_a}{R}\left(\frac{1}{783\ K} - \frac{1}{743\ K}\right)$$

$$\ln(0.108) = \frac{E_a}{(0.008134 kJ / mol \cdot K)}\left(0.001277\ K^{-1} - 0.001346\ K^{-1}\right)$$

$$-2.227 = \frac{E_a}{(0.008134 kJ / mol \cdot K)}\left(-0.000069\ K^{-1}\right)$$

$$-2.227 = E_a\ (-0.0083\ mol/kJ)$$

$$E_a = 270\ kJ$$

(b) Use the same equation to determine the rate constant at 500. °C.

500.0 °C + 273.15 = 773.2 K

$$\ln\left(\frac{1.10 \times 10^{-4}\ s^{-1}}{k_2}\right) = \frac{E_a}{R}\left(\frac{1}{773.2\ K} - \frac{1}{743.2\ K}\right)$$

$$\ln\left(\frac{1.10 \times 10^{-4}\ s^{-1}}{k_2}\right) = \frac{(270\ kJ)}{(0.008134\ kJ / mol \cdot K)}\left(0.001293\ K^{-1} - 0.001346\ K^{-1}\right)$$

$$\ln\left(\frac{1.10 \times 10^{-4}\ s^{-1}}{k_2}\right) = -1.7$$

$$\frac{1.10 \times 10^{-4}\ s^{-1}}{e^{-1.7}} = k_2 \quad \textit{(with one sig fig from exponent of e)}$$

$$6 \times 10^{-4}\ s^{-1} = k_2 \quad \textit{(round to one sig fig)}$$

The rate constants are first-order units, so use first order integrated rate law to determine the time:

$$kt = \ln[A]_0 - \ln[A]_t = \ln(0.10) - \ln(0.023) = 1.47$$

$$t = \frac{1.47}{k} = \frac{1.47}{6 \times 10^{-4}\ s^{-1}}$$

$$t = 2467\ s \simeq 2 \times 10^3\ s$$

Applying Concepts

102. *Result:* **Curve A represents [H_2O(g)] increase with time, Curve B represents [O_2(g)] increase with time, and Curve C represents [H_2O_2(g)] decrease with time.**

Explanation: This Question uses the plan described in the solution to Question 11.

Curve A represents the increase in the concentration of the product H_2O(g) with time. Curve B represents the increase in the concentration of product O_2 with time. The O_2 curve is half as steep as the H_2O line because the stoichiometric relationship between them is 1:2. Curve C represents the decrease in the concentration of reactant $H_2O_2(g)$ with time.

104. *Result:* **Snapshot (b); products form more slowly at lower temperatures, so choose the snapshot with fewer HI molecules**

Explanation: Products form more slowly at low temperatures, so (b) the snapshot with fewer HI molecules is the one corresponding to a lower temperature.

106. *Result:* **rate = $k[A]^3[B][C]^2$**

Analyze: Given a set of pictures showing combinations of A, B, and C atoms, determine the rate law.

Plan: Use the methods described in Section 11-2, and similar to those in the solution to Question 26.

Execute: The reactants are A, B and C, so the rate law looks like this:

$$\text{rate} = k[A]^i[B]^j[C]^h \quad \text{where k, i, j, and h are currently unknown.}$$

Picture 1 has 2A, 2B, and 2C.

Picture 2 has 4A, 2B, and 2C.

Picture 3 has 4A, 4B, and 4C.

Picture 4 has 4A, 2B, and 6C.

Between Pictures 1 and 2, the initial concentration of A doubles (2 A → 4 A), and the initial concentrations of B and C stay constant (2 of each), the initial rate changes by a factor of (12/1.5 =) eight. The rate change is the cube of the concentration change ($8 = 2^3$), which suggests that the order with respect to A is three.

Between Pictures 2 and 3, the initial concentration of B doubles (2 B → 4 B), and the initial concentrations of A and C stay constant (4 A and 4 C), the initial rate changes by a factor of (23/12 =) two. The rate change is equal to the concentration change, which suggests that the order with respect to B is one.

Between Pictures 2 and 4, the initial concentration of C triples (2 C → 6 C), and the initial concentrations of A and B stay constant (4A and 2B), the initial rate changes by a factor of nine. The rate change is the square of the concentration change ($9 = 3^2$), which suggests that the order with respect to C is two.

So, the rate law looks like this: **rate = $k[A]^3[B][C]^2$**

107. *Result:* **(a) Three, because each hill in the reaction energy diagram represents one elementary reaction (b) Exothermic, because products have lower energy than reactants**

Explanation:

(a) Each hill in the reaction energy diagram represents one elementary reaction. So, this mechanism has three steps.

(b) The products are lower in energy than the reactants, so forming the products from the reactants causes a release of energy, making the reaction exothermic.

108. *Result:* **E_a is approximately zero.**

Explanation: The Arrhenius equation described in Section 11-5 describes the temperature dependence of the rate constant.

$$k = A\,e^{-E_a/RT}$$

If a radioactive decay reaction is independent of temperature, that says the activation energy E_a is indistinguishable from zero, and $k = A\,e^{-0/RT} = Ae^0 = A(1) = A$, with no functional dependence on T.

110. *Result:* **29.6 s, 94.7 s**

Analyze and Plan: Given time required for specific concentration changes, calculate the half-life and the time required to reach a new concentration. Use the method described in the solution to Question 39.

Execute:

$$t = t_2 - t_1 = 45.0 \text{ s} - 30.5 \text{ s} = 14.5 \text{ s}$$

$$kt = \ln[A]_1 - \ln[A]_2$$

$$k(14.5 \text{ s}) = \ln(0.0451 \text{ M}) - \ln(0.0321 \text{ M}) = 0.340$$

$$k = = \frac{0.340}{14.5 \text{ s}} = 0.0235 \text{ s}^{-1}$$

$$t_{1/2} = \frac{\ln 2}{k} = \frac{\ln 2}{0.0235 \text{ s}^{-1}} = 29.6 \text{ s}$$

$$t = t_3 - t_1 = t_3 - 30.5 \text{ s}$$

$$kt = \ln[A]_1 - \ln[A]_2$$

$$0.0235 \text{ s}^{-1}(t_3 - 30.5 \text{ s}) = \ln(0.0451 \text{ M}) - \ln(0.0100 \text{ M})$$

$$t_3 - 30.5 \text{ s} = 64.2 \text{ s}$$

$$t_3 = 94.7 \text{ s}$$

More Challenging Questions

113. *Result:* **(a) rate = 2.4 × 10^{-7}mol L^{-1}s^{-1} + k[BSC][F$^-$] (b) 0.3 L mol^{-1}s^{-1}**

Analyze: Given the initial concentrations and initial rates of a reaction at several different experimental conditions, determine the rate law and rate constant for the reaction.

Plan: In Section 11-2, the method of finding the rate law from initial rates is described for getting the orders. However, before we compare the experimental rates, we need to subtract the residual rate. Then try comparing pairs of experiments where only one of the concentrations is different and relating that to the changes in the rate. If the result of the pair-wise comparison is ambiguous, make a linear graph. Once the orders are determined, plug the data into the rate law to determine the value of k.

Execute: Benzenesulfonyl chloride (abbreviated BSC) has a known effect on the rate (first-order), but it has a constant concentration in this experiment, 2 × 10^{-4} M. As a result, we can't say how it is involved in the constant term representing the residual rate, so we will seek a rate law that looks like this:

rate = 2.4 × 10^{-7}mol L^{-1}s^{-1} + k[BSC][F$^-$]^i where k and i are currently unknown.

To study how [F$^-$] affects the rate of the reaction, subtract the residual rate from the observed rate.

Experiment	[F$^-$] × 10^2 (mol/L)	Initial rate × 10^7 (mol L^{-1}s^{-1})	Adjusted initial rate × 10^7 (mol L^{-1}s^{-1})
1	0	2.4	2.4 – 2.4 = 0.0
2	0.5	5.4	5.4 – 2.4 = 3.0
3	1.0	7.9	7.9 – 2.4 = 5.5
4	2.0	13.9	13.9 – 2.4 = 11.5
5	3.0	20.2	20.2 – 2.4 = 17.8
6	4.0	25.2	25.2 – 2.4 = 22.8
7	5.0	32.0	32.0 – 2.4 = 29.6

(a) The adjusted rate law has the functional form: adjusted rate = k[BSC][F$^-$]^i

To get the order of F$^-$, compare experiments 2 to 3, 3 to 4, and 4 to 6. They each show a concentration increase of a factor of two. The ratio of the adjusted initial rate in each of these instances is:

$$2 \text{ to } 3: \frac{5.5}{3.0} = 1.8, \quad 3 \text{ to } 4: \frac{11.5}{5.5} = 2.1, \quad 4 \text{ to } 6: \frac{22.8}{11.5} = 2.0$$

It looks like the adjusted rate change is approximately the same as the concentration change, suggesting that the adjusted rate is proportional to the concentration of fluoride ions. The relationship is only approximate, though, so let's prepare a graph that uses all of the data. By taking the log of the rate law, the order becomes the slope of a linear graph.

$$\log(\text{adjusted rate}) = \log(k[\text{BSC}]) + \log([\text{F}^-]^i)$$

$$\log(\text{adjusted rate}) = i \log[\text{F}^-] + \log(k[\text{BSC}])$$

$\log[\text{F}^-]$	log(adjusted initial rate) ($\text{mol L}^{-1}\text{s}^{-1}$)	$\log[\text{F}^-]$	log(adjusted initial rate) ($\text{mol L}^{-1}\text{s}^{-1}$)
–2.3	–6.52	–1.52	–5.750
–2.00	–6.26	–1.40	–5.642
–1.70	–5.939	–1.30	–5.529

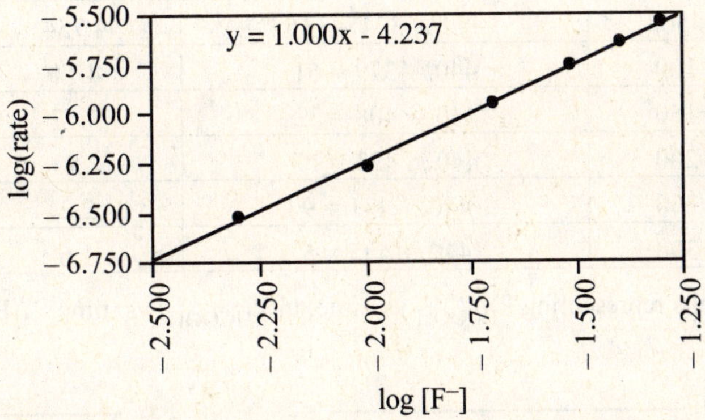

This confirms that the order of the adjusted rate law with respect to fluoride (i) is one, and the most complete rate law we can write for the reaction looks like this:

$$\text{rate} = 2.4 \times 10^{-7} \text{mol L}^{-1}\text{s}^{-1} + k[\text{BSC}][\text{F}^-]$$

(b) Solve the adjusted rate law for k: $k = \dfrac{\text{adjusted rate}}{[\text{BSC}][\text{F}^-]}$

Plug in each set of data. Here is a sample of a calculation for the first experiment with nonzero $[\text{F}^-]$:

$$k_2 = \frac{3.0 \times 10^{-7} \text{ mol L}^{-1}\text{s}^{-1}}{(2 \times 10^{-4} \text{ M})(0.5 \times 10^{-2} \text{ mol/L})} = 0.3 \text{ M}^{-1}\text{s}^{-1}$$

Experiment	$[\text{F}^-] \times 10^2$ (mol/L)	Adjusted initial rate × 10^7 ($\text{mol L}^{-1}\text{s}^{-1}$)	Adjusted rate constant ($\text{M}^{-1}\text{s}^{-1}$)
2	0.5	3.0	0.3
3	1.0	5.5	0.3
4	2.0	11.5	0.3
5	3.0	17.8	0.3
6	4.0	22.8	0.3
7	5.0	29.6	0.3

The average of these seven rate constants is 0.3 $\text{L mol}^{-1}\text{s}^{-1}$

☑ *Reasonable Result Check:* It is satisfying to get the same rate constant for each data set.

115. *Result:* **$0.0127\ s^{-1}$, 54.6 s**

Analyze: The chemical reaction takes one mole of gas and makes it two, so the partial pressure rises as a result of the formation of products. The total pressure P_{TOT} initially is entirely due to the pressure exerted by HCOOH(g), 220 torr. As the reaction proceeds the reduction in the partial pressure of HCOOH(g) causes an increase in the partial pressure of products CO_2(g) and H_2(g). The stoichiometry is 1:1:1. So, at any given time, the partial pressure of HCOOH(g), $P_{HCOOH} = 220.\ \text{torr} - x$, The partial pressures of CO_2(g) and H_2(g) increase to $P_{CO_2} = P_{H_2} = x$. Dalton's law of partial pressures (Chapter 8 Section 8-6) indicates that the sum of the partial pressures is equal to the total pressure.

$$P_{TOT} = P_{HCOOH} + P_{CO_2} + P_{H_2} = (220.\ \text{torr} - x) + x + x = 220.\ \text{torr} + x$$

$$P_{TOT} - 220.\ \text{torr} = x$$

$$P_{HCOOH} = 220.\ \text{torr} - (P_{TOT} - 220.\ \text{torr}) = 440.\ \text{torr} - P_{TOT}$$

Time (s)	P_{HCOOH} (torr)	$\ln(P_{HCOOH})$	$1/P_{HCOOH}$
0	220.	5.394	0.00455
50	440. – 324 = 116	4.754	0.00862
100	440. – 379 = 61	4.11	0.016
150	440. – 408 = 32	3.47	0.031
200	440. – 423 = 17	2.83	0.059
250	440. – 431 = 9	2.2	0.1
300	440. – 435 = 5	1.6	0.2

Create three graphs representing P_{HCOOH} vs. time, $\ln(P_{HCOOH})$ vs. time, $1/(P_{HCOOH})$ vs. time:

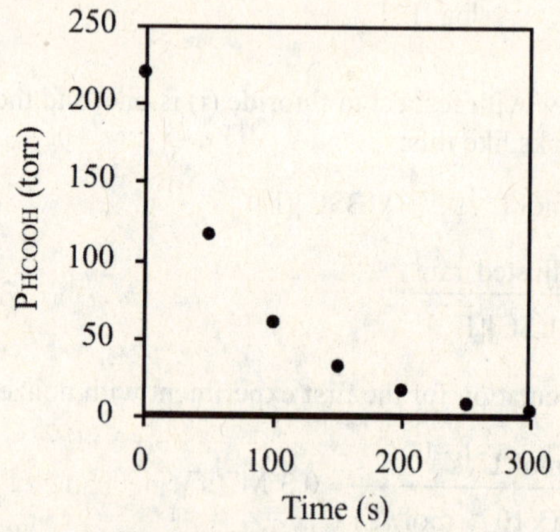

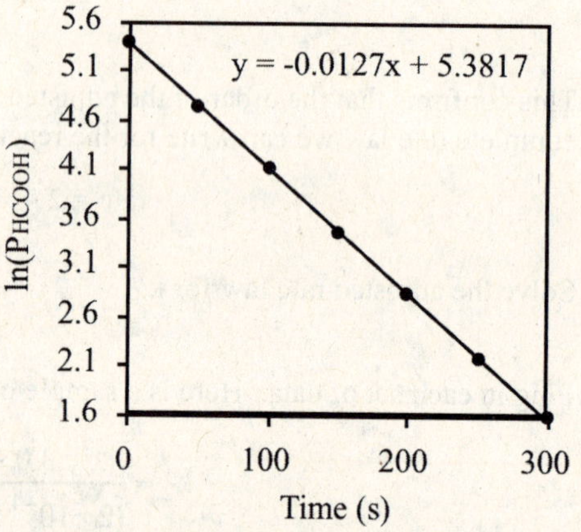

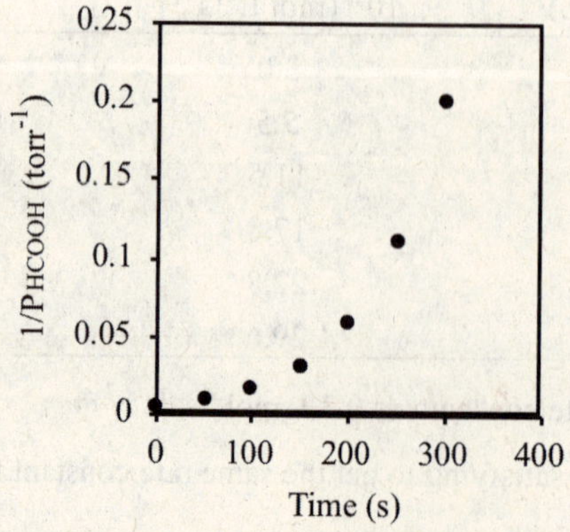

The linear graph is the second one, which represents a first-order graph. The slope of the first-order graph is –k; therefore, $k = 0.0127\ s^{-1}$. The first-order half-life equation looks like this:

$$t_{1/2} = \frac{\ln 2}{k} = \frac{\ln 2}{0.0127\ s^{-1}} = 54.6\ s$$

☑ *Reasonable Result Check:* It makes sense that the half life is a little more than 50 s since half the initial pressure of HCOOH is 110 torr, and the pressure at t = 50 pressure is 116 torr.

117. *Result:* **See mechanisms below (There can be other correct answers.)**

Analyze: For each of a given set of chemical equations and rate laws, propose a reasonable mechanism. People may come up with correct answers to this Question that are NOT the same as the answers given here. In all cases, a proposed mechanism must have three things: First, it must be composed of first or second order elementary reactions. Second, it must recreate the given overall net reaction, including the elimination of all intermediates and the recreation of any catalysts. Third, it must recreate the observed rate law.

Plan: If the rate law is first or second order, try making a mechanism with the first step the slow step, and use the chemical(s) in the rate law as reactants. If the rate law is more complicated, start with a first step that is fast, and include some of the chemicals from the rate law in that reaction, then use the product of that reaction as a reactant in the next reaction. In each case, once you have written the mechanism, confirm that its mechanism matches the observed (given) mechanism.

Execute:

(a)

$CH_3CO_2CH_3 + H^+ \longrightarrow CH_3COHOCH_3^+$ slow

$CH_3COHOCH_3^+ + H_2O \longrightarrow CH_3COH(OH_2)OCH_3^+$ fast

$CH_3COH(OH_2)OCH_3^+ \longrightarrow CH_3C(OH)_2^+ + CH_3OH$ fast

$CH_3C(OH)_2^+ \longrightarrow CH_3COOH + H^+$ fast

Check this mechanism:

Each reaction is unimolecular or bimolecular.

Check the overall reaction:

$CH_3CO_2CH_3 + \cancel{H^+} \longrightarrow \cancel{CH_3COHOCH_3^+}$

$\cancel{CH_3COHOCH_3^+} + H_2O \longrightarrow \cancel{CH_3COH(OH_2)OCH_3^+}$

$\cancel{CH_3COH(OH_2)OCH_3^+} \longrightarrow \cancel{CH_3C(OH)_2^+} + CH_3OH$

$\cancel{CH_3C(OH)_2^+} \longrightarrow CH_3COOH + \cancel{H^+}$

$CH_3CO_2CH_3 + H_2O \longrightarrow CH_3COOH + CH_3OH$ This is the net equation.

The rate law of the slow first elementary reaction is the rate law for the whole mechanism. As described in Section 11-6, the stoichiometric coefficient of a reactant in an elementary reaction is the reaction order for that reactant.

$CH_3CO_2CH_3 + H^+ \longrightarrow$ products rate = $k_1[CH_3CO_2CH_3][H^+]$

This rate law matches the reported rate law for the reaction. Therefore, the proposed mechanism is plausible.

(b) $H_2 + I_2 \longrightarrow 2\ HI$ slow

Check this mechanism: The reaction is bimolecular and represents the overall reaction.

As described in (a), check the mechanism's rate law and compare to the experimental rate law.

$H_2 + I_2 \longrightarrow$ products rate = $k[H_2][I_2]$

This rate law matches the reported rate law for the reaction. The proposed mechanism is plausible.

(c) The presence of I_2 and Pt in the rate law suggests that they are catalysts in this reaction:

$$H_2 + Pt(s) \longrightarrow PtH_2 \qquad \text{fast}$$
$$PtH_2 + I_2 \longrightarrow PtH_2I_2 \qquad \text{slow}$$
$$PtH_2I_2 + O_2 \longrightarrow PtI_2O + H_2O \qquad \text{fast}$$
$$PtI_2O + H_2 \longrightarrow Pt(s) + I_2 + H_2O \qquad \text{fast}$$

Check this mechanism:

Each reaction is unimolecular or bimolecular.

Overall:

$$H_2 + \cancel{Pt} \longrightarrow \cancel{PtH_2}$$
$$\cancel{PtH_2} + \cancel{I_2} \longrightarrow \cancel{PtH_2I_2}$$
$$\cancel{PtH_2I_2} + O_2 \longrightarrow \cancel{PtI_2O} + H_2O$$
$$\cancel{PtI_2O} + H_2 \longrightarrow \cancel{Pt} + \cancel{I_2} + H_2O$$

$$2\,H_2 + O_2 \longrightarrow 2\,H_2O$$ This is the overall equation.

Determine the rate law for this mechanism.

The slow step is step 2, so the rate of the reaction is equal to the rate of this rate-determining step. Because this reaction is elementary, its rate law is related to the stoichiometry of its reactants:

$$\text{rate of reaction} = \text{rate}_{\text{reaction 2}} = k_2[PtH_2][I_2]$$

PtH_2 is an intermediate, so we need to eliminate it from the proposed mechanism. Look for all the ways this intermediate is created and destroyed during the mechanism. It is only created in the forward reaction of step 1. It is destroyed in the reverse reaction of step 1 and also in the forward reaction of step 2. Because Pt is a solid, its surface area shows up in the rate law. Because all three of these reactions are elementary, use their reactants' stoichiometric coefficients to write rate laws for each of these three reactions:

$$\text{rate}_{\text{forward reaction 1}} = k_1[H_2](\text{area of Pt surface})$$
$$\text{rate}_{\text{reverse reaction 1}} = k_{-1}[PtH_2]^1$$
$$\text{rate}_{\text{forward reaction 2}} = k_2[PtH_2][I_2]$$

The rate of PtH_2 creation is equal to the rate of its destruction once the steady state condition is reached:

$$\text{rate}_{\text{forward reaction 1}} = \text{rate}_{\text{reverse reaction 1}} + \text{rate}_{\text{forward reaction 2}}$$
$$k_1[H_2](\text{area of Pt surface}) = k_{-1}[PtH_2]^1 + k_2[PtH_2][I_2]$$

Because the rate of step 2 is presumed to be much smaller than the rate of step 1, the second term is expected to be negligibly small compared to the first term:

$$\text{rate}_{\text{forward reaction 1}} \cong \text{rate}_{\text{reverse reaction 1}}$$
$$k_1[H_2](\text{area of Pt surface}) = k_{-1}[PtH_2]^1$$
$$[PtH_2] = \frac{k_1}{k_{-1}}[H_2](\text{area of Pt surface})$$

Substitute the equality just derived into the rate law: $\text{rate} = k[PtH_2][I_2]$

$$\text{rate} = k_2\frac{k_1}{k_{-1}}[H_2](\text{area of Pt surface})[I_2]$$

We will define the new group of rate constants by one variable, $k' = k_2 \frac{k_1}{k_{-1}}$.

$$\text{rate} = k'[H_2][I_2](\text{area of Pt surface})$$

This matches the reported rate law for the reaction. Therefore, the proposed mechanism is a plausible one.

(d) Whenever a rate law has a substance's concentration with a square root, it's typically due to the decomposition of that substance into two identical atoms or fragments, then using only one of those atoms or fragments in the slow step:

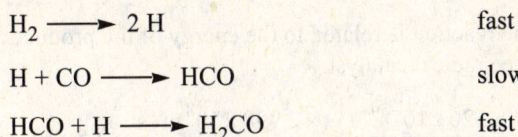

$H_2 \longrightarrow 2\,H$	fast
$H + CO \longrightarrow HCO$	slow
$HCO + H \longrightarrow H_2CO$	fast

Check this mechanism:

Each reaction is unimolecular or bimolecular.

$$H_2 \longrightarrow 2\cancel{H}$$
$$\cancel{H} + CO \longrightarrow \cancel{HCO}$$
$$\cancel{HCO} + \cancel{H} \longrightarrow H_2CO$$

$H_2 + CO \longrightarrow H_2CO$ This is the overall equation.

Determine the rate law for this mechanism.

The slow step is step 2, so the rate of the reaction is equal to the rate of this rate-determining step. Because this reaction is elementary, its rate law is related to the stoichiometry of its reactants:

$$\text{rate of reaction} = \text{rate}_{\text{reaction 2}} = k_2[H][CO]$$

H is an intermediate, so we need to eliminate it from the proposed mechanism. Look for all the ways this intermediate is created and destroyed during the mechanism. It is only created in the forward reaction of step 1. It is destroyed in the reverse reaction of step 1 and also in the reaction of step 2. Because all three of these reactions are elementary, use their reactants' stoichiometric coefficients to write rate laws for each of these three reactions:

$$\text{rate}_{\text{forward reaction 1}} = k_1[H_2]$$
$$\text{rate}_{\text{reverse reaction 1}} = k_{-1}[H]^2$$
$$\text{rate}_{\text{reaction 2}} = k_2[H][CO]$$

The rate of H creation is equal to the rate of its destruction once the steady state condition is reached:

$$\text{rate}_{\text{forward reaction 1}} = \text{rate}_{\text{reverse reaction 1}} + \text{rate}_{\text{reaction 2}}$$
$$k_1[H_2] = k_{-1}[H]^2 + k_2[H][CO]$$

Because the rate of step 2 is presumed to be much smaller than the rate of step 1, the second term is expected to be negligibly small compared to the first term:

$$\text{rate}_{\text{forward reaction 1}} \cong \text{rate}_{\text{reverse reaction 1}}$$
$$k_1[H_2] = k_{-1}[H]^2$$
$$[H]^2 = \frac{k_1}{k_{-1}}[H_2]$$
$$[H] = \left(\frac{k_1}{k_{-1}}[H_2]\right)^{1/2}$$

Substitute the equality just derived into the rate law: rate = k_2[H][CO]

$$\text{rate} = k_2\left(\frac{k_1}{k_{-1}}\right)^{1/2}[H_2]^{1/2}$$

We will define the new group of rate constants by one variable, $k' = k_2\left(\frac{k_1}{k_{-1}}\right)^{1/2}$.

$$\text{rate} = k'[H_2]^{1/2}[CO]$$

119. *Result/Explanation:* The speed of the reaction is related to the energy of the production of a photon that can break a bond in iodine and make a free radical catalyst.

$$E_{photon} = \frac{hc}{\lambda} = \frac{6.626\times10^{-34}\ J\cdot s \times 2.998\times10^{8}\ m/s}{800\ nm \times \dfrac{1\times10^{-9}\ m}{1\ nm}} = 2.5\times10^{-19}\ J$$

$$\text{Energy per mole of bonds} = \frac{2.5\times10^{-19}\ J\times}{1\ photon}\times\frac{6.022\times10^{23}\ photons}{1\ mol\ photons}\times\frac{1\ kJ}{1000\ J} = 150\ \frac{kJ}{mol}$$

This is approximately the bond enthalpy of I_2. Wavelengths longer than 800 nm have photons with energy less than the bond enthalpy of I_2.

121. *Result/Explanation:* Use Section 11-3 and the margin note in Section 11-3a associated with Equation 11.5 to assist in the solution to this Question.

(i) Define the reaction rate in terms of [A]: $\text{Rate} = -\frac{\Delta[A]}{\Delta t}$

(ii) Write the rate law in terms of [A], k and t: $-\frac{\Delta[A]}{\Delta t} = k[A]$

Calculus uses "d" to replace "Δ" to describe very small changes: $-\frac{d[A]}{dt} = k[A]$

(iii) Separate the variables by putting everything with [A] on one side and everything with t on the other side:

$$\frac{d[A]}{[A]} = -\,kdt$$

(iv) Integrate each side, using calculus: $\int\frac{dx}{x} = \ln x$ and $\int dx = x$

and use the initial and final conditions to resolve the definite integral:

$$\int_0^t \frac{d[A]}{[A]} = -k\int_0^t dt$$

$$\ln[A]_t - \ln[A]_0 = -\,k(t_t - t_0)$$

(v) If the reaction starts at time t=0 and goes to time t, then $t_t - t_0$ is the elapsed time, t.

$$\ln[A]_t - \ln[A]_0 = -\,kt$$

$$\ln[A]_t = -\,kt + \ln[A]_0$$

This equation is in the form of y = mx + b, where variable y = $\ln[A]_t$, m = – k (a constant), variable x = t, and b = $\ln[A]_0$ (also a constant)

(vi) Half life is the elapsed time for $[A]_0$ to drop by half, so:

$$[A]_t = \frac{1}{2}[A]_0 \text{, when } t = t_{1/2}$$

$$\ln\left(\frac{1}{2}[A]_0\right) - \ln[A]_0 = -kt_{1/2}$$

$$\ln[A]_0 - \ln 2 - \ln[A]_0 = -kt_{1/2}$$

$$-\ln 2 = -kt_{1/2}$$

$$t_{1/2} = \frac{-\ln 2}{-k} = \frac{\ln 2}{k}$$

Chapter 12: Chemical Equilibrium

Solutions for Red-Numbered Questions for Review and Thought

Topical Questions

Characteristics of Chemical Equilibrium (Section 12-1)

8. *Result/Explanation:* There are many answers to this Question. One possible answer is given later in Question 115. Prepare a sample of N_2O_4 in which the N atoms are the heavier isotopes ^{15}N. Introduce the heavy isotope of N_2O_4 into an equilibrium mixture of N_2O_4 and NO_2. Use spectroscopic methods, such as infrared spectroscopy to observe the distribution of the radioisotope among the reactants and products.

10. *Result:* (a) 0 °C (b) Dynamic equilibrium (see explanation below)

Explanation:

(a) The temperature of an equilibrium mixture of ice and water is the melting point of water: 0 °C.

(b) This is a dynamic equilibrium. Molecules are not smart enough to stay in a particular phase. Some molecules at the interface between the water and the ice detach from the ice and enter the liquid phase or attach to the solid phase leaving the liquid phase.

The Equilibrium Constant (Section 12-2)

12. *Result:* **See drawings below**

Explanation: Draw reactant concentration vs. time as a downward curve to level off flat at the equilibrium concentration. Draw product concentration vs. time curve as an upward curve to level off flat at the equilibrium concentration. The slope of [NO] increase is steeper than slope of $[N_2]$ decrease, and the equilibrium concentration of NO is equal to half of the equilibrium concentration of N_2 or O_2.

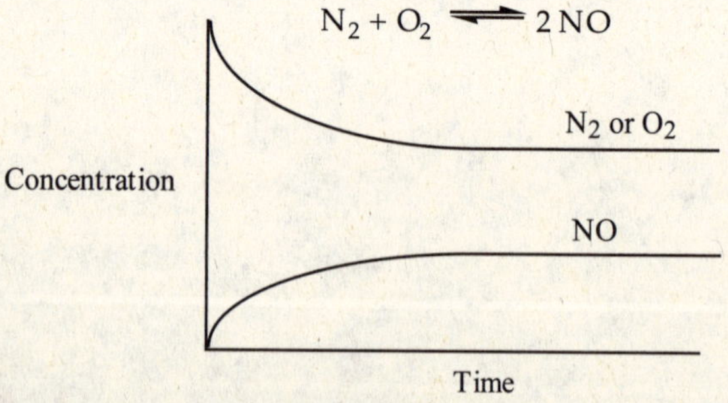

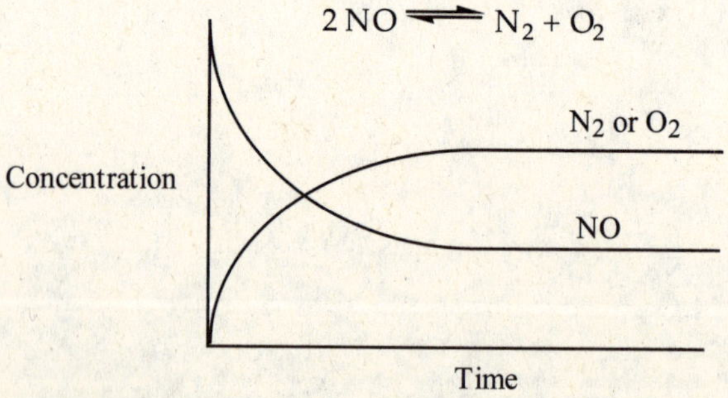

14. *Result:* **See equations below**

Analyze: Given chemical equations, write equilibrium constant expressions.

Plan: For the expression of the equilibrium constant, use this form: $K_c = \frac{[\text{products}]}{[\text{reactants}]}$; if a stoichiometric coefficient precedes a species, that number is used as the mathematical power of the concentration in the expression. Remember that the equilibrium state is not affected by the relative quantity of any solids or liquids, so those materials do not show up in the equilibrium expression.

Execute: The answer for (b) does not have Fe(s) in the expression, the answer for (c) does not have
$(NH_4)_2CO_3$ (s) in the expression, and the answer for (d) does not have Ag_2SO_4 (s) in the expression.

(a) $K_c = \frac{[O_3]^2}{[O_2]^3}$ (b) $K_c = \frac{[Fe(CO)_5]}{[CO]^5}$

(c) $K_c = [NH_3]^2[CO_2][H_2O]$ (d) $K_c = [Ag^+]^2[SO_4^{2-}]$

16. *Result:* **See equations below**

Analyze and Plan: Given chemical equations, write equilibrium constant expressions. Follow the method described in Question 14.

Execute: The answer for (c) does not have SiO_2(s), C(s), or SiC(s) in the expression, and the answer for (d) does not have S_8(s) in the expression.

(a) $K_c = \frac{[H_2O]^2[O_2]}{[H_2O_2]^2}$ (b) $K_c = \frac{[PCl_5]}{[PCl_3][Cl_2]}$

(c) $K_c = [CO]^2$ (d) $K_c = \frac{[H_2S]}{[H_2]}$

18. *Result:* **See equations below**

Analyze and Plan: Given chemical equations, write equilibrium constant expressions. Follow the method described in Question 14.

Execute: The answer for (a) does not include solid PCl_5. The answer for (b) does not include $H_2O(\ell)$ in the expression, however (d) does include the gaseous H_2O(g).

(a) $K_c = [PCl_3][Cl_2]$ (b) $K_c = \frac{[CoCl_4^{2-}]}{[Co(H_2O)_6^{2+}][Cl^-]^4}$

(c) $K_c = \frac{[CH_3COO^-][H^+]}{[CH_3COOH]}$ (d) $K_c = \frac{[OF_2][HF]^2}{[F_2]^2[H_2O]}$

20. *Result:* **See equations below**

Analyze and Plan: Given chemical equations, write equilibrium constant expressions. Follow the method described in Question 14.

Execute: The answer for (a) does not have $CaSO_4 \cdot 5H_2O$(s) nor $CaSO_4 \cdot 3H_2O$(s) in the expression, the answer for (b) does not have SiO_2(s) in the expression, and the answer for (c) does not have $LaCl_3$(s) nor LaClO(s) in the expression.

(a) $K_c = [H_2O]^2$ (b) $K_c = \frac{[HF]^4}{[SiF_4][H_2O]^2}$ (c) $K_c = \frac{[HCl]^2}{[H_2O]}$

22. *Result:* **See equations below**

Analyze and Plan: Given chemical equations, write equilibrium constant expressions. Reverse the instructions given in the solution to Question 14. Identify solid and liquid reactants and products to complete balanced equations.

Execute: The Br_2 product is in the liquid state in (c), which explains its absence from the equilibrium constant expression.

(a) $2\ SO_2(g) + 2\ H_2O(g) \rightleftharpoons 2\ H_2S(g) + 3\ O_2(g)$

(b) $IF(g) \rightleftharpoons \frac{1}{2}F_2\ (g) + \frac{1}{2}I_2(g)$

(c) $Cl_2(g)\ +\ 2Br^- \longrightarrow Br_2(\ell)\ +\ 2\ Cl^-(aq)$

24. *Result:* **Equation (e)**

Analyze: The two equations are related, so we will use the information in Section 12-2d to identify how their equilibrium constants are related.

Plan and Execute: Multiplying the first equation by a constant factor of 2 and then reversing that equation gives the second equation. The first change means we need to raise the K from the first reaction to the power of 2. The second change means we need to take the reciprocal of the K after the first change. Therefore the second equation's equilibrium constant is related to the first equation's equilibrium constant as represented in (e), $K_{c_2} = 1/K_{c_1}{}^2$.

26. *Result:* **(a) 0.87 (b) 1.3**

Analyze: Given a chemical equation and its K_c, calculate the K_c of a related chemical equation.

Plan: The two equations are related, so we will use the information in Section 12-2d to identify how their equilibrium constants are related.

Execute:

(a) The synthesis of 1 mol sulfur trioxide gas is represented by the formation reaction:

$$SO_2(g) + \tfrac{1}{2}O_2(g) \rightleftharpoons SO_3(g)$$

Multiplying the given equation by a constant factor of $\frac{1}{2}$ gives the second equation. This change means we need to raise the K_c from the first reaction to the power of $\frac{1}{2}$.

$$K_{(a)} = K_c{}^{1/2} = (0.76)^{1/2} = 0.87$$

(b) The decomposition of 2 mol sulfur trioxide gas is represented by the formation reaction:

$$2\ SO_3(g) \rightleftharpoons 2\ SO_2(g) + O_2(g)$$

This reaction is the reverse of the given reaction, so we need to take the reciprocal of the K_c:

$$K_{(b)} = 1/K_c = 1/(0.76) = 1.3$$

28. *Result:* **See equations below**

Analyze: Given chemical equations, write their equilibrium constant expressions, K_p.

Plan: For the expression of the equilibrium constant, use this form: $K_p = \frac{P_{products}}{P_{reactants}}$; if a stoichiometric coefficient precedes a species, that number is used as the mathematical power of the pressure in the expression. Remember that the equilibrium is not affected by the quantity of any solids or liquids, so those materials do not show up in the equilibrium expression.

Execute: The answer for (c) does not have SiO_2(s), C(s), or SiC(s) in the expression, and the answer for (d) does not have S_8(s) in the expression.

(a) $K_p = \frac{P^2_{H_2O} P_{O_2}}{P^2_{H_2O_2}}$

(b) $K_p = \frac{P_{PCl_5}}{P_{PCl_3} P_{Cl_2}}$

(c) $K_p = P^2_{CO}$

(d) $K_p = \frac{P_{H_2S}}{P_{H_2}}$

30. *Result:* **2.6 × 10^{11}**

Analyze: Given 3 chemical equations and their equilibrium constants, calculate the K for a reaction described.

Plan and Execute: Use the naming system described in Section 2-8 for binary compounds. Dinitrogen oxide is N_2O and dinitrogen tetraoxide is N_2O_4. Write equation for the desired equation.

$$N_2O(g) + {}^3\!/_2\, O_2(g) \rightleftharpoons N_2O_4\,(g) \qquad K = ?$$

Follow the method shown in Problem-Solving Example 12.2, look at the desired equation and ascertain which of the given equations can provide reactants and products, then arrange them in a way that when added together produce the desired equation as described for Hess' Law in Chapter 4. Use the information in Section 12.2 to identify how the equilibrium constants are affected by manipulating these equations. The reactant $N_2O(g)$ is only found in the first equation, but in that equation it is a product. We need to reverse and take half of the first reaction and take the reciprocal of the square-root of the K value:

$$1\ N_2O(g) \rightleftharpoons 1\ N_2(g) + \tfrac{1}{2}\ O_2(g) \qquad K = 1/\sqrt{1.2\times10^{-35}}$$

The product $N_2O_4(g)$ is only found in the second equation, but in that equation it is a reactant. We need to reverse the second reaction and take the reciprocal of the K value:

$$2\ NO_2(g) \rightleftharpoons N_2O_4(g) \qquad K = 1/(4.6 \times 10^{-3})$$

The reactant in the equation above uses $NO_2(g)$, but that compound is not in the desired equation, so we need to recreate that compound. $NO_2(g)$ is only found in the third equation, but there's only one equation and we need two. We need to multiply the third reaction by two and square the K value:

$$N_2(g) + 2\,O_2(g) \rightleftharpoons 2\,NO_2(g) \qquad K = (4.1 \times 10^{-9})^2$$

The sum of the three equations, results in the mathematical product of their K values:

$$N_2O(g) \rightleftharpoons 2\,N_2(g) + O_2(g) \qquad K = 1/\sqrt{1.2\times10^{-35}}$$

$$2\,NO_2(g) \rightleftharpoons N_2O_4(g) \qquad K = 1/(4.6 \times 10^{-3})$$

$$+ \quad N_2(g) + 2\,O_2(g) \rightleftharpoons 2\,NO_2(g) \qquad K = (4.1 \times 10^{-9})^2$$

$$N_2O(g) + O_2(g) \rightleftharpoons N_2O_4\,(g) \qquad K = \frac{\left(4.1\times10^{-9}\right)^2}{\sqrt{1.2\times10^{-35}}\left(4.6\times10^{-3}\right)} = 2.6 \times 10^{11}$$

32. *Result:* **K_c = 0.0161**

Analyze: Given a phase change reaction and the vapor pressure of the gaseous product, find the value of K_c.

Write the K_p expression for the reaction and plug in the known value. Then use the relationship between K_p and K_c (Section 12-2f), with $R = 0.08206\ \frac{L\cdot atm}{mol\cdot K}$, and Δn = change in the number of moles of gas in the reaction, to get K_c.

$$H_2O(\ell) \rightleftharpoons H_2O(g) \qquad K_p = P_{H_2O(g)} = 0.467\text{ atm}$$

(Assume temperature is known within ± 1 °C.) T = 80. °C + 273.15 = 353 K

$$\Delta n = 1\text{ mol } H_2O\text{ gas product} - 0\text{ mol gas reactants} = 1$$

$$K_p = K_c(RT)^{\Delta n} \qquad K_p(RT)^{-\Delta n} = K_c$$

$$K_c = (0.467\text{ atm}) \times \left\{\left(0.08206\ \frac{L\cdot atm}{mol\cdot K}\right)\times\left(353\text{ K}\right)\right\}^{-1} = 0.0161$$

✓ *Reasonable Result Check:* This $(RT)^{-1}$ factor makes K_c smaller than K_p.

Determining Equilibrium Constants (Section 12-3)

34. *Result:* **$K_c = 2.6 \times 10^{-9}$**

Analyze: Given a chemical equation and the equilibrium concentrations of reactants and products, calculate K_c.

Plan: Write the K_c expression and plug the concentrations into the expression to get the value of K_c.

Execute:

$$K_c = \frac{[Br_2][F_2]^5}{[BrCl_5]^2} = \frac{(0.0018)\times(0.0090)^5}{(0.0064)^2} = 2.6 \times 10^{-9}$$

36. *Result:* **$K_p = 1.6$**

Analyze: Given an equation for a reaction and the moles of the gaseous reactants and products at equilibrium in a known volume, find the value of K_p.

Plan: Write the K_c expression for the reaction, calculate the concentrations of the gases, and plug them into the expression to get the value of K_c. Use the relationship between K_p and K_c (also described in the solution to Question 32) to get K_p

$$K_p = K_c(RT)^{\Delta n}$$

Execute:

$$H_2(g) + CO_2(g) \rightleftharpoons H_2O(g) + CO(g)$$

$$K_c = \frac{[H_2O][CO]}{[H_2][CO_2]} = \frac{\left(\frac{0.11\text{ mol}}{1.0\text{ L}}\right)\times\left(\frac{0.11\text{ mol}}{1.0\text{ L}}\right)}{\left(\frac{0.087\text{ mol}}{1.0\text{ L}}\right)\times\left(\frac{0.087\text{ mol}}{1.0\text{ L}}\right)} = 1.6$$

$$T = 986\ °C + 273.15 = 1259\text{ K}$$

$$\Delta n = 1\text{ mol } H_2O(g) + 1\text{ mol } CO(g) - 1\text{ mol } H_2(g) - 1\text{ mol } CO_2(g) = 0$$

$$K_p = (1.6) \times \left\{\left(0.08206\ \frac{\text{L}\cdot\text{atm}}{\text{mol}\cdot\text{K}}\right)\times(1259\text{ K})\right\}^0$$

$$K_p = 1.6$$

☑ *Reasonable Result Check:* The slightly larger moles of products suggest this reaction is slightly product-favored, so it makes sense that the K_c is larger than 1. The equal moles of gas-phase reactants and gas-phase products, which is responsible for the power in the $(RT)^0$ factor, makes K_p the same as K_c.

38. *Result:* **(a) [CO] = 0.0071 mol/L, [$COCl_2$] = 0.00308 mol/L (b) 1.4×10^2**

Analyze: Given an equation for a reaction, the initial concentration of the gaseous reactants and the equilibrium concentration of a gaseous product, find the value of K_c.

Plan and Execute: Following the procedure given in Section 12.3, write the equation and construct an ICE reaction table. No products are initially present.

	$CO(g)$	+	$Cl_2(g)$	$\rightleftharpoons$	$COCl_2(g)$
initial conc. (mol/L)	0.0102		0.00609		0
change as reaction occurs (mol/L)	______		______		______
equilibrium conc. (mol/L)	______		0.00301		______

Describe the stoichiometric changes in terms of one variable, x.

	$CO(g)$	$+ Cl_2(g)$ ⇌	$COCl_2(g)$
initial conc. (mol/L)	0.0102	0.00609	0
change as reaction occurs (mol/L)	– x	– x	+ x
equilibrium conc. (mol/L)	0.0102 – x	0.00609 – x	x

(a) Calculate the concentrations of the gases at equilibrium using equilibrium $[COCl_2]$

$$[Cl_2] = 0.00609 - x = 0.00301 \text{ mol/L}$$

$$x = 0.00308 \text{ mol/L}$$

$$[CO] = 0.0102 - x = 0.0102 \text{ mol/L} - 0.00308 \text{ mol/L} = [CO] = 0.0071 \text{ mol/L}$$

$$[COCl_2] = x = 0.00308 \text{ mol/L}$$

(b) Then write the K_c expression and plug the concentrations into the expression to get the value of K_c.

$$K_c = \frac{[COCl_2]}{[CO][Cl_2]} = \frac{(0.00308)}{(0.0071)(0.00301)} = 1.4 \times 10^2 = 0.080$$

☑ *Reasonable Result Check:* The larger product concentrations suggest this reaction is product-favored, so it makes sense that the K_c is greater than 1.

40. *Result:* $\mathbf{K_c = 0.075}$

Analyze: Given an equation for a reaction, the initial moles of the gaseous reactant in a known volume, and the equilibrium concentration of the gaseous reactant, find the value of K_c.

Plan and Execute: Following the procedure given in Section 12.3, write the equation and construct an ICE reaction table. Put all known concentrations in the table. There is no $NO_2(g)$ present, initially. The equilibrium $[N_2O_4] = 0.00090$ M

$$(\text{conc. } N_2O_4) = 0.010 \text{ mol } N_2O_4/2.0 \text{ L} = 0.0050 \text{ M}$$

	$N_2O_4(g)$ ⇌	$2\ NO_2(g)$
initial conc. (M)	0.0050	0
change as reaction occurs (M)	______	______
equilibrium conc. (M)	0.00090	

Describe the stoichiometric changes in terms of one variable, x.

	$N_2O_4(g)$ ⇌	$2\ NO_2(g)$
initial conc. (M)	0.0050	0
change (M)	– x	+ 2x
equilibrium conc. (M)	0.00090 = 0.0050 – x	0 + 2x

Calculate the concentrations of the gases at equilibrium using what you know about the equilibrium concentration of the reactant.

Use equilibrium $[N_2O_4]$ to solve for x: 0.00090 M = 0.0050 M – x

$$x = 0.0050\text{ M} - 0.00090\text{ M} = 0.0041\text{ M}$$

Use the value of x to find the equilibrium $[NO_2]$: 2x = 2(0.0041 M) = 0.0082 M

Then write the K_c expression and plug the concentrations into the expression to get the value of K_c.

$$K_c = \frac{[NO_2]^2}{[N_2O_4]} = \frac{(2x)^2}{0.0050 - x} = \frac{(0.0082)^2}{0.00090} = 0.075$$

☑ *Reasonable Result Check:* The smaller product concentration suggests this reaction is reactant-favored, so it makes sense that the K_c is smaller than 1.

42. *Result:* **3.9×10^{-4}**

Analyze: Given an equation for a reaction, the initial moles of the gaseous reactant and the percent dissociation at equilibrium, find the value of K_c.

Plan and Execute: Following the procedure given in Section 12.3, write the equation and construct an ICE reaction table. No products are present, initially.

$$(\text{conc. } Br_2) = 0.086\text{ mol } Br_2/1.26\text{ L} = 0.068\text{ mol/L}$$

And describe the stoichiometric changes in terms of one variable, x.

	$Br_2(g)$ ⇌	$2\,Br(g)$
initial conc. (M)	0.068	0
change as reaction occurs (M)	– x	+ 2x
equilibrium conc. (M)	0.068 – x	2x

Calculate the concentrations of the Br_2 gas at equilibrium using what you know about the percent dissociation. Given that 3.7% Br_2 dissociates, x = 0.037 × (0.068 M) = 0.0025

Write the expression for K_c and plug in the concentrations from part (a)

$$K_c = \frac{[Br]^2}{[Br_2]} = \frac{(2x)^2}{(0.068 - x)} = \frac{(2 \times 0.0025)^2}{(0.068 - 0.0025)} = 3.9 \times 10^{-4}$$

☑ *Reasonable Result Check:* The small percent ionization causes smaller product concentrations. Both of these are consistent with a reactant-favored reaction. It also makes sense that the K_c is smaller than 1.

44. *Result:* **1.4×10^3**

Analyze: Given an equation for a reaction, the initial concentrations of the gaseous reactants and the concentration of one reactant at equilibrium, find the value of K_c.

Plan and Execute: Following the procedure given in Section 12.3, write the equation and construct an ICE reaction table. No products are present, initially.

Describe the stoichiometric changes in terms of one variable, x.

	$2\ SO_2(g)$ +	$O_2(g)$ $\rightleftharpoons$	$2\ SO_3(g)$
initial conc. (M)	0.0076	0.0036	0
change as reaction occurs (M)	– 2x	– x	+2x
equilibrium conc. (M)	0.0076 – 2x	0.0036 – x	2x

Use equilibrium $[SO_2]$, to find the value of x, then use x to determine the other equilibrium concentrations.

$$[SO_2] = 0.0032 = 0.0076 - 2x$$

$$2x = 0.0044$$

$$x = 0.0022\ M$$

$$[O_2] = 0.0036\ M - 0.002\ M = 0.0014\ M$$

$$[SO_3] = 0.0044\ M$$

Write the expression for K_c and plug in the concentrations from part (a)

$$K_c = \frac{[SO_3]^2}{[SO_2]^2[O_2]} = \frac{(0.0044)^2}{(0.0032)^2 \times (0.0014)} = 1.4 \times 10^3$$

☑ *Reasonable Result Check:* The reduction in the concentration of a reactant causes relatively large product concentrations. Both of these are consistent with a product-favored reaction. It also makes sense that the K_c is larger than 1.

The Meaning of the Equilibrium Constant (Section 12-4)

45. *Result:* **Reactions (b) and (c) are product-favored. Most reactant-favored (a), then (b), then (c).**

Analyze: Given Table 12.1 with values of K_c and K_p and a set of chemical equations, predict which reaction is product-favored, and order the members of a set or equations from most reactant-favored to most product-favored.

Plan: Look up the given chemical equation or a related chemical equation, determine the size of its K_c, using techniques described at the end of Section 12.2, as needed. When K_c is larger than 1 is product-favored. A smaller K_c has a more reactant-favored reaction. A larger K_c has a more product-favored reaction. Order the equations from smallest K_c to largest K_c.

Execute:

(a) $2\ NH_3(g) \rightleftharpoons N_2(g) + 3\ H_2(g)$

This reaction is the reverse of the third reaction in Table 12.1, so $K_{c,(a)} = 1/K_c = 1/(3.5 \times 10^8) = 2.9 \times 10^{-9}$.

K_c is smaller than 1, so the reaction is not product-favored.

(b) $NH_4^+(aq) + OH^-(aq) \rightleftharpoons NH_3(aq) + H_2O(\ell)$

This reaction is the reverse of the 12th reaction in Table 12.1, so $K_{c,(b)} = 1/K_c = 1/(1.8 \times 10^{-5}) = 5.6 \times 10^4$.

K_c is larger than 1, so the reaction is product-favored.

(c) $2\ NO(g) \rightleftharpoons N_2(g) + O_2(g)$ is the reverse of the fourth reaction in Table 12.1, so $K_{c,(c)} = 1/K_c = 1/(4.5 \times 10^{-31}) = 2.2 \times 10^{30}$. K_c is larger than 1, so the reaction is product-favored.

Therefore, the order is (a) 2.9×10^{-9}, then (b) 5.6×10^4, then (c) 2.2×10^{30}.

☑ *Reasonable Result Check:* Reversing a reaction takes the products and makes them the reactants and vice versa, so the reverse of a product-favored reaction will be a reactant-favored reaction and vice versa. The values of K_c make sense. Comparing the values of K to determine which is the most reactant-favored and product-favored among reaction with different stoichiometric relationships is not always legitimate. Here, while the denominators all have two concentration values, we find four concentration values in the numerator of (a), only two in (c), and only one in (b). These differences will affect the size of K_c dramatically. In general, one should compare the values of K_c only among reactions with the same stoichiometric relationships (i.e., with the same number of reactants and products.)

47. *Result:* **(a), (b), and (c)**

Analyze: Given chemical equations and their associated K value, determine which of them favors reactants.

A reaction favors reactants if the value of K is less than 1, because the concentrations of the reactant would be larger than those of the products, making the fraction [prod]/[react] < 1. The reactions in (a), (b) and (c) all have K < 1, so they favor reactants.

48. *Result:* **(a) See reactions and equations below (b) Ag_2SO_4; larger K_c (c) Ag_2S; smaller K_c**

Analyze: Given the equilibrium constants for the dissolving of two compounds, write the dissolving equations, the equilibrium constant expression, and explain which is more soluble.

Plan: Dissolving ionic compounds produces aqueous solutions of ions as described in Chapters 2 and 3. Water is not explicitly included in the dissolving equation, because its only function is as the solvent to stabilize the aqueous products via intermolecular forces. Write equilibrium expressions as was done in the solution to Question 14. Assess solubility by comparing the size of K_c as was done in the solution to Question 45.

Execute

(a) $Ag_2SO_4(s) \rightleftharpoons 2\ Ag^+(aq) + SO_4^{2-}(aq)$ $\qquad K_c = [Ag^+]^2[SO_4^{2-}] = 1.7 \times 10^{-5}$

$Ag_2S(s) \rightleftharpoons 2\ Ag^+(aq) + S^{2-}(aq)$ $\qquad K_c = [Ag^+]^2[S^{2-}] = 6 \times 10^{-30}$

(b) When the dissociation stoichiometry is comparable, as it is here, the compound that is more soluble has the **larger K_c**, and that is **$Ag_2SO_4(s)$**.

(c) When the dissociation stoichiometry is comparable, as it is here, the compound that is least soluble has the **smaller K_c**, and that is **$Ag_2S(s)$**.

Using Equilibrium Constants (Section 12-5)

50. *Result:* **(a) See table below (b)** $\frac{0.100 + x}{0.100 - x} = 2.5$**, x = 0.043 (c) [2-methylpropane] = 0.024 M, [butane] = 0.010 M**

Analyze: Given an equation for a reaction, the initial concentration or moles of the gaseous reactant in a known volume, and the value of the equilibrium constant, determine the change in the concentrations and the equilibrium concentrations of the reactants and products.

Follow the procedure given in Section 12.5. When necessary, find Q to determine direction of the reaction. Then write the equation and construct an ICE reaction table. Describe the stoichiometric changes in terms of one variable, x, the change in the concentration of butane. Write the K_c expression for the reaction in terms of x. Solve that equation to get x, and use it to calculate the concentrations of the gases at equilibrium.

$$\text{butane (g)} \rightleftharpoons \text{2-methylpropane (g)}$$

(a) (conc. butane) = (conc. 2-methylpropane) = 0.100 M

$$Q_c = \frac{\text{(conc. 2-methylpropane)}}{\text{(conc. butane)}} = \frac{0.100\ \text{M}}{0.100\ \text{M}} = 1.00 < 2.5$$

$Q_c < K_c$, so reaction goes from reactants toward products:

	butane (g) ⇌	2-methylpropane (g)
initial conc. (mol/L)	0.100	0.100
change as reaction occurs (mol/L)	– x	+ x
equilibrium conc. (mol/L)	0.100 – x	0.100 + x

(b) $$K_c = \frac{[\text{2-methylpropane}]}{[\text{butane}]} = 2.5$$

At equilibrium $$\frac{0.100 + x}{0.100 - x} = 2.5$$

$$0.100 + x = 2.5(0.100 - x) = 0.25 - 2.5x$$

$$x + 2.5x = 0.25 - 0.100$$

$$3.5x = 0.15$$

$$x = 0.043$$

(c) (conc. butane) = 0.017 mol butane/0.50 L = 0.034 M

	butane (g) ⇌	2-methylpropane (g)
initial conc. (M)	0.034	0
change as reaction occurs (M)	– x	+ x
equilibrium conc. (M)	0.034 – x	x

At equilibrium $$\frac{x}{0.034 - x} = 2.5$$

$$x = 2.5(0.034 - x) = 0.085 - 2.5x$$

$$x + 2.5x = 0.085$$

$$3.5x = 0.085$$

$$x = 0.024 \text{ M} = [\text{2-methylpropane}]$$

$$[\text{butane}] = 0.034 \text{ M} - 0.024 \text{ M} = 0.010 \text{ M}$$

☑ *Reasonable Result Check:* Substituting the equilibrium concentrations into the equilibrium expression will reproduce K_c:

Part (b), [butane] = 0.100 – 0.043 = 0.057 M, [2-methylpropane] = 0.100 + 0.043 = 0.143 M. $K_c = \frac{0.143}{0.057} = 2.5$. (c) $K_c = \frac{0.024}{0.010} = 2.4$. These are both right, within the uncertainty of the data.

52. *Result:* **3.39 g**

Analyze: Given an equation for a reaction, the initial mass of the gaseous reactant in a known volume, and the value of the equilibrium constant, determine the mass of the reactant present at equilibrium.

Plan: Convert grams to moles using the molar mass. Then follow the method described in the solution to Question 50 to get the equilibrium concentration of the reactant gas. Then use the volume and molar mass to determine the grams.

Execute: cyclohexane (g) $\rightleftharpoons$ methylcyclopentane (g)

$$(\text{conc. cyclohexane}) = \frac{3.79 \text{ g cyclohexane}}{2.80 \text{ L}} \times \frac{1 \text{ mol cyclohexane}}{84.15 \text{ g cyclohexane}} = 0.0161 \text{ M}$$

	cyclohexane (g) $\rightleftharpoons$	methylcyclopentane (g)
initial conc. (M)	0.0161	0
change as reaction occurs (M)	– x	+ x
equilibrium conc. (M)	0.0161 – x	x

At equilibrium $$K_c = \frac{[\text{methylcyclopentane}]}{[\text{cyclohexane}]} = 0.12$$

$$\frac{x}{0.0161 - x} = 0.12$$

$$x = 0.12(0.0161 - x) = 0.0019 - 0.12x$$

$$x + 0.12x = 0.0019$$

$$1.12x = 0.0019$$

$$x = 0.0017 \text{ M} = [\text{methylcyclopentane}]$$

$$[\text{cyclohexane}] = 0.0161 \text{ M} - 0.0017 \text{ M} = 0.0144 \text{ M}$$

$$2.80 \text{ L} \times \frac{0.0144 \text{ mol cyclohexane}}{1 \text{ L}} \times \frac{84.15 \text{ g cyclohexane}}{1 \text{ mol cyclohexane}} = 3.39 \text{ g}$$

☑ *Reasonable Result Check:* The quantity present at equilibrium is less than the initial quantity. The equilibrium concentrations can be used to recreate the value of the equilibrium constant: $K_c = \frac{0.0017}{0.0144} = 0.12$.

54. *Result:* **$[Br_2] = [F_2] = 0.047$ M, $[BrF] = 0.347$ M**

Analyze and Plan: Given a chemical equation, K_c, initial concentrations, and constant volume, calculate the equilibrium concentrations for all three substances. Use a method similar to that described in the solution to Question 51(b).

Execute:

	$Br_2(g)$	$+ F_2(g)$	$\rightleftharpoons$ 2 BrF(g)
initial conc. (mol/L)	0.220	0.220	0
change (mol/L)	– x	– x	+ 2x
equilibrium conc. (mol/L)	0.220 – x	0.220 –x	2x

At equilibrium $$K_c = \frac{[BrF]^2}{[Br_2][F_2]} = \frac{(2x)^2}{(0.220-x)^2} = 55.3$$

Take the square root of each side: $$\frac{(2x)}{(0.220-x)} = \sqrt{55.3} = 7.44$$

Solve for x: $$7.44(0.220 - x) = 2x$$

$$1.64 = 2x + 7.44x = (2 + 7.44)x = 9.44x$$

$$1.64 = 9.44x$$

$$x = 0.173 \text{ M}$$

$$0.220 - x = 0.220 - 0.173 = 0.047 \text{ M} = [Br_2] = [F_2]$$

$$[BrF] = 2 \times (0.173) = 0.347 \text{ M}$$

☑ *Reasonable Result Check:* The large concentration of product is consistent with a K greater than 1.

55. *Result:* **(a) 1.94 mol (b) 1.92 mol (c) 1.98 mol**

Analyze and Plan: Given a chemical equation, K_c, quantity (in moles), volume, and temperature, calculate the equilibrium concentrations for all three substances. Use a method similar to that described in the solution to Question 52.

Execute:

(a) $$(\text{conc. } I_2) = 1.00 \text{ mol}/10.00 \text{ L} = 0.100 \text{ M}$$

$$(\text{conc. } H_2) = 3.00 \text{ mol}/10.00 \text{ L} = 0.300 \text{ M}$$

	H_2 (g)	$+ I_2$ (g)	$\rightleftharpoons$ 2 HI (g)
initial conc. (M)	0.300	0.100	0
change (M)	– x	– x	+ 2x
equilibrium conc. (M)	0.300 – x	0.100 – x	2x

At equilibrium $$K_c = \frac{[HI]^2}{[H_2][I_2]} = \frac{(2x)^2}{(0.300 - x)(0.100 - x)} = 50.0$$

Set up to use the quadratic equation (see Appendix A, Section A.7, page A.13):

$$50.0\{(0.300)(0.100) - (0.300 + 0.100)x + x^2\} = 4x^2$$

$$1.50 - 20.0x + 50.0x^2 = 4x^2$$

$$1.50 - 20.0x + 46.0x^2 = 0$$

Plug into the quadratic equation:

$$x = \frac{-b \pm \sqrt{b^2 - 4ac}}{2a} = \frac{20.0 \pm \sqrt{(-20.0)^2 - 4(46.0)(1.50)}}{2(46.0)}$$

The two roots are:

$$x = \frac{20.0 - 11.1}{2(46.0)} = \frac{8.9}{92.0} = 0.097$$

$$x = \frac{20.0 + 11.1}{2(46.0)} = \frac{31.1}{92.0} = 0.338$$

Notice: The second of these two roots must be discarded because it is too large and produces a nonsensical result; the equilibrium concentration of I_2 (0.100 – x) must be a positive number, but using the second root produces a negative answer, 0.100 – 0.338 = –0.238.

$$x = 0.097\text{ M}$$

$$2x = 2(0.097\text{ M}) = 0.194\text{ M} = [HI]$$

$$10.00\text{ L} \times \frac{0.194\text{ mol}}{\text{L}} = 1.94\text{ mol}$$

$$[H_2] = 0.300 - x = 0.300\text{ M} - (0.097\text{ M}) = 0.203\text{ M}$$

$$[I_2] = 0.100 - x = 0.100\text{ M} - (0.097\text{ M}) = 0.003\text{ M}$$

☑ *Reasonable Result Check:* (a) The larger concentration of product is consistent with a K greater than 1. Substituting the equilibrium concentrations into the equilibrium expression will reproduce K_c within the

precision of the data: $K_c = \frac{(0.194)^2}{(0.203) \times (0.003)} = 6 \times 10^1 \cong 50.0$ with 1 sig fig or ±10.

(b) $(\text{conc. } I_2) = 1.00\text{ mol}/5.00\text{ L} = 0.200\text{ M}$

$(\text{conc. } H_2) = 3.00\text{ mol}/5.00\text{ L} = 0.600\text{ M}$

	H_2 (g)	+ I_2 (g)	⇌ 2 HI (g)
initial conc. (M)	0.600	0.200	0
change as reaction occurs (M)	– x	– x	+ 2x
equilibrium conc. (M)	0.600 – x	0.200 – x	2x

At equilibrium $K_c = \dfrac{[HI]^2}{[H_2][I_2]} = \dfrac{(2x)^2}{(0.600-x)(0.200-x)} = 50.0$

Set up to use the quadratic equation (Appendix A, Section A.7, page A.13):

$$50.0\{(0.600)(0.200) - (0.600 + 0.200)x + x^2\} = 4x^2$$

$$6.00 - 40.0x + 46.0x^2 = 0$$

Plug into the quadratic equation:

$$x = \frac{-b \pm \sqrt{b^2 - 4ac}}{2a} = \frac{40.0 \pm \sqrt{(-40.0)^2 - 4(46.0)(6.00)}}{2(46.0)} = \frac{17.7}{92.0} = 0.192$$

$$2x = 0.384 \text{ M} = [HI]$$

$$5.00 \text{ L} \times \frac{0.384 \text{ mol}}{\text{L}} = 1.92 \text{ mol}$$

☑ *Reasonable Result Check:* (b) The larger concentration of product is consistent with a K greater than 1. Substituting the equilibrium concentrations into the equilibrium expression will reproduce K_c within the precision of the data:

$$[H_2] = 0.600 - x = 0.600 \text{ M} - (0.192 \text{ M}) = 0.408 \text{ M}$$

$$[I_2] = 0.200 - x = 0.200 \text{ M} - (0.192 \text{ M}) = 0.008 \text{ M}$$

$K_c = \dfrac{(0.384)^2}{(0.408) \times (0.008)} = 5 \times 10^1 \cong 50.0$. The answers for (a) and (b) should be the same, because the volume dependence cancels due to the fact that the number of products and reactants are the same. Within the cited uncertainty, ±10, (a) and (b) give the same results.

(c)

$$(\text{conc. } I_2) = 1.00 \text{ mol}/10.00 \text{ L} = 0.100 \text{ M}$$

$$(\text{conc. } H_2) = (3.00 \text{ mol} + 3.00 \text{ mol})/10.00 \text{ L} = 0.600 \text{ M}$$

	H_2 (g)	+ I_2 (g)	⇌ 2 HI (g)
initial conc. (M)	0.600	0.100	0
change as reaction occurs (M)	– x	– x	+ 2x
equilibrium conc. (M)	0.600 – x	0.100 – x	2x

At equilibrium $K_c = \dfrac{[HI]^2}{[H_2][I_2]} = \dfrac{(2x)^2}{(0.600-x)(0.100-x)} = 50.0$

Set up to use the quadratic equation (Appendix A, Section A.7, page A.13):

$$50.0\{(0.600)(0.100) - (0.600 + 0.100)x + x^2\} = 4x^2$$

$$3.00 - 35.0x + 46.0x^2 = 0$$

Plug into the quadratic equation:

$$x = \frac{-b \pm \sqrt{b^2 - 4ac}}{2a} = \frac{35.0 \pm \sqrt{(-35.0)^2 - 4(46.0)(3.00)}}{2(46.0)} = \frac{9.1}{92.0} = 0.099$$

$$2x = 0.198 \text{ M} = [HI]$$

$$10.00 \text{ L} \times \frac{0.198 \text{ mol}}{\text{L}} = 1.98 \text{ mol}$$

☑ *Reasonable Result Check:* (c) The larger concentration of product is consistent with a K greater than 1. Substituting the equilibrium concentrations into the equilibrium expression will reproduce K_c within the precision of the data:

$$[H_2] = 0.300 - x = 0.600 \text{ M} - (0.099 \text{ M}) = 0.501 \text{ M}$$

$$[I_2] = 0.100 - x = 0.100 \text{ M} - (0.099 \text{ M}) = 0.001 \text{ M}$$

$K_c = \dfrac{(0.198)^2}{(0.501) \times (0.001)} = 7 \times 10^1 \cong 50.0$. With only one significant figure, round off errors have started to dramatically affect the smallest number, $[I_2]$. These results still look reasonable.

57. *Result:* **(a) [CO] = [H_2O] = 0.95 mol/L, [CO_2] = [H_2] = 0.0489 mol/L (b) [CO] = [H_2O] = 1.90 mol/L, [CO_2] = [H_2] = 0.0977 mol/L**

Analyze and Plan: Given the chemical equation, K_c, temperature, quantity (in moles) if reactants, volume, calculate the final concentrations of all species. After the addition of greater quantities, calculate the final concentrations of all species. Use a method similar to that described in the solution to Question 52.

Execute:

(a) (conc. CO) = 1.00 mol/1.00 L = 1.00 **mol/L**

(conc. H_2O) = 1.00 mol/1.00 L = 1.00 **mol/L**

	CO (g)	+	H_2O(g)	⇌ CO_2 (g)	+ H_2 (g)
initial conc. (**mol/L**)	1.00		1.00	0	0
change in conc. (**mol/L**)	– x		– x	+ x	+ x
equilibrium conc. (**mol/L**)	1.00 – x		1.00 – x	x	x

At equilibrium $K_c = \dfrac{[CO_2][H_2]}{[CO][H_2O]} = \dfrac{(x)(x)}{(1.00-x)(1.00-x)} = 2.64 \times 10^{-3}$

Take the square root of each side: $\dfrac{(x)}{(1.00-x)} = 0.0514$

$$x = 0.0514 - 0.0514x$$

$$x + 0.0514x = 0.0514$$

$$(1.0514)x = 0.0514$$

$$x = 0.0489 \ \textbf{mol/L} = [CO_2] = [H_2]$$

$$[CO] = [H_2O] = 1.00 \text{ M} - x = 1.00 \ \textbf{mol/L} - 0.0489 \ \textbf{mol/L} = 0.95 \ \textbf{mol/L}$$

☑ *Reasonable Result Check:* The smaller concentration of products is consistent with a K smaller than 1. Substituting the equilibrium concentrations into the equilibrium expression will reproduce K_c within the precision of the data: $K_c = \frac{(0.0489)\times(0.0489)}{(0.95)\times(0.95)} = 2.6\times10^{-3}$.

(b) The initial quantities can be added together, since the final equilibrium state will be the same as if they were added to the equilibrium concentrations from part (a).

$$(\text{conc. CO}) = (1.00\text{ mol} + 1.00\text{ mol})/1.00\text{ L} = 2.00\ \mathbf{mol/L}$$

$$(\text{conc. }H_2O) = (1.00\text{ mol} + 1.00\text{ mol})/1.00\text{ L} = 2.00\ \mathbf{mol/L}$$

	$CO(g)$ +	$H_2O(g)$ ⇌	$CO_2(g)$ +	$H_2(g)$
initial conc. (**mol/L**)	2.00	2.00	0	0
change in conc. (**mol/L**)	– x	– x	+ x	+ x
equilibrium conc. (**mol/L**)	2.00 – x	2.00 – x	x	x

At equilibrium $$K_c = \frac{[CO_2][H_2]}{[CO][H_2O]} = \frac{(x)(x)}{(2.00-x)(2.00-x)} = 2.64\times10^{-3}$$

Take the square root of each side: $$\frac{(x)}{(2.00-x)} = 0.0514$$

Solve for x $$x = 2.00(0.0514) - 0.0514x$$

$$x + 0.0514x = 0.103$$

$$(1.0514)x = 0.103$$

$$x = 0.0977\ \mathbf{mol/L} = [CO_2] = [H_2]$$

$$[CO] = [H_2O] = 2.00\ \mathbf{mol/L} - x = 2.00\ \mathbf{mol/L} - 0.0977\ \mathbf{mol/L} = 1.90\ \mathbf{mol/L}$$

☑ *Reasonable Result Check:* The smaller concentration of products is consistent with a K smaller than 1. Substituting the equilibrium concentrations into the equilibrium expression will reproduce K_c within the precision of the data: $K_c = \frac{(0.0977)\times(0.0977)}{(1.90)\times(1.90)} = 2.64\times10^{-3}$

59. *Result:* **(a) No (b) Proceeds toward products**

Analyze: Given an equation for a reaction, the initial moles of the gaseous reactants and products in a known volume, and the value of the equilibrium constant, determine if the system is at equilibrium and, if it is not, determine which direction it must proceed to reach equilibrium.

Plan: Follow the procedure given in Section 12-5. Calculate Q and compare it with K. If Q = K, then the system is at equilibrium. If Q < K, the reaction proceeds toward products to reach equilibrium; if Q > K, the reaction proceeds toward reactants to reach equilibrium.

Execute: $K_c = 3.58\times10^{-3}$

$$(\text{conc. }SO_3) = 0.15\text{ mol}/10.0\text{ L} = 0.015\text{ M}$$

$$(\text{conc. }SO_2) = 0.015\text{ mol}/10.0\text{ L} = 0.0015\text{ M}$$

$$(\text{conc. }O_2) = 0.0075\text{ mol}/10.0\text{ L} = 0.00075\text{ M}$$

$$Q_c = \frac{(\text{conc. }SO_2)^2(\text{conc. }O_2)}{(\text{conc. }SO_3)^2} = \frac{(0.0015)^2(0.00075)}{(0.015)^2} = 7.5\times10^{-6}$$

(a) $Q_c \neq K_c$, so the reaction is not at equilibrium.

(b) $Q_c < K_c$, so the reaction must proceeds toward products to reach equilibrium.

☑ *Reasonable Result Check:* Considering the powers of 10, the size of Q makes sense:

$$\frac{(10^{-3})^2(10^{-4})}{(10^{-2})^2} = \frac{(10^{-6})(10^{-4})}{(10^{-4})} = 10^{-6}$$

61. *Result:* **(a) No (b) proceed toward reactants (c) $[N_2] = [O_2] = 0.025$ M; [NO] = 0.0010 M**

Analyze: Given an equation for a reaction, the initial moles of the gaseous reactant in a known volume, and the value of the equilibrium constant, determine if the system is at equilibrium and, if it is not, determine which direction it must proceed to reach equilibrium, and calculate the equilibrium concentrations.

Start with the procedure given in Section 12-5. Calculate Q and compare it to K to see if the system is at equilibrium, and, as needed, which direction the reaction proceeds in order to reach equilibrium. If the system is not already at equilibrium, use a method similar to that described in the solution to Question 52 to find the equilibrium concentrations.

Execute: $K_c = 1.7 \times 10^{-3}$

(conc. NO) = 0.015 mol/10.0 L = 0.0015 M

(conc. N_2) = 0.25 mol/10.0 L = 0.025 M (conc. O_2) = 0.25 mol/10.0 L = 0.025 M

$$Q_c = \frac{(\text{conc. NO})^2}{(\text{conc. } N_2)(\text{conc. } O_2)} = \frac{(0.0015)^2}{(0.025)(0.025)} = 3.6 \times 10^{-3}$$

(a) $Q_c \neq K_c$, so the reaction is not at equilibrium.

(b) $Q_c > K_c$, so the reaction must proceeds toward reactants to reach equilibrium.

(c)

	$N_2(g)$	+ $O_2(g)$	$\rightleftharpoons$ 2 NO(g)
initial conc. (M)	0.025	0.025	0.0015
change as reaction occurs (M)	+ x	+ x	– 2x
equilibrium conc. (M)	0.025 + x	0.025 + x	0.0015 – 2x

At equilibrium $$K_c = \frac{[NO]^2}{[N_2][O_2]} = \frac{(0.0015-2x)^2}{(0.025+x)(0.025+x)} = 1.7 \times 10^{-3}$$

Take the square root of each side: $$\frac{(0.0015-2x)}{(0.025+x)} = \sqrt{1.7 \times 10^{-3}} = 0.041$$

Solve for x: $$0.0015 - 2x = 0.041(0.025 + x)$$

$$0.0015 - 2x = 0.001025 + 0.041x$$

$$0.0015 - 0.001025 = (2 + 0.041)x$$

$$x = 2.3 \times 10^{-4} \text{ M}$$

$$[N_2] = [O_2] = 0.025 - x = 0.025 - 2.3 \times 10^{-4} \text{ M} = 0.025 \text{ M}$$

$$[NO] = 0.0015 - 2x = 0.0015 \text{ M} - 2 \times (2.3 \times 10^{-4} \text{ M}) = 0.0010 \text{ M}$$

☑ *Reasonable Result Check:* The smaller concentrations of products are consistent with a Q > K and a K less than 1.

Shifting a Chemical Equilibrium: Le Chatelier's Principle (Section 12-6)

64. *Result:* **(a) Left (b) Left (c) Left (d) Right**

Analyze: Given Kp, a chemical equation, and several changes, some at constant volume, predict the effect of each on the position of equilibrium.

Plan: As described in Section 12-6: Changing concentrations or pressures of reactants and products or changing the available energy in the system will all take the system out of equilibrium. The response to that change will be a shift away from the increase. Because this reaction has a positive $\Delta_r H$, heat energy is absorbed as the reaction goes from reactants to products.

Execute:

(a) Adding more Br_2 increases the concentration of a product, so the equilibrium responds by shifting to the **left** to reduce the product concentration.

(b) Removing NOBr decreases the concentration of a reactant, so the equilibrium responds by shifting to the **left** to increase the reactant concentration.

(c) Decreasing the temperature takes energy away from the system, so the equilibrium responds by shifting to the **left** to release some energy.

(d) Increasing the volume of the container, lowers the concentrations of every gas-phase reactant. To raise the concentrations, the reaction shifts to the **right** because combining two reactant molecules makes three product molecules.

66. *Result:* **(a) reverse reaction (b) forward reaction (c) forward reaction**

Analyze: Given a chemical equation for a reaction at equilibrium and some changes made, explain whether the forward or reverse reactions rate is faster immediately after the change.

Plan: Changing concentrations or pressures of reactants and products will take the system out of equilibrium. The response to that change will be a shift away from the increase, because the reaction rate will be temporarily faster for the reaction with the newly increased concentrations.

Execute:

(a) Adding more of the product F_2 causes the **reverse reaction** rate to be faster than the forward reaction immediately after the change.

(b) Decreasing the amount of the product ClF_3 causes the reverse reaction rate to be slower than the forward reaction, so the **forward reaction** rate will be faster immediately after the change.

(c) Doubling the total volume of the system lowers all the concentrations by half. The rate of the forward reaction is first order, rate = $k[ClF_5]$. The rate of the reverse reaction is second order, rate = $k[ClF_3][F_2]$. Therefore the reverse reaction will be slowed more than the forward reaction when the concentrations drop and the **forward reaction** rate will be faster immediately after the change.

68. *Result:* **See chart below**

Analyze: Given a thermochemical expression for a reaction at equilibrium and some changes made, explain how the changes affect the indicated quantities.

Plan: As described in Section 12-6: Changing concentrations or pressures of reactants and products or changing the available energy in the system will take the system out of equilibrium. The response to that change will be a shift away from the increase. Changing the amounts of solids or liquids will not affect the equilibrium position. Only temperature can affect the value of the equilibrium constant. Because this reaction is exothermic, energy is released.

Execute:

$$H_2(g) + Br_2(g) \rightleftharpoons 2\,HBr(g) \qquad \Delta H = -103.7 \text{ kJ}$$

Change	$[Br_2]$	[HBr]	K_c	K_p
Some H_2 *(a reactant)* is added to the container	decrease	increase	no change	no change
The temperature of the gases in the container is increased. *(increase in energy, energy released in the reaction)*	increase	decrease	decrease	decrease
The pressure of HBr (*a product*) is increased.	increase	increase (since extra is added)	no change	no change

70. *Result:* **(a) no change (b) left (c) left**

Analyze and Plan: Given a chemical equation for a reaction at equilibrium, Δ_rH, K_c, and some changes made at constant volume, explain whether the forward or reverse reactions rate is faster immediately after the change.

Analyze: As described in Section 12-6: Changing concentrations or pressures of reactants and products or changing the available energy in the system will take the system out of equilibrium. The response to that change will be to shift away from an increase (or shift towards a decrease). Changing the amounts of solids or liquids will not affect the equilibrium position. Because the reaction is endothermic, energy is absorbed.

Execute:

$$2\,NOBr(g) \rightleftharpoons 2\,NO(g) + Br_2(\ell) \qquad \Delta_rH = 16.1 \text{ kJ}$$

(a) The equilibrium is not affected by the addition of a pure liquid, so adding more liquid Br_2 will cause **no change.**

(b) Removing reactant, NOBr, shifts the reaction toward the reactants (**left**) to increase the [NOBr].

(c) Decreasing the temperature removes energy. The reaction shifts toward the reactants (**left**), to evolve more energy.

72. *Result:* **(a) (i) (b) (ii) (c) (i) (d) (iii) (e) (iii)**

Analyze: Given a balanced chemical equation and the fact that heat causes decomposition to occur and some changes, predict the Use the plan from the solution to Question 68.

(a) Adding solid $BaCO_3$ causes no shift, so the answer is **(i)**.

(b) Adding the product, CO_2, causes a shift to the left, so the answer is **(ii)**.

(c) Adding solid BaO causes no shift, so the answer is **(i)**.

(d) Raising the temperature increases the heat energy, which leads to decomposition, so the answer is **(iii)**.

(e) Increasing the volume decreases the pressure, which leads to a lower concentration of the gaseous product CO_2, causing a shift to the right, so the answer is **(iii)**.

73. *Result:* **(a) Decrease (b) See sketch below**

Explanation: Given a chemical equation determine how one products's concentration changes when adding another product ion. Make a graph like Figure 12.6 to illustrate what happens to the concentrations after addition.

(a) Use the methods described in Section 12-6.

$$PbCl_2(s) \rightleftharpoons Pb^{2+}(aq) + 2\,Cl^-(aq)$$

Adding the soluble ionic compound, NaCl, produces a solution with more of the product, Cl^-. The reaction shifts toward the reactants (left) and **decreases** the $[Pb^{2+}]$.

(b) Each time NaCl is added, the chloride concentration rises, then the reaction shifts toward reactants producing more solid and dropping the concentrations of both reactants until a new equilibrium condition is achieved.

The following qualitative sketch (similar to that of Figure 12.6) illustrates these changes.

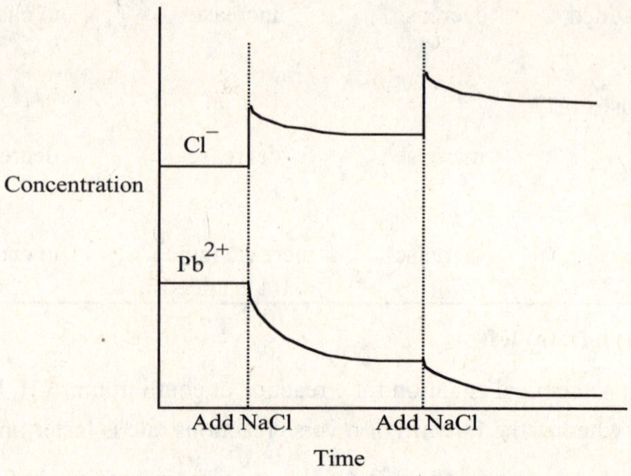

75. *Result:* **(a) (i) increase (ii) increase (iii) no shift (iv) increase (v) increase (b) No change increases K; Change (v) decreases K.**

Analyze and Plan: Given a thermochemical expression and some changes, determine how the amount if a reactant will be affected by those changes. Use the plan from the solution to Question 68.

Execute:

(a) (i) Removing reactant O_2 causes a shift to the **left** to replace some of the missing O_2. The shift toward reactants increases the amount of NH_3.

(ii) Adding product N_2 causes a shift to the **left** to remove some of the new O_2. The shift toward reactants increases the amount of NH_3.

(iii) Adding liquid water causes **no shift** in the position of equilibrium, so there is no change in the amount of NH_3.

(iv) Expanding the volume of the container decreases the concentration of all the gases. The reactants have seven moles of gas and the products have two moles of gas, so the decrease in concentration will be more prominent on the reactant side, causing a shift to the **left** to increase the concentration. The shift toward reactants increases the amount of NH_3.

(v) Increasing the temperature on an exothermic reaction drives the reactions to the **left**. The shift toward reactants increases the amount of NH_3.

(b) Only changes in the temperature affect the value of K. The change described in **change (5)** affects the value of K. None of the changes increase the value of K. Section 4-6 explains, for an exothermic reaction, an increase in temperature always means a decrease in K_c. The reaction will become less product-favored at higher temperatures in Change (5) of part (a).

76. *Result:* **(a) Left (b) $[PCl_5] = 0.0198$ M, $[PCl_3] = 0.0231$M, $[Cl_2] = 0.0403$ M**

Analyze: Given an equation for a reaction, the equilibrium mass of the gaseous reactants and products in a known volume and a mass of product subsequently added, determine how the addition of the product will affect the equilibrium and the equilibrium concentrations when equilibrium is re-established.

Plan and Execute:

(a) Adding a product will shift the equilibrium to the **left**.

(b) First, calculate the equilibrium concentrations from the masses and the volume.

$$\frac{3.120 \text{ g } PCl_5}{1.00 \text{ L}} \times \frac{1 \text{ mol } PCl_5}{208.2388 \text{ g } PCl_5} = 0.0150 \text{ M } PCl_5$$

$$\frac{3.845 \text{ g } PCl_3}{1.00 \text{ L}} \times \frac{1 \text{ mol } PCl_3}{137.3328 \text{ g } PCl_3} = 0.0280 \text{ M } PCl_3$$

$$\frac{1.787 \text{ g } Cl_2}{1.00 \text{ L}} \times \frac{1 \text{ mol } Cl_2}{70.906 \text{ g } Cl_2} = 0.0252 \text{ M } Cl_2$$

Adapt the plan from the solution to Question 26 and Question 35 to calculate the equilibrium constant.

$$K_c = \frac{[PCl_3][Cl_2]}{[PCl_5]} = \frac{(0.0280 \text{ M}) \times (0.0252 \text{ M})}{(0.0150 \text{ M})} = 0.0471$$

Adapt the plan from the solution to Question 52 to find the equilibrium concentrations.

Total mass of Cl_2 = 1.418 g + 1.787 g = 3.205 g Cl_2

$$(\text{conc. } Cl_2) = \frac{3.205 \text{ g } Cl_2}{1.00 \text{ L}} \times \frac{1 \text{ mol } Cl_2}{70.906 \text{ g } Cl_2} = 0.0452 \text{ M } Cl_2$$

	$PCl_5(g)$ ⇌	$PCl_3(g)$	+ $Cl_2(g)$
initial conc. (M)	0.0150	0.0280	0.0452
change as reaction occurs (M)	+ x	− x	− x
equilibrium conc. (M)	0.0150 + x	0.0280 − x	0.0452 − x

At equilibrium $$K_c = \frac{[PCl_3][Cl_2]}{[PCl_5]} = \frac{(0.0280 - x)(0.0452 - x)}{(0.0150 + x)} = 0.0471$$

Set up to use the quadratic equation (Appendix A, Section A.7, page A.13):

$$(0.0280 - x)(0.0452 - x) = 0.0471(0.0150 + x)$$

$$(0.0280)(0.0452) - (0.0280 + 0.0452)x + x^2 = (0.0471)(0.0150) + 0.0471x$$

$$0.00127 - 0.07320x + x^2 = 0.000706 + 0.0471x$$

$$x^2 - 0.1203x - 0.00056 = 0$$

Plug into the quadratic equation:

$$x = \frac{-b \pm \sqrt{b^2 - 4ac}}{2a} = \frac{-(-0.1203) \pm \sqrt{(-0.1203)^2 - 4(1)(-0.00056)}}{2(1)} = \frac{0.1203 - 0.1106}{2} = 0.00485$$

$$[PCl_5] = 0.0150 \text{ M} + x = 0.0150 \text{ M} + 0.00485 \text{ M} = 0.0198 \text{ M}$$

$$[PCl_3] = 0.00280 \text{ M} - x = 0.0280 \text{ M} - 0.00485 \text{ M} = 0.0231\text{M}$$

$$[Cl_2] = 0.0452 \text{ M} - x = 0.0452 \text{ M} - 0.00485 \text{ M} = 0.0403 \text{ M}$$

☑ *Reasonable Result Check:* Substituting the equilibrium concentrations into the equilibrium expression will reproduce K_c within the uncertainty of the significant figures:

$$K_c = = \frac{(0.0231)(0.0403)}{(0.0198)} = 0.0470$$

Equilibrium at the Nanoscale (Section 12-7)

78. *Result:* **(a) energy effect (b) entropy effect (c) neither**

Analyze: Given three thermochemical expressions, indicate whether the entropy effect, the energy effect, both, or neither favors the reaction.

Plan: The entropy effect is related to randomness and probability. The more random a system is, the more probable it is. In many of the chemical reactions we study, we can get a qualitative idea of the entropy change by looking at the physical phases of the reactants and products. In general, gases have more entropy than liquids, which have more entropy than solids. Aqueous substances have more entropy than solids, but less than gases.

The energy effect is related to the relative stability of the reactants and products. If the products are more stable than the reactants (exothermic, ΔH = negative), then the products are favored energetically. If the reactants are more stable than the products (endothermic, ΔH = positive), then the reactants are favored energetically.

Execute:

(a) 4 mol gas $\rightleftharpoons$ 2 mol gas Entropy effect favors reactants, not products.

Reaction is exothermic, so energy effect favors products.

Reaction (a) (in the forward direction) is favored only by the energy effect.

(b) 1 mol gas $\rightleftharpoons$ 2 mol gas Entropy effect favors products.

Reaction is endothermic, so energy effect favors reactants, not products.

Reaction (b) (in the forward direction) is favored only by the entropy effect.

(c) 4 mol gas $\rightleftharpoons$ 2 mol gas Entropy effect favors reactants, not products.

Reaction is endothermic, so energy effect favors reactants, not products.

Reaction (c) (in the forward direction) is not favored by either effect.

80. *Result:* **(a) Insufficient information is available (b) Greater than 1; product-favored; (c) Less than 1; reactant-favored.**

Analyze: Given three thermochemical expressions, predict the size of the equilibrium constant.

Plan: We will use methods described in the solution to Question 78 and in Section 12-7. If the reaction is favored by both effects, it is product-favored and K will be greater than 1. If the reaction is favored by neither effect, it is reactant-favored and K will be less than 1. If the reaction is only favored by one of the two effects, we have insufficient information available.

(a) 3 mol gas $\rightleftharpoons$ 2 mol gas Entropy effect favors reactants, not products.

Reaction is exothermic, so energy effect favors products.

We have insufficient information to judge the size of K for reaction (a) or which side of the reaction is favored at equilibrium.

(b) 2 mol gas $\rightleftharpoons$ 3 mol gas Entropy effect favors products.

Reaction is exothermic, so energy effect favors products.

Reaction (b) (in the forward direction) is favored by both effects, so it is product-favored and its K will be greater than 1.

(c) 4 mol gas $\rightleftharpoons$ 2 mol gas Entropy effect favors reactants, not products.

Reaction is endothermic, so energy effect favors reactants, not products.

Reaction (c) (in the forward direction) is not favored by either effect, so it is reactant-favored and K will be less than 1.

Controlling Chemical Reactions: The Haber-Bosch Process (Section 12-8)

82. *Result/Explanation:* A reaction will only go significantly towards products if it is product-favored. As is described in Section 12-7 and 12-8, the change in entropy and the change in enthalpy both affect the whether a reaction is product-favored. An increase in entropy favors products and having products lower in energy favors products. In some instances, these two quantities have the opposite signs, in which cases the reaction is either always product-favored or never product-favored:

Sign of entropy change	Sign of enthalpy change	Product favored?
+	–	Yes
–	+	No

In some instances, these two quantities have the same signs, in which cases the reaction is product-favored only at some temperatures. If the reaction is favored only by an entropy effect, then it needs high temperatures to assist the endothermic process. If the reaction is favored only by an energy effect, then it needs low temperatures to keep the randomness at a minimum:

Sign of entropy change	Sign of enthalpy change	Product favored?
+	+	Only at high temperatures
–	–	Only at low temperatures

84. *Result:* **(a) Step 1: –296.830 kJ/mol, Step 2: –197.78 kJ/mol, Step 3: –132.44 kJ/mol (b) all three steps are exothermic (c) None have entropy increase, Steps 2 and 3 have entropy decrease, the entropy in Step 1 stays about the same. (d) All three steps have products favored more at low temperatures.**

Analyze: Given a series of three chemical equations, calculate $\Delta_r H°$ for each; identify which are exothermic and which are endothermic; which has an increasing entropy, which have decreasing entropy which stay the same; and which has products favored at low temperature.

Plan and Execute:

(a) Follow the method described in Chapter 4. Look up the $\Delta_f H°$ values in Appendix J.

Step 1: $\Delta_r H° = (1 \text{ mol}) \times \Delta_f H°\{SO_2(g)\} - (1 \text{ mol}) \times \Delta_f H°\{S(s)\} - (1 \text{ mol}) \times \Delta_f H°\{O_2(g)\}$

$\Delta_r H° = (1 \text{ mol}) \times (-296.830 \text{ kJ/mol}) - (1 \text{ mol}) \times (0 \text{ kJ/mol}) - (1 \text{ mol}) \times (0 \text{ kJ/mol}) = -296.830 \text{ kJ/mol}$

Step 2: $\Delta_r H° = (2 \text{ mol}) \times \Delta_f H°\{SO_3(g)\} - (2 \text{ mol}) \times \Delta_f H°\{SO_2(g)\} - (1 \text{ mol}) \times \Delta_f H°\{O_2(g)\}$

$\Delta_r H° = (2 \text{ mol}) \times (-395.72 \text{ kJ/mol}) - (2 \text{ mol}) \times (-296.830 \text{ kJ/mol}) - (1 \text{ mol}) \times (0 \text{ kJ/mol})$

$= -197.78 \text{ kJ/mol}$

Step 3: $\Delta_r H° = (1 \text{ mol}) \times \Delta_f H°\{H_2SO_4(\ell)\} - (1 \text{ mol}) \times \Delta_f H°\{SO_3(g)\} - (1 \text{ mol}) \times \Delta_f H°\{H_2O(\ell)\}$

$\Delta_r H° = (1 \text{ mol}) \times (-813.989 \text{ kJ/mol}) - (1 \text{ mol}) \times (-395.72 \text{ kJ/mol})$

$- (1 \text{ mol}) \times (-285.830 \text{ kJ/mol}) = -132.44 \text{ kJ/mol}$

(b) All three reactions are exothermic ($\Delta_r H$ = negative)

(c) Follow the method described in the solution to Question 78.

Step 1:	1 mol gas $\rightleftharpoons$ 1 mol gas	No large entropy change.
Step 2:	3 mol gas $\rightleftharpoons$ 2 mol gas	Entropy decreases.
Step 3:	1 mol gas $\rightleftharpoons$ 0 mol gas	Entropy decreases.

(d) Heat is produced in all three steps. A low temperature in each step would serve to reduce the energy and drive the reaction toward products. All three steps would favor products more at lower temperatures.

General Questions

86. *Result:* **First Reaction:** $K_c = [H^+][OH^-]$

Second Reaction: $K_c = \dfrac{[CH_3COO^-][H^+]}{[CH_3COOH]}$

Third Reaction: $K_c = \dfrac{[NH_3]^2}{[N_2][H_2]^3}$

Fourth Reaction: $K_c = \dfrac{[CO_2][H_2]}{[CO][H_2O]}$

For answers to (a), (b) and (c) for each reaction, see below.

Analyze: Given several chemical equations, their K_c, the initial concentrations of the reactants and products, write the equilibrium constant expression, set up an ICE reaction table with –x on the first reactant, write the equilibrium constant expression in terms of x, identify if any of them yields a quadratic equation, and identify how you would go about solving the ones that are not quadratic.

Plan: For each reaction, follow the directions given in the question: Assume that all gases and solutes have initial concentrations of 1.0 mol/L. Then let the *first* reagent in each reaction changes its concentration by –x. *(Notice: This is not always the best approach, but those are the instructions for this question. Additional details are given afterward.*

First reaction:	$H_2O\ (\ell)$	$\rightleftharpoons$	$H^+\ (aq)$	+	$OH^-\ (aq)$
initial conc. (mol/L)	liquid		1.0		1.0
change as reaction occurs (mol/L)	– x		+ x		+ x
equilibrium conc. (mol/L)	liquid		1.0 + x		1.0 + x

(a) $K_c = [H^+][OH^-] = (1.0 + x)(1.0 + x) = 1.0 \times 10^{-14}$

(b) This equation is a quadratic equation.

(c) Not applicable.

Second reaction:	$CH_3COOH\ (aq)$	$\rightleftharpoons$	$CH_3COO^-\ (aq)$	+	$H^+\ (aq)$
initial conc. (mol/L)	1.0		1.0		1.0
change as reaction occurs (mol/L)	– x		+ x		+ x
equilibrium conc. (mol/L)	1.0 – x		1.0 + x		1.0 + x

(a) $K_c = \dfrac{[CH_3COO^-][H^+]}{[CH_3COOH]} = \dfrac{(1.0+x)(1.0+x)}{(1.0-x)} = 1.8\times10^{-5}$

(b) This equation is a quadratic equation.

(c) Not applicable.

Third reaction:	$N_2(g)$	+	$3\ H_2(g)$	$\rightleftharpoons$	$2\ NH_3\ (g)$
initial conc. (mol/L)	1.0		1.0		1.0
change as reaction occurs (mol/L)	– x		– 3x		+2x
equilibrium conc. (mol/L)	1.0 – x		1.0 – 3x		1.0 + 2x

(a) $K_c = \dfrac{[NH_3]^2}{[N_2][H_2]^3} = \dfrac{(1.0+2x)^2}{(1.0-x)(1.0-3x)^3} = 3.5\times10^8$

(b) This equation is not a quadratic equation.

(c) This K is so large that it might be worth starting with a limiting reactant problem first and running the reaction as far to the right as possible using stoichiometry, before defining an appropriate change variable:

	$N_2(g)$	$+ \; 3\,H_2(g)$	$\rightleftharpoons 2\,NH_3(g)$
initial conc. (mol/L)	1.0	1.0	1.0
change as forward reaction occurs (mol/L)	– 0.33	– 1.0	+ 0.67
final conc. (mol/L)	0.7	0.0	1.7
change as reverse reaction occurs (mol/L)	+ x	+ 3x	– 2x
equilibrium conc. (mol/L)	0.7 + x	+ 3x	1.7 – 2x

Since this x is assumed to be very small compared to 0.7 or 1.7, we can ignore subtraction of x from 0.7 and addition of 2x from 1.7:

$$K_c = \frac{[NH_3]^2}{[N_2][H_2]^3} = \frac{(1.7-2x)^2}{(0.7+x)(3x)^3} \cong \frac{(1.7)^2}{(0.7)(3x)^3} = 4 \times 10^8 \text{, then solve for x.}$$

$x = 0.0008$ mol/L<<1.7 or 0.7

Fourth reaction:	$CO(g)$	$+ \; H_2O(g)$	$\rightleftharpoons CO_2(g)$	$+ \; H_2(g)$
initial conc. (M)	1.0	1.0	1.0	1.0
change as reaction occurs (M)	– x	– x	+ x	+ x
equilibrium conc. (M)	1.0 – x	1.0 – x	1.0 + x	1.0 + x

(a) $$K_c = \frac{[CO_2][H_2]}{[CO][H_2O]} = \frac{(1.0+x)(1.0+x)}{(1.00-x)(1.00-x)} = 4.00$$

(b) This equation is a quadratic equation.

(c) Not applicable.

IMPORTANT NOTICE FOR THIS QUESTION: If we were actually doing this question to get the answers, it would be a very good idea to calculate Q to determine the direction of reaction before deciding which side of the reaction experiences a decrease in concentration (– x) and which side experiences an increase in concentration (+ x). The side where concentration decreases should get the – x, so that we can be certain that the value of x is a positive quantity. Doing that significantly simplifies the mathematics, especially when dealing with quadratic equations and other higher order polynomials. For the reactions in this question, here are the Q calculations.

First reaction: $Q = (1.0M)(1.0M) = 1.0 > K_c = 1.0 \times 10^{-14}$

Reaction goes toward reactants, not products. The x defined earlier for the first reaction will be negative!

Second reaction: $Q = \frac{(1.0)(1.0)}{(1.0)} = 1.0 > K_c = 1.8 \times 10^{-5}$

Reaction goes toward reactants, not products. The x defined earlier for the second reaction will be negative!

Third reaction: $Q = \frac{(1.0)^2}{(1.0)(1.0)^3} = 1.0 < K_c = 3.5 \times 10^8$

Reaction goes toward products. The x defined above will be positive.

Fourth reaction: $Q = \frac{(1.0)(1.0)}{(1.0)(1.0)} = 1.0 < K_c = 4.00$

Reaction goes toward products. The x defined above will be positive.

Notice: The fourth reaction has a large K value and is product-favored, so it might be worth starting with a limiting reactant problem first and running the reaction as far forward as possible using stoichiometry then running the reverse reaction (a reactant-favored reaction) to establish the equilibrium, so that x will be small.

88. *Result:* **$K_p = 0.108$**

Analyze: Given an equation for a reaction, the initial mass of solid reactant, the partial pressure of both products, and the percentage increase of a product at equilibrium, find the value of K_P.

Plan and Execute: Construct an ICE reaction table. Put all known pressures in the table:

	$NH_4HS(s)$ $\rightleftharpoons$	$NH_3(g)$	+ $H_2S(g)$
initial pressure (atm)	solid	0.692	0.0532
change as reaction occurs (atm)	____	____	____
equilibrium pressure (atm)	____	____	____

Describe the stoichiometric changes in terms of one variable, x.

	$NH_4HS(s)$ $\rightleftharpoons$	$NH_3(g)$	+ $H_2S(g)$
initial conc. (mol/L)	solid	0.692	0.0532
change as reaction occurs (mol/L)	– x	+x	+x
equilibrium conc. (mol/L)	solid	0.692 + x	0.0532 + x

Use equilibrium change in pressure to solve for x. The partial pressure of ammonia increases by 12.4%.

$$\frac{12.4 \text{ atm}}{100 \text{ atm}} \times 0.692 \text{ atm} = 0.0858 \text{ atm} = x$$

Use the value of x to find the equilibrium concentrations of the other substances:

$$P_{NH_3} = 0.692 + x = 0.692 + 0.0858 = 0.778 \text{ atm}$$

$$P_{H_2S} = 0.0532 + x = 0.0532 + 0.0858 = 0.1390 \text{ atm}$$

Then write the K_P expression and plug the concentrations into the expression to get the value of K_P.

$$K_P = P_{NH_3}P_{H_2S} = (0.778)(0.1390) = 0.108$$

☑ *Reasonable Result Check:* The small pressures suggest this reaction is reactant-favored, so it makes sense that the K_P is smaller than 1.

90. *Result:* **(a)**

Species	**Br_2**	**Cl_2**	**F_2**	**H_2**	**N_2**	**O_2**
[E] (mol/L)	**0.28**	**0.057**	**1.44**	**1.76×10^{-5}**	**4×10^{-14}**	**4.0×10^{-6}**

(b) F_2; see explanation below.

Analyze and Plan: Given a general equation for the decomposition of diatomic molecules and Kc for each decomposition reactions, calculate the equilibrium concentration of the atomic form for each and predict which has the lowest bond energy, and compare with thermochemical calculations and Lewis structures. Use a method similar to that described in the solution to Question 52.

Execute:

(a) (conc. E_2) = 1.00 mol E_2/1.0 L = 1.0 M

	E_2 (g)	$\rightleftharpoons$ 2 E (g)
initial conc. (M)	1.0	0
change as reaction occurs (M)	– x	+ 2x
equilibrium conc. (M)	1.0 – x	2x

At equilibrium $K_c = \dfrac{[E]^2}{[E_2]} = \dfrac{(2x)^2}{1.0-x}$

Set up to use the quadratic equation (Appendix A, Section A.7, page A.13):

$$(2x)^2 = K_c(1.0 - x) = K_c - K_c x$$

$$4x^2 + K_c x - K_c = 0$$

Plug into the quadratic equation:

$$x = \frac{-b \pm \sqrt{b^2 - 4ac}}{2a} = \frac{-K_c \pm \sqrt{K_c^2 - 4(K_c)(-K_c)}}{2(4)}$$

$$[E] = 2x$$

Here is an example of the calculation for the first species: E = Br, $K_c = 8.9 \times 10^{-2}$.

$$x = \frac{-(8.9\times10^{-2}) - \sqrt{(8.9\times10^{-2})^2 - 4(8.9\times10^{-2})(-8.9\times10^{-2})}}{2(4)} = \frac{1.1}{8} = 0.14$$

$$[Br] = 2\times(0.14\ M) = 0.28\ M$$

Similar calculations give these results:

E species	K_c	x (M)	[E] (M)
Br	8.9×10^{-2}	0.14	0.28
Cl	3.4×10^{-3}	0.029	0.057
F	7.4	0.72	1.44
H	3.1×10^{-10}	8.8×10^{-6}	1.76×10^{-5}
N	1×10^{-27}	2×10^{-14}	4×10^{-14}
O	1.6×10^{-11}	2.0×10^{-6}	4.0×10^{-6}

☑ *Reasonable Result Check:* [E] is small when K is small. [E] is large when K is large. The equilibrium concentrations combine to reproduce $K_c = \dfrac{(2x)^2}{1.0-x}$.

(b) At this temperature, the lowest bond enthalpy is predicted from the reaction that gives the most products, so F_2 is predicted to have the lowest bond dissociation enthalpy.

Comparing these results to the bond enthalpy values in Table 6.2, the product production decreases as the bond enthalpy increases: 158 kJ F_2, 193 kJ Br_2, 242 kJ Cl_2, 436 kJ H_2, 498 kJ O_2, 946 kJ N_2.

Lewis structures of F_2, Br_2, Cl_2, H_2 have a single bonds and these compounds have more product atoms than do the compounds with multiple bonds: O_2 with a double bond and N_2 with a triple bond.

94. *Result:* **(a) 0.927 atm (b) 0.0420 moles**

Analyze and Plan: Given K_p and the equation for the decomposition of a solid into two gases, the mass of the reactant, and the volume of the container, determine the total pressure of gas in the flask at equilibrium and how much of the solid decomposed. Use a method similar to that described in the solution to Question 52. Use Dalton's law and the ideal gas law from Chapter 8 for total pressure and quantity calculations.

Execute:

Use the molar mass of NH_4I to determine how many moles of the solid are present initially:

$$15.0 \text{ g } NH_4I \times \frac{1 \text{ mol } NH_4I}{145.9507 \text{ g } NH_4I} = 0.1028 \text{ mol } NH_4I$$

Remember, the quantity of solids does not affect the equilibrium state.

	$NH_4I(s)$ ⇌	NH_3 (g)	+ HI (g)
initial press. (atm)	solid	0	0
change as reaction occurs (atm)	- x	+ x	+ x
equilibrium press. (atm)	solid - x	x	x

At equilibrium $$K_p = P_{NH_3}P_{HI} = x^2 = 0.215$$

$$x = 0.464$$

$$P_{NH_3} = P_{HI} = x = 0.46\underline{3}7 \text{ atm}$$

(a) Use Dalton's law (Section 8-6): $P_{tot} = P_{NH_3} + P_{HI} = 0.46\underline{3}7 \text{ atm} + 0.46\underline{3}7 \text{ atm} = 0.927 \text{ atm}$

(b) First, calculate the moles of $NH_3(g)$ at equilibrium, using the ideal gas law, PV = nRT (Section 8-4):

$$P_{NH_3}V = n_{NH_3}RT$$

$$n_{NH_3} = \frac{P_{NH_3}V}{RT} = \frac{(0.4637 \text{ atm})(5.00 \text{ L})}{\left(0.08206 \frac{\text{atm}\cdot\text{L}}{\text{K}\cdot\text{mol}}\right)(673 \text{ K})} = 0.0420 \text{ mol } NH_3$$

The stoichiometric relationship between production of NH_3 and the decomposition of NH_4I is 1:1, so 0.0420 moles of ammonium iodide decomposes.

☑ *Reasonable Result Check:* (a) Substituting the equilibrium pressures into the equilibrium expression will reproduce K_p:

$$K_p = P_{NH_3}P_{HI} = 0.464$$

Less than half of the solid decomposes, which makes sense because the K_p is slightly smaller than 1.

95. *Result:* **Cases c and d will have increased [HI]; Cases a and b will have decreased [HI]**

Analyze and Plan: Given the initial amounts (in moles) of reactants and products, the volume, the chemical equation, and the equilibrium constant, determine which will have an increase in the reactant concentration to attain equilibrium and which will have a decrease in the reactant concentration to attain equilibrium. Use some of the methods described in the solution to Question 45.

Execute:

$$2 \text{ HI(g)} \rightleftharpoons H_2(g) + I_2(g) \qquad K_c = \frac{[H_2][I_2]}{[HI]^2} = 0.0200$$

Construct Q_c: $$Q_c = \frac{(\text{conc. } H_2)(\text{conc. } I_2)}{(\text{conc. HI})^2}$$

Then compare to K_c.

Case a: $$Q_c = \frac{\left(\frac{0.10 \text{ mol } H_2}{10.00 \text{ L}}\right)\left(\frac{0.10 \text{ mol } I_2}{10.00 \text{ L}}\right)}{\left(\frac{1.0 \text{ mol HI}}{10.00 \text{ L}}\right)^2} = 0.010 < 0.0200$$

Reaction goes toward products, and the concentration of HI decreases.

Case b: $$Q_c = \frac{\left(\frac{1.0 \text{ mol } H_2}{10.00 \text{ L}}\right)\left(\frac{1.0 \text{ mol } I_2}{10.00 \text{ L}}\right)}{\left(\frac{10. \text{ mol HI}}{10.00 \text{ L}}\right)^2} = 0.010 < 0.0200$$

Reaction goes toward products, and the concentration of HI decreases.

Case c: $$Q_c = \frac{\left(\frac{10. \text{ mol } H_2}{10.00 \text{ L}}\right)\left(\frac{1.0 \text{ mol } I_2}{10.00 \text{ L}}\right)}{\left(\frac{10. \text{ mol HI}}{10.00 \text{ L}}\right)^2} = 0.10 > 0.0200$$

Reaction goes toward reactants, and the concentration of HI increases.

Case d: $$Q_c = \frac{\left(\frac{0.381 \text{ mol } H_2}{10.00 \text{ L}}\right)\left(\frac{1.75 \text{ mol } I_2}{10.00 \text{ L}}\right)}{\left(\frac{5.62 \text{ mol HI}}{10.00 \text{ L}}\right)^2} = 0.0211 < 0.0200$$

The reaction goes toward reactants a very small amount. The concentration of HI increases a very small amount.

In conclusion, the HI concentration increases in case c and very slightly in case d. The HI concentration decreases in cases a and b.

97. *Result:* **(a) 0.18 (b) P_{CO} =1.02 atm, P_{Br_2}= 0.52 atm, P_{COBr_2} =0.10 atm**

Analyze: Given the partial pressures of three gases at equilibrium for a given equation, determine the value of K_p. Given the reduced partial pressure of one of the gases (due to partial condensation), determine the equilibrium partial pressures of all the gases after equilibrium is re-established.

Plan and Execute:

(a) Set up the K_p expression and plug in the equilibrium partial pressures:

$$K_p = \frac{P_{COBr_2}}{P_{CO}P_{Br_2}} = \frac{(0.12)}{(1.00)(0.65)} = 0.1846 = 0.18 \text{ (2 sig figs)}$$

(b) LeChatlier's principle tells us that when a reactant is removed, the equilibrium will shift to the left to make more reactant. So, use signs on the change variable, x, that indicate the reaction going to the left. This way we can assure that the value of x we are seeking is positive.

	$CO(s)$	$+ Br_2(g)$ ⇌	$+ COBr_2(g)$
initial press. (atm)	1.00	0.50	0.12
change as reaction occurs (atm)	+ x	+ x	– x
equilibrium press. (atm)	1.00 + x	0.50 + x	0.12 – x

At equilibrium $$K_p = \frac{P_{COBr_2}}{P_{CO}P_{Br_2}} = \frac{(0.12 - x)}{(1.00 + x)(0.50 + x)} = 0.1846$$

Set up to use the quadratic equation (Appendix A, Section A.7, page A.13):

$$(0.1\underline{8}46)(1.00)(0.50) + (0.1\underline{8}46)(1.00)x + (0.1\underline{8}46)(0.50)x + (0.1\underline{8}46)x^2 = 0.12 - x$$

$$(0.09\underline{2}31) + (0.1\underline{8}46 + 0.09231)x + (0.1\underline{8}46)x^2 = 0.12 - x$$

$$0.1\underline{8}46x^2 + 1.2\underline{7}69x - 0.0\underline{2}769 = 0$$

Plug into the quadratic equation:

$$x = \frac{-b \pm \sqrt{b^2 - 4ac}}{2a} = \frac{-1.2769 \pm \sqrt{(1.2769)^2 - 4(0.1846)(-0.0276)}}{2(0.1846)}$$

$$x = 0.02\underline{1}6$$

$$P_{CO} = 1.00 + x = 1.02 \text{ atm}$$

$$P_{Br_2} = 0.50 + x = 0.52 \text{ atm}$$

$$P_{COBr_2} = 0.12 - x = 0.10 \text{ atm}$$

☑ *Reasonable Result Check:* The equilibrium has greater pressures of reactants, so it makes sense that the value of K_p is less than 1. Plugging the equilibrium pressures into the K_p expression, recreates the value calculated in (a) within the limits of uncertainty:

$$K_p = \frac{P_{COBr_2}}{P_{CO}P_{Br_2}} = \frac{(0.10)}{(1.02)(0.52)} = 0.18$$

99. *Result:* **35.7 g $CaCO_3$**

Analyze: Use a method similar to that described in the solution to Question 62.

Remember, the quantity of solids does not affect the equilibrium state.

	$CaCO_3(s)$ ⇌	$CaO(s)$	$+ CO_2(g)$
initial press. (atm)	solid	0	0
change as reaction occurs (atm)	– x	+ x	+ x
equilibrium press. (atm)	solid – some	solid	x

At equilibrium $$K_p = P_{CO_2} = 3.87 \text{ atm}$$

Now, calculate the moles of $CO_2(g)$ at equilibrium, using the ideal gas law, PV = nRT (Section 8.4) and T = 1000.°C + 273.15 = 1273 K:

$$P_{CO_2}V = n_{CO_2}RT$$

$$n_{CO_2} = \frac{P_{CO_2}V}{RT} = \frac{(3.87 \text{ atm})(5.00 \text{ L})}{\left(0.08206 \frac{\text{atm} \cdot \text{L}}{\text{K} \cdot \text{mol}}\right)(1273 \text{ K})} = 0.185 \text{ mol } CO_2$$

The equation gives the mole ratio between production of CO_2 and the decomposition of $CaCO_3$. Use the mole ratio and the molar mass of the solid to determine the mass decomposed.

$$0.185 \text{ mol } CO_2 \times \frac{1 \text{ mol } CaCO_3}{1 \text{ mol } CO_2} \times \frac{192.9144 \text{ g } CaCO_3}{1 \text{ mol } CaCO_3} = 35.7 \text{ g } CaCO_3 \text{ decomposed}$$

☑ *Reasonable Result Check:* The size of K is large, so it is not surprising that a relatively large mass of the solid must decompose to reach equilibrium.

Applying Concepts

102. *Result:* **Yes, the system is at equilibrium at 600. K; no further experiments would be needed.**

Explanation: For the system to be at equilibrium, it must contain some quantity of both reactants and products, the concentrations of all the reactants and products must be constant over time, and the same relative concentrations are established regardless of which direction the equilibrium was approached. The 42% *cis* concentration was achieved in two different experiments; therefore, the system is at equilibrium. No further experiments are needed.

105. *Result/Explanation:* In the warmer sample, the molecules would be moving faster, and more NO_2 molecules would be seen. In the cooler sample, the molecules would be moving slower, and fewer NO_2 molecules would be seen. In both samples, the molecules are moving very fast. The average speed of gas molecules is commonly hundreds of miles per hour. In both samples, one would see some N_2O_4 molecules decomposing and some NO_2 molecules reacting with each other, at equal rates.

107. *Result:* **Diagrams (b), (c), and (d)**

Analyze: Given a chemical equation and a set of diagrams, determine which diagrams represents equilibrium mixtures.

Plan and Execute: The equilibrium constant expression for this reaction is: $K_c = \dfrac{[AB]^2}{[A_2][B_2]}$.

We need to find diagrams that fit this range: $10^2 >= \dfrac{[AB]^2}{[A_2][B_2]} > 0.1$

Diagram (a) has only reactants, A_2 and B_2. No AB molecules are found in the mixture. That means the value of $K_c = 0$. Diagram (e) has only product molecules, AB. No A_2 or B_2 are found in the mixture. That means the value of $K_c = \infty$. Neither of these K_c values is within the stated range.

In the other three diagrams, we notice that the number of A_2 molecules and the number of B_2 molecules are equal.

$$K = \frac{[AB]^2}{[A_2][B_2]} = \frac{[(\text{Number AB}) \times N_0 / V]^2}{(\text{Number } A_2) \times N_0 / V \times (\text{Number } B_2) \times N_0 / V}$$

$$= \frac{(\text{Number AB})^2}{(\text{Number } A_2) \times (\text{Number } B_2)} = \frac{(\text{Number AB})^2}{(\text{Number } A_2)^2} = \left(\frac{\text{Number AB}}{\text{Number } A_2}\right)^2$$

That means we can simplify the search range by taking the square-root of the range variables and compare them to the ratio of the numbers:

$$\sqrt{10^2} = 10 >= \frac{\text{Number of AB}}{\text{Number of } A_2} > = \sqrt{0.1} = 0.3$$

Diagram (b) has 4 AB and 2 A_2 = 2 B_2: $\dfrac{\text{Number of AB}}{\text{Number of } A_2} = \dfrac{4}{2} = 2$

Diagram (c) has 6 AB and 1 A_2 = 1 B_2 : $\frac{\text{Number of AB}}{\text{Number of } A_2} = \frac{6}{1} = 6$

Diagram (d) has 2 AB and 3 A_2 = 3 B_2 : $\frac{\text{Number of AB}}{\text{Number of } A_2} = \frac{2}{3} = 0.667$

Therefore, all three of these remaining diagrams, (b), (c), and (d), represent equilibrium mixtures where $10^2 > K_c > 0.1$.

109. *Result:* **See diagram below (many other answers are possible)**

Analyze: The nanoscale-level diagram will have CO, H_2O, CO_2, and H_2 gas molecules in relative proportions dictated by the relationship described by the equilibrium constant expression.

$$K_c = 4.00 = \frac{[CO_2][H_2]}{[CO][H_2O]}$$

Assume the container has a volume of V and contains one CO molecule and one H_2O molecule. We can use the equilibrium expression to calculate how many product molecules to put in the container at equilibrium with these reactants.

The concentrations of each of the reactants is: [CO] = [H_2O] = 1 molecule / V = 1/V. If we let x represent the number of CO_2 and H_2 molecule, their concentrations are: [CO_2] = [H_2] = x molecules / V. Substitute the concentrations into the equilibrium expression and solve for x:

$$4.00 = \frac{(\text{Number of } CO_2 / V)(\text{Number of } H_2 / V)}{(\text{Number of } CO / V)(\text{Number of } H_2O / V)} = \frac{(x/V)^2}{(1/V)^2}$$

$$4.00 = x^2$$

$$2.00 = x = \text{number of } CO_2 \text{ molecules} = \text{number of } H_2 \text{ molecules}$$

To represent an equilibrium state, we need to put two CO_2 molecules and two H_2 molecules in the box with one CO molecule and one H_2O molecule.

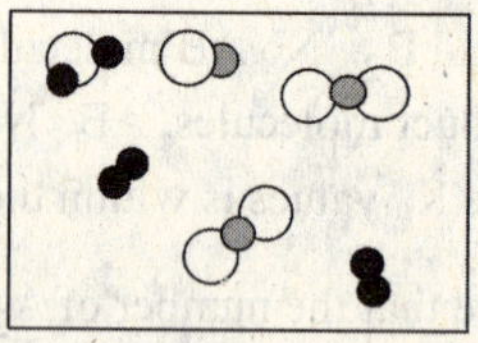

Many other answers are possible as long as K_c is satisfied, such as: four CO_2 molecules and one H_2 molecules in the box with one CO molecule and one H_2O molecule $\frac{(4/V)(1/V)}{(1/V)(1/V)} = 4$, or: four CO_2 molecules and two H_2 molecules in the box with one CO molecule and two H_2O molecule $\frac{(4/V)(2/V)}{(1/V)(2/V)} = 4$, etc.

☑ *Reasonable Result Check:* The equilibrium constant can be reproduced by plugging in the concentrations of the reactants and products: $\frac{(2/V)(2/V)}{(1/V)(1/V)} = 4$

111. *Result:* **Diagram (b)**

Analyze: Given K_c, a chemical equation, and a series of diagrams, determine which diagram represents an equilibrium mixture.

Plan and Execute: Count the molecules of each kind in each diagram. Divide the number of molecules by volume, V, to get concentrations, and combine them in an expression matching the K_c:

$$K_c = \frac{[CH_4][CCl_4]}{[CH_2Cl_2]} = \frac{\left(\frac{N_{CH_4}}{V}\right)\left(\frac{N_{CCl_4}}{V}\right)}{\left(\frac{N_{CH_2Cl_2}}{V}\right)^2} = \frac{\left(N_{CH_4}\right)\left(N_{CCl_4}\right)}{N_{CH_2Cl_2}^{\ \ 2}} = 1.05$$

The diagram that produces a K_c closest to 1.05 is the answer.

Diagram (a) has four CCl_4, four CH_4, and one CH_2Cl_2.

$$K_c = \frac{(4)(4)}{(1)^2} = 16$$

Diagram (b) has three CCl_4, three CH_4, and three CH_2Cl_2.

$$K_c = \frac{(3)(3)}{(3)^2} = 1.0$$

Diagram (c) has two CCl_4, two CH_4, and five CH_2Cl_2.

$$K_c = \frac{(2)(2)}{(5)^2} = 0.16$$

Diagram (b) has an equilibrium constant closest to the given value.

113. *Result/Explanation:* Dynamic equilibria with small values of K introduce a small amount of D^+ ions in place of H^+ ions in the acidic hydrogen.

$$H_2O(\ell) \rightleftharpoons H^+(aq) + OH^-(aq)$$

$$D_2O(\ell) \rightleftharpoons D^+(aq) + OH^-(aq)$$

$$C_6H_5COOH\ (s) \rightleftharpoons C_6H_5COOH\ (aq) \rightleftharpoons H^+(aq) + C_6H_5COO^-(aq)$$

$$C_6H_5COOD\ (s) \rightleftharpoons C_6H_5COOD\ (aq) \rightleftharpoons D^+(aq) + C_6H_5COO^-(aq)$$

$$D^+(aq) + C_6H_5COO^-(aq) \rightleftharpoons C_6H_5COOD\ (aq) \rightleftharpoons C_6H_5COOD\ (s)$$

115. *Result/Explanation:* Dynamic equilibria representing the decomposition of the dimer $N_2O_4(g)$ produces NO_2 (g) and N^*O_2 (g), which will occasionally recombine into the mixed dimer, $O_2N^*\text{–}NO_2$ (g)

$$O_2N\text{–}NO_2\ (g) \rightleftharpoons 2\ NO_2\ (g)$$

$$O_2N^*\text{–}N^*O_2\ (g) \rightleftharpoons 2\ N^*O_2\ (g)$$

$$O_2N^*\text{–}NO_2\ (g) \rightleftharpoons N^*O_2\ (g) + NO_2\ (g)$$

No $O_2N^*\text{–}N^*O_2$ is observed since the chance of two N^*O_2 molecules meeting is vanishingly small.

More Challenging Questions

117. *Result:* **1.34 atm**

Analyze: Given the mass of a mixture of two compounds in equilibrium, the volume and temperature of the container, the equation and equilibrium constant, K_p, for the reaction, determine the total pressure of gas in the container.

Plan and Execute: There are a few ways to solve this question. Below we focus on quantities (in moles).

Let x = moles of NO_2 and y = moles of N_2O_4. Now, we'll set up two equations that relate these quantities. One is the equilibrium expression relating concentration, in moles per liter. The second is the mass relationship identified for the mixture.

First, find K_c, adapting the method described in the solution to Question 33. T = 300. K

$$\Delta n = 1 \text{ mol } N_2O_4(g) - 2 \text{ mol } NO_2(g) = -1$$

$$K_c = K_p(RT)^{-\Delta n}$$

$$K_c = (6.67) \times \left\{\left(0.08206 \frac{L \cdot atm}{mol \cdot K}\right) \times (300.\ K)\right\}^{-(-1)} = 164$$

Next, set up the equilibrium constant expression to relate the concentrations of the reactants and products in terms of x and y.

$$K_c = \frac{[N_2O_4]}{[NO_2]^2} = \frac{\left(\frac{y}{15.0\ L}\right)}{\left(\frac{x}{15.0\ L}\right)^2} = \frac{y}{x^2} \times 15.0$$

$$164 = = \frac{y}{x^2} \times 15.0$$

$$y = 10.\underline{9}468\ x^2$$

Now, relate the relative masses of the two substances to the total mass, in terms of x and y.

$$\text{Mass total} = \text{mass of } NO_2 + \text{mass of } N_2O_4$$

$$64.4 \text{ g total} = x\left(\frac{46.0055 \text{ g } NO_2}{1 \text{ mol } NO_2}\right) + y\left(\frac{92.0110 \text{ g } N_2O_4}{1 \text{ mol } N_2O_4}\right)$$

Plug in the relationship for y, which was derived in the equilibrium equation:

$$64.4 = 46.0055\ x + 92.0110\ y = 46.0055\ x + 92.0110\ (10.\underline{9}468\ x^2)$$

Set up to use the quadratic equation (Appendix A, Section A.7, page A.13):

$$-64.4 + 46.0055\ x + 10\underline{0}7.226\ x^2 = 0$$

Plug into the quadratic equation:

$$x = \frac{-b \pm \sqrt{b^2 - 4ac}}{2a} = \frac{-46.0055 \pm \sqrt{(46.0055)^2 - 4(1007.226)(-64.4)}}{2(1007.226)}$$

$$x = 0.231 \text{ mol}$$

$$y = 10.\underline{9}468\ x^2 = 10.\underline{9}468\ (0.231)^2 = 0.584 \text{ mol}$$

Now, use a combination of Dalton's law of partial pressures (Section 8-7) and the ideal gas law, PV = nRT (Section 8-4), to calculate the total pressure.

$$n_{tot} = 0.231 \text{ mol } NO_2 + 0.584 \text{ mol } N_2O_4 = 0.815 \text{ mol total}$$

$$P_{tot}V = n_{tot}RT$$

$$P_{tot} = \frac{n_{tot}RT}{V} = \frac{(0.815 \text{ mol}) \times \left(0.08206 \frac{atm \cdot L}{K \cdot mol}\right) \times (300.\ K)}{(15.0\ L)} = 1.34 \text{ atm}$$

☑ *Reasonable Result Check:* The values of x and y need to simultaneously satisfy the equilibrium constant equation and the total mass equation. Use the resulting moles and total pressure to determine the partial pressures and recreate the given K_p.

$$P_{N_2O_4} = Ptot\left(\frac{n_{NO_2}}{n_{tot}}\right) = 1.34 \text{ atm} \times \left(\frac{0.231 \text{ mol}}{0.815 \text{ mol}}\right) = 0.379 \text{ atm}$$

$$P_{NO_2} = Ptot\left(\frac{n_{N_2O_4}}{n_{tot}}\right) = 1.34 \text{ atm} \times \left(\frac{0.584 \text{ mol}}{0.815 \text{ mol}}\right) = 0.959 \text{ atm}$$

$$K_p = \frac{P_{N_2O_4}}{P_{NO_2}{}^2} = \frac{0.959}{0.379^2} = 6.67$$

Use the calculated moles and the molar masses to recreate the given total mass:

$$0.231 \text{ mol}\left(\frac{46.0055 \text{ g } NO_2}{1 \text{ mol } NO_2}\right) + 0.584 \text{ mol}\left(\frac{92.0110 \text{ g } N_2O_4}{1 \text{ mol } N_2O_4}\right) = 64.4 \text{ g total}$$

119. *Result/Explanation:* Look first at the relationship between the equilibrium constant and the activation energies of the forward and reverse reactions.

$$K_c = \frac{A_f e^{(-E_{a,f}/RT)}}{A_r e^{(-E_{a,r}/RT)}} = \frac{A_f}{A_r} e^{(-E_{a,f}/RT)} e^{(E_{a,r}/RT)} = \frac{A_f}{A_r} e^{[(-E_{a,f}/RT)+(E_{a,r}/RT)]}$$

$$K_c = \frac{A_f}{A_r} e^{[(E_{a,r}-E_{a,f})/RT]}$$

When a catalyst is added, the activation energies of both the forward and reverse reactions get smaller and the equation that describes the K_c for the catalyzed reaction is:

$$K_{c,cat} = \frac{A_{f,cat}}{A_{r,cat}} e^{[(E_{a,r,cat}-E_{a,f,cat})/RT]}$$

We were told to assume that the frequency factors did not change. The activation energies are reduced, and because the reaction has to conserve energy, they must each be reduced by the same amount. Let us call that quantity of energy X, here.

$$A_{f,\,cat} = A_f \qquad E_{a,f,\,cat} = E_{a,f} - X$$

$$A_{r,\,cat} = A_r \qquad E_{a,r,\,cat} = E_{a,r} - X$$

Notice that $E_{a,f,\,cat} - E_{a,r,\,cat} = (E_{a,f} - X) - (E_{a,r} - X) = E_{a,f} - X - E_{a,r} + X) = E_{a,f} - E_{a,r}$

Substituting into the $K_{c,\,cat}$ equation gives: $K_{c,cat} = \frac{A_f}{A_r} e^{[(E_{a,r}-E_{a,f})/RT]} = K_c$

Therefore, the equilibrium state, as described quantitatively by the equilibrium constant, does not change with the addition of a catalyst.

122. *Result:* **(a) $K_p = 0.03126$; $K_c = 0.0128$ (b) $K_p = 1$ (c) $K_c = 1/(RT_{bp})$**

Analyze: Given the vapor pressure of water at a specific temperature and a chemical equation, calculate K_p and K_c at that temperature, calculate K_p at a different temperature, and suggest a general rule for calculating K_p for any liquid in equilibrium with its vapor at its normal boiling point.

(a) *Plan:* Use the method described in the solution to Question 28.
Execute:

$$H_2O(\ell) \rightleftharpoons H_2O(g) \qquad K_p = P_{H_2O(g)} = 0.03126 \text{ atm}$$

$$T = 25.\ ^\circ C + 273.15 = 298\ K$$

$$\Delta n = 1 \text{ mol } H_2O \text{ gas product} - 0 \text{ mol gas reactants} = 1$$

$$K_p = K_c(RT)^{\Delta n} \qquad\qquad K_p(RT)^{-\Delta n} = K_c$$

$$K_c = (0.03126 \text{ atm}) \times \left\{\left(0.08206\ \frac{L \cdot atm}{mol \cdot K}\right) \times (298\ K)\right\}^{-1} = 0.0128$$

(b) *Explanation:* At the normal boiling point, the vapor pressure of the gas above the liquid is exactly 1 atmosphere. Because $K_p = P_{H_2O(g)}$, then $K_p = 1$, under these conditions.

(c) *Explanation:* Δn for a pure liquid's vaporization equation will always be 1. Also, when the temperature is the boiling point $T = T_{bp}$ and $K_p = P_{(g)} = 1$. So the equation, $K_c = K_p(RT)^{-\Delta n}$, can be simplified to:

$$K_c = 1/(RT_{bp})$$

124. *Result:* **1.15%**

Analyze and Plan: Given a chemical equation, K_c, temperature, and quantity of reactants, volume, calculate the percentage of a reactant converted to product. Use a method similar to that described in the solution to Question 52.

Execute: (conc. N_2) = (conc. H_2) = 1.00 mol/10.00 L = 0.100 M

	N_2 (g)	+ $3\ H_2$ (g)	⇌ $2\ NH_3$ (g)
initial conc. (M)	0.100	0.100	0
change as reaction occurs (M)	− x	− 3x	+ 2x
equilibrium conc. (M)	0.100 − x	0.100 − 3x	2x

At equilibrium
$$K_c = \frac{[NH_3]^2}{[N_2][H_2]^3} = \frac{(2x)^2}{(0.100 - x)(0.100 - 3x)^3} = 5.97 \times 10^{-2}$$

The reactant-favored reaction will have larger reactant concentrations than product concentrations, so let's assume x is small, such that subtraction from the reactant concentrations is negligible: $0.100 - x \cong 0.100$, and $0.100 - 3x \cong 0.100$.

$$\frac{(2x)^2}{(0.100)(0.100)^3} = 5.97 \times 10^{-2}$$

$$x = \sqrt{\frac{5.97 \times 10^{-2}(0.100)(0.100)^3}{4}} = 0.00122$$

However, this value of x is not very small, so the assumption may not be a good one. Using the method described in Appendix A, Section A.7, plug this approximate value of x back into the equation in the places where we ignored it, to obtain a more accurate value. Repeat the procedure until the value of x stops changing:

$$x = \sqrt{\frac{5.97 \times 10^{-2}(0.100 - 0.00122)(0.100 - 3(0.00122))^3}{4}} = 0.00115$$

$$x = \sqrt{\frac{5.97 \times 10^{-2}(0.100 - 0.00115)(0.100 - 3(0.00115))^3}{4}} = 0.00115$$

$$\text{Percentage } N_2 \text{ converted} = \frac{\text{amount of } N_2 \text{ converted}}{\text{initial amount of } N_2} \times 100\ \%$$

$$= \frac{0.00115\ \text{M}}{0.100} \times 100\ \% = 1.15\ \%$$

☑ *Reasonable Result Check:* The equilibrium concentrations should combine to reproduce K_c:

$[N_2] = 0.100 - x = 0.099$ M, $[H_2] = 0.100 - 3x = 0.097$ M, $[NH_3] = 2x = 0.00230$ M

$$K_c = \frac{(0.00230)^2}{(0.099)(0.097)^3} = 5.9 \times 10^{-2}$$

Chapter 13: The Chemistry of Solutes and Solutions

Solutions for Red-Numbered Questions for Review and Thought

Topical Questions

Solubility and Intermolecular Forces (Section 13-1)

13. *Result/Explanation:* If the solid interacts with the solvent using similar or stronger intermolecular forces, it will dissolve readily. If the solute interacts with the solvent using different intermolecular forces than those experienced in the solvent, it will be almost insoluble. For example, consider dissolving an ionic solid in water and oil. The interactions between the ions in the solid and water are very strong, since ions would be attracted to the highly polar water molecule; hence the solid would have a relatively high solubility. However, the ions in the solid interact with each other much more strongly than the London dispersion forces experienced between the nonpolar hydrocarbons in the oil; hence, the solid would have a low solubility.

15. *Result:* **Beaker (c)**

    Analyze: Benzene is a nonpolar organic compound that interacts with other molecules via London dispersion forces. Water molecules are very polar and are capable of experiencing hydrogen bonding interactions. The two are not miscible. Water has a higher density than benzene, so benzene will sit on top of the water. Beaker (c) best represents the results of their mixing.

17. *Result/Explanation:* When an organic acid has a large (nonpolar) piece, it interacts primarily using London dispersion intermolecular forces. Since water interacts via hydrogen bonding intermolecular forces, it would rather interact with itself than with the acid. Hence, the solubility of the large organic acids drops, and some are completely insoluble.

Solubility and Equilibrium (Section 13-2)

18. *Result:* **Put more than 143 g oxalic acid into 1 L of water, agitate/heat until it dissolves, then cool to 25°C. Keep some un-dissolved solid in the bottom of the container when stored.**

    Analyze: Given the solubility of a solid, describe how you would prepare a fixed volume of a saturated solution.

    Plan: The calculation involves simple unit conversions. Start with the sample: targeted volume of 1 L of the solution. Use the density of water and the target mass ratio of the solute to solvent in the solution to determine how much oxalic acid to use.

    Execute:
    $$1\text{ L} \times \frac{1000\text{ mL}}{1\text{ L}} \times \frac{1.00\text{ g}}{1\text{ mL}} \times \frac{1\text{ g oxalic acid}}{7\text{ g water}} = 143\text{ g oxalic acid}$$

    Put more than 143 grams of oxalic acid into 1.00 L of water and wait until most of it dissolves. Heating the solution will make it dissolve faster. To be certain it is completely saturated, some solid oxalic acid can be left in the bottom of the solution when it is stored.

20. *Result:* **(a) supersaturated (b) unsaturated**

    Analyze: Given the masses of solutes and solvents in two solutions and their temperature, and a graph of solubility (g/100g) vs. temperature (Figure 13.4), determine whether the solutions are unsaturated, saturated, or supersaturated.

    Plan: Calculate the solubility of each solution in grams solute per 100. grams of solvent, which is the same as mass percent. Then draw a horizontal line from the concentration axis and a vertical line from the temperature axis at 25°C. If they cross at a point on the red curve, the solution is saturated. If they cross at a point below the curve, then the solution unsaturated. If they cross at a point above the curve, then the solution is supersaturated.

    Execute:

(a) $\dfrac{25.0 \text{ g } NH_4Cl}{50.0 \text{ g water}} 100\% = 50.0\%$. The point at a concentration of 50.0 g/100g and temperature of 25°C is above the red curve on the graph, so the solution is supersaturated.

(b) $\dfrac{25.0 \text{ g } NH_4Cl}{75.0 \text{ g water}} 100\% = 33.3\%$. The point at a concentration of 33.3 g/100g and temperature of 25°C is below the red curve on the graph, so the solution is unsaturated.

Enthalpy, Entropy, and Dissolving Solutes (Section 13-3)

22. *Answer:* **endothermic**

Analyze: Given the lattice energy and the total hydration enthalpy, determine whether the process of dissolving is exothermic or endothermic.

Plan: The relationship among the processes is described in Section 13-3. Combine reactions with Hess's Law:

Reverse lattice formation:	$CaCl_2(s) \longrightarrow \cancel{Ca^{2+}(g) + 2\,Cl^-(g)}$	$\Delta_{latt}H = -(-\,2258 \text{ kJ/mol})$
Hydration:	$+\ \cancel{Ca^{2+}(g) + 2\,Cl^-(g)} \longrightarrow Ca^{2+}(aq) + 2\,Cl^-(aq)$	$+\ \Delta_{hyd}H = -\,2175 \text{ kJ/mol}$
Hess's law sum:	$CaCl_2(s) \longrightarrow Ca^{2+}(aq) + 2\,Cl^-(aq)$	$\Delta_{latt}H + \Delta_{hyd}H = +\,83 \text{ kJ/mol}$
Solution equation:	$CaCl_2(s) \longrightarrow Ca^{2+}(aq) + 2\,Cl^-(aq)$	$\Delta_{soln}H = +\,83 \text{ kJ/mol}$

Because $\Delta_{soln}H$ is positive, the dissolving reaction is **endothermic**.

24. *Answer:* **+ 157 kJ/mol**

Analyze: Given the lattice energy and the total hydration enthalpy, calculate the solution enthalpy.

Plan: The relationship between these processes is described in Section 13-3. Combine the reactions using Hess's Law:

Reverse lattice formation:	$CaBr_2(s) \longrightarrow \cancel{Ca^{2+}(g) + 2\,Br^-(g)}$	$\Delta_{latt}H = -(-\,1984 \text{ kJ/mol})$
Hydration:	$+\ \cancel{Ca^{2+}(g) + 2\,Br^-(g)} \longrightarrow Ca^{2+}(aq) + 2\,Br^-(aq)$	$+\ \Delta_{hyd}H = -\,1827 \text{ kJ/mol}$
Hess's law sum:	$CaBr_2(s) \longrightarrow Ca^{2+}(aq) + 2\,Br^-(aq)$	$\Delta_{latt}H + \Delta_{hyd}H = +\,157 \text{ kJ/mol}$
Solution equation:	$CaBr_2(s) \longrightarrow Ca^{2+}(aq) + 2\,Br^-(aq)$	$\Delta_{soln}H =$ **+ 157 kJ/mol**

26. *Result/Explanation:* The positive side (H side) of the very polar water molecule interacts with the negative ions. The negative side of the very polar water molecule (O side) interacts with the positive ions.

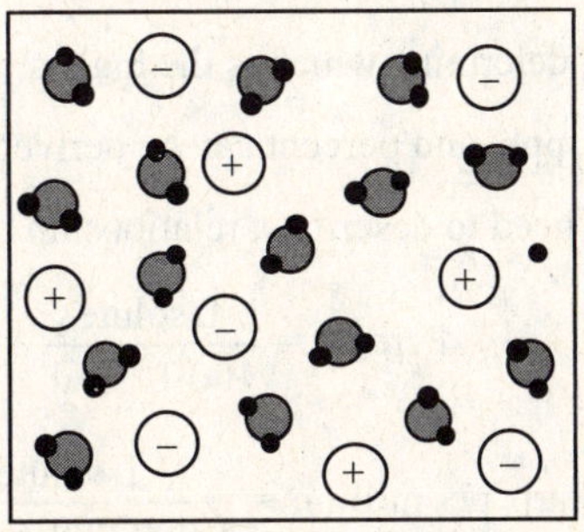

Temperature and Solubility (Section 13-4)

27. *Result:* **(a) No, 65 g < 90 g (b) Supersaturated, 95 g > 90 g (c) Unsaturated, 25 g < 30 g**

Analyze: Given masses of solute and solvent and a graph of solubility (in g/100g water) vs. temperature (Figure 13.10) and temperature, determine if a saturated solution occurs or if a solution is saturated, unsaturated, or supersaturated.

Plan: Adapt the plans used in Questions 20.

Execute: At 40°C, the LiCl curve on the graph shows the concentration of the saturated solution is 90. g/100g.

(a) The concentration is 65.0 g/100g. Because 65 g is less than 90. g, this solution is **not saturated**.

(b) The concentration is 95.0 g/100g. Because 95.0g more than 90. g, this solution is **supersaturated**.

(c) At 50°C, the Li_2SO_4 curve on the graph shows the concentration of the saturated solution is 30. g/100g.

The concentration is $\frac{50.\text{ g}}{200.\text{ g water}} 100\% = 25\%$, which means the solubility is 25 g/100g.

Because 25 g < 30. G, the solution is **unsaturated**.

29. *Result/Explanation:* The dissolving process was endothermic, so the temperature dropped as more solute was added. The solubility of the solid at the lower temperature is lower, so some of the solid did not dissolve. As the solution warmed up, however, the solubility increased again. What remained of the solid dissolved. The solution was saturated at the lower temperature, but is no longer saturated at the current temperature.

Henry's Law (Section 13-5)

31. *Result:* **1.2×10^{-3} M N_2**

Analyze: Given the partial pressure of a gas, the temperature, and the Henry's law constant, determine the concentration of the gas dissolved in the water.

Plan: Find the partial pressure of the gas in units of mm Hg, then use Henry's law described in Section 13-5: $S_g = k_H P_g$ Where s_g is the solubility of the gas, k_H is the Henry's law constant, and P_g is the partial pressure of a gas.

Execute: 78% by volume of gaseous N_2 in air is directly proportional to the percentage by mole (Avogadro's law, Section 8-4). The mole fraction is directly proportional to the pressure (Section 8-6), so we get 0.78 atm N_2 in every 1.00 atm air.

$$2.5 \text{ atm air} \times \frac{0.78 \text{ atm } N_2}{1 \text{ atm air}} \times \frac{760 \text{ mmHg } N_2}{1 \text{ atm } N_2} = 1.5 \times 10^3 \text{ mmHg } N_2$$

For N_2, $k_H = 8.42 \times 10^{-7}$ mol L^{-1}mm Hg^{-1} (Table 13.3)

$$S_g = (8.42 \times 10^{-7} \text{ mol L}^{-1}\text{mm Hg}^{-1})(1.5 \times 10^3 \text{ mm Hg}) = 1.2 \times 10^{-3} \text{ mol L}^{-1} = 1.2 \times 10^{-3} \text{ M}$$

☑ *Reasonable Result Check:* The solubility of nonpolar N_2 should be small. Always double check to be sure that the gas does not react with the water, since Henry's law does not apply to those gases.

Expressing Solution Composition (Section 13-6)

33. *Result:* 0.05%

Analyze: Given three concentrations, determine which is the highest.

Plan: The relationship between ppm, ppb, and percent can be derived:

To accomplish these conversions, we need to describe a relationship between percent and ppm and ppb:

$$\text{Percent} = \frac{\text{L solute}}{100 \text{ L total}}$$

$$\text{Parts per million} = \frac{\text{L solute}}{1{,}000{,}000 \text{ L total}}$$

So, 1% = 10,000 ppm. Multiply the number in units of percent by 10,000 to get ppm.

$$\text{Parts per billion} = \frac{\text{L solute}}{1{,}000{,}000{,}000 \text{ L total}}$$

So, 1ppm = 1,000 ppb. Multiply the number in units of ppm by 1,000 to get ppm.

Calculate these three concentrations with the same units so we can compare them.

Execute: 500 ppb

$$50 \text{ ppm} \times 1{,}000 = 50{,}000 \text{ ppb}$$

$$0.05\% \times 1{,}000 = 500 \text{ ppm}$$

$$500 \text{ ppm} \times 1{,}000 = 500{,}000 \text{ ppb}$$

Thus, 0.05% is the highest concentration. (500,000 ppb > 50,000 ppb > 500 ppb)

35. *Result:* **7.32×10^{-3}%**

Analyze: Convert a concentration in ppm to weight percent

Plan: The relationship between ppm and percent was derived in the solution to Question 33: 1% = 10,000 ppm.

Execute:

$$73.2 \text{ ppm} \times \frac{1\%}{10000 \text{ ppm}} = 7.32 \times 10^{-3}\,\%$$

37. *Result/Explanation:* The definition of parts per billion by mass can be expressed by this ratio:

$$1 \text{ ppb} = \frac{1 \text{ g part}}{10^9 \text{ g whole}}$$

$$\text{Therefore, } 1\text{ppb} = \frac{1 \text{ g part}}{10^9 \text{ g whole}} \times \frac{1 \text{ μg part}}{10^{-6} \text{ g part}} \times \frac{1000 \text{ g whole}}{1 \text{ kg whole}} = \frac{1 \text{ μg part}}{1 \text{ kg whole}}$$

39. *Result:* **90. g ethanol**

Analyze: Given the volume, percent, and density of a solution, calculate the mass of the solute.

Plan: Convert volume to mass with density, then use percent to calculate mass of solute.

Execute: The density of the solution is assumed to be the same as that of water, 1.0 g/mL.

$$750 \text{ mL solution} \times \frac{1.0 \text{ g solution}}{1 \text{ mL solution}} \times \frac{12 \text{ g ethanol}}{100 \text{ g solution}} = 90. \text{ g ethanol}$$

42. *Result:* **1.6×10^{-6} g lead**

Analyze: Given ppm lead in paint, square area of a paint sample, density of paint, volume of paint, and whole area of wall, determine the mass of lead in the sample.

Plan: Convert the sample area into square feet, then determine the volume of paint in the sample, then use the density to get the mass of paint in the sample, then use the concentration to determine the mass of lead in the sample.

Execute: (Here, we assume that "1 gallon" of paint is measured with at least three significant figures.)

$$1.0 \text{ cm}^2 \times \left(\frac{1 \text{ in}}{2.54 \text{ cm}}\right)^2 \times \left(\frac{1 \text{ ft}}{12 \text{ in}}\right)^2 \times \frac{1 \text{ gal paint}}{500. \text{ ft}^2} \times \frac{10.0 \text{ lb paint}}{1 \text{ gal paint}} \times \frac{453.6 \text{ g paint}}{1 \text{ lb pa int}} \times \frac{200. \text{ g Pb}}{10^6 \text{ g paint}} = 2.0 \times 10^{-6} \text{ g Pb}$$

44. *Result:* **(a) 160. g NH_4Cl (b) 83.9 g KCl (c) 7.46 g Na_2SO_4**

Analyze: Given volume and molarity of solution samples, calculate the mass of solute required to prepare the sample.

Plan: Calculate quantity (in moles) of solute using volume and molarity, then use the molar mass to determine mass of solute.

Execute:

$$\text{(a) } 750. \text{ mL} \times \frac{1 \text{ L}}{1000 \text{ mL}} \times \frac{4.00 \text{ mol NH}_4\text{Cl}}{1 \text{ L}} \times \frac{53.491 \text{ g NH}_4\text{Cl}}{1 \text{ mol NH}_4\text{Cl}} = 160. \text{ g NH}_4\text{Cl}$$

(b) $1.50\text{ L} \times \frac{0.750\text{ mol KCl}}{1\text{ L}} \times \frac{74.5513\text{ g KCl}}{1\text{ mol KCl}} = 83.9\text{ g KCl}$

(c) $150.\text{ mL} \times \frac{1\text{ L}}{1000\text{ mL}} \times \frac{0.350\text{ mol Na}_2\text{SO}_4}{1\text{ L}} \times \frac{142.0422\text{ g Na}_2\text{SO}_4}{1\text{ mol Na}_2\text{SO}_4} = 7.46\text{ g Na}_2\text{SO}_4$

46. *Result:* **(a) 0.762 M KCl (b) 0.174 M K_2CrO_4 (c) 0.0126 M $KMnO_4$ (d) 0.167 M $C_6H_{12}O_6$**

Analyze: Given mass of solute and volume of solution samples, calculate the molarity of the solute in the sample.

Plan: Using molar mass, calculate quantity of solute (in moles), then get molarity by dividing the moles by the volume of the solution in liters.

Execute:

(a)
$$\frac{14.2\text{ g KCl} \times \frac{1\text{ mol KCl}}{74.5513\text{ g KCl}}}{250.\text{ mL} \times \frac{1\text{ L}}{1000\text{ mL}}} = 0.762\text{ M KCl}$$

The calculation in (a) can be written more concisely this way:

$$\frac{14.2\text{ g KCl}}{250.\text{ mL}} \times \frac{1\text{ mol KCl}}{74.5513\text{ g KCl}} \times \frac{1000\text{ mL}}{1\text{ L}} = 0.762\text{ M KCl}$$

(b) $\frac{5.08\text{ g K}_2\text{CrO}_4}{150.\text{ mL}} \times \frac{1\text{ mol K}_2\text{CrO}_4}{194.1903\text{ g K}_2\text{CrO}_4} \times \frac{1000\text{ mL}}{1\text{ L}} = 0.174\text{ M K}_2\text{CrO}_4$

(c) $\frac{0.799\text{ g KMnO}_4}{400.\text{ mL}} \times \frac{1\text{ mol KMnO}_4}{158.0339\text{ g KMnO}_4} \times \frac{1000\text{ mL}}{1\text{ L}} = 0.0126\text{ M KMnO}_4$

(d) $\frac{15.0\text{ g C}_6\text{H}_{12}\text{O}_6}{500.\text{ mL}} \times \frac{1\text{ mol C}_6\text{H}_{12}\text{O}_6}{180.1554\text{ g C}_6\text{H}_{12}\text{O}_6} \times \frac{1000\text{ mL}}{1\text{ L}} = 0.167\text{ M C}_6\text{H}_{12}\text{O}_6$

48. *Result:* **96% H_2SO_4**

Analyze: Given density and molarity, calculate the weight percent of the solute.

Plan: Start with the molarity in moles solute per liter solution. Using density, determine the mass of the solution from the volume (1L). Using molar mass, determine the mass of the solute, then multiply by 100% to get the weight percent.

Execute:

$$\frac{18\text{ mol H}_2\text{SO}_4}{1\text{ L solution}} \times \frac{1\text{ L solution}}{1000\text{ cm}^3\text{ solution}} \times \frac{1\text{ cm}^3\text{ solution}}{1.84\text{ g solution}} \times \frac{98.078\text{ g H}_2\text{SO}_4}{1\text{ mol H}_2\text{SO}_4} \times 100\ \% = 96\%\text{ H}_2\text{SO}_4$$

50. *Result:* **0.1 M NaCl**

Analyze: Given volume and mass of solute, calculate the molarity of the solute.

Plan: Using molar mass, calculate quantity of solute (in moles) then divide by volume of solution (in liters).

Execute:

$$\frac{4.\text{mg NaCl} \times \frac{1\text{ g}}{1000\text{ mg}} \times \frac{1\text{ mol NaCl}}{58.4428\text{ g NaCl}}}{0.6\text{ mL} \times \frac{1\text{ L}}{1000\text{ mL}}} = 0.1\text{ M NaCl}$$

52. *Result:* **59 g $C_2H_4(OH)_2$**

Analyze: Given concentration (mol/kg) and the mass of water, calculate the mass of solute.

Plan: Assume that mol/kg means moles of solute per kilogram of water. Start with the mass of water, use the concentration to calculate quantity (in moles) then the molar mass to calculate mass in grams.

Execute:

$$950.\text{ g } H_2O \times \frac{1\text{ kg } H_2O}{1000\text{ g } H_2O} \times \frac{1.0\text{ mol } C_2H_4(OH)_2}{1\text{ kg } H_2O} \times \frac{62.07\text{ g } C_2H_4(OH)_2}{1\text{ mol } C_2H_4(OH)_2} = 59\text{ g } C_2H_4(OH)_2$$

54. *Result:* **0.764 M sucrose**

Analyze: Given percent by weight and density, calculate the molarity of solute in the solution.

Plan: Start with the percent in grams solute per 100 g solution. Using density, determine the volume of the solution. Convert milliliters to liters. Using molar mass, determine the moles of the solute.

Execute:

The formula for sucrose ($C_{12}H_{22}O_{11}$) is given in Exercise 13.14 and Question 45(b).

$$\frac{23.2\text{ g sucrose}}{100\text{ g solution}} \times \frac{1.127\text{ g solution}}{1\text{ mL solution}} \times \frac{1000\text{ mL solution}}{1\text{ L solution}} \times \frac{1\text{ mol sucrose}}{342.2956\text{ g sucrose}} = 0.764\text{ M sucrose}$$

56. *Result:* **(a) 108 ppm (b) 5.64 × 10^{-4} M (c) 5.58 × 10^{-4} mol/kg**

Analyze: Given volume of solution, the mass of the solute, the molar mass of the solute, and the density, determine the solute concentration in units of ppm, molarity and molality of the solute.

(a) As specified in Question 38, 1 ppm = 1 mg/1 kg.

$$355\text{ mL solution} \times \frac{1.01\text{ g solution}}{1\text{ mL solution}} \frac{1\text{ kg}}{1000\text{ g}} = 0.359\text{ kg solution}$$

$$\frac{38.9\text{ mg caffeine}}{0.359\text{ kg water}} = 108\text{ ppm}$$

(b) Molarity, M, is moles solute per liter solution. The molar mass of caffeine is 194.2 g/mol.

$$38.9\text{ mg caffeine} \times \frac{1\text{ g}}{1000\text{ mg}} \times \frac{1\text{ mol caffeine}}{194.2\text{ g caffeine}} = 2.00 \times 10^{-4}\text{ mol caffeine}$$

$$355\text{ mL solution} \times \frac{1\text{ L}}{1000\text{ mL}} = 0.355\text{ mL solution}$$

$$\frac{2.00 \times 10^{-4}\text{ mol caffeine}}{0.355\text{ L}} = 5.64 \times 10^{-4}\text{ M}$$

(c) To get molality, *m*, divide the moles of solute by the mass of solvent in kilograms.

$$\frac{2.00 \times 10^{-4}\text{ mol caffeine}}{0.359\text{ kg water}} = 5.58 \times 10^{-4}\ m$$

58. *Result:* **(a) 0.106, 10.6% (b) 0.127, 12.7% (c) 0.308, 30.8% (d) 2.49 × 10^{-3}, 0.249%**

Strategy: Given masses of solutes and solvents for several solutions, calculate the mass fraction and weight percent.

Plan: Divide the mass of solute by the sum of the two masses to calculate the fraction. Multiply the fraction by 100% to calculate percent.

Execute:

(a) $\dfrac{20.7 \text{ g NaCl}}{20.7 \text{ g NaCl} + 175 \text{ g } H_2O} = 0.106 \text{ NaCl}$ and $0.106 \times 100\% = 10.6 \text{ \% NaCl}$

(b) $\dfrac{1.45 \text{ g ethanol}}{1.45 \text{ g ethanol} + 10.0 \text{ g } H_2O} = 0.127 \text{ ethanol}$ and $0.127 \times 100\% = 12.7 \text{ \% NaCl}$

(c) $\dfrac{20.0 \text{ g } CS_2}{20.0 \text{ g } CS_2 + 45.0 \text{ g } CHCl_3} = 0.308 \text{ } CS_2$ and $0.308 \times 100\% = 30.8 \text{ \% } CS_2$

(d) $4.00 \text{ mL benzene} \times \dfrac{0.877 \text{ g}}{1 \text{ mL}} = 3.51 \text{ g benzene}$

$$\frac{3.51 \text{ g benzene}}{3.51 \text{ g benzene} + 120. \text{ g diethyl ether}} = 2.49 \times 10^{-3} \text{ benzene} \text{ and } 2.49 \times 10^{-3} \times 100\% = 0.249 \text{ \% } CS_2$$

Colligative Properties (Section 13-7)

61. *Result/Explanation*: The addition of a solute to a solvent lowers the vapor pressure of the solvent, so the upper curve (a) is for benzene and the lower curve (b) is for the solution.

62. *Result:* **100.26 °C**

Analyze: Given the moles of solute, the mass of solvent and the normal boiling point of the solvent, determine the boiling point of the solution.

Plan: Calculate the molality of the solute. Use molality to get ΔT_b with the molal boiling-point-elevation constant of the solvent, given in Table 13.5 as 0.512 °C kg mol^{-1}. Then add ΔT_b to the normal boiling point (exactly 100 °C) to get the boiling point of the solution.

Execute:

$$m_{\text{solute}} = \frac{15.0 \text{ g } (NH_2)_2CO}{0.500 \text{ kg water}} \times \frac{1 \text{ mol } (NH_2)_2CO}{60.06 \text{ g} (NH_2)_2CO} = 0.500 \text{ mol/kg}$$

$$\Delta T_b = (0.512 \text{ °C kg mol}^{-1}) \times (0.500 \text{ mol/kg}) = 0.256 \text{ °C}$$

$$T_b(\text{solution}) = 100.000 \text{ °C} + 0.256 \text{ °C} = 100.256 \text{ °C}$$

☑ *Reasonable Result Check:* The boiling point is elevated by the presence of the solute.

64. *Result:* **Freezing point of (a) > freezing point of (d) > freezing point of (b) > freezing point of (c)**

Analyze: Given the concentration (mol/kg) of four substances, list them in order of decreasing freezing point.

Plan: In Section 13-7c, we learn that the freezing point decreases as the molality of solute particles increases. As described in Section 13-7d, solutes that ionize provide a larger number of particles in the solution than solutes that do not ionize. So, we must calculate the concentration of particles in each solution and order the solutions from the smallest concentration (highest freezing point) to the highest concentration (lowest freezing point) .

Execute:

(a) Each methanol (a covalent molecule) provides one particle. = 0.10 mol/kg particles

(b) Each KCl ionizes into K^+ and Cl^-, two particles.

$$0.10 \text{ mol/kg KCl} \times (2 \text{ particles/mol KCl}) = 0.20 \text{ mol/kg particles}$$

(c) Each $BaCl_2$ ionizes into Ba^{2+} and 2 Cl^-, three particles.

$$0.080 \text{ mol/kg } BaCl_2 \times (3 \text{ particles/mol } BaCl_2) = 0.24 \text{ mol/kg particles}$$

(d) Each Na_2SO_4 ionizes into 2 Na^+ and SO_4^{2-}, three particles.

$$0.040 \text{ mol/kg } Na_2SO_4 \times (3 \text{ particles/mol } Na_2SO_4) = 0.12 \text{ mol/kg particles}$$

Order of increasing concentration, (a) < (d) < (b) < (c), gives order of decreasing freezing points: (a) > (d) > (b) > (c)

66. *Result:* $\mathbf{T_f = -1.65\ °C,\ T_b = 100.46\ °C}$

Analyze: Given mass of solute and mass of solvent, calculate the freezing and boiling points of the solution.

Plan: Follow the method described in the solution to Question 63 for the boiling point calculation and a similar method for the freezing point calculation, as described in Section 13-7c:

$$T_f(\text{solution}) = T_f(\text{solvent}) + \Delta T_f$$

$$\Delta T_f = K_f m_{\text{solute}}$$

Execute: Molal boiling-point-elevation and freezing-point-lowering constants for water are given Table 13.5.

$$m_{\text{solute}} = \frac{4.00\text{ g }CO(NH_2)_2}{75.0\text{ g water}} \times \frac{1\text{ mol }CO(NH_2)_2}{60.06\text{ g }CO(NH_2)_2} \times \frac{1000\text{ g water}}{1\text{ kg water}} = 0.888\text{ mol/kg}$$

$$\Delta T_b = (0.512\ °C\text{ kg mol}^{-1}) \times (0.888\text{ mol/kg}) = 0.455\ °C$$

$$T_b(\text{solution}) = 100.000\ °C + 0.455\ °C = 100.455\ °C$$

$$\Delta T_f = (-1.86\ °C\text{ kg mol}^{-1}) \times (0.888\text{ mol/kg}) = -1.65\ °C$$

$$T_f(\text{solution}) = 0.00\ °C + (-1.65\ °C) = -1.65\ °C$$

68. *Result:* $\mathbf{X_{H_2O} = 0.79999;}$ **712 g sucrose**

Analyze: Given the vapor pressure of the solvent at a given temperature, the vapor pressure of a solution at the same temperature and the mass of solvent, determine the mole fraction of the solvent and the mass of solute in the solution.

Plan: Use Raoult's law, described in Section 13-7a. $P_1 = X_1 P_1^0$

Where P_1 is the vapor pressure over the solution, P_1^0 is the vapor pressure over the pure solvent, and X_1 is the mole fraction of the solvent (described in Section 8-6). $X_1 = X_{\text{solvent}} = \dfrac{n_{\text{solvent}}}{n_{\text{tot}}}$

Calculate the mole fraction from the vapor pressures. Calculate moles of solvent from the grams, then use it and the mole fraction to find moles of solute. Then convert to mass.

Execute:

$$X_{H_2O} = \frac{P_{H_2O}}{P^0_{H_2O}} = \frac{119.55\text{ mmHg}}{149.44\text{ mmHg}} = 0.79999$$

$$150.\text{ g }H_2O \times \frac{1\text{ mol }H_2O}{18.02\text{ g }H_2O} = 8.32\text{ mol }H_2O$$

$$X_{H_2O} = \frac{n_{H_2O}}{n_{H_2O} + n_{\text{sucrose}}}$$

Solve for n_{sucrose}:

$$X_{H_2O}(n_{H_2O} + n_{\text{sucrose}}) = X_{H_2O}n_{H_2O} + X_{H_2O}n_{\text{sucrose}} = n_{H_2O}$$

$$X_{H_2O}n_{\text{sucrose}} = n_{H_2O} - X_{H_2O}n_{H_2O}$$

$$n_{\text{sucrose}} = \frac{(1 - X_{H_2O})n_{H_2O}}{X_{H_2O}}$$

$$n_{sucrose} = \frac{(1-0.79999)(8.32\text{ mol})}{0.79999} = \frac{(0.20001)(8.32\text{ mol})}{0.79999} = 2.08\text{ mol}$$

$$2.08\text{ mol } C_{12}H_{22}O_{11} \times \frac{342.297\text{ g } C_{12}H_{22}O_{11}}{1\text{ mol } C_{12}H_{22}O_{11}} = 712\text{ g } C_{12}H_{22}O_{11}$$

☑ *Reasonable Result Check:* The mole fraction of sucrose = 2.08 mol/(8.32 mol + 2.08 mol) = 0.200. The sum of all the mole fractions (0.79999 + 0.200) is one.

70. *Result:* **190 g/mol**

Analyze: Given the boiling point increase, the mass of solute, the mass of solvent, calculate the approximate molar mass of the solute.

Plan: Adapt the method described in the solution to Question 62 to find moles of solute. Divide mass by moles to get molar mass.

Execute: Table 13.5 shows K_b for benzene is 2.53 °C kg mol^{-1}.

$$\Delta T_b = 0.65\ °C$$

$$\Delta T_b = K_b m_{unknown}$$

$$m_{unknown} = \frac{\Delta T_b}{K_b} = \frac{0.65\ °C}{2.53\ °C\text{ kg mol}^{-1}} = 0.26\text{ mol/kg}$$

$$100.\text{ g benzene} \times \frac{1\text{ kg benzene}}{1000\text{ g benzene}} \times \frac{0.26\text{ mol unknown}}{1\text{ kg benzene}} = 0.026\text{ mol unknown}$$

$$\text{Molar mass} = \frac{5.0\text{ g}}{0.026\text{ mol}} = 190\text{ g/mol}$$

72. *Result:* **3.6 × 10^2 g/mol; $C_{20}H_{16}Fe_2$**

Analyze: Given a solute mass, its empirical formula, the solvent mass, and the boiling point of the solution, determine the molar mass and molecular formula of the solute.

Plan: Adapt the method described in the solution to Question 62 to find moles of solute, then divide mass by moles for molar mass. Use methods described in Section 2-12 to find molecular formula.

Execute: Table 13.5 shows K_b for benzene is 2.53 °C kg mol^{-1}.

$$\Delta T_b = T_b(\text{solution}) - T_b(\text{solvent}) = 80.26\ °C - 80.10\ °C = 0.16\ °C$$

$$\Delta T_b = K_b m_{C_{10}H_8Fe}$$

$$m_{C_{10}H_8Fe} = \frac{\Delta T_b}{K_b} = \frac{0.16\ °C}{2.53\ °C\text{ kg mol}^{-1}} = 0.063\text{ mol/kg}$$

$$11.12\text{ g benzene} \times \frac{1\text{ kg benzene}}{1000\text{ g benzene}} \times \frac{0.063\text{ mol } C_{10}H_8Fe}{1\text{ kg benzene}} = 0.00070\text{ mol } C_{10}H_8Fe$$

$$\text{Molar mass of the compound} = \frac{0.255\text{ g}}{0.00070\text{ mol}} = 3.6 \times 10^2\text{ g/mol}$$

The molecular formula is a multiple of the empirical formula: $(C_{10}H_8Fe)_n$

The molar mass of the empirical formula $C_{10}H_8Fe$ is 184.01 g/mol

$$n = \frac{3.6 \times 10^2\text{ g/mol}}{184.01\text{ g/mol}} = 2.0 \cong 2$$

The molecular formula is: $C_{20}H_{16}Fe_2$

75. *Result:* **1.8×10^2 g/mol; $C_{14}H_{10}$**

Analyze: Given the formula of a solute, the mass of solute, the mass of solvent, and the boiling point of the solution, calculate the molar mass of the solute.

Plan: Adapt the method described in the solution to Question 70 to find moles of solute, then divide mass by moles for molar mass. Use methods described in Section 2.12 to find the molecular formula.

Table 13.5 gives the boiling point and K_b for benzene.

Execute:

$$\Delta T_b = T_b(\text{solution}) - T_b(\text{solvent})$$

$$\Delta T_b = 80.34\ °C - 80.10\ °C = 0.24\ °C$$

$$\Delta T_b = K_b m_{C_7H_5}$$

$$m_{C_7H_5} = \frac{\Delta T_b}{K_b} = \frac{0.24\ °C}{2.53\ °C\ \text{kg mol}^{-1}} = 0.095\ \text{mol/kg}\ \textit{(two sig figs)}$$

$$30.0\ \text{g benzene} \times \frac{1\ \text{kg benzene}}{1000\ \text{g benzene}} \times \frac{0.095\ \text{mol } C_7H_5}{1\ \text{kg benzene}} = 0.0029\ \text{mol } C_7H_5\ \textit{(two sig figs)}$$

$$\text{Molar mass of the compound} = \frac{0.500\ \text{g}}{0.0029\ \text{mol}} = 1.8 \times 10^2\ \text{g/mol}\ \textit{(two sig figs)}$$

The molecular formula is a multiple of the empirical formula: $(C_7H_5)_n$

The molar mass of the empirical formula C_7H_5 is 89.117 g/mol

$$n = \frac{1.8 \times 10^2\ \text{g/mol}}{89.117\ \text{g/mol}} = 2.0 \cong 2$$

The molecular formula is: $C_{14}H_{10}$

☑ *Reasonable Result Check:* The value of n is very close to an integer value.

77. *Result:* **(a) 2500 g $C_2H_6O_2$ (b) 104.2 °C**

Analyze: Given the formula of a solute, the mass of the solvent, and the freezing point of the solution, , calculate the mass of solute needed to make the solution and the boiling point.

Plan: Adapt the plan described in the solution to Question 66 to calculate the mass, then adapt the plan from the solution to Question 62 to calculate the boiling point.

Execute:

(a)

$$\Delta T_f = T_f(\text{solution}) - T_f(\text{solvent}) = -15.0\ °C - 0.0°C = -15.0\ °C$$

$$m_{C_2H_6O_2} = \frac{\Delta T_f}{K_f} = \frac{-15.0\ °C}{-1.86\ °C\ \text{kg mol}^{-1}} = 8.06\ \text{mol/kg}$$

$$5.0\ \text{kg water} \times \frac{8.06\ \text{mol } C_2H_6O_2}{1\ \text{kg water}} \times \frac{62.07\ \text{g } C_2H_6O_2}{1\ \text{mol } C_2H_6O_2} = 2500\ \text{g} C_2H_6O_2$$

You must add 2500 g of ethylene glycol to 5.0 kg of water for this much freezing protection.

(b)

$$\Delta T_b = (0.512\ °C\ \text{kg mol}^{-1}) \times (8.06\ \text{mol/kg}) = 4.13\ °C$$

$$T_b(\text{solution}) = 100.000\ °C + 4.13\ °C = 104.13\ °C$$

79. *Result:* **29 atm**

Analyze: Given the freezing point of a solution, the temperature, and an assumption that the density is the same as pure water, determine the osmotic pressure.

Plan: Adapt the method described in Problem-Solving Example 13.13 to find the molal concentration of the solution. With the concentration (in units of molality) and the density of the solution, determine the molar concentration.

Then use the equation described in Section 13-7c to get the osmotic pressure: $\Pi = cRT$ where Π is the osmotic pressure, c is the concentration of the solute in the solution, R is the familiar gas constant with liter and atmosphere units, 0.08206 L·atm/mol·K, and T is the absolute temperature in Kelvin units.

Execute:

$$\Delta T_f = T_f(\text{solution}) - T_f(\text{solvent}) = -2.3\ °C - 0.0\ °C = -2.3\ °C$$

$$m_{\text{solute}} = \frac{\Delta T_f}{K_f} = \frac{-2.3\ °C}{-1.86\ °C\ \text{kg mol}^{-1}} = 1.2\ \text{mol/kg}$$

Assume the mass of the solvent is essentially equal to the mass of the solution: $T = 20.0\ °C + 273.15 = 293.2\ K$

$$\frac{1.2\ \text{mol solute}}{1\ \text{kg solvent}} \times \frac{1\ \text{kg solvent}}{1\ \text{kg solution}} \times \frac{1.00\ \text{kg solution}}{1\ \text{L solution}} = 1.2\ \text{mol/L}$$

$$\Pi = (1.2\ \text{mol/L})(0.08206\ \text{L·atm/mol·K})(293.2\ \text{K}) = 29\ \text{atm}$$

☑ *Reasonable Result Check:* The solute concentration of seawater is fairly high, so it makes sense that the osmotic pressure is high, also.

80. *Result:* **1.2×10^4 g/mol**

Analyze: Given the osmotic pressure, the mass of solute, the volume of solution, and the temperature, calculate the molar mass of the solute.

Plan: Use the equation described in Section 13-7c to get the osmotic pressure: $\Pi = cRT$ as described in the plan for the solution to Question 79. Assume that the addition of solute does not change the volume of the solution, so the volume of the solvent is equal to the volume of the solution, then calculate the moles of solute. Divide the moles by the mass to get the molar mass.

Execute:

$$c = \frac{\Pi}{RT} = \frac{7.6\ \text{mmHg} \times \left(\frac{1\ \text{atm}}{760\ \text{mmHg}}\right)}{\left(0.08206\ \frac{\text{L·atm}}{\text{mol·K}}\right)(298.15\ \text{K})} = 4.1 \times 10^{-4}\ \text{mol/L}\ \textit{(two sig figs)}$$

Assuming that the addition of solute does not change the volume of the solution, the volume of the solvent is equal to the volume of the solution:

$$1.0\ \text{L water} \times \frac{1\ \text{L solution}}{1\ \text{L water}} \times \frac{4.1 \times 10^{-4}\ \text{mol solute}}{1\ \text{L solution}} = 4.1 \times 10^{-4}\ \text{mol solute}\ \textit{(two sig figs)}$$

$$\text{Molar mass of the compound} = \frac{5.0\ \text{g}}{4.1 \times 10^{-4}\ \text{mol}} = 1.2 \times 10^{4}\ \text{g/mol}\ \textit{(two sig figs)}$$

☑ *Reasonable Result Check:* Polymers are large molecules, so a large molar mass is expected.

Colloids (Section 13-8)

82. *Result/Explanation:* Colloids are now understood to be mixtures in which **particles** from 10 to nearly 1000 times the size of a small molecule, the **dispersed phase**, are **distributed uniformly** throughout a solvent-like medium, the continuous phase, or dispersing medium. Consult Table 13.7.

(a) Soap suds are a form of foam with a gas dispersed phase and a liquid continuous phase.

(b) Milk is an emulsion with a liquid dispersed phase and a liquid continuous phase.

(c) Airborne pollen grains are a form of an aerosol with a solid dispersed phase and a gas continuous phase.

(d) Margarine is a gel with a liquid dispersed phase and a solid continuous phase.

Surfactants (Section 13-9)

86. *Result/Explanation:* These terms are described in Section 13-9. A soap molecule looks like this:

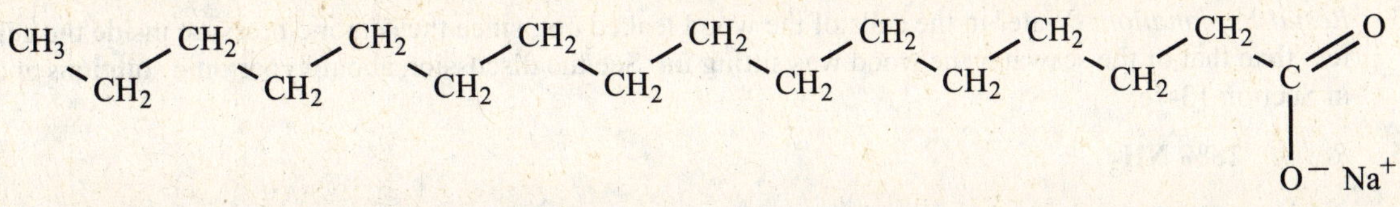

It has a long nonpolar hydrocarbon chain that interacts with other molecules using London dispersion forces and a carboxylate anion salt at one end that interacts with water molecules using ion-dipole forces.

88. *Result/Explanation:* See Figure 13.24: The hydrophobic ends of the detergent molecules (which have greater attraction for hydrocarbons than for water) are embedded in oils, forming colloidal particles called micelles. The micelles are stabilized in solution by the detergent molecules' hydrophilic ends, which are strongly attracted to water molecules. The detergent-oil micelles are then rinsed away.

Water: Natural, Clean, and Otherwise (Section 13-10)

90. *Result/Explanation:* No, water drawn from a well does not have the same type and concentration as those on the surface water. Surface water will have higher concentrations of atmospheric gases (O_2, CO_2, etc.) than well water. Rock and soil deeper in the ground will have different mineral content; for example, some reduced forms of metal ions such as Fe^{2+} may exist in well water and not in surface water.

92. *Result:* **1×10^{-4} g As**

Analyze: Given the number of glasses of water drunk per day, the volume of a glass, and the concentration (ppm) of arsenic in water, calculate how much arsenic you would consume in one week.

Plan: This is a standard conversion factor problem. Start with time, convert time to glasses, convert glasses to volume, then use density to determine mass of solution and use ppm solute to get mass of solute.

Execute: $1\text{ week} \times \frac{7\text{ days}}{1\text{ week}} \times \frac{6\text{ glasses}}{1\text{ day}} \times \frac{8\text{ oz}}{1\text{ glass}} \times \frac{1\text{ quart}}{32\text{ oz}} \times \frac{1\text{ L}}{1.0567\text{ quart}} \times \frac{1000\text{ mL}}{1\text{ L}}$

$$\times \frac{1.00\text{ g water}}{1\text{ mL}} \times \frac{0.010\text{ g As}}{10^6\text{ g water}} = 1 \times 10^{-4}\text{ g As}$$

94. *Result:* **No**

Analyze: Given the maximum contamination level (MCL) for a substance in ppm, and the concentration (in ppb) of a sample, determine if the sample is within the MCL.

Plan: Use the relationship between ppm and ppb derived in the solution to Question 33.

Execute: MCL = 0.002 ppm. 1 ppm = 1,000 ppb

$$5\text{ ppb chlorodane} \times \frac{1\text{ ppm chlorodane}}{1000\text{ ppb chlorodane}} = 0.005\text{ ppm}$$

This concentration is higher than the MCL of 0.002 ppm, so it is not within the MCL for chlordane.

96. *Result/Explanation:* The lime-soda process relies on the precipitation of insoluble compounds to remove the "hard water" ions. The ion exchange process relies on the high charge of the "hard water" ions to attract them to an ion exchange resin, thereby removing them from the water.

98. *Result/Explanation:* Water is sprayed into the air to oxidize organic substances dissolved in it.

General Questions

101. *Result:* **4%**

Analyze: Given the mass of solvent and the masses of two solutes, calculate the weight percent of one of the solutes.

Plan: This is a standard percent problem: find the total mass and divide mass of A by total mass times 100%, *Assume all masses are known ± 0.1 g.*

Total mass = 5.0 g + 0.2 g + 0.3 g = 5.5 g   Weight percent of A = $\frac{0.2 \text{ g A}}{5.5 \text{ g total}} \times 100\% = 4\%$

103. *Result/Explanation:* Water in the cells of the wood leaked out, since the osmotic pressure inside the cells was less than that of the seawater the wood was sitting in. See the discussion about hypertonic solutions and cells in Section 13-7e.

105. *Result:* **28% NH_3**

Analyze: Given concentration (M) and density, calculate the weight percent.

Plan: Start with the concentration in moles per liter, then use the density to calculate mass of solution and molar mass to calculate mass of solute, then multiply by 100%.

Execute:

$$\frac{14.8 \text{ mol } NH_3}{1 \text{ L solution}} \times \frac{1 \text{ L solution}}{1000 \text{ cm}^3 \text{ solution}} \times \frac{1 \text{ cm}^3 \text{ solution}}{0.90 \text{ g solution}} \times \frac{17.03 \text{ g } NH_3}{1 \text{ mol } NH_3} \times 100\% = 28\% \; NH_3$$

107. *Result:* **0.982 *m* $C_4H_8N_2O_2$; 10.2% $C_4H_8N_2O_2$**

Analyze: Given the mass of solute, the volume of solvent, and the density of the solvent, calculate the molality (mol/kg) and weight percent of the solute in this solution.

Plan: Use density to calculate mass of solvent and molar mass to calculate quantity of solute (in moles), then divide moles of solute by kilograms of solvent for molality. Calculate the total mass, then divide mass of solute by total mass and multiply by 100% to get the weight percent.

Execute:

$$500. \text{ mL } CH_3OH \times \frac{0.7893 \text{ g } CH_3OH}{1 \text{ mL } CH_3OH} = 395. \text{ g } CH_3OH$$

$$45.0 \text{ g } C_4H_8N_2O_2 \times \frac{1 \text{ mol } C_4H_8N_2O_2}{116.13 \text{ g } C_4H_8N_2O_2} = 0.387 \text{ mol } C_4H_8N_2O_2$$

$$\text{molality} = \frac{0.387 \text{ mol } C_4H_8N_2O_2}{395 \text{ g ethanol}} \times \frac{1000 \text{ g}}{1 \text{ kg}} = 0.982 \text{ mol/kg} = 0.982 \; m \; C_4H_8N_2O_2$$

$$\text{Total mass} = 45.0 \text{ g} + 395. \text{ g} = 440. \text{ g}$$

$$\text{Weight percent of } C_4H_8N_2O_2 = \frac{45.0 \text{ g } C_4H_8N_2O_2}{440. \text{ g total}} \times 100\% = 10.2\% \; C_4H_8N_2O_2$$

108. *Result:* **freezing point of (a) = freezing point of (c) > freezing point of (d) > freezing point of (b)**

Analyze and Plan: Given the concentration in mol/kg of several solutes, arrange the solutions in order of decreasing freezing point. Adapt the plan from the solution to Question 64.

Execute:

(a) Each ethylene glycol (a covalent molecule) provides one particle, so the actual concentration is 0.20 mol/kg particles.

(b) Each Na_2SO_4 ionizes into 2 Na^+ and SO_4^{2-}, three particles.

$$0.12 \text{ mol/kg } Na_2SO_4 \times (3 \text{ particles/mol } Na_2SO_4) = 0.36 \text{ mol/kg particles}$$

(c) Each NaBr ionizes into Na^+ and Br^-, two particles.

$$0.10 \text{ mol/kg NaBr} \times (2 \text{ particles/mol NaBr}) = 0.20 \text{ mol/kg particles}$$

(d) Each KI ionizes into K^+ and I^-, two particles.

$$0.12 \text{ mol/kg KI} \times (2 \text{ particles/mol KI}) = 0.24 \text{ mol/kg particles}$$

Order of increasing concentrations, (a) = (c) < (d) < (b), gives the order for decreasing freezing points: (a) = (c) > (d) > (b)

111. *Result:* **1.77 g $[(C_4H_9)_4N][ClO_4]$**

Analyze: Given a salt that dissolved into two ions, the mass of solvent, the boiling points of the solution and the solvent, and Kb, calculate the mass of solute that must be added to make the solution.

Plan: Adapt the method described in the solution to Question 62 and use the boiling-point-elevation equation from Sections 13.7d to calculate the quantity of solute (in moles). Use the molar mass to calculate the mass.

Execute:

$$\Delta T_b = T_b(\text{solution}) - T_b(\text{solvent}) = 63.20\ °C - 61.70\ °C = 1.50\ °C$$

$$\Delta T_b = K_b m_{solute} i_{solute}$$

When 100% ionized, one $[(C_4H_9)_4N][ClO_4]$ ionizes into $(C_4H_9)_4N^+$ and ClO_4^-, two particles, so $i_{solute} = 2$.

$$m_{solute} = \frac{\Delta T_b}{K_b i} = \frac{1.50\ °C}{(3.63\ °C\ \text{kg mol}^{-1})(2)} = 0.207\ \text{mol/kg}$$

$$25.0\ \text{g chloroform} \times \frac{1\ \text{kg chloroform}}{1000\ \text{g chloroform}} \times \frac{0.207\ \text{mol}\ [(C_4H_9)_4N][ClO_4]}{1\ \text{kg chloroform}}$$

$$\times \frac{341.90\ \text{g}\ [(C_4H_9)_4N][ClO_4]}{1\ \text{mol}\ [(C_4H_9)_4N][ClO_4]} = 1.77\ \text{g}\ [(C_4H_9)_4N][ClO_4]$$

113. *Result:* **(a) See equation below (b) 15.0 mL $Ba(NO_3)_2$ (b) 12.0 mL Na_2SO_4**

Analyze: Given the concentration and names of two reactants, and the mass and name of a product, write a balanced equation, calculate the volume of each reactant used.

Plan and Execute: Use the methods from Chapter 2 and 3. Balance the equation then use molar mass, mole ratio from the balanced equation, and the concentration (in molarity) to calculate the volume.

Execute:

Use the names to get the formulas. Sodium sulfate is: Na_2SO_4, barium nitrate is: $Ba(NO_3)_2$, and barium sulfate is: $BaSO_4$.

(a) The balanced molecular equation has two sodium ions, two nitrate ions, one barium ion and one sulfate ion on each side of the equation:

$$Na_2SO_4(aq) + Ba(NO_3)_2(aq) \longrightarrow 2\ NaNO_3(aq) + BaSO_4(s)$$

(b) $$0.700\ \text{g}\ BaSO_4 \times \frac{1\ \text{mol}\ BaSO_4}{233.390\ \text{g}\ BaSO_4} \times \frac{1\ \text{mol}\ Ba(NO_3)_2}{1\ \text{mol}\ BaSO_4} \times \frac{1\ \text{L}\ Ba(NO_3)_2}{0.200\ \text{mol}\ Ba(NO_3)_2}$$

$$\times \frac{1000\ \text{mL}}{1\ \text{L}} = 15.0\ \text{mL}\ Ba(NO_3)_2$$

(c) $$0.700\ \text{g}\ BaSO_4 \times \frac{1\ \text{mol}\ BaSO_4}{233.390\ \text{g}\ BaSO_4} \times \frac{1\ \text{mol}\ Na_2SO_4}{1\ \text{mol}\ BaSO_4} \times \frac{1\ \text{L}\ Na_2SO_4}{0.250\ \text{mol}\ Na_2SO_4}$$

$$\times \frac{1000\ \text{mL}}{1\ \text{L}} = 12.0\ \text{mL}\ Na_2SO_4$$

114. *Result:* **(a) No (b) 108.9 °C**

Analyze: Given the definition of "proof", the density of the solute, the proof of a solution, and the temperature, calculate whether the solution will freeze and calculate the boiling point. Adapt the plan described in the solution to Question 66.

Execute:

(a) Calculate the molality starting from the proof:

$$\frac{100 \text{ proof ethanol}}{2} = 50\% \text{ by volume ethanol}$$

$$\frac{50.0 \text{ mL } C_2H_5OH}{50.0 \text{ mL water}} \times \frac{0.789 \text{ g } C_2H_5OH}{1 \text{ mL } C_2H_5OH} \times \frac{1 \text{ mL } H_2O}{1.00 \text{ g } H_2O} \times \frac{1000 \text{ g}}{1 \text{ kg}}$$

$$\times \frac{1 \text{ mol } C_2H_5OH}{46.068 \text{ g } C_2H_5OH} = \frac{17.13 \text{ mol } C_2H_5OH}{\text{kg } H_2O}$$

$$\Delta T_f = m_{C_2H_6O_2} K_f = (17.13 \text{ mol/kg})(-1.86 \text{ °C kg mol}^{-1}) = -31.9 \text{ °C}$$

$$T_f(50\% \text{ solution}) = T_f(H_2O) + \Delta T_f = 0.0 \text{ °C} + (-31.9 \text{ °C}) = -31.9 \text{ °C}$$

So, the vodka will not freeze at a temperature of –15°C.

(b)

$$\Delta T_b = m_{C_2H_6O_2} K_b = (17.13 \text{ mol/kg})(0.512 \text{ °C kg mol}^{-1}) = 8.77 \text{ °C}$$

$$T_b(50\% \text{ solution}) = T_b(H_2O) + \Delta T_b = 100.00 \text{ °C} + 8.9 \text{ °C} = 108.9 \text{ °C}$$

(Notice: This answer assumes that the dissolved alcohol will not evaporate from the solution as the temperature is raised, which is probably not true. If the alcohol does evaporate, it will lower the concentration of the solute, and the boiling temperature will be closer to that of pure water than that calculated here.)

Applying Concepts

117. *Result:* **(a)**

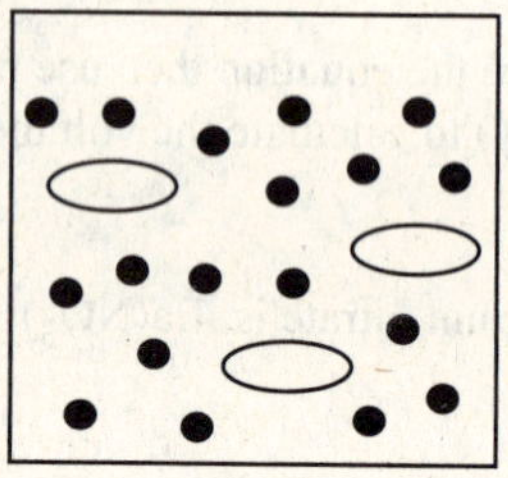

(b)

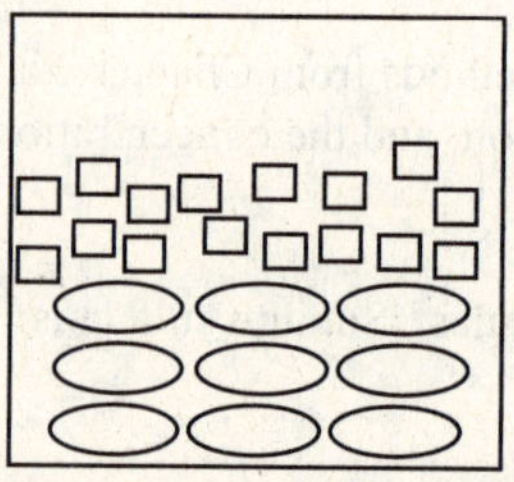

Explanation:

(a) Sugar and water interact with the same hydrogen bonding attractive forces, so they will commingle.

(b) Carbon tetrachloride and sugar interact with very different interactive forces, so they will remain separate phases.

119. *Result:* **(a) unsaturated (b) supersaturated (c) supersaturated**

Analyze: Given the masses of solute and solvent for several solutions, the temperature and curves showing solubility versus temperature (Figure 13.10), determine if the solutions are unsaturated, saturated, or supersaturated.

Plan: Drawn a vertical line at the given concentration and a horizontal line at the given temperature. If these two lines intersect on the curve, the solution is saturated. If they intersect above the curve, the solution is supersaturated. If they intersect below the curve at the given temperature, the solution is unsaturated.

Execute:

(a) 40. g NH_4Cl/100. g H_2O at 80 °C: This point is below the curve, so the solution is unsaturated.

(b) 100. g LiCl/100 g H_2O at 30. °C: This point is above the curve, so the solution is supersaturated.

(c) 120. g $NaNO_3$/100 g H_2O at 40. °C: This point is above the curve, so the solution is supersaturated.

121. *Result:* **See table below**

Analyze: Given an incomplete table of solute compound, masses of solutes and water, mass fraction, weight percent and ppm of solute, complete the table.

Plan: Mass fraction is calculated by dividing the mass of the substance by the total mass of the solution. Calculate weight percent by multiplying mass fraction by 100%. Calculate ppm, using the equality 1% = 10,000 ppm. Solve the mass fraction equation for mass of compound and mass of water:

$$\text{mass fraction} = \frac{m_{acetylene}}{m_{acetylene} + m_{water}}$$

$$(\text{mass fraction}) \times (m_{compound} + m_{water}) = m_{compound}$$

$$(\text{mass fraction}) \times m_{water} = m_{compound} - (\text{mass fraction}) \times m_{compound}$$

$$(\text{mass fraction}) \times m_{water} = [1 - (\text{mass fraction})] \times m_{compound}$$

$$m_{water} = \frac{[1-(\text{mass fraction})] \times m_{compound}}{(\text{mass fraction})}$$

$$m_{compound} = \frac{(\text{mass fraction}) \times m_{water}}{1-(\text{mass fraction})}$$

Execute:

For lye: weight percent = 0.0375 × 100% = **37.5%** and ppm = 37.5 × 10,000 ppm = **3.75 × 10⁵ ppm**

$$m_{lye} = \frac{0.375 \times 125}{1-0.375} = 75.0 \text{ g}$$

For glycerol: mass total = 33 g + 200. g = 233 g, Mass fraction = $\frac{33 \text{ g}}{233 \text{ g}} = 0.14$

weight percent = 0.14 × 100% = **14%** and ppm = 14 × 10,000 ppm =**1.4 × 10⁵**

For acetylene: Mass fraction = 0.0009/100 = **0.000009** and ppm = 37.5 × 10,000 = **9 ppm**

$$m_{water} = \frac{(1-0.000009) \times 0.0015}{0.000009} = 166 \approx \mathbf{2 \times 10^2\ g}\ \textit{(1 sig fig)}$$

Compound	Mass of compound	Mass of water	Mass fraction	Weight percent	ppm of solute
Lye	**75.0 g**	125 g	0.375	**37.5%**	**3.75×10^5**
Glycerol	33 g	200. g	**0.14**	**14%**	**1.4×10^5**
Acetylene	0.0015 g	**2×10^2 g**	**0.000009**	0.0009%	**9**

123. *Result/Explanation:* Molecules slow down and move less at lower temperatures. The reduced motion prevents them from randomly moving as they had in the liquid state. As a result, the intermolecular forces between one molecule and the next begin to align them into a crystal form. The presence of a nonvolatile solute disrupts the regular formation of the crystal. Its size and shape will be different from that of the solute. Intermolecular forces between the solute and solvent are also different from those of solvent molecules with each other. To form the crystalline solid, the solute has to be excluded. If the ice in an iceberg is in regular crystalline form, the water will be pure. Only if the ice is crushed or dirty will other particles be included. So, melting an iceberg will produce relatively pure water.

125. *Result/Explanation:*

(a) Sea water contains more dissolved solutes than fresh water. The presence of a solute lowers the freezing point. That means a lower temperature is required to freeze the sea water than to freeze fresh water.

(b) Salt added to a mixture of ice and water will lower the freezing point of the water. If the ice cream is mixed at a lower temperature, its temperature will drop faster; hence, it will freeze faster.

More Challenging Questions

128. *Result:* **(a) Empirical formula is $C_{18}H_{24}Cr$. (b) Molecular formula is $C_{18}H_{24}Cr$.**

Analyze: Given a partial empirical formula, the percent by mass of two elements in the compound, the temperature, the mass of a sample, the osmotic pressure, calculate the empirical and molecular formulas.

Plan and Execute: This is a combination of the problem-solving from several different chapters.

(a) From Section 3-9: 100.00 g compound – 73.94 g C – 8.27 g H = 17.79 g Cr

$$73.94 \text{ g C} \times \frac{1 \text{ mol C}}{12.0107 \text{ g C}} = 6.156 \text{ mol C} \qquad 8.27 \text{ g H} \times \frac{1 \text{ mol H}}{1.0079 \text{ g H}} = 8.21 \text{ mol H}$$

$$17.79 \text{ g Cr} \times \frac{1 \text{ mol Cr}}{51.996 \text{ g Cr}} = 0.342 \text{ mol Cr}$$

Mole Ratio: 6.156 mol C : 8.21 mol H : 0.342 mol Cr

Simplify the ratio: 18 C : 24 H : 1 Cr Empirical formula is $C_{18}H_{24}Cr$.

(b) Adapt the plan from Question 80.

$$T = 25\ °C + 273.15 = 298 \text{ K}$$

$$c = \frac{\Pi}{RT} = \frac{3.17 \text{ mmHg}\left(\dfrac{1 \text{ atm}}{760 \text{ mmHg}}\right)}{\left(0.08206 \dfrac{\text{L}\cdot\text{atm}}{\text{mol}\cdot\text{K}}\right)(298 \text{ K})} = 1.71\times10^{-4} \text{ mol/L}$$

Assuming that the addition of solute does not change the volume of the solution, so that the volume of the solvent is equal to the volume of the solution:

$$100. \text{ mL chloroform} \times \frac{1 \text{ L chloroform}}{1000 \text{ mL chloroform}} \times \frac{1 \text{ L solution}}{1 \text{L chloroform}}$$

$$\times \frac{1.71\times10^{-4} \text{ mol solute}}{1 \text{ L solution}} = 1.71\times10^{-5} \text{ mol solute}$$

$$\text{Molar mass of the compound} = \frac{5.00 \text{ mg}}{1.71\times10^{-5} \text{ mol}} \times \frac{1 \text{ g}}{1000 \text{ mg}} = 292 \text{ g/mol}$$

The molecular formula is a multiple of the empirical formula: $(C_{18}H_{24}Cr)_n$

The molar mass of the empirical formula is 292.37 g/mol, so the molecular formula is $C_{18}H_{24}Cr$.

130. *Result:* **28 m**

Analyze: Given the concentration of the non-electrolyte solute present in tree sap and a relationship between pressure and density, determine the height of the sap in a tree.

Plan and Execute: Use methods described in Section 13-7e to determine the osmotic pressure (in mm Hg) for a nonelectrolyte related to the difference in the concentration of the sap inside and outside of the tree, then compare the measure of liquid height by relating their densities (see Table 1.1 of the textbook for these densities). Assume the temperature is 25.00°C and the density of a dilute aqueous solution is the same as water.

$$\Pi = cRT$$

$$\Pi = (0.13 \text{ M} - 0.020 \text{ M}) \times \left(0.08206 \frac{\text{L}\cdot\text{atm}}{\text{mol}\cdot\text{K}}\right) \times (298.15 \text{ K}) \times \left(\frac{760 \text{ mmHg}}{1 \text{ atm}}\right) = 2.0\times10^{3} \text{ mmHg } \textit{(two sig figs)}$$

Since pressure is proportional to density of the liquid, 1 mm Hg is directly related to the density of mercury with the same proportionality constant as 1 mm sap is related to the density of sap.

$$2.0 \times 10^3 \text{ mmHg} \times \frac{13.55 \text{ g/mL}}{1 \text{ mmHg}} \times \frac{1 \text{ mm sap}}{0.998 \text{ g/mL}} \times \frac{1 \text{ m}}{1000 \text{ mm}} = 28 \text{ m sap}$$

131. *Result:* **0.300 mol/L**

Analyze and Plan: Given the osmotic pressure of blood at a given temperature, calculate the concentration (M) of a glucose solution with the same osmotic pressure (to make it safe to administer intravenously). Adapt method in the solution to Question 81.

Execute: T = 37 °C + 273.15 = 310. K

$$c = \frac{\Pi}{RTi} = \frac{7.63\text{atm}}{\left(0.08206 \frac{\text{L} \cdot \text{atm}}{\text{mol} \cdot \text{K}}\right)(310. \text{ K})(1.00)} = 0.300 \text{ mol/L}$$

133. *Result:* **(a) 12.3 M (b) 44.1 *m***

Analyze: Given a series of concentrations (mass percent) of ethanol and the boiling points, plot these data and use the graph to determine the boiling point of a solution with a different concentration, then calculate the molarity and molality of one of the solutions in the series.

Plan and Execute:

(a) A graph of mass percent of ethanol (on the x-axis) and boiling point (on the y-axis).

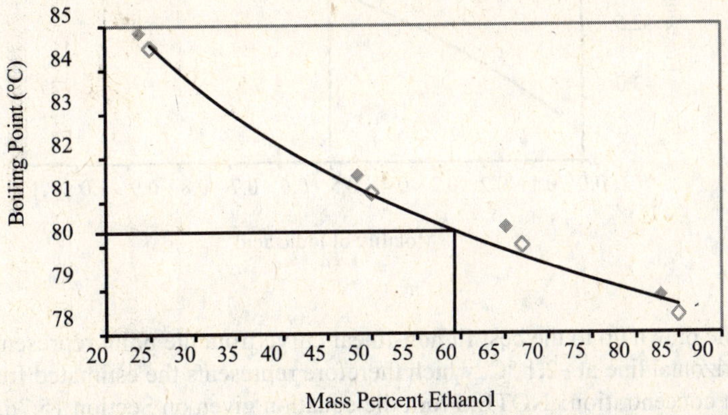

The vertical line drawn up to the best fit non-linear curve from the point representing 59% ethanol by mass produces a horizontal line at 80.4 °C, which therefore represents the estimated boiling point of a solution at that concentration.

(b) In exactly 100 grams of the solution that is 67.0% by mass ethanol, there are 67.0 grams of ethanol and 33.0 grams of water.

Calculate molarity: We need the density of the solution, which we will assume is a weighted average of the densities of the two liquids:

$$67.0 \text{ g ethanol} \times \frac{1 \text{ mL}}{0.789 \text{ g ethanol}} + 33.0 \text{ g water} \times \frac{1 \text{ mL}}{0.998 \text{ g water}} = 117.9 \text{ mL in exactly 100 grams}$$

$$d_{\text{soln}} = \frac{100 \text{ g}}{117.9 \text{ mL}} = 0.848 \text{ g/mL}$$

$$\frac{67.0 \text{ g ethanol}}{100 \text{ g solution}} \times \frac{0.848 \text{ g solution}}{1 \text{ mL solution}} \times \frac{1000 \text{ mL}}{1 \text{ L}} \times \frac{1 \text{ mole ethanol}}{46.0682 \text{ ethanol}} = 12.3 \text{ M}$$

Find molality: We need the mass of solvent = 100 g – 67 g ethanol = 33 g water:

$$\frac{67.0 \text{ g ethanol}}{33.0 \text{ g solvent}} \times \frac{1000 \text{ g}}{1 \text{ kg}} \times \frac{1 \text{ mol ethanol}}{46.0682 \text{ g ethanol}} = 44.1 \text{ mol/kg} = 44.1\ m$$

134. *NOTE: This question involves an aqueous solution that is not characterized well by the simple equations describing colligative properties given in the textbook. Applying those equations to answer this question does not produce answers that will make sense to students. If you assign this question to students, be prepared to answer many questions as they try to make sense of the results.*

Result: **(a) see graph below; –2.1 °C (b) 34, 7.3, 0.92 (c) 3300%, 630%, –8%**

Analyze: Given some concentrations (mol/kg) and the freezing points, plot these data and use the graph to determine the freezing point of a solution with a different concentration, then calculate the van't Hoff factor and the percent dissociation of each solution.

Plan and Execute:

(a) A graph of molality of iodic acid (on the x-axis) and freezing point (on the y-axis).

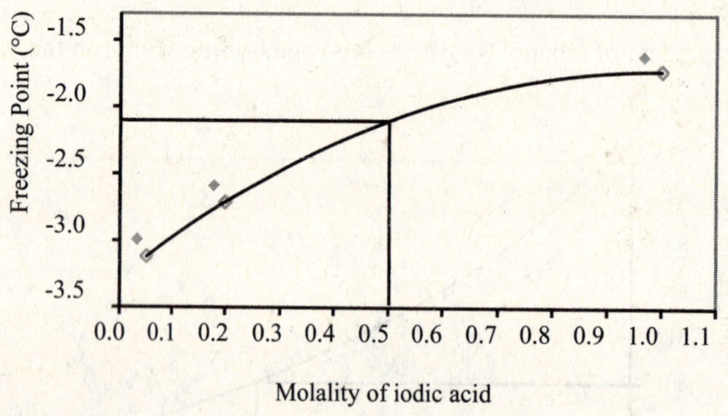

The vertical line drawn up to the best fit non-linear curve from the point representing 0.50 *m* iodic acid produces a horizontal line at –2.1 °C, which therefore represents the estimated freezing point of a solution at that concentration. NOTE: From the equation given on Section 15.7d, the degree of ionization is the only thing that could explain the non-linear shape of this graph, assuming that the solution is an ideal solution.

(b) Adapt the method described in the solution to Question 76. If iodic acid does not dissociate at all, the value of i would be 1. If it dissociates completely it would form two ions, IO_3^- and H_3O^+, and theoretically i would be 2. As a result, the expected range of i values is from 1 (at 0% ionized) to 2 (at 100%).

First solution: $\Delta T_f = T_f(\text{solution}) - T_f(\text{solvent}) = -3.12\ °C - 0.00\ °C = -3.12\ °C$

$$i = \frac{\Delta T_f}{K_f m_{\text{solute}}} = \frac{-3.12\ °C}{(-1.86\ °C\ \text{kg mol}^{-1})(0.050\ \text{mol/kg})} = 34 \text{ (two sig figs)}$$

Second solution: $\Delta T_f = 0.00\ °C - 2.71\ °C = -2.71\ °C$

$$i = \frac{-2.71\ °C}{(-1.86\ °C\ \text{kg mol}^{-1})(0.20\ \text{mol/kg})} = 7.3$$

Third solution: $\Delta T_f = 0.00\ °C - 1.72\ °C = -1.72\ °C$

$$i = \frac{-1.72\ °C}{(-1.86\ °C\ \text{kg mol}^{-1})(1.0\ \text{mol/kg})} = 0.92$$

(c) In general: $\text{percent dissociation} = \dfrac{i_{expt} - i_{0\%}}{i_{100\%} - i_{0\%}} \times 100\%$

First solution: $\text{percent} = \dfrac{34 - 1.00}{2.00 - 1.00} \times 100\% = 3300\%$ *(two sig figs)*

Second solution: $\text{percent} = \dfrac{7.3 - 1.00}{2.00 - 1.00} \times 100\% = 630\%$

Third solution: $\text{percent} = \dfrac{0.92 - 1.00}{2.00 - 1.00} \times 100\% = -8\%$ *(one sig fig after subtraction)*

	i	percent dissociation
First Solution	34	3300%
Second Solution	7.3	630%
Third Solution	0.92	–8%

Except for the fact that the percent ionization is higher at lower concentrations, these calculated results are NOT in the expected range at all, suggesting that this is NOT an ideal solution.

135. *NOTE: This question involves an aqueous solution that is not characterized well by the simple equations describing colligative properties given in the textbook. Using those equations to solve this question does not produce sensible answers. Plan to describe non-ideal solutions before assigning this question.*

Result: **(a) See graph below; –3.36 °C (b) 390, 95, 9.1, 1.8 (c) 39000%, 9400%, 810%, 80%**

Analyze: Given some concentrations (mol/kg) and the freezing points, plot these data and use the graph to determine the freezing point of a solution with a different concentration, then calculate the van't Hoff factor and the percent dissociation of each solution.

Plan and Execute:

(a) A graph of concentration (molality) of ammonium chloride (on the x-axis) and freezing point (on the y-axis).

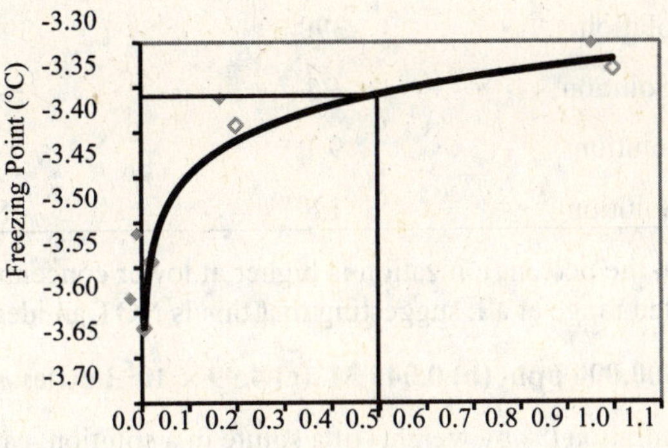

The vertical line drawn down to the best fit non-linear curve from the point representing 0.50 *m* ammonium chloride produces a horizontal line at –3.36 °C, which therefore represents the estimated freezing point of a solution at that concentration.

(b) Adapt the method described in the solution to Question 76. If ammonium chloride does not dissociate at all, the value of i would be 1. If it dissociates completely it would form NH_4^+ and Cl^-, and theoretically i would be 2. As a result, the expected range of i values is from 1 (at 0% ionized) to 2 (at 100%).

First solution: $\Delta T_f = T_f(\text{solution}) - T_f(\text{solvent}) = (-3.617\ °C) - 0.000\ °C + = -3.617\ °C$

$$i = \frac{\Delta T_f}{K_f m_{solute}} = \frac{-3.617\ ^\circ C}{(-1.86\ ^\circ C\ kg\ mol^{-1})(0.0050\ mol/kg)} = 390\ \textit{(two sig figs)}$$

Second solution: $\Delta T_f = T_f(\text{solution}) - T_f(\text{solvent}) = -3.544\ ^\circ C - 0.000\ ^\circ C = -3.544\ ^\circ C$

$$i = \frac{-3.544\ ^\circ C}{(-1.86\ ^\circ C\ kg\ mol^{-1})(0.020\ mol/kg)} = 95$$

Third solution: $\Delta T_f = T_f(\text{solution}) - T_f(\text{solvent}) = -3.392\ ^\circ C - 0.000\ ^\circ C = -3.392\ ^\circ C$

$$i = \frac{-3.392\ ^\circ C}{(-1.86\ ^\circ C\ kg\ mol^{-1})(0.20\ mol/kg)} = 9.1$$

Fourth solution: $\Delta T_f = T_f(\text{solution}) - T_f(\text{solvent}) = -3.33\ ^\circ C - 0.000\ ^\circ C = -3.33\ ^\circ C$

$$i = \frac{-3.33\ ^\circ C}{(-1.86\ ^\circ C\ kg\ mol^{-1})(1.0\ mol/kg)} = 1.8$$

(c) In general: $\text{percent dissociation} = \frac{i_{expt} - i_{0\%}}{i_{100\%} - i_{0\%}} \times 100\%$

First solution: $\text{percent} = \frac{390 - 1.00}{2.00 - 1.00} \times 100\% = 39{,}000\%$ *(two sig figs)*

Second solution: $\text{percent} = \frac{95 - 1.00}{2.00 - 1.00} \times 100\% = 9{,}400\%$ *(two sig figs)*

Third solution: $\text{percent} = \frac{9.1 - 1.0}{2.0 - 1.0} \times 100\% = 810\%$

Fourth solution: $\text{percent} = \frac{1.8 - 1.0}{2.0 - 1.0} \times 100\% = 80\%$

	i	percent dissociation
First Solution	390	39000%
Second Solution	95	9400%
Third Solution	9.1	810%
Fourth Solution	1.8	80%

Except for the fact that the percent ionization is higher at lower concentrations, these calculated results are NOT in the expected range at all, suggesting that this is NOT an ideal solution.

137. *Result:* **(a) 6300 ppm, 6,300,000 ppb (b) 0.040 M (c) 4.99 × 10^5 bottles**

Analyze: Given the concentration (% by weight) of a solute in a solution, calculate the concentration in ppm, ppb, and molarity. Given a reactant, balanced chemical equation and its percent yield, the description of another reaction and its percent yield, the volume of one bottle, and the total mass of one reactant, calculate how many bottles could be prepared

Plan and Execute: A sample of exactly 100 g of solution contains 0.63 g SnF_2 and 99.37 g water.

(a) The units ppm and ppb are discussed in the solution to Question 33: 1% = 10,000 ppm.

$$0.63\ \%\ SnF_2 \times \frac{10{,}000\ ppm}{1\ \%} = 6300\ ppm\ SnF_2$$

$$6300\ ppm\ SnF_2 \times \frac{1000\ ppb\ SnF_2}{1\ ppm\ SnF_2} = 6{,}300{,}000\ ppb\ SnF_2$$

(b) $\frac{0.63 \text{ g SnF}_2}{10^2 \text{ g solution}} \times \frac{0.998 \text{ g solution}}{1 \text{ mL}} \times \frac{1 \text{ mol SnF}_2}{156.707 \text{ g SnF}_2} \times \frac{1000 \text{ mL}}{1 \text{ L}} = 0.040 \text{ M SnF}_2$

(c) In exactly one metric ton, there are exactly 10^6 grams.

$$10^6 \text{ g SnO}_2 \times \frac{1 \text{ mol SnO}_2}{150.709 \text{ g SnO}_2} \times \frac{1 \text{ mol Sn}}{1 \text{ mol SnO}_2} \times \frac{118.710 \text{ g Sn}}{1 \text{ mol Sn}} = 7.88 \times 10^5 \text{ g Sn theoretical}$$

$$7.88 \times 10^5 \text{ g Sn theoretical} \times \frac{80 \text{ g Sn actual}}{100 \text{ g Sn theoretical}} = 6.30 \times 10^5 \text{ g Sn actual}$$

$$6.30 \times 10^5 \text{ g Sn} \times \frac{1 \text{ mol Sn}}{118.710 \text{ g Sn}} \times \frac{1 \text{ mol SnF}_2}{1 \text{ mol Sn}} \times \frac{156.707 \text{ g SnF}_2}{1 \text{ mol SnF}_2} = 8.32 \times 10^5 \text{ g SnF}_2 \text{ theoretical}$$

$$8.32 \times 10^5 \text{ g SnF}_2 \text{ theoretical} \times \frac{94 \text{ g SnF}_2 \text{ actual}}{100 \text{ g SnF}_2 \text{ theoretical}} = 7.82 \times 10^5 \text{ g SnF}_2 \text{ actual}$$

Use the answer from (b), in units of molarity to determine volume, then use the bottle volume to calculate the number of bottles.

$$7.82 \times 10^5 \text{ g SnF}_2 \times \frac{1 \text{ mol SnF}_2}{156.707 \text{ g SnF}_2} \times \frac{1 \text{ L}}{0.040 \text{ mol SnF}_2} \times \frac{1000 \text{ mL}}{1 \text{ L}} \times \frac{1 \text{ bottle}}{250. \text{ mL}} = 4.99 \times 10^5 \text{ bottles}$$

Chapter 14: Acids and Bases

Solutions for Red-Numbered Questions for Review and Thought

Topical Questions

Brønsted-Lowry Acids and Bases (Sections 14-1, 14-2)

9. *Result:* **See equations below**

Analyze: Given formulas of acids, write chemical equations for proton-transfer reactions.

Plan: An H^+ ion is transferred from the acid to the water molecule to make $H_3O^+(aq)$ and the conjugate base of the acid.

Execute:

(a) $HCO_3^-(aq) + H_2O(\ell) \rightleftharpoons H_3O^+(aq) + CO_3^{2-}(aq)$

(b) $HCl(aq) + H_2O(\ell) \rightleftharpoons H_3O^+(aq) + Cl^-(aq)$

(c) $CH_3COOH(aq) + H_2O(\ell) \rightleftharpoons H_3O^+(aq) + CH_3COO^-(aq)$

(d) $HCN(aq) + H_2O(\ell) \rightleftharpoons H_3O^+(aq) + CN^-(aq)$

Notice: These reactions will proceed toward products to a variable extent depending on their strength.

11. *Result:* **See equations below**

Analyze: Given formulas of bases, write chemical equations for proton-transfer reactions.

Plan: An $H^+(aq)$ ion is transferred from the water molecule to the base to make the conjugate acid of the base and $OH^-(aq)$.

Execute:

(a) $HSO_4^-(aq) + H_2O(\ell) \rightleftharpoons H_2SO_4(aq) + OH^-(aq)$

(b) $CH_3NH_2(aq) + H_2O(\ell) \rightleftharpoons CH_3NH_3^+(aq) + OH^-(aq)$

(c) $I^-(aq) + H_2O(\ell) \rightleftharpoons HI(aq) + OH^-(aq)$

(d) $H_2PO_4^-(aq) + H_2O(\ell) \rightleftharpoons H_3PO_4(aq) + OH^-(aq)$

Notice: These reactions will proceed toward products to a variable extent depending on their strength. In particular, the two bases in (a) and (c) I^- are very weak bases. The reaction that is actually observed between HSO_4^- and water is $HSO_4^-(aq) + H_2O(\ell) \rightleftharpoons H_3O^+(aq) + SO_4^{2-}(aq)$. I^- ion is not observed to react with water.

13. *Result:* **(a) I^-, iodide ion (b) HNO_3, nitric acid (c) HCO_3^-, hydrogen carbonate ion (d) HCO_3^-, hydrogen carbonate ion (e) conjugate acid is sulfuric acid, and conjugate base is SO_4^{2-}, sulfate ion (f) HSO_3^{2-}, hydrogen sulfite ion**

Analyze: Given formulas of acids and bases, write the formula and name of the conjugate partner.

Plan: Conjugate acid-base pairs differ by one H^+ ion. The conjugate partner of an acid is a base with one fewer H^+ than the acid has. The conjugate partner of a base is an acid with one more H^+ than the base has.

Execute:

(a) HI is a Brønsted-Lowry acid. Its conjugate base is I^-, **iodide ion.**

(b) NO_3^- is a Brønsted-Lowry base. Its conjugate acid is $\mathbf{HNO_3}$, **nitric acid.**

(c) CO_3^{2-} is a Brønsted-Lowry base. Its conjugate acid is $\mathbf{HCO_3^-}$, **hydrogen carbonate ion.**

(d) H_2CO_3 is a Brønsted-Lowry acid. Its conjugate base is $\mathbf{HCO_3^-}$, **hydrogen carbonate ion.**

(e) HSO_4^- as a Brønsted-Lowry base, has a conjugate acid of $\mathbf{H_2SO_4}$, **sulfuric acid.** HSO_4^- as a Brønsted-Lowry acid, has a conjugate base of $\mathbf{SO_4^{2-}}$, **sulfate ion.**

(f) SO_3^{2-} is a Brønsted-Lowry base. Its conjugate acid is $\mathbf{HSO_3^-}$, **hydrogen sulfite ion.**

15. *Result:* **pairs (b), (c), and (d)**

Analyze: Given several pairs, determine which are acid-base conjugate pairs.

Plan: Conjugate acid-base pairs must differ by only one H^+ ion.

Execute:

(a) NH_2^- is the conjugate base of NH_3, not NH_4^+. NH_3 is the conjugate base of NH_4^+.

(b) NH_3 is the conjugate acid of NH_2^-.

(c) H_3O^+ is the conjugate acid of H_2O.

(d) OH^- is the conjugate acid of O^{2-}.

(e) H_3O^+ is the conjugate acid of H_2O, not OH^-. H_2O is the conjugate acid of OH^-.

Of the choices given, proper conjugate acid-base pairs are described in (b), (c), and (d).

17. *Result:* **See identifications below**

Analyze: Given several chemical equations for proton-transfer reactions, identify the acid reactant, base reactant, conjugate acid product, and conjugate base product.

Plan: The conjugate base of an acid has one less H^+ than its acid partner; the conjugate acid of a base has one more H^+ than its base partner. For further assistance, consult Section 14-1b and Figure 14.1.

Execute:

(a) $HS^-(aq) + H_2O(\ell) \rightleftharpoons H_2S(aq) + OH^-(aq)$

reactant base	**reactant acid**	**conj. acid of $\mathbf{HS^-}$**	**conj. base of $\mathbf{H_2O}$**

H_2O/OH^- and H_2S/HS^- are the two acid-base conjugate pairs.

(b) $S^{2-}(aq) + NH_4^+(aq) \rightleftharpoons NH_3(aq) + HS^-(aq)$

reactant base	**reactant acid**	**conj. acid of $\mathbf{NH_4^+}$**	**conj. base of $\mathbf{S^{2-}}$**

NH_4^+/NH_3 and H_2S/HS^- are the two acid-base conjugate pairs.

(c) $HCO_3^-(aq) + HSO_4^-(aq) \rightleftharpoons H_2CO_3(aq) + SO_4^{2-}(aq)$

reactant base	**reactant acid**	**conj. acid of $\mathbf{HCO_3^-}$**	**conj. base of $\mathbf{HSO_4^-}$**

HSO_4^-/SO_4^{2-} and H_2CO_3/HCO_3^- are the two acid-base conjugate pairs.

(d) $NH_3(aq)$	+ $NH_2^-(aq)$	$\rightleftharpoons$	$NH_2^-(aq)$	+ $NH_3(aq)$
reactant acid	**reactant base**		**conj. base of NH_3**	**conj. acid of NH_2^-**

NH_3/NH_2^- and NH_3/NH_2^- are the two acid-base conjugate pairs.

pH and Autoionization of Water (Sections 14-3, 14-4)

The following is a summary of relationships between H_3O^+ concentration, OH^-concentration, pH, and pOH.

$$[H_3O^+][OH^-] = 10^{-14}$$

When $[H_3O^+] > 10^{-7}$ M and $[OH^-] < 10^{-7}$ M, the solution is **acidic**.

When $[H_3O^+] = [OH^-] = 10^{-7}$ M, the solution is **neutral**

When $[H_3O^+] < 10^{-7}$ M and $[OH^-] > 10^{-7}$ M, the solution is **basic**.

To get pH or pOH from $[H_3O^+]$ or $[OH^-]$ concentrations, use the following relationships:

$$pH = -\log[H_3O^+]$$

$$pOH = -\log[OH^-]$$

$$14 = pH + pOH$$

To get $[H_3O^+]$ or $[OH^-]$ concentrations from pH or pOH, use the following relationships:

$$[H_3O^+] = 10^{-pH}$$

$$[OH^-] = 10^{-pOH}$$

When pH < 7 and pOH > 7, the solution is **acidic**.

When pH = 7 = pOH, the solution is **neutral**.

When pH > 7 and pOH < 7, the solution is **basic**.

19. *Result:* **(a) D (b) E (c) E**

Analyze: Given concentrations of H_3O^+ or OH^- for four solutions, determine which has the highest concentration of H_3O^+, which has the highest concentration of OH^-, and which is closest to neutral.

Plan: Use the relationship shown in the summary above, $[H_3O^+][OH^-] = 10^{-14}$ to calculate the missing concentrations.

$$[OH^-] = \frac{10^{-14}}{[H_3O^+]} \text{ and } [H_3O^+] = \frac{10^{-14}}{[OH^-]}$$

Execute:

Solution D: $$[OH^-] = \frac{10^{-14}}{2\times10^{-3}} = 5\times10^{-12}\text{ M}$$

Solution E: $$[H_3O^+] = \frac{10^{-14}}{2\times10^{-7}} = 5\times10^{-8}\text{ M}$$

Solution F: $$[OH^-] = \frac{10^{-14}}{4\times10^{-5}} = 3\times10^{-10}\text{ M}$$

Solution G: $$[H_3O^+] = \frac{10^{-14}}{5\times10^{-11}} = 2\times10^{-4}\ M$$

Order the four solutions in order of decreasing $[H_3O^+]$ and increasing $[OH^-]$:

Solution	$[H_3O^+]$ (M)	$[OH^-]$ (M)
D	2×10^{-3}	5×10^{-12}
G	2×10^{-4}	5×10^{-11}
F	4×10^{-5}	3×10^{-10}
E	5×10^{-8}	2×10^{-7}

(a) **Solution D** has the highest concentration of H_3O^+

(b) **Solution E** has the highest concentration of OH^-

(c) The solution with $[H_3O^+]$ and $[OH^-]$ closest to 10^{-7} M is the solution closest to neutral. **Solution E** is the closest to neutral.

21. *Result/Explanation:* $C_5H_5N(aq) + H_2O(\ell) \rightleftharpoons C_5H_5NH^+(aq) + OH^-(aq)$

The aqueous solution is basic because some $OH^-(aq)$ is produced.

23. *Result/Explanation:* $CH_3COCOOH(aq) + H_2O(\ell) \rightleftharpoons H_3O^+(aq) + CH_3COCOO^-(aq)$

The aqueous solution is acidic because some $H_3O^+(aq)$ is produced.

25. *Result:* **3×10^{-11} M; basic**

Analyze: Given pH, calculate the concentration of H_3O^+ and determine whether the solution is acidic or basic.

Plan: Use appropriate relationships from the summary given before the solution to Question 19.

Execute: pH = 10.5

$$[H_3O^+] = 10^{-pH} = 10^{-10.5} = \mathbf{3 \times 10^{-11}\ M}$$

Since pH > 7, the antacid is basic.

27. *Result:* **pH = 12.40, pOH = 1.60**

Analyze: Given the concentration of an aqueous base, calculate the pH and pOH.

Plan: Use appropriate relationships from the summary given before the solution to Question 19.

Execute: NaOH is a soluble hydroxide compound and a strong base, so it ionizes to form Na^+ and OH^-:

$$[OH^-] = 0.025\ M$$

$$pOH = -\log[OH^-] = -\log(0.025) = \mathbf{1.60}$$

$$pH = 14.00 - pOH = 14.00 - 1.60 = \mathbf{12.40}$$

29. *Result:* **pOH = 12.51**

Analyze: Given the concentration of H_3O^+, calculate the pOH.

Plan: Use appropriate relationships from the summary given before the solution to Question 19.

Execute: $[H_3O^+] = 0.032$ M

$$pH = -\log[H_3O^+] = -\log(0.032) = 1.49$$

$$pOH = 14.00 - pH = 14.00 - 1.49 = 12.51$$

31. *Result:* **2 g HCl**

Analyze: Given the volume and pH of an HCl solution, calculate the mass of HCl dissolved in the solution.

Plan: Use appropriate relationships from the summary given <u>before</u> the solution to Question 19 and Chapter 3.

Execute: pH = 1.3

$$[H_3O^+] = 10^{-pH} = 10^{-1.3} = 5 \times 10^{-2}\ M.$$

HCl is a strong acid: $HCl(aq) + H_2O(\ell) \longrightarrow H_3O^+(aq) + Cl^-(aq)$

One mole of HCl completely ionizes to form one mole of H_3O^+ and one mole Cl^- ions.

$$1000.\ \text{mL solution} \times \frac{1\ \text{L}}{1000\ \text{mL}} \times \frac{5\times10^{-2}\ \text{mol}\ H_3O^+}{1\ \text{L solution}} \times \frac{1\ \text{mol HCl}}{1\ \text{mol}\ H_3O^+} \times \frac{36.4609\ \text{g HCl}}{1\ \text{mol HCl}} = 2\ \text{g HCl}$$

33. *Result:*

	pH	$[H_3O^+]$ (M)	$[OH^-]$ (M)	acidic or basic
(a)	-	**1.0×10^{-1}**	**1.0×10^{-13}**	**acidic**
(b)	-	**3×10^{-11}**	**3×10^{-4}**	**basic**
(c)	**3.74**	-	**5.6×10^{-11}**	**acidic**
(d)	**9.36**	**4.3×10^{-10}**	-	**basic**

Analyze: Given one of these: pH, $[H_3O^+]$, or $[OH^-]$ for several solutions, calculate other two.

Plan: Use appropriate relationships from the summary given <u>before</u> the solution to Question 19 and Chapter 3.

Execute:

(a) pH = 1.00

$$[H_3O^+] = 10^{-pH} = 1.0 \times 10^{-1}\ M \quad \text{acidic solution (pH} < 10^{-7})$$

$$pOH = 14 - pH = 14 - 1.00 = 13.00$$

$$[OH^-] = 10^{-pOH} = 10^{-13.00} = 1.0 \times 10^{-13}\ M$$

(b) pH = 10.5

$$[H_3O^+] = 10^{-pH} = 10^{-10.5} = 3 \times 10^{-11}\ M$$

Basic solution (pH > 10^{-7})

$$pOH = 14 - pH = 14 - 10.5 = 3.5$$

$$[OH^-] = 10^{-pOH} = 10^{-3.5} = 3 \times 10^{-4}\ M$$

(c) $[H_3O^+] = 1.8 \times 10^{-4}$ M

$$pH = -\log[H_3O^+] = -\log(1.8 \times 10^{-4}) = 3.74$$

Acidic (pH < 10^{-7})

$$pOH = 14 - pH = 14 - 3.74 = 10.26$$

$$[OH^-] = 10^{-pOH} = 10^{-10.26} = 5.6 \times 10^{-11}\ M$$

(d) $[OH^-] = 2.3 \times 10^{-5}$ M

$$pOH = -\log[OH^-] = 4.64$$

$$pH = 14 - pOH = 14 - 4.64 = 9.36$$

$$[H_3O^+] = 10^{-pH} = 10^{-9.36} = 4.3 \times 10^{-10}\ M$$

Basic solution (pH > 10^{-7})

(b) $[OH^-] = 2.2 \times 10^{-9}$ M

$$pOH = -\log[OH^-] = -\log(2.2 \times 10^{-9}) = 8.66$$

$$pH = 14 - pOH = 14 - 8.66 = 5.34$$

$$[H_3O^+] = 10^{-pH} = 10^{-5.34} = 4.5 \times 10^{-6}\ M$$

Acidic (pH < 10^{-7})

(c) pH = 4.67

$$[H_3O^+] = 10^{-pH} = 10^{-4.67} = 2.1 \times 10^{-5}\ M$$

Acidic (pH < 10^{-7})

$$pOH = 14 - pH = 14 - 4.67 = 9.33$$

$$[OH^-] = 10^{-pOH} = 10^{-9.33} = 4.7 \times 10^{-10}\ M$$

(d) $[H_3O^+] = 2.5 \times 10^{-2}$ M

$$pH = -\log[H_3O^+] = -\log(2.5 \times 10^{-2}) = 1.60$$

Acidic (pH < 10^{-7})

$$pOH = 14 - pH = 14 - 1.60 = 12.40$$

$$[OH^-] = 10^{-pOH} = = 10^{-12.40} = 4.0 \times 10^{-13}\ M$$

(e) pH = 9.12

$$[H_3O^+] = 10^{-pH} = 10^{-9.12} = 7.6 \times 10^{-10}\ M$$

Basic (pH > 10^{-7})

$$pOH = 14 - pH = 14 - 9.12 = 4.88$$

$$[OH^-] = 10^{-pOH} = 1.3 \times 10^{-5}\ M$$

35. *Result:* **(a) six times more acidic (b) three times more basic (c) three times more basic (d) 10,000,000 times more acidic**

Analyze: Given the pH of some common solutions, determine how many times more acidic or basic they are compared to neutral.

Plan: As we see from Figure 14.3, the pH scale is logarithmic. Each time the pH unit changes by one, the concentrations of H_3O^+ and OH^- change by a factor of ten. $[H_3O^+] = 10^{-pH}$

Execute:

(a) Milk has a pH about 6.2. $[H_3O^+] = 6 \times 10^{-7}$ M, which is about six times more acidic than neutral.

(b) Seawater has a pH about 7.5. $[H_3O^+] = 3 \times 10^{-8}$ M, which is about three times more basic than neutral.

(c) Blood has a pH about 7.5. $[H_3O^+] = 3 \times 10^{-8}$ M, which is about three times more basic than neutral.

(d) Battery acid has a pH about 0.0. $[H_3O^+] = 1$ M, which is about 10,000,000 times more acidic than neutral.

37. *Result:* **(a) 5.0 × 10^{-9} M (b) basic**

Analyze: Given a solution's pH, calculate the $[H_3O^+]$ and identify if it is acidic or basic.

Plan: Analyze: Use appropriate relationships from the summary given <u>before</u> the solution to Question 19.

Execute:

(a) When pH = 8.30: $[H_3O^+] = 10^{-pH} = 10^{-8.30} = 5.0 \times 10^{-9}$ M

(b) For pH > 7, the solution is basic.

Acid-Base Strengths (Sections 14-5, 14-6)

38. *Result:* (a) A (b) X (c) Y (d) A; see explanations below

Analyze: Given K_a for two acids and K_b for two bases, explain which acid is stronger, which base is stronger, which base has the stronger conjugate acid, and which acid has the weaker conjugate base.

Plan: The acid with the larger K_a is the stronger acid. The base with the larger K_b is the stronger base. The weaker base has a stronger conjugate acid. The weaker acid has a stronger conjugate base.

Execute:

(a) $K_{a,A} = 1 \times 10^{-5}$ is larger than $K_{a,Z} = 5 \times 10^{-6}$, so **A is the stronger acid.**

(b) $K_{b,X} = 1 \times 10^{-4}$ is larger than $K_{b,Y} = 4 \times 10^{-5}$, so **X is the stronger base.**

(c) As shown in (b), X is the stronger base, so Z is the weaker base, and the **conjugate acid of Z** is a stronger acid.

(d) As shown in (a), A is the stronger acid, so the **conjugate base of A** is the weaker base.

39. *Result:* **See Lewis structure below**

Analyze: Given the structural formula of an amino acid, write the Lewis structure for the zwitterion form.

Plan: Zwitterions are explained in Section 14-6d. Remove the H+ from the —COOH group, and add it to the —NH_2 group.

Execute:

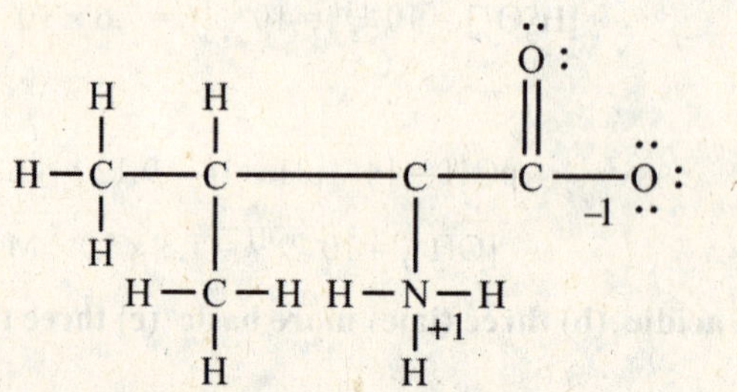

41. *Result:* **(a) See Lewis structures below (b) negative form (c) positive form**

Analyze: Given the structural formula of an amino acid, write the Lewis structure for one positive ion form and one negative ion form that could be formed, and determine which form will exist at high and low pH.

Plan: Add a H^+ to the —NH_2 group to make the cation. Remove a H+ from the —COOH group to make an anion.

Execute:

(a) The positive ion has a —NH_3^+ group:

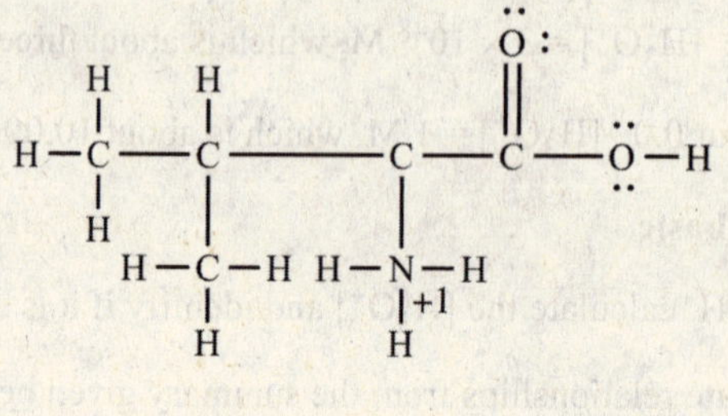

The negative ion has a $—COO^-$ group:

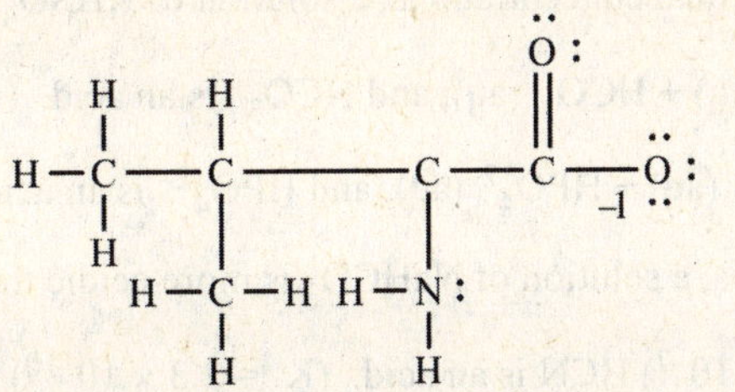

(b) At high pH the concentration of OH^- is high and all the acidic groups will have their acidic H^+ removed, so the net-negative form exists.

(c) At low pH the concentration of H_3O^+ is high and all the acidic groups will have their acidic H^+, so the net-positive form exists.

43. *Result:* **See chemical equations and equilibrium expressions below**

Analyze: Given acids and bases, write ionization equations and ionization constants for them.

Plan: In acid ionization reactions, the acid donates H^+ to a water molecule to make H_3O^+ and the conjugate base. In base ionization reactions, the base receives H^+ from a water molecule to make OH^- and the conjugate acid.

Execute:

(a) $CH_3COOH(aq) + H_2O(\ell) \rightleftharpoons H_3O^+(aq) + CH_3COO^-(aq)$ $\quad K = \dfrac{[H_3O^+][CH_3COO^-]}{[CH_3COOH]}$

(b) $HCN(aq) + H_2O(\ell) \rightleftharpoons H_3O^+(aq) + CN^-(aq)$ $\quad K = \dfrac{[H_3O^+][CN^-]}{[HCN]}$

(c) $SO_3^{2-}(aq) + H_2O(\ell) \rightleftharpoons HSO_3^-(aq) + OH^-(aq)$ $\quad K = \dfrac{[HSO_3^-][OH^-]}{[SO_3^{2-}]}$

(d) $PO_4^{3-}(aq) + H_2O(\ell) \rightleftharpoons HPO_4^{2-}(aq) + OH^-(aq)$ $\quad K = \dfrac{[HPO_4^{2-}][OH^-]}{[PO_4^{3-}]}$

(e) $NH_4^+(aq) + H_2O(\ell) \rightleftharpoons H_3O^+(aq) + NH_3(aq)$ $\quad K = \dfrac{[H_3O^+][NH_3]}{[NH_4^+]}$

(f) $H_2SO_4(aq) + H_2O(\ell) \rightleftharpoons H_3O^+(aq) + HSO_4^-(aq)$ $\quad K = \dfrac{[H_3O^+][HSO_4^-]}{[H_2SO_4]}$

45. *Result:* **(a) 0.10 M H_2CO_3 (b) 0.10 M $KHSO_4$ (c) 0.1 M $NaHCO_3$ (d) 0.1 M H_2S**

Analyze: Given the concentration and identity of solutes in pairs of solutions, determine which is more acidic.

Plan: In all these comparisons the concentrations are the same, so we can use the ionization constants from Appendix G or Table 14.2 to compare their strengths. The larger the K_a, the more acidic the solution. Ionic compounds provide cations or anions that may be acids, so watch for them.

Execute:

(a) H_2CO_3 is an acid. ($K_a = 4.3 \times 10^{-7}$) $NH_4Cl(s) \longrightarrow NH_4^+(aq) + Cl^-(aq)$, and NH_4^+ is an acid. ($K_a = 5.6 \times 10^{-10}$) Given equal concentrations, a solution of H_2CO_3 is more acidic than a solution of NH_4Cl.

(b) HF is an acid. ($K_a = 6.8 \times 10^{-4}$) $KHSO_4(s) \longrightarrow K^+(aq) + HSO_4^-(aq)$, and HSO_4^- is an acid. ($K_a = 1.1 \times 10^{-2}$) Given equal concentrations, a solution of $KHSO_4$ is more acidic than a solution of HF.

(c) $NaHCO_3(s) \longrightarrow Na^+(aq) + HCO_3^-(aq)$, and HCO_3^- is an acid. ($K_a = 4.7 \times 10^{-11}$)

$Na_2HPO_4(s) \longrightarrow 2\ Na^+(aq) + HPO_4^{2-}(aq)$, and HPO_4^{2-} is an acid. ($K_a = 4.6 \times 10^{-13}$)

Given equal concentrations, a solution of $NaHCO_3$ is more acidic than a solution of Na_2HPO_4.

(d) H_2S is an acid. ($K_a = 1 \times 10^{-7}$) HCN is an acid. ($K_a = 3.3 \times 10^{-10}$)

Given equal concentrations, a solution of H_2S is more acidic than a solution of HCN.

48. *Result:* **(a) H_2SO_4 (b) HNO_3 (c) $HClO_4$ (d) $HClO_3$ (e) H_2SO_4**

Analyze: Given formulas of pairs of acids, determine which is a stronger acid.

Plan: Two relationships between acids can be used to help us decide: If the central atoms are the same, then the acid with more O atoms is stronger. If the numbers of O atoms are the same, then the acid with the more-electronegative central atom is stronger. (Electronegativities are found in Figure 6.8.)

Execute:

(a) H_2SO_4 is stronger than H_2SO_3 because it has more O atoms, and H_2SO_3 is comparable in strength to H_2CO_3 because the electronegativities of C and S are almost identical ($EN_S = 2.3$, $EN_C = 2.4$), therefore H_2SO_4 is stronger than H_2CO_3.

(b) HNO_3 is stronger than HNO_2 because it has more O atoms.

(c) $HClO_4$ is stronger than H_2SO_4 because it has a more electronegative atom ($EN_{Cl} > EN_S$).

(d) $HClO_3$ is stronger than H_3PO_4 because it has a more electronegative atom ($EN_{Cl} > EN_P$).

(e) H_2SO_4 is stronger than H_2SO_3 because it has more O atoms.

51. *Result:* **See equations below**

Analyze: Given the formulas of an acid and two bases, write stepwise chemical equations showing the step-wise proteniation (of bases) or deprotonation (of the acid).

Plan: For the acid, sequentially transfer H^+ from the acid to water. For bases, sequentially transfer H^+ from water to the base. When protonating the anioninc form of an amino acid, form the zwitter ion, as explained in Section 14-6d.

Execute:

(a) CO_3^{2-} is the anion of a diprotic acid, H_2CO_3, so write two protonation equations for this base.

$$CO_3^{2-}(aq) + H_2O(\ell) \rightleftharpoons HCO_3^-(aq) + OH^-(aq)$$

$$HCO_3^-(aq) + H_2O(\ell) \rightleftharpoons H_2CO_3(aq) + OH^-(aq)$$

(b) $H_3AsO_4(aq)$ is a triprotic acid, write three deprotonation equations for this acid.

$$H_3AsO_4(aq) + H_2O(\ell) \rightleftharpoons H_3O^+(aq) + H_2AsO_4^-(aq)$$

$$H_2AsO_4^-(aq) + H_2O(\ell) \rightleftharpoons H_3O^+(aq) + HAsO_4^{2-}(aq)$$

$$HAsO_4^{2-}(aq) + H_2O(\ell) \rightleftharpoons H_3O^+(aq) + AsO_4^{3-}(aq)$$

(c) $NH_2CH_3COO^-(aq)$ can be protonated in two places, at the N atom and at the O^- atom, so write two protonation equations for this base. Protonate the —NH_2 group first to form the zwitter ion, as explained in Section 14-6d.

$$NH_2CH_2COO^-(aq) + H_2O(\ell) \rightleftharpoons {}^+NH_3CH_2COO^-(aq) + OH^-(aq)$$

$${}^+NH_3CH_2COO^-(aq) + H_2O(\ell) \rightleftharpoons {}^+NH_3CH_2COOH + OH^-(aq)$$

Problem Solving Using K_a and K_b (Section 14-7)

54. *Result:* **(a) 2.09 is between 2 and 6 (b) 5.13 is between 2 and 6 (c) 8.09 is between 8 and 12 (d) 9.02 is between 8 and 12 (e) 13.30 pH 12 (or higher) (f) 1.55 is close to being between 2 and 6 (g) 9.68 is between 8 and 12 (h) 7.00 is between 6 and 8**

Analyze: Given the concentation of various acids bases, calculate the pH for the solutions.

Plan: Follow the procedure described in Problem-Solving Example 14.9 in Section 14-7, adapting the methods described in the solution to Question 12.50.

Execute: Given in Question 14.47, the initial concentration of the solute is 0.10 M.

(a) The equation for the equilibrium and the equilibrium expression are:

$$HNO_2(aq) + H_2O(\ell) \rightleftharpoons H_3O^+(aq) + NO_2^-(aq) \qquad K_a = \frac{[H_3O^+][NO_2^-]}{[HNO_2]}$$

As the reactants decompose, the concentrations of the products increase stoichiometrically, until they reach equilibrium concentrations. Set up a reaction (ICE) table:

	$HNO_2(aq)$	$H_3O^+(aq)$	$NO_2^-(aq)$
initial conc. (M)	0.10	1.0×10^{-7}*	0
change as reaction occurs (M)	$-$ x	+ x	+ x
equilibrium conc. (M)	0.10 – x	x	x

* 1.0×10^{-7} M H_3O^+ from autoionization of pure water. We assume it is small compared to the change.

At equilibrium $$K_a = \frac{(x)(x)}{(0.10 - x)} = 7.4 \times 10^{-4}$$

$$x^2 = (7.4 \times 10^{-4})(0.10 - x)$$

$$x^2 + 7.4 \times 10^{-4}x - 7.4 \times 10^{-5} = 0$$

Use the quadratic equation: (see Appendix A, Section A.7, page A.13)

$$x = 8.2 \times 10^{-3}\ M = [H_3O^+] \qquad pH = -\log[H_3O^+] = -\log(8.2 \times 10^{-3}) = 2.09$$

If you did Question 47, compare this pH to your qualitative prediction. You should have predicted between 2 and 6.

(b) The equation for the equilibrium and the equilibrium expression are:

$$NH_4^+(aq) + H_2O(\ell) \rightleftharpoons H_3O^+(aq) + NH_3(aq) \qquad K_a = \frac{[H_3O^+][NH_3]}{[NH_4^+]}$$

As the reactants decompose, the concentrations of the products increase stoichiometrically, until they reach equilibrium concentrations. Set up a reaction (ICE) table:

	$NH_4^+(aq)$	$H_3O^+(aq)$	$NH_3(aq)$
initial conc. (M)	0.10	1.0×10^{-7}*	0
change as reaction occurs (M)	$-$ x	+ x	+ x
equilibrium conc. (M)	0.10 – x	x	x

* 1.0×10^{-7} M H_3O^+ from autoionization of pure water. We assume it is small compared to the change.

At equilibrium $$K_a = \frac{(x)(x)}{(0.10 - x)} = 5.6 \times 10^{-10}$$

The small-K, reactant-favored reaction will have a larger reactant concentration than product concentrations, so assume x is very small, such that subtraction from the reactant concentration is negligible: $0.10 - x \cong 0.10$.

$$x^2 = (5.6 \times 10^{-10})(0.10)$$

$$x = 7.5 \times 10^{-6}\text{ M} = [H_3O^+]$$

$$pH = -\log[H_3O^+] = -\log(7.5 \times 10^{-6}) = 5.13$$

If you did Question 47, compare this pH to your qualitative prediction. You should have predicted between 2 and 6.

(c) The equation for the equilibrium and the equilibrium expression are:

$$F^-(aq) + H_2O(\ell) \rightleftharpoons HF(aq) + OH^-(aq) \qquad K_b = \frac{[HF][OH^-]}{[F^-]}$$

As the reactants decompose, the concentrations of the products increase stoichiometrically, until they reach equilibrium concentrations. Set up a reaction (ICE) table:

	$F^-(aq)$	$HF(aq)$	$OH^-(aq)$
initial conc. (M)	0.10	0	1.0×10^{-7}*
change as reaction occurs (M)	– x	+ x	+ x
equilibrium conc. (M)	0.10 – x	x	x

* 1.0×10^{-7} M H_3O^+ from autoionization of pure water. We assume it is small compared to the change.

At equilibrium $$K_b = \frac{(x)(x)}{(0.10 - x)} = 1.5 \times 10^{-11}$$

For the same reasons as in (b), assume x is very small and: $0.10 - x \cong 0.10$.

$$1.5 \times 10^{-11} = \frac{x^2}{(0.10)}$$

$$x^2 = (1.5 \times 10^{-11})(0.10)$$

$$x = 1.2 \times 10^{-6}\text{ M} = [OH^-]$$

$$pOH = -\log[OH^-] = -\log(1.2 \times 10^{-6}) = 5.91$$

$$pH = 14.00 - pOH = 8.09$$

If you did Question 47, compare this pH to your qualitative prediction. You should have predicted between pH 8 and 12.

(d) $$(\text{conc. } CH_3COO^-) = 0.10\text{ M } Ba(CH_3COO)_2 \times \frac{2\text{ mol } CH_3COO^-}{1\text{ mol } Ba(CH_3COO)_2} = 0.20\text{ M}$$

The equation for the equilibrium and the equilibrium expression are:

$$CH_3COO^-(aq) + H_2O(\ell) \rightleftharpoons CH_3COOH(aq) + OH^-(aq) \qquad K_b = \frac{[CH_3COOH][OH^-]}{[CH_3COO^-]}$$

As the reactants decompose, the concentrations of the products increase stoichiometrically, until they reach equilibrium concentrations. Set up a reaction (ICE) table:

	$CH_3COO^-(aq)$	$CH_3COOH(aq)$	$OH^-(aq)$
initial conc. (M)	0.20	0	$1.0 \times 10^{-7*}$
change as reaction occurs (M)	– x	+ x	+ x
equilibrium conc. (M)	0.20 – x	x	x

* 1.0×10^{-7} M OH^- from autoionization of pure water. We assume it is small compared to the change.

At equilibrium $K_b = \dfrac{(x)(x)}{(0.20 - x)} = 5.6 \times 10^{-10}$

For the same reasons as in (b), assume x is very small and: $0.20 - x \cong 0.20$.

$$5.6 \times 10^{-10} = \frac{x^2}{(0.20)}$$

$$x^2 = (5.6 \times 10^{-10})(0.20)$$

$$x = 1.1 \times 10^{-5} \text{ M} = [OH^-]$$

$$pOH = -\log[OH^-] = -\log(1.1 \times 10^{-5}) = 4.98$$

$$pH = 14.00 - pOH = 14.00 - 4.98 = 9.02$$

If you did Question 47, compare this pH to your qualitative prediction. You should have predicted between pH 8 and 12.

(e) $O^{2-}(aq) + H_2O(\ell) \longrightarrow 2\ OH^-(aq)$

This must have a very large K. 100% reaction.

$$[OH^-] = 0.10 \text{ M } O^{2-} \times \frac{2 \text{ mol } OH^-}{1 \text{ mol } O^{2-}} = 0.20 \text{ M}$$

$$pOH = -\log[OH^-] = -\log(0.20) = 0.70$$

$$pH = 14.00 - pOH = 14.00 - 0.70 = 13.30$$

If you did Question 47, compare this pH to your qualitative prediction. You should have predicted pH 12 (or higher).

(f) $HSO_4^-(aq) + H_2O(aq) \rightleftharpoons H_3O^+(aq) + SO_4^{2-}(aq)$ $\quad K_a = \dfrac{[H_3O^+][SO_4^{2-}]}{[HSO_4^-]} = 1.1 \times 10^{-2}$

$HSO_4^-(aq) + H_2O(aq) \rightleftharpoons H_2SO_4(aq) + OH^-(aq)$ $\quad K_b = \dfrac{[H_2SO_4][OH^-]}{[HSO_4^-]} = \text{very small}$

We will use the first reaction since it has the largest K value, and produces more products.

As the reactants decompose, the concentrations of the products increase stoichiometrically, until they reach equilibrium concentrations. Set up a reaction (ICE) table:

	$HSO_4^-(aq)$	$H_3O^+(aq)$	$SO_4^{2-}(aq)$
initial conc. (M)	0.10	$1.0 \times 10^{-7*}$	0
change as reaction occurs (M)	– x	+ x	+ x
equilibrium conc. (M)	0.10 – x	x	x

* 1.0×10^{-7} M H_3O^+ from autoionization of pure water. We assume it is small compared to the change.

At equilibrium $$K_a = \frac{(x)(x)}{(0.10 - x)} = 1.1 \times 10^{-2}$$

$$x^2 = (1.1 \times 10^{-2})(0.10 - x)$$

$$x^2 + 1.1 \times 10^{-2}x - 1.1 \times 10^{-3} = 0$$

Use the quadratic equation: (see Appendix A, Section A.7, page A.13)

$$2.8 \times 10^{-2}\text{ M} = [H_3O^+] \qquad pH = -\log[H_3O^+] = -\log(2.8 \times 10^{-2}) = 1.55$$

If you did Question 47, compare this pH to your qualitative prediction. You should have predicted between 2 and 6. While this calculated pH doesn't fall in that range, it is still greater than the pH of 1 we would predict for a 0.10 M strong acid.

(g) $HCO_3^-(aq) + H_2O(aq) \rightleftharpoons H_3O^+(aq) + CO_3^{2-}(aq)$ $$K_a = \frac{[H_3O^+][CO_3^{2-}]}{[HCO_3^-]} = 4.7 \times 10^{-11}$$

$HCO_3^-(aq) + H_2O(aq) \rightleftharpoons H_2CO_3(aq) + OH^-(aq)$ $$K_b = \frac{[H_2CO_3][OH^-]}{[HCO_3^-]} = 2.3 \times 10^{-8}$$

We will use the second reaction since it has the largest K value, and produces more products.

As the reactants decompose, the concentrations of the products increase stoichiometrically, until they reach equilibrium concentrations. Set up a reaction (ICE) table:

	$HCO_3^-(aq)$	$HCO_3^-(aq)$	$OH^-(aq)$
initial conc. (M)	0.10	0	1.0×10^{-7}*
change as reaction occurs (M)	$-x$	$+x$	$+x$
equilibrium conc. (M)	$0.10 - x$	x	x

* 1.0×10^{-7} M OH^- from autoionization of pure water. We assume it is small compared to the change.

At equilibrium $$K_b = \frac{(x)(x)}{(0.10 - x)} = 2.3 \times 10^{-8}$$

For the same reasons as in (b), assume x is very small and: $0.10 - x \cong 0.10$.

$$2.3 \times 10^{-8} = \frac{x^2}{(0.10)} \qquad x^2 = (2.3 \times 10^{-8})(0.10) \qquad x = 4.8 \times 10^{-5}\text{ M} = [OH^-]$$

$$pOH = -\log[OH^-] = -\log(4.8 \times 10^{-5}) = 4.32$$

$$pH = 14.00 - pOH = 14.00 - 4.32 = 9.68$$

If you did Question 47, compare this pH to your qualitative prediction. You should have predicted between pH 8 and 12.

(h) $BaCl_2 \longrightarrow Ba^{2+} + 2\,Cl^-$ Neither of these ions affect the pH of a water solution, because Ba^{2+} is the cation of a soluble ionic hydroxide compound, $Ba(OH)_2$ and Cl^- is a weak base (K_b = very small), so pH 7.00 and no calculation is needed. If you did Question 47, you should have predicted pH 7, which is between 6 and 8.

55. *Result:* **3.6×10^{-4}**

Analyze: Given the concentration and pH of an acid, calculate its ionization constant.

Plan: Follow the plan from Problem-Solving Example 14.8.

Execute:

HCNO is cyanic acid. The equation for the equilibrium and the equilibrium expression are:

$$HCNO(aq) + H_2O(\ell) \rightleftharpoons H_3O^+(aq) + CNO^-(aq) \qquad K_a = \frac{[H_3O^+][CNO^-]}{[HCNO]}$$

As the reactants decompose, the concentrations of the products increase stoichiometrically, until they reach equilibrium concentrations.

At equilibrium, pH = 2.67, so

$$[H_3O^+] = 10^{-pH} = 10^{-2.67} = 2.1 \times 10^{-3}\ M$$

Since the H_3O^+ ions are produced from the decomposition of HCNO, an equal quantity of CNO^- is formed, and the concentrations change. Set up a reaction (ICE) table:

	HCNO(aq)	$H_3O^+(aq)$	$CNO^-(aq)$
initial conc. (M)	0.015	1.0×10^{-7}*	0
change as reaction occurs (M)	-2.1×10^{-3}	$+2.1 \times 10^{-3}$	$+2.1 \times 10^{-3}$
equilibrium conc. (M)	$0.015 - 2.1 \times 10^{-3}$	2.1×10^{-3}	2.1×10^{-3}

* 1.0×10^{-7} M H_3O^+ from autoionization of pure water. We assume it is small compared to the change.

$$K_a = \frac{(2.1\times10^{-3})(2.1\times10^{-3})}{(0.015-2.1\times10^{-3})} = 3.6 \times 10^{-4}$$

57. *Result:* **1.4×10^{-5}**

Analyze: Given the concentration and pH of an acid, calculate its ionization constant.

Plan: Follow the plan from Problem-Solving Example 14.8.

Execute:

The equation for the equilibrium and the equilibrium expression are:

$$CH_3CH_2COOH(aq) + H_2O(\ell) \rightleftharpoons H_3O^+(aq) + CH_3CH_2COO^-(aq) \qquad K_a = \frac{[H_3O^+][CH_3CH_2COO^-]}{[CH_3CH_2COOH]}$$

At equilibrium, pH = 2.93, so $[H_3O^+] = 10^{-pH} = 10^{-2.93} = 1.2 \times 10^{-3}$ M. Since the H_3O^+ ions are produced from the decomposition of C_6H_5COOH, an equal quantity of $C_6H_5COO^-$ is formed, and the concentrations change in the following way:

	$C_6H_5COOH(aq)$	$H_3O^+(aq)$	$C_6H_5COO^-(aq)$
initial conc. (M)	0.15	1.0×10^{-7}*	0
change as reaction occurs (M)	-1.2×10^{-3}	$+1.2 \times 10^{-3}$	$+1.2 \times 10^{-3}$
equilibrium conc. (M)	$0.015 - 1.2 \times 10^{-3}$	1.2×10^{-3}	1.2×10^{-3}

* 1.0×10^{-7} M H_3O^+ from autoionization of pure water. We assume it is small compared to the change.

$$K_a = \frac{(1.2\times10^{-3})(1.2\times10^{-3})}{(0.015-1.2\times10^{-3})} = 1.4 \times 10^{-5}$$

58. *Result:* $[H_3O^+] = 1.9 \times 10^{-3}$ M, $[CH_3COO^-] = 1.9 \times 10^{-3}$ M, $[CH_3COOH] = 0.20$ M

Analyze: Given the initial concentrations of an acid, calculate the equilibrium concentrations.

Plan: Combine the information described before Question 19 and the plan from the solution to Question 54 and Problem-Solving Example 14.9.

Execute:

The equation for the equilibrium and the equilibrium expression are:

$$CH_3COOH(aq) + H_2O(\ell) \rightleftharpoons H_3O^+(aq) + CH_3COO^-(aq) \qquad K_a = \frac{[H_3O^+][CH_3COO^-]}{[CH_3COOH]}$$

As the reactants decompose, the concentrations of the products increase stoichiometrically, until they reach equilibrium concentrations. Set up a reaction (ICE) table:

	$CH_3COOH(aq)$	$H_3O^+(aq)$	$CH_3COO^-(aq)$
initial conc. (M)	0.20	1.0×10^{-7}*	0
change as reaction occurs (M)	– x	+ x	+ x
equilibrium conc. (M)	0.20 – x	x	x

* 1.0×10^{-7} M H_3O^+ from autoionization of pure water. We assume it is small compared to the change.

At equilibrium $\quad K_a = \dfrac{(x)(x)}{(0.20 - x)} = 1.8 \times 10^{-5}$

Assume x is very small and: $0.20 - x \cong 0.20$.

$$1.8 \times 10^{-5} = \frac{x^2}{(0.20)}$$

$$x^2 = (1.8 \times 10^{-5})(0.20)$$

$$x = 1.9 \times 10^{-3}\ M = [H_3O^+] = [CH_3COO^-]$$

$$[CH_3COOH] = 0.20\ M - x = 0.20\ M - 1.9 \times 10^{-3}\ M = 0.20\ M \text{ (as assumed)}$$

60. *Result:* **8.84**

Analyze: Given the initial concentration of a base and its K_b, calculate the pH.

Plan: Adapt the plan from Problem-Solving Example 14.10.

Execute: The equation for the equilibrium and the equilibrium expression are:

$$C_6H_5NH_2(aq) + H_2O(\ell) \rightleftharpoons C_6H_5NH_3^+(aq) + OH^-(aq) \qquad K_b = \frac{[C_6H_5NH_3^+][OH^-]}{[C_6H_5NH_2]}$$

As the reactants decompose, the concentrations of the products increase stoichiometrically, until they reach equilibrium concentrations. Set up a reaction (ICE) table:

	$C_6H_5NH_2(aq)$	$C_6H_5NH_3^+(aq)$	$OH^-(aq)$
initial conc. (M)	0.12	0	1.0×10^{-7}*
change as reaction occurs (M)	– x	+ x	+ x
equilibrium conc. (M)	0.12 – x	x	x

* 1.0×10^{-7} M OH^- from autoionization of pure water. We assume it is small compared to the change.

At equilibrium $\quad K_b = \dfrac{(x)(x)}{(0.12 - x)} = 3.9 \times 10^{-10}$

Assume x is very small and: $0.12 - x \cong 0.12$.

$$x^2 = (3.9 \times 10^{-10})(0.12)$$

$$x = 6.8 \times 10^{-6}\ M = [\ OH^-\ ]$$

$$pOH = -\log[OH^-] = -\log(6.8 \times 10^{-6}) = 5.16$$

$$pH = 14.00 - pOH = 14.00 - 5.16 = 8.84$$

62. *Result:* **10.47**

Analyze: Given the initial concentration of a base and its K_b, calculate the $[OH^-]$ and pH.

Plan: Adapt the plan from Problem-Solving Example 14.10.

Execute: The equation for the equilibrium is:

$$C_{10}H_{15}NH_2(aq) + H_2O(\ell) \rightleftharpoons C_{10}H_{15}NH_3^+(aq) + OH^-(aq)$$

The K_a given is for the dissociation of $C_{10}H_{15}NH^+(aq)$. We need to find the K_b from the K_a:

$$K_b = \frac{K_w}{K_a} = \frac{1.00 \times 10^{-14}}{7.9 \times 10^{-11}} = 1.3 \times 10^{-4}$$

The reactants decompose, the concentrations of the products increase stoichiometrically, until they reach equilibrium concentrations. Set up a reaction (ICE) table:

	$C_{10}H_{15}N(aq)$	$C_{10}H_{15}NH^+(aq)$	$OH^-(aq)$
initial conc. (M)	0.0010	0	1.0×10^{-7}*
change as reaction occurs (M)	$-x$	$+x$	$+x$
equilibrium conc. (M)	$0.0010 - x$	x	x

* 1.0×10^{-7} M OH^- from autoionization of pure water. We assume it is small compared to the change.

At equilibrium $$K_b = \frac{[C_{10}H_{15}NH_3^+][OH^-]}{[C_{10}H_{15}NH_2]} = \frac{(x)(x)}{(0.0010 - x)} = 1.3 \times 10^{-4}$$

$$x^2 = (1.3 \times 10^{-4})(0.0010 - x)$$

$$x^2 + 1.3 \times 10^{-4}x - 1.5 \times 10^{-7} = 0$$

Use the quadratic equation: (see Appendix A, Section A.7, page A.13)

$$x = 3.0 \times 10^{-4}\text{ M} = [OH^-]$$

$$pOH = -\log[OH^-] = -\log(3.0 \times 10^{-4}) = 3.53$$

$$pH = 14.00 - pOH = 14.00 - 3.53 = 10.47$$

64. *Result:* **3.28**

Analyze: Given the mass of solute, volume of solution, and K_a, calculate the pH.

Plan: First use methods from Chapter 3 to calculate the initial concentration of the $C_3H_6O_3$. Then adapt the plan from Problem-Solving Example 14.9.

Execute: $$\frac{56\text{ mg } C_3H_6O_3}{250\text{ mL soln}} \times \frac{1\text{ g}}{1000\text{ mg}} \times \frac{1\text{ mol } C_3H_6O_3}{90.08\text{ g } C_3H_6O_3} \times \frac{1000\text{ mL}}{1\text{ L}} = 0.0025\text{ M}$$

According to the structure shown in Section 14-2, lactic acid is a monoprotic acid, so we'll write the formula as: $HC_3H_5O_3$. The equation for the equilibrium and the equilibrium expression are:

$$HC_3H_5O_3(aq) + H_2O(\ell) \rightleftharpoons H_3O^+(aq) + C_3H_5O_3^-(aq) \qquad K_a = \frac{[H_3O^+][C_3H_5O_3^-]}{[HC_3H_5O_3]}$$

As the reactants decompose, the concentrations of the products increase stoichiometrically, until they reach equilibrium concentrations. Set up a reaction (ICE) table:

	$HC_3H_5O_3(aq)$	$H_3O^+(aq)$	$C_3H_5O_3^-(aq)$
initial conc. (M)	0.0025	1.0×10^{-7}*	0
change as reaction occurs (M)	$-x$	$+x$	$+x$
equilibrium conc. (M)	$0.0025 - x$	x	x

* 1.0×10^{-7} M H_3O^+ from autoionization of pure water. We assume it is small compared to the change.

At equilibrium $$K_a = \frac{(x)(x)}{(0.0025 - x)} = 1.4 \times 10^{-4}$$

$$x^2 = (1.4 \times 10^{-4})(0.0025 - x)$$

$$x^2 + 1.4 \times 10^{-4}x - 3.5 \times 10^{-7} = 0$$

Use the quadratic equation: (see Appendix A, Section A.7, page A.13)

$x = 5.2 \times 10^{-4}\text{ M} = [H_3O^+]$ $\qquad$ $pH = -\log[H_3O^+] = -\log(5.2 \times 10^{-4}) = 3.28$

Acid-Base Reactions of Salts (Section 14-8)

66. *Result:* **(a) NH_4^+, reactant-favored, weaker reactants (b) CH_3COO^-, product-favored, weaker products (c) NH_2^-, product-favored, weaker products**

Analyze: Given incomplete equations for chemical reactions, determine the missing reactant or product and explain the prediction of whether each reaction is product-favored or reactant=favored.

Plan: Adapt the plan described in the solutions to Question 13, 17, and 44. Compare the reactant acid to the product acid and identify which is stronger and which is weaker. Do the same with the bases. Equilibrium favors the weaker species in the reaction.

Execute:

(a) $\mathbf{NH_4^+(aq)}$ + $Br^-(aq)$ $\rightleftharpoons$ $NH_3(aq)$ + $HBr(aq)$

weaker acid — weaker base — stronger base — stronger acid

Since the relatively weaker species are in the reactants, the reaction is reactant-favored.

(b) $CH_3COOH(aq)$ + $CN^-(aq)$ $\rightleftharpoons$ $\mathbf{CH_3COO^-(aq)}$ + $HCN(aq)$

stronger acid — stronger base — weaker base — weaker acid

Since the relatively weaker species are in the products, the reaction is product-favored.

(c) $\mathbf{NH_2^-(aq)}$ + $H_2O(\ell)$ $\rightleftharpoons$ $NH_3(aq)$ + $OH^-(aq)$

stronger base — stronger acid — weaker acid — weaker base

Since the relatively weaker species are in the products, the reaction is product-favored.

68. *Result:* **(a) and (c) are product-favored; (b) and (d) are reactant-favored; see equations below**

Analyze: Given reactants for chemical equations, determine the products and explain the prediction of whether each reaction is product-favored or reactant-favored.

Plan: Adapt the plan described in the solutions to Question 13, 17, and 44. Compare the reactant acid to the product acid and identify which is stronger and which is weaker. Do the same with the bases. Equilibrium favors the weaker species in the reaction.

Execute:

(a) $H_2O(\ell)$ + $HNO_3(aq)$ $\rightleftharpoons$ $\mathbf{H_3O^+(aq)}$ + $\mathbf{NO_3^-(aq)}$

stronger base — stronger acid — weaker acid — weaker base

The reaction is product-favored.

(b) $H_3PO_4(aq)$ + $H_2O(\ell)$ ⇌ **$H_3O^+(aq)$** + **$HPO_4^-(aq)$**

weaker acid — weaker base — stronger acid — stronger base

The reaction is reactant-favored.

(c) $CN^-(aq)$ + $HCl(aq)$ ⇌ **HCN(aq)** + **$Cl^-(aq)$**

stronger base — stronger acid — weaker acid — weaker base

The reaction is product-favored.

(d) $NH_4^+(aq)$ + $F^-(aq)$ ⇌ **$NH_3(aq)$** + **HF(aq)**

weaker acid — weaker base — stronger base — stronger acid

The reaction is reactant-favored.

70. *Result:* **(a) less than 7 (b) greater than 7 (c) equal to 7; see explanations below**

Analyze: Given three salts, predict whether their aqueous solutions will have pH less than, greater than or equal to 7.

Plan: Adapt the plan from the solutions to Question 43-45.

Execute:

(a) $AlCl_3(s) \longrightarrow Al^{3+}(aq) + 3\ Cl^-(aq)$

The Cl^- ions do not affect the pH of a water solution, because HCl is a strong acid. However, the hydrated aqueous aluminum ion is formed:

$$Al^{3+} + 6\ H_2O(aq) \longrightarrow [Al(H_2O)_6]^{3+}$$

It is a weak acid, so we predict pH less than 7.

(b) $Na_2S(s) \longrightarrow 2\ Na^+(aq) + S^{2-}(aq)$

The Na^+ ions do not affect the pH of a water solution. S^{2-} is a strong base, so we predict pH greater than 7.

(c) $NaNO_3(s) \longrightarrow Na^+(aq) + NO_3^-(aq)$

Neither of these ions affects the pH of a water solution, because NaOH is a strong base and HNO_3 is a strong acid, so we predict pH equal to 7.

73. *Result:* **(a), (b), (c), (d), and (e) have greater solubility at pH 2 than at pH 7**

Analyze: Given the chemical formulas of five substances, determine which among them will have greater solubility at pH = 2 than at pH = 7.

Plan: All of these solids have the same cation, whose aqueous form, $Cu(H_2O)_6^{2+}$, is an acid with $K_a = 1.6 \times 10^{-7}$. The presence of extra H_3O^+ will not affect the cation concentration from one solution to the next. Therefore, we will compare the strength of the anionic bases to judge variations in the solubility in acid solutions. The solid will be more soluble at pH 2 than pH 7 if its anion reacts with H_3O^+ to form a weak conjugate acid.

Execute:

(a) $Cu(OH)_2(s) \longrightarrow Cu^{2+}(aq) + 2\ OH^-(aq)$

The conjugate of OH^- is H_2O, a very weak acid. This salt would be more soluble at pH 2.

(b) $CuSO_4(s) \longrightarrow Cu^{2+}(aq) + SO_4^{2-}(aq)$

The conjugate of SO_4^{2-} is HSO_4^-, a slightly weak acid. This salt would be slightly more soluble at pH 2.

(c) $CuCO_3(s) \longrightarrow Cu^{2+}(aq) + CO_3^{2-}(aq)$

The conjugate of CO_3^{2-} is HCO_3^-, a weak acid. This salt would be more soluble at pH 2.

(d) $CuS(s) \longrightarrow Cu^{2+}(aq) + S^{2-}(aq)$

The conjugate of S^{2-} is HS^-, a weak acid. This salt would be more soluble at pH 2.

(e) $CuS(s) \longrightarrow 3\ Cu^{2+}(aq) + 2\ PO_4^{3-}(aq)$

The conjugate of PO_4^{3-} is HPO_4^{2-}, a weak acid. This salt would be more soluble at pH 2.

In conclusion, all of these salts would be more soluble at pH 2 than at pH 7.

Lewis Acids and Bases (Section 14-9)

74. *Result:* **See equation below**

Analyze: Given the formula of a compound and the fact that it acts as an acid, write a chemical equation that illustrates acid ionization

Plan: The only hydrogen atoms in the formula are the ones on the water molecules. Write an wuation snowing the transfer of one H^+ from the compound to water.

Execute: $\mathbf{[Ni(H_2O)_6]^{2+} + H_2O(\ell) \longrightarrow [Ni(H_2O)_5(OH)]^+ + H_3O^+(aq)}$

76. *Result:* **Lewis acid: (b); Lewis bases: (a), (b), and (c)**

Analyze: Given three chemical formulas, determine which is a Lewis acid and which is a Lewis base.

Plan: The Lewis model focuses on electron pairs. The substance capable of donating the electron pair to form a new bond is called a Lewis base. The substance capable of accepting an electron pair is a Lewis acid.

Execute:

(a) O^{2-} has a lone pair of electrons that can form a new bond, so O^{2-} can be a Lewis base. It cannot accept any more electrons, so it is not a Lewis acid.

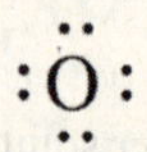

(b) CO_2 has a lone pair of electrons on the O atoms that can form a new bond, so CO_2 can be a Lewis base. Its central C atom can interact with lone pairs on other Lewis bases, so CO_2 can also be a Lewis acid.

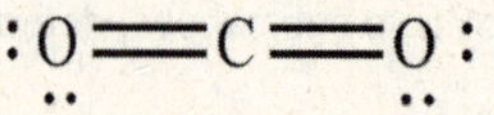

(c) H^- has a lone pair of electrons that can form a new bond, so H^- can be a Lewis base. It cannot accept any more electrons, so it is not a Lewis acid.

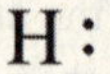

78. *Result:* **Lewis acids: (a), (b) and (c); Lewis base: (b)**

Analyze: Given three chemical formulas, determine which is a Lewis acid and which is a Lewis base.

Plan: Follow the plan from the solution to Question 76.

Execute:

(a) Cr^{3+} can interact with lone pairs on Lewis bases, so Cr^{3+} can also be a Lewis acid. It has no valence electrons that can form a new bond, so it cannot be a Lewis base.

(b) The S atoms in SO_3 could interact with lone pairs on other Lewis bases, so SO_3 can also be a Lewis acid.

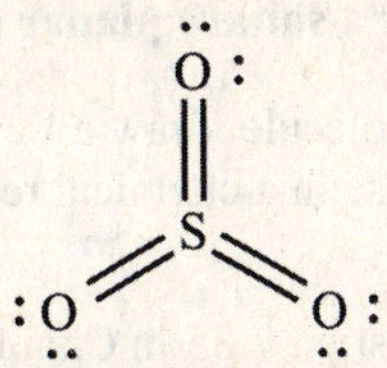

(c) CH_3NH_2 has a lone pair of electrons on N that can form a new bond, so CH_3NH_2 can be a Lewis base. The H atoms polar-covalently bonded to N could interact with other Lewis bases and be removed, so it could function as a Lewis acid; however, that reaction requires a very strong Lewis base.

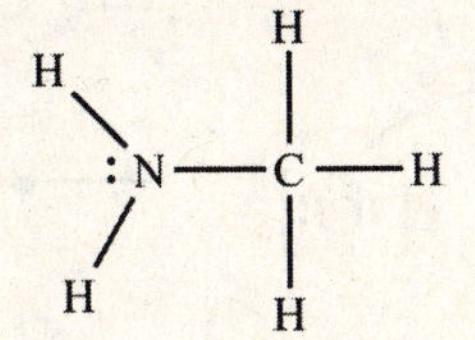

80. *Result:* **(a) Lewis acid: SO_2; Lewis base: H_2O (b) Lewis acid: H_3BO_3; Lewis base: OH^- (c) Lewis acid: Cu^{2+}; Lewis base: NH_3 (d) Lewis acid: Sn^{2+}; Lewis base: Cl^-**

Analyze: Given chemical equations, identify the Lewis acid and the Lewis base in each.

Plan: Draw Lewis structures and identify the reactant donating the electrons (the Lewis base) and the reactant accepting them (the Lewis acid) to make a new bond.

Execute:

(a) The curved arrow shows how the lone pair on O becomes a new bond between O atom and S atom. (Notice that after the initial Lewis acid-base reaction, one of the H atoms migrates to a nearby O atom to balance the positive and negative charges.)

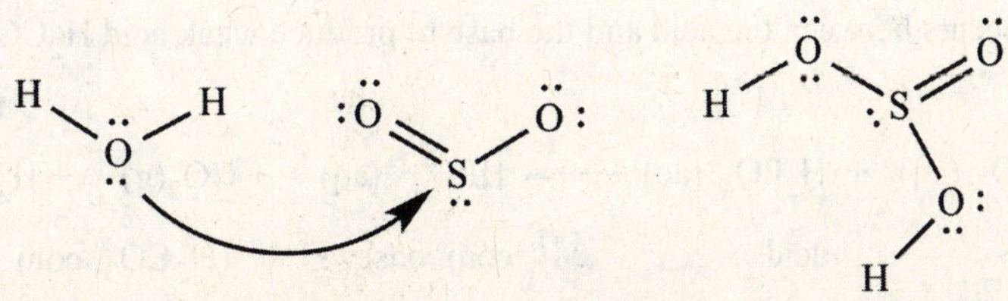

Therefore, SO_2 is the Lewis acid and H_2O is the Lewis base.

(b) The lone pair on OH^- becomes a new bond between the O atom and B atom.

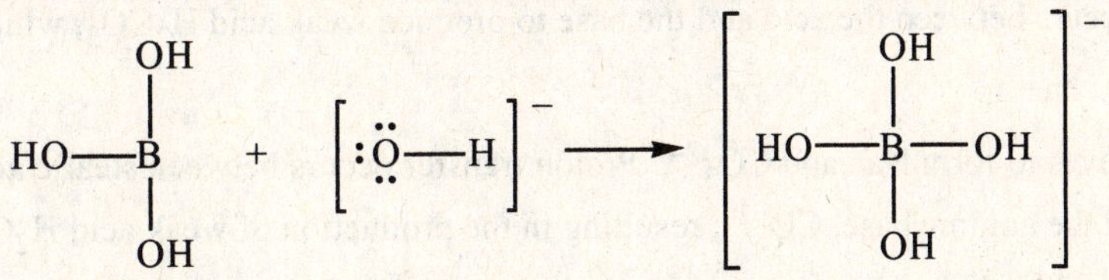

Therefore, H_3BO_3 is the Lewis acid and OH^- is the Lewis base.

(c) Cu^{2+} ions have no valence electrons. The lone pair electrons on the NH_3 molecules make new bonds with Cu^{2+}, so Cu^{2+} is the Lewis acid and NH_3 is the Lewis base.

(d) As in (c), the Sn^{2+} metal ion accepts electrons from the Cl^- ions to form new bonds between Sn and Cl atoms, so Sn^{2+} is the Lewis acid and Cl^- is the Lewis base.

83. *Result:* **See Lewis structure below; ICl_3 is T-shaped; ICl_3 functions as the Lewis acid to form ICl_4^-; see below for Lewis structure of ICl_4^-; it has a square planar geometry.**

Analyze: Given a chemical formula of a molecule, draw a Lewis structures and predict the shape, and identify if it functions as an Lewis acid or Lewis base in a chemical reaction with an ion. Determine the structure of the product.

Plan and Execute: See the solution to Question 7.85 in Chapter 7 for the shape determinations. The ICl_3 molecule has AB_3E_2 form, so the shape is T-shaped. The curved arrow shows how the lone pair on Cl^- becomes a new bond between the Cl atom and I atom. Therefore, ICl_3 is the Lewis acid in the reaction with chloride ion. The ICl_4^- ion has AB_4E_2 form, so the shape is square planar.

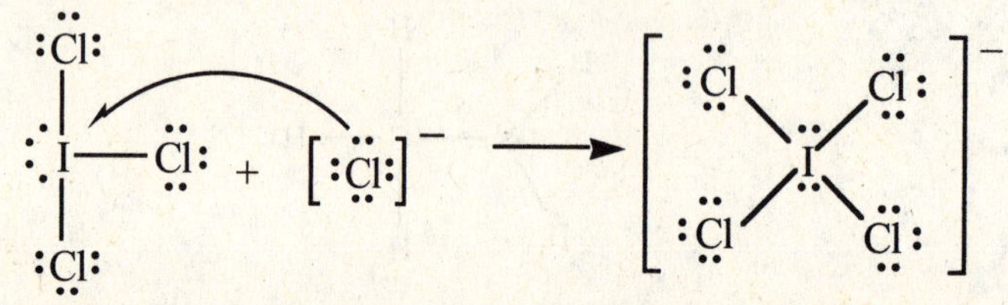

84. *Result/ Explanation:* Conjugates in an acid-base pair must differ by only one H^+ ion. The conjugate acid of the $[Zn(OH)_4]^{2-}$ ion will have one more H^+ ion, so attached one H^+ to one of the OH^- ions oxygen atoms:

$$[Zn(H_2O)(OH)_3]^-$$

Additional Applied Acid-Base Chemistry (Section 14-10)

86. *Result:* **$HCO_3^-(aq) + H_2PO_4^-(aq) \longrightarrow HPO_4^{2-}(aq) + CO_2(g) + H_2O(\ell)$; base: HCO_3^-; acid: $H_2PO_4^-$**

Analyze: Given a description of reactants and products, write a balanced chemical equation and determine which anion is the acid and which is the base.

Plan: Proton transfer occurs between the acid and the base to produce weak acid H_2CO_3, which decomposes into $CO_2(g)$ and water.

Execute:

	$HCO_3^-(aq)$	$+ H_2PO_4^-(aq) \longrightarrow$	$HPO_4^{2-}(aq)$	$+ CO_2(g)$	$+ H_2O(\ell)$
	base	acid	conj. base	(H_2CO_3, conj. acid)	

87. *Result:* **$Na_2CO_3 + 2\ CH_3(CH_2)_{16}COOH \longrightarrow 2\ CH_3(CH_2)_{16}COONa + H_2O + CO_2$**

Analyze: Given a description of reactants and products, write a balanced chemical equation.

Plan: Proton transfer occurs between the acid and the base to produce weak acid H_2CO_3, which decomposes into $CO_2(g)$ and water.

Execute: Na_2CO_3 dissolves to form Na^+ and CO_3^{2-}. Proton transfer occurs between stearic acid, $CH_3(CH_2)_{16}COOH$, and the anionic base, CO_3^{2-}, resulting in the production of weak acid H_2CO_3, which decomposes into $CO_2(g)$ and water.

89. *Result/Explanation:* Dishwasher detergent is very basic, and should not be used to wash anything by hand, including a car. If it gets into the engine area, it can also dissolve automobile grease and oil, which could prevent the engine from running correctly.

91. *Result/Explanation:* Lemon juice contains citric acid. The acid protonates the basic amines. The conjugate acids formed from the neutralized bases are ions and not volatile.

General Questions

92. *Result:* **(a) CH_3COOH is a weak acid. (b) Na_2O, contains O^{2-}, which is a strong base. (c) H_2SO_4 is a strong acid. (d) NH_3 is a weak base. (e) $Ba(OH)_2$, contains OH^-, which is a strong base. (f) $H_2PO_4^-$ is amphiprotic.**

Analyze: Given several chemical formulas, classify them as stong acid, weak acid, stong base, weak base, amphiprotic or neigher acid nor base.

Plan: The term "amphiprotic" describes a species that can function both as a Bronsted base and as a Bronsted acid. In general, any species with an H can donate H^+ ions, though sometimes that will not happen in the common water solvent. In addition, any species with one or more lone pairs can be a base, since the electrons could make a bond to a proton. However, because this question asks for a judgment of "weak" and "strong," we will restrict our designations to the reactions of these species in water and use Table 14.2 to assist us. Notice that Na_2O contains O^{2-}.

(a) CH_3COOH is a weak acid.

(b) Na_2O, contains O^{2-}, which is a strong base.

(c) H_2SO_4 is a strong acid.

(d) NH_3 is a weak base.

(e) $Ba(OH)_2$, contains OH^-, which is a strong base.

(f) $H_2PO_4^-$ is amphiprotic.

95. *Result:* **(a) less than 7 (b) equal to 7 (c) greater than 7**

Explanation: If equal molar amounts of an acid and a base are combined, the stronger of them will dictate the pH of the solution.

(a) A weak base and a strong acid will have an acidic pH, less than 7.

(b) A strong base and a strong acid will have a neutral pH, equal to 7.

(c) A strong base and a weak acid will have a basic pH, greater than 7.

96. *Result:* **2.85**

Analyze: Adapt the plan described in the solution to Questions 64.

$$\frac{5.0 \text{ mg } C_6H_8O_6}{1 \text{ mL soln}} \times \frac{1 \text{ g}}{1000 \text{ mg}} \times \frac{1 \text{ mol } C_6H_8O_6}{176 \text{ g } C_6H_8O_6} \times \frac{1000 \text{ mL}}{1 \text{ L}} = 0.028 \text{ M}$$

Ascorbic acid, $C_6H_8O_6$, is diprotic acid, so give it a formula of H_2A. The equation for the equilibrium and the equilibrium expression are:

$$H_2A(aq) + H_2O(\ell) \rightleftharpoons H_3O^+(aq) + HA^-(aq) \qquad K_a = \frac{[H_3O^+][HA^-]}{[H_2A]}$$

We are told to assume only the first ionization is important to determining the pH.

As the reactants decompose, the concentrations of the products increase stoichiometrically, until they reach equilibrium concentrations. Set up a reaction (ICE) table:

	$H_2A(aq)$	$H_3O^+(aq)$	$HA^-(aq)$
initial conc. (M)	0.028	1.0×10^{-7}*	0
change as reaction occurs (M)	– x	+ x	+ x
equilibrium conc. (M)	0.028 – x	x	x

* 1.0×10^{-7} M H_3O^+ from autoionization of pure water. We assume it is small compared to the change.

At equilibrium $$K_a = \frac{(x)(x)}{(0.028 - x)} = 7.9 \times 10^{-5}$$

$$x^2 = (7.9 \times 10^{-5})(0.028 - x)$$

If we assume that x is small, and $0.028 - x \cong 0.028$

$$x^2 = (7.9 \times 10^{-5})(0.028)$$

$$x = 1.5 \times 10^{-3}\ M = [H_3O^+]$$

$$pH = -\log[H_3O^+] = -\log(1.5 \times 10^{-3}) = 2.82$$

If we use the quadratic equation: (see Appendix A, Section A.7, page A.13)

$$x^2 + 7.9 \times 10^{-5}x - 2.2 \times 10^{-6} = 0$$

$$x = 1.4 \times 10^{-3}\ M = [H_3O^+]$$

$$pH = -\log[H_3O^+] = -\log(1.4 \times 10^{-3}) = 2.85$$

97. *Result:* **(a) decreases (b) increases (c) stays the same**

Analyze: Given the description of adding a solid compound to an aqueous solution, determine if the pH increases, decreases or stays the same.

Plan: Adding acids to bases decreases the pH. Adding bases to acids increases the pH. Adding weak acids or bases to strong acids or bases, respectively, does not change the pH significantly.

Execute:

(a) NH_4Cl contains NH_4^+, a weak acid. The acid is added to a base, NH_3, so the pH decreases.

(b) CH_3COONa contains CH_3COO^-. The base is added to an acid, CH_3COOH, so the pH increases.

(c) Adding a neutral salt, NaCl, will not change the pH of any solution, so the pH stays the same.

100. *Result:* **9.59**

Analyze: Given the concentration of an aqueous compound that contains a base and its K_b, calculate the pH of the solution.

Plan: Adapt the plan from the solution to Question 60.

Execute: The solution resulting from adding NaOCl to water is basic because the OCl^- anion reacts with water according to the equation below:

$$OCl^-(aq) + H_2O(\ell) \rightleftharpoons HOCl(aq) + OH^-(aq) \qquad K_b = \frac{[HOCl][OH^-]}{[OCl^-]}$$

As the reactants decompose, the concentrations of the products increase stoichiometrically, until they reach equilibrium concentrations. Set up a reaction (ICE) table:

	$OCl^-(aq)$	$HOCl(aq)$	$OH^-(aq)$
initial conc. (M)	0.010	0	1.0×10^{-7}*
change as reaction occurs (M)	– x	+ x	+ x
equilibrium conc. (M)	0.010 – x	x	x

* 1.0×10^{-7} M OH^- from autoionization of pure water. We assume it is small compared to the change.

At equilibrium $$K_b = \frac{(x)(x)}{(0.010 - x)} = 1.5 \times 10^{-7}$$

Assume x is very small and: $0.010 - x \cong 0.010$.

$$x^2 = (1.5 \times 10^{-7})(0.010)$$

$$x = 3.9 \times 10^{-5}\ M = [OH^-]$$

$$pOH = -\log[OH^-] = -\log(3.9 \times 10^{-5}) = 4.41$$

$$pH = 14.00 - pOH = 14.00 - 4.41 = 9.59$$

Applying Concepts

101. *Result:* **See equation below**

Analyze: Given the names of reactants in a reaction and the fact that a gas forms, write balanced chemical equations for the reactions that created the gas and explain the role of water.

Plan: We know that the neutralization of carbonate produces carbon dioxide. We also know that some hydrated metals are amphiprotic. Let the hydrate of aluminum be the source of acid for the neutralization of hydrogen carbonate.

Execute: Aluminum sulfate is soluble, producing Al^{3+} and sulfate ions.

$$Al^{3+}(aq) + 6\ H_2O(\ell) \rightleftharpoons Al(H_2O)_6^{\ 3+}(aq)$$

$$Al(H_2O)_6^{\ 3+}(aq) + H_2O(\ell) \rightleftharpoons Al(OH)_5(OH)^{2+}(aq) + H_3O^+(aq)$$

$$H_3O^+(aq) + HCO_3^-(aq) \rightleftharpoons CO_2(g) + 2\ H_2O(\ell)$$

The role of **water is to hydrate the aluminum ion,** permitting it to serve as the source of acidity needed to activate the production of carbon dioxide gase from the neutralization of hydrogen carbonate ion

106. *Result:* **7.27, 6.998, 6.631; all three solutions are neutral.**

Analyze: Using a table of K_w values at different temperatures, calculate what the pH will be at three different temperatures and classify the water at these temperatures as either acidic, neutral, or basic.

Plan and Execute: Table 14.1 provides the ionization constant for water at various temperatures. The equation representing the autoionization of water and the expression for its equilibrium constant are:

$$2\ H_2O(\ell) \rightleftharpoons H_3O^+(aq) + OH^-(aq) \qquad K_w = [H_3O^+][OH^-]$$

The concentration of products increase stoichiometrically, until they reach equilibrium concentrations. Set up a reaction (ICE) table:

	$H_3O^+(aq)$	$OH^-(aq)$
initial conc. (M)	0	0
change as reaction occurs (M)	+ x	+ x
equilibrium conc. (M)	x	x

So, at equilibrium $K_w = x^2$

At 10 °C $x = \sqrt{0.29 \times 10^{-14}} = 5.4 \times 10^{-8}\ M = [H_3O^+]$

$$pH = -\log[H_3O^+] = -\log(5.4 \times 10^{-8})$$

$$pH = 7.27$$

At 25 °C $x = \sqrt{1.01 \times 10^{-14}} = 1.00 \times 10^{-7}\ M = [H_3O^+]$

$$pH = -\log[H_3O^+] = -\log(1.00 \times 10^{-7})$$

$$pH = 6.998$$

At 50 °C $x = \sqrt{5.48 \times 10^{-14}} = 2.34 \times 10^{-7}\ M = [H_3O^+]$

$$pH = -\log[H_3O^+] = -\log(2.34 \times 10^{-7})$$

$$pH = 6.631$$

These solutions are all still neutral, since $[H_3O^+] = [OH^-]$ in each of them.

108. *Result:* **(a) $H_2O > H_3O^+ = Cl^- >> OH^-$ (b) $H_2O > Na^+ = ClO_4^- >> H_3O^+ = OH^-$ (c) $H_2O > HNO_2 > H_3O^+ = NO_2^- >> OH^-$ (d) $H_2O > Na^+ \cong ClO^- > OH^- = HClO >> H_3O^+$ (e) $H_2O > NH_4^+ \cong Cl^- > H_3O^+ = NH_3 >> OH^-$ (f) $H_2O > Na^+ = OH^- >> H_3O^+$**

Analyze: Given several chemical formulas, predict what ions and molecules will be present and list them in order of decreasing concentration.

Plan: Determine if the compound is a soluble ionic compound, acid or base. If it is an acid or base, determine if it is weak or struong. If the compound is a soluble ionic compound, a strong acid, or strong base, the compound ionizes making aqueous ions. If the compound is an acid, the concentration of H_3O^+ will be much larger than the concentration of OH≂. If the compound is a base, the concentration of OH^- will be much larger than the concentration of H_3O^+. If the substance is a soluble ionic compond that does not react as an acid or a base, then the concentrations of H_3O^+ and OH^- will be equal and small. Water has a concentration in dilute solutions of 55.5 M, so it is presumed to have the highest concentration in the aqueous solution.

Execute:

(a) HCl is a strong acid, resulting in a solution of H_3O^+ and Cl^-. The solution contains the following molecules and ions in order of decreasing concentrations:

$$H_2O > H_3O^+ = Cl^- >> OH^-$$

(b) $NaClO_4$ is a soluble ionic compound, resulting in a solution of Na^+ and ClO_4^-. Neither of these ions reacts with water. The solution contains the following molecules and ions in order of decreasing concentrations:

$$H_2O > Na^+ = ClO_4^- >> H_3O^+ = OH^-$$

(c) HNO_2 is a weak acid, resulting in a solution of a large proportion of HNO_2 and a small proportion of H_3O^+ and NO_2^-. The solution contains the following molecules and ions in order of decreasing concentrations:

$$H_2O > HNO_2 > H_3O^+ = NO_2^- >> OH^-$$

(d) NaClO is a soluble ionic compound, resulting in a solution of Na^+ and ClO^-. The anion is a weak base and reacts with water to a small extent to make a small proportion of HClO and OH^-. The solution contains the following molecules and ions in order of decreasing concentrations:

$$H_2O > Na^+ \cong ClO^- > OH^- = HClO >> H_3O^+$$

(e) NH_4Cl is a soluble ionic compound, resulting in a solution of NH_4^+ and Cl^-. The cation is a weak acid and reacts with water to a small extent to make a small proportion of NH_3 and H_3O^+. The solution contains the following molecules and ions in order of decreasing concentrations:

$$H_2O > NH_4^+ \cong Cl^- > H_3O^+ = NH_3 >> OH^-$$

(f) NaOH is a strong base, resulting in a solution of Na^+ and OH^-. The solution contains the following molecules and ions in order of decreasing concentrations:

$$H_2O > Na^+ = OH^- >> H_3O^+$$

110. *Result/Explanation:* Conjugates in an acid-base pair must differ by only one H^+ ion. The acids and bases were identified correctly, but the conjugates were not. HCO_3^- is the conjugate of H_2CO_3 and HSO_4^- is the conjugate of SO_4^{2-}.

112. *Result:* **(a) 2.01 (b) 23 times**

Analyze: Given the formula of an acid and its adic dissocuation constant, calculate the pH of the solution and determine how many times greater the H_3O^+ concentation is than a given concentration of another acid.

Plan: Adapt the plans described in Problem-Solving Example 14.9 as shown in the solution to Question 54.

(a) The equation for the equilibrium and the equilibrium expression are:

$$CCl_3COOH(aq) + H_2O(\ell) \rightleftharpoons H_3O^+(aq) + CCl_3COO^-(aq) \qquad K_a = \frac{[H_3O^+][CCl_3COO^-]}{[CCl_3COOH]}$$

The concentrations of products increase stoichiometrically, until equilibrium is reached. Set up a reaction (ICE) table:

	$CCl_3COOH(aq)$	$H_3O^+(aq)$	$CCl_3COO^-(aq)$
initial conc. (M)	0.010	1.0×10^{-7}*	0
change as reaction occurs (M)	– x	+ x	+ x
equilibrium conc. (M)	0.010 – x	x	x

* 1.0×10^{-7} M H_3O^+ from autoionization of pure water. We assume it is small compared to the change.

At equilibrium $$K_a = \frac{(x)(x)}{(0.010 - x)} = 3.0 \times 10^{-1}$$

$$x^2 = (3.0 \times 10^{-1})(0.010 - x)$$

$$x^2 + 3.0 \times 10^{-1}x - 3.0 \times 10^{-5} = 0$$

Use the quadratic equation: (see Appendix A, Section A.7, page A.13)

$$x = 9.7 \times 10^{-3}\ M = [H_3O^+]$$

$$pH = -\log[H_3O^+] = -\log(9.7 \times 10^{-3}) = 2.01$$

(b) The equation for the equilibrium and the equilibrium expression are:

$$CH_3COOH(aq) + H_2O(\ell) \rightleftharpoons H_3O^+(aq) + CH_3COO^-(aq) \qquad K_a = \frac{[H_3O^+][CH_3COO^-]}{[CH_3COOH]}$$

The concentrations of products increase stoichiometrically, until equilibrium is reached. Set up a reaction (ICE) table:

	$CH_3COOH(aq)$	$H_3O^+(aq)$	$CH_3COO^-(aq)$
initial conc. (M)	0.010	1.0×10^{-7}*	0
change as reaction occurs (M)	– x	+ x	+ x
equilibrium conc. (M)	0.010 – x	x	x

* 1.0×10^{-7} M H_3O^+ from autoionization of pure water. We assume it is small compared to the change.

At equilibrium $$K_a = \frac{(x)(x)}{(0.010 - x)} = 1.8 \times 10^{-5}$$

Assume x is very small and: $0.20 - x \cong 0.20$.

$$1.8 \times 10^{-5} = \frac{x^2}{(0.010)}$$

$$x^2 = (1.8 \times 10^{-5})(0.010)$$

$$x = 4.2 \times 10^{-4}\text{ M} = [H_3O^+]$$

To determine the number of times more hydronium ions in the trichloroacetic acid versus the acetic acid, divide the hydronium ion concentration in the trichloroacetic acid solution by that in the acetic acid solution.

$$\frac{9.7 \times 10^{-3}\text{ M}}{4.2 \times 10^{-4}\text{ M}} = 23 \text{ times more } H_3O^+ \text{ in } CCl_3COOH(aq)$$

115. *Result:* **(a) HY; see explanation below (b) Z^-; see explanation below**

Analyze: Given a series of illustrations of nanoscale representations of aqueous solutions and three unknown acids, determine which acid had the largest K_a and which conjugate base has the largest K_b value.

Plan and Execute:

(a) For each box, count the number of each species in the box and dividing by the box volume, V, to get the concentration. Then calculate the K_a value by putting the concentrations into the form of the reaction quotient. (A = X, Y, or Z)

$$HA(aq) + H_2O(\ell) \rightleftharpoons H_3O^+(aq) + A^-(aq) \qquad K_a = \frac{[H_3O^+][A^-]}{[HA]}$$

In the first box, there are eight HX molecules, four X^-, and four H_3O^+:

$$K_a = \frac{\left(\frac{4}{V}\right)\left(\frac{4}{V}\right)}{\left(\frac{8}{V}\right)} = \frac{2}{V}$$

In the second box, there are six HY molecules, six Y^-, and six H_3O^+:

$$K_a = \frac{\left(\frac{6}{V}\right)\left(\frac{6}{V}\right)}{\left(\frac{6}{V}\right)} = \frac{6}{V}$$

In the third box, there are six HZ molecules, two Z^-, and two H_3O^+:

$$K_a = \frac{\left(\frac{2}{V}\right)\left(\frac{2}{V}\right)}{\left(\frac{6}{V}\right)} = \frac{4}{6V}$$

Assuming the boxes are all the same volume, **HY** has the largest K_a value.

(b) The weaker the acid, the stronger the conjugate base. The acid with the smallest K_a value, HZ, is the weakest of these three acid, so its conjugate base, Z^-, is the strongest conjugate base.

118. *Result:* **(a) Boxes C, F, and G (b) Box I (c) Boxes A and H (d) Boxes D and E (e) Box B (f) Box I (g) Boxes D and E**

Analyze: Give a series of lettered boxes and a series of items, match the items to the correct lettered box.

Plan and Execute:

(a) The pH > 7.0 in basic solutions, so look for solutions with $[OH^-] > 10^{-7}$ M or $[H_3O^+] < 10^{-7}$ M. There are four boxes with concentrations given, Boxes C, F, G, and I; examine each one.

Box C has $[OH^-]$ 10^{-5} M > 10^{-7} M

Box F has $[OH^-] = 0.10$ M > 10^{-7} M

Box G has $[H_3O^+] = 10^{-8}$ M < 10^{-7} M

Box I has $[H_3O^+] = 0.01$ M > 10^{-7} M

The boxes that answer this item are **Boxes C, F, and G**.

(b) $[H_3O^+] > 1.0 \times 10^{-7}$ M in acidic solutions, so look for ones with $[OH^-] < 10^{-7}$ M or $[H_3O^+] > 10^{-7}$ M. There are four boxes with concentrations given, Boxes C, F, G and I; examine each one as done in (a).

The only box that answers this item is **Box I**.

(c) A basic salt is a salt that contains an anion that is the conjugate acid of a weak acid. There are two boxes that contain salts, Boxes A and H; examine each one.

Box A contains a salt with the anion, CO_3^{2-}. The acid H_2CO_3 is a weak acid

Box H contains a salt with the anion, NO_2^-. The acid HNO_2 is a weak acid

The boxes that answer this item are **Boxes A and H**.

(d) The item is limited to the species in the grid that can function as a base and as an acid. Those are found in Boxes D and E; examine each one.

Box D, H_2SO_3, is a fairly strong weak acid. The compound does still have lone pairs of electrons on the O atoms, so it can also be a base, but it would be a very weak base. $K_a > K_b$

Box E, $H_2PO_4^-$, contains an acid anion that can be found in Table 14.2. $K_a = 10^{-8}$ and $K_b = 10^{-12}$, so $K_a > K_b$

(Note: Boxes C, F, G, and I also contain acids or bases that could be considered, but because they are given with specific concentration values, they will not be considered in the answer to this question. If they were, the acids in Boxes G and I, H_3O^+ and HNO_2, would also qualify as answers to this question.)

The boxes that answer this item are **Boxes D and E**.

(e) A Lewis acid is an electron-pair acceptor. The metallic cation Cu^{2+} ion can function as a Lewis acid, by accepting electron pairs. The answer is **Box B**.

(f) A solution with a pH = 2.00, has $[H_3O^+] = 10^{-2.00}$ M = 0.01 M. Because $HClO_4$ is a strong acid, the solution identified in Box I has the right concentration for pH 2.00. The answer is **Box I**.

(g) A polyprotic acid has more than one acidic hydrogen atom in its formula. There are two acids listed in the grid, in Boxes D and E.

Box D, H_2SO_3, is a diprotic acid.

Box E, $H_2PO_4^-$, is a diprotic acid.

(Note: Boxes G and I also contain acids that could be considered, but because they are given with specific concentration values, they will not be considered in the answer to this question. If they were, the acids in Boxes G, triprotic H_3O^+, would also qualify as an answer to this question.)

The boxes that answer this item are **Boxes D and E**.

More Challenging Questions

121. *Result:* (a) 10.46 (b) 10.45 (c) 10.45

Analyze: Given the initial concentration of an aqueous compound that contains a conjungate acid cation and a conjugate base anion, calculate the pH of the solution using the small change approximation, the quadratic equation, and successive approximations.

Plan: Adapt the plan from the solution to Question 100.

Execute: The solution contains the cation NH_4^+ and the anion $[B(OH)_4]^-$. Table 14.2 gives

K_a for NH_4^+ to be 5.6×10^{-10} and K_b for $B(OH)_4^-$ to be 1.7×10^{-5}. Because the base has a much larger K_b value, the solution is basic, because $B(OH)_4^-$ reacts with water according to the equation below:

$$B(OH)_4^-(aq) + H_2O(\ell) \rightleftharpoons B(OH)_3H_2O(aq) + OH^-(aq) \qquad K_b = \frac{[B(OH)_3(H_2O)][OH^-]}{[B(OH)_3^-]}$$

As the reactants decompose, the concentrations of the products increase stoichiometrically, until they reach equilibrium concentrations. Set up a reaction (ICE) table:

	$B(OH)_4^-(aq)$	$B(OH)_3H_2O(aq)$	$OH^-(aq)$
initial conc. (M)	0.00500	0	1.0×10^{-7}*
change as reaction occurs (M)	– x	+ x	+ x
equilibrium conc. (M)	0.00500 – x	x	x

* 1.0×10^{-7} M OH^- from autoionization of pure water. We assume it is small compared to the change.

At equilibrium $$K_b = \frac{(x)(x)}{(0.00500 - x)} = 1.7 \times 10^{-5}$$

(a) Assume x is very small and: $0.00500 - x \cong 0.00500$

$$x^2 = (1.7 \times 10^{-5})(0.00500)$$

$$x = 2.9 \times 10^{-4}\text{ M} = [OH^-]$$

$$pOH = -\log[OH^-] = -\log(2.9 \times 10^{-4}) = 3.54$$

$$pH = 14.00 - pOH = 14.00 - 3.54 = 10.46$$

(a) Algebraically rearrange the equation to the form of ax + bx + c = 0

$$x^2 = (1.7 \times 10^{-5})(0.00500 - x)$$

$$x^2 = (1.7 \times 10^{-5})(0.00500) - (1.7 \times 10^{-5})x$$

$$x^2 + (1.7 \times 10^{-5})x - 8.5 \times 10^{-8} = 0$$

Use the quadratic equation: (see Appendix A, Section A.7, page A.13)

$$x = 2.8 \times 10^{-4}$$

$$pOH = -\log[OH^-] = -\log(2.8 \times 10^{-4}) = 3.55$$

$$pH = 14.00 - pOH = 14.00 - 3.55 = 10.45$$

(c) Use the method of successive approximations explained in Appendix A, Section A.7, page A.13-A.14

Use the answer from part (a) to refine $0.00500 - x = 0.00500 - 2.9 \times 10^{-4}\text{ M} = 0.00471$ M

Then use 0.00471 in place of 0.00500 in the equation $x^2 = (1.7 \times 10^{-5})(0.00500)$ and solve for x again:

$$x^2 = (1.7 \times 10^{-5})(0.00471)$$

$$x = 2.8 \times 10^{-4}\ M = [OH^-]$$

$$pOH = -\log[OH^-] = -\log(2.8 \times 10^{-4}) = 3.55$$

$$pH = 14.00 - pOH = 14.00 - 3.55 = 10.45$$

124. *Result:* **lactic acid sample**

Explanation: This question involves some interpretation and research. Of the three materials described, two are solutions containing acids: Vinegar contains acetic acid (CH_3COOH, $K_a = 1.8 \times 10^{-5}$) at a concentration of approximately 5% and lemon juice contains triprotic citric acid ($H_3C_5H_5O_7$, Table 14.3) at a concentration of approximately 1%. The third material is lactic acid ($C_3H_6O_3$, $K_a = 1.4 \times 10^{-4}$, from Question 64), a 100% pure molecular acid that is sometimes found in milk.

The ion being neutralized is $HCO_3^-(aq)$.

$$H_3O^+(aq) + HCO_3^-(aq) \rightleftharpoons H_2CO_3(aq) + H_2O(\ell) \qquad K = \frac{1}{K_{a,H_2CO_3}} = \frac{1}{4.3\times10^{-7}} = 2.3 \times 10^6$$

$$HA(aq) + H_2O(\ell) \rightleftharpoons H_3O^+(aq) + A^-(aq) \qquad K_{a,HA}$$

Because lactic acid, acetic acid, and citric acid each presumably have $K_a > 10^{-7}$, the three neutralization reactions we are studying here will all be product-favored. ($K > 1$). So, all we need to do is determine which sample provides the largest number of H_3O^+ ions.

Using the molar mass of lactic acid, we can determine the moles of H_3O^+ ions available if one gram of lactic acid reacts:

$$1\ g\ C_3H_6O_3 \times \frac{1\ mol\ C_3H_6O_3}{90.08\ g\ C_3H_6O_3} \times \frac{1\ mol\ H_3O^+}{1\ mol\,C_3H_6O_3} = 0.01\ mol\ H_3O^+$$

Using the percentages of acids in solution and the molar masses, determine the H_3O^+ ions available in the vinegar and lemon juice samples.

$$1\ g\ vinegar \times \frac{0.05\ g\ CH_3COOH}{1\ g\ vinegar} \times \frac{1\ mol\ CH_3COOH}{60.0518\ g\ CH_3COOH} \times \frac{1\ mol\ H_3O^+}{1\ mol\ CH_3COOH} = 0.0008\ mol\ H_3O^+$$

$$1\ g\ lemon\ juice \times \frac{0.01\ g\ H_3C_5H_3O_7}{1\ g\ lemon\ juice} \times \frac{1\ mol\ H_3C_5H_3O_7}{178.097\ g\ H_3C_5H_3O_7} \times \frac{3\ mol\ H_3O^+}{1\ mol\ H_3C_5H_3O_7} = 0.0002\ mol\ H_3O^+$$

With this analysis, we find that the 1gram sample of pure lactic acid will make more CO_2, even though the citric acid in the lemon juice is the strongest acid.

125. *Result:* **0.76 L; no**

Analyze: Determine the volume of gas produced with a specific amount of a compound and if that volume exceeds the volume of a person's stomach.

Plan and Execute: This problem can be solved by methods described in Chapter 3 and 8. Use the chemical equation in Table 14.8. The product-favored reaction goes essentially to completion in the presence of sufficient acid.

$$2.5\ g\ NaHCO_3 \times \frac{1\ mol\ NaHCO_3}{84.0066\ g\ NaHCO_3} \times \frac{1\ mol\ CO_2(g)}{1\ mol\ NaHCO_3} = 0.030\ mol\ CO_2(g)$$

$$V_{CO_2} = \frac{n_{CO_2}RT}{P} = \frac{(0.030\ mol) \times \left(0.08206 \frac{L\cdot atm}{mol\cdot K}\right) \times (37+273)K}{1\ atm} = 0.76\ L$$

This, by itself, is an insufficient volume of CO_2 to rupture a 1-L stomach, though it would be enough to be uncomfortable. To be able to predict a rupture we'd need to know how much volume the food occupied and whether this added CO_2 would exceed the capacity of his stomach. We don't know what pressure of gas the stomach at maximum volume can take, nor do we know if some other involuntary action such as burping or vomiting would occur to prevent the excess gas from causing damage. One would need to be a gastroenterologist, to accurately answer this part of the question.

126. *Result:* **Yes, pH increases**

Analyze: Given physical changes in a carbonated beverage, determine if the pH changes.

Plan and Execute: When the can is initially pressurized with CO_2, the following chemical equilibrium expression is shifted toward the formation of products. The beverage becomes acidic.

$$CO_2(aq) + 2\ H_2O(\ell) \rightleftharpoons H_3O^+(aq) + HCO_3^-(aq)$$

When the can is opened and warmed, carbon dioxide escapes from the carbonated beverage. The loss of $CO_2(aq)$ shifts the above equilibrium to the left. Hydronium ion concentration is decreased and the solution pH increases (i.e. becomes more basic).

128. *Result/Explanation:* Br has a higher electronegativity than H. The bromine withdraws electron density from the nitrogen atom, reducing the nitrogen's ability to bind the positively charged proton. The K_b value is smaller for $BrNH_2$ than for NH_3. $ClNH_2$ is a weaker base than $BrNH_2$ because Cl is more electronegative than Br.

129. *Result:* **HM > HQ > HZ ; HZ, $K_{a,HZ} = 1 \times 10^{-5}$; HQ, $K_{a,HQ} = 1 \times 10^{-3}$; HM, $K_{a,HM} = 1 \times 10^{-1}$ or larger**

Analyze: Given the pH values of three sodium salt solutions, arrange the conjugate acids in order of decreasing strength.

Plan and Execute: Stronger acids result in weaker conjugate bases. The stronger base has the larger pH. So, the salt solution with the highest pH contains the anion of the weakest acid.

Smallest pH: NaM > NaQ > NaZ :Largest pH

Strongest: HM > HQ > HZ :Weakest

The general equation for the equilibrium and the equilibrium expression are:

$$A^-(aq) + H_2O(\ell) \rightleftharpoons HA(aq) + OH^-(aq) \qquad K_b = \frac{[HA][OH^-]}{[A^-]}$$

(where A = M, Q, or Z)

The initial $[OH^-]$ of neutral water is 1×10^{-7} M. The initial concentration of the base is 0.1 M. The concentrations of products increase stoichiometrically, until they reach equilibrium concentrations. Set up a reaction (ICE) table:

In general:	$A^-(aq)$	$HA(aq)$	$OH^-(aq)$
initial conc. (M)	0.1	0	1×10^{-7}*
change as reaction occurs (M)	– x	+ x	+ x
equilibrium conc. (M)	0.1 – x	x	10^{-7} + x

* 1.0×10^{-7} M OH^- from autoionization of pure water. We assume it is small compared to the change.

At equilibrium $$K_b = \frac{(x)(1\times10^{-7}+x)}{(0.1-x)}$$

$$1 \times 10^{-7} + x = [OH^-] = 10^{-pOH} = 10^{-(14.00-\,pH)} = 10^{(pH-\,14.00)}$$

$$x = 10^{(pH-\,14.00)} - 1 \times 10^{-7}$$

For NaZ, A^- above is Z^- and pH = 9.0, so

$$x = 10^{(9.0-14.00)} - 1 \times 10^{-7} = 1 \times 10^{-5} - 1 \times 10^{-7} = 1 \times 10^{-5}, \text{ and}$$

$$K_b = \frac{(1\times10^{-5})(1\times10^{-7}+1\times10^{-5})}{(0.1-1\times10^{-5})} = 1\times10^{-9}$$

For NaQ, A^- above is Q^- and pH = 8.0, so

$$x = 10^{(8.0-14.00)} - 1 \times 10^{-7} = 1 \times 10^{-6} - 1 \times 10^{-7} = 1 \times 10^{-6}, \text{ and}$$

$$K_b = \frac{(1\times10^{-6})(1\times10^{-7}+1\times10^{-6})}{(0.1-1\times10^{-6})} = 1\times10^{-11}$$

For NaM, A^- above is M^- and pH = 7.0, so $x = 10^{(7.0-14.00)} - 1 \times 10^{-7}$

$x = 1 \times 10^{-7} - 1 \times 10^{-7} = 0$, suggesting that the reaction with water does not go toward products at all. That means K_b of M^- may be so small that the pH of the solution is accounted for exclusively by the ionization of water. If indeed the base does provide all the OH^- ions to make the solution's pH: $[HM] = [OH^-] = 10^{-7.0} = 1 \times 10^{-7}$:

$$K_b = \frac{(1\times10^{-7})(1\times10^{-7})}{(0.1)} = 1\times10^{-13}$$

A relationship between K_a and K_b is described in Section 14-7a.

For HZ, $K_a = K_w/K_b = (1.0 \times 10^{-14})/(1 \times 10^{-9}) = 1 \times 10^{-5}$

For HQ, $K_a = K_w/K_b = (1.0 \times 10^{-14})/(1 \times 10^{-11}) = 1 \times 10^{-3}$

For HM, $K_a = K_w/K_b = (1.0 \times 10^{-14})/(1 \times 10^{-13}) = 1 \times 10^{-1}$ or larger

130. *Result:* **(a) weak (b) weak (c) conjugate acid-base pair (d) 6.29**

Analyze: Given the K_a and K_b for a compound, determine if it is a strong or weak acid and if its conjugate base is a strong or weak base. Identify the relationship between the acid and the conjugate base. Given the concentration of a solution of the compound, calculate the pH.

Plan and Execute:

(a) The acid ionization constant for hydrogen is very small. H_2O_2 is a weak acid.

(b) K_b for OOH^- is larger than K_a, but it is still small. OOH^- is a weak base.

(c) H_2O_2 and OOH^- are a conjugate acid and base.

(d) Hydrogen peroxide will ionize in solution according to the following equation with the following equilibrium expression:

$$HOOH(aq) + H_2O(\ell) \rightleftharpoons H_3O^+(aq) + HOO^-(aq) \qquad K_b = \frac{[H_3O^+][HOO^-]}{[HOOH]}$$

Set up a reaction (ICE) table:

	HOOH	$H_3O^+(aq)$	$HOO^-(aq)$
initial conc. (M)	0.100	1.0×10^{-7}	0
change as reaction occurs (M)	– x	+ x	+ x
equilibrium conc. (M)	0.100 – x	1.0×10^{-7} + x	x

* 1.0×10^{-7} M H_3O^+ from autoionization of pure water. We assume it is small compared to the change.

$$K_b = \frac{(1.0 \times 10^{-7} + x)(x)}{(0.100 - x)} = 2.1 \times 10^{-12}$$

$$(1.0 \times 10^{-7} + x)x = (2.1 \times 10^{-12})(0.100 - x)$$

If we assume that x is small, and $0.100 - x \cong 0.100$

$$(1.0 \times 10^{-7} + x)x = (2.1 \times 10^{-12})(0.100)$$

$$x^2 + x(1.0 \times 10^{-7}) - 2.1 \times 10^{-13} = 0$$

Use the quadratic equation: (see Appendix A, Section A.7, page A.13)

$$x = 4.1 \times 10^{-7}$$

$$[H_3O^+] = 1.0 \times 10^{-7} + 4.1 \times 10^{-7} = 5.1 \times 10^{-7}\text{ M}$$

$$pH = -\log[H_3O^+] = -\log(5.1 \times 10^{-7}) = 6.29$$

133. *Result:* **(a) $H_2NCH_2CH_2CH_2CH_2CH_2NH_2$ (b) $^+H_3NCH_2CH_2CH_2CH_2CH_2NH_3^+$ (c) 9.13**

Analyze: Given a structural formula, determine the formula of the molecule and the protonated form. Given some pK_as, determine which one it applies to.

Plan and Execute:

(a) The formula for cadaverine is $H_2NCH_2CH_2CH_2CH_2CH_2NH_2$.

(b) The formula for protonated cadaverine is $^+H_3NCH_2CH_2CH_2CH_2CH_2NH_3^+$.

(c) The smaller pK_a (i.e. the larger K_a), 9.13, refers to the first deprotonation of cadaverine.

The larger pK_a (i.e. the smaller K_a), 10.25, refers to the second deprotonation of cadaverine.

135. *Result/Explanation:* Lysine is given in Question 133. It has three functional groups, a carboxylic acid and two amine groups. All three of these have acid-base chemistry. The most acidic functional group is the carboxylic acid, so it must account for the pK_a of 2.18. The amine groups are bases, which means that their pK_b's would be less than 7; hence their pK_a's will be greater than 7. The amine group closest to the carboxylic acid accounts for the pK_a of 8.95. The second amine function accounts for the pK_a of 10.53.

Chapter 15: Additional Aqueous Equilibria
Solutions for Red-Numbered Questions for Review and Thought

Topical Questions

Buffer Solutions (Section 15-1)

12. *Result:* **(b) and (c)**

Explanation: A weak acid-base conjugate pair must be in the solution. The weak base can neutralize added acid making more of the weak base, and the weak acid can neutralize added base making more of the weak acid; hence, the pair together makes a buffer solution capable of resisting pH.

(a) A strong acid, HCl, and a weak acid, CH_3COOH, together in a system are able to neutralize added base, but are not able to neutralize added acid, so this pair does **not** form a buffer.

(b) The dissolved NaH_2PO_4 salt contains the weak acid $H_2PO_4^-$, and the dissolved Na_2HPO_4 salt contains its conjugate weak base, HPO_4^{2-}, so this pair **would form a buffer**.

(c) The solution contains the weak acid H_2CO_3 and a dissolved salt, $NaHCO_3$, containing its conjugate weak base, HCO_3^-, so this pair **would form a buffer**.

14. *Result:* **Combination (b) because the pK_a for the acid in that solution is closest to 7**

Analyze and Plan: To determine the pH of a buffer, look up the pK_a (Table 15.1) or look up the K_a (Table 14.2 or Appendix G) and calculate the pK_a ($pK_a = -\log K_a$). The pK_a closest to the desired pH is the best buffer, since close to equal quantities of the acid and base would be used, giving the solution approximately equal ability to neutralize added acid or added base.

Execute:

(a) The acid in H_3PO_4/NaH_2PO_4 is H_3PO_4. It has $K_a = 7.2 \times 10^{-3}$.

$$pK_a = -\log(7.2 \times 10^{-3}) = 2.14$$

(b) The NaH_2PO_4/Na_2HPO_4 buffer system has $pK_a = 7.20$.

(c) The acid in Na_2HPO_4/Na_3PO_4 is HPO_4^{2-}. It has $K_a = 4.6 \times 10^{-13}$.

$$pK_a = -\log(4.6 \times 10^{-13}) = 12.34$$

The combination that would make the best pH 7 buffer system is (b) NaH_2PO_4/Na_2HPO_4 because the pK_a for the acid in that solution (7.20) is closest to 7 than the pK_a for the acids in (a) or (c).

16. *Result:* **(a) 3.2 (b) 9.25 (c) 3.4**

Analyze and Plan: To answer this question quantitatively we need the value of pK_a, since that is equal to the pH in an equimolar buffer solution. In some cases, that value is not in the textbook. Without doing any calculations, we must estimate the pK_a from the K_a. We will look at the size of the K_a value and estimate its power of ten. We can also use Table 15.1 to give us advice about the fractional part of the pK_a.

Execute:

(a) The acid of the pair is HNO_2, with $K_a = 7.2 \times 10^{-4}$. Here the K_a is between 10^{-3} and 10^{-4}. In Table 15.1, the hypochlorous acid K_a has a similar number multiplying its power of ten, 6.8, and its pK_a has a fractional component of .17, so we will estimate the $HNO_2/NaNO_2$ buffer system will have a pH of approximately 3.2.

(b) The NH_3/NH_4Cl buffer system has $pK_a = 9.25$ so pH = 9.25, given in Table 15.1.

(c) The acid of the pair is HCOOH, with $K_a = 3.0 \times 10^{-4}$. Here the K_a is between 10^{-3} and 10^{-4}. In Table 15.1, the carbonic acid K_a has a slightly higher number multiplying its power of ten, 4.3, and its pK_a has a fractional component of .37, so we will estimate the HCOOH/NaHCOO buffer system will have a fractional component slight more than .4, so we estimate a pH of about 3.5.

☑ *Reasonable Result Check:* To check the answers, we will now disobey the instructions and do a calculation to check estimates in (a) and (c): (a) $pK_a = -\log(7.2 \times 10^{-4}) = 3.14$. (c) $pK_a = -\log(3.0 \times 10^{-4}) = 3.52$.

18. *Result:* **(a) Lactic acid/lactate ion (b) Dihydrogen phosphate ion/hydrogen phosphate ion (c) Acetic acid/acetate ion (d) Hydrogen carbonate ion/carbonate ion; see explanations below**

Analyze and Plan: We will compare the $[H_3O^+]$ to the values of K_a in Table 15.1, since that is equal to the $[H_3O^+]$ in an equimolar buffer solution. The K_a closest to the desired $[H_3O^+]$ is most suitable.

Execute:

(a) $[H_3O^+] = 4.5 \times 10^{-3}$ M, needs a lactic acid/lactate ion buffer ($K_a = 1.4 \times 10^{-4}$).

(b) $[H_3O^+] = 5.2 \times 10^{-8}$ M, needs dihydrogen phosphate ion/hydrogen phosphate ion buffer ($K_a = 6.3 \times 10^{-8}$).

(c) $[H_3O^+] = 8.3 \times 10^{-6}$ M, needs an acetic acid/acetate ion buffer ($K_a = 1.8 \times 10^{-5}$).

(d) $[H_3O^+] = 9.7 \times 10^{-11}$ M, needs a hydrogen carbonate/carbonate ion buffer ($K_a = 4.7 \times 10^{-11}$).

20. *Result:* **4.2 g $NaCH_3COO$**

Analyze and Plan: This question uses methods learned in several previous Chapters. First use the Henderson-Hasselbalch equation and Table 15.1 to find the concentration of the CH_3COO^- present in the equilibrium solution.

$$pH = pK_a + \log\left(\frac{[\text{conj. base}]}{[\text{conj. acid}]}\right)$$

In such calculations, we assume that the concentrations of conjugate acid and conjugate base are large enough to not be changed as the reaction equilibrium is established. This assumption is valid when the H_3O^+ concentration is small compared to the concentrations of conjugate acid and conjugate base.

Execute: Table 15.1 gives pK_a of acetic acid/acetate buffer as 4.74. The pH is 4.57. The initial concentration of CH_3COOH, the conjugate acid, is 0.150 M. The conjugate base, CH_3COO^-, is being added, but its initial concentration is unknown, so use the variable "c" to identify that quantity.

$$4.57 = 4.74 + \log\left(\frac{c}{0.150}\right)$$

Solve for c: $c = 0.150\ \text{M} \times 10^{(4.57 - 4.74)} = 0.10\ \text{M}$

Then adapt the methods from Chapter 3 to find the mass of the salt.

$$500.\ \text{mL} \times \frac{1\ \text{L}}{1000\ \text{mL}} \times \frac{0.10\ \text{mol CH}_3\text{COO}^-}{1\ \text{L}} \times \frac{1\ \text{mol NaCH}_3\text{COO}}{1\ \text{mol CH}_3\text{COO}^-} \times \frac{82.0337\ \text{g NaCH}_3\text{COO}}{1\ \text{mol NaCH}_3\text{COO}}$$

$$= 4.2\ \text{g NaCH}_3\text{COO}$$

22. *Result:* **13 g C_6H_5COOH per liter**

Analyze and Plan: Adapt the method described in the solution to Question 21.

Use methods from Chapter 3 to find the initial concentration of $C_6H_5COO^-$, the conjugate base:

$$(\text{conc. } C_6H_5COO^-) = \frac{14.4 \text{ g } C_6H_5COONa}{1 \text{ L}} \times \frac{1 \text{ mol } C_6H_5COONa}{144.1030 \text{ g } C_6H_5COONa} \times \frac{1 \text{ mol } C_6H_5COO^-}{1 \text{ mol } C_6H_5COONa}$$

$$= 0.100 \text{ M } C_6H_5COO^-$$

In Table 14.2, the K_a of benzoic acid is given: 1.2×10^{-4}. As described in Section 15.1, calculate the pK_a:

$$pK_a = -\log K_a = -\log(1.2 \times 10^{-4}) = 3.92$$

Execute: The pH is 3.88. The conjugate acid, C_6H_5COOH, is being added, but its initial concentration is unknown, so use the variable "c" to identify that quantity.

$$pH = pK_a + \log\left(\frac{[\text{conj. base}]}{[\text{conj. acid}]}\right)$$

$$3.88 = 3.92 + \log\left(\frac{0.100}{c}\right)$$

Solve for c: $c = 0.100 \text{ M} \times 10^{-(3.88-3.92)} = 0.11 \text{ M}$

Calculate the mass using methods from Chapter 3:

$$\frac{0.11 \text{ mol } C_6H_5COOH}{1 \text{ L}} \times \frac{122.1211 \text{ g } C_6H_5COOH}{1 \text{ mol } C_6H_5COOH} = 13 \text{ g } C_6H_5COOH \text{ per liter}$$

24. *Result:* **7.23**

Analyze: Adapt the methods from Chapter 14 and Question 21.

Plan and Execute: Find the initial concentration of OH^- ions from the NaOH, after it is added to the solution but before it reacts:

$$(\text{conc. } OH^-) = \frac{0.425 \text{ g NaOH}}{2.00 \text{ L}} \times \frac{1 \text{ mol NaOH}}{39.9971 \text{ g NaOH}} \times \frac{1 \text{ mol } OH^-}{1 \text{ mol NaOH}} = 0.00531 \text{ M } OH^-$$

The base neutralizes the acid, via a product-favored neutralization reaction, as shown in this equation:

$$H_2PO_4^-(aq) + OH^-(aq) \longrightarrow HPO_4^{2-}(aq) + H_2O(\ell)$$

This reaction is product-favored, so we will first make as many products as possible. Using the method of limiting reactants, run the reaction towards products until one of the reactants runs out.

	$H_2PO_4^-(aq)$	$OH^-(aq)$	$HPO_4^{2-}(aq)$
initial conc. (M)	0.132	0.00531	0.132
change as reaction occurs (M)	– 0.00531	– 0.00531	+ 0.00531
final conc. (M)	0.127	0	0.137

The solution is a buffer solution, containing a weak-acid/weak-base conjugate pair, so we can find the pH using the Henderson-Hasselbalch equation. In Table 15.1, the pK_a of dihygydrogen phosphate is given: 7.20.

$$pH = pK_a + \log\left(\frac{[\text{conj. base}]}{[\text{conj. acid}]}\right)$$

$$pH = 7.20 + \log\left(\frac{0.137}{0.127}\right) = 7.23$$

☑ *Reasonable Result Check:* An equimolar concentration buffer has a pH of 7.20. After added some strong base, the solution would be slightly more basic, so a higher pH makes sense.

26. *Result:* **Samples (a) and (b) are buffers; they both contain a conjugate acid-base pair. (c) is not a buffer because there is no base. (d) is not a buffer because too much base was added.**

Analyze and Plan: A 1-L solution of 0.20 M CH_3COOH provides the weak acid for the buffer solution. The added solution must provide the conjugate base, CH_3COO^-, either directly or by partial neutralization.

Execute:

(a) Adding any amount of the soluble salt, $NaCH_3COO$, will form a buffer solution, **containing a conjugate acid-base pair**.

(b) Adding 0.10 mol NaOH to the 1-L solution makes a 0.10 M OH^- solution. The strong base neutralizes the acid in the solution, CH_3COOH.

$$CH_3COOH(aq) + OH^-(aq) \longrightarrow CH_3COO^-(aq) + H_2O(\ell)$$

This reaction is product-favored, so we will first make as many products as possible. Using the method of limiting reactants, run the reaction towards products until one of the reactants runs out.

	$CH_3COOH(aq)$	$OH^-(aq)$	$CH_3COO^-(aq)$
initial conc. (M)	0.20	0.10	0
change as reaction occurs (M)	– 0.10	– 0.10	+ 0.10
final conc. (M)	0.10	0	0.10

The solution produced is a buffer solution, **containing a conjugate acid-base pair.**

(c) **Adding a strong acid to a weak acid** will **not** produce a buffer.

(d) Adding 0.30 mol NaOH to the 1-L solution makes a 0.30 M OH^- solution. The strong base neutralizes the acid in the solution, as in (b). This reaction is product-favored, so we will first make as many products as possible. Using the method of limiting reactants, run the reaction towards products until one of the reactants runs out.

	$CH_3COOH(aq)$	$OH^-(aq)$	$CH_3COO^-(aq)$
initial conc. (M)	0.20	0.30	0
change as reaction occurs (M)	– 0.20	– 0.20	+ 0.20
final conc. (M)	0	0.10	0.20

The solution produced is a **not** buffer solution. **Too much base has been added** and all the CH_3COOH has been neutralized.

Therefore, of the four substances added to the acid, only samples (a) or (b) produce buffer solutions.

29. *Result:* **(a) ΔpH = 0.1 (b) ΔpH = 3.8 (c) ΔpH = 7.25**

Analyze and Plan: Calculate the initial pH by finding the pK_a. Calculate initial OH^- concentration using the methods from Chapter 3. Then adapt the method used in the solution to Question 26.

Equimolar buffer solutions have pH = pK_a.

Execute: Calculate the initial diluted concentration of OH^- using the total volume of the solution after the addition.

$$V = 0.100\text{ L} \times \frac{1000\text{ mL}}{1\text{ L}} + 1.0\text{ mL}$$

$$(\text{conc. OH}^-) = \frac{1.0\text{ mL}}{101\text{ mL}} \times \frac{1.0\text{ mol OH}^-}{1\text{ L}} = 0.0099\text{ M OH}^-$$

The strong base neutralizes the acid in the solution, CH_3COOH.

$$CH_3COOH(aq) + OH^-(aq) \longrightarrow CH_3COO^-(aq) + H_2O(\ell)$$

This reaction is product-favored, so we will first make as many products as possible. Using the method of limiting reactants, run the reaction towards products until one of the reactants runs out.

(a)

	$CH_3COOH(aq)$	$OH^-(aq)$	$CH_3COO^-(aq)$
initial conc. (M)	0.10	0.0099	0.10
change as reaction occurs (M)	– 0.0099	– 0.0099	+ 0.0099
final conc. (M)	0.09	0	0.11

The solution is still a buffer solution, containing an acid-base conjugate pair, so we can find the pH using the Henderson-Hasselbalch equation. In Table 15.1, the pK_a is 4.74.

$$pH = pK_a + \log\left(\frac{[\text{conj. base}]}{[\text{conj. acid}]}\right)$$

$$pH = 4.74 + \log\left(\frac{0.11}{0.09}\right) = 4.8 \quad \textit{(1 decimal place)}$$

$$\Delta pH = 4.8 - 4.74 = 0.1 \quad \textit{(1 decimal place)}$$

(b)

	$CH_3COOH(aq)$	$OH^-(aq)$	$CH_3COO^-(aq)$
initial conc. (M)	0.010	0.0099	0.010
change as reaction occurs (M)	– 0.0099	– 0.0099	+ 0.0099
final conc. (M)	0.000	0.0000	0.020

Within known significant figures (three decimal places), the solution is no longer a buffer solution. It contains only the conjugate base. We must find the pH using K_b of the base, as was done in Chapter 14.

$$CH_3COO^-(aq) + H_2O(\ell) \rightleftharpoons CH_3COOH(aq) + OH^-(aq) \qquad K_b = \frac{[CH_3COOH][OH^-]}{[CH_3COO^-]}$$

As the reactants react, the concentrations of the products increase stoichiometrically, until they reach equilibrium concentrations.

	$CH_3COO^-(aq)$	$CH_3COOH(aq)$	$OH^-(aq)$
initial conc. (M)	0.020	0	0
change as reaction occurs (M)	– x	+ x	+ x
equilibrium conc. (M)	0.020 – x	x	x

At equilibrium $$K_b = \frac{(x)(x)}{(0.020 - x)} = 5.6 \times 10^{-10}$$

Assume x is very small and does not affect the difference.

$$5.6 \times 10^{-10} = \frac{x^2}{(0.020)}$$

$$x^2 = (5.6 \times 10^{-10})(0.20)$$

$$x = 3.3 \times 10^{-6}\ M = [OH^-]$$

$$pOH = -\log[OH^-] = -\log(3.3 \times 10^{-6}) = 5.48$$

$$pH = 14.00 - pOH = 8.52$$

$$\Delta pH = 8.52 - 4.74 = 3.8$$

Notice: Do not use the Henderson-Hasselbalch equation for very low acid concentration.

(c) In this situation, the acetic acid is the limiting reactant:

	$CH_3COOH(aq)$	$OH^-(aq)$	$CH_3COO^-(aq)$
initial conc. (M)	0.0010	0.0099	0.0010
change as reaction occurs (M)	– 0.0010	– 0.0010	+ 0.0010
final conc. (M)	0.0000	0.0089	0.0020

The solution is no longer a buffer solution. This time, some OH^- ions are left over. The minor amount produced by the weak base reaction will not change this value, so:

$$pH = 14.00 + \log[OH^-] = 14.00 + \log(0.0089) = 11.95$$

$$\Delta pH = 11.95 - 4.74 = 7.25$$

31. *Result:* **(a) 5.02 (b) 4.99 (c) 4.06**

Analyze and Plan: Adapt the method described in the solutions to Question 23, 27, and 30.

Execute:

Use the abbreviation HProp for the monoprotic propanoic acid. The salt sodium propanoate contains the $Prop^-$ ion.

(a) Find the pH using Henderson-Hasselbalch. The K_a of propanoic acid is given: 1.4×10^{-5}. As described in Section 15-1c, calculate the pK_a:

$$pK_a = -\log K_a = -\log(1.4 \times 10^{-5}) = 4.85$$

The concentration of the conjugate acid, HProp, is 0.20 M; the concentration of the conjugate base, $Prop^-$, is 0.30 M.

$$pH = pK_a + \log\left(\frac{[\text{conj. base}]}{[\text{conj. acid}]}\right)$$

$$pH = 4.85 + \log\left(\frac{0.30}{0.20}\right) = 4.85 + 0.18 = 5.02$$

(b) Calculate the new concentration of dilute hydronium ion, using the total volume of the solution after the addition, but before the reaction.

$$V = 0.010\ L \times \frac{1000\ mL}{1\ L} + 1.0\ mL = 11\ mL$$

$$(\text{conc. } H_3O^+) = \frac{1.0\ mL}{11\ mL} \times \frac{1.0\ mol\ H_3O^+}{1\ L} = 0.0091\ M\ H_3O^+$$

The strong acid neutralizes the base in the solution, CH_3COO^-.

$$Prop^-(aq) + H_3O^+(aq) \longrightarrow HProp + H_2O(\ell)$$

This reaction is product-favored, so we will first make as many products as possible. Using the method of limiting reactants, run the reaction towards products until one of the reactants runs out.

	$Prop^-$ (aq)	H_3O^+(aq)	HProp(aq)
initial conc. (M)	0.30	0.0091	0.20
change as reaction occurs (M)	– 0.0091	– 0.0091	+ 0.0091
final conc. (M)	0.29	0	0.21

The solution is still a buffer solution, containing an acid-base conjugate pair, so determine the pH as in (a).

$$pH = pK_a + \log\left(\frac{[\text{conj. base}]}{[\text{conj. acid}]}\right)$$

$$pH = 4.85 + \log\left(\frac{0.29}{0.21}\right) = 4.85 + 0.14 = 4.99$$

(c) Calculate the concentration of hydronium ion, using the total volume of the solution after the addition, but before the reaction.

$$V = 0.010\ \text{L} \times \frac{1000\ \text{mL}}{1\ \text{L}} + 3.0\ \text{mL} = 13\ \text{mL}$$

$$(\text{conc. } H_3O^+) = \frac{3.0\ \text{mL}}{13\ \text{mL}} \times \frac{1.0\ \text{mol } H_3O^+}{1\ \text{L}} = 0.23\ \text{M } H_3O^+$$

The strong acid neutralizes the base in the solution, $Prop^-$.

$$Prop^-\ (aq) + H_3O^+(aq) \longrightarrow HProp + H_2O(\ell)$$

This reaction is product-favored, so we will first make as many products as possible. Using the method of limiting reactants, run the reaction towards products until one of the reactants runs out.

	$Prop^-$ (aq)	H_3O^+(aq)	HProp(aq)
initial conc. (M)	0.30	0.23	0.20
change as reaction occurs (M)	– 0.23	– 0.23	+ 0.23
final conc. (M)	0.07	0	0.43

The solution is still a buffer solution, containing an acid-base conjugate pair, so we can find pH as in (a).

$$pH = pK_a + \log\left(\frac{[\text{conj. base}]}{[\text{conj. acid}]}\right)$$

$$pH = 4.85 + \log\left(\frac{0.07}{0.43}\right) = 4.85 - 0.79 = 4.06$$

☑ *Reasonable Result Check:* Adding strong acid makes the solution more acidic, so it makes sense that the pH goes down.

Acid-Base Titrations and Titration Curves (Section 15-2)

32. *Result:* **(a) Acid 2; see explanation below (b) Acid 1: pH ≈ 7; acid 2: pH ≈ 9.5 (c) See explanation below (d) See explanation below (e) The best choice for acid 1 is bromthymol blue; phenolphthalein should be used for Acid 2; see explanation below**

Explanation:

(a) The effects of acid strength on the shape of the titration curve are shown in Figures 15.8. The curve for Acid 2 is the curve for the weak acid.

(b) The equivalence point for Acid 1 is about pH 7, and the equivalence point for Acid 2 is about pH 9.5.

(c) At the equivalence point of a strong acid and a strong base, the solution contains the cation of the strong base and the anion of the strong acid, both of which are too weak to change the pH of the water solution. Thus the solution is neutral, and pH = 7. At the equivalence point of a weak acid and a strong base, the solution contains the cation of the strong base and the anion of the weak acid. The cation is too weak to change the pH of the water solution, but the anion of a weak acid is a weak base. Thus the solution is basic, and pH > 7.

(d) The pH of the weak acid starts at a higher value because of a smaller degree of ionization of the weak acid compared to the ionization of a strong acid.

(e) Bromthymol blue should be used for Acid 1 and phenolphthalein should be used for Acid 2, because their color changes are near the respective equivalence points.

33. *Result/Explanation:* At the equivalence point, the solution contains the conjugate base of the weak acid. Weaker acids have stronger conjugate bases. Stronger bases have higher pH, so the titration of a weaker acid will have a more alkaline (i.e., more basic) equivalence point.

34. *Result:* **See titration curve below; (a) 1.000 (b) 1.477 (c) < 7 (d) basic**

Analyze and Plan: Equal volumes of acid and base are required to reach the equivalence point, because the molarities of the two solutions are the same. As the weak base titrant is added, it will be completely neutralized by the strong acid, and the volume will increase.

Execute:

(a) Prior to the titration, only the strong acid is present. The pH = – log[H_3O^+] = – log(0.100) = 1.000.

(b) When half the required titrant has been added, half of the acid has been neutralized and the total volume has been increased by 50%. That means the concentration of H_3O^+ at this point is equal to 0.500 × (0.100 M)/1.50 = 0.0333 M. The pH = – log[H_3O^+] = – log(0.0333) = 1.477 (assuming that the H_3O^+ present due to BH^+ concentration is negligible).

(c) At the equivalence point, 20.0 mL of the base has been added. The acid has exactly neutralized the base, leaving the solution with conjugate acid of the weak base. The equivalence point will be acidic, but without information about the K_a for this acid, all we can say about the equivalence point is that the pH is less than 7.

(d) At 10 mL beyond the equivalence point, the solution will definitely be above the equivalence point pH and will probably be basic. The solution will again be a buffer solution because the strong acid will have been completely neutralized forming the conjugate acid of the base titrant, then more weak base titrant is added. That is all we can say without more information about the K_b for the weak base.

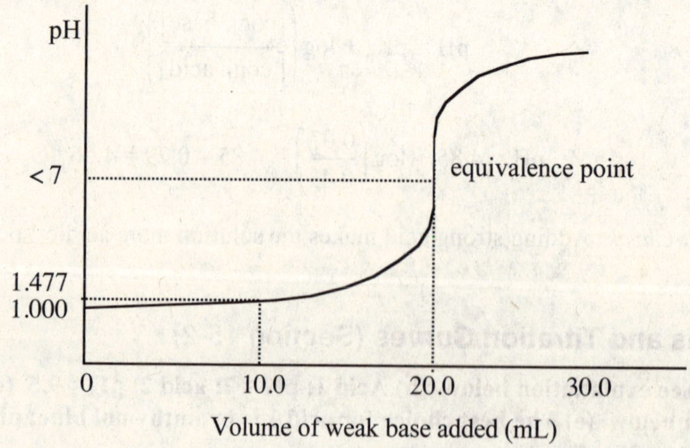

35. *Result:* **Best choices: (a) bromthymol blue (b) phenolphthalein (c) methyl red (d) bromthymol blue; see explanations below**

Analyze and Plan: The color change needs to be near the pH of the equivalence point.

Execute:

(a) The strong base, NaOH, titrated with a strong acid, $HClO_4$, has a neutral equivalence point. It would be best to choose bromthymol blue, which is shown changing color in Figure 15.6 at a pH near 7. In practice, any of them would be suitable, because of the extreme change in the pH of the solution very close to the equivalence point.

(b) The weak acid, CH_3COOH, titrated with a strong base, KOH, has a basic equivalence point, due to the presence of the weak base CH_3COO^- in the solution. It would be best to choose phenolphthalein, which is shown changing color in Figure 15.7 at a pH near 9.

(c) The weak base, NH_3, titrated with a strong acid, HBr, has an acidic equivalence point, due to the presence of the weak acid NH_4^+ in the solution. It would be best to choose methyl red, which is shown changing color in Figure 15.9 at a pH near 5.

(d) The strong base, KOH, titrated with a strong acid, HNO_3, has a neutral equivalence point. It would be best to choose bromthymol blue, which is shown changing color in Figure 15.9 at a pH near 7. In practice, any of them would be suitable, because of the extreme change in the pH of the solution very close to the equivalence point.

37. *Result:* **0.0253 M HCl**

Analyze, Plan, and Execute: Use the methods described in Sections 3-11 and 3.12.

$$\frac{22.6\text{ mL Ba(OH)}_2}{25.00\text{ mL HCl}} \times \frac{1\text{ L Ba(OH)}_2}{1000\text{ mL Ba(OH)}_2} \times \frac{1000\text{ mL HCl}}{1\text{ L HCl}}$$

$$\times \frac{0.0140\text{ mol Ba(OH)}_2}{1\text{ L Ba(OH)}_2} \times \frac{2\text{ mol OH}^-}{1\text{ mol Ba(OH)}_2} \times \frac{1\text{ mol HCl}}{1\text{ mol OH}^-} = 0.0253\ \frac{\text{mol HCl}}{\text{L HCl}} = 0.0253\text{ M HCl}$$

39. *Result:* **93.6%**

Analyze, Plan, and Execute: Use the methods described in Sections 3-11 and 3.12.

$$24.4\text{ mL NaOH} \times \frac{1\text{ L}}{1000\text{ mL}} \times \frac{0.110\text{ mol NaOH}}{1\text{ L NaOH}} \times \frac{1\text{ mol C}_6\text{H}_8\text{O}_6}{1\text{ mol NaOH}} \times \frac{176.1238\text{ g C}_6\text{H}_8\text{O}_6}{1\text{ mol C}_6\text{H}_8\text{O}_6} = 0.473\text{ g C}_6\text{H}_8\text{O}_6$$

$$\frac{0.473\text{ g C}_6\text{H}_8\text{O}_6}{0.505\text{ g capsule}} \times 100\% = 93.6\%$$

41. *Result:* **(a) 29.2 mL HCl (b) 600. mL HCl (c) 1.20 L HCl (d) 2.7 mL HCl**

Analyze and Plan: Use the methods described in Sections 3-11 and 3.12.

Execute:

(a) $$25.0\text{ mL KOH} \times \frac{0.175\text{ mol KOH}}{1\text{ L KOH}} \times \frac{1\text{ mol HCl}}{1\text{ mol KOH}} \times \frac{1\text{ L HCl}}{0.150\text{ mol HCl}} = 29.2\text{ mL HCl}$$

(b) $$15.0\text{ mL NH}_3 \times \frac{6.00\text{ mol NH}_3}{1\text{ L NH}_3} \times \frac{1\text{ mol HCl}}{1\text{ mol NH}_3} \times \frac{1\text{ L HCl}}{0.150\text{ mol HCl}} = 600.\text{ mL HCl}$$

(c) $$15.0\text{ mL C}_3\text{H}_7\text{NH}_2 \times \frac{0.712\text{ g C}_3\text{H}_7\text{NH}_2}{1\text{ mL C}_3\text{H}_7\text{NH}_2} \times \frac{1\text{ mol C}_3\text{H}_7\text{NH}_2}{59.1099\text{ g C}_3\text{H}_7\text{NH}_2}$$

$$\times \frac{1\text{ mol HCl}}{1\text{ mol C}_3\text{H}_7\text{NH}_2} \times \frac{1\text{ L HCl}}{0.150\text{ mol HCl}} = 1.20\text{ L HCl}$$

(d) $$40.0\text{ mL Ba(OH)}_2 \times \frac{0.0050\text{ mol Ba(OH)}_2}{1\text{ L Ba(OH)}_2} \times \frac{2\text{ mol OH}^-}{1\text{ mol Ba(OH)}_2}$$

$$\times \frac{1\text{ mol HCl}}{1\text{ mol OH}^-} \times \frac{1\text{ L HCl}}{0.150\text{ mol HCl}} = 2.7\text{ mL HCl}$$

43. *Result:* **(a) 3.62 (b) 8.31 (c) 12.15**

Analyze and Plan: Adapt the method used in Problem-Solving Example 15.7.

The chemical formula for benzoic acid is given in Table 14.2, C_6H_5COOH. Benzoate anion has the formula $C_6H_5COO^-$.

Plan and Execute:

(a) Calculate the moles of hydroxide and the benzoic acid.

$$\text{mol } OH^- = 10.00 \text{ mL} \times \frac{1 \text{ L}}{1000 \text{ mL}} \times \frac{0.100 \text{ mol } OH^-}{1 \text{ L}} = 0.00100 \text{ mol } OH^-$$

$$\text{mol } C_6H_5COOH = 30.00 \text{ mL} \times \frac{1 \text{ L}}{1000 \text{ mL}} \times \frac{0.100 \text{ mol } C_6H_5COOH}{1 \text{ L}} = 0.00300 \text{ mol } C_6H_5COOH$$

The strong base neutralizes the acid in the solution:

C_6H_5COOH	+ $OH^-(aq)$	$\longrightarrow$	$C_6H_5COO^-(aq)$	+ $H_2O(\ell)$
0.00300 mol in acid soln	0.00100 mol added		0.00100 mol formed	

After neutralization, mol C_6H_5COOH = 0.00300 mol – 0.00100 mol = 0.00200 mol, and the total volume is 30.0 mL + 10.0 mL = 40.0 mL or 0.0400 L.

Use the K_a expression associated with the benzoic acid ionization equilibrium to calculate $[H_3O^+]$, then calculate the pH:

$$K_a = \frac{[H_3O^+][C_6H_5COO^-]}{[C_6H_5COOH]} = \frac{[H_3O^+](0.00100 \text{ mol}/0.0400 \text{ L})}{(0.00200 \text{ mol}/0.0400 \text{ L})} = 1.2 \times 10^{-4}$$

$$[H_3O^+] = (1.2 \times 10^{-4}) \times 2.00 = 2.4 \times 10^{-4} \text{ M}$$

$$pH = -\log[H_3O^+] = -\log(2.4 \times 10^{-4}) = 3.62$$

(b) The initial moles of benzoic acid are unchanged, but we need to calculate the new moles of hydroxide.

$$\text{mol } OH^- = 30.00 \text{ mL} \times \frac{1 \text{ L}}{1000 \text{ mL}} \times \frac{0.100 \text{ mol } OH^-}{1 \text{ L}} = 0.00300 \text{ mol } OH^-$$

C_6H_5COOH	+ $OH^-(aq)$	$\longrightarrow$	$C_6H_5COO^-(aq)$	+ $H_2O(\ell)$
0.00300 mol in acid soln	0.00300 mol added		0.00300 mol formed	

The total volume is 30.0 mL + 30.0 mL = 60.0 mL or 0.0600 L. At this point in the neutralization, all the C_6H_5COOH has been neutralized and the solution's pH is governed by the hydrolysis of conjugate base, $C_6H_5COO^-$. Use K_a given for benzoic acid to find the K_b for the conjugate base hydrolysis reaction.

$$K_b = \frac{K_w}{K_a} = \frac{1.00 \times 10^{-14}}{1.2 \times 10^{-4}} = 8.3 \times 10^{-11}$$

Then use the K_b expression for the hydrolysis reaction to calculate OH^-, then calculate the pH.

$$C_6H_5COO^- = \frac{0.00300 \text{ mol } C_6H_5COO^-}{0.06000 \text{ L}} = 0.0500 \text{ mol } C_6H_5COO^-$$

Substituting into the K_b expression, we let $x = [C_6H_5COOH] = [OH^-]$:

$$K_b = \frac{[C_6H_5COOH][OH^-]}{[C_6H_5COO^-]} = 8.3 \times 10^{-11} = \frac{x^2}{0.0500}$$

$$x^2 = (8.3 \times 10^{-11})(0.0500)$$

$$x = 2.0 \times 10^{-6}\ M = [OH^-]$$

$$pOH = -\log[OH^-] = -\log(2.0 \times 10^{-6}) = 5.69$$

$$pH = 14.00 - pOH = 14.00 - 5.69 = 8.31$$

(c) The initial moles of benzoic acid are unchanged, but we need to calculate the moles of hydroxide.

$$\text{mol } OH^- = 40.00\ \text{mL} \times \frac{1\ \text{L}}{1000\ \text{mL}} \times \frac{0.100\ \text{mol } OH^-}{1\ \text{L}} = 0.00400\ \text{mol } OH^-$$

Neutralization occurs, as in (a), but this time benzoic acid runs out first:

C_6H_5COOH	+ $OH^-(aq)$	⟶	$C_6H_5COO^-(aq)$	+ $H_2O(\ell)$
0.00300 mol in acid soln	0.00400 mol added		0.00300 mol formed	

After neutralization, mol OH^- = 0.00400 mol – 0.00300 mol = 0.00100 mol.

The total volume in the solution after the two solutions are mixed is 30.00 mL + 40.00 mL = 70.00 mL or 0.07000 L.

The pH of the solution now is governed by the presence of excess strong base. We calculate the final concentration of OH^- after neutralization, then determine the pH.

$$\text{Final } OH^- \text{ conc.} = \frac{0.00100\ \text{mol } OH^-}{0.07000\ \text{L}} = 0.0143\ M\ OH^-$$

$$pOH = -\log[OH^-] = -\log(0.0143) = 1.85$$

$$pH = 14.00 - pOH = 14.00 - 1.85 = 12.15$$

44. *Result:* **(a) 13.176 (b) 12.699 (c) 10.2 (d) 7.000 (e) 3.8 (f) 1.52; see titration curve and identification of the equivalence point below**

Analyze and Plan: Adapt the method shown in Problem-Solving Example 15.6. For each of the points, determine the limiting reactant and use the concentration of the excess reactant to calculate the pH.

Plan and Execute:

$$\text{Total volume of base in liters} = 50.00\ \text{mL} \times \frac{1\ \text{L}}{1000\ \text{mL}} = 0.05000\ \text{L}$$

$$\text{Original mol } OH^- \text{ added} = 0.05000\ \text{L} \times \frac{0.150\ \text{mol } OH^-}{1\ \text{L}} = 0.00750\ \text{mol } OH^-$$

(a) Solution contains only 0.150 M NaOH, so:

$$pOH = \log[OH^-] = -\log(0.150) = 0.824$$

$$pH = 14.00 - pOH = 14.00 - 0.824 = 13.176$$

(b) After the addition of the titrant, adapt Equation 15.1 to determine the equilibrium concentration of OH^- in this titration:

$$[OH^-] = \frac{\text{original moles base} - \text{total moles acid added}}{\text{volume of acid (L)} + \text{volume of base (L)}}$$

$$\text{Volume of acid in liters} = 25.00 \text{ mL} \times \frac{1 \text{ L}}{1000 \text{ mL}} = 0.02500 \text{ L}$$

$$\text{Total mol } H_3O^+ \text{ added} = 0.02500 \text{ L} \times \frac{0.150 \text{ mol } H_3O^+}{1 \text{ L}} = 0.00375 \text{ mol } H_3O^+$$

$$[OH^-] = \frac{0.00750 \text{ mol} - 0.00375 \text{ mol}}{0.05000 \text{ L} + 0.02500 \text{ L}} = 0.0500 \text{ M}$$

$$pOH = -\log[OH^-] = -\log(0.0500) = 1.301$$

$$pH = 14.00 - pOH = 14.00 - 1.301 = 12.699$$

(c) Use the equation given in (b) again.

$$\text{Volume of acid (L)} = 49.9 \text{ mL} \times \frac{1 \text{ L}}{1000 \text{ mL}} = 0.0499 \text{ L}$$

Total mol H_3O^+ added =

$$0.0499 \text{ L} \times \frac{0.150 \text{ mol } H_3O^+}{1 \text{ L}} = 0.007485 \text{ mol } H_3O^+ \approx 0.00749 \text{ mol } H_3O^+ \quad \textit{(3 sig fig)}$$

$$[OH^-] = \frac{0.00750 \text{ mol} - 0.007485 \text{ mol}}{0.05000 \text{ L} + 0.0499 \text{ L}} = \frac{0.000015 \text{ mol}}{0.0999 \text{ L}} = 0.00015 \text{ M} \approx 0.0002 \text{ M}$$

(1 sig fig after subtraction)

$$pOH = -\log[OH^-] = -\log(0.00015) = 3.8 \quad \textit{(1 decimal place)}$$

$$pH = 14.00 - pOH = 14.00 - 3.8 = 10.2$$

(d) Calculate volume and total moles of acid:

$$\text{Volume of acid (L)} = 50.00 \text{ mL} \times \frac{1 \text{ L}}{1000 \text{ mL}} = 0.05000 \text{ L}$$

$$\text{Total mol } H_3O^+ \text{ added} = 0.05000 \text{ L} \times \frac{0.150 \text{ mol } H_3O^+}{1 \text{ L}} = 0.00750 \text{ mol } H_3O^+$$

Total moles of H_3O^+ added is equal to the original moles OH^-, so the solution is neutral with pH = 7.000.

(e) At this point, the moles of acid begin to exceed the moles of base, so adapt Equation 15.1 for a solution with excess acid:

$$[H_3O^+] = \frac{\text{total moles acid added} - \text{original moles base}}{\text{volume of acid (L)} + \text{volume of base (L)}}$$

$$\text{Volume of acid (L)} = 50.1 \text{ mL} \times \frac{1 \text{ L}}{1000 \text{ mL}} = 0.0501 \text{ L}$$

$$\text{Total mol } H_3O^+ \text{ added} = 0.0501 \text{ L} \times \frac{0.150 \text{ mol } H_3O^+}{1 \text{ L}} = 0.007515 \text{ mol } H_3O^+ \approx 0.00752 \text{ mol } H_3O^+$$

$$[H_3O^+] = \frac{0.007515 \text{ mol} - 0.00750 \text{ mol}}{0.05000 \text{ L} + 0.0501 \text{ L}} = \frac{0.000015 \text{ mol}}{0.1001 \text{ L}} = 0.00015 \text{ M} \approx 0.0002 \text{ M}$$ *(1 sig fig after subtraction)*

$$pH = -\log[H_3O^+] = -\log(0.00015) = 3.8 \quad \textit{(1 decimal place)}$$

(f) Use the equation given in (e) again.

$$\text{Volume of acid (L)} = 75.00\text{ mL} \times \frac{1\text{ L}}{1000\text{ mL}} = 0.0750\text{ L}$$

$$\text{Total mol }H_3O^+\text{ added} = 0.0750\text{ L} \times \frac{0.150\text{ mol }H_3O^+}{1\text{ L}} = 0.0113\text{ mol }H_3O^+$$

$$[H_3O^+] = \frac{0.0113\text{ mol} - 0.00750\text{ mol}}{0.05000\text{ L} + 0.0750\text{ L}} = \frac{0.00038\text{ mol}}{0.1250\text{ L}} = 0.0030\text{ M}$$

$$\text{pH} = -\log[H_3O^+] = -\log(0.030) = 1.52$$

Combine these results in a pH *verses* volume of titrant to make a titration curve, including equivalence point:

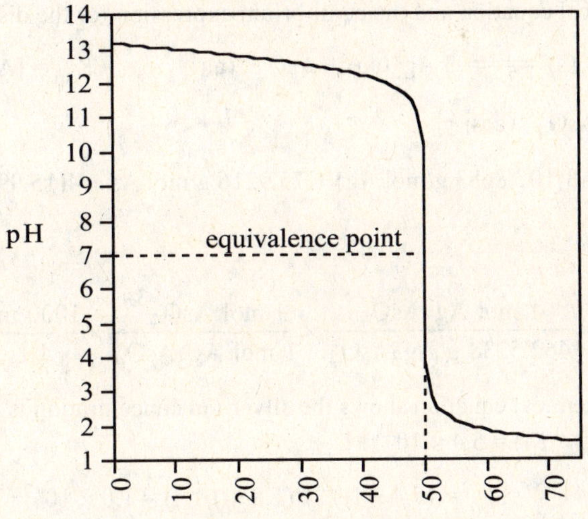

Acid Rain (Section 15-3)

47. *Result:* **NO_2 and SO_3; see equations below**

Explanation: Two oxides that are key producers of acid rain are NO_2 and SO_3. NO_2 reacts with water in the air to make nitric acid and nitrous acid:

$$2\,NO_2(g) + H_2O(\ell) \longrightarrow HNO_3(aq) + HNO_2(aq)$$

SO_2 reacts with oxygen in the air to make SO_3 and SO_3 reacts water in the air to make sulfuric acid:

$$2\,SO_2(g) + O_2(g) \longrightarrow 2\,SO_3(g)$$

$$SO_3(g) + H_2O(\ell) \longrightarrow H_2SO_4(aq)$$

49. *Result:* **$CaCO_3(s) + 2\,H^+(aq) \longrightarrow Ca^{2+}(aq) + CO_2(g) + H_2O(\ell)$**

Analyze: Limestone is made of calcium carbonate. The anion of this compound is a base and reacts with acid in rainwater to neutralize it by a gas-forming exchange reaction (Section 3-4e).

Solubility Product (Section 15-4)

50. **See equations and expressions below**

Analyze and Plan: Adapt the method developed in Problem-Solving Example 15.8.

Execute:

(a) $Ag_3AsO_4(s) \rightleftharpoons 3\ Ag^+(aq) + AsO_4^{3-}(aq)$ $K_{sp} = [Ag^+]^3[AsO_4^{3-}]$

(b) $Ag_2SO_4(s) \rightleftharpoons 2\ Ag^+(aq) + SO_4^{2-}(aq)$ $K_{sp} = [Ag^+]^2[SO_4^{2-}]$

(c) $Ca_3(PO_4)_2(s) \rightleftharpoons 3\ Ca^{2+}(aq) + 2\ PO_4^{3-}(aq)$ $K_{sp} = [Ca^{2+}]^3[PO_4^{3-}]^2$

(d) $Mn(OH)_3(s) \rightleftharpoons Mn^{3+}(aq) + 3\ OH^-(aq)$ $K_{sp} = [Mn^{3+}][OH^-]^3$

(e) $FeCO_3(s) \rightleftharpoons Fe^{2+}(aq) + CO_3^{2-}(aq)$ $K_{sp} = [Fe^{2+}][CO_3^{2-}]$

52. *Result:* $\mathbf{K_{sp} = 1.0 \times 10^{-7}}$

Analyze and Plan: Adapt the method developed in Problem-Solving Example 15.10.

Execute: Write the chemical equation and the equilibrium expression for the dissociation of the solute:

$$Ag_3AsO_4(s) \rightleftharpoons 3\ Ag^+(aq) + AsO_4^{3-}(aq) \qquad K_{sp} = [Ag^+]^3[AsO_4^{3-}]$$

Calculate the molarity of $AsO_4^{3-}(aq)$:

Molar mass of Ag_3AsO_4 = 3(107.8683 g/mol Ag) + 75.9216 g/mol As + 4(15.9994 g/mol O)

= 462.5238 g/mol Ag_3AsO_4

$$\frac{8.5\times10^{-6}\text{ g } Ag_3AsO_4}{1\text{ mL}}\times\frac{1\text{ mol } Ag_3AsO_4}{462.5238\text{ g } Ag_3AsO_4}\times\frac{1\text{ mol } AsO_4^{3-}}{1\text{ mol } Ag_3AsO_4}\times\frac{1000\text{ mL}}{1\text{ L}} = 1.8\times10^{-4}\text{ M } AsO_4^{3-}$$

The mole ratio from the chemical equation shows the silver ion concentration is three times the arsenate ion concentration, $3 \times (1.8 \times 10^{-4}\text{ M}) = 5.4 \times 10^{-4}$ M.

$$K_{sp} = (5.4 \times 10^{-4})^3(1.8 \times 10^{-5}) = 1.0 \times 10^{-7}$$

54. *Result:* $\mathbf{K_{sp} = 2.22 \times 10^{-4}}$

Analyze and Plan: Adapt the method developed in Problem-Solving Example 15.10.

Execute:

Write the chemical equation and the equilibrium expression for the dissociation of the solute:

$$CaSO_4(s) \rightleftharpoons Ca^{2+}(aq) + SO_4^{2-}(aq) \qquad K_{sp} = [Ca^{2+}][SO_4^{2-}]$$

At equilibrium, the moles of solid that dissolve per liter is:

$$\frac{2.03\text{ g } CaSO_4}{1\text{ L}}\times\frac{1\text{ mol } CaSO_4}{136.14\text{ g } CaSO_4} = 0.0149\text{ M}$$

The stoichiometry of the equation shows that the concentrations of calcium ion and sulfate ion are both the same as the moles of solid that dissolve per liter, 0.0149 M.

$$K_{sp} = [Ca^{2+}][SO_4^{2-}] = (0.0149)(0.0149) = 2.22 \times 10^{-4}$$

56. *Result:* $\mathbf{K_{sp} = 2.2 \times 10^{-12}}$

Analyze and Plan: Adapt the method developed in Problem-Solving Example 15.10.

Execute: Write the chemical equation and the equilibrium expression for the dissociation of the solute:

$$Ag_2CrO_4(s) \rightleftharpoons 2\ Ag^+(aq) + CrO_4^{2-}(aq) \qquad K_{sp} = [Ag^+]^2[CrO_4^{2-}]$$

At equilibrium, the moles of solid that dissolve per liter is:

$$\frac{2.7\times10^{-3}\text{ g }Ag_2CrO_4}{100.\text{ mL}}\times\frac{1\text{ mol }Ag_2CrO_4}{331.7301\text{ g }Ag_2CrO_4}\times\frac{1\text{ mol }CrO_4^{\;2-}}{1\text{ mol }Ag_2CrO_4}\times\frac{1000\text{ mL}}{1\text{ L}}=8.1\times10^{-5}\text{ M }CrO_4^{\;2-}$$

The mole ratio of the chemical equation shows that the silver ion concentration is twice the concentration of chromate ion, $2 \times (8.1 \times 10^{-5}\text{ M}) = 1.6 \times 10^{-4}\text{ M}$.

$$K_{sp} = (1.6 \times 10^{-4})^2(8.1 \times 10^{-5}) = 2.2 \times 10^{-12}$$

58. *Result:* $\mathbf{K_{sp} = 1.7 \times 10^{-5}}$

Analyze and Plan: Adapt the method developed in Problem-Solving Example 15.10.

Plan and Execute: Write the chemical equation and the equilibrium expression for the dissociation of the solute:

$$PbCl_2(s) \rightleftharpoons Pb^{2+} + 2\,Cl^-(aq) \qquad K_{sp} = [Pb^{2+}][Cl^-]^2$$

At equilibrium, the moles of solid that dissolve per liter is given as the solubility 1.62×10^{-2} M. The mole ratios of the chemical equation show that the concentration of lead ion is the same as the moles of solid that dissolve per liter, 1.62×10^{-2} M, and the chloride ion concentration is two times that value, $2 \times (1.62 \times 10^{-2}\text{ M}) = 3.24 \times 10^{-2}\text{ M}$.

$$K_{sp} = (1.62 \times 10^{-2})(3.24 \times 10^{-2})^2 = 1.70 \times 10^{-5}$$

Factors Affecting Solubility (Section 15-5)

60. *Result:* $\mathbf{1.2 \times 10^{-9}}$ **M or lower**

Analyze and Plan: Follow the method developed Problem-Solving Example 15.11 and use the relationships provided in Chapter 14 to determine the hydroxide concentration from the pH.

Plan and Execute: Given the pH, we can determine the concentration of the aqueous ion, OH^-. The precipitation of sparingly soluble $Zn(OH)_2$ will occur above a certain concentration of Zn^{2+}.

$$[OH^-] = 10^{-pOH} = 10^{pH - 14.00} = 10^{pH - 14.00} = 10^{10.00 - 14.00} = 1.0 \times 10^{-4}\text{ M}$$

$$K_{sp} = [Zn^{2+}][OH^-]^2 = 1.2 \times 10^{-17} \text{ (from Appendix H)}$$

$$1.2 \times 10^{-17} = [Zn^{2+}](1.0 \times 10^{-4})^2$$

$$[Zn^{2+}] = 1.2 \times 10^{-9}\text{ M}$$

An equilibrium $[Zn^{2+}]$ of 1.2×10^{-9} M or lower can exist in a solution with pH 10.00. Above that concentration, $Zn(OH)_2$ will precipitate. So, the maximum $[Zn^{2+}] = 1.2 \times 10^{-9}$ M.

62. *Result:* **9.16**

Analyze and Plan: Adapt the method developed in the solution to Question 60. Get the K_{sp} for $Mg(OH)_2$ from Appendix H: $K_{sp} = 1.8 \times 10^{-11}$.

Plan and Execute:

The acid added affects the concentration of the hydroxide. So, if all of the solid is dissolved it must be acidic enough to hold all the magnesium ions without precipitation. So, we will calculate the resulting magnesium ion concentration:

$$[Mg^{2+}] = \frac{5.00\text{ g }Mg(OH)_2}{1\text{ L}}\times\frac{1\text{ mol }Mg(OH)_2}{58.32\text{ g }Mg(OH)_2}\times\frac{1\text{ mol }Mg^{2+}}{1\text{ mol }Mg(OH)_2}=0.0857\text{ M}$$

$$K_{sp} = [Mg^{2+}][OH^-]^2$$

$$1.8 \times 10^{-11} = (0.0857)[OH^-]^2$$

$$[OH^-] = 1.4 \times 10^{-5}\ M$$

$$pOH = -\log[OH^-] = -\log(1.4 \times 10^{-5}) = 4.84$$

$$pH = 14.00 - pOH = 14.00 - 4.84 = 9.16$$

Enough acid must be added to drop the pH to 9.16, before all the solid will dissolve.

Common Ion Effect (Section 15-5b)

64. *Result:* **(a) Precipitation of $PbCl_2$ (b) More precipitate forms. Some $PbCl_2$ solid dissolves, but more AgCl forms. (c) Precipitation of $PbCl_2$**

Analyze: Beginning with a saturated solution of $PbCl_2$, identify what would be observed after specific changes are made to separate samples of the solution.

Plan: The solution initially contains the ions $Pb^{2+}(aq)$ and $Cl^-(aq)$. Using LeChatlier's Principle, determine how the change affects the solubility reaction

$$PbCl_2(s) \rightleftharpoons Pb^{2+}(aq) + 2\ Cl^-(aq)$$

Execute:

(a) $Pb(NO_3)_2$ is a source of $Pb^{2+}(aq)$. Increasing in the concentration of the reactant $Pb^{2+}(aq)$, causes more product $PbCl_2(s)$ to form, so $PbCl_2(s)$ will be observed to precipitate.

(b) $AgNO_3$ is a source of $Ag^+(aq)$. Increasing in the concentration of $Ag^+(aq)$, causes the reaction: $Ag^+(aq) + Cl^-(aq) \rightleftharpoons AgCl(s)$ and AgCl will precipitate. Some $PbCl_2(s)$ will dissolve as the $Ag^+(aq)$ consumes the reactant $Cl^-(aq)$, but the K_{sp} for AgCl is smaller than the K_{sp} for $PbCl_2$, so there will be a net increase in the solid precipitated.

(c) NaCl is a source of $Cl^-(aq)$. Increasing in the concentration of reactant $Cl^-(aq)$, causes more product $PbCl_2(s)$ to form, so $PbCl_2(s)$ will be observed to precipitate.

66. *Result:* **6.2×10^{-11} mol/L**

Analyze and Plan: We will use the equilibrium expression for K_{sp} and the known values to determine the unknown value. We will get the value of K_{sp} from Appendix H.

Execute:

The soluble Na_2CO_3 salt produces carbonate ions in the solution.

The anion base hydrolysis reaction should be considered because its K is larger than the K_{sp} above. The equation for that reaction is: $CO_3^{2-} + H_2O \rightleftharpoons HCO_3^- + OH^-$, some of the carbonate (–x) reacts to form hydrogen carbonate (x) and hydroxide (x). $K_b = \dfrac{x^2}{0.25 - x} = 2.1 \times 10^{-4}$. Solving for x (0.0072 M) and calculating the carbonate concentration, we find that it is slightly but measurably lower:

$$[CO_3^{2-}] = 0.25\ M - x = 0.25\ M - 0.0072\ M = 0.24\ M$$

$$K_{sp} = [Zn^{2+}][CO_3^{2-}] = [Zn^{2+}] \times (0.24) = 1.5 \times 10^{-11},\ [Zn^{2+}] = 6.2 \times 10^{-11}\ mol/L$$

68. *Result:* **(a) 1.8×10^{-11} (b) 9.2×10^{-3} M or higher**

Analyze and Plan: Adapt the methods developed in the solutions to Problem-Solving Example 15.10 and Question 66.

Plan and Execute:

(a) Write the chemical equation and the equilibrium expression for the dissociation of the solute:

$$Mg(OH)_2(s) \rightleftharpoons Mg^{2+}(aq) + 2\ OH^-(aq) \qquad K_{sp} = [Mg^{2+}][OH^-]^2$$

At equilibrium, the moles of solid that dissolve per liter are:

$$\frac{9.6 \text{ mg Mg(OH)}_2}{1 \text{ L}} \times \frac{1 \text{ g}}{1000 \text{ mg}} \times \frac{1 \text{ mol Mg(OH)}_2}{58.32 \text{ g Mg(OH)}_2} = 1.65 \times 10^{-4} \text{ M} \cong 1.6 \times 10^{-4} \text{ M}$$

The stoichiometry of the equation shows that the concentration of magnesium ion is the same as the moles of solid that dissolve per liter, 1.6×10^{-4} M, and the hydroxide ion concentration is two times that value, $2 \times (1.6 \times 10^{-4} \text{ M}) = 3.3 \times 10^{-4}$ M.

$$K_{sp} = (1.6 \times 10^{-4})(3.3 \times 10^{-4})^2 = 1.8 \times 10^{-11}$$

(b) First we determine the concentration of the magnesium ion in a 1.0 μg Mg^{2+} ion solution:

$$[Mg^{2+}] = \frac{5.0 \text{ μg } Mg^{2+}}{1 \text{ L}} \times \frac{10^{-6} \text{ g}}{1 \text{ μg}} \times \frac{1 \text{ mol } Mg^{2+}}{24.305 \text{ g } Mg^{2+}} = 2.1 \times 10^{-7} \text{ M}$$

Next, we use the calculated K_{sp} from (a) to calculate the hydroxide ion concentration in this solution:

$$K_{sp} = [Mg^{2+}][OH^-]^2$$

$$1.8 \times 10^{-11} = (2.1 \times 10^{-7})[OH^-]^2$$

$$[OH^-] = 9.2 \times 10^{-3} \text{ M}$$

An equilibrium $[OH^-]$ of 9.2×10^{-3} M or higher will keep the $[Mg^{2+}]$ at or below 1.0 μg/L

70. *Result:* **(a) 3.7×10^{-6} mol/L (b) 2.8×10^{-10} mol/L (c) 2.8×10^{-10} mol/L**

Analyze: Adapt the methods described in Problem-Solving Examples 15.9 and 15.11.

Plan and Execute: Insert variables describing the concentrations of the ions present from another sources as initial concentrations in the equilibrium calculation.

	$ZnCO_3(s) \rightleftharpoons$	$Zn^{2+}(aq)$	$+ \ CO_3^{2-}(aq)$
conc. initial (M)	solid	c_{zinc}	c_{carb}
change as reaction occurs (M)	solid	$+S$	$+S$
equilibrium conc. (M)	solid	$c_{zinc} + S$	$c_{carb} + S$

At equilibrium $\quad K_{sp} = [Zn^{2+}][CO_3^{2-}] = (c_{zinc} + S)(c_{carb} + S) = 1.4 \times 10^{-11}$

(a) In water, no other source of Zn^{2+} or CO_3^{2-} is present so, before dissociation, $c_{zinc} = c_{carb} = 0$.

$$(S)(S) = 1.4 \times 10^{-11}$$

$$S = 3.7 \times 10^{-6} \text{ mol/L}$$

(b) In $Zn(NO_3)_2$, an external source of Zn^{2+} is present, so $c_{zinc} = 0.050$ M. No external sources of CO_3^{2-} are present so $c_{carb} = 0$.

$$(0.050 + S)(S) = 1.4 \times 10^{-11}$$

Assuming S is small compared to 0.050, ignore its addition.

$$(0.050)(S) = 1.4 \times 10^{-11}$$

$$S = 2.8 \times 10^{-10} \text{ mol/L}$$

(c) In K_2CO_3, an external source of CO_3^{2-} is present. Neglecting the reaction of carbonate as a base, $c_{carb} = 0.050$ M. No external sources of Zn^{2+} are present so, before dissociation, $c_{zinc} = 0$.

$$(S)(0.050 + S) = 1.4 \times 10^{-11}$$

Assuming S is small compared to 0.050, ignore its addition.

$$(S)(0.050) = 1.4 \times 10^{-11}$$

$$S = 2.8 \times 10^{-10}\ \text{mol/L}$$

(Notice the assumption that S is small is a good one.)

Complex Ion Formation (Section 15-5c)

73. *Result:* **See equations and expressions below**

Analyze and Plan: The charge of the reactant metal ion is determined by subtracting the Lewis base's charge(s), if any, from the complex ion charge.

Execute:

(a) $Ag^+ + 2\ CN^- \rightleftharpoons [Ag(CN)_2]^-$ $\qquad K_f = \dfrac{[[Ag(CN)_2]^-]}{[Ag^+][CN^-]^2}$

(b) $Cd^{2+} + 4\ NH_3 \rightleftharpoons [Cd(NH_3)_4]^{2+}$ $\qquad K_f = \dfrac{[[Cd(NH_3)_4]^{2+}]}{[Cd^{2+}][NH_3]^4}$

75. *Result:* **6.3×10^{-3} mol or more**

Analyze: Adapt the method described in the solution to Question 62, and Problem-Solving Example 15.13 in Section 15-5c.

Plan: The $Na_2S_2O_3$ salt is a source of $S_2O_3^{2-}$, a Lewis base capable of forming a complex ion with the silver ion. If all the solid is dissolved, it must have enough $S_2O_3^{2-}$ to complex enough the silver ions and prevent the precipitation of AgBr in the resulting Br^- solution. First, we get the balanced equation for the reaction from Problem-Solving Example 15.13 and determine the value of its equilibrium constant using Tables 15.2 and 15.3

Execute:

$$AgBr(s) + 2\ S_2O_3^{2-}(aq) \rightleftharpoons [Ag(S_2O_3)_2]^{3-}(aq) + Br^-(aq) \qquad K = \frac{[[Ag(S_2O_3)_2]^{3-}][Br^-]}{[S_2O_3^{2-}]^2}$$

$$K = K_{sp} \times K_f = [Ag^+][Br^-] \times \frac{[[Ag(S_2O_3)_2]^{3-}]}{[Ag^+][S_2O_3^{2-}]^2}$$

$$K = \frac{[[Ag(S_2O_3)_2]^{3-}][Br^-]}{[S_2O_3^{2-}]^2}$$

$$K = K_{sp} \times K_f = (5.0 \times 10^{-13}) \times (2.0 \times 10^{13}) = 10.$$

Now, calculate the $[Br^-]$ and $[[Ag(S_2O_3)_2]^{3-}]$, once all the AgBr has dissolved:

$$[Br^-] = \frac{0.020\ \text{mol AgBr}}{1.0\ \text{L}} \times \frac{1\ \text{mol Br}^-}{1\ \text{mol AgBr}} = 0.020\ \text{M},$$

Similarly, $[[Ag(S_2O_3)_2]^{3-}] = 0.020$ M

Now, we can calculate the necessary $[S_2O_3^{2-}]$:

$$10. = \frac{(0.020)(0.020)}{[S_2O_3^{2-}]^2}$$

$$[S_2O_3^{2-}] = 6.3 \times 10^{-3}\ \text{M}$$

One mole of $S_2O_3^{2-}$ is found in each mole of $Na_2S_2O_3$, so we must add 6.3×10^{-3} mol of $Na_2S_2O_3$ to this 1.0 L solution.

77. *Result:* **See equations below**

Analyze and Plan: The hydroxide anion of the solid is neutralized in excess acid, and the aqueous complex ion is formed in excess base.

Execute:

(a)
$$Zn(OH)_2(s) + 2\ H_3O^+(aq) \longrightarrow Zn^{2+}(aq) + 4\ H_2O(\ell)$$
$$Zn(OH)_2(s) + 2\ OH^-(aq) \longrightarrow [Zn(OH)_4]^{2-}(aq)$$

(b)
$$Sb(OH)_3(s) + 3\ H_3O^+(aq) \longrightarrow Sb^{3+}(aq) + 6\ H_2O(\ell)$$
$$Sb(OH)_3(s) + OH^-(aq) \longrightarrow [Sb(OH)_4]^-(aq)$$

General Questions

80. *Result:* **(a) $BaSO_4$ (b) 2.5×10^{-4} M**

Analyze: Given the concentrations of two solutions whose anions, phosphate and sulfate, form a precipitate when combined with barium ion and the slow addition of a barium compound, determine which solid precipitates first and calculate the concentration of the first anion when the second anion species starts to precipitate.

Plan: Adapt method used in Problem-Solving Example 15.15

Execute:

(a) $Ba_3(PO_4)_2(s) \rightleftharpoons 3\ Ba^{2+}(aq) + 2\ PO_4^{3-}(aq)$ $\quad K_{sp} = [Ba^{2+}]^3[PO_4^{3-}]^2 = 3.4 \times 10^{-23}$

$$[Ba^{2+}] = \sqrt[3]{\frac{3.4 \times 10^{-23}}{[PO_4^{3-}]^2}} = \sqrt[3]{\frac{3.4 \times 10^{-23}}{(0.020\ M)^2}} = 4.4 \times 10^{-7}\ M$$

$BaSO_4(s) \rightleftharpoons Ba^{2+}(aq) + SO_4^{2-}(aq)$ $\quad K_{sp} = [Ba^{2+}]^2[SO_4^{2-}] = 1.1 \times 10^{-10}$

$$[Ba^{2+}] = \frac{1.1 \times 10^{-10}}{[SO_4^{2-}]} = \frac{1.1 \times 10^{-10}}{0.050\ M} = 2.2 \times 10^{-9}\ M$$

2.2×10^{-9} M < 4.4×10^{-7} M, so $BaSO_4$ precipitates first.

(b) The $[Ba^{2+}] = 4.4 \times 10^{-7}$ M, when the second solid, $Ba_3(PO_4)_2$ begins to precipitate.

Use the K_{sp} for $BaSO_4$ to calculate the SO_4^{2-} when the $[Ba^{2+}] = 4.4 \times 10^{-7}$ M:

$$[SO_4^{2-}] = \frac{1.1 \times 10^{-10}}{[Ba^{2+}]} = \frac{1.1 \times 10^{-10}}{4.4 \times 10^{-7}\ M} = 2.5 \times 10^{-4}\ M$$

☑ *Reasonable Result Check:* The concentration of the SO_4^{2-} should be very small when the second precipitation begins.

82. *Result:* **(a) CaF_2 (b) 79%**

Analyze: Given the concentrations of two solutions whose cations, Pb^{2+} and Ca^{2+}, form a precipitate when combined with fluoride ion and the slow addition of a fluoride compound, determine which solid precipitates first and calculate the percentage of Pb^{2+} and Ca^{2+} when the second compound begins to precipitate.

Plan: Adapt the method used in Question 80 and Problem-Solving Example 15.15

Execute:

(a) $PbF_2(s) \rightleftharpoons Pb^{2+}(aq) + 2\ F^-(aq)$ $K_{sp} = [Pb^{2+}][F^-]^2 = 2.7 \times 10^{-8}$

$$[F^-] = \sqrt{\frac{2.7\times10^{-8}}{[Pb^{2+}]}} = \sqrt[3]{\frac{2.7\times10^{-8}}{0.010\ M}} = 1.6\times10^{-3}\ M$$

$CaF_2(s) \rightleftharpoons Ca^{2+}(aq) + 2\ F^-(aq)$ $K_{sp} = [Ca^{2+}][F^-]^2 = 5.3 \times 10^{-9}$

$$[F^-] = \sqrt{\frac{5.3\times10^{-9}}{[Ca^{2+}]}} = \sqrt{\frac{5.3\times10^{-9}}{0.0100\ M}} = 7.3\times10^{-4}\ M$$

7.3×10^{-4} M < 1.6×10^{-3} M, so CaF_2 precipitates first.

(a) The $[F^-] = 1.6 \times 10^{-3}$ M, when the second solid, PbF_2 begins to precipitate.

Use the K_{sp} for CaF_2 to calculate the $[Ca^{2+}]$ when the $[F^-] = 1.6 \times 10^{-3}$ M:

$$[Ca^{2+}] = \frac{5.3\times10^{-9}}{[F^-]^2} = \frac{5.3\times10^{-9}}{(1.6\times10^{-3}\ M)^2} = 2.1\times10^{-3}\ M$$

$$\text{Percent } Ca^{2+} = \frac{\text{final M } Ca^{2+}}{\text{initial M } Ca^{2+}}\times100\% = \frac{0.01000-0.0021\ M\ Ca^{2+}}{0.0100\ M\ Ca^{2+}}\times100\% = 79\%\ \text{Cl by mass}$$

☑ *Reasonable Result Check:* The percent of Ca^{2+} precipitated will be large, but not very close to 100% because the two K_{sp}'s are very close together.

85. *Result:* **(a) H_2O, CH_3COO^-, Na^+, CH_3COOH, H_3O^+, OH^- (b) 4.95 (c) 5.05**
(d) $CH_3COOH(aq) + H_2O(\ell) \rightleftharpoons H_3O^+(aq) + CH_3COO^-(aq)$

(a) *Explanation:*

The solution contains an abundance of water, more sodium acetate than acetic acid, a small amount of H_3O^+ since the solution is an acidic buffer, and an even smaller amount of OH^-. Therefore, the ions and molecules in solution from the largest concentration to the smallest is: H_2O, CH_3COO^-, Na^+, CH_3COOH, H_3O^+, OH^-.

(b) *Analyze and Plan:* Adapt the methods used in the solutions to Question 29.

Plan and Execute:

The equation for the equilibrium and the equilibrium expression are:

$$CH_3COOH(aq) + H_2O(\ell) \rightleftharpoons H_3O^+(aq) + CH_3COO^-(aq) \qquad K_a = \frac{[H_3O^+][CH_3COO^-]}{[CH_3COOH]}$$

$$\frac{4.95\ \text{g NaCH}_3\text{COO}}{250.\ \text{mL}}\times\frac{1\ \text{mol NaCH}_3\text{COO}}{82.0337\ \text{g NaCH}_3\text{COO}}\times\frac{1000\ \text{mL}}{1\ \text{L}}\times\frac{1\ \text{mol CH}_3\text{COO}^-}{1\ \text{mol NaCH}_3\text{COO}} = 0.241\ \text{M CH}_3\text{COO}^-$$

As the reactants react, the concentrations change stoichiometrically, until they reach equilibrium.

	$CH_3COOH(aq)$	$H_3O^+(aq)$	$CH_3COO^-(aq)$
initial conc. (M)	0.150	0	0.241
change as reaction occurs (M)	– x	+ x	+ x
equilibrium conc. (M)	0.150 – x	x	0.241 + x

At equilibrium $$K_a = \frac{(x)(0.241+x)}{(0.150-x)} = 1.8 \times 10^{-5}$$

Assume x is very small and does not affect the sum or the difference.

$$1.8 \times 10^{-5} = \frac{(x)(0.241)}{(0.150)}$$

$$x = 1.1 \times 10^{-5}\ M = [H_3O^+]$$

$$pH = -\log[H_3O^+]$$

$$pH = -\log(1.1 \times 10^{-5}) = 4.95$$

(c) The strong base neutralizes the acid in the solution, CH_3COOH.

$$CH_3COOH(aq) + OH^-(aq) \longrightarrow CH_3COO^-(aq) + H_2O(\ell)$$

$$(\text{conc. } OH^-) = \frac{80.\ \text{mg NaOH}}{100.\ \text{mL}} \times \frac{1000\ \text{mL}}{1\ \text{L}} \times \frac{1\ \text{g}}{1000\ \text{mg}} \times \frac{1\ \text{mol NaOH}}{40.00\ \text{g NaOH}} \times \frac{1\ \text{mol } OH^-}{1\ \text{mol NaOH}} = 0.020\ M$$

We take it as far to the products as possible, using the method of limiting reactants.

	$CH_3COOH(aq)$	$OH^-(aq)$	$CH_3COO^-(aq)$
initial conc. (M)	0.150	0.020	0.241
change as reaction occurs (M)	– 0.020	– 0.020	+ 0.020
conc. final (M)	0.130	0	0.261

The solution is still a buffer solution, containing an acid-base conjugate pair, so we can find the pH using K_a of the acid.

$$CH_3COOH(aq) + H_2O(\ell) \rightleftharpoons H_3O^+(aq) + CH_3COO^-(aq) \qquad K_a = \frac{[H_3O^+][CH_3COO^-]}{[CH_3COOH]}$$

As the reactants react, the concentrations change stoichiometrically, until they reach equilibrium.

	$CH_3COOH(aq)$	$H_3O^+(aq)$	$CH_3COO^-(aq)$
initial conc. (M)	0.130	0	0.261
change as reaction occurs (M)	– x	+ x	+ x
equilibrium conc. (M)	0.130 – x	x	0.261 + x

At equilibrium $$K_a = \frac{(x)(0.261 + x)}{(0.130 - x)} = 1.8 \times 10^{-5}$$

Assume x is very small and does not affect the sum or the difference.

$$1.8 \times 10^{-5} = \frac{(x)(0.261)}{(0.130)} \qquad x = 9.0 \times 10^{-6}\ M = [H_3O^+]$$

$$pH = -\log[H_3O^+] = -\log(9.0 \times 10^{-6}) = 5.05$$

(d) $CH_3COOH(aq) + H_2O(\ell) \rightleftharpoons H_3O^+(aq) + CH_3COO^-(aq)$

87. *Result:* [*o*-ethylbenzoic acid] must be approximately 0.63 times the [potassium *o*-ethylbenzoate]

Analyze and Plan: Adapt the methods used in the solution to Question 21.

$$pH = pK_a + \log\left(\frac{[o\text{-ethylbenzoate}]}{[o\text{-ethylbenzoic acid}]}\right)$$

Plan and Execute:

At equilibrium, pH = 4.0 and $pK_a = 3.79$

$$pH - pK_a = \log\left(\frac{[o\text{-ethylbenzoate}]}{[o\text{-ethylbenzoic acid}]}\right)$$

$$4.0 - 3.79 = 0.2 = \log\left(\frac{[o\text{-ethylbenzoate}]}{[o\text{-ethylbenzoic acid}]}\right)$$

$$\frac{[o\text{-ethylbenzoate}]}{[o\text{-ethylbenzoic acid}]} = 10^{0.2}$$

$$\frac{[o\text{-ethylbenzoic acid}]}{[o\text{-ethylbenzoate}]} = 10^{-0.2} = 0.63$$

The [*o*-ethylbenzoic acid] must be approximately 0.63 times the [potassium *o*-ethylbenzoate].

89. *Result:* **0.012 mol HNO_2**

Analyze and Plan: Adapt the method used in the solution to Question 21.

Plan and Execute:

$$[NO_2^-] = \frac{7.50 \text{ g } KNO_2}{1 \text{ L}} \times \frac{1 \text{ mol } KNO_2}{85.11 \text{ g } KNO_2} \times \frac{1 \text{ mol } NO_2^-}{1 \text{ mol } KNO_2} = 0.0881 \text{ M}$$

In Table 14.2, the K_a of nitrous acid is given: 7.1×10^{-4}. As described in Section 15-1, calculate the pK_a:

$$pK_a = -\log K_a = -\log(7.1 \times 10^{-4}) = 3.15$$

The pH is 4.00. The initial concentration of NO_2^-, the conjugate base, is 0.0881 M. The conjugate acid, HNO_2, is being added, but its initial concentration is unknown, so use the variable "c" to identify that quantity.

$$pH = pK_a + \log\left(\frac{[\text{conj. base}]}{[\text{conj. acid}]}\right)$$

$$4.00 = 3.15 + \log\left(\frac{0.0881}{c}\right)$$

Solve for c: $c = 0.0881 \text{ M} \times 10^{-(4.00-3.35)} = 0.012 \text{ M}$

Add 0.012 mol HNO_2 to one liter of solution to make this buffer.

91. *Result:* **(a) 2.78 (b) 5.39**

(a) *Analyze and Plan:* Use the method similar to that used in the solution to Question 14.54.

Plan and Execute:

The equation for the equilibrium and the equilibrium expression are:

$$CH_3COOH(aq) + H_2O(\ell) \rightleftharpoons H_3O^+(aq) + CH_3COO^-(aq) \qquad K_a = \frac{[CH_3COO^-][H_3O^+]}{[CH_3COOH]}$$

As the reactants react, the concentrations change stoichiometrically, until they reach equilibrium.

	$CH_3COOH(aq)$	$H_3O^+(aq)$	$CH_3COO^-(aq)$
initial conc. (M)	0.15	0	0
change as reaction occurs (M)	– x	+ x	+ x
equilibrium conc. (M)	0.15 – x	x	x

At equilibrium $K_a = \frac{(x)(x)}{(0.15 - x)} = 1.8 \times 10^{-5}$

Assume x is very small and does not affect the difference.

$$x^2 = (1.8 \times 10^{-5})(0.15) \qquad x = 1.6 \times 10^{-3}\ M = [H_3O^+]$$

$$pH = -\log[H_3O^+] = -\log(1.6 \times 10^{-3}) = 2.78$$

(b) *Analyze and Plan:* Use the methods described in the solutions to Questions 26(a) and 29.

Plan and Execute:

The solution is a buffer solution, containing an acid-base conjugate pair, so we can find the pH using K_a of the acid described in (a).

$$[CH_3COO^-] = \frac{83\ g\ NaCH_3COO}{1.50\ L} \times \frac{1\ mol\ NaCH_3COO}{82.03\ g\ NaCH_3COO} \times \frac{1\ mol\ CH_3COO^-}{1\ mol\ NaCH_3COO} = 0.67\ M\ CH_3COO^-$$

The solution is now a buffer solution, containing an acid-base conjugate pair, so we can find the pH using the Henderson-Hasselbalch equation. In Table 15.1, the pK_a is 4.74.

$$pH = pK_a + \log\left(\frac{[\text{conj. base}]}{[\text{conj. acid}]}\right)$$

$$pH = 4.74 + \log\left(\frac{0.67}{0.15}\right) = 4.74 + 0.65 = 5.39$$

93. *Result:* **3.5×10^{-6}**

Analyze and Plan: Adapt the methods described in the solution to Question 43. We will use the equivalence point data to find the total moles of the weak acid, HA in the solution.

$$35.00\ mL\ NaOH \times \frac{1\ L}{1000\ mL} \times \frac{0.100\ mol\ NaOH}{1\ L\ NaOH} \times \frac{1\ mol\ HA}{1\ mol\ NaOH} = 0.00350\ mol\ HA$$

Plan and Execute:

Calculate the moles of hydroxide added, after 20.00 mL of titrant have been added.

$$mol\ OH^- = 20.00\ mL \times \frac{1\ L}{1000\ mL} \times \frac{0.100\ mol\ OH^-}{1\ L} = 0.00200\ mol\ OH^-$$

The strong base neutralizes the acid, HA, in the solution.

$$HA + OH^-(aq) \longrightarrow A^-(aq) + H_2O(\ell)$$

We take it as far to the products as possible, using the method of limiting reactants.

	Initial	After Reaction of HA with NaOH
mol A^-	0	0 + 0.00200 = 0.00200
mol HA	0.00350	0.00300 – 0.00200 = 0.00100

The equation for the equilibrium and the equilibrium expression are:

$$HA(aq) + H_2O(\ell) \rightleftharpoons H_3O^+(aq) + A^-(aq) \qquad K_a = \frac{[H_3O^+][A^-]}{[HA]}$$

We calculate the concentrations using the total volume after 20.0 mL of titrant has been added:

20.0 mL + 40.0 mL = 60.0 mL, which is 0.0600 L.

$$(\text{conc. } A^-) = \frac{0.00200 \text{ mol}}{0.0600 \text{ L}} = 0.0333 \text{ M} \qquad (\text{conc. HA}) = \frac{0.00100 \text{ mol}}{0.0600 \text{ L}} = 0.0167 \text{ M}$$

The solution is now a buffer solution, so use the Henderson-Hasselbalch equation to calculate pK_a. The concentration of the conjugate acid, HA, is 0.0167 M; the concentration of the conjugate base, A^-, is 0.0333 M. The pH is 5.75.

$$pH = pK_a + \log\left(\frac{[\text{conj. base}]}{[\text{conj. acid}]}\right)$$

$$pK_a = 5.75 - \log\left(\frac{0.0333}{0.0167}\right) = 5.75 - 0.30 = 5.45$$

$$pK_a = -\log K_a, \text{ so } K_a = 10^{-pK_a} = 10^{-5.45} = 3.5 \times 10^{-6}$$

95. *Result:* **No effect**

Explanation: There is no effect on the equilibrium state if more solid is added, since the quantity of that solid present does not affect the equilibrium state.

Applying Concepts

96. *Result:* **(a) 5.004 (b) 6.84 (c) Pure water; the pH changed drastically**

Analyze and Plan: Adapt the method described in the solutions to Question 31 and 29.

Plan and Execute: Calculate the concentration of new hydronium ion, using the total volume of the solutions after the addition, but before any reaction.

$$V = 1.00 \text{ L} + 10.0 \text{ mL} \times \frac{1 \text{ L}}{1000 \text{ mL}} = 1.01 \text{ L} = 11 \text{ mL}$$

$$(\text{conc. } H_3O^+) = \frac{10.0 \text{ mL}}{1010 \text{ mL}} \times \frac{0.0010 \text{ mol HCl}}{1 \text{ L}} \times \frac{1 \text{ mol } H_3O^+}{1 \text{ mol HCl}} = 9.90 \times 10^{-6} \text{ M } H_3O^+$$

(a) In pure water, the pH drops due to the presence of HCl:

$$pH = -\log[H_3O^+] = -\log[9.90 \times 10^{-6} \text{ M}] = 5.004$$

(b) In the lake water, the strong acid neutralizes the conjugate base in the solution, HCO_3^-.

$$HCO_3^-(aq) + H_3O^+(aq) \longrightarrow H_2CO_3 + H_2O(\ell)$$

This reaction is product-favored, so we will first make as many products as possible. Using the method of limiting reactants, run the reaction towards products until one of the reactants runs out.

	$HCO_3^-(aq)$	$H_3O^+(aq)$	$H_2CO_3(aq)$
initial conc. (M)	1.6×10^{-4}	9.90×10^{-6}	4.0×10^{-5}
change as reaction occurs (M)	-9.90×10^{-6}	-9.90×10^{-6}	$+9.90 \times 10^{-6}$
final conc. (M)	1.5×10^{-4}	0	5.0×10^{-5}

The solution is still a buffer solution, but the concentrations are very small, so we will find the pH using K_a of the acid, as was done in Chapter 14 using K_a for H_2CO_3 from Table 14.2.

	$H_2CO_3(aq)$ $\rightleftharpoons$	$HCO_3^-(aq)$	$H_3O^+(aq)$
initial conc. (M)	5.0×10^{-5}	1.5×10^{-4}	0
change as reaction occurs (M)	$-x$	$+x$	$+x$
equilibrium conc. (M)	$5.0 \times 10^{-5} - x$	$1.5 \times 10^{-4} + x$	x

At equilibrium $$K_a = \frac{(1.5\times10^{-4}+x)(x)}{(5.0\times10^{-5}-x)} = 4.3 \times 10^{-7}$$

Assume x is small and does not affect the sum or difference, then check those assumptions.

$$\frac{(1.5\times10^{-4})(x)}{(5.0\times10^{-5})} = 4.3 \times 10^{-7}$$

$$x = 1.4 \times 10^{-7}\ M = [H_3O^+]$$

Check assumptions: $1.5 \times 10^{-4}\ M + 1.4 \times 10^{-7}\ M = 1.5 \times 10^{-4}\ M$

$$5.0 \times 10^{-5}\ M - 1.4 \times 10^{-7}\ M = 5.0 \times 10^{-5}\ M$$

Adding and subtracting x, here, is indeed too small to change these concentrations.

Now, calculate the new pH:

$$pH = -\log[H_3O^+] = \log(1.4 \times 10^{-7}\ M) = 6.84$$

(c) The pH change after adding HCl to pure water: $\Delta pH = 7.00 - 5.004 = -2.00$

The pH change after adding HCl to lake water: $\Delta pH = 6.96 - 6.84 = -0.12$

The pure water did not act as a buffer, as seen by the large decrease in its pH when the HCl was added. The pH for the lake water did decrease when HCl was added, but not as drastically as that of pure water.

☑ *Reasonable Result Check:* Adding strong acid makes the solution more acidic, so it makes sense that the pH goes down. It also makes sense that if a solution contains a buffer, as the lakewater does, it can resist changes in pH.

99. *Result/Explanation:* The tiny amount of base (CH_3COO^-) present is insufficient to prevent the pH from changing dramatically if a strong acid is introduced into the solution.

101. *Result:* **2.3 x 10⁻⁴**

Analyze and Plan: When exactly half of the acid in the original solution has been neutralized, that means that equal quantities of the acid and base are present in the solution. As described in the solution to Question 29, when equimolar quantities are present, $pH = pK_a$.

Execute: $$pH = 3.64 = pK_a$$

$$K_a = 10^{-pK_a} = 10^{-3.64} = 2.3 \times 10^{-4}$$

103. *Result/Explanation:* The extra dissolved CO_2 causes an increase in H_2CO_3, a weak acid, the excess acid raises the $[H_3O^+]$ decreasing the blood pH, causing acidosis, which is the term used to describe acidification of the blood.

105. *Result/Explanation:* Given: $Ca_5(PO_4)_3OH(s) \rightleftharpoons 5\ Ca^{2+}(aq) + 3\ PO_4^{3-}(aq) + OH^-(aq)$

(a) Apatite is a relatively insoluble compound. Drinking milk containing calcium ion increases the concentration of the product Ca^{2+}. According to LeChatelier's principle, this change drives the equilibrium shown above toward the formation of apatite.

(b) The presence of lactic acid will react with OH^-. The subsequent decrease in OH^- concentration will cause apatite to dissolve as equilibrium is re-established.

107. *Result/Explanation:* During a recent television medical drama, a person went into cardiac arrest and stopped breathing. A doctor quickly injected sodium hydrogen carbonate into the heart. This would indicate that cardiac arrest leads to **acidosis** and that the sodium hydrogen carbonate helps to **increase** the pH. Explanation: When the patient stopped breathing, his blood pH decreases because of the increase in H_2CO_3. This leads to acidosis, an acidification of the blood. The pH of the blood drops too low. To counteract the extra acidity, the weak base sodium hydrogen carbonate is introduced to increase the pH back to normal.

109. *Result:* **(a) Boxes C (b) Boxes C and D (c) Box E (d) Box D (e) Box A (f) Box B (g) Box E**

Explanation:

(a) The amphiprotic $Zn(OH)_2$ will dissolve in sufficient acid (enough to neutralize the OH^-) and sufficient base, to form the complex ion, $Zn(OH)_4^{2-}$. So, **Boxes C** can be correct in answering this prompt.

(b) HPO_4^{2-} can be partially neutralized with NaOH to form a HPO_4^{2-}/PO_4^{3-} buffer. It can also be partially protonated with an increase in the solution's acidity by decreasing the pH to form $H_2PO_4^-/HPO_4^{2-}$. Therefore, **Boxes C and D** both serve as correct answers for this prompt.

(c) At halfway to the equivalence point in a titration of a weak monoprotic acid, the solution has equal concentrations of acid and conjugate base:

$$pH = pK_a + \log\left(\frac{[base]}{[acid]}\right)$$

Because, log(1) = 0, pH = pK_a. This answer is given in **Box E.**

(d) When SO_2 is present in the atmosphere, precipitation becomes acidic. Decreasing pH is in **Box D.**

(e) A Lewis acid-base reaction occurs between Cr^{3+} ion (a Lewis acid) and the four OH^- ions (Lewis bases) to form $[Cr(OH)_4]^-$. The answer is **Box A.**

(f) The reaction quotient must be greater than the solubility product for a precipitation reaction to be assured of no precipitation. $Q > K_{sp}$ is in **Box B.**

(g) The buffer solution has equal concentrations of acid and conjugate base, so as described in (c) pH = pK_a, given in **Box E.**

111. *Result:* **(a) Box III (b) Box IV (c) Box II (d) Box I**

Explanation: The solution starts with an aqueous acidic solution of strong acid, HA, because the acid is strong, the solution will contain an equal quantity (16) of H_3O^+ and A^- ions present.

(a) After a very small amount of base has been added, a very small amount of H_3O^+ ions will be neutralized, leaving slightly fewer H_3O^+ ions (15) than A^- ions (16). This is shown in **Box III.**

(b) When the solution is halfway to the equivalence point, there will be half as many H_3O^+ molecules as there were to begin with. Eight are shown in **Box IV**.

(c) When enough titrant has been added to take the solution to just past the equivalence point, there will be no H_3O^+ ions left and some leftover OH^-. This is shown in **Box II.**

(d) At the equivalence point, there will be only A^- ions. This is shown in **Box I.**

113. *Result:* **5.64**

Analyze and Plan: Adapt the method in the solution to Question 24.

Plan and Execute:

Find the initial moles of H_3O^+ and C_5H_5N, after the solutions are combined:

$$25.0 \text{ mL} \times \frac{1 \text{ L}}{1000 \text{ mL}} \times \frac{0.085 \text{ mol } C_5H_5N}{1 \text{ L}} = 0.00212 \text{ mol } C_5H_5N$$

$$5.5 \text{ mL} \times \frac{1 \text{ L}}{1000 \text{ mL}} \times \frac{0.102 \text{ mol HCl}}{1 \text{ L}} \times \frac{1 \text{ mol } H_3O^+}{1 \text{ mol HCl}} = 0.00056 \text{ mol } H_3O^+$$

The acid neutralizes the base, via product-favored neutralization reaction.

$$H_3O^+(aq) + C_5H_5N(aq) \longrightarrow C_5H_5NH^+(aq) + H_2O(\ell)$$

We take it as far to the products as possible, using the method of limiting reactants.

	$H_3O^+(aq)$	$C_5H_5N(aq)$	$C_5H_5NH^+(aq)$
init. mol	0.00056	0.00212	0
change in mol	-0.00056	-0.00056	+0.00056
after rxn (mol)	0	0.00156	0.00056

The solution is a buffer solution, containing an acid-base conjugate pair, so we can find the pH using Henderson-Hasselbalch equation. In the question, we are given K_b for $C_5H_5N = 1.6x10^{-9}$. We will use that to find the K_a for the dissociation of the pyradinium cation, $C_5H_5NH^+(aq)$:

$$K_a = \frac{K_w}{K_b} = \frac{1.00\times10^{-14}}{1.6\times10^{-9}} = 6.3\times10^{-6}$$

Determine the post-neutralization concentrations of $C_5H_5NH^+$ and C_5H_5N, using moles and the combined volumes of solutions: 25.0 mL + 5.5 mL = 30.5 mL, which is 0.0305 L

$$[C_5H_5NH^+] = \frac{0.00056 \text{ mol } C_5H_5NH^+}{0.03050 \text{ L}} = 0.0184 \text{ M } C_5H_5N^+$$

$$[C_5H_5N] = \frac{0.00156 \text{ mol } C_5H_5N}{0.03050 \text{ L}} = 0.0511 \text{ M } C_5H_5N$$

As described in Section 15.1, calculate the pK_a: $pK_a = -\log K_a = -\log(6.3 \times 10^{-6}) = 5.20$

$$pH = 5.20 + \log\left(\frac{0.0511}{0.0184}\right) = 5.20 + 0.44 = 5.64$$

115. *Result:* **3.22**

Analyze, Plan, and Execute: Distilled water with a pH of 5.6, has an H_3O^+ concentration.

$$[H_3O^+] = 10^{-pH} = 10^{-5.6} = 2.5 \times 10^{-6} \text{ M}$$

The acidic pH is a result of the following equlibria:

$$CO_2(aq) + H_2O(\ell) \rightleftharpoons H_2CO_3(aq)$$

$$H_2CO_3(aq) + H_2O(\ell) \rightleftharpoons H_3O^+(aq) + HCO_3^-(aq)$$

HCl is strong acid. It ionizes completely to produce H_3O^+. As a result, the equilibria reactions above shift the left in order to re-establish a new equilibrium state. That means the $H_3O^+(aq)$ concentration *resulting from the ionization of the weak acid* will be even less than that calculated, above.

Determine the $H_3O^+(aq)$ concentration resulting from the ionization of the HCl after one drop of concentrated HCl solution is added to the 1.0 L of water. Notice that the drop adds negligible volume to the 1.0 L solution.

$$\frac{1 \text{ drop}}{1.0 \text{ L}} \times \frac{1 \text{ mL}}{20 \text{ drop}} \times \frac{1 \text{ L}}{1000 \text{ mL}} \times \frac{12 \text{ mol HCl}}{1 \text{ L}} = 6.0 \times 10^{-4} \text{ M HCl}$$

It is clear that the addition of HCl produces so much H_3O^+ that the presence of aqueous CO_2 as an acid source is negligible. Ultimately, the pH depends only on the molarity of the HCl.

$$pH = -\log[H_3O^+] = -\log(6.0 \times 10^{-4}) = 3.22$$

117. *Result:* **(a) 1.1×10^{-9} M (b) 0.010 M**

Analyze: Determine the barium ion concentration in the presence of each these anions at equilibrium.

Plan and Execute:

(a) $BaSO_4$ equilibrium: $BaSO_4(s) \rightleftharpoons Ba^{2+}(aq) + SO_4^{2-}(aq)$

$$K_{sp} = [Ba^{2+}][SO_4^{2-}] = [Ba^{2+}](0.010) = 1.1 \times 10^{-10}$$

$$[Ba^{2+}] = 1.1 \times 10^{-9}\text{ M}$$

The $[Ba^{2+}]$ will be 1.1×10^{-9} M when the $BaSO_4$ begins to precipitate.

(b) BaF_2 equilibrium: $BaF_2(s) \rightleftharpoons Ba^{2+}(aq) + 2\,F^{-}(aq)$

$$K_{sp} = [Ba^{2+}][F^{-}]^2 = [Ba^{2+}](0.010)^2 = 1.0 \times 10^{-6}$$

$$[Ba^{2+}] = 0.010\text{ M}$$

The $[Ba^{2+}]$ will be 0.010 M when the BaF_2 begins to precipitate.

122. *Result:* **(a) 13.00 (b) 5.0 × 10^{-4}; T is different from 25°C.**

Analyze, Plan, and Execute: This question uses the following:

$$Ca(OH)_2(s) \rightleftharpoons Ca^{2+}(aq) + 2\,OH^{-}(aq) \qquad K_{sp} = [Ca^{2+}][OH^{-}]^2$$

(a) First stoichiometrically determine the concentration of OH^- created in the dissolution of $Ca(OH)_2$.

$$\frac{0.0050\text{ mol }Ca(OH)_2}{0.100\text{ L}} \times \frac{2\text{ mol }OH^{-}}{1\text{ mol }Ca(OH)_2} = 0.10\text{ M }OH^{-}$$

$$pOH = -\log(0.10) = 1.00$$

$$pH = 14 - pOH = 13.00$$

(b) Stoichiometrically, the concentration of calcium ion is one half that of the hydroxide ion.

$$[Ca^{2+}] = \frac{1}{2}(0.10\text{ M}) = 0.050\text{ M}$$

$$K_{sp} = [Ca^{2+}][OH^{-}]^2 = (0.050)(0.10)^2$$

$$K_{sp} = 5.0 \times 10^{-4}$$

The K_{sp} differs from the value given in Appendix H because the T is different from 25°C.

123. *Result:* **Sample A: $NaHCO_3$; Sample B: NaOH; Sample C: NaOH and $NaHCO_3$; Sample D: Na_2CO_3 (other answers can also be correct)**

Analyze, Plan, and Execute: We must assume that any of these substances, if present, are present in "reasonable" concentrations.

Sample A:

Phenolphthalein is colorless, suggesting the pH is below 8.3. The most acidic of the choices is $NaHCO_3$, and that is a likely guess for Sample A. (Notice: any of these compounds in solution, even NaOH, can have a pH below 8.3, if it is sufficiently dilute, that's why we have to assume reasonable concentrations.)

Sample B:

We interpret the evidence to say that the methyl orange changed to its acidic color as soon as it was added. That means the solution became acidic as soon as the phenolphthalein end point was reached. That suggests the titration of a strong base with a very dramatic decline to acidic pH values after the equivalence point. In addition, at very low pH values (below pH = 3.01), bubbles would have been observed as the carbonate reacted to form $CO_2(g)$. All of these interpretations lead us to believe that Sample B is probably NaOH.

Sample C:

Presumably, both indicators changed color rapidly, suggesting that they were true endpoints, and not just a result of pH changes. Because two different acidic end points were reached with different indicators, this sample must be a mixture of excess strong base, NaOH, and one or more of the salts, either Na_2CO_3 or

$NaHCO_3$. It is not possible to distinguish which salt is present, due to the immediate reaction of the strong base with $NaHCO_3$ to make the same product Na_2CO_3.

Sample D:

This sample has a two endpoints the second at exactly twice the volume of the first, suggesting that two neutralizations occurred. That means the sample may have been pure Na_2CO_3, which undergoes two sequential neutralization reactions during the titration. Alternative scenarios, equally plausible, would be a mixture of equimolar quantities of NaOH and $NaHCO_3$ or a mixture of NaOH, Na_2CO_3, and $NaHCO_3$ with the following proportions:

$$(\text{conc. NaOH}) = (\text{conc. NaHCO}_3) = \frac{1}{2}(\text{conc. Na}_2\text{CO}_3)$$

Other proportions are also possible. Without some more data, or other limits on the quantities, no more definite answers are possible.

Chapter 16: Thermodynamics: Directionality of Chemical Reactions

Solutions for Red-Numbered Questions for Review and Thought

Topical Questions

Reactant-Favored and Product-Favored Processes (Section 16-1)

13. *Result:* **(a) 2 $H_2O(\ell) \longrightarrow 2\ H_2(g) + O_2(g)$; reactant-favored (b) $C_8H_{18}(\ell) \longrightarrow C_8H_{18}(g)$; product-favored (c) $C_{12}H_{22}O_{11}(s) \longrightarrow C_{12}H_{22}O_{11}(aq)$; product-favored**

Analyze and Plan: Think about the process and whether you have observed it happening spontaneously. If it has, it will be classified as product-favored. If it has not, it will be classified as reactant-favored.

Execute:

(a) Water decomposing into its elements. $2\ H_2O(\ell) \longrightarrow 2\ H_2(g) + O_2(g)$

The decomposition of water has not been observed to occur spontaneously at normal temperature and pressure, thus this reaction is classified as reactant-favored.

(b) Gasoline spills and evaporates, changing from liquid to gas phase: $C_8H_{18}(\ell) \longrightarrow C_8H_{18}(g)$

The process has been observed to occur spontaneously at normal temperature and pressure, thus this reaction is classified as product-favored.

(c) Dissolving sugar at room temperature, changing it from solid to aqueous phase:

$$C_{12}H_{22}O_{11}(s) \longrightarrow C_{12}H_{22}O_{11}(aq)$$

The process has been observed to occur spontaneously at normal temperature and pressure, thus this reaction is classified as product-favored.

Chemical Reactions and Dispersal of Energy (Section 16-2)

15. *Result:* **(a) $\frac{1}{2}$ (b) $\frac{1}{2}$ (c) 50 "heads" and 50 "tails"**

Analyze and Plan: Determine the total number of possible outcomes and use that to build a fraction describing the probability of obtaining one outcome.

Execute:

(a) "Heads" is one of two possible outcomes, so the probability is $\frac{1}{2}$.

(b) "Tails" is one of two possible outcomes, so the probability is $\frac{1}{2}$.

(c) If you flip a coin 100 times, you get 100 outcomes—each outcome has an equal probability, $\frac{1}{2}$, of being "heads" or "tails". The most likely number of "heads" or "tails" would be half of the total number of flips, or 50 "heads" and 50 "tails".

17. *Result:* **(a) $\frac{1}{2}$ in A ; $\frac{1}{2}$ in B (b) 50 molecules in flask A and 50 molecules in flask B is most probably arrangement and has the highest entropy.**

Analyze and Plan: Determine the total number of possible outcomes and use that to build a fraction describing the probability of obtaining one outcome. We will assume that the connecting tube between the two flasks is so short than there is no probability that the molecules will actually be found there.

Execute:

(a) The probability of the molecule being in flask A is one of two possible outcomes, so the probability is $\frac{1}{2}$.

The probability of the molecule being in flask B is one of two possible outcomes, so the probability is $\frac{1}{2}$.

(b) If you put 100 molecules in the two-flask system, the most likely arrangement of the molecules in flask A and flask B would be half of the molecules in each, or 50 in flask A and 50 in flask B. This 50-50 arrangement has the highest entropy, because the molecules have a higher disorder than any other state that would crowd more than 50 atoms into one flask or the other.

Measuring Dispersal of Energy: Entropy (Section 16-3)

19. *Result:* **(a) Negative (b) Positive (c) Positive**

Analyze and Plan: Use the qualitative guidelines for entropy changes described in Section 16-3.

Execute:

(a) $H_2O(g) \longrightarrow H_2O(s)$

The solid product has lower entropy than the gas-phase reactant, so the entropy change is negative.

(b) $CO_2(aq) \longrightarrow CO_2(g)$

The gas-phase product has higher entropy than the aqueous reactant, so the entropy change is positive.

(c) $\text{glass}(s) \longrightarrow \text{glass}(\ell)$

The liquid product has higher entropy than the solid reactant, so the entropy change is positive.

21. *Result:* **(a) Positive (b) Positive (c) Positive**

Analyze and Plan: Use the qualitative guidelines for entropy changes described in Section 16-3.

Execute:

(a) $2\ H_2O(\ell) \longrightarrow 2\ H_2(g) + O_2(g)$

The gas-phase products have higher entropy than the liquid reactant, so the entropy change is positive.

(b) $C_8H_{18}(\ell) \longrightarrow C_8H_{18}(g)$

The gas-phase product has higher entropy than the liquid reactant, so the entropy change is positive.

(c) $C_{12}H_{22}O_{11}(s) \longrightarrow C_{12}H_{22}O_{11}(aq)$

The aqueous product has higher entropy than the solid reactant, so the entropy change is positive.

23. *Result:* **(a) Item 2 (b) Item 2 (c) Item 2**

Analyze and Plan: Use the qualitative guidelines for entropy changes described in Section 16-3.

Execute:

(a) Item 2 has higher entropy since it is identical to item 1 except that its temperature is higher. Molecules at higher temperature have higher entropy.

(b) Item 2, dissolved sugar, has higher entropy than item 1, solid sugar, because solute molecules are more random than those in a solid crystal.

(c) Item 2, the mixture of water and alcohol together has higher entropy than item 1, water and alcohol separate. Mixing makes the molecules more random.

25. *Result:* **(a) NaCl (b) P_4 (c) $NH_4NO_3(aq)$; see explanations below.**

Analyze and Plan: Use the methods described in Problem-Solving Example 16.2 and the qualitative guidelines for entropy changes described in Section 16-3.

Execute:

(a) Comparing NaCl and CaO, we find that the biggest difference between these two ionic solids is the attractions due to the charges on the ions. According to Coulomb's law, the **Ca^{2+} and O^{2-} ions have greater interaction than the Na^{+} and Cl^{-} ions**, so NaCl has a larger entropy/mol than CaO.

(b) **P_4 molecules have more atoms than Cl_2 molecules**, so P_4 has a larger entropy per mol than Cl_2.

(c) The **solid NH_4NO_3 crystal is more ordered than the aqueous NH_4^+ and NO_3^- ions**, so the aqueous NH_4NO_3 has a larger entropy per mol than the solid NH_4NO_3.

27. *Result:* **(a) Negative (b) Positive (c) Negative (d) Positive**

Analyze and Plan: Use the qualitative guidelines for entropy changes described in Section 16-3.

Execute:

(a) The reaction has more gas-phase reactants (2 mol) than gas-phase products (1 mol), so the entropy change will be negative.

(b) The reaction has fewer gas-phase reactants ($\frac{3}{2}$ mol) than gas-phase products (3 mol), so the entropy change will be positive.

(c) The reaction has more gas-phase reactants (4 mol) than gas-phase products (2 mol), so the entropy change will be negative.

(d) The reaction has fewer gas-phase reactants (0 mol) than gas-phase products (1 mol), so the entropy change will be positive.

29. *Result:* **(a) Negative (b) Negative (c) Positive**

Analyze and Plan: Use the qualitative guidelines for entropy changes described in Section 16-3.

Execute:

(a) The reaction has more gas-phase reactants (3 mol) than gas-phase products (2 mol), so the entropy change will be negative.

(b) The reaction has more gas-phase reactants (3 mol) than gas-phase products (0 mol), so the entropy change will be negative.

(c) The reaction has fewer gas-phase reactants (2 mol) than gas-phase products (3 mol), so the entropy change will be positive.

31. *Result:* **112 J K^{-1}mol^{-1}**

Analyze and Plan: Use $\Delta_r G = 0$ and rearrange Equation 16.4 to solve for $\Delta_r S°$: $T = \dfrac{\Delta_r H°}{\Delta_r S°}$

Execute:

Use the enthalpy of vaporization and the normal boiling point of ethanol:

$$\Delta_r S° = \frac{\Delta_r H°}{T} = \frac{\left(39.3\ \frac{\text{kJ}}{\text{mol}}\right)\times\left(\frac{1000\ \text{J}}{1\ \text{kJ}}\right)}{(78.3+273.15)\text{K}} = 112\ \text{J K}^{-1}\text{mol}^{-1}$$

33. *Result:* **(a) 2.63 J K^{-1}mol^{-1} (b) 1000 J K^{-1}mol^{-1} (c) 2.45 J K^{-1}mol^{-1}**

Analyze and Plan: Adapt the method described in Problem-Solving Example 16.1. Use Equation 16.1:

$$\Delta_r S = \frac{q_{rev}}{T}.$$

Execute:

For example, the calculation for (a) looks like this:

$$\Delta_r S^\circ = \frac{q_{rev}}{T} = \frac{\left(0.775\ \frac{kJ}{mol}\right) \times \left(\frac{1000\ J}{1\ kJ}\right)}{295\ K} = 2.63\ J\ K^{-1} mol^{-1}$$

	q°_{rev} (kJ/mol)	T (K)	$\Delta_r S^\circ$ ($J\ K^{-1}$)
(a)	0.775	295	2.63
(b)	500	500.	1000
(c)	2.45	1000.	2.45

Calculating Entropy Changes (Section 16-4)

35. *Result:* **(a) –120.64 $J\ K^{-1} mol^{-1}$; negative $\Delta_r S$ prediction confirmed (b) 156.9 $J\ K^{-1} mol^{-1}$; positive $\Delta_r S$ prediction confirmed (c) –198.76 $J\ K^{-1} mol^{-1}$; negative $\Delta_r S$ prediction confirmed (d) 160.6 $J\ K^{-1} mol^{-1}$; positive $\Delta_r S$ prediction confirmed**

Analyze and Plan: Adapt the method described in Problem-Solving Example 16.3. Use the balanced equation to adapt Equation 16.2:

$$\Delta_r S^\circ = \sum\left[(\text{coefficient of product}) \times S^\circ(\text{product})\right] - \sum\left[(\text{coefficient of reactant}) \times S^\circ(\text{reactant})\right]$$

Execute:

(a) $\Delta S^\circ = S^\circ\{C_2H_6(g)\} - S^\circ\{C_2H_4(g)\} - S^\circ\{H_2(g)\}$

$= (229.60\ J\ K^{-1} mol^{-1}) - (219.56\ J\ K^{-1} mol^{-1}) - (130.684\ J\ K^{-1} mol^{-1}) = -120.64\ J\ K^{-1} mol^{-1}$

This confirms the negative $\Delta_r S$ prediction from Question 27.

(b) $\Delta_r S^\circ = S^\circ\{CO_2(g)\} + 2\ S^\circ\{H_2O(g)\} - S^\circ\{CH_3OH(\ell)\} - \frac{3}{2}\ S^\circ\{O_2(g)\}$

$= (213.74\ J\ K^{-1} mol^{-1}) + 2\ (188.825\ J\ K^{-1} mol^{-1})$

$- (126.8\ J\ K^{-1} mol^{-1}) - \frac{3}{2}\ (205.138\ J\ K^{-1} mol^{-1}) = 156.9\ J\ K^{-1} mol^{-1}$

This confirms the positive $\Delta_r S$ prediction from Question 27.

(c) $\Delta_r S^\circ = 2\ S^\circ\{NH_3(g)\} - S^\circ\{N_2(g)\} - 3\ S^\circ\{H_2(g)\}$

$= 2\ (192.45\ J\ K^{-1} mol^{-1}) - (191.61\ J\ K^{-1} mol^{-1}) - 3\ (130.684\ J\ K^{-1} mol^{-1}) = -198.76\ J\ K^{-1} mol^{-1}$

This confirms the negative $\Delta_r S$ prediction from Question 27.

(d) $\Delta_r S^\circ = S^\circ\{CaO(s)\} + S^\circ\{CO_2(g)\} - S^\circ\{CaCO_3(s)\}$

$= (39.75\ J\ K^{-1} mol^{-1}) + (213.74\ J\ K^{-1} mol^{-1}) - (92.9\ J\ K^{-1} mol^{-1}) = 160.6\ J\ K^{-1} mol^{-1}$

This confirms the positive $\Delta_r S$ prediction from Question 27.

37. *Result:* **(a) –173.01 $J\ K^{-1} mol^{-1}$; negative $\Delta_r S$ prediction confirmed (b) –326.69 $J\ K^{-1} mol^{-1}$; negative $\Delta_r S$ prediction confirmed (c) 137.55 $J\ K^{-1} mol^{-1}$; positive $\Delta_r S$ prediction confirmed**

Analyze and Plan: Adapt the method described in the solution to Question 35.

Execute:

(a) $\Delta_r S° = 2\ S°\{CO_2(g)\} - 2\ S°\{CO(g)\} - S°\{O_2(g)\}$

$= 2\ (213.74\ J\ K^{-1}mol^{-1}) - 2\ (197.674\ J\ K^{-1}mol^{-1}) - (205.138\ J\ K^{-1}mol^{-1}) = -173.01\ J\ K^{-1}mol^{-1}$

This confirms the negative $\Delta_r S$ prediction from Question 29.

(b) $\Delta_r S° = 2\ S°\{H_2O(\ell)\} - 2\ S°\{H_2(g)\} - S°\{O_2(g)\}$

$= 2\ (69.91\ J\ K^{-1}mol^{-1}) - 2\ (130.684\ J\ K^{-1}mol^{-1}) - (205.138\ J\ K^{-1}mol^{-1}) = -326.69\ J\ K^{-1}mol^{-1}$

This confirms the negative $\Delta_r S$ prediction from Question 29.

(c) $\Delta_r S° = 3\ S°\{O_2(g)\} - 2\ S°\{O_3(g)\}$

$= 3\ (205.138\ J\ K^{-1}mol^{-1}) - 2\ (238.93\ J\ K^{-1}mol^{-1}) = 137.55\ J\ K^{-1}mol^{-1}$

This confirms the positive $\Delta_r S$ prediction from Question 29.

Entropy and the Second Law of Thermodynamics (Section 16-5)

39. *Result:* **No, we need $\Delta_r H°$ also; the reaction is product-favored.**

Analyze and Plan: Adapt the method described in the solution to Question 35.

Execute:

$\Delta_r S° = S°\{C_2H_5OH(\ell)\} - S°\{C_2H_4(g)\} - S°\{H_2O(g)\}$

$= (160.7\ J\ K^{-1}mol^{-1}) - (219.56\ J\ K^{-1}mol^{-1}) - (188.825\ J\ K^{-1}mol^{-1}) = -247.7\ J\ K^{-1}mol^{-1}$

We cannot tell from the results of this calculation whether this reaction is product-favored. We need the $\Delta_r H°$, also. Use Equation 6.10:

$$\Delta_r H° = \sum\left[(\text{coefficient of product}) \times \Delta H_f^\circ(\text{product})\right] - \sum\left[(\text{coefficient of reactant}) \times \Delta H_f^\circ(\text{reactant})\right]$$

$\Delta_r H° = \Delta_f H°\{C_2H_5OH(\ell)\} - \Delta_f H°\{C_2H_4(g)\} - \Delta_f H°\{H_2O(g)\}$

$= (-277.69\ kJ/mol) - (52.26\ kJ/mol) - (-241.818\ kJ/mol) = -88.13\ kJ/mol$

Since, both $\Delta_r H°$ and $\Delta_r S°$ are negative, the product-favored nature of the reaction depends on temperature.

$$T = \frac{\left(-88.13\frac{kJ}{mol}\right) \times \left(\frac{1000\ J}{1\ kJ}\right)}{-247.7\ J\ K^{-1}mol^{-1}} = 355.8\ K$$

When 355.8 K or lower, the reaction is product-favored. The temperature here is 25 °C + 273.15 = 298 K, so the reaction is product-favored.

41. *Result:* **Low temperatures**

Analyze and Plan: Adapt the method described in the solution to Question 31, then consult Table 16.2.

Execute: The reaction has more gas-phase reactants (1 mol) than gas-phase products (0 mol), so the entropy change will be negative. When $\Delta_r H°$ and $\Delta_r S°$ are both negative, the reaction is product-favored only at low temperatures. The exothermicity is sufficient to favor products, if the temperature is low enough to overcome the decrease in entropy.

43. *Result/Explanation:* Exothermic reactions with an increase in disorder, exhibited by a larger number of gas-phase products (7 mol) than gas-phase reactants (5 mol), never need help from the surroundings to favor products.

45. *Result:* **(a) $\Delta_r S°$ is positive; $\Delta_r H°$ is negative. (b) $\Delta_r H° = -184.28$ kJ/mol, which is negative as predicted; $\Delta_r S° = -7.7\ J\ K^{-1}mol^{-1}$ which is not positive as predicted, but it is very small.**

(a) *Explanation:* The products of this reaction have a larger number of gas-phase products (½ mol) than gas-phase reactants (0 mol), so we predict the entropy change, ΔS°, to be positive. Sodium reacts violently, suggesting that a great amount of heat is produced in the reaction. We will predict the enthalpy change, $\Delta_r H°$, is negative.

(b) *Analyze and Plan:* Adapt the method described in the solution to Question 35.

Execute:

$$\Delta_r H° = \Delta_f H°\{NaOH(s)\} + (½\ mol)\times\Delta_f H°\{H_2(g)\} - \Delta_f H°\{Na(s)\} - \Delta_f H°\{H_2O(\ell)\}$$

$$= (-470.114\ kJ/mol) + (½\ mol)\ (0\ kJ/mol) - (0\ kJ/mol) - (-285.83\ kJ/mol) = -184.28\ kJ/mol$$

$$\Delta_r S° = S°\{NaOH(s)\} + (½\ mol) \times S°\{H_2(g)\} - \Delta S°\{Na(s)\} - \Delta S°\{H_2O(\ell)\}$$

$$= (48.1\ J\ K^{-1}mol^{-1}) + (½\ mol)\ (130.684\ J\ K^{-1}mol^{-1})$$

$$- (51.21\ J\ K^{-1}mol^{-1}) - (69.91\ J\ K^{-1}mol^{-1}) = -7.7\ J\ K^{-1}mol^{-1}$$

The enthalpy change is negative, as predicted in (a); however, the entropy change is negative, not as predicted in (a). Looking at the values of S°, the aqueous solute has lower entropy than pure water, providing sufficient order to compensate for the higher disorder of the gas. The value of $-7.7\ J\ K^{-1}$ is pretty small.

49. *Result:* **(a) –851.5 kJ/mol; $-37.52\ J\ K^{-1}mol^{-1}$; product-favored at low temperatures (b) 66.36 kJ/mol; $-21.77\ J\ K^{-1}mol^{-1}$; never product-favored**

Analyze and Plan: Adapt the method described in the solution to Question 39 and use Table 16.2.

Execute:

(a) $$\Delta_r H° = 2\ \Delta_f H°\{Fe(s)\} + \Delta_f H°\{Al_2O_3(s)\} - \Delta_f H°\{Fe_2O_3(s)\} - 2\ \Delta_f H°\{Al(s)\}$$

$$= 2\ (0\ kJ/mol) + (-1675.7\ kJ/mol) - (-824.2\ kJ/mol) - 2\ (0\ kJ/mol) = -851.5\ kJ/mol$$

$$\Delta_r S° = 2\ S°\{Fe(s)\} + S°\{Al_2O_3(s)\} - S°\{Fe_2O_3(s)\} - 2\ S°\{Al(s)\}$$

$$= 2\ (27.78\ J\ K^{-1}mol^{-1}) + (50.92\ J\ K^{-1}mol^{-1})$$

$$- (87.40\ J\ K^{-1}mol^{-1}) - 2\ (28.3\ J\ K^{-1}mol^{-1}) = -37.52\ J\ K^{-1}mol^{-1}$$

The reaction is exothermic, but its entropy change is negative, so it would stop being product-favored above a specific temperature. (Calculated from Equation 16.4, this temperature is 22,700 K.)

(b) $$\Delta_r H° = 2\ \Delta_f H°\{NO_2(g)\} - \Delta_f H°\{N_2(g)\} - 2\ \Delta_f H°\{O_2(g)\}$$

$$= 2\ (33.18\ kJ/mol) - (0\ kJ/mol) - 2\ (0\ kJ/mol) = 66.36\ kJ/mol$$

$$\Delta_r S° = 2\ S°\{NO_2(g)\} - S°\{N_2(g)\} - 2\ S°\{O_2(g)\}$$

$$= 2\ (240.06\ J\ K^{-1}mol^{-1}) - (191.61\ J\ K^{-1}mol^{-1}) - 2\ (205.138\ J\ K^{-1}mol^{-1}) = -21.77\ J\ K^{-1}mol^{-1}$$

The reaction is endothermic; the entropy change is negative, so it will never be product-favored.

Gibbs Free Energy (Section 16-6)

51. *Result:* **(a) $\Delta_r S_{universe} = 4.92 \times 10^3\ J\ K^{-1}mol^{-1}$ (b) $\Delta_r G°_{system} = -1.47 \times 10^3$ kJ/mol and $\Delta_r G°_{system} = -T\Delta_r S_{universe}$ (c) Yes, since ethane is used as a fuel.**

Analyze and Plan: Adapt the method described in Problem-Solving Example 16.4, using these three equations:

From Section 16-5: $$\Delta S_{universe} = -\frac{\Delta H°_{system}}{T} + \Delta S°_{system}$$

Equation 16.4 $\Delta_r G^\circ_{system} = \Delta_r H^\circ_{system} - T\Delta_r S^\circ_{system}$

and Equation 16.5, as described in Problem-Solving Example 16.5.

$$\Delta_r G^\circ = \sum\left[(\text{coefficient of product}) \times \Delta G_f^\circ(\text{product})\right] - \sum\left[(\text{coefficient of reactant}) \times \Delta G_f^\circ(\text{reactant})\right]$$

Execute: $T(K) = T(°C) + 273.15 = 25\ °C + 273.15 = 298\ K$

(a) $\Delta_r H^\circ_{system} = 2\ \Delta_f H^\circ\{CO_2(g)\} + 3\ \Delta_f H^\circ\{H_2O(\ell)\} - \Delta_f H^\circ\{C_2H_6(g)\} - \frac{7}{2}\ \Delta_f H^\circ\{O_2(g)\}$

$= 2\ (-393.509\ kJ/mol) + 3\ (-285.83\ kJ/mol) - (-84.68\ kJ/mol) - \frac{7}{2}\ (0\ kJ/mol) = -1559.83\ kJ/mol$

$\Delta_r S^\circ_{system} = 2\ S^\circ\{CO_2(g)\} + 3\ S^\circ\{H_2O(\ell)) - S^\circ\{C_2H_6(g)\} - \frac{7}{2}\ S^\circ\{O_2(g)\}$

$= 2\ (213.74\ J\ K^{-1}mol^{-1}) + 3\ (69.91\ J\ K^{-1}mol^{-1})$

$- (229.60\ J\ K^{-1}mol^{-1}) - \frac{7}{2}\ (205.138\ J\ K^{-1}mol^{-1}) = -310.37\ J\ K^{-1}mol^{-1}$

$$\Delta_r S_{universe} = -\frac{(-1559.83\ kJ/mol)}{(298\ K)} \times \frac{1000\ J}{1\ kJ} + (-310.37\ J\ K^{-1}mol^{-1}) = 4.92 \times 10^3\ J\ K^{-1}mol^{-1}$$

(b) $\Delta_r G^\circ_{system} = \Delta_r H^\circ_{system} - T\Delta_r S^\circ_{system}$

$$= (-1559.83\ kJ/mol) - (298\ K)(-310.37\ J\ K^{-1}mol^{-1}) \times \frac{1\ kJ}{1000\ J} = -1467.3\ kJ/mol$$

Independently, we can get $\Delta_r G^\circ_{system}$ using Equation 16.5:

$\Delta_r G^\circ_{system} = 2\ \Delta_f G^\circ\{CO_2(g)\} + (3\ mol) \times \Delta_f G^\circ\{H_2O(\ell)\} - \Delta_f G^\circ\{C_2H_6(g)\} - (\frac{7}{2}\ mol) \times \Delta_f G^\circ\{O_2(g)\}$

$= 2\ (-394.359\ kJ/mol) + 3\ (-237.129\ kJ/mol)$

$- (-32.82\ kJ/mol) - \frac{7}{2}\ (0\ kJ/mol) = -1467.28\ kJ/mol$

A negative $\Delta_r G^\circ_{system}$ is consistent with a positive $\Delta S_{universe}$. In addition:

$$-T\Delta_r S_{universe} = (298\ K)(4.92 \times 10^3\ J\ K^{-1}mol^{-1}) \times \frac{1\ kJ}{1000\ J} = -1.47 \times 10^3\ kJ/mol = \Delta_r G^\circ_{system}.$$

(c) Yes. Ethane is used as a fuel; hence, we would expect its combustion reaction to be product–favored.

53. *Result/Explanation:* Adapt the method described in the solution to Question 51 using Equation 16.4:

$$\Delta_r G^\circ = \Delta_r H^\circ - T\Delta_r S^\circ$$

Here: (sign of $\Delta_r H^\circ$) = negative, and (sign of $\Delta_r S^\circ$) = positive. The Kelvin temperature is always positive.

$$\Delta_r G^\circ = (\text{negative } \Delta_r H^\circ) - (\text{positive Kelvin temperature})\ (\text{positive } \Delta_r S^\circ)$$

Take the absolute values of the each term and then include the resulting signs explicitly:

$$\Delta_r G^\circ = -|\Delta_r H^\circ| - |T\Delta_r S^\circ|$$

$$\Delta_r G^\circ = -(|\Delta_r H^\circ| + |T\Delta_r S^\circ|)$$

$$(\text{sign of } \Delta_r G^\circ) = -(\text{positive}) = \text{negative}$$

So, $\Delta G^\circ < 0$ for all temperature values.

55. *Result:* **$\Delta_r G° = 28.63$ kJ/mol, so reaction is reactant-favored.**

Analyze and Plan: Adapt the method described in the solution to Question 51(b).

Execute:

$$\Delta_r G° = \Delta_r H° - T\Delta_r S° = (41.17 \text{ kJ/mol}) - (298 \text{ K})(42.08 \text{ J K}^{-1}\text{mol}^{-1}) \times \frac{1 \text{ kJ}}{1000 \text{ J}} = 28.63 \text{ kJ/mol}$$

The sign of $\Delta_r G°$ is positive, so the reaction is reactant-favored.

57. *Result:* **$\Delta_r G° = 462.28$ kJ/mol, so reaction is not a good way to make pure Si.**

Analyze and Plan: Adapt the method described in the solution to Question 51(b).

Execute:

$$\Delta_r G° = \Delta_f G°\{Si(s)\} + \Delta_f G°\{CO_2(g)\} - \Delta_f G°\{SiO_2(s)\} - \Delta_f G°\{C(s)\}$$

$$= (0 \text{ kJ/mol}) + (-394.359 \text{ kJ/mol}) - (-856.64 \text{ kJ/mol}) - (0 \text{ kJ/mol}) = 462.28 \text{ kJ/mol}$$

The reaction is not product-favored, so this would not be a good way to make pure silicon.

59. *Result:* **(a) 385.7 K (b) 835.1 K**

Analyze and Plan: Adapt the method described in the solution to Question 35.

Execute:

(a) $$\Delta_r H° = \Delta_f H°\{CH_3OH(\ell)\} + \Delta_f H°\{CO(g)\} - 2\,\Delta_f H°\{H_2(g)\}$$

$$= (-238.66 \text{ kJ/mol}) + (-110.525 \text{ kJ/mol}) - 2\,(0 \text{ kJ/mol}) = -128.14 \text{ kJ/mol}$$

$$\Delta_r S° = S°\{CH_3OH(\ell)\} + S°\{CO(g)\} - 2\,S°\{H_2(g)\}$$

$$= (126.8 \text{ J K}^{-1}\text{mol}^{-1}) + (197.674 \text{ J K}^{-1}\text{mol}^{-1}) - 2\,(130.684 \text{ J K}^{-1}\text{mol}^{-1}) = -332.2 \text{ J K}^{-1}\text{mol}^{-1}$$

$$T = \frac{\left(-128.14 \dfrac{\text{kJ}}{\text{mol}}\right) \times \left(\dfrac{1000 \text{ J}}{1 \text{ kJ}}\right)}{-332.2 \text{ J K}^{-1}\text{mol}^{-1}} = 385.7 \text{ K}$$

(b) $$\Delta_r H° = 4\,\Delta_f H°\{Fe(s)\} + 3\,\Delta_f H°\{CO_2(g)\} - 2\,\Delta_f H°\{Fe_2O_3(s)\} - 3\,\Delta_f H°\{C(s)\}$$

$$= 4\,(0 \text{ kJ/mol}) + 3\,(-393.509 \text{ kJ/mol}) - 2\,(-824.2 \text{ kJ/mol}) - 3\,(0 \text{ kJ/mol}) = 467.9 \text{ kJ/mol}$$

$$\Delta_r S° = 4\,S°\{Fe(s)\} + 3\,S°\{CO_2(g)\} - 2\,S°\{Fe_2O_3(s)\} - 3\,S°\{C(s)\}$$

$$= 4\,(27.78 \text{ J K}^{-1}\text{mol}^{-1}) + 3\,(213.74 \text{ J K}^{-1}\text{mol}^{-1})$$

$$- 2\,(87.40 \text{ J K}^{-1}\text{mol}^{-1}) - 3\,(5.740 \text{ J K}^{-1}\text{mol}^{-1}) = 560.32 \text{ J K}^{-1}\text{mol}^{-1}$$

$$T = \frac{\left(467.9 \dfrac{\text{kJ}}{\text{mol}}\right) \times \left(\dfrac{1000 \text{ J}}{1 \text{ kJ}}\right)}{560.32 \text{ J K}^{-1}\text{mol}^{-1}} = 835.1 \text{ K}$$

61. *Result:* **(a) 49.7 kJ/mol (b) –178.2 kJ/mol (c) –1267.5 kJ/mol**

Analyze and Plan: Adapt the method described in the solution to Question 51(b).

The values of $\Delta_r H°$ and $\Delta_r S°$ were calculated for the reactions in the solution to Question 59. We will assume $\Delta_r H°$ and $\Delta_r S°$ are approximately independent of temperature:

Execute:

(a) $\Delta_r H° = -174.47$ kJ/mol and $\Delta_r S° = -112.1$ J $K^{-1}mol^{-1}$

$$\Delta_r G° = \Delta_r H° - T\Delta_r S° = (-174.47 \text{ kJ/mol}) - (2000.)(-112.1 \text{ J K}^{-1}\text{mol}^{-1}) \times \frac{1 \text{ kJ}}{1000 \text{ J}} = 49.7 \text{ kJ/mol}$$

(b) $\Delta_r H = 197.78$ kJ/mol and $\Delta_r S° = 188.0$ J $K^{-1}mol^{-1}$

$$\Delta_r G° = \Delta_r H° - T\Delta_r S° = (197.78 \text{ kJ/mol}) - (2000.)(188.0 \text{ J K}^{-1}\text{mol}^{-1}) \times \frac{1 \text{ kJ}}{1000 \text{ J}} = -178.2 \text{ kJ/mol}$$

(c) $\Delta_r H° = -905.47$ kJ/mol and $\Delta_r S° = 181.0$ J $K^{-1}mol^{-1}$

$$\Delta_r G° = \Delta_r H° - T\Delta_r S° = (-905.47 \text{ kJ/mol}) - (2000.)(181.0 \text{ J K}^{-1}\text{mol}^{-1}) \times \frac{1 \text{ kJ}}{1000 \text{ J}} = -1267.5 \text{ kJ/mol}$$

63. *Result:* **(a) $\Delta_r H° = 178.32$ kJ/mol; $\Delta_r S° = 160.6$ J $K^{-1}mol^{-1}$; $\Delta_r G° = 130.5$ kJ/mol (b) Reactant-favored (c) It is not product-favored at all temperatures. (d) 1110. K**

Analyze and Plan: Adapt methods described in the solution to Questions 39 and 49.

Execute:

(a) $\Delta_r H° = (1 \text{ mol}) \times \Delta_f H°\{CaO(s)\} + \Delta_f H°\{CO_2(g)\} - \Delta_f H°\{CaCO_3(s)\}$

$= (-635.09 \text{ kJ/mol}) + (-393.509 \text{ kJ/mol}) - (-1206.92 \text{ kJ/mol}) = 178.32 \text{ kJ/mol}$

$\Delta_r S° = S°\{CaO(s)\} + S°\{CO_2(g)\} - S°\{CaCO_3(s)\}$

$= (39.75 \text{ J K}^{-1}\text{mol}^{-1}) + (213.74 \text{ J K}^{-1}\text{mol}^{-1}) - (92.9 \text{ J K}^{-1}\text{mol}^{-1}) = 160.6 \text{ J K}^{-1}\text{mol}^{-1}$

$$\Delta_r G° = \Delta_r H° - T\Delta_r S° = (178.32 \text{ kJ/mol}) - (298 \text{ K})(160.6 \text{ J K}^{-1}\text{mol}^{-1}) \times \frac{1 \text{ kJ}}{1000 \text{ J}} = 130.5 \text{ kJ/mol}$$

(b) The change in Gibbs free energy is positive, so the reaction is reactant-favored.

(c) Both $\Delta_r H°$ and $\Delta_r S°$ are positive, so the reaction is only product-favored at high temperatures.

(d)
$$T = \frac{\left(178.32 \frac{\text{kJ}}{\text{mol}}\right) \times \left(\frac{1000 \text{ J}}{1 \text{ kJ}}\right)}{160.6 \text{ J K}^{-1}\text{mol}^{-1}} = 1110. \text{ K}$$

65. *Result:* **–867.8 kJ/mol**

Analyze and Plan: Rearrange Equation 16.5 to determine the value of $\Delta_f G°$ sought.

Execute:

$\Delta_r G° = \Delta_f G°\{C_2H_2(g)\} + \Delta_f G°\{Ca(OH)_2(s)\} - \Delta_f G°\{CaC_2(s)\} - 2\,\Delta_f G°\{H_2O(\ell)\}$

$$\Delta_f G°\{Ca(OH)_2(s)\} = -\frac{\Delta_r G°}{(1 \text{ mol})} - \Delta_f G°\{C_2H_2(g)\} + \Delta_f G°\{CaC_2(s)\} + 2 \times \Delta_f G°\{H_2O(\ell)\}$$

$= (-119.282 \text{ kJ/mol}) - (209.20 \text{ kJ/mol}) - (-64.9 \text{ kJ/mol}) - 2\,(-237.129 \text{ kJ/mol}) = -867.8 \text{ kJ/mol}$

Gibbs Free Energy Changes and Equilibrium Constants (Section 16-7)

67. *Result:* **(a) $K_p = 4 \times 10^{-34}$ (b) $K_p = 5 \times 10^{-31}$**

Analyze and Plan: Use the method shown in the solution to Question 51(b), then apply Equation 16.9 to calculate the equilibrium constant: K°, also called, K_p, for the gas phase reaction, here.

$$\Delta_r G° = -RT\ln K_p$$

Execute:

(a) $\Delta_r G° = \Delta_f G°\{H_2(g)\} + \Delta_f G°\{Cl_2(g)\} - 2\,\Delta_f G°\{HCl(g)\}$

$= (0\text{ kJ/mol}) + (0\text{ kJ/mol}) - 2\,(-95.299\text{ kJ/mol}) = 190.598\text{ kJ/mol}$

$$K_p = e^{(-\Delta_r G°/RT)} = e^{-\left(\frac{190.598\text{ kJ/mol}}{(0.008314\text{ kJ/mol·K})\times(298\text{ K})}\right)} = e^{-76.9} = 4 \times 10^{-34}$$

(b) $\Delta_r G° = 2\,\Delta_f G°\{NO(g)\} - \Delta_f G°\{N_2(g)\} - \Delta_f G°\{O_2(g)\}$

$= 2\,(86.55\text{ kJ/mol}) - (0\text{ kJ/mol}) - (0\text{ kJ/mol}) = 173.10\text{ kJ/mol}$

$$K_p = e^{(-\Delta_r G°/RT)} = e^{-\left(\frac{173.10\text{ kJ/mol}}{(0.008314\text{ kJ/mol·K})\times(298\text{ K})}\right)} = e^{-69.9} = 5 \times 10^{-31}$$

69. *Result:* **(a) –100.97 kJ/mol; product-favored (b) 5×10^{17}; since $\Delta_r G°$ is negative, K is greater than 1.**

Analyze and Plan: Use methods described in the solution to Question 67:

Execute:

(a) $\Delta_r G° = \Delta_f G°\{H_3C{-}CH_3(g)\} - \Delta_f G°\{H_2C{=}CH_2(g)\} - \Delta_f G°\{H_2(g)\}$

$= (-32.82\text{ kJ/mol}) - (-68.15\text{ kJ/mol}) - (0\text{ kJ/mol}) = -100.97\text{ kJ/mol}$

The negative sign of $\Delta_r G°$ indicates that the reaction is product-favored.

(b) $$K_p = e^{(-\Delta_r G°/RT)} = e^{-\left(\frac{-100.97\text{ kJ/mol}}{(0.008314\text{ kJ/mol·K})\times(298\text{ K})}\right)} = e^{+40.8} = 5 \times 10^{17}$$

When $\Delta_r G°$ is positive, K is less than 1. When $\Delta_r G°$ is negative, K is greater than 1. The latter is the case.

71. *Result:* **(a) $K°_{800} = 8.7 \times 10^{26}$, product-favored (b) $K°_{500} = 7 \times 10^{10}$, product-favored**

(c) $K°_{2000} = 1.3 \times 10^{-15}$, reactant-favored

Analyze and Plan: Adapt the methods described in the solutions to Questions 39 and 67.

Execute:

(a) $\Delta_r H° = 2\,\Delta_f H°\{H_2O(g)\} - 2\,\Delta_f H°\{H_2(g)\} - \Delta_f H°\{O_2(g)\}$

$= 2\,(-241.818\text{ kJ/mol}) - 2\,(0\text{ kJ/mol}) - (0\text{ kJ/mol}) = -483.636\text{ kJ/mol}$

$\Delta_r S° = 2\,S°\{H_2O(g)\} - 2\,S°\{H_2(g)\} - S°\{O_2(g)\}$

$= 2\,(188.825\text{ J K}^{-1}\text{mol}^{-1}) - 2\,(130.684\text{ J K}^{-1}\text{mol}^{-1}) - (205.138\text{ J K}^{-1}\text{mol}^{-1}) = -88.856\text{ J K}^{-1}\text{mol}^{-1}$

$$\Delta_r G°_{800} = \Delta_r H° - T\Delta_r S° = (-483.636\text{ kJ/mol}) - (800.\text{ K})\,(-88.856\text{ J K}^{-1}\text{mol}^{-1}) \times \frac{1\text{ kJ}}{1000\text{ J}} = -412.6\text{ kJ/mol}$$

$$K°_{800} = e^{(-\Delta_r G°/RT)} = e^{-\left(\frac{-412.6\text{ kJ/mol}}{(0.008314\text{ kJ/mol·K})\times(800.\text{ K})}\right)} = e^{62.0} = 8.7 \times 10^{26}$$ K>1, product-favored

(b) $\Delta_r H° = 2\,\Delta_f H°\{SO_3(g)\} - 2\,\Delta_f H°\{SO_2(g)\} - \Delta_f H°\{O_2(g)\}$

$= 2\,(-395.72\text{ kJ/mol}) - 2\,(-296.830\text{ kJ/mol}) - (0\text{ kJ/mol}) = -197.78\text{ kJ/mol}$

$\Delta_r S° = 2\,S°\{SO_3(g)\} - 2\,S°\{SO_2(g)\} - S°\{O_2(g)\}$

$= 2\,(256.76\text{ J K}^{-1}\text{mol}^{-1}) - 2\,(248.22\text{ J K}^{-1}\text{mol}^{-1}) - (205.138\text{ J K}^{-1}\text{mol}^{-1}) = -188.06\text{ J K}^{-1}\text{mol}^{-1}$

$$\Delta_r G°_{500} = \Delta_r H° - T\Delta_r S° = (-197.78\text{ kJ/mol}) - (500.\text{ K})\,(-188.06\text{ J K}^{-1}\text{mol}^{-1}) \times \frac{1\text{ kJ}}{1000\text{ J}} = -103.8\text{ kJ/mol}$$

$$K^\circ_{500} = e^{(-\Delta_r G^\circ/RT)} = e^{-\left(\frac{-103.8 \text{ kJ/mol}}{(0.008314 \text{ kJ/mol}\cdot\text{K})\times(500.\text{ K})}\right)} = e^{25.0} = 7 \times 10^{10} \qquad K>1\text{, product-favored}$$

(c) $\Delta_r H^\circ = \Delta_f H^\circ\{H_2(g)\} + (1 \text{ mol}) \times \Delta_f H^\circ\{F_2(g)\} - 2\,\Delta_f H^\circ\{HF(g)\}$

$= (0 \text{ kJ/mol}) + (0 \text{ kJ/mol}) - 2\,(-271.1 \text{ kJ/mol}) = 542.2 \text{ kJ/mol}$

$\Delta_r S^\circ = S^\circ\{H_2(g)\} + S^\circ\{F_2(g)\} - 2\,S^\circ\{HF(g)\}$

$= (130.684 \text{ J K}^{-1}\text{mol}^{-1}) + (202.78 \text{ J K}^{-1}\text{mol}^{-1}) - 2\,(173.779 \text{ J K}^{-1}\text{mol}^{-1}) = -14.09 \text{ J K}^{-1}\text{mol}^{-1}$

$$\Delta_r G^\circ_{2000} = \Delta_r H^\circ - T\Delta_r S^\circ = (542.2 \text{ kJ/mol}) - (2000.\text{ K})(-14.09 \text{ J K}^{-1}\text{mol}^{-1}) \times \frac{1 \text{ kJ}}{1000 \text{ J}} = 570.4 \text{ kJ/mol}$$

$$K^\circ_{2000} = e^{(-\Delta_r G^\circ/RT)} = e^{-\left(\frac{570.4 \text{ kJ/mol}}{(0.008314 \text{ kJ/mol}\cdot\text{K})\times(2000.\text{ K})}\right)} = e^{-34.30} = 1.3 \times 10^{-15} \qquad K<1\text{, reactant-favored}$$

73. *Result:* **(a) –106 kJ/mol (b) 8.55 kJ/mol (c) –33.8 kJ/mol**

Analyze and Plan: Adapt the strategy described in the solution to Question 67. In part (c), we will also need to use the equation that follows Equation 12.4 to get K_p from K_c. Use the appropriate value of R in each equation. To convert K_p to K_c, keep in mind that standard pressure is 1 bar. 1 atm is 1.01325 bar, so R = (0.08206 L·atm/mol·K)(1atm / 1.01325 bar) = 0.083147 L·bar/mol·K

Execute:

(a) $\Delta_r G^\circ = -RT\ln K_p = -(0.008314 \text{ kJ/mol}\cdot\text{K})(298 \text{ K}) \times \ln(4.4 \times 10^{18}) = -106 \text{ kJ/mol}$

(b) $\Delta_r G^\circ = -RT\ln K_p = -(0.008314 \text{ kJ/mol}\cdot\text{K})(298 \text{ K}) \times \ln(3.17 \times 10^{-2}) = 8.55 \text{ kJ/mol}$

(c) Get K_p from K_c: $K_p = K_c(RT)^{\Delta_r n}$ In this equation use R = 0.083147 L·bar/mol·K

$\Delta_r n = 2$ mol product gas – 4 mol reactant gas = –2

$\Delta_r G^\circ = -RT\ln K_p = -RT\ln[K_c(RT)^{-2}] = -(0.008314 \text{ kJ/mol}\cdot\text{K})(298 \text{ K}) \times$

$\ln[(0.083147 \text{ L}\cdot\text{bar/mol}\cdot\text{K} \times 298 \text{ K})^{-2}(3.5 \times 10^8)] = -33.8 \text{ kJ/mol}$

Gibbs Free Energy, Maximum Work, and Energy Resources (Section 16-8)

75. *Result:* **Reaction (a) can be used to do useful work; reactions (b) and (c) require work to be done.**

Analyze and Plan: In Section 4-5(b) and Section 8-1, we were told that for a gaseous substance the standard thermodynamic properties are given for a gas pressure of 1 bar. If we also assume that 298 K is close enough to 298.15 K, then we can use the standard thermodynamic table in Appendix J. We will calculate the $\Delta_r G^\circ$, the Gibbs free energy change, which is a measure of maximum useful work. If $\Delta_r G^\circ$ is negative, the reaction can be harnessed to do useful work. If $\Delta_r G^\circ$ is positive, the reaction cannot be harnessed to do useful work.

Execute:

(a) $\Delta_r G^\circ = (12 \text{ mol}) \times \Delta_f G^\circ\{CO_2(g)\} + (6 \text{ mol}) \times \Delta_f G^\circ\{H_2O(g)\}$

$- 2\,\Delta_f G^\circ\{C_6H_6(\ell)\} - (15 \text{ mol}) \times \Delta_f G^\circ\{O_2(g)\}$

$= 12\,(-394.359 \text{ kJ/mol}) + 6\,(-228.572 \text{ kJ/mol}) - 2\,(124.5 \text{ kJ/mol}) - 15\,(0 \text{ kJ/mol}) = -6352.7 \text{ kJ/mol}$

The reaction can be harnessed to do useful work

(b) $\Delta_r G^\circ = \Delta_f G^\circ\{N_2(g)\} + 3\,\Delta_f G^\circ\{F_2(g)\} - 2\,\Delta_f G^\circ\{NF_3(g)\}$

$= (0 \text{ kJ/mol}) + 3\,(0 \text{ kJ/mol}) - 2\,(-83.2 \text{ kJ/mol}) = 166.4 \text{ kJ/mol}$

The reaction requires work to be done.

(c) $\Delta_rG° = \Delta_fG°\{Ti(s)\} + 3\ \Delta_fG°\{O_2(g)\} - 2\ \Delta_fG°\{TiO_2(s)\}$

$= (0\ kJ/mol) + 3\ (0\ kJ/mol) - (-884.5\ kJ/mol) = 884.5\ kJ/mol$

The reaction requires work to be done.

77. *Result:* **Reaction (b) needs 5.068 g. Reactions (c) needs 26.94 g.**

Analyze and Plan: Find the Gibbs free energy of the graphite reaction and use that as a thermochemical conversion factor to supply the Gibbs free energy of the reactions that require work be done in Question 75 items (b) and (c).

Execute:

$$C(graphite) + O_2(g) \longrightarrow CO_2(g)$$

This equation is the formation reaction for CO_2, so $\Delta_rG° = \Delta_fG° = -394.359$ kJ/mol

The calculation for 75(b) looks like this:

$$166.4\ \text{kJ endergonic reaction} \times \frac{1\ \text{mol C}}{394.359\ \text{kJ provided}} \times \frac{12.011\ \text{g C}}{1\ \text{mol C}} = 5.068\ \text{g C}$$

The rest of the results are described in the table below.

Endergonic Reaction Reference	Calculated $\Delta_rG°$ (X) (kJ/mol)	Mass of graphite needed (g)
75(b)	166.4 kJ	5.068
75(c)	884.5 kJ	26.94

79. *Result:* **(a) $2\ CuO(s) \longrightarrow 2\ Cu(s) + O_2(g)$; $CuO(s) + C(graphite) \longrightarrow Cu(s) + CO(g)$; $\Delta_rG° = -7.5$ kJ/mol (b) $2\ Ag_2O(s) \longrightarrow 4\ Ag(s) + O_2(g)$; $Ag_2O(s) + C(graphite) \longrightarrow 2\ Ag(s) + CO(g)$; $\Delta_rG° = -125.97$ kJ/mol (c) $2\ HgO(s) \longrightarrow 2\ Hg(\ell) + O_2(g)$; $HgO(s) + C(graphite) \longrightarrow Hg(\ell) + CO(g)$; $\Delta_rG° = -125.97$ kJ/mol (d) $2\ MgO(s) \longrightarrow 2\ Mg(s) + O_2(g)$; $MgO(s) + C(graphite) \longrightarrow Mg(s) + CO(g)$; $\Delta_rG° = 432.26$ kJ/mol (e) $2\ PbO(s) \longrightarrow 2\ Pb(s) + O_2(g)$; $PbO(s) + C(graphite) \longrightarrow Pb(s) + CO(g)$; $\Delta_rG° = 50.72$ kJ/mol; Cu, Ag, and Hg can be obtained by this method.**

Analyze and Plan: Balance the equations, then use the method in the solution to Question 51(b).

Execute:

(a) Decomposition $\quad 2\ CuO(s) \longrightarrow 2\ Cu(s) + O_2(g)$

Coupled $\quad CuO(s) + C(graphite) \longrightarrow Cu(s) + CO(g)$

$\Delta_rG°_{coupled} = \Delta_fG°\{Cu(s)\} + \Delta_fG°\{CO(g)\} - \Delta_fG°\{CuO(s)\} - \Delta_fG°\{C(s)\}$

$= (0\ kJ/mol) + (-137.168\ kJ/mol) - (-129.7\ kJ/mol) - (0\ kJ/mol) = -7.5\ kJ/mol$

(b) Decomposition $\quad 2\ Ag_2O(s) \longrightarrow 4\ Ag(s) + O_2(g)$

Coupled $\quad Ag_2O(s) + C(graphite) \longrightarrow 2\ Ag(s) + CO(g)$

$\Delta_rG°_{coupled} = 2\ \Delta_fG°\{Ag(s)\} + \Delta_fG°\{CO(g)\} - \Delta_fG°\{Ag_2O(s)\} - \Delta_fG°\{C(s)\}$

$= 2\ (0\ kJ/mol) + (-137.168\ kJ/mol) - (-11.20\ kJ/mol) - (0\ kJ/mol) = -125.97\ kJ/mol$

(c) Decomposition $\quad 2\ HgO(s) \longrightarrow 2\ Hg(\ell) + O_2(g)$

Coupled $\quad HgO(s) + C(graphite) \longrightarrow Hg(\ell) + CO(g)$

$\Delta_rG° = \Delta_fG°\{Hg(\ell)\} + \Delta_fG°\{CO(g)\} - \Delta_fG°\{HgO(s)\} - \Delta_fG°\{C(s)\}$

$= (0\ kJ/mol) + (-137.168\ kJ/mol) - (-58.539\ kJ/mol) - (0\ kJ/mol) = -78.63\ kJ/mol$

(d) Decomposition $2\ MgO(s) \longrightarrow 2\ Mg(s) + O_2(g)$

Coupled $MgO(s) + C(graphite) \longrightarrow Mg(s) + CO(g)$

$\Delta_rG° = \Delta_fG°\{Mg(s)\} + \Delta_fG°\{CO(g)\} - \Delta_fG°\{MgO(s)\} - \Delta_fG°\{C(s)\}$

$= (0\ kJ/mol) + (-137.168\ kJ/mol) - (-569.43\ kJ/mol) - (0\ kJ/mol) = 432.26\ kJ/mol$

(e) Decomposition $2\ PbO(s) \longrightarrow 2\ Pb(s) + O_2(g)$

Coupled $PbO(s) + C(graphite) \longrightarrow Pb(s) + CO(g)$

$\Delta_rG° = \Delta_fG°\{Pb(s)\} + \Delta_fG°\{CO(g)\} - \Delta_fG°\{PbO(s)\} - \Delta_fG°\{C(s)\}$

$= (0\ kJ/mol) + (-137.168\ kJ/mol) - (-187.89\ kJ/mol) - (0\ kJ/mol) = 50.72\ kJ$

The coupled reactions that have negative $\Delta_rG°$ can be used to produce the respective metals, so Cu, Ag, and Hg can be obtained by this method at 25 °C.

80. *Result:* **CuO, Ag_2O, HgO, and PbO**

Analyze and Plan: Adapt the method used in the solution to Question 71. ($X = \Delta_fH$ or S)

Execute:

Kelvin temperature = 800 + 273.15 = 1.1×10^3 K

(a) $\Delta_rX° = X°\{Cu(s)\} + X°\{CO(g)\} - X°\{CuO(s)\} - X°\{C(s)\}$

$\Delta_rH° = (0\ kJ/mol) + (-110.525\ kJ/mol - (-157.3\ kJ/mol) - (0\ kJ/mol) = 46.8\ kJ/mol$

$\Delta_rS° = (33.15\ J\ K^{-1}mol^{-1}) + (197.674\ J\ K^{-1}mol^{-1})$

$- (42.63\ J\ K^{-1}mol^{-1}) - (5.740\ J\ K^{-1}mol^{-1}) = 182.45\ J\ K^{-1}mol^{-1}$

$\Delta_rG°_{800} = \Delta_rH° - T\Delta_rS° = (46.8\ kJ/mol) - (1.1 \times 10^3\ K)(182.45\ J\ K^{-1}mol^{-1}) \times \frac{1\ kJ}{1000\ J}$

$= 46.8\ kJ - 2.0 \times 10^2\ kJ/mol = -1.5 \times 10^2\ kJ/mol$

(b) $\Delta_rX° = 2\ X°\{Ag(s)\} + X°\{CO(g)\} - X°\{Ag_2O(s)\} - X°\{C(s)\}$

$\Delta_rH° = 2\ (0\ kJ/mol) + (-110.525\ kJ/mol) - (-31.05\ kJ/mol) - (0\ kJ/mol) = -79.47\ kJ/mol$

$\Delta_rS° = 2\ (42.55\ J\ K^{-1}mol^{-1}) + (197.674\ J\ K^{-1}mol^{-1})$

$- (121.3\ J\ K^{-1}mol^{-1}) - (5.740\ J\ K^{-1}mol^{-1}) = 155.7\ J\ K^{-1}mol^{-1}$

This is an exothermic reaction ($\Delta_rH°$ = negative) with a positive entropy change, so it will be spontaneous at any temperature, including 800. °C.

(c) $\Delta_rX° = X°\{Hg(\ell)\} + X°\{CO(g)\} - X°\{HgO(s)\} - X°\{C(s)\}$

$\Delta_rH° = (0\ kJ/mol) + (-110.525\ kJ/mol) - (-90.83\ kJ/mol) - (0\ kJ/mol) = -19.70\ kJ/mol$

$\Delta_rS° = (29.87\ J\ K^{-1}mol^{-1}) + (197.674\ J\ K^{-1}mol^{-1})$

$- (70.29\ J\ K^{-1}mol^{-1}) - (5.740\ J\ K^{-1}mol^{-1}) = 151.51\ J\ K^{-1}mol^{-1}$

This is an exothermic reaction ($\Delta_rH°$ = negative) with a positive entropy change, so it will be spontaneous at any temperature, including 800. °C.

(d) $\Delta_rX° = X°\{Mg(s)\} + X°\{CO(g)\} - X°\{MgO(s)\} - X°\{C(s)\}$

$\Delta_rH° = (0\ kJ/mol) + (-110.525\ kJ/mol) - (-601.70\ kJ/mol) - (0\ kJ/mol) = 491.18\ kJ/mol$

$\Delta_r S° = (32.68 \text{ J K}^{-1}\text{mol}^{-1}) + (197.674 \text{ J K}^{-1}\text{mol}^{-1})$

$- (26.94 \text{ J K}^{-1}\text{mol}^{-1}) - (5.740 \text{ J K}^{-1}\text{mol}^{-1}) = 197.67 \text{ J K}^{-1}\text{mol}^{-1}$

$$\Delta_r G°_{800} = \Delta_r H° - T\Delta_r S° = (491.18 \text{ kJ/mol}) - (1.1 \times 10^3 \text{ K})(197.67 \text{ J K}^{-1}\text{mol}^{-1}) \times \frac{1 \text{ kJ}}{1000 \text{ J}}$$

$= 491.18 \text{ kJ/mol} - 2.1 \times 10^2 \text{ kJ/mol} = 2.8 \times 10^2 \text{ kJ/mol}$

(e) $\Delta_r X° = X°\{Pb(s)\} + X°\{CO(g)\} - X°\{PbO(s)\} - X°\{C(s)\}$

$\Delta_r H° = (0 \text{ kJ/mol}) + (-110.525 \text{ kJ/mol}) - (-217.32 \text{ kJ/mol}) - (0 \text{ kJ/mol}) = 106.80 \text{ kJ/mol}$

$\Delta_r S° = (64.81 \text{ J K}^{-1}\text{mol}^{-1}) + (197.674 \text{ J K}^{-1}\text{mol}^{-1})$

$- (68.7 \text{ J K}^{-1}\text{mol}^{-1}) - (5.740 \text{ J K}^{-1}\text{mol}^{-1}) = 188.0 \text{ J K}^{-1}\text{mol}^{-1}$

$$\Delta_r G°_{800} = \Delta_r H° - T\Delta_r S° = (106.80 \text{ kJ/mol}) - (1.1 \times 10^3 \text{ K})(188.0 \text{ J K}^{-1}\text{mol}^{-1}) \times \frac{1 \text{ kJ}}{1000 \text{ J}}$$

$= 106.80 \text{ kJ/mol} - 2.0 \times 10^2 \text{ kJ/mol} = -1 \times 10^2 \text{ kJ/mol}$

The coupled reactions that have negative $\Delta_r G°$ can be used to produce the respective metals, so CuO, Ag_2O, HgO, and PbO can be used to obtain Cu, Ag, Hg, and Pb, respectively, by this method at 800 °C.

Gibbs Free Energy in Biological Systems (Section 16-9)

82. *Result:* **(a) $\Delta_r H° \cong -2873$ kJ/mol; Broken: 5 O–H, 7 C–O, 7 C–H, 5 C–C, 6 O=O; formed: 12 C=O, 12 O–H (b) It is close; intermolecular forces in solid glucose and liquid water are being neglected.**

Analyze and Plan: Adapt the method described in Section 6-6.

Execute:

(a) Looking at the given ball-and-stick model structure we see that we must break five mol of O–H bonds, seven mol of C–O bonds, seven mol of C–H bonds, and five mol of C–C bonds in glucose. We must also break six mol of O=O bonds. Two C=O bonds in each of six mol of CO_2 and two O–H bonds in each of six mol of H_2O are formed.

The enthalpy of a reaction is approximated by adding the energy required to break one mole of bonds of each type of bond broken, described by the bond energies (D), to the energy required to form one mole of bonds of each type of bonds produced in the reactants, described by the negative of the bond energies (–D).

$\Delta_r H° \cong$ (bonds broken in glucose) + (bonds broken in O_2)

+ (bonds formed in CO_2) + (bonds formed in H_2O)

$= (5 \text{ mol} \times D_{O-H} + 7 \text{ mol} \times D_{C-O} + 7 \text{ mol} \times D_{C-H} + 5 \text{ mol} \times D_{C-C}) + 6 \text{ mol} \times D_{O=O}$

$- 6 (2 \text{ mol} \times D_{C=O}) - 6 (2 \text{ mol} \times D_{O-H})$

Notice that the number of moles multiplied by the energy per mol (in kJ/mol) gives the result for each term in kJ.

$\Delta_r H° \cong 5 D_{O-H} + 7 D_{C-O} + 7 D_{C-H} + 5 D_{C-C} + 6 D_{O=O} - 12 D_{C=O} - 12 D_{O-H}$

$= 7 D_{C-O} + 7 D_{C-H} + 5 D_{C-C} + 6 D_{O=O} - 12 D_{C=O} - 7 D_{O-H}$

Now use the data in Table 6.2.

$= 7 (336 \text{ kJ/mol}) + 7 (416 \text{ kJ/mol}) + 5 (356 \text{ kJ/mol}) + 6 (498 \text{ kJ/mol})$

$- 12 (803\text{kJ/mol}) - 7 (467 \text{ kJ/mol}) = -2873 \text{ kJ/mol} \cong \Delta_r H°$

(b) The actual $\Delta_rH°$ (–2816 kJ/mol) is close to the estimated value in (a). Intermolecular forces in condensed phases (solid glucose and liquid water) are being neglected in this calculation, which could explain the discrepancy. Also, the bond enthalpies given in Table 6.2 are average values; the actual enthalpy of a particular bond in a particular molecule may be slightly greater or less depending upon other factors.

84. *Result:* **(a) 6.46 mol ATP/mol glucose (b) –106 kJ/mol (c) product-favored**

Analyze and Plan: The calculations here are solved with conversion factors.

Execute:

(a) $$\frac{197 \text{ kJ produced}}{1 \text{ mol glucose}} \times \frac{1 \text{ mol ATP}}{30.5 \text{ kJ needed}} = 6.46 \text{ mol ATP per mol of glucose}$$

(b) $$\frac{3 \text{ mol ATP produced}}{1 \text{ mol glucose}} \times \frac{30.5 \text{ kJ needed}}{1 \text{ mol ATP}} = 91.5 \text{ kJ/mol of glucose}$$

The actual reaction must be less exergonic than the given reaction to produce fewer ATP.

$$\Delta_rG°_{\text{overall reaction}} = \Delta_rG°_{\text{conversion}} + \Delta_rG°_{\text{ATP}}$$

$$\Delta_rG°_{\text{overall reaction}} = -197 \text{ kJ/mol} + 91.5 \text{ kJ/mol} = -106 \text{ kJ/mol}$$

(c) The overall reaction in part (b) has a negative Gibbs free energy change, so it is product-favored.

Conservation of Gibbs Free Energy (Section 16-10)

86. *Result/Explanation:* Food we eat provides us with a supply of Gibbs free energy. Coal, petroleum, and natural gas are the most common fuel sources used to supply Gibbs free energy by combustion. We also use solar and nuclear energy, as well as the kinetic energy of wind and water. Consult Section 4-11 for other details.

87. *Result/Explanation:* Most of the activities that involve the intake of Gibbs free energy in our bodies are eating activities, which involve nutrient metabolism. The sun or some other heat energy source usually warms us during the day. If we drive an automobile, chemical energy in the gasoline is consumed during the course of transporting us places. There are many other answers to this question, depending on what people do in the day.

Thermodynamic and Kinetic Stability (Section 16-11)

88. *Result:* **(a) $\Delta_rG°$ = –86.5 kJ so the reaction is product-favored (b) $\Delta_rG°$ = –873.1 kJ (c) No (d) Yes**

Analyze and Plan: Balance the equation, when not provided, and adapt the method in Question 51(b).

Execute:

(a) $$\Delta_rG° = \Delta_fG°\{CH_3COOH(\ell)\} - \Delta_fG°\{CH_3OH(\ell)\} - \Delta_fG°\{CO(g)\}$$

$$= (-389.9 \text{ kJ/mol}) - (-166.27 \text{ kJ/mol}) - (-137.168 \text{ kJ/mol}) = -86.5 \text{ kJ/mol}$$

Because $\Delta_rG°$ is negative, the reaction is product-favored at 298 K.

(b) $$CH_3COOH(\ell) + 2\,O_2(g) \longrightarrow 2\,CO_2(g) + 2\,H_2O(\ell)$$

$$\Delta_rG° = 2\,\Delta_fG°\{CO_2(g)\} + 2\,\Delta_fG°\{H_2O(\ell)\} - \Delta_fG°\{CH_3COOH(\ell)\} - 2\,\Delta_fG°\{(O_2(g)\}$$

$$= 2\,(-394.359 \text{ kJ/mol}) + 2\,(-237.129 \text{ kJ/mol})$$

$$- (-389.9 \text{ kJ/mol}) - 2\,(0 \text{ kJ/mol}) = -873.1 \text{ kJ/mol at } 298.15 \text{ K}$$

(c) Based on the answer to (b), the products of oxidation are more stable, so acetic acid is not thermodynamically stable at 298 K.

(d) Acetic acid can be kept both in liquid form and in solution form if stored properly. In the presence of air, it does not explode, so we can classify it as kinetically stable.

90. *Result:* **$CH_4(g) + 2\ O_2(g) \longrightarrow CO_2(g) + 2\ H_2O(\ell)$; $\Delta_cG° = -817.90$ kJ/mol, product-favored; $C_6H_6(g) + \frac{15}{2}\ O_2(g) \longrightarrow 6\ CO_2(g) + 3\ H_2O(\ell)$; $\Delta_cG° = -3202.0$ kJ/mol, product-favored; $CH_3OH(\ell) + \frac{3}{2}\ O_2(g) \longrightarrow CO_2(g) + 2\ H_2O(\ell)$; $\Delta_cG° = -702.34$ kJ/mol, product-favored; complex molecules require significant rearrangement of atoms, indicating kinetic stability is likely.**

Analyze and Plan: Balance the combustion equations, and adapt methods from the solution to Question 51(b).

Execute: $CH_4(g) + 2\ O_2(g) \longrightarrow CO_2(g) + 2\ H_2O(\ell)$

$$\Delta_cG° = \Delta_fG°\{CO_2(g)\} + 2\ \Delta_fG°\{H_2O(\ell)\} - \Delta_fG°\{CH_4(g)\} - 2\ \Delta_fG°\{O_2(g)\}$$

$$= (-394.359\text{ kJ/mol}) + 2\ (-237.129\text{ kJ/mol})$$

$$- (-50.72\text{ kJ/mol}) - 2\ (0\text{ kJ/mol}) = -817.90\text{ kJ/mol for } CH_4(g) \text{ at } 298\text{ K}$$

Therefore the $CH_4(g)$ combustion reaction is product-favored.

$$C_6H_6(g) + \frac{15}{2}\ O_2(g) \longrightarrow 6\ CO_2(g) + 3\ H_2O(\ell)$$

$$\Delta_cG° = (6\text{ mol}) \times \Delta_fG°\{CO_2(g)\} + 3\ \Delta_fG°\{H_2O(\ell)\} - \Delta_fG°\{CH_4(g)\} - (\tfrac{15}{2}\text{ mol}) \times \Delta_fG°\{O_2(g)\}$$

$$= (6\text{ mol})\ (-394.359\text{ kJ/mol}) + 3\ (-237.129\text{ kJ/mol})$$

$$- (124.5\text{ kJ/mol}) - (\tfrac{15}{2}\text{ mol})\ (0\text{ kJ/mol}) = -3202.0\text{ kJ/mol for } C_6H_6(g) \text{ at } 298\text{ K}$$

Therefore the $C_6H_6(g)$ combustion reaction is product-favored at 298 K.

$$CH_3OH(\ell) + \frac{3}{2}\ O_2(g) \longrightarrow CO_2(g) + 2\ H_2O(\ell)$$

$$\Delta_cG° = \Delta_fG°\{CO_2(g)\} + 2\ \Delta_fG°\{H_2O(\ell)\} - \Delta_fG°\{CH_3OH\} - (\tfrac{3}{2}\text{ mol}) \times \Delta_fG°\{O_2(g)\}$$

$$= (-394.359\text{ kJ/mol}) + 2\ (-237.129\text{ kJ/mol})$$

$$- (-166.27\text{ kJ/mol}) - (\tfrac{3}{2}\text{ mol})\ (0\text{ kJ/mol}) = -702.34\text{ kJ/mol for } CH_3OH(\ell) \text{ at } 298\text{ K}$$

Therefore the $CH_3OH(\ell)$ combustion reaction is product-favored at 298 K.

Organic compounds are complex molecular systems that require significant rearrangement of atoms and bonds to undergo combustion. This makes them likely candidates for being kinetically stable.

General Questions

92. *Result:* **(a) 113.0 J $K^{-1}mol^{-1}$ (b) 38.17 kJ/mol**

(a) *Analyze and Plan:* Adapt the method described in Problem-Solving Example 16.3. Write the balanced chemical equation and use Equation 16.2:

$$\Delta_rS° = \sum\Big[(\text{coefficient of product}) \times S°(\text{product})\Big] - \sum\Big[(\text{coefficient of reactant}) \times S°(\text{reactant})\Big]$$

Execute: Vaporization equation: $CH_3OH(\ell) \longrightarrow CH_3OH(g)$

$$\Delta_rS° = (1\text{ mole}) \times S°\{CH_3OH(g)\} - (1\text{ mole}) \times S°\{CH_3OH(\ell)\}$$

$$\Delta_rS° = (1\text{ mole})\ (239.81\text{ J K}^{-1}\text{mol}^{-1}) - (1\text{ mole})\ (126.8\text{ J K}^{-1}\text{mol}^{-1}) = 113.0\text{ J K}^{-1}\text{mol}^{-1}$$

(b) *Analyze and Plan:* Adapt the method described in the solution to Question 31.

Execute: $\Delta_r G° = 0$ kJ/mol

$$\Delta_r H° = T\Delta_r S° = (64.6 + 273.15)\text{ K }(113.0\text{ J K}^{-1}\text{J K}^{-1}\text{mol}^{-1}) \times \frac{1\text{ kJ}}{1000\text{ J}} = 38.17\text{ kJ/mol}$$

94. *Result:* **(a) 3.5×10^{18} J/yr (b) 9.5×10^{15} J/day (c) 1.1×10^{11} J/s (d) 1.1×10^{11} W (e) 3×10^{2} W/person**

Analyze and Plan: Use standard conversion factors to solve this question.

Execute: Agriculture, mining and construction industries are represented on the graph with approximately 3.3 quadrillion BTUs per year. One quadrillion is a thousand times more than a trillion, or 10^{15}. One BTU (British thermal unit) is defined in Appendix B in Table B.4. as 1055.06 J.

(a) $$\frac{3.3\times10^{15}\text{ BTU}}{\text{yr}}\times\frac{1055.06\text{ J}}{1\text{ BTU}}=\frac{3.5\times10^{18}\text{ J}}{\text{yr}}$$

(b) $$\frac{3.5\times10^{18}\text{ J}}{\text{yr}}\times\frac{1\text{ year}}{365\text{ day}}=\frac{9.5\times10^{15}\text{ J}}{\text{day}}$$

(c) $$\frac{9.5\times10^{15}\text{ J}}{\text{day}}\times\frac{1\text{ day}}{24\text{ hr}}\times\frac{1\text{ hr}}{3600\text{ s}}=\frac{1.1\times10^{11}\text{ J}}{\text{s}}$$

(d) $$\frac{1.1\times10^{11}\text{ J}}{\text{s}}\times\frac{1\text{ W}}{1\text{ J/s}}=1.1\times10^{11}\text{ W}$$

(e) $$\frac{1.1\times10^{11}\text{ W}}{317\times10^{6}\text{ persons}}=3\times10^{2}\ \frac{\text{W}}{\text{person}}$$

96. *Result:* **(a) Reaction 2 (b) Reactions 1 & 5 (c) Reaction 2 (d) Reactions 2 & 3 (e) None**

Analyze and Plan: Adapt the method described in the solution to Question 73(c) and concepts described in Sections 12-2 and 12-5.

Execute:

(a) Because of the relationship: $K_p = K_c(RT)^{\Delta_r n}$, all we need to do is check $\Delta_r n$ for each reaction. K_p is larger than K_c when $\Delta_r n$ is positive, as long as RT >1(RT <1 only below T = 12 K!)

Reaction	$\Delta_r n$	$K_p > K_c$
1	2 mol product gas – 2 mol reactant gas = 0	No
2	1 mol product gas – 0 mol reactant gas = 1	Yes
3	2 mol product gas – 2 mol reactant gas = 0	No
4	1 mol product gas – 3 mol reactant gas = –2	No
5	2 mol product gas – 2 mol reactant gas = 0	No

Only reaction 2 has K_p larger than K_c.

(b) Now find the value of K_p, because $K_p = K° > 1$ for a gas-phase reaction that is product-favored.

$$(RT) = (0.08206\text{ L atm K}^{-1}\text{mol}^{-1})(298\text{ K}) = 24.5$$

$$K_p = K_c(RT)^{\Delta_r n}$$

Reaction	$(RT)^{\Delta_r n}$	K_c	K_p	Product-favored
1	1	3.6×10^{20}	3.6×10^{20}	Yes
2	$(24.5)^1$	1.24×10^{-5}	3.03×10^{-4}	No
3	1	9.5×10^{-13}	9.5×10^{-13}	No
4	$(24.5)^{-2}$	3.76	6.29×10^{-3}	No
5	1	1.9×10^{24}	1.9×10^{24}	Yes

Only reactions 1 and 5 are product-favored.

(c) Every gas contributes a concentration to the K_c expression, as described in Section 12-2. Only one reaction given has just one gas-phase component: Reaction 2.

(d) The product concentrations get higher when the value of K_c increases. In Section 12-6, we learned that the K_c increases with an increase in temperature when the reaction is endothermic. So, reactions 2 and 3 will have an increase in the product concentrations when the temperature increases.

(e) Only two reactions (1 and 2) have a product that is H_2O. The difference between $\Delta_f G°\{H_2O(\ell)\}$ and $\Delta_f G°\{H_2O(g)\}$ is 8.557 kJ/mol, so the switch would make the reaction $\Delta_r G°$ less negative by 8.557 kJ/mol. If these reactions have positive $\Delta_r G°$ between 8.557 kJ and zero, then the sign would switch. If these reactions have negative $\Delta_r G°$ values with gas-phase water, then the $\Delta_r G°$ will still be negative with liquid water.

Reaction	K_p	$\Delta_r G° = -RT\ln K°$ (kJ/mol)	$\Delta_r G°$ sign affected by (ℓ) to (g)?
1	3.6×10^{20}	– 117	No
2	3.03×10^{-4}	20.1	No

None of these has a $\Delta_r G°$ whose sign is affected by a switch from $H_2O(g)$ to $H_2O(\ell)$.

98. *Result:* **(a) $\Delta_r G° = 31.8$ kJ/mol (b) $K_p = P_{Hg(g)}$ (c) $K° = 2.7 \times 10^{-6}$ (d) 2.7×10^{-6} atm (e) 450 K**

Analyze and Plan: Use the method described in the solution to Question 67. Then use the equation derived in the solution to Question 123.

Execute:

(a) $\Delta_r G° = \Delta_f G°\{Hg(g)\} - \Delta_f G°\{Hg(\ell)\} = (31.8 \text{ kJ/mol}) - (0 \text{ kJ/mol}) = 31.8 \text{ kJ/mol}$

(b) $K° = K_p = P_{Hg(\ell)}$

(c) $K° = e^{(-\Delta_r G°/RT)} = e^{-\left(\frac{31.8 \text{ kJ/mol}}{(0.008314 \text{ kJ/mol·K})\times(298 \text{ K})}\right)} = e^{-12.8} = 10^{-5.56} = 2.7 \times 10^{-6}$

(d) $K° = K_p = P_{Hg(\ell)} = 2.7 \times 10^{-6}$ atm

(e) $K_1° = 2.7 \times 10^{-6}$ atm at $T_1 = 298$ K, $K_2° = P_{Hg(\ell)} = 10$ mm Hg at T_2

$\Delta_r H° = \Delta_f H°\{Hg(g)\} - \Delta_f H°\{Hg(\ell)\}$

$= (61.4 \text{ kJ/mol}) - (0 \text{ kJ/mol}) = 61.4 \text{ kJ}$

$$\ln\left(\frac{K_1^\circ}{K_2^\circ}\right) = \frac{\Delta H^\circ}{R}\left(\frac{1}{T_2} - \frac{1}{T_1}\right)$$

$$\ln\left(\frac{2.7\times10^{-6}\ \text{atm}}{10\ \text{mmHg}\times\dfrac{1\ \text{atm}}{760\ \text{mmHg}}}\right) = \frac{61.4\text{kJ/mol}}{0.008314\text{kJ/mol}\cdot\text{K}}\left(\frac{1}{T_2} - \frac{1}{298\ \text{K}}\right)$$

Solve for T_2:

$$T_2 = 450\ \text{K}$$

Applying Concepts

100. *Result/Explanation:* If you don't know the Humpty Dumpty poem, it should be possible to look it up at http://en.wikipedia.org/wiki/Humpty_Dumpty. Humpty Dumpty is a fictional character who was an egg. A scrambled egg is a very disordered state for an egg. The second law of thermodynamics says that the more disordered state is the more probable state. Putting the delicate tissues and fluids back where they were before the scrambling occurred would take a great deal of energy. Humpty Dumpty fell off a wall. A very probable result of that fall is for the egg to become scrambled. The story goes on to tell that all the energy of the king's horses and men was not sufficient to put Humpty together again.

101. *Result/Explanation:* It is possible to define conditions for which the entropy of a substance has its lowest possible value, namely zero at T = 0 K, so absolute entropies can be determined. It is not possible to define conditions for a specific minimum value for internal energy, enthalpy, or Gibbs free energy of a substance, so relative quantities must be used.

103. *Result:* **exothermic; product-favored, very stable ions are formed.**

Explanation: Oxides of alkali metal ions are not provided in Appendix J, another source of formation enthalpies shows the oxides of sodium and potassium, Na_2O and K_2O, have $\Delta_fH°$= – 414.2 kJ/mol and – 363 kJ/mol, respectively. Oxides of alkaline earth metals barium and magnesium, BaO and MgO, have $\Delta_fH°$= – 553.4 kJ/mol and –601.70 kJ/mol. An oxide of aluminum, corumdum, Al_2O_3 has $\Delta_fH°$= –1675.7 kJ/mol. The oxides of some transition metals for chromium(III), copper(II), iron(II), and iron(III), Cr_2O_3, CuO, FeO, Fe_2O_3, have $\Delta_fH°$= –1139.7 kJ/mol, –157.3 kJ/mol, –266.27 kJ/mol, and –824.2 kJ/mol, respectively. These oxides all have negative enthalpies of reaction, which means their oxidations are exothermic. These are probably product-favored reactions. The formation of stable metal ions along with the very stable oxide ion would very likely favor the products.

105. *Result:* **(a) false (b) false (c) true (d) true (e) true**

Analyze and Plan: Use Equation 16.4 ($\Delta_rG° = \Delta_rH° - T\Delta_rS°$) and Equation 16.9 ($\Delta_rG° = -RT\ln K°$) to relate equilibrium constant to thermodynamic values. When K = 1.0, then lnK = ln(1.0) = 0.0, so $\Delta_rG° = 0.0$ and $\Delta_rH° - T\Delta_rS° = 0.0$

Execute:

(a) It is false to say that a chemical reaction with K = 1.0 has $\Delta_rH° = 0$, unless $\Delta_rS° = 0$, also (this is unlikely to be true, except in the trivial cases where the reaction has the same reactants as products).

(b) It is false to say that a chemical reaction with K = 1.0 has $\Delta_rS° = 0$, unless $\Delta_rH° = 0$, also (this is unlikely to be true, except in the trivial cases where the reaction has the same reactants as products).

(c) It is true to say that a chemical reaction with K = 1.0 has $\Delta_rG° = 0$.

(d) It is true to say that $\Delta_rH°$ and $\Delta_rS°$ have equal sign, since $\Delta_rH°$ must have the same sign as $T\Delta_rS°$ and T is always positive.

(e) It is true to say that $\Delta_rH°/T = \Delta_rS°$ at the temperature T. The relationship is obtained when we solve this equation: $\Delta_rH° - T\Delta_rS° = 0$ for $\Delta_rS°$.

107. *Result/Explanation:* Living organisms have a high degree of order, which results in their having relatively low entropy. As a result, they must contain a large amount of free energy to maintain this low-probability state. During metabolism, the energy obtained from nutrients is stored as ATP. Conversion of ATP to ADP is an exergonic reaction. The energy from this reaction is used to drive the endergonic reactions needed to maintain the organized state of the organism. The original source of the energy needed to synthesize the sugars was sunlight used by plants to produce the sugars and other carbohydrates.

109. *Result/Explanation:* $\Delta_r G < 0$ means products are favored; however, the equilibrium state will always have some reactants present, too. To get all the reactants to go away requires the removal of the products from the reactants, so that the reaction continues forward.

111. *Result:* **(a) Case (iii)** **(b) Case (i)**

Analyze and Plan: We are given the same molar mass and the same number of molecules (five).

Execute:

(a) The most complex of the molecules (with the most atoms bonded together in one molecule) will have the greatest opportunity for complex motion, so **Case (iii)** has the highest entropy.

(b) The least complex of the species shown are the diatomic molecules. Therefore, **Case (i)** has the lowest entropy.

113. *Result:* **(a) $\Delta_r S°$ =58.78 J $K^{-1}mol^{-1}$ (b) $\Delta_r S°$ = –53.29 J $K^{-1}mol^{-1}$ (c) $\Delta_r S°$ = –173.93 J $K^{-1}mol^{-1}$; Adding more H atoms decreases the $\Delta_r S$ (i.e., makes it more negative).**

Analyze and Plan: Adapt the method described in the solution to Question 92. Formation is defined as producing one mol of a substance from standard state elements.

Execute:

(a) $2\,C(graphite) + H_2(g) \longrightarrow C_2H_2(g)$

$\Delta_r S° = S°\{C_2H_2(g)\} - 2\,S°\{C(s)\} - S°\{H_2(g)\}$

$= (200.94\ J\ K^{-1}mol^{-1}) - 2\,(5.740\ J\ K^{-1}mol^{-1}) - (130.684\ J\ K^{-1}mol^{-1}) = 58.78\ J\ K^{-1}mol^{-1}$

(b) $2\,C(graphite) + 2\,H_2(g) \longrightarrow C_2H_4(g)$

$\Delta_r S° = S°\{C_2H_4(g)\} - 2\,S°\{C(s)\} - 2\,S°\{H_2(g)\}$

$= (219.56\ J\ K^{-1}mol^{-1}) - 2\,(5.740\ J\ K^{-1}mol^{-1}) - 2\,(130.684\ J\ K^{-1}mol^{-1}) = -53.29\ J\ K^{-1}mol^{-1}$

(c) $2\,C(graphite) + 3\,H_2(g) \longrightarrow C_2H_6(g)$

$\Delta_r S° = S°\{C_2H_6(g)\} - 2\,S°\{C(s)\} - 3\,S°\{H_2(g)\}$

$= (229.60\ J\ K^{-1}mol^{-1}) - 2\,(5.740\ J\ K^{-1}mol^{-1}) - 3\,(130.684\ J\ K^{-1}mol^{-1}) = -173.93\ J\ K^{-1}mol^{-1}$

Adding more H atoms decreases the $\Delta_r S$ (i.e., makes it more negative).

More Challenging Questions

116. *Result:* **(a) 331.51 K (b) 371 K**

Analyze and Plan: Adapt methods described in the solution to Question 92.

Execute:

(a) $Br_2(\ell) \longrightarrow Br_2(g)$

$\Delta_r H° = (1\ mol) \times \Delta_f H°\{Br_2(g)\} - \Delta_f H°\{Br_2(\ell)\}$

$= (30.907\ kJ/mol) - (0\ kJ/mol) = 30.907\ kJ/mol$

$\Delta_r S° = S°\{Br_2(g)\} - S°\{Br_2(\ell)\}$

$= (245.463\ J\ K^{-1}mol^{-1}) - (152.231\ K^{-1}mol^{-1}) = 93.232\ J\ K^{-1}mol^{-1}$

$$T = \frac{\left(30.907 \frac{\text{kJ}}{\text{mol}}\right) \times \left(\frac{1000 \text{ J}}{1 \text{ kJ}}\right)}{93.232 \text{ J K}^{-1}\text{mol}^{-1}} = 331.51 \text{ K}$$

(b) $SnCl_4(\ell) \longrightarrow SnCl_4(g)$

$\Delta_r H° = (1 \text{ mol}) \times \Delta_f H°\{SnCl_4(g)\} - \Delta_f H°\{SnCl_4(\ell)\}$

$= (-471.5 \text{ kJ/mol}) - (-511.3 \text{ kJ/mol}) = 39.8 \text{ kJ/mol}$

$\Delta_r S° = S°\{SnCl_4(g)\} - S°\{SnCl_4(\ell)\}$

$= (365.8 \text{ J K}^{-1}\text{mol}^{-1}) - (258.6 \text{ K}^{-1}\text{mol}^{-1}) = 107.2 \text{ J K}^{-1}\text{mol}^{-1}$

$$T = \frac{\left(39.8 \frac{\text{kJ}}{\text{mol}}\right) \times \left(\frac{1000 \text{ J}}{1 \text{ kJ}}\right)}{107.2 \text{ J K}^{-1}\text{mol}^{-1}} = 371 \text{ K}$$

118. *Result:* **(a) $K_p = 1.5 \times 10^7$ (b) product-favored (c) $K_c = 3.7 \times 10^8$**

Analyze and Plan: Adapt methods described in the solution to Questions 67 and 73.

Execute:

(a) $\Delta_r G° = 2\, \Delta_f G°\{NOCl(g)\} + 2\, \Delta_f G°\{NO(g)\} - \Delta_f G°\{Cl_2(g)\}$

$= 2\,(66.08 \text{ kJ/mol}) + 2\,(86.55 \text{ kJ/mol}) - (0 \text{ kJ/mol}) = -40.94 \text{ kJ/mol}$

$$K_p = e^{(-\Delta_r G°/RT)} = e^{-\left(\frac{-40.94 \text{ kJ/mol}}{(0.008314 \text{ kJ/mol K}) \times (298 \text{ K})}\right)} = e^{16.5} = 10^{7.16} = 1.5 \times 10^7$$

(b) The negative Gibbs free energy change indicates that the reaction is product-favored.

(c) Get K_c from K_p: $K_p = K_c(RT)^{\Delta_r n}$

$\Delta_r n = 2 \text{ mol product gas} - 3 \text{ mol reactant gas} = -1$

$K_c = K_p(RT)^{-\Delta_r n} = (1.5 \times 10^7)\,(0.08206 \text{ L·atm/mol·K} \times 298 \text{ K})^{-(-1)} = 3.7 \times 10^8$

120. *Result:* **(a) See equations below (b) C_2H_6 reaction: $\Delta_r H° = 101.02$ kJ/mol; $\Delta_r S° = -72.71$ J K^{-1}mol^{-1}; $\Delta_r G° = 122.72$ kJ/mol; C_3H_8 reaction: $\Delta_r H° = 76.5$ kJ/mol; $\Delta_r S° = -86.6$ J K^{-1}mol^{-1} $\Delta_r G° = 102.08$ kJ/mol; CH_3OH reaction: $\Delta_r H° = 96.44$ kJ/mol; $\Delta_r S° = -305.2$ J K^{-1}mol^{-1}; $\Delta_r G° = 187.39$ kJ/mol; None are feasible.**

Analyze and Plan: Balance the equations, then use the method described in Question 51.

(a) $7\,C(s) + 6\,H_2O(g) \longrightarrow 2\,C_2H_6(g) + 3\,CO_2(g)$

$5\,C(s) + 4\,H_2O(g) \longrightarrow C_3H_8(g) + 2\,CO_2(g)$

$3\,C(s) + 4\,H_2O(g) \longrightarrow 2\,CH_3OH(\ell) + CO_2(g)$

(b) $X = \Delta_f H$, S, or $\Delta_f G$, here:

C_2H_6 reaction:

$\Delta_r X° = 2\,X°\{C_2H_6(g)\} + 3\,X°\{CO_2(g)\} - (7 \text{ mol}) \times X°\{C(s)\} - (6 \text{ mol}) \times X°\{H_2O(g)\}$

$\Delta_r H° = 2\,(-84.68 \text{ kJ/mol}) + 3\,(-393.509 \text{ kJ/mol}) - 7\,(0 \text{ kJ/mol}) - 6\,(-241.818 \text{ kJ/mol}) = 101.02 \text{ kJ/mol}$

$\Delta_r S° = 2\,(229.60 \text{ J K}^{-1}\text{mol}^{-1}) + 3\,(213.74 \text{ J K}^{-1}\text{mol}^{-1})$

$- 6\,(5.740 \text{ J K}^{-1}\text{mol}^{-1}) - 6\,(188.825 \text{ J K}^{-1}\text{mol}^{-1}) = -72.71 \text{ J K}^{-1}\text{mol}^{-1}$

$\Delta_rG° = 2\ (-32.82\ kJ/mol) + 3\ (-394.359\ kJ/mol) - 7\ (0\ kJ/mol) - 6\ (-228.572\ kJ/mol) = 122.72\ kJ/mol$

C_3H_8 reaction:

$\Delta X° = X°\{C_3H_8(g)\} + 2\ X°\{CO_2(g)\} - 5\ X°\{C(s)\} - 4\ X°\{H_2O(g)\}$

$\Delta_rH° = (-103.8\ kJ/mol) + 2\ (-393.509\ kJ/mol) - 5\ (0\ kJ/mol) - 4\ (-241.818\ kJ/mol) = 76.5\ kJ/mol$

$\Delta_rS° = (269.9\ J\ K^{-1}mol^{-1}) + 2\ (213.74\ J\ K^{-1}mol^{-1})$

$- 5\ (5.740\ J\ K^{-1}mol^{-1}) - 4\ (188.825\ J\ K^{-1}mol^{-1}) = -86.6\ J\ K^{-1}mol^{-1}$

$\Delta_rG° = (-23.49\ kJ/mol) + 2\ (-394.359\ kJ/mol) - 5\ (0\ kJ/mol) - 4\ (-228.572\ kJ/mol) = 102.08\ kJ/mol$

CH_3OH reaction:

$\Delta_rX° = 2\ X°\{CH_3OH(\ell)\} + X°\{CO_2(g)\} - 3\ X°\{C(s)\} - 4\ X°\{H_2O(g)\}$

$\Delta_rH° = 2\ (-238.66\ kJ/mol) + (-393.509\ kJ/mol) - 3\ (0\ kJ/mol) - 4\ (-241.818\ kJ/mol) = 96.44\ kJ/mol$

$\Delta_rS° = 2\ (126.8\ J\ K^{-1}mol^{-1}) + (213.74\ J\ K^{-1}mol^{-1})$

$- 3\ (5.740\ J\ K^{-1}mol^{-1}) - 4\ (188.825\ J\ K^{-1}mol^{-1}) = -305.2\ J\ K^{-1}mol^{-1}$

$\Delta_rG° = 2\ (-166.27\ kJ/mol) + (-394.359\ kJ/mol) - 3\ (0\ kJ/mol) - 4\ (-228.572\ kJ/mol) = 187.39\ kJ/mol$

None of these is feasible. $\Delta_rG°$ is positive. In addition, $\Delta_rH°$ is positive and $\Delta_rS°$ is negative, suggesting that there is no temperature at which the products would be favored.

122. *Result/Explanation:* Use Equation 16.9 and Equation 16.4: $\Delta_rG° = -RT\ln K = \Delta_rH° - T\Delta_rS°$

Divide everything by – RT: $$\ln K = -\frac{\Delta_rH°}{RT} + \frac{\Delta_rS°}{R}$$

If lnK is plotted against 1/T, the slope would be $-\Delta_rH°/R$, and the y-intercept would be $\Delta_rS°/R$. A straight line on this graph proves that the quantities of $\Delta_rH°$ and $\Delta_rS°$ are independent of temperature.

124. *Result:* **(i) Reaction (b) (ii) Reactions (a) & (c) (iii) None of them (iv) None of them**

Explanation: Determine the qualitative changes in $\Delta_rH°$ and $\Delta_rS°$, then refer to Table 16.2.

Execute:

(a) Using the hint at the end of the question, we look up bond enthalpies in Chapter 6 Section 6.7. We find that forming a bond is exothermic ($\Delta_rH°$ = negative). Two mol of gas-phase reactants form one mol of gas-phase products, so the entropy decreases ($\Delta_rS°$ = negative). That means this reaction is (ii) product-favored at low temperatures but not at high temperatures.

(b) A combustion reaction for a hydrocarbon is exothermic ($\Delta_rH°$ = negative). Nine mol of gas-phase reactants form eleven mol of gas-phase products, so the entropy increases ($\Delta_rS°$ = positive). That means this reaction is (i) always product-favored.

(c) Going farther with the hint at the end of the question, we can look up actual bond energies in Chapter 8 (Table 8.2). We conclude that forming very strong P–F bonds (490 kJ/mol) produces more energy than is used breaking the weak P–P (209 kJ/mol) and F–F (158 kJ/mol) bonds. Hence, we will predict that the reaction is exothermic ($\Delta_rH°$ = negative). Eleven mol of gas-phase reactants form four mol of gas-phase products, so the entropy decreases ($\Delta_rS°$ = negative). That means this reaction is (ii) product-favored at low temperatures but not at high temperatures.

126. *Result:* **205.2 J K^{-1}mol^{-1}**

Analyze and Plan: Calculate the $\Delta_rH°$ for the reaction, using $\Delta_fH°$ given and Hess' Law. Determine $\Delta_rS°$, from $\Delta_rG°$ and $\Delta_rH°$. Use Hess' Law to calculate $S°\{O_2(g)\}$.

Execute:

$$\Delta_r H^\circ = 2\,\Delta_f H^\circ\{H_2O(\ell)\} + \Delta_f H^\circ\{CO_2(g)\} - \Delta_f H^\circ\{CH_3OH(\ell)\} - \frac{3}{2}\,\Delta_f H^\circ\{O_2(g)\}$$

$$= 2\,(-285.83\text{ kJ/mol}) + (-393.509\text{ kJ/mol}) - (-238.66\text{ kJ/mol}) - \frac{3}{2}\,(0\text{ kJ/mol}) = -726.51\text{ kJ/mol}$$

$$\Delta_r S^\circ = \frac{\Delta H^\circ - \Delta G^\circ}{T} = \frac{-726.51\text{ kJ / mol} - (-702.35\text{ kJ / mol})}{298.15\text{ K}} \times \frac{1000\text{ J}}{1\text{ kJ}} = -81.17\text{ J K}^{-1}\text{mol}^{-1}$$

$$\Delta_r S^\circ = 2\,S^\circ\{H_2O(\ell)\} + S^\circ\{CO_2(g)\} - S^\circ\{CH_3OH(\ell)\} - \frac{3}{2}\,S^\circ\{O_2(g)\}$$

$$\frac{3}{2}\,S^\circ\{O_2(g)\} = 2\,S^\circ\{H_2O(\ell)\} + S^\circ\{CO_2(g)\} - S^\circ\{CH_3OH(\ell)\} - \Delta_r S^\circ$$

$$S^\circ\{O_2(g)\} = \frac{2}{3}\Big[\,2\,S^\circ\{H_2O(\ell)\} + S^\circ\{CO_2(g)\} - S^\circ\{CH_3OH(\ell)\} - \Delta_r S^\circ\,\Big]$$

$$S^\circ\{O_2(g)\} = \frac{2}{3}\Big[\,2(69.91\text{ J K}^{-1}\text{mol}^{-1}) + (213.74\text{ J K}^{-1}\text{mol}^{-1}) - (126.8\text{ J K}^{-1}\text{mol}^{-1}) - (-81.17\text{ J K}^{-1}\text{mol}^{-1})\,\Big]$$

$$S^\circ\{O_2(g)\} = 205.2\text{ J K}^{-1}\text{mol}^{-1}$$

Chapter 17: Electrochemistry and Its Applications

Solutions for Red-Numbered Questions for Review and Thought

Topical Questions

Redox Reactions (Section 17-1)

6. *Result:* **In reactants: Ox. # H = +1, Ox. # Fe = +2, Ox. # Mn = +7, and Ox. O = –2. In products, Ox. # Mn = +2, Ox. # Fe = +3, Ox. O = –2, Ox. # H = +1. Fe^{2+} is oxidized and MnO_4^- is reduced. MnO_4^- is the oxidizing agent and Fe^{2+} is reducing agent.**

   Analyze, Plan, and Execute: Follow the methods described in Chapter 3.

   The oxidation number of atoms monatomic ions is their charge, so:

   H^+ Ox. # H = +1

   Fe^{2+} Ox. # Fe = +2

   Mn^{2+} Ox. # Mn = +2

   Fe^{3+} Ox. # Fe = +3

   Ox. O = –2, here, since O is bonded to Mn and H.

   In compounds and polyatomic ions, the sum of the oxidation numbers of the atoms is equal to the net charge.

   In MnO_4^- Ox. # Mn + 4(Ox.# O) = –1 so: Ox. # Mn = –1 – 4(–2) = +7

   In H_2O Ox. # H + 2(Ox.# O) = 0 so: Ox. # H = – ½ (–2) = +1

   The substance containing the element that loses electrons is oxidized and represents the reducing agent. Here, that is Fe^{2+}.

   The substance containing the element that that gains electrons is reduced and represents the oxidizing agent. Here, that is MnO_4^-.

   ☑ *Reasonable Result Check:* Oxidation numbers increase from reactant to products for atoms oxidized and reducing agents; oxidation numbers decrease from reactant to products for atoms reduced and oxidizing agents.

Half-Reactions and Redox Reactions (Section 17-2)

10. *Result:* **(a) $Cd(s) \longrightarrow Cd^{2+}(aq) + 2\,e^-$ (b) $Fe^{3+}(aq) + 3\,e^- \longrightarrow Fe(s)$ (c) $Sn^{4+}(aq) + 2\,e^- \longrightarrow Sn^{2+}(aq)$ (d) $Cl_2 + 2\,e^- \longrightarrow 2\,Cl^-$ (e) $2\,H_2O(\ell) + SO_2(g) \longrightarrow SO_4^{2-}(aq) + 4\,H^+(aq) + 2\,e^-$**

    Analyze and Plan: Follow the methods described in Appendix F Problem-Solving Example F.1 Step 3.

    Execute:

    (a) Balance Cd atoms, then balance charge with electrons:

    $$Cd(s) \longrightarrow Cd^{2+}(aq) + 2\,e^- \qquad \text{(Check: 1 Cd, zero net charge)}$$

    (b) Balance Fe atoms, then balance charge with electrons:

    $$Fe^{3+}(aq) + 3\,e^- \longrightarrow Fe(s) \qquad \text{(Check: 1 Fe, zero net charge)}$$

    (c) Balance Sn atoms, then balance charge with electrons:

    $$Sn^{4+}(aq) + 2\,e^- \longrightarrow Sn^{2+}(aq) \qquad \text{(Check: 1 Sn, +2 net charge)}$$

(d) Balance Cl atoms, then balance charge with electrons:

$$Cl_2(g) + 2\ e^- \longrightarrow 2\ Cl^-(aq) \qquad \text{(Check: 2 Cl, –2 net charge)}$$

(e) Balance S atoms, then balance O atoms with H_2O, then balance H atoms with H^+, then balance charge with electrons:

$$2\ H_2O(\ell) + SO_2(g) \longrightarrow SO_4^{2-}(aq) + 4\ H^+(aq) + 2\ e^- \quad \text{(Check: 4 H, 4 O, 1 S, zero net charge)}$$

12. *Result:* **$Fe^{2+}(aq) \longrightarrow Fe^{3+}(aq) + e^-$; $MnO_4^-(aq) + 8\ H^+(aq) + 5\ e^- \longrightarrow Mn^{2+}(aq) + 4\ H_2O(\ell)$**

Analyze and Plan: Follow methods described in Appendix F Problem-Solving Examples F.1 and 17.2

The substances for oxidation all go in one half-reaction and the substances for reduction all go in the other half-reaction. We do not have to determine where the H^+ ions and H_2O molecules go because they will show up where they belong as we follow Step 3 to balance the half-reactions.

Execute:

Put Fe^{2+} and Fe^{3+} in one half-reaction. Put MnO_4^- and Mn^{2+} in the other. Balance the atoms and charge.

$$Fe^{2+}(aq) \longrightarrow Fe^{3+}(aq) + e^- \qquad \text{(Check: 1 Fe, +2)}$$

$$MnO_4^-(aq) + 8\ H_3O^+(aq) + 5\ e^- \longrightarrow Mn^{2+}(aq) + 12\ H_2O(\ell) \qquad \text{(Check: 1 Mn, 12 O, 24 H, +2)}$$

14. *Result/Explanation:* Use the half-reactions balanced in Question 12.

Multiply each half-reactions by constant values that equalizes the number of electrons, then add the oxidation and reduction reaction, cancelling anything that appears on both sides of the net equation.

The first reaction gets multiplied by 4, and nothing but the electrons cancel.

$$5\ Fe^{2+}(aq) \longrightarrow 5\ Fe^{3+}(aq) + 5\ e^-$$

$$+\ \ 8\ H^+(aq) + MnO_4^-(aq) + 5\ e^- \longrightarrow Mn^{2+}(aq) + 4\ H_2O(\ell)$$

$$\mathbf{8\ H^+(aq) + MnO_4^-(aq) + 5\ Fe^{2+}(aq) \longrightarrow 5\ Fe^{3+}(aq) + Mn^{2+}(aq) + 12\ H_2O(\ell)}$$

(Check: 8 H, 1 Mn, 4 O, 5 Fe, +17 net charge)

16. *Result:* **(a) $3\ CO(g) + O_3(g) \longrightarrow 3\ CO_2(g)$; oxidizing agent: O_3; reducing agent: CO**

(b) $H_2(g) + Cl_2(g) \longrightarrow 2\ HCl(g)$; oxidizing agent: Cl_2; reducing agent: H_2
(c) $2\ H^+(aq) + H_2O_2(aq) + Ti^{2+}(aq) \longrightarrow 2\ H_2O(\ell) + Ti^{4+}(aq)$; ox. agent: H_2O_2; red. agent: Ti^{2+}
(d) $2\ MnO_4^-(aq) + 6\ Cl^-(aq) + 8\ H^+ \longrightarrow 2\ MnO_2(s) + 3\ Cl_2(g) + 4\ H_2O(\ell)$; oxidizing agent: MnO_4^-; reducing agent: Cl^- (e) $4\ FeS_2(s) + 11\ O_2(g) \longrightarrow 2\ Fe_2O_3(s) + 8\ SO_2(g)$; oxidizing agent: O_2; reducing agent: FeS_2 (f) $O_3(g) + NO(g) \longrightarrow O_2(g) + NO_2(g)$; oxidizing agent: O_3; reducing agent: NO

(g) $Zn(s) + HgO(s) + H_2O(\ell) \longrightarrow Zn(OH)_2(s) + Hg(\ell)$; oxidizing agent: HgO; reducing agent: Zn(s)

Analyze and Plan: Follow the methods described in the solution to Questions 10, 12 and Appendix F Problem-Solving Examples F.1, and F.2. After the half-reactions are separated and balanced, equalize the electrons with appropriate multipliers and add the two half-reactions.

Execute:

(a) Put CO and CO_2 in one half-reaction. Put O_3 in the other. Balance each:

$$H_2O(\ell) + CO(g) \longrightarrow CO_2(g) + 2\ H^+(aq) + 2\ e^-$$

$$6\ e^- + 6\ H^+(aq) + O_3(g) \longrightarrow 3\ H_2O(\ell)$$

Multiply each half-reaction by a constant to get the same number of electrons:

$$3 \times [H_2O(\ell) + CO(g) \longrightarrow CO_2(g) + 2\,H^+(aq) + 2\,e^-]$$

$$6\,e^- + 6\,H^+(aq) + O_3(g) \longrightarrow 3\,H_2O(\ell)$$

Now add them:

$$3\,H_2O(\ell) + 3\,CO(g) \longrightarrow 3\,CO_2(g) + 6\,H^+(aq) + 6\,e^-$$

$$+\ 6\,e^- + 6\,H^+(aq) + O_3(g) \longrightarrow 3\,H_2O(\ell)$$

$$3\,CO(g) + O_3(g) \longrightarrow 3\,CO_2(g)$$

(Check: 3 C, 6 O, no net charge)

O_3 is the oxidizing agent. CO is the reducing agent.

(b) This reaction is easy to balance the old-fashioned way: $H_2(g) + Cl_2(g) \longrightarrow 2\,HCl(g)$

The Ox. # Cl changes from 0 to –1, Cl_2 is reduced and serves as the oxidizing agent. The Ox. # H changes from 0 to +1, so H_2 is oxidized and serves as the reducing agent.

(c) Put H_2O_2 in one half-reaction. Put Ti^{2+} and Ti^{4+} in the other. Balance each, then add.

$$2\,e^- + 2\,H^+(aq) + H_2O_2(aq) \longrightarrow 2\,H_2O(\ell)$$

$$+\ Ti^{2+}(aq) \longrightarrow Ti^{4+}(aq) + 2\,e^-$$

$$2\,H^+(aq) + H_2O_2(aq) + Ti^{2+}(aq) \longrightarrow 2\,H_2O(\ell) + Ti^{4+}(aq)$$

(Check: 4 H, 2 O, 1 Ti, +4 net charge)

H_2O_2 is the oxidizing agent. Ti^{2+} is the reducing agent.

(d) Put MnO_4^- and MnO_2 in one half-reaction. Put Cl^- and Cl_2 in the other.

$$3\,e^- + MnO_4^-(aq) + 4\,H^+(aq) \longrightarrow MnO_2(s) + 2\,H_2O(\ell)$$

$$2\,Cl^-(aq) \longrightarrow Cl_2(g) + 2\,e^-$$

Multiply each half-reaction by a constant to get the same number of electrons:

$$2 \times [3\,e^- + MnO_4^-(aq) + 4\,H^+(aq) \longrightarrow MnO_2(s) + 2\,H_2O(\ell)]$$

$$3 \times [2\,Cl^-(aq) \longrightarrow Cl_2(g) + 2\,e^-]$$

Now add them:

$$6\,e^- + 2\,MnO_4^-(aq) + 8\,H^+(aq) \longrightarrow 2\,MnO_2(s) + 4\,H_2O(\ell)$$

$$+\ 6\,Cl^-(aq) \longrightarrow 3\,Cl_2(g) + 6\,e^-$$

$$2\,MnO_4^-(aq) + 6\,Cl^-(aq) + 8\,H^+(aq) \longrightarrow 2\,MnO_2(s) + 3\,Cl_2(g) + 4\,H_2O(\ell)$$

(Check: 2 Mn, 8 O, 8 H, zero net charge)

MnO_4^- is the oxidizing agent. Cl^- is the reducing agent.

(e) Put FeS_2, Fe_2O_3 and SO_2 in one half-reaction. Put O_2 in the other.

$$11\,H_2O(\ell) + 2\,FeS_2(s) \longrightarrow Fe_2O_3(s) + 4\,SO_2(g) + 22\,H^+(aq) + 22\,e^-$$

$$4\,e^- + 4\,H^+(aq) + O_2(g) \longrightarrow 2\,H_2O(\ell)$$

Multiply each half-reaction by a constant to get the same number of electrons:

$$2 \times [11\ H_2O(\ell) + 2\ FeS_2(s) \longrightarrow Fe_2O_3(s) + 4\ SO_2(g) + 22\ H^+(aq) + 22\ e^-]$$

$$11 \times [4\ e^- + 4\ H^+(aq) + O_2(g) \longrightarrow 2\ H_2O(\ell)]$$

Now add them:

$$22\ H_2O(\ell) + 4\ FeS_2(s) \longrightarrow 2\ Fe_2O_3(s) + 8\ SO_2(g) + 44\ H^+(aq) + 44\ e^-]$$

$$+\ 44\ e^- + 44\ H^+(aq) + 11\ O_2(g) \longrightarrow 22\ H_2O(\ell)$$

$$4\ FeS_2(s) + 11\ O_2(g) \longrightarrow 2\ Fe_2O_3(s) + 8\ SO_2(g)$$

(Check: 4 Fe, 8 S, 22 O, zero net charge)

O_2 is the oxidizing agent. FeS_2 is the reducing agent.

(f) This reaction is already balanced, but let's separate the half reactions to elucidate what is oxidized and what is reduced.

$$2\ e^- + O_3(g) + 2\ H^+(aq) \longrightarrow O_2(g) + H_2O(\ell)$$

$$+\ NO(g) + H_2O(\ell) \longrightarrow NO_2(g) + 2\ H^+(aq) + 2\ e^-$$

$$O_3(g) + NO(g) \longrightarrow O_2(g) + NO_2(g)$$

O_3 is the oxidizing agent. NO is the reducing agent.

(g) Put Zn and $Zn(OH)_2$ in one half-reaction. Put HgO and Hg in the other. Balance each, then add them

$$Zn(s) + 2\ H_2O(\ell) \longrightarrow Zn(OH)_2(s) + 2\ H^+(aq) + 2\ e^-$$

$$+\ 2\ e^- + HgO(s) + 2\ H^+(aq) \longrightarrow Hg(\ell) + H_2O(\ell)$$

$$Zn(s) + HgO(s) + H_2O(\ell) \longrightarrow Zn(OH)_2(s) + Hg(\ell)$$

(Check: 1 Zn. 1Hg, 2 O, 2 H, zero net charge)

Because there is no remaining $H^+(aq)$ in the net reaction, nothing else needs to be done to balance this in basic solution.

Hg is the oxidizing agent. Zn(Hg) is the reducing agent.

Voltaic Cells (Section 17-3)

18. *Result/Explanation:* The generation of electricity occurs when electrons are transmitted through a wire from the metal to the cation. Here, the transfer of electrons would occur directly from the metal to the cation and the electrons would not flow through any wire.

21. *Result:* **(a) $Zn + Pb^{2+} \longrightarrow Zn^{2+} + Pb$ (b) Anode: $Zn(s) \longrightarrow Zn^{2+} + 2\ e^-$ Cathode: $2\ e^- + Pb^{2+} \longrightarrow Pb(s)$ (c) see diagram below**

Explanation:

(a) $Zn(s) + Pb^{2+}(aq) \longrightarrow Zn^{2+}(aq) + Pb(s)$

(b) Oxidation of zinc atoms occurs at the anode, which is metallic zinc:

$$Zn(s) \longrightarrow Zn^{2+}(aq) + 2\ e^-$$

The reduction of lead(II) ions occurs at the cathode, which is metallic lead:

$$2\ e^- + Pb^{2+}(aq) \longrightarrow Pb(s)$$

(c)

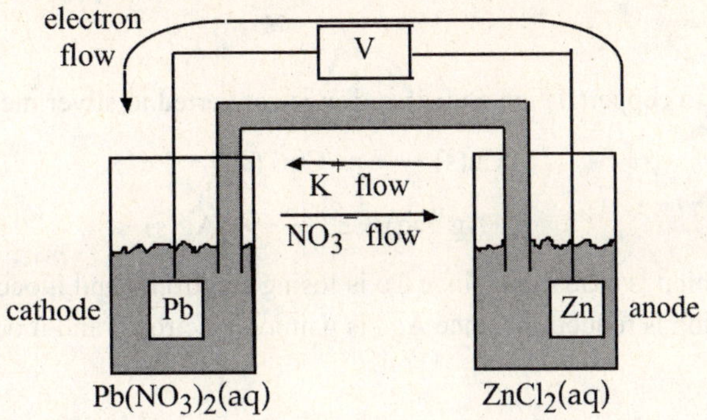

23. *Result*: **See diagrams below**

Explanation: anode half-cell || cathode half-cell

(a)

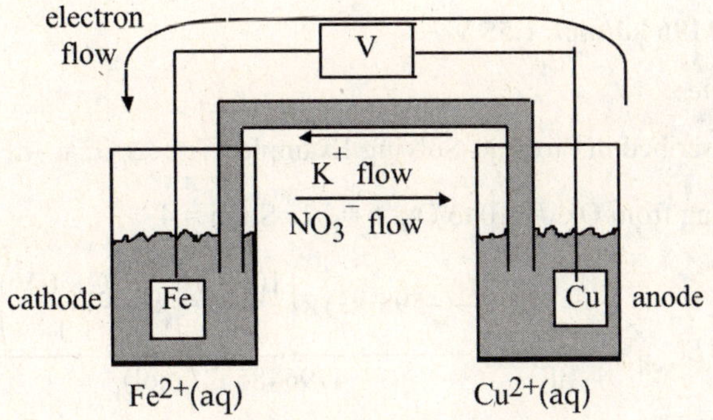

(b)

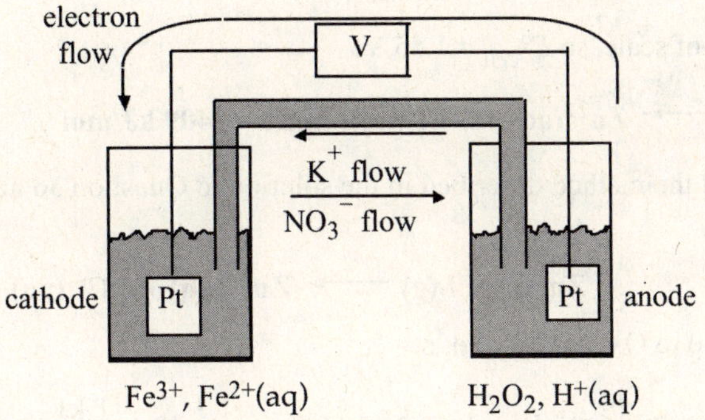

Voltaic Cells and Cell Potential (Section 17-4)

24. *Result:* **9.0×10^4 J; 7500 C**

Analyze and Plan: This is a standard conversion factor problem. Divide the time by the voltage and use the definition of "watt" given in the Question to show that 25 J are transferred per second.

Execute: 1 J = 1C × 1V

$$1.0 \text{ hr} \times \frac{60 \text{ min}}{1 \text{ hr}} \times \frac{60 \text{ s}}{1 \text{ min}} \times \frac{25 \text{ J}}{1 \text{ s}} = 9.0 \times 10^4 \text{ J}$$

$$\frac{9.0 \times 10^4 \text{ J}}{12 \text{ V}} \times \frac{1 \text{ C} \times 1 \text{ V}}{1 \text{ J}} = 7500 \text{ C}$$

25. *Result:* **(a) Cu $\longrightarrow$ Cu^{2+} + 2 e^-; Ag^+ + e^- $\longrightarrow$ Ag (b) Oxidation: Cu half-reaction; Reduction: Ag half-reaction; Anode compartment: Cu half-reaction; Cathode compartment: Ag half-reaction**

Explanation:

(a) Copper is converted to copper(II) ion and silver ion is converted to silver metal.

$$Cu(s) \longrightarrow Cu^{2+}(aq) + 2\ e^-$$

$$Ag^+(aq) + e^- \longrightarrow Ag(s)$$

(b) The copper half-reaction is oxidation (since Cu is losing electrons) and it occurs in the anode compartment. The silver half-reaction is reduction (since Ag^+ is gaining electrons) and it occurs in the cathode compartment.

Using Standard Half-Cell Potentials (Section 17-5)

35. *Result:* **Greater; less**

Explanation: In a product-favored chemical reaction, the standard cell potential, $E°$, is **greater** than zero and the Gibbs free energy, $\Delta_r G°$, is **less** than zero.

36. *Result:* **(a) 1.55 V (b) –1196 kJ/mol, 1.55 V**

Analyze, Plan, and Execute:

(a) Adapt the method described in Problem-Solving Example 17.6: $\Delta_r G° = -nFE°_{cell}$

Two O atoms are going from Ox. # = 0 to Ox. # = –2. So, n = 4.

$$E°_{cell} = \frac{-\Delta G°}{nF} = \frac{-(-598\ \text{kJ}) \times \left(\frac{1000\ \text{J}}{1\ \text{kJ}}\right) \times \left(\frac{1\ \text{C} \times 1\ \text{V}}{1\ \text{J}}\right)}{4\ (96485\ \text{C/mol})} = 1.55\ \text{V}$$

(b) When the reaction is written with all the coefficients doubled:

$$\Delta_r G°_{double} = \Delta_r G°_{original} + \Delta_r G°_{original} = -598\ \text{kJ/mol} + (-598\ \text{kJ/mol}) = -1196\ \text{kJ/mol}$$

$E°_{cell}$ is independent of scale, so $E°_{cell}$ = 1.55 V.

38. *Result:* **Zn(s) + Cl_2(g) $\longrightarrow$ Zn^{2+}(aq) + 2 Cl^-(aq); $\Delta_r G°$ = –409 kJ/mol**

Analyze and Plan: Adapt the method described in the solution to Question 36 and Problem-Solving Example 17.6.

Execute: **Zn(s) + Cl_2(g) $\longrightarrow$ Zn^{2+}(aq) + 2 Cl^-(aq)**

Zn is going from Ox. # = 0 to Ox. # = +2. So, n = 2

$$\Delta_r G° = -2\ (96485\ \text{C/mol}) \times (2.12\ \text{V}) \times \frac{1\ \text{J}}{1\ \text{C} \times 1\ \text{V}} \times \frac{1\ \text{kJ}}{1000} = \mathbf{-409\ kJ/mol}$$

40. *Result:* **$K_c = 1 \times 10^{25}$, $\Delta_r G°$ = –143 kJ/mol**

Analyze and Plan: Adapt the method described in the solution to Question 38 and Problem-Solving Example 17.7.

Execute:

$$Cd(s) \longrightarrow Cd^{2+}(aq) + 2\ e^- \qquad E°_{anode} = -0.403\ \text{V}$$

$$Cu^{2+}(aq) + 2\ e^- \longrightarrow Cu(s) \qquad E°_{cathode} = 0.340\ \text{V}$$

$$Cd(s) + Cu^{2+}(aq) \longrightarrow Cd^{2+}(aq) + Cu(s) \qquad E°_{cell} = (0.340\ \text{V}) - (-0.403\ \text{V}) = 0.743\ \text{V}$$

$$\ln K° = \frac{nFE°_{cell}}{RT} = \frac{(2\text{ mol}) \times \left(\frac{96,485\text{ C}}{1\text{ mol e}^-}\right) \times (0.743\text{ V})}{(8.314\text{J K}^{-1}\text{mol}^{-1})(298\text{ K})} = 57.9$$

$$K° = e^{57.9} = 1 \times 10^{25} = \mathbf{K_c}$$

$$\Delta_r G° = -2\ (96485\text{ C/mol}) \times (0.743\text{ V}) \times \frac{1\text{ J}}{1\text{ C} \times 1\text{ V}} \times \frac{1\text{ kJ}}{1000} = -143\text{ kJ/mol}$$

42. *Result:* $\mathbf{K_c = 2 \times 10^{-53}}$, $\mathbf{\Delta_r G° = 301.5}$ **kJ/mol**

Analyze and Plan: Adapt the method described in the solution to Question 41.

Execute:

$2\ Ag(s) \longrightarrow 2\ Ag^+(aq) + 2\ e^-$ oxidation half-reaction $E°_{anode} = +\ 0.7991$ V

$Zn^{2+}(aq) + 2\ e^- \longrightarrow Zn(s)$ reduction half-reaction $E°_{cathode} = -0.763$ V

$Zn^{2+}(aq) + 2\ Ag^+(aq) \longrightarrow Sn^{4+}(aq) + 2\ Ag(s)$ $E°_{cell} = +(-0.763\text{ V}) - (+\ 0.7991\text{ V}) = -1.562$ V

$$\ln K° = \frac{nFE°_{cell}}{RT} = \frac{(2\text{ mol}) \times \left(\frac{96,485\text{ C}}{1\text{ mol e}^-}\right) \times (-1.562\text{ V})}{(8.314\text{J K}^{-1}\text{mol}^{-1})(298.15\text{ K})} = -121.7$$

$$K° = e^{-121.7} = 2 \times 10^{-53} = \mathbf{K_c}$$

$$\Delta_r G° = -2\ (96485\text{ C/mol}) \times (-1.562\text{ V}) \times \frac{1\text{ J}}{1\text{ C} \times 1\text{ V}} \times \frac{1\text{ kJ}}{1000} = 301.5\text{ kJ/mol}$$

45. *Result:* $\mathbf{K_c = 0.12}$

Analyze and Plan: Adapt the method described in the solution to Question 36 and Problem-Solving Example 19.7.

$$\ln K° = \frac{nFE°_{cell}}{RT}$$

Execute:

$E°_{cell}$ is given to be 0.046 V and the number of moles of electrons passed to take 2+ ions to neutral is 2.

$$\log K° = = \frac{(2\text{ mol}) \times \left(\frac{96,485\text{ C}}{1\text{ mol e}^-}\right) \times (-0.027\text{ V})}{(8.314\text{J K}^{-1}\text{mol}^{-1})(298.15\text{ K})} = -2.1$$

$$K° = e^{-2.1} = 0.12 = K_c$$

Effect of Concentration on Cell Potential: The Nernst Equation (Section 17-7)

46. *Result:* **(a)** $\mathbf{E°_{cell} = 0.360}$ **V (b) (conc.** $\mathbf{Zn^{2+}}$**) = 0.20 M (c)** $\mathbf{E_{cell} = 0.33}$ **V**

Analyze and Plan: Adapt the method described in the solution to Question 31. Use the method described in Problem-Solving Example 17.6, then use the Nernst equation at T = 298 K, by adapting the method described in the Problem-Solving Example 17.8.

$$E_{cell} = E°_{cell} - \frac{RT}{nF} \ln Q$$

Execute:

(a) $Zn(s) \longrightarrow Zn^{2+}(aq) + 2\,e^-$ $\quad E^\circ_{anode} = -0.763$ V

$Cd^{2+}(aq) + 2\,e^- \longrightarrow Cd(s)$ $\quad E^\circ_{cathode} = -0.403$ V

$Zn(s) + Cd^{2+}(aq) \longrightarrow Zn^{2+}(aq) + Cd(s)$ $\quad E^\circ_{cell} = (-0.403\text{ V}) - (-0.763\text{ V}) = 0.360$ V

For this reaction, $Q = \dfrac{(\text{conc } Zn^{2+})}{(\text{conc } Cd^{2+})}$ and n = 2

(b)
$$\ln Q = \ln\left(\frac{(\text{conc } Zn^{2+})}{(\text{conc } Cd^{2+})}\right) = \frac{nF}{RT}(E^\circ_{cell} - E_{cell})$$

$$\ln\left(\frac{(\text{conc } Zn^{2+})}{(\text{conc } Cd^{2+})}\right) = \frac{(2)\times\left(\dfrac{96,485\text{ C}}{1\text{ mol e}^-}\right)}{(8.314\text{J K}^{-1}\text{mol}^{-1})(298\text{ K})} \times (0.360\text{ V} - 0.390\text{ V})$$

$$\ln\left(\frac{(\text{conc } Zn^{2+})}{(\text{conc } Cd^{2+})}\right) = \frac{2}{0.0592\text{ V}} \times (-0.030\text{ V}) = -2.3$$

$$\ln\left(\frac{(\text{conc } Zn^{2+})}{2.00\text{ M}}\right) = -1.0$$

$$\frac{(\text{conc } Zn^{2+})}{2.00\text{ M}} = 0.10$$

$$(\text{conc. } Zn^{2+}) = 0.20\text{ M}$$

(c)
$$E_{cell} = E^\circ_{cell} - \frac{0.0592\text{ V}}{n}\log\left(\frac{(\text{conc. } Zn^{2+})}{(\text{conc. } Cd^{2+})}\right) = 0.36\text{ V} - \frac{0.0592\text{ V}}{n}\log\left(\frac{1.00\text{ M}}{0.068\text{ M}}\right)$$

$$E_{cell} = 0.36\text{ V} - \frac{0.0592\text{ V}}{2}\log(15)$$

$$E_{cell} = 0.36\text{ V} - \frac{0.0592\text{ V}}{2} \times 1.2$$

$$E_{cell} = 0.36\text{ V} - 0.03\text{ V}$$

$$E_{cell} = 0.33\text{ V}$$

49. *Result:* $\boldsymbol{E_{cell} = -0.378}$ **V**

Analyze and Plan: Adapt the method described in the solution to Question 46 and the methods described in Chapter 14 for determining the concentration of H_3O^+.

Execute: $H_2(g) \longrightarrow 2\,H^+(aq) + 2\,e^-$ $\quad E^\circ_{anode} = 0.00$ V

$2\,H^+(aq) + 2\,e^- \longrightarrow H_2(g)$ $\quad E^\circ_{cathode} = 0.00$ V

$H_2(g)_{anode} + 2\,H^+(aq)_{cathode} \longrightarrow 2\,H^+(aq)_{anode} + H_2(g)_{cathode}$ $\quad E^\circ_{cell} = (0.00\text{ V}) - (0.00\text{ V}) = 0.00$ V

For this reaction, $Q = \dfrac{(P_{H_2})_{cathode}(\text{conc. } H^+)^2_{anode}}{(P_{H_2})_{anode}(\text{conc. } H^+)^2_{cathode}}$ and n = 2

The cathode has pH = 7.8, so

$(\text{conc. H}^+)_{\text{cathode}} = (\text{conc. H}_3\text{O}^+)_{\text{cathode}} = 10^{-\text{pH}} = 10^{-7.8} = 2 \times 10^{-8}$ M *(must round to 1 sig fig)*

$$E_{\text{cell}} = 0.00\text{ V} - \frac{0.0592\text{ V}}{2}\log\frac{(1.00\text{ bar})(0.05\text{ M})^2}{(1.00\text{ bar})(2\times10^{-8}\text{ M})^2}$$

$$E_{\text{cell}} = 0.00\text{ V} - \frac{0.0592\text{ V}}{2}\log(1\times10^{13})$$

$$E_{\text{cell}} = 0.00\text{ V} - \frac{0.0592\text{ V}}{2}\times(13.0)$$

$$E_{\text{cell}} = 0.00\text{ V} - (0.378\text{ V})$$

$$E_{\text{cell}} = -0.378\text{ V}$$

51. *Result:* $\dfrac{\textbf{conc. Pb}^{2+}}{\textbf{conc. Sn}^{2+}} = \mathbf{0.38}$

Analyze and Plan: Adapt the method described in the solution to Question 46.

Execute:

$$\text{Pb(s)} \longrightarrow \text{Pb}^{2+} + 2e^- \qquad E°_{\text{anode}} = -0.125\text{ V}$$

$$+ \quad \text{Sn}^{2+} + 2e^- \longrightarrow \text{Sn(s)} \qquad E°_{\text{cathode}} = -0.1375\text{ V}$$

$$\text{Pb(s)} + \text{Sn}^{2+} \longrightarrow \text{Pb}^{2+} + \text{Sn(s)} \qquad E°_{\text{cell}} = -0.0125\text{ V}$$

$= -0.013$ V *(must round to three decimal places)*

$$n = 2$$

$$E_{\text{cell}} = E°_{\text{cell}} - \frac{0.0592\text{ V}}{n}\log\left(\frac{\text{conc. Pb}^{2+}}{\text{conc. Sn}^{2+}}\right)$$

$$0\text{ V} = -0.013\text{ V} - \frac{0.0592\text{ V}}{2}\log\left(\frac{\text{conc. Pb}^{2+}}{\text{conc. Sn}^{2+}}\right)$$

$$-0.42 = \log\left(\frac{\text{conc Pb}^{2+}}{\text{conc Sn}^{2+}}\right)$$

$$0.38 = \frac{\text{conc Pb}^{2+}}{\text{conc Sn}^{2+}}$$

Common Batteries (Section 17-8)

52. *Result:* Advantages: rechargeable, durable, inexpensive, reliable, and simple. Disadvantages: heavy, dangerous (can be corrosive and might explode if sparked), contains lead, a known environmental danger.

Explanation: The advantages of the lead-acid storage battery are: It is rechargeable, durable, inexpensive, reliable, and simple.

The disadvantages of the lead-acid storage battery are: It is heavy. It can be dangerous (sulfuric acid is corrosive; the side-reaction hydrolysis of water in the recharge stage creates hydrogen and oxygen gases, which can explode if sparked.) It contains lead, which is a known environmental danger. If the battery is improperly stored or disposed of, the air, soil, and ground water could become contaminated.

54. *Result:* **(a) $Ni^{2+} + Cd \longrightarrow Ni + Cd^{2+}$ (b) oxidized: Cd; reduced: Ni^{2+}; oxidizing agent: Ni^{2+}; reducing agent: Cd (c) anode: Cd; cathode: Ni (d) $E°_{cell} = 0.15$ V (e) from the Cd electrode to the Ni electrode (f) toward the anode compartment**

Explanation: Apply the methods described in the solutions to Questions 21, 25, and 32.

(a) Nickel ion reacts with cadmium metal to form Nickel metal and cadmium ion.

$$Ni^{2+}(aq) + Cd(s) \longrightarrow Ni(s) + Cd^{2+}(aq)$$

(b)

Substance oxidized	Substance reduced	Oxidizing agent	Reducing agent
Cd	Ni^{2+}	Ni^{2+}	Cd

(c) Metals in the half-reaction will serve as the electrodes. Metallic Cd is the anode and metallic Ni is the cathode.

(d) $Cd(s) \longrightarrow Cd^{2+}(aq) + 2\,e^-$ $\quad E°_{anode} = -\,0.403$ V

$Ni^{2+}(aq) + 2\,e^- \longrightarrow Ni(s)$ $\quad E°_{cathode} = -\,0.25$ V

$Cd(s) + Ni^{2+}(aq) \longrightarrow Cd^{2+}(aq) + Ni(s)$ $\quad E°_{cell} = (-\,0.25\text{ V}) - (-\,0.403\text{ V}) = 0.15$ V

(e) The half-reactions above show that electrons flow spontaneously from the Cd electrode to the Ni electrode.

(f) The NO_3^- anions in the salt bridge flow toward the anode compartment to replenish the negative charges to neutralize the Cd^{2+} ions being formed.

Fuel Cells (Section 17-9)

56. *Result/Explanation:* A fuel cell has a continuous supply of reactants, and will be useable for as long as the reactants are supplied. A battery contains all the reactants of the reaction. Once the reactants are gone, the battery is no longer useable.

58. *Result:* **(a) Anode: N_2H_4 oxidation; cathode: O_2 reduction (b) $N_2H_4(g) + O_2(g) \longrightarrow N_2(g) + 2H_2O(\ell)$ (c) 7.5 g N_2H_4 (d) 7.5 g O_2**

Analyze and Plan: Adapt the methods described in the solutions to Question 47, Chapter 2, and Problem-Solving Example 17.11.

Execute:

(a) The N_2H_4 oxidation occurs at the anode. The O_2 reduction occurs at the cathode.

(b) Adding the two half-reactions gives: $N_2H_4(g) + O_2(g) \longrightarrow N_2(g) + 2H_2O(\ell)$

(c) $$50.0\text{ hr} \times 0.50\text{ A} \times \frac{3600\text{ s}}{1\text{ hr}} \times \frac{1\text{ C}}{1\text{ A}\cdot 1\text{ s}} \times \frac{1\text{ mol e}^-}{96485\text{ C}} \times \frac{1\text{ mol }N_2H_4}{4\text{ mol e}^-} \times \frac{32.05\text{ g }N_2H_4}{1\text{ mol }N_2H_4} = 7.5\text{ g }N_2H_4$$

(d) $$7.5\text{ g }N_2H_4 \times \frac{1\text{ mol }N_2H_4}{32.05\text{ g }N_2H_4} \times \frac{1\text{ mol }O_2}{1\text{ mol }N_2H_4} \times \frac{32.00\text{ g }O_2}{1\text{ mol }O_2} = 7.5\text{ g }O_2$$

Electrolysis—Causing Reactant-Favored Redox Reactions to Occur (Section 17-10)

59. *Result:* **Anode O_2; cathode H_2; 2 moles of H_2 produced per mole O_2**

Analyze and Plan: Electrolysis of water in sulfuric acid involves the oxidation of water to form oxygen gas and the reduction of the strong acid to form hydrogen gas.

Execute: Anode - oxidation: $\quad 6\,H_2O(\ell) \longrightarrow O_2(g) + 4\,H_3O^+(aq) + 4\,e^-$

Cathode - reduction: $\quad 4\,H_3O^+(aq) + 4\,e^- \longrightarrow 2\,H_2(g) + 4\,H_2O(\ell)$

$$6\,H_2O(\ell) + 4\,H_3O^+(aq) \longrightarrow O_2(g) + 2\,H_2(g) + 4\,H_3O^+(aq) + 4\,H_2O(\ell)$$

Simplify: $2\ H_2O(\ell) + 4\ H_3O^+(aq) \longrightarrow O_2(g) + 2\ H_2(g) + 4\ H_3O^+(aq)$

Gaseous O_2 is produced is at the anode, gaseous H_2 is produced at the cathode, and 2 moles of $H_2(g)$ are produced for each mole of $O_2(g)$.

61. *Result:* **Au^{3+}, Hg^{2+}, Ag^+, Hg_2^{2+}, Fe^{3+}, Cu^{2+}, Sn^{4+}, Sn^{2+}, Ni^{2+}, Cd^{2+}, Fe^{2+}, Zn^{2+}**

Explanation: All the metals with a reduction potential more positive than the half-reaction with water as a reactant (– 0.8277 V) can be electrolyzed from their aqueous ions to the corresponding metals. The metals in Table 17.1 that qualify are listed above.

63. *Result:* **$H_2(g)$, $Br_2(\ell)$, and 2 $OH^-(aq)$ are formed; H_2O, Na^+, OH^- (and small amounts of dissolved Br_2 and H_3O^+) are in the solution. H_2 is formed at the cathode. Br_2 is formed at the anode.**

Explanation: Electrolysis of NaBr involves the oxidation of the anion Br^- to Br_2 and the reduction of the water.

Anode - oxidation: $2\ Br^-(aq) \longrightarrow Br_2(\ell) + 2\ e^-$

Cathode - reduction: $2\ H_2O(\ell) + 2\ e^- \longrightarrow H_2(g) + 2\ OH^-(aq)$

$$2\ H_2O(\ell) + 2\ Br^-(\ell) \longrightarrow H_2(g) + Br_2(\ell) + 2\ OH^-(aq)$$

H_2 and Br_2 are produced in a basic solution. After the reaction is complete, Br^- is used up and the solution contains Na^+, OH^-, a small amount of dissolved Br_2 (though it has low solubility in water), and a very small amount of H_3O^+. H_2 is formed in the reduction reaction at the cathode. Br_2 is formed in the oxidation reaction at the anode.

Counting Electrons (Section 17-11)

65. *Result:* **0.16 g Ag**

Analyze and Plan: Follow the method described in the solutions to Question 56 and Problem-Solving Example 17-11.

Execute: $Ag^+(aq) + e^- \longrightarrow Ag(s)$

$$155\text{ min} \times 0.015\text{ A} \times \frac{60\text{ s}}{1\text{ min}} \times \frac{1\text{ C}}{1\text{ A}\cdot 1\text{ s}} \times \frac{1\text{ mol e}^-}{96485\text{ C}} \times \frac{1\text{ mol Ag}}{1\text{ mol e}^-} \times \frac{107.9\text{ g Ag}}{1\text{ mol Ag}} = 0.16\text{ g Ag}$$

67. *Result:* **5.93 g Cu**

Analyze and Plan: Adapt the methods described in the solution to Question 65.

Execute: $Cu^{2+}(aq) + 2\ e^- \longrightarrow Cu(s)$

$$2.00\text{ hr} \times 2.50\text{ A} \times \frac{3600\text{ s}}{1\text{ hr}} \times \frac{1\text{ C}}{1\text{ A}\cdot 1\text{ s}} \times \frac{1\text{ mol e}^-}{96485\text{ C}} \times \frac{1\text{ mol Cu}}{2\text{ mol e}^-} \times \frac{63.55\text{ g Cu}}{1\text{ mol Cu}} = 5.93\text{ g Cu}$$

69. *Result:* **2.7×10^5 g Al**

Analyze and Plan: Follow the method described in the solution to Question 65.

Execute:

$$8.0\text{ hr} \times (1.0 \times 10^5\text{ A}) \times \frac{3600\text{ s}}{1\text{ hr}} \times \frac{1\text{ C}}{1\text{ A}\cdot 1\text{ s}} \times \frac{1\text{ mol e}^-}{96485\text{ C}} \times \frac{1\text{ mol Al}}{3\text{ mol e}^-} \times \frac{26.98\text{ g Al}}{1\text{ mol Al}} = 2.7 \times 10^5\text{ g Al}$$

71. *Result:* **1.9×10^2 g Pb**

Analyze and Plan: Follow the method described in the solution to Question 65. Equations for chemical reactions are found in Section 17-8a.

Execute:

$$50.\text{ hr} \times 1.0\text{ A} \times \frac{3600\text{ s}}{1\text{ hr}} \times \frac{1\text{ C}}{1\text{ A}\cdot 1\text{ s}} \times \frac{1\text{ mol e}^-}{96485\text{ C}} \times \frac{1\text{ mol Pb}}{2\text{ mol e}^-} \times \frac{207.2\text{ g Pb}}{1\text{ mol Pb}} = 1.9 \times 10^2\text{ g Pb}$$

73. *Result:* **The lithium battery uses 0.043 g Li, while the lead battery uses 0.64 g Pb.**

Analyze and Plan: Adapt the method described in the solutions to Question 65.

Execute:

$$10.\text{ min} \times 1.0\text{ A} \times \frac{60\text{ s}}{1\text{ min}} \times \frac{1\text{ C}}{1\text{ A}\cdot 1\text{ s}} \times \frac{1\text{ mol e}^-}{96485\text{ C}} = 6.2 \times 10^{-3}\text{ mol e}^-$$

$$6.2 \times 10^{-3}\text{ mol e}^- \times \frac{1\text{ mol Li}}{1\text{ mol e}^-} \times \frac{6.941\text{ g Li}}{1\text{ mol Li}} = 0.043\text{ g Li}$$

$$6.2 \times 10^{-3}\text{ mol e}^- \times \frac{1\text{ mol Pb}}{2\text{ mol e}^-} \times \frac{207.2\text{ g Pb}}{1\text{ mol Pb}} = 0.64\text{ g Pb}$$

The lithium battery consumes only 0.043 g Li, while the lead-acid storage battery consumes 0.64 g Pb.

75. *Result:* **6.85 min**

Analyze and Plan: Adapt the method described in the solution to Question 65.

Execute:

$$\frac{0.500\text{ g Ni}}{4.00\text{ A}} \times \frac{1\text{ mol Ni}}{58.6934\text{ g Ni}} \times \frac{2\text{ mol e}^-}{1\text{ mol Ni}} \times \frac{96485\text{ C}}{1\text{ mol e}^-} \times \frac{1\text{A}\cdot 1\text{s}}{1\text{C}} \times \frac{1\text{ min}}{60\text{ s}} = 6.85\text{ min}$$

Corrosion: Undesirable Product-Favored Redox Reactions (Section 17-12)

78. *Result/Explanation:* As described in Section 17-12, one requirement for corrosion is an electrolyte in contact with both the anode and the cathode. The presence of sodium and chloride ions in salt water increases the electrolytic capacity of the solution.

80. *Result/Explanation:* Chromium is highly resistant to corrosion and protects the more active iron metal in the steel from oxidizing.

General Questions

84. *Result:* **0.00689 g Cu; 0.00195 g Al**

Analyze and Plan: Adapt the method described in the solution to Question 65.

Execute:

$Ag^+(aq) + e^- \longrightarrow Ag(s)$, $\quad 0.0234\text{ g Ag} \times \frac{1\text{ mol Ag}}{107.9\text{ g Ag}} \times \frac{1\text{ mol e}^-}{1\text{ mol Ag}} = 2.17 \times 10^{-4}\text{ mol e}^-$

$Cu^{2+}(aq) + 2\,e^- \longrightarrow Cu(s)$, $\quad 2.17 \times 10^{-4}\text{ mol e}^- \times \frac{1\text{ mol Cu}}{2\text{ mol e}^-} \times \frac{63.55\text{ g Cu}}{1\text{ mol Cu}} = 0.00689\text{ g Cu}$

$Al^{3+}(aq) + 3\,e^- \longrightarrow Al(s)$, $\quad 2.17 \times 10^{-4}\text{ mol e}^- \times \frac{1\text{ mol Al}}{3\text{ mol e}^-} \times \frac{26.98\text{ g Al}}{1\text{ mol Al}} = 0.00195\text{ g Al}$

86. *Result:* **(a) 9.5×10^6 g HF (b) 1.7×10^3 kWh**

Analyze and Plan: Perform a standard stoichiometry problem (Chapter 2) and then adapt the method described in the solution to Question 65 and in Section 17-11b.

Execute:

(a) $$9.0 \text{ metric tons} \times \frac{1000 \text{ kg}}{1 \text{ metric ton}} \times \frac{1000 \text{ g}}{1 \text{ kg}} \times \frac{1 \text{ mol } F_2}{38.00 \text{ g } F_2} \times \frac{2 \text{ mol HF}}{1 \text{ mol } F_2} \times \frac{20.01 \text{ g HF}}{1 \text{ mol HF}} = 9.5 \times 10^6 \text{ g HF}$$

(b) $$24. \text{ hr} \times (6.0 \times 10^3 \text{ A}) \times \frac{3600 \text{ s}}{1 \text{ hr}} \times \frac{1 \text{ C}}{1 \text{ A} \cdot 1 \text{ s}} \times 12 \text{ V} \times \frac{1 \text{ J}}{1 \text{ C} \cdot 1 \text{ V}} \times \frac{1 \text{ kWh}}{3.60 \times 10^6 \text{ J}} = 1.7 \times 10^3 \text{ kWh}$$

88. *Result:* **4+**

Analyze and Plan: Adapt methods described in the solution to Question 65 and in Section 17.11.

Execute:

$$3.00 \text{hr} \times 2.00 \text{ A} \times \frac{1\text{C}}{1\text{A} \cdot 1\text{s}} \times \frac{3600 \text{ s}}{1 \text{ hr}} \times \frac{1 \text{ mol } e^-}{96485 \text{ C}} = 0.224 \text{mol } e^-$$

$$10.9 \text{ g Pt} \times \frac{1 \text{ mol Pt}}{195.08 \text{ g Pt}} = 0.0559 \text{ mol Pt}$$

$$\frac{0.224 \text{ mole } e^-}{0.0559 \text{ mol Pt}} = 4.01$$

The platinum ion is a 4+ ion, Pt^{4+}.

89. *Result:* **75 s**

Analyze and Plan: Use the methods described in the solution to Question 75 and in Section 17-11.

Execute:

$$V = 1200 \text{ mm}^2 \times 1.0 \text{ μm} \times \left(\frac{1 \text{ m}}{1000 \text{ mm}} \times \frac{100 \text{ cm}}{1 \text{ m}}\right)^2 \times \left(\frac{10^{-6} \text{ m}}{1 \text{ μm}} \times \frac{100 \text{ cm}}{1 \text{ m}}\right) = 0.0012 \text{ cm}^3$$

$$0.0012 \text{ cm}^3 \times \frac{10.5 \text{ g}}{1 \text{ cm}^3} \times \frac{1 \text{ mol}}{107.8682 \text{ g}} = 1.2 \times 10^{-4} \text{ mol Ag}$$

$$1.2 \times 10^{-4} \text{ mol Ag} \times \frac{1 \text{ mol } Ag^+}{1 \text{ mol Ag}} \times \frac{1 \text{ mol } e^-}{1 \text{ mol } Ag^+} \times \frac{96485 \text{ C}}{1 \text{ mol } e^-} = 11 \text{ C}$$

$$\frac{11 \text{ C}}{150.0 \text{ mA}} \times \frac{1000 \text{ mA}}{1 \text{ A}} \times \frac{1\text{A} \cdot 1\text{s}}{1 \text{ C}} = 75 \text{ s}$$

Applying Concepts

91. *Result:* **B < D < A < C**

Explanation: The strongest reducing agent is the most reactive metal. From the information in (b), we find that metal C reacts with all the other metals' ions, so it will be last on the list. From the information in (a), we find that metals A and C are more reactive than the other metals, so A will precede C in the list. From the information in (c), we find that metal D reacts with the ions of metal B, so B will be first on the list.

93. *Result:* **(a) Oxidized: B(s); reduced: A^{2+} (b) Oxidizing agent: A^{2+}; reducing agent: B(s) (c) Anode: B(s); cathode: A(s) (d) See equations below (e) A(s) (f) From B(s) to A(s) (g) Towards the A^{2+} solution**

Explanation: The solution labeled "A^{2+}" is getting lighter and the solution labeled "B^{2+}" is getting darker. If we assume that means A^{2+} is getting less concentrated and that B^{2+} is getting more concentrated we can make the following conclusions:

(a) B(s) is being oxidized to B^{2+} and A^{2+} is being reduced to A(s).

(b) A^{2+} is the oxidizing agent since it is being reduced, and B(s) is the reducing agent since it is being oxidized.

(c) B(s) is the anode, the solid at the site of oxidation, and A(s) is the cathode, the solid at the site of reduction.

(d) $A^{2+} + 2\,e^- \longrightarrow A(s)$

$B(s) \longrightarrow B^{2+} + 2\,e^-$

(e) The A(s) metal gains mass.

(f) Electrons flow from the B(s) electrode to the A(s) electrode.

(g) K^+ ions in the salt bridge will migrate towards the A^{2+} solution to replace the cations that plated out as A(s).

95. *Result:* **(a) The reaction of Co^{3+} with H_2O is spontaneous. (b) The reaction of Fe^{2+} with O_2 is spontaneous.**

Analyze and Plan: The spontaneous direction of electron flow has a positive value of $E°_{cell}$. Determine if $E°_{cell}$ is positive.

Execute:

(a)

$$\begin{array}{lll} & 2\,H_2O \longrightarrow O_2 + 4\,H^+ + 4\,e^- & E°_{anode} = 1.229\text{ V} \\ + & 4\,Co^{3+} + 4\,e^- \longrightarrow 4\,Co^{2+} & E°_{cathode} = -1.92\text{ V} \\ \hline & 4\,Co^{3+} + 2\,H_2O \longrightarrow O_2 + 4\,H^+ + 4\,Co^{2+} & E°_{cell} = 0.69\text{ V} \end{array}$$

Since the cell potential is positive, the reaction is spontaneous, and Co^{3+} is not stable in water.

(b)

$$\begin{array}{lll} & O_2 + 4\,H^+ + 4\,e^- \longrightarrow 4\,H_2O & E°_{cathode} = 1.229\text{ V} \\ + & 4\,Fe^{2+} \longrightarrow 4\,Fe^{3+} + 4\,e^- & E°_{anode} = -\,0.771\text{ V} \\ \hline & O_2 + 4\,H^+ + 4\,Fe^{2+} \longrightarrow 4\,Fe^{3+} + 2\,H_2O & E°_{cell} = 0.458\text{ V} \end{array}$$

Since the cell potential is positive, the reaction is spontaneous, and Fe^{2+} is not stable in air.

More Challenging Questions

101. *Result:* **(a) 1.0×10^{-2} M (b) 9.5×10^{-17}**

Analyze: To measure the Ag^+ concentration and the K_{sp} for AgI, a titration is done using KI and the E_{cell} is measured against a standard hydrogen electrode.

Plan: Use standard conversion factors to calculate the concentration of silver ion in the solution. Then, set up the half-reactions, balance the redox equation, determine the $E°_{cell}$, then using the Nernst equation to calculate the Q. Use Q to determine silver ion concentration. Use the volume and concentration of the KI solution, the mole ratio of the balanced precipitation equation, and an equilibrium calculation to determine K_{sp} as described in Chapter 15.

Execute:

(a)

$$\frac{16.7\text{ mL KI}}{25.00\text{ mL Ag}^+} \times \frac{0.015\text{ mol KI}}{1000\text{ mL KI}} \times \frac{1\text{ mol Ag}^+}{1\text{ mol KI}} \times \frac{1000\text{ mL Ag}^+}{1\text{ L Ag}^+} = 1.0 \times 10^{-2}\text{ M Ag}^+$$

(b)

$$\begin{array}{lll} & 2\,Ag^+(aq) + 2\,e^- \longrightarrow 2\,Ag(s) & E°_{cathode} = 0.7991\text{ V} \\ + & H_2(g) \longrightarrow 2\,H^+(aq) + 2\,e^- & E°_{anode} = 0.000\text{ V} \\ \hline & 2\,Ag^+(aq) + H_2(g) \longrightarrow 2\,H^+(aq) + 2\,Ag(s) & E°_{cell} = 0.7991\text{ V} \end{array}$$

$$E_{cell} = E°_{cell} - \frac{RT}{nF}\ln Q \qquad Q = \frac{(\text{conc. H}^+)^2}{(\text{conc. Ag}^+)^2 P_{H_2}} \qquad n = 2$$

$$0.325\text{ V} = 0.7991\text{ V} - \frac{\left(8.314\text{Jmol}^{-1}\text{K}^{-1}\right)(298\text{ K})}{(2)(96485\text{ C mol}^{-1})\left(\frac{1\text{ J}}{1\text{ C }1\text{ V}}\right)}\ln\left(\frac{1^2}{(\text{conc. Ag}^+)^2(1)}\right)$$

$$(\text{conc. Ag}^+) = 9.7 \times 10^{-9}\text{ M} = [\text{Ag}^+] \text{ in equilibrium with } \text{I}^-$$

At the equivalence point, the moles of titrant, I^-, are equal to the moles of reactant, Ag^+. So, $[I^-] = [Ag^+]$

For AgI, $K_{sp} = [Ag^+][I^-] = (9.7 \times 10^{-9}\text{ M})(9.7 \times 10^{-9}\text{ M}) = 9.5 \times 10^{-17}$

☑ *Reasonable Result Check:* This is close to the value listed in Appendix H (8.3×10^{-17})

103. *Result:* **Calculate pH from the cell potential of the test solution, pH = 6.69**

Analyze: Use method described in Problem-Solving Example 17.9.

Plan: Write the half-reaction equations and an overall equation. From the overall equation determine n and write the expression for Q. Substitute into the Nernst equation and solve for the unknown concentration of use [H⁺] to calculate pH.

Execute:

$$\tfrac{1}{4}\,O_2(g) + H^+(aq) + e^- \longrightarrow \tfrac{1}{2}\,H_2O(\ell) \qquad E°_{cathode} = 0.401\text{ V}$$

$$+ \quad \tfrac{1}{2}\,H_2O(\ell) \longrightarrow \tfrac{1}{4}\,O_2(g) + H^+(aq) + e^- \qquad E°_{anode} = 0.401\text{ V}$$

$$\tfrac{1}{4}\,O_2(g)_{cat} + H^+(aq)_{cat} + \tfrac{1}{2}\,H_2O(\ell) \longrightarrow \tfrac{1}{4}\,O_2(g)_{an} + H^+(aq)_{an} + \tfrac{1}{2}\,H_2O(\ell) \qquad E°_{cell} = 0.000\text{ V}$$

$$E = E° - \frac{RT}{nF}\ln Q \qquad Q = \frac{(\text{conc. H}^+)_{standard}\left(P_{O_2,\text{ standard}}\right)^{1/4}}{(\text{conc. H}^+)_{unknown}\left(P_{O_2,\text{ unknown}}\right)^{1/4}} \qquad n = 1$$

Pressure of $O_2(g)$ is constant.

$$\text{Standard pH} = 9.40 \qquad (\text{conc. H}^+)_{standard} = 10^{-pH} = 10^{-9.40} = 4.0 \times 10^{-10}\text{ M}$$

$$0.22\text{ V} = 0.060\text{ V} - \frac{\left(8.314\text{Jmol}^{-1}\text{K}^{-1}\right)(298\text{ K})}{(1)(96485\text{ C mol}^{-1})\left(\frac{1\text{ J}}{1\text{ C }1\text{ V}}\right)}\ln\left(\frac{4.0\times10^{-10}}{[\text{H}^+]_{unknown}}\right)$$

$$0.025678\text{ V}\ln\left(\frac{4.0\times10^{-10}}{[\text{H}^+]_{unknown}}\right) = 0.060\text{ V} - 0.22\text{ V}$$

$$\ln\left(\frac{4.0\times10^{-10}}{[\text{H}^+]_{unknown}}\right) = -6.23$$

$$\frac{4.0\times10^{-10}}{[\text{H}^+]_{unknown}} = e^{-6.23} = 0.00197$$

$$[\text{H}^+]_{unknown} = \frac{4.0\times10^{-10}}{0.00197} = 2.02 \times 10^{-7}\text{ M}$$

$$\text{pH} = -\log[\text{H}^+] = -\log(2.02 \times 10^{-7}) = 6.69$$

106. *Result:* **(a) 48.1 min (b) O_2, 1.93 L**

Analyze and Plan: Adapt the method described in the solution to Question 65, then use the mole ratio to determine moles of gas, and the ideal gas law to calculate volume of gas.

Execute:

(a) $$\frac{9.50\text{ g Cui}}{10.0\text{ A}} \times \frac{1\text{ mol Cu}}{63.546\text{ g Cu}} \times \frac{2\text{ mol e}^-}{1\text{ mol Cu}} = 0.299\text{ mol e}^-$$

$$0.299\text{ mol e}^- \times \frac{96485\text{ C}}{1\text{ mol e}^-} \times \frac{1\text{A}\cdot 1\text{s}}{1\text{C}} \times \frac{1\text{ min}}{60\text{ s}} = 48.1\text{ min}$$

(b) $2\ H_2O(\ell) \longrightarrow 4\ H^+(aq) + O_2(g) + 4\ e^-$ $\quad E^\circ_{anode} = 1.229\text{ V}$

Oxygen gas, $\mathbf{O_2(g)}$, is produced.

$$0.299\text{ mol e}^- \times \frac{1\text{ mol } O_2(g)}{1\text{ mol e}^-} = 0.0747\text{ mol } O_2(g)$$

$$PV = nRT$$

$$V = \frac{nRT}{P} = \frac{(0.0747\text{ mol})\left(0.08206\frac{\text{L atm}}{\text{mol K}}\right)(298\text{ K})}{0.945\text{ atm}} = \mathbf{1.93\text{ L}}$$

108. *Result:* $\mathbf{Cl_2 + 2\ Br^- \longrightarrow Br_2 + 2\ Cl^-}$

Explanation: $Br_2 + 2\ e^- \longrightarrow 2\ Br^-$ $\quad +1.066\text{ V}$

$Cl_2 + 2\ e^- \longrightarrow 2\ Cl^-$ $\quad +1.358\text{ V}$

Reverse the first half-reaction to make a spontaneous reaction (with a positive cell potential).

$$\begin{array}{llr} & 2\ Br^- \longrightarrow Br_2 + 2\ e^- & E^\circ_{anode} = +1.066\text{ V} \\ + & Cl_2 + 2\ e^- \longrightarrow 2\ Cl^- & E^\circ_{cathode} = +1.358\text{ V} \\ \hline & Cl_2 + 2\ Br^- \longrightarrow Br_2 + 2\ Cl^- & E^\circ_{cell} = +\ 0.292\text{ V} \end{array}$$

111. *Result:* **conc. $\mathbf{Cu^{2+} = 4 \times 10^{-6}}$ M**

Analyze and Plan: Use the method shown in the solution to Question 47 involving the Nernst equation.

Execute:

$$\begin{array}{llrr} & Cu(s) \longrightarrow Cu^{2+} + 2e^- & E^\circ_{anode} = 0.340\text{ V} & \\ + & 2(Ag^+ + e^- \longrightarrow Ag(s)) & E^\circ_{cathode} = 0.7991\text{ V} & \\ \hline & Cu(s) + 2\ Ag^+ \longrightarrow Cu^{2+} + 2\ Ag(s) & E^\circ_{cell} = 0.459\text{ V} & n = 2 \end{array}$$

$$E_{cell} = E^\circ_{cell} - \frac{0.0592\text{ V}}{n}\log\left(\frac{\text{conc. } Cu^{2+}}{(\text{conc. } Ag^+)^2}\right)$$

$$0.62\text{ V} = 0.459\text{ V} - \frac{0.0592\text{ V}}{2}\log\left(\frac{\text{conc } Cu^{2+}}{(1.00\text{ M})^2}\right)$$

$$\log(\text{conc. } Cu^{2+}) = -5.4$$

$(\text{conc. } Cu^{2+}) = 10^{-5.4} = 4 \times 10^{-6}\text{ M}$ *(round to 1 sig fig)*

Chapter 18: Nuclear Chemistry

Topical Questions

The Nature of Radioactivity (Section 18-1)

10. *Result/Explanation:* Recap Table 18.1:

	Symbol	Mass	Charge
α particle	${}^{4}_{2}\text{He}$	6.65×10^{-24} g/particle	+ 2
β particle	${}^{0}_{-1}\text{e}$	9.11×10^{-28} g/particle	– 1
γ radiation	${}^{0}_{0}\gamma$	0 g	0

Nuclear Reactions (Section 18-2)

12. *Result:* **(a) Alpha emission (b) Beta emission (c) Electron capture or positron emission (d) Beta emission**

Analyze and Plan: The mass numbers and atomic numbers must balance. Use that information to determine the mass number and the atomic number of the decay particle. Use the periodic table and the atomic number to get the symbol of an element. Use the identity of the decay particle to identify what transformation that occurs.

Execute:

(a) ${}^{230}_{90}\text{Th} \longrightarrow {}^{226}_{88}\text{Ra} + \underline{{}^{4}_{2}\text{He}}$ $\quad 230 - 226 = 4,\ 90 - 88 = 2$

Thorium-230 decays by **alpha emission** to form radium-226.

(b) ${}^{137}_{55}\text{Cs} \longrightarrow {}^{137}_{56}\text{Ba} + \underline{{}^{0}_{-1}\text{e}}$ $\quad 137 - 137 = 0,\ 55 - 56 = -1$

Cesium-137 decays by **beta emission** to form barium-137.

(c) ${}^{38}_{19}\text{K} + \underline{{}^{0}_{-1}\text{e}} \longrightarrow {}^{38}_{18}\text{Ar}$ $\quad 38 - 38 = 0,\ 19 + x = 18,\ x = -1$

${}^{38}_{19}\text{K} \longrightarrow {}^{38}_{18}\text{Ar} + \underline{{}^{0}_{+1}\text{e}}$ $\quad 38 - 38 = 0,\ 19 - 18 = 1$

Potassium-38 decays by **electron capture or positron emission** to form argon-38.

(d) ${}^{97}_{40}\text{Zr} \longrightarrow {}^{97}_{41}\text{Nb} + \underline{{}^{0}_{-1}\text{e}}$ $\quad 97 - 97 = 0,\ 41 - 40 = -1$

Zirconium-97 decays by **beta emission** to form niobiom-137.

☑ *Reasonable Result Check:* Each reaction matches a known radioactive decay process, atomic numbers and mass numbers are conserved.

14. *Result:* **(a) ${}^{238}_{92}\text{U}$ (b) ${}^{32}_{15}\text{P}$ (c) ${}^{10}_{5}\text{B}$ (d) ${}^{0}_{-1}\text{e}$ (e) ${}^{15}_{7}\text{N}$**

Analyze and Plan: The mass numbers and atomic numbers must balance. Use that information to determine the mass number and the atomic number of the missing entry. Use the periodic table and the atomic number to get the symbol of an element.

Execute:

(a) ${}^{242}_{94}\text{Pu} \longrightarrow {}^{4}_{2}\text{He} + \boxed{{}^{238}_{92}\text{U}}$ $\quad 242 - 4 = 238,\ 94 - 2 = 92$, Element 92 is U.

(b) $\boxed{{}^{32}_{15}\text{P}} \longrightarrow {}^{32}_{16}\text{S} + {}^{0}_{-1}\text{e}$ $\quad 32 + 0 = 32,\ 16 + (-1) = 15$, Element 15 is P.

(c) ${}^{252}_{98}\text{Cf} + \boxed{{}^{10}_{5}\text{B}} \longrightarrow 3\ {}^{1}_{0}\text{n} + {}^{259}_{103}\text{Lr}$ $\quad 3 \times (1) - 259 - 252 = 10$

$3 \times (0) - 103 - 98 = 5$, Element 5 is B.

(d) $^{55}_{26}Fe + \boxed{^{0}_{-1}e} \longrightarrow ^{55}_{25}Mn$ 55 – 55 = 0, 25– 26 = – 1, electron captured $^{0}_{-1}e$.

(e) $^{15}_{8}O \longrightarrow \boxed{^{15}_{7}N} + ^{0}_{+1}e$ 15 – 0 = 15, 8 – 1 = 7, Element 7 is N.

☑ *Reasonable Result Check:* Atomic numbers and mass numbers are conserved.

16. *Result:* **(a) $^{28}_{12}Mg \longrightarrow ^{28}_{13}Al + ^{0}_{-1}e$ (b) $^{238}_{92}U + ^{12}_{6}C \longrightarrow 4\ ^{1}_{0}n + ^{246}_{98}Cf$ (c) $^{2}_{1}H + ^{3}_{2}He \longrightarrow ^{4}_{2}He + ^{1}_{1}H$ (d) $^{38}_{19}K \longrightarrow ^{38}_{18}Ar + ^{0}_{+1}e$ (e) $^{175}_{78}Pt \longrightarrow ^{4}_{2}He + ^{171}_{76}Os$**

Analyze and Plan: Interpret the statement by identifying the nuclear symbol(s) for the given reactant and/or product isotope(s) and identifying the details of the radioactive decay process. Then follow the balancing method described in the solution to Question 12.

Execute:

(a) Magnesium-28 is $^{28}_{12}Mg$ and β emission is the production of $^{0}_{-1}e$: $^{28}_{12}Mg \longrightarrow ^{28}_{13}Al + ^{0}_{-1}e$

(b) Uranium-238 is $^{238}_{92}U$, carbon-12 is $^{12}_{6}C$, and the neutron symbol is $^{1}_{0}n$:

$$^{238}_{92}U + ^{12}_{6}C \longrightarrow 4\ ^{1}_{0}n + ^{246}_{98}Cf$$

(c) Hydrogen-2 is $^{2}_{1}H$, helium-3 is $^{3}_{2}He$, and helium-4 is : $^{2}_{1}H + ^{3}_{2}He \longrightarrow ^{4}_{2}He + ^{1}_{1}H$

(d) Argon-38 is $^{38}_{18}Ar$ and positron emission is the production of $^{0}_{+1}e$: $^{38}_{19}K \longrightarrow ^{38}_{18}Ar + ^{0}_{+1}e$

(e) Platinum-175 is $^{175}_{78}Pt$, and osmium-171 is $^{171}_{76}Os$: $^{175}_{78}Pt \longrightarrow ^{4}_{2}He + ^{171}_{76}Os$

☑ *Reasonable Result Check:* Atomic numbers (Zs) and mass numbers (As) are balanced.

18. *Result:* **$^{231}_{90}Th$, $^{231}_{91}Pa$, $^{227}_{89}Ac$, $^{227}_{90}Th$, and $^{223}_{88}Ra$**

Analyze and Plan: The first five steps of the decay series are: α, β, α, β, α. Therefore, start with uranium-235 undergoing an α decay reaction. Then take the radioisotope produced and make it the reactant of the second β decay reaction. Repeat this process for the remaining three steps undergoing α, then β, then α.

Execute:

$$^{235}_{92}U \longrightarrow ^{4}_{2}He + ^{231}_{90}Th$$

$$^{231}_{90}Th \longrightarrow ^{0}_{-1}e + ^{231}_{91}Pa$$

$$^{231}_{91}Pa \longrightarrow ^{4}_{2}He + ^{227}_{89}Ac$$

$$^{227}_{89}Ac \longrightarrow ^{0}_{-1}e + ^{227}_{90}Th$$

$$^{227}_{90}Th \longrightarrow ^{4}_{2}He + ^{223}_{88}Ra$$

So, the radioisotopes produced in the first five steps are: $^{231}_{90}Th$, $^{231}_{91}Pa$, $^{227}_{89}Ac$, $^{227}_{90}Th$, and $^{223}_{88}Ra$.

☑ *Reasonable Result Check:* Atomic numbers (Zs) and mass numbers (As) are balanced.

20. *Result:* **$^{222}_{86}Rn \longrightarrow ^{4}_{2}He + ^{218}_{84}Po$, $^{218}_{84}Po \longrightarrow ^{4}_{2}He + ^{214}_{82}Pb$, $^{214}_{82}Pb \longrightarrow ^{0}_{-1}e + ^{214}_{83}Bi$, $^{214}_{83}Bi \longrightarrow ^{0}_{-1}e + ^{214}_{84}Po$, $^{214}_{84}Po \longrightarrow ^{4}_{2}He + ^{210}_{82}Pb$, $^{210}_{82}Pb \longrightarrow ^{0}_{-1}e + ^{210}_{83}Bi$, $^{210}_{83}Bi \longrightarrow ^{0}_{-1}e + ^{210}_{84}Po$, $^{210}_{84}Po \longrightarrow ^{4}_{2}He + ^{206}_{82}Pb$**

Analyze and Plan: The decay series of radon-222 has eight steps: α, α, β, β, α, β, β, α. Start with radon-222 undergoing an α decay reaction. Then take the radioisotope produced and make it the reactant of the second α decay reaction. Repeat this process for the remaining six steps.

Execute:

$$^{222}_{86}Rn \longrightarrow {}^{4}_{2}He + {}^{218}_{84}Po$$

$$^{218}_{84}Po \longrightarrow {}^{4}_{2}He + {}^{214}_{82}Pb$$

$$^{214}_{82}Pb \longrightarrow {}^{0}_{-1}e + {}^{214}_{83}Bi$$

$$^{214}_{83}Bi \longrightarrow {}^{0}_{-1}e + {}^{214}_{84}Po$$

$$^{214}_{84}Po \longrightarrow {}^{4}_{2}He + {}^{210}_{82}Pb$$

$$^{210}_{82}Pb \longrightarrow {}^{0}_{-1}e + {}^{210}_{83}Bi$$

$$^{210}_{83}Bi \longrightarrow {}^{0}_{-1}e + {}^{210}_{84}Po$$

$$^{210}_{84}Po \longrightarrow {}^{4}_{2}He + {}^{206}_{82}Pb \quad \text{Lead-206 is stable.}$$

☑ *Reasonable Result Check:* Atomic numbers (Zs) and mass numbers (As) are balanced. The final product is a stable isotope of lead.

Stability of Atomic Nuclei (Section 18-3)

21. *Result:* **(a)** $^{19}_{10}\mathbf{Ne} \longrightarrow {}^{19}_{9}\mathbf{F} + {}^{0}_{+1}\mathbf{e}$ **(b)** $^{230}_{90}\mathbf{Th} \longrightarrow {}^{0}_{-1}\mathbf{e} + {}^{230}_{91}\mathbf{Pa}$ **(c)** $^{82}_{35}\mathbf{Br} \longrightarrow {}^{0}_{-1}\mathbf{e} + {}^{82}_{36}\mathbf{Kr}$ **(d)** $^{212}_{84}\mathbf{Po} \longrightarrow {}^{4}_{2}\mathbf{He} + {}^{208}_{82}\mathbf{Pb}$

Analyze and Plan: Identify which type of radioactive decay is most likely for the isotope, by identifying the N/Z ratio and seeing where it falls on Figure 18.2.

Execute:

(a) The neon-19 isotope has 9 neutrons and 10 protons, giving an N/Z ratio of 0.90. That means it has too few neutrons and undergoes positron emission or electron capture. Neon is relatively small, so we will predict that it undergoes positron emission:

$$^{19}_{10}Ne \longrightarrow {}^{19}_{9}F + {}^{0}_{+1}e$$

(b) The thorium-230 isotope has 140 neutrons and 90 protons, giving an N/Z ratio of 1.56. Stable isotopes in this range of the graph have an N/Z ratio of about 1.52. That means it has too many neutrons, so it undergoes β emission:

$$^{230}_{90}Th \longrightarrow {}^{0}_{-1}e + {}^{230}_{91}Pa$$

(c) The bromine-82 isotope has 47 neutrons and 35 protons, giving an N/Z ratio of 1.34. Stable isotopes in this range of the graph have an N/Z ratio of about 1.20. That means it has too many neutrons, so it undergoes β emission:

$$^{82}_{35}Br \longrightarrow {}^{0}_{-1}e + {}^{82}_{36}Kr$$

(d) The lead-212 isotope has 128 neutrons and 84 protons, more than the threshold limit of 126 neutrons and 83 protons above which all isotopes undergo α decay:

$$^{212}_{84}Po \longrightarrow {}^{4}_{2}He + {}^{208}_{82}Pb$$

☑ *Reasonable Result Check:* Each reaction is a well-known radioactive decay process. Atomic numbers (Zs) and mass numbers (As) are balanced.

23. *Result:* **6.001×10^8 kJ/mol nucleons ^{10}B; 6.460×10^8 kJ/ mol nucleon ^{11}B; ^{11}B is more stable than ^{10}B**

Analyze and Plan: Use the method described in Section 18-3b. We want to compare the binding energy per nucleon. Calculate the change in mass (Δm) when the reactant nucleus is formed from combining hydrogen atoms (^1_1H = 1.007276 g/mol) and neutrons (^1_0n = 1.008665 g/mol). *(Note: these values are more precise than the ones given in the question.)* Use Einstein's equation: $E = (\Delta m)c^2$ to calculate the total energy generated, then divide that number by the number of nucleons to determine the binding energy per nucleon.

Execute:

$$\Delta m = [(5 \times 1.007276 \text{ g/mol } ^1_1H) + (5 \times 1.008665 \text{ g/mol } ^1_0n)] - 10.01294 \text{ g } ^{10}B = 0.06677 \text{ g/mol}$$

The nuclear binding energy = $E_b = (\Delta m)c^2$

$$E_b = (\Delta m)c^2 = \left(\frac{0.06677 \text{ g}}{1 \text{ mol}} \times \frac{1 \text{ kg}}{1000 \text{ g}}\right) \times (2.99792 \times 10^8 \text{ m/s})^2 \times \frac{1 \text{ J}}{1 \text{ kg m}^2 \text{ s}^{-2}} \times \frac{1 \text{ kJ}}{1000 \text{ J}} = 6.001 \times 10^9 \text{ kJ/mol}$$

As described in Section 18-3b, "Binding energy per nucleon" is really E_b per mol nucleons.

One mol of boron-10 nuclei has 5 mol of protons and 5 mol of neutrons, or a total of 10 mol of nucleons:

$$E_b \text{ per mol nucleon} = \frac{6.001 \times 10^9 \text{ kJ}}{10 \text{ mol nucleons}} = 6.001 \times 10^8 \text{ kJ/mol nucleons}$$

$$\Delta m = [(5 \times 1.007276 \text{ g/mol } ^1_1H) + (6 \times 1.008665 \text{ g/mol } ^1_0n)] - 11.00931 \text{ g/mol } ^{11}B = 0.07906 \text{ g/mol}$$

$$E_b = (\Delta m)c^2 = \left(\frac{0.07906 \text{ g}}{1 \text{ mol}} \times \frac{1 \text{ kg}}{1000 \text{ g}}\right) \times (2.99792 \times 10^8 \text{ m/s})^2 \times \frac{1 \text{ J}}{1 \text{ kg m}^2 \text{ s}^{-2}} \times \frac{1 \text{ kJ}}{1000 \text{ J}} = 7.106 \times 10^9 \text{ kJ/mol}$$

One mol of boron-11 nuclei has 5 mol of protons and 6 mol of neutrons, or a total of 11 mol of nucleons:

$$E_b \text{ per nucleon} = \frac{7.106 \times 10^9 \text{ kJ}}{11 \text{ mol nucleons}} = 6.460 \times 10^8 \text{ kJ/ mol nucleon}$$

Boron-11 has a larger E_b per mol nucleon than boron-10, so ^{11}B is more stable than ^{10}B.

☑ *Reasonable Result Check:* The average atomic mass of boron (10.811) is closer to 11 than 10, consistent with the calculation that boron-11 is more stable than boron-10.

25. *Result:* **1.6930 × 10^{11} kJ; 7.1133 × 10^{8} kJ/mol nucleon ^{238}U**

Analyze and Plan: Use the method described in Question 23. One mol of uranium-238 nuclei has 92 mol of protons and 146 mol of neutrons, or a total of 238 mol of nucleons:

Execute:

$$\Delta m = [(92 \times 1.007276 \text{ g/mol } ^1_1H) + (146 \times 1.008665 \text{ g/mol } ^1_0n)] - 238.0508 \text{ g/mol } ^{238}U = 1.8837 \text{ g/mol}$$

$$E_b = (\Delta m)c^2 = \left(\frac{1.8837 \text{ g}}{1 \text{ mol}} \times \frac{1 \text{ kg}}{1000 \text{ g}}\right) \times (2.99792 \times 10^8 \text{ m/s})^2 \times \frac{1 \text{ J}}{1 \text{ kg m}^2 \text{ s}^{-2}} \times \frac{1 \text{ kJ}}{1000 \text{ J}} = 1.6930 \times 10^{11} \text{ kJ/mol}$$

"Binding energy per nucleon" is E_b per mol nucleon.

$$E_b \text{ per mol nucleon} = \frac{1.6930 \times 10^{11} \text{ kJ}}{238 \text{ mol nucleons}} = 7.1133 \times 10^8 \text{ kJ/mol nucleon}$$

Rates of Disintegration Reactions (Section 18-4)

28. *Result:* **5 mg**

Analyze and Plan: Because of the special circumstances where we are given an amount of time that is a multiple of the half-life, we can use the method shown in Problem-Solving Example 18.4:

Execute: $t = 1\ d \times \left(\frac{24\ h}{1\ d}\right) + 6\ h = 30.\ h$ *(Assume, in this context, that 1 d is exact.)*

$$30.\ h \times \frac{1\ \text{half - life}}{15\ h} = 2.0\ \text{half-lives}$$

The time of 30. hours is 2.0 times the half-life of 15 hours, so the sample mass is reduced by half twice:

$$\frac{1}{2} \times \frac{1}{2} \times 20\ mg = 5\ mg.$$

It is only legitimate to use the quick method described above when the elapsed time is a whole-number multiple of the half-life.

If the elapsed time were not a whole-number multiple of the half-life, it would be necessary to adapt the methods described in other Problem-Solving Examples in Section 18-4 and use these equations:

$$A = kN \qquad \ln\left(\frac{N}{N_0}\right) = -kt \qquad \ln\left(\frac{A}{A_0}\right) = -kt \qquad t_{1/2} = \frac{\ln 2}{k}$$

(Notice: ln2 is used here in the last equation instead of 0.693 for three reasons: it is faster to type into a calculator—requiring only two buttons be pushed instead of four, provides less opportunity to type the wrong button or have a transcription error, and it is more precise than its 3 sig. fig. approximation.)

Use the last equation to determine the value of k:

$$k = \frac{\ln 2}{t_{1/2}} = \frac{\ln 2}{15\ h} = 4.6 \times 10^{-3}\ h^{-1}$$

The mass (m) of a sample of a pure isotopic substance is directly proportional to the number of atoms (N) because its atomic mass is a constant value and the number of atoms in a mole is also a constant. Therefore, the atom ratio is the same as the mass ratio:

$$\frac{N}{N_0} = \frac{m}{m_0}$$

Now, use the second equation to determine the new mass:

$$\ln\left(\frac{N}{N_0}\right) = \ln\left(\frac{m}{m_0}\right) = -kt = -(4.6 \times 10^{-3}\ h^{-1}) \times (30.\ h) = -1.4$$

$$m = m_0 e^{-1.4} = (20\ mg) \times e^{-1.4} = (20\ mg) \times (0.25) = 5\ mg$$

☑ *Reasonable Result Check:* The quick method and the actual calculation produce the same answer, which is a small fraction of the initial mass.

30. *Result:* **(a)** $^{131}_{53}I \longrightarrow {}^{0}_{-1}e + {}^{131}_{54}Xe$ **(b) 1.56 mg**

Analyze and Plan: Adapt the methods described in the solutions to Questions 16 and 28.

Execute:

(a) The reaction's equation is: $^{131}_{53}I \longrightarrow {}^{0}_{-1}e + {}^{131}_{54}Xe$ $\qquad 131 = 0 + 131;\ 53 = -1 + 54.$

(b) $$32.2\ d \times \frac{1\ \text{half - life}}{8.04\ d} = 4.02\ \text{half-lives}$$

Since the elapsed time is not exactly a whole-number multiple of the half-life, it is necessary to do the actual calculations:

$$k = \frac{\ln 2}{t_{1/2}} = \frac{\ln 2}{8.04\ d} = 8.62 \times 10^{-2}\ d^{-1}$$

The mass (m) of a sample of a compound is directly proportional to the number of radioactive atoms (N) because there is a fixed percentage of that atom in the compound, so the ratios are interchangeable, as described in the solution to Question 28:

$$\ln\left(\frac{m}{m_0}\right) = -kt = -(8.62 \times 10^{-2}\ d^{-1}) \times (32.2\ d) = -2.78$$

$$m = m_0 e^{-kt} = (25.0\ mg) \times e^{-2.78} = (25.0\ mg) \times (0.0623) = 1.56\ mg$$

☑ *Reasonable Result Check:* 32.2 days is very close to exactly four half-lives. After four half lives, the sample mass is reduced by half four times:

$$\frac{1}{2} \times \frac{1}{2} \times \frac{1}{2} \times \frac{1}{2} \times 25.0\ mg = 1.56\ mg$$

Unsurprisingly, this mass matches the calculated sample mass of 1.56 mg.

32. *Result:* **4.0 yr**

Analyze and Plan: Adapt the methods described in the solution to Question 28.

We can adapt the quick method by applying the definition of half-life.

Execute: Initially, the radioactivity measured is defined as 100%.

After one half-life, the radioactivity will drop by half to 50.0% of the original.

After the second half-life, the radioactivity will drop by half again to 25.0% of the original.

After the third half-life, the radioactivity will drop by half again to 12.5% of the original

Stated more concisely, $\frac{1}{2} \times \frac{1}{2} \times \frac{1}{2} \times 100\% = 12.5\%$

We are told that it takes 12 years to get to 12.5% of the original radioactivity; therefore, 12 years must represent three half-lives:

$$t_{1/2} = \frac{1}{3} \times (12\ yr) = 4.0\ yr$$

If the given percentage did not correspond exactly to one of the subsequent reductions of 100% by half, it would be necessary to do the actual calculations: $A_0 = 100.0\%$ and $A = 12.5\%$

$$\ln\left(\frac{A}{A_0}\right) = -kt$$

$$\ln\left(\frac{12.5\%}{100.0\%}\right) = -2.079 = -k \times (12\ yr)$$

$$k = 0.17\ yr^{-1}$$

$$t_{1/2} = \frac{\ln 2}{k} = \frac{\ln 2}{0.17\ yr^{-1}} = 4.0\ yr$$

☑ *Reasonable Result Check:* The quick method and the actual calculation produce the same answer. The half-life must be less than the elapsed time, since the remaining radioactivity is less than half of the original radioactivity.

33. *Result:* **0.5 h**

Analyze and Plan: Adapt the methods described in the solution to Question 32.

Execute: It is possible to note that 16 is a multiple of 2: $\frac{1}{16} = \frac{1}{2} \times \frac{1}{2} \times \frac{1}{2} \times \frac{1}{2}$

Therefore, the elapsed time, 2 h, represents four half-lives:

$$\frac{1}{4} \times (2\text{ h}) = 0.5\text{ h}$$

If the fraction remaining was not a multiple of $\frac{1}{2}$, we would need to do the actual calculations: $A = \frac{1}{16}A_0$

$$\ln\left(\frac{\frac{1}{16}A_0}{A_0}\right) = -2.7726 = -k \times (2\text{ h}) \quad \textit{(assuming 1/16 and 2 h are exact)}$$

$$k = 1.386\text{ h}^{-1}$$

$$t_{1/2} = \frac{\ln 2}{k} = \frac{\ln 2}{1.386\text{ hr}^{-1}} = 0.5\text{ hr}$$

☑ *Reasonable Result Check:* The quick method and the actual calculation produce the same answer. The half-life must be less than the elapsed time, since the remaining radioactivity is less than half of the original radioactivity.

34. *Result:* **34.8 d**

Analyze and Plan: Adapt the methods described in the solutions to Questions 28, 30 and 32.

Execute: The percentage decrease given (5.0%) does not correspond exactly to one of the sequential reductions of 100% by half, so we must set up the time calculation using the equation given in Question 30:

$$\ln\left(\frac{A}{A_0}\right) = -kt$$

In the solution to Question 28, we calculated $k = 8.62 \times 10^{-2}\text{ d}^{-1}$.

$A_0 = 100.0\%$ and $A = 5.0\%$ $\qquad \ln\left(\frac{5.0\%}{100.0\%}\right) = -3.00$

$$-3.00 = -(8.62 \times 10^{-2}\text{ d}^{-1})t$$

$$t = 34.8\text{ d}$$

☑ *Reasonable Result Check:* The elapsed time (34.8 d) is much longer than the half-life (8.05 d) and is consistent with the small remaining percent (5.0%).

36. *Result:* **2.58 × 10³ y**

Analyze and Plan: Adapt the methods described in the solution to Question 34 and Problem-Solving Example 18.5.

Execute: $\qquad k = \frac{\ln 2}{5.73 \times 10^3\text{ yr}} = 1.21 \times 10^{-4}\text{ y}^{-1}$

$A_0 = 15.3\text{ d min}^{-1}\text{g}^{-1}$ and $A = 11.2\text{ d min}^{-1}\text{g}^{-1}$ $\qquad \ln\left(\frac{11.2\text{ d min}^{-1}\text{ g}^{-1}}{15.3\text{ d min}^{-1}\text{ g}^{-1}}\right) = -0.312$

$$-0.312 = -(1.21 \times 10^{-4}\text{ yr}^{-1})t$$

$$t = 2.58 \times 10^3\text{ yr}$$

☑ *Reasonable Result Check:* The elapsed time is shorter than the half-life and is consistent with the small reduction in the measured activity.

38. *Result:* **5.0 × 10¹² s**

Analyze and Plan: Adapt the methods described in the solution to Question 34.

Execute: The percentage decrease (10.0%) does not correspond exactly to one of the sequential reductions of 100% by half, so we must set up the calculations for k and then solve for time using: $\ln\left(\frac{A}{A_0}\right) = -kt$

$$k = \frac{\ln 2}{2.4 \times 10^4 \text{ yr}} = 2.9 \times 10^{-5} \text{ yr}^{-1}$$

$N_0 = 100.\%$ and $N = 1\%$ $\qquad \ln\left(\frac{1\%}{100\%}\right) = -4.605$ *(Assume 1% is exact, here.)*

$$-4.605 = -(2.9 \times 10^{-5} \text{ yr}^{-1})t$$

$$t = 1.6 \times 10^5 \text{ yr}$$

$$1.6 \times 10^5 \text{ yr} \times \frac{365.25 \text{ days}}{1 \text{ yr}} \times \frac{24 \text{ hr}}{1 \text{ day}} \times \frac{3600 \text{ s}}{1 \text{ hr}} = 5.0 \times 10^{12} \text{ s}$$

☑ *Reasonable Result Check:* The elapsed time is much longer than the half-life and is consistent with the small remaining percent.

Artificial Transmutation (Section 18-5)

40. *Result:* ${}^{18}_{8}\text{O} + {}^{1}_{1}\text{p} \longrightarrow {}^{18}_{9}\text{F} + {}^{1}_{0}\text{n}$

Analyze and Plan: As described in the solutions to Questions 12 and 14, the mass numbers and atomic numbers must balance. Use that information to determine the mass number and the atomic number of the missing entry. Use the periodic table and the identity of the element to get the atomic number.

Execute: Fluorine has Z = 9. Oxygen has Z = 8

${}^{18}_{8}\text{O} + {}^{1}_{1}\text{p} \longrightarrow {}^{18}_{9}\text{F} + {}^{1}_{0}\text{n}$ $\qquad 18 + 1 = 19 = 18 + 1$ $\qquad 8 + 1 = 9 = 9 + 0$

42. *Result:* ${}^{209}_{83}\text{Bi} + {}^{1}_{0}\text{n} \longrightarrow {}^{210}_{84}\text{Po} + {}^{0}_{-1}\text{e}$

Analyze and Plan: Adapt the method described in the solution to Question 40. Use the Mass Number balance and Atomic Number balance to determine the second product.

Execute: Polonium has Z = 84. Bismuth has Z = 83

${}^{209}_{83}\text{Bi} + {}^{1}_{0}\text{n} \longrightarrow {}^{210}_{84}\text{Po} + ?$ $\qquad 209 + 1 = 210 = 210 + Z$ $\qquad 83 + 0 = 83 = 84 + A$

To balance the reaction, a particle with Z = –1 and A = 0 must form, thus a beta particle also forms.

${}^{209}_{83}\text{Bi} + {}^{1}_{0}\text{n} \longrightarrow {}^{210}_{84}\text{Po} + {}^{0}_{-1}\text{e}$ $\qquad 209 + 1 = 210 = 210 + 0$ $\qquad 83 + 0 = 83 = 84 + (-1)$

45. *Result:* $\mathbf{{}^{239}_{94}Pu + 2\,{}^{1}_{0}n \longrightarrow {}^{0}_{-1}e + {}^{241}_{95}Am}$

Analyze and Plan: Use the method described in the solution to Question 40.

Execute: ${}^{239}_{94}\text{Pu} + 2\,{}^{1}_{0}\text{n} \longrightarrow {}^{0}_{-1}\text{e} + {}^{241}_{95}\text{Am}$ $\qquad 239 + 2(1) = 0 + 241$ $\qquad 94 + 2(0) = -1 + 95$

47. *Result:* $\mathbf{{}^{12}_{6}C}$

Analyze and Plan: Use the method described in the solution to Question 14.

Execute: ${}^{238}_{92}\text{U} + {}^{x}_{y}? \longrightarrow 4\,{}^{1}_{0}\text{n} + {}^{246}_{98}\text{Cf}$ $\qquad 238 + x = 4(1) + 246 = 250$ $\qquad 92 + y = 4(0) + 98$

x = 12, and y = 6, so the isotope is: ${}^{12}_{6}\text{C}$

So, the atomic number of the element that would be formed should be between 116, 118, or 119

Nuclear Fission and Fusion (Sections 18-6, 18-7)

49. *Result:* $^{4}_{2}He + ^{4}_{2}He \longrightarrow ^{8}_{4}Be$

Analyze and Plan: Adapt the method described in the solution to Question 12.

Execute: Helium has Z = 2. Beryllium has Z = 4

$^{4}_{2}He + ^{4}_{2}He \longrightarrow ^{8}_{4}Be$ 4 + 4 = 8 = Z 2 + 2 = 8 = A

51. *Result:* $^{12}_{6}C + ^{16}_{8}O \longrightarrow ^{28}_{14}Si$

Analyze and Plan: Adapt the method described in the solution to Question 12.

Execute: Carbon has Z = 6. Oxygen has Z = 8. Silicon has Z = 14

$^{12}_{6}C + ^{16}_{8}O \longrightarrow ^{28}_{14}Si$ 12 + 16 = 28 6 + 8 = 14

53. *Result/Explanation:* Three components represent the fundamental parts of a nuclear fission reactor. Cadmium rods are used as a neutron absorber to control the rate of the fission reaction. Uranium rods are the source of fuel, since uranium is a reactant in the nuclear equation. Water is used for cooling by removing excess heat energy. It is also used in the form of steam in the steam/water cycle to produce the turning torque for the generator.

55. *Result:* **(a)** $\mathbf{^{140}_{54}Xe}$ **(b)** $\mathbf{^{104}_{41}Nb}$ **(c)** $\mathbf{^{92}_{36}Kr}$

Analyze and Plan: Use the method described in Question 40.

Execute:

(a) $^{235}_{92}U + ^{1}_{0}n \longrightarrow ^{A}_{Z}? + ^{93}_{38}Sr + 3\ ^{1}_{0}n$

235 + 1 = A + 93 + 3(1) = 236 92 + 0 = Z + 38 + 3(0)

A = 140, and Z = 54, so the isotope is: $^{140}_{54}Xe$

(b) $^{235}_{92}U + ^{1}_{0}n \longrightarrow ^{A}_{Z}? + ^{132}_{51}Sb + 3\ ^{1}_{0}n$

235 + 1 = A + 132 + 3(1) = 236 92 + 0 = Z + 51 + 3(0)

A = 101, and Z = 41, so the isotope is: $^{101}_{41}Nb$

(c) $^{235}_{92}U + ^{1}_{0}n \longrightarrow ^{A}_{Z}? + ^{141}_{56}Ba + 3\ ^{1}_{0}n$

235 + 1 = A + 141 + 3×(1) = 236 92 + 0 = Z + 56 + 3×(0)

A = 92 and Z = 36, so the isotope is: $^{92}_{36}Kr$

57. *Result:* **6.9 × 10³ barrels**

Analyze and Plan: This is a typical conversion factor question.

Execute: $1.0 \text{ lb } ^{235}U \times \dfrac{453.6 \text{ g}}{1 \text{ lb}} \times \dfrac{1 \text{ mol } ^{235}U}{235 \text{ g U}} \times \dfrac{2.1\times10^{10} \text{ kJ}}{1 \text{ mol } ^{235}U} \times \dfrac{1 \text{ barrel oil}}{5.9\times10^{6} \text{ kJ}} = 6.9 \times 10^3 \text{ tons of coal}$

Nuclear Radiation: Effects and Units (Section 18-8)

59. *Results*: (a) 400 rad (b) 4 Gy (c) 7.3 × 10⁶ mrem

Analyze: Given the radiation reading in sieverts and the quality factor, determine rads, grays, and mrem.

Plan: Use appropriate conversion factors and definitions.

Execute:

(a) $$73\text{ Sv} \times \frac{100\text{ rem}}{1\text{ Sv}} = 7300\text{ rem}$$

$$\frac{7300\text{ rem}}{20\text{ Q}} = 365\text{ rad} \equiv 400\text{ rad } \textit{(1 sig fig)}$$

(b) $$365\text{ rad} \times \frac{1\text{ Gy}}{100\text{ rad}} = 3.65\text{ Gy} \equiv 4\text{ Gy } \textit{(1 sig fig)}$$

(c) $$7300\text{ rem} \times \frac{1000\text{ mrem}}{1\text{ rem}} = 7.3 \times 10^6\text{ mrem}$$

61. *Result/Explanation:* The unit "rad" is the measure of the amount of radiation absorbed. The unit "rem" includes a quality factor that better describes the biological impact of a radiation dose. The unit rem would be more appropriate when talking about the effects of an atomic bomb on humans. The unit gray (Gy) is 100 rad.

63. *Result/Explanation:* Since most elements have some proportion of unstable isotopes that decay and we are composed of these elements (e.g., ^{14}C), our bodies emit radiation particles.

Applications of Radioisotopes (Section 18-9)

65. *Result/Explanation:* The gamma ray is a high-energy photon. Its interaction with matter is most likely just going to be imparting large quantities of energy. The alpha and beta particles are charged particles of matter, which could interact and possibly react with and alter the matter composing the food. Gamma rays penetrate most matter readily, passing through unaffected. Beta particles are slower and penetration is stopped by dense tissues, such as bones. Alpha particles are stopped by skin.

67. *Result:* **0.13 L**

Analyze and Plan: Adapt the methods described in the solution to Questions 28 and Chapter 3.

Execute:

$$A = kN \propto \text{mol}$$

$$\textit{Molarity}(\text{conc}) = \text{mol/L} \propto A/L \propto A/\text{mL}$$

$$\textit{Molarity}(\text{conc}) \times V(\text{conc}) = \textit{Molarity}(\text{dil}) \times V(\text{dil})$$

$$A(\text{conc})/\text{mL} \times V(\text{conc}) = A(\text{dil})/\text{mL} \times V(\text{dil})$$

$$V(\text{dil}) = \frac{A/\text{mL(conc)} \times V(\text{conc})}{A/\text{mL}} = \frac{\left(2.0 \times 10^6\text{ dps/mL conc}\right) \times (1.0\text{ mL conc}) \times \frac{1\text{ L}}{1000\text{ mL}}}{(1.5 \times 10^4\text{ dps/mL dil})} = 0.13\text{ L}$$

General Questions

69. *Result:* (a) $^{0}_{-1}e$ (b) $^{4}_{2}He$ (c) $^{87}_{35}Br$ (d) $^{216}_{84}Po$ (e) $^{68}_{31}Ga$

Analyze and Plan: Follow the method described in the solutions to Questions 14 and 16.

Execute:

(a) $^{214}_{83}Bi \longrightarrow \underline{^{0}_{-1}e} + ^{214}_{84}Po$ $214 = \underline{0} + 214;\ 83 = \underline{-1} + 84$

(b) $4\,^{1}_{1}H \longrightarrow \underline{^{4}_{2}He} + 2\,^{0}_{+1}e$ $4\times(1) = \underline{4} + 2\times(0);\ 4\times(1) = \underline{2} + 2\times(0)$

(c) $^{249}_{99}Es + ^{1}_{0}n \longrightarrow 2\,^{1}_{0}n + \underline{^{87}_{35}Br} + ^{161}_{64}Gd$ $249 + 1 = 2\times(1) + \underline{87} + 161;\ 99 + 0 = 2\times(0) + \underline{35} + 64$

(d) $^{220}_{86}Rn \longrightarrow \underline{^{216}_{84}Po} + ^{4}_{2}He$ $220 = \underline{216} + 4;\ 86 = \underline{84} + 2$

(e) $^{68}_{32}Ge + ^{0}_{-1}e \longrightarrow \underline{^{68}_{31}Ga}$ $68 + 0 = \underline{68};\ 32 + (-1) = \underline{31}$

71. *Result:* **2 mg**

Analyze and Plan: Use the methods described in the solution to Question 28.

Execute: $$t = (1\ \text{h}) \times \left(\frac{60\ \text{min}}{1\ \text{h}}\right) = 60\ \text{min}\ \textit{(1 sig fig)}$$

The elapsed time (1 hour = 60 min) is six times the half-life (10. min), so the quantity remaining will be reduced by half six times:

$$\frac{1}{2}\times\frac{1}{2}\times\frac{1}{2}\times\frac{1}{2}\times\frac{1}{2}\times\frac{1}{2}\times 96\ \text{mg} = 1.5\ \text{mg} \cong 2\ \text{mg}\ \textit{(1 sig fig)}$$

If the time elapsed were not a whole-number multiple of the half-life, it would be necessary to do the calculations:

$$k = \frac{\ln 2}{10.\ \text{min}} = 6.9 \times 10^{-2}\ \text{min}^{-1}$$

$$\ln\left(\frac{m}{m_0}\right) = -kt = -(6.9 \times 10^{-2}\ \text{min}^{-1}) \times (60\ \text{min}) = -4.2$$

$$m = m_0 e^{-4.2} = (96\ \text{mg}) \times e^{-4} = (96\ \text{mg}) \times (0.016) = 1.5\ \text{mg} \cong 2\ \text{mg}\ \textit{(1 sig fig)}$$

73. *Result:* **3.6×10^9 yr**

Analyze and Plan: Adapt the methods described in the solution to Question 36.

Execute: Because the quantity ratio is not a multiple of $\frac{1}{2}$, we will set up the calculations for k and t:

$$k = \frac{\ln 2}{4.9 \times 10^{10}\ \text{yr}} = 1.4 \times 10^{-11}\ \text{yr}^{-1}$$

$$\ln\left(\frac{N}{N_0}\right) = -kt$$

$$\frac{N}{N_0} = 0.951$$

$$\ln(0.951) = -0.0502 = -(1.4 \times 10^{-11}\ \text{yr}^{-1})t$$

$$t = 3.6 \times 10^9\ \text{yr}$$

74. *Result:* **(a) $^{247}_{99}$Es (b) $^{16}_{8}$O (c) $^{4}_{2}$He (d) $^{12}_{6}$C (e) $^{10}_{5}$B**

Analyze and Plan: Follow the method described in the solution to Question 40.

Execute:

(a) $^{238}_{92}\text{U} + ^{14}_{7}\text{N} \longrightarrow \boxed{^{247}_{99}\text{Es}} + 5\ ^{1}_{0}\text{n}$ $\quad 238 + 14 = 252 = \boxed{247} + 5(1) \quad 92 + 7 = 99 = \boxed{99} + 5(0)$

(b) $^{238}_{92}\text{U} + \boxed{^{16}_{8}\text{O}} \longrightarrow ^{249}_{100}\text{Fm} + 5\ ^{1}_{0}\text{n}$ $\quad 238 + \boxed{16} = 254 = 249 + 5(1) \quad 92 + \boxed{8} = 100 = 100 + 5(0)$

(c) $^{253}_{99}\text{Es} + \boxed{^{4}_{2}\text{He}} \longrightarrow ^{256}_{101}\text{Md} + ^{1}_{0}\text{n}$ $\quad 253 + \boxed{4} = 257 = 256 + 1 \quad 99 + \boxed{2} = 101 = 101 + 0$

(d) $^{246}_{96}\text{Cm} + \boxed{^{12}_{6}\text{C}} \longrightarrow ^{254}_{102}\text{No} + 4\ ^{1}_{0}\text{n}$ $\quad 246 + \boxed{12} = 258 = 254 + 4(1) \quad 96 + \boxed{6} = 102 = 102 + 4(0)$

(e) $^{252}_{98}\text{Cf} + \boxed{^{10}_{5}\text{B}} \longrightarrow ^{257}_{103}\text{Lr} + 5\ ^{1}_{0}\text{n}$ $\quad 252 + \boxed{10} = 262 = 257 + 5(1) \quad 98 + \boxed{5} = 103 = 103 + 5(0)$

76. *Result:* **(a) See equations below (b) See equations below (c) See equations below (d) Group 7A (e) Halogens**

Analyze and Plan: Follow the method described in the solution to Questions 14 and 40.

Execute:

(a) $^{48}_{20}Ca + ^{249}_{97}Bk \longrightarrow ^{294}_{117}Uus + 3\ ^{1}_{0}n$

$48 + 249 = 297 = 294 + 3(1) \qquad 20 + 97 = 117 = 117 + 3(0)$

$^{48}_{20}Ca + ^{249}_{97}Bk \longrightarrow ^{293}_{117}Uus + 4\ ^{1}_{0}n$

$48 + 249 = 297 = 293 + 4(1) \qquad 20 + 97 = 117 = 117 + 4(0)$

(b) $^{293}_{117}Uus \longrightarrow ^{289}_{115}Uup + ^{4}_{2}He \qquad 293 = 289 + 4 \qquad 117 = 115 + 2$

$^{289}_{115}Uup \longrightarrow ^{285}_{113}Uut + ^{4}_{2}He \qquad 289 = 285 + 4 \qquad 115 = 113 + 2$

$^{285}_{113}Uut \longrightarrow ^{281}_{111}Rg + ^{4}_{2}He \qquad 285 = 281 + 4 \qquad 113 = 111 + 2$

(c) $^{294}_{117}Uus \longrightarrow ^{290}_{115}Uup + ^{4}_{2}He \qquad 294 = 290 + 4 \qquad 117 = 115 + 2$

$^{290}_{115}Uup \longrightarrow ^{286}_{113}Uut + ^{4}_{2}He \qquad 290 = 286 + 4 \qquad 115 = 113 + 2$

$^{286}_{113}Uut \longrightarrow ^{282}_{111}Rg + ^{4}_{2}He \qquad 286 = 282 + 4 \qquad 113 = 111 + 2$

$^{282}_{111}Rg \longrightarrow ^{278}_{109}Mt + ^{4}_{2}He \qquad 282 = 278 + 4 \qquad 111 = 109 + 2$

$^{278}_{109}Mt \longrightarrow ^{274}_{107}Bh + ^{4}_{2}He \qquad 278 = 274 + 4 \qquad 109 = 107 + 2$

$^{274}_{107}Bh \longrightarrow ^{270}_{105}Db + ^{4}_{2}He \qquad 274 = 270 + 4 \qquad 107 = 105 + 2$

(d) Looking at the periodic table, element 117 will be found directly under astatine, At, in **group 7A.**

(e) Group 7A is called the **halogens**.

78. *Result:* **1.70×10^8 kJ/g (b) 4.798×10^{28} kJ (c) 9.8×10^{-10} %**

(a) *Analyze and Plan:* Use the method described in Question 23 and Section 18-3b.

Execute:

$$^{2}_{1}H + ^{1}_{1}H \longrightarrow ^{3}_{2}He$$

Calculate the change in mass (Δm) during the reaction using the molar masses of $^{1}_{1}H$ = 1.007276 g/mol) and $^{2}_{1}H$ = 1.008665 g/mol (from Section 18-3b), then use Einstein's equation: $\Delta E = (\Delta m)c^2$ to calculate the total energy (in kJ/mol) generated and divide by the molar mass to determine the binding energy in joules per gram.

$\Delta m = 2.013553\text{g/mol}\ ^{2}_{1}H + 1.007276\text{ g/mol}\ ^{1}_{1}H - 3.0160297\text{ g/mol}\ ^{3}_{2}He = 0.004799\text{ g/mol}$

The nuclear binding energy = $E_b = (\Delta m)c^2$

$$\Delta E_b = (\Delta m)c^2 = \left(\frac{0.004799\text{ g}}{1\text{ mol}} \times \frac{1\text{ kg}}{1000\text{ g}}\right) \times (2.99792 \times 10^8\text{ m/s})^2 \times \frac{1\text{ J}}{1\text{ kg m}^2\text{ s}^{-2}} \times \frac{1\text{ kJ}}{1000\text{ J}}$$

$$\times \frac{1\text{ mol}}{2.013553\text{ g }^{2}_{1}H} = \mathbf{2.142 \times 10^8\ kJ/g\ ^{2}_{1}H}$$

(b) *Analyze and Plan:* Use conversion factors to determine the mass of deuterium in the oceans and use the results from part (a) to determine the kJ:

Execute:
$$1.4 \times 10^{24} \text{ g sea water} \times \frac{0.016 \text{ g } {}^2_1\text{H}}{100 \text{ g sea water}} \times \frac{2.142 \times 10^8 \text{ kJ}}{\text{g } {}^2_1\text{H}} = \mathbf{4.798 \times 10^{28}\ kJ}$$

(b) *Analyze and Plan:* Energy percent is equal to mass percent:

Execute:
$$\frac{4.7 \times 10^{17} \text{ kJ}}{4.799 \times 10^{28} \text{ kJ}} \times 100\ \% = \mathbf{9.8 \times 10^{-10}\ \%}$$

Applying Concepts

83. *Result/Explanation:* Isotopes that could be used as a tracer are ^{14}C, ^{18}O, and ^{3}H. For example, one can take acetic acid made using ^{1}H and place it in water made using ^{3}H, or take acetic acid made using ^{3}H with and react it with water made using ^{1}H. In either case, by analyzing the acetic acid, we see the acidic hydrogen would be replaced with hydrogen from the water until the ^{3}H is evenly distributed between the two reactants.

85. *Result/Explanation:* Alpha and beta radiation decay particles are charged (${}^4_2He^{2+}$ and ${}^{0}_{-1}e^{-}$), so they are better able to interact with and ionize tissues, disrupting the function of the cancer cells. Gamma radiation, like X-rays, goes through soft tissue without much being absorbed. This radiation can exit the body and be detected, thereby locating the cancerous cells.

87. *Result/Explanation:* The ^{20}Ne isotope is stable. The ^{17}Ne isotope is likely to decay by positron emission, to increase the ratio of neutrons to protons. The ^{23}Ne isotope is likely to decay by beta emission to decrease the ratio of neutrons to protons. For more details, refer to the discussion in Section 18-3 and examine Figure 18.3.

89. *Result/Explanation:* A nuclear reaction occurred, making products. Therefore, some of the lost mass is found in the decay particles, if the decay is alpha or beta decay, and almost all the rest is found in the element produced by the reaction.

More Challenging Questions

92. *Result/Explanation:* All radioactive decays are first order because, to occur, there needs to be one and only one reactant. The reaction involves only the nucleus of an atom.

94. *Result:* **75.1 yr**

Analyze and Plan: Adapt the methods described in the solution to Question 34.

Execute:
$$N_0 = 100\%,\ N = 1.45\%$$

The ratio of percentage activity is not a multiple of $\frac{1}{2}$, so we can't use the quick method. Therefore, set up the calculation for k and t.

$$k = \frac{\ln 2}{12.3 \text{ yr}} = 0.0564 \text{ yr}^{-1}$$

$$\ln\left(\frac{1.45\ \%}{100\ \%}\right) = -4.234 = -(0.0564 \text{ yr}^{-1})t$$

$$t = 75.1 \text{ yr}$$

96. *Result:* **3.92 × 10³ yr**

Analyze and Plan: Use conversion factors to determine the mass of carbon, then calculate the disintegrations per minute per gram (d min^{-1} g^{-1}) of carbon, and use the method described in the solution to Question 36.

Execute:
$$1.14 \text{ g CaCO}_3 \times \frac{1 \text{ mole CaCO}_3}{100.087 \text{ g CaCO}_3} \times \frac{1 \text{ mole C}}{1 \text{ mole CaCO}_3} \times \frac{12.0107 \text{ g C}}{1 \text{ mole C}} = 0.137 \text{ g C}$$

$$k = \frac{\ln 2}{5730\ \text{yr}} = 1.21 \times 10^{-4}\ \text{yr}^{-1}$$

$$2.17 \times 10^{-2}\ \text{Bq} = 2.17 \times 10^{-2}\ \text{dis}\cdot\text{s}^{-1}$$

$$A = \frac{2.17 \times 10^{-2}\ \text{dis}\cdot\text{s}^{-1} \times \frac{60\,\text{s}}{1\,\text{min}}}{0.137\ \text{g C}} = 9.52\ \text{min}^{-1}\ \text{g}^{-1} \quad \text{and} \quad A_0 = 15.3\ \text{dis min}^{-1}\ \text{g}^{-1}$$

$$\ln\left(\frac{9.52\ \text{dis min}^{-1}\ \text{g}^{-1}}{15.3\ \text{dis min}^{-1}\ \text{g}^{-1}}\right) = -0.474 = -(1.21 \times 10^{-4}\ \text{yr}^{-1})t$$

$$t = 3.92 \times 10^{3}\ \text{yr}$$

Chapter 19: The Chemistry of the Main Group Elements

Solutions for Red-Numbered Questions for Review and Thought

Topical Questions

Formation of Elements (Section 19-1)

13. *Result:* (a) $^{16}_{8}O + ^{4}_{2}He \longrightarrow ^{20}_{10}Ne$ (b) $^{12}_{6}C + ^{12}_{6}C \longrightarrow ^{20}_{10}Ne + ^{4}_{2}He$

Analyze and Plan: Balance the atomic numbers and the mass numbers to determine what radiation particles are involved

Execute:

(a) Oxygen has Z = 8. Neon has Z = 10

$$^{16}_{8}O + \square \longrightarrow ^{20}_{10}Ne \qquad 16 + A = 20 \qquad 8 + Z = 10$$

A = 4, Z = 2, so $^{4}_{2}He$ reacts with oxygen-16.

$$^{16}_{8}O + ^{4}_{2}He \longrightarrow ^{20}_{10}Ne \qquad 16 + 4 = 20 \qquad 8 + 2 = 10$$

(b) Oxygen has Z = 8. Neon has Z = 10

$$^{12}_{6}C + ^{12}_{6}C \longrightarrow ^{20}_{10}Ne + \square \qquad 12 + 12 = 24 = 20 + A \qquad 6 + 6 = 12 = 10 + Z$$

A = 4, Z = 2, so $^{4}_{2}He$ is produced along with neon-20.

$$^{12}_{6}C + ^{12}_{6}C \longrightarrow ^{20}_{10}Ne + ^{4}_{2}He\ ^{20}_{10}Ne \qquad 12 + 12 = 24 = 20 + 4 \qquad 6 + 6 = 12 = 10 + 2$$

15. *Result/Explanation:* Heavier atoms are formed from lighter elements by "burning" and capture of neutrons described in Section 19-1b and Figure 19.1.

Terrestrial Elements (Section 19-2)

17. *Result/Explanation:* Several elements are found in nature in uncombined form. Some examples are: Nitrogen, oxygen, sulfur, argon, platinum, gold (other answers are also correct). See Figure 19.3 for other examples.

19. *Result/Explanation:* Toxicity is determined by the persistence in human tissue. Amphibole asbestos is composed of long, thin, straight fibers that can penetrate narrow lung passages, so it is relatively insoluble and persists in human tissue. In contrast, serpentine asbestos fibers are curly, so they ball up like yarn and are more easily rejected by the body; they tend to be soluble and disappear in human tissue. See Section 19-2a for more details.

22. *Result/Explanation:* Amphiboles are formed by joining two pyroxene chains (made of two SiO_3^{2-}) such that they share one oxygen atom to give the total formula $Si_4O_{11}^{6-}$. The amphibole mineral is discussed in detail in Section 19-2a and the combining of pyroxene chains into an amphibole structure is shown in Figure 19.5.

Extraction by Physical Means, N_2, O_2, S (Section 19-3)

23. *Result/Explanation:* The physical properties that allow for the separation of nitrogen, oxygen and argon in fractional distillation of air are their boiling points. As shown in Figure 19.7, N_2 boils first at –195.8°C, then O_2 boils at –183°C. As described in 19-3a, Ar doesn't boil until –189°C.

25. *Result:* **(a) Nitrogen (b) Sulfur**

Explanation:

(a) The Linde process is described near the end of Section 19-3a. It produces nitrogen.

(b) The Frasch process is decribed near the beginning of Section 19-3b. It produces sulfur.

Extraction by Electrolysis:, Na, Cl_2, Mg, Al (Section 19-4)

27. *Result:* **(a) Sodium metal and chlorine gas; see equation below (b) Sodium hydroxide; see equation below (c) Magnesium metal; see equation below (d) Aluminum metal; see equation below.**

Analyze and Plan: Given names of several commercial processes, identify the substance or substances produced by each and write a balanced equation for the main reaction of the process. Look for details in Section 19-4.

Execute:

(a) The Downs cell is described in Section 19-4a and Figure 19.8. It produces sodium metal and chlorine gas. The equation representing the main reaction is:

$$2\ NaCl(\ell) \longrightarrow 2\ Na(\ell) + Cl_2(g)$$

(b) The chlor-alkali process is described in Section 19-4b and Figure 19.9. It produces sodium hydroxide, chlorine and hydrogen. The chemical equation representing the main reaction is:

$$2\ NaCl\ (aq) + 2\ H_2O(\ell) \longrightarrow Cl_2(g) + 2\ NaOH(aq) + H_2(g)$$

(c) The Dow process is described in Section 19-4c and Figure 19.11. It produces magnesium metal. The chemical equation representing the main reaction is:

$$Mg^{2+}(\text{in melt}) + 2\ Cl^-(\text{in melt}) \longrightarrow Mg(\text{in melt}) + Cl_2(g)$$

(d) The Hall-Heroult process is described in Section 19-4d and Figure 19.12. It produces aluminum metal. The chemical equation representing the main reaction is:

$$4\ Al^{3+}(\text{in melt}) + 3\ C(s) + 6\ O^{2-}(\text{in melt}) \longrightarrow 4\ Al(\text{in melt}) + 3\ CO_2(g)$$

29. *Result:* **(a) $2\ Cl^-(aq) + 2\ H_2O(\ell) \longrightarrow Cl_2(g) + 2\ OH^-(aq) + H_2(g)$ (b) The 2002 Cl_2/NaOH ratio (1.27) is higher than the balanced equation ratio (0.88639). Electrolysis of brine may not be the only method used by industry to produce Cl_2 and NaOH. (Other answers are possible to explain the difference.)**

Analyze: The membrane cell pictured in Figure 19.9 shows what reactions occur in the electrolysis of brine (aqueous NaCl). Chloride ion is oxidized to chlorine gas (Cl_2) at the anode, and water is reduced to hydrogen gas (H_2) and hydroxide ion (OH^-) at the cathode.

Plan and Execute:

(a) Using the method of half-reactions (introduced in Section 17.2) we can write the following half-reactions. The sum of the two balanced half-reactions provides the net-cell reaction.

Notice that aqueous sodium ion is a spectator ion in the reaction has been omitted from these equations.

$$2\ Cl^-(aq) \longrightarrow Cl_2(g) + 2e^-$$

$$2\ H_2O(\ell) + 2e^- \longrightarrow 2\ OH^-(aq) + H_2(g)$$

$$2\ Cl^-(aq) + 2\ H_2O(\ell) \longrightarrow Cl_2(g) + 2\ OH^-(aq) + H_2(g)$$

(b) Calculate the masses of one mol of Cl_2 and two mol of NaOH, and compare the mass ratios.

$$2\ \text{mol OH}^- \times \frac{1\ \text{mol NaOH}}{1\ \text{mol OH}^-} \times \frac{39.9971\ \text{g NaOH}}{1\ \text{mol NaOH}} = 79.9942\ \text{g NaOH}$$

$$1 \text{ mol Cl}_2 \times \frac{70.906 \text{ g Cl}_2}{1 \text{ mol Cl}_2} = 70.906 \text{ g Cl}_2$$

The mass ratio of Cl_2 to NaOH from this reaction is: $\frac{70.906 \text{ g Cl}_2}{79.9942 \text{ g NaOH}} = 0.88639$

The mass ratio of Cl_2 to NaOH produced in 2002: $\frac{1.14 \times 10^{10} \text{ kg Cl}_2}{8.98 \times 10^{9} \text{ kg NaOH}} = 1.27$

The ratio produced in 2002 (0.88639) is substantially higher than that calculated based on the balanced equation (1.27). The difference suggests that the electrolysis of brine may not be the only method used by industry to produce these materials.

Other answers may also provide a legitimate suggestion for why the ratio is different.

31. *Result:* **(a) cathode (b) Cl_2(g) (c) 2.027 × 10^9 C (d) 1.6 × 10^3 kJ/mol**

Analyze and Plan: This electrolysis process is described in Section 19-4c.

Execute:

(a) The magnesium ion is reduced to magnesium metal. This occurs at the cathode.

(b) Chlorine gas, Cl_2(g), is formed at the other electrode.

(c) For each mol of Mg and Cl_2 produced, two mol of electrons are transferred. One mol of electrons is one Faraday. One Faraday is equal to the charge of 96485 C. These types of problems are discussed in Section 17-11.

$$1000. \text{ kg MgCl}_2 \times \frac{1000 \text{ g}}{1 \text{ kg}} \times \frac{1 \text{ mol MgCl}_2}{95.211 \text{ g MgCl}_2} \times \frac{2 \text{ mol e}^-}{1 \text{ mol MgCl}_2} \times \frac{1 \text{ Faraday}}{1 \text{ mol e}^-} = 2.101 \times 10^4 \text{ Faradays}$$

$$2.101 \times 10^4 \text{ Faradays} \times \frac{96485 \text{ C}}{1 \text{ Faraday}} = 2.027 \times 10^9 \text{ C}$$

(d) The relationship 1 kWh = 3.60 × 10^6 J/kWh is given in a margin note in Section 17-11a; this conversion factor is also used in the Estimation Box in Section 17-12.

$$\frac{8.4 \text{ kwh}}{\text{lb Mg}} \times \frac{3.60 \times 10^6 \text{ J}}{1 \text{ kwh}} \times \frac{1 \text{ kJ}}{1000 \text{ J}} \times \frac{1 \text{ lb Mg}}{453.6 \text{ g Mg}} \times \frac{24.3050 \text{ g Mg}}{1 \text{ mol Mg}} = 1.6 \times 10^3 \frac{\text{kJ}}{\text{mol Mg}}$$

33. *Result:* **7 × 10^3 kWh**

Analyze and Plan: Adapt the method described in the solution to Question 17.65. The relationship 1 kWh = 3.60 × 10^6 J/kWh is given in a margin note in Section 17-11a; this conversion factor is also used in the Estimation Box in Section 17-12.

Notice: it is not necessary to use the amperes given in the question.

Execute:

$$1 \text{ ton Na} \times \frac{2000 \text{ lb}}{1 \text{ ton}} \times \frac{453.6 \text{ g}}{1 \text{ lb}} \times \frac{1 \text{ mol Na}}{22.9898 \text{ g Na}} \times \frac{2 \text{ mol e}^-}{2 \text{ mol Na}} \times \frac{96485 \text{ C}}{1 \text{ mol e}^-}$$

$$\times 7.0 \text{ V} \times \frac{1 \text{ J}}{1 \text{ C} \cdot 1 \text{ V}} \times \frac{1 \text{ kWh}}{3.60 \times 10^6 \text{ J}} = 7 \times 10^3 \text{ kWh}$$

35. *Result:* **67.1 g Al**

Analyze and Plan: Adapt the method described in the solution to Question 33.

Execute:

$$2.00 \text{ hr} \times 100. \text{ A} \times \frac{3600 \text{ s}}{1 \text{ hr}} \times \frac{1 \text{ C}}{1 \text{ A} \cdot 1 \text{ s}} \times \frac{1 \text{ mol e}^-}{96485 \text{ C}} \times \frac{1 \text{ mol Al}}{3 \text{ mol e}^-} \times \frac{26.9815 \text{ g Al}}{1 \text{ mol Al}} = 67.1 \text{ g Al}$$

Extraction by Chemical Oxidation-Reduction, P, Br_2, I_2 (Section 19-5)

36. *Result/Explanation:* PH_3 has a lower boiling point because its molecules do not form strong hydrogen bonds, whereas the molecules of NH_3 do.

38. *Result:* **Product-favored**

Analyze and Plan: Balance the redox equation as described in Section 17-2. Use the method described in Section 17-4a to determine the $E°_{cell}$ using $E°_{cell} = E°_{cathode} - E°_{anode}$. If the $E°_{cell}$ is positive, the reaction is product-favored. If it is negative, it is not product favored.

Execute:

$$Br_2(g) + 2\ e^- \longrightarrow 2\ Br^-(aq) \qquad E°_{cathode} = 1.07\ V$$

$$2\ I^-(aq) \longrightarrow I_2(g) + 2\ e^- \qquad E°_{anode} = 0.54\ V$$

$$Br_2(g) + 2\ I^-(aq) \longrightarrow I_2(g) + 2Br^-(aq) \qquad E°_{cell} = (1.07\ V) - (0.54\ V) = 0.53\ V$$

$E°_{cell} =$ 0.53 V, so the reaction is **product-favored** at standard state.

Main-Group Elements (Section 19-6)

40. Result: **(a) nitric acid; see equation below (b) Sodium hydroxide; see equation below (c) Magnesium metal; see equation below (d) Aluminum metal; see equation below.**

Analyze and Plan: Given names of several commercial processes, identify the substance or substances produced by each and write a balanced equation for the main reaction of the process. Look for details in Section 19-4.

Execute:

(a) The Ostwald process is described in Section 19-4e. It produces nitric acid. The equations representing the main reactions are:

$$4\ NH_3(g) + 5\ O_2(g) \longrightarrow 4\ NO(g) + 6\ H_2O(\ell)$$

$$2\ NO(g) + O_2(g) \longrightarrow NO_2(g)$$

$$3\ NO_2(g) + 2\ H_2O(\ell) \longrightarrow HNO_3(aq) + NO(g)$$

(b) The contact process is described in Section 19-6f. It produces sulfuric acid. The chemical equations representing the main reactions are:

$$S_8(s) + 8\ O_2(g) \longrightarrow 8\ SO_2(g)$$

$$2\ SO_2(g) + O_2(g) \longrightarrow 2\ SO_3(g)$$

$$SO_3(g) + H_2O(\ell) \longrightarrow H_2SO_4(aq)$$

This final step is actually accomplished using this pair of reactions:

$$SO_3(g) + H_2SO_4(\ell) \longrightarrow H_2S_2O_7(\ell)$$

$$H_2S_2O_7(\ell) + H_2O(\ell) \longrightarrow 2\ H_2SO_4(aq)$$

(c) The Haber-Bosch process is described in Sections 12-8 and 19-4e. It produces ammonia. The chemical equation representing the main reaction is:

$$N_2(g) + 3\ H_2(g) \longrightarrow 2\ NH_3(g)$$

42. *Result:* **0.2 - 0.4 g Li**

Analyze and Plan: Use conversion factors with molar mass and mole ratios, to determine the mass of lithium in each treatment dose of a given mass range.

Execute:

$$1\ g\ Li_2CO_3 \times \frac{1\ mol\ Li_2CO_3}{66.950\ g\ Li_2CO_3} \times \frac{2\ mol\ Li}{1\ mol\ Li_2CO_3} \times \frac{6.941\ g\ Li}{1\ mol\ Li} = 0.2\ g$$

$$2 \text{ g Li}_2\text{CO}_3 \times \frac{1 \text{ mol Li}_2\text{CO}_3}{66.950 \text{ g Li}_2\text{CO}_3} \times \frac{2 \text{ mol Li}}{1 \text{ mol Li}_2\text{CO}_3} \times \frac{6.941 \text{ g Li}}{1 \text{ mol Li}} = 0.4 \text{ g}$$

The dose contains **0.2 - 0.4 grams** of lithium.

44. *Result:* Oxidizing agent is SnO_2; reducing agent is C.

Analyze and Plan: The oxidizing agent is the reactant of reduction. The reducing agent is the reactant of oxidation. Identify oxidation numbers for each element in the reaction and determine which element is gaining electrons (thus undergoing reduction) and which is losing electrons (thus undergoing oxidation).

Execute:

SnO_2	Ox. # Sn + 2(–2) = 0	Ox. # S = 4
C	Ox. # C = 0	
CO	Ox. # C + (–2) = 0	Ox. # C = 2
Sn	Ox. # Sn = 0	

Ox. # Sn atom goes from +4 to 0, so Sn is gaining electrons and represents to reactant of reduction.

Ox. # C atom goes from 0 to +2, so C is losing electrons and represents to reactant of oxidation.

Therefore, SnO_2 is the oxidizing agent and C is the reducing agent.

46. *Results:* **(a) No (b) Yes (c) Yes (d) Yes (e) No**

Analyze and Plan: Given the formula of a liquid, determine which of the given substances are soluble in that liquid. Determine the molecular shape and identify noncovalent interactions. Those molecules that interact in with similar type and strength of forces as are experienced in the solvent will be soluble. Those that have very different noncovalent interactions will not be soluble.

Execute: CS_2(s) is a linear molecule with non-polar bonds, so it is a non-polar molecule and interacts with other substances using dispersion forces.

(a) Water, H_2O, is bent and has polar bonds, so it is polar and (due to O—H bonds) can interact with other molecules using hydrogen bonding. Because its non-covalent interactions are very different from that of CS_2, water will not be soluble in CS_2.

(b) Benzene, is a symmetrical cyclic aromatic hydrocarbon

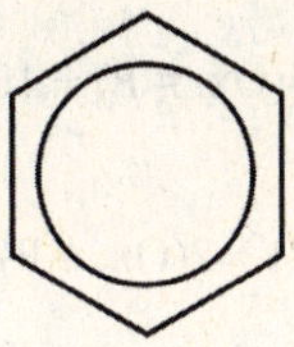

The symmetry and the tiny bond poles make this molecule largely non-polar, so it interacts with other molecules with dispersion forces. Since these are the same types of noncovalent interactions used by CS_2, benzene will be soluble in CS_2.

(c) Elemental white phosphorus is symmetrical and non-polar, , so it interacts with other molecules with dispersion forces. Since these are the same types of noncovalent interactions used by CS_2, elemental white phosphorus will be soluble in CS_2.

(d) Elemental iodine, I_2, is symmetrical and non-polar, so it interacts with other molecules with dispersion forces. Since these are the same types of noncovalent interactions used by CS_2, elemental iodine will be soluble in CS_2.

(e) Ethanol, CH_3CH_2OH, has low symmetry and one very polar bond, so the molecule is polar and (due to O—H bond) can interact with other molecules using hydrogen bonding. Because it's noncovalent interactions are very different from that of CS_2, ethanol will not be soluble in CS_2.

In general, the molecules that are polar and especially those that interact primarily using H-bonding do not dissolve. The molecules that are nonpolar like nonpolar CS_2 do dissolve.

General Questions

48. *Result:*

Formula	Name	Oxidation state of phosphorus
$\mathbf{P_4}$	Phosphorus	**0**
$(NH_4)_2HPO_4$	**Ammonium hydrogen phosphate**	**+5**
H_3PO_4	Phosphoric acid	**+5**
$\mathbf{P_4O_{10}}$	Tetraphosphorus decaoxide	**+5**
$Ca_3(PO_4)_2$	**Calcium phosphate**	**+5**
$\mathbf{Ca(H_2PO_4)_2}$	Calcium dihydrogen phosphate	**+5**

Analyze and Plan: Use the methods described in Chapter 2 for names and formulas: Section 2-4 and 2-5, including Table 2.2 and Chapter 3, Table 3.2. For oxidation states refer to Chapter 3, Section 3-4. Here, we use Ox. # H = +1, Ox. # O = –2, and the sum of the oxidation numbers in a compound add up to the net charge.

Execute:

Formula, name	Calculation of oxidation state	Oxidation state of phosphorus
P_4, phosphorus	4(Ox. # P) = 0	Ox. # P = 0
$(NH_4)_2HPO_4$, ammonium hydrogen phosphate	Ammonium, NH_4^+ hydrogen phosphate, HPO_4^{2-} (+1) + Ox. # P + 4(–2) = –1	Ox. # P = +5
H_3PO_4, phosphoric acid	3(+1) + Ox. # P + 4(–2) = 0	Ox. # P = +5
P_4O_{10}, tetraphosphorus decaoxide	4(Ox. # P) + 10(–2) = 0	Ox. # P = +5
$Ca_3(PO_4)_2$, calcium phosphate	3(+2) + 2(Ox. # P) + 8(–2) = 0	Ox. # P = +5
$Ca(H_2PO_4)_2$, calcium dihydrogen phosphate	(+2) + 4(+1) + 2(Ox. # P) + 8(–2) = 0	Ox. # P = +5

50. *Result:* **1.7×10^7 tons**

Analyze and Plan: Given the percent mass of a compound in an ore, determine how much ore is required to make a given mass of an element contained in the compound. This is a conversion factor question as described in Chapter 2.

Execute:

$$5.0 \times 10^6 \text{ tons Al} \times \frac{2000 \text{ lb}}{\text{ton}} \times \frac{453.6 \text{ g}}{1 \text{ lb}} \times \frac{1 \text{ mol Al}}{26.9815 \text{ g Al}} \times \frac{1 \text{ mol Al}_2\text{O}_3}{2 \text{ mol Al}}$$

$$\times \frac{101.961 \text{ g Al}_2\text{O}_3}{1 \text{ mol Al}_2\text{O}_3} \times \frac{100 \text{ g bauxite}}{55 \text{ g Al}_2\text{O}_3} \times \frac{1 \text{ lb}}{453.6 \text{ g}} \times \frac{1 \text{ ton}}{2000 \text{ lb}} = 1.7 \times 10^7 \text{ tons bauxite}$$

52. *Result:* **See structure below**

Analyze and Plan: Use the methods described in Chapter 6.

Execute: In the solution to Question 6.15(d), we found that phosphate, PO_4^{3-}, has 4 O atoms single-bonded to the P atom. Use that pattern to develop a three-dimensional version of the P_4O_{10} molecule: Each P atom forms four single bonds to O atoms. Three of the four O atoms have single bonds to two different P atoms.

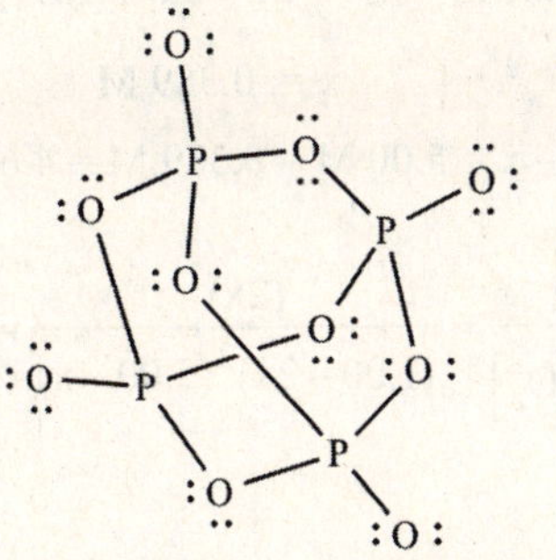

Other Lewis structures are possible, but the structure above is the most plausible one that follows the octet rule.

54. *Result:* **962 K**

Analyze and Plan: Use the methods described in the solution to Question 16.50.

Execute:

$$T = \frac{(-17.6\ \text{kJ/mol}) \times \left(\frac{1000\ \text{J}}{1\ \text{kJ}}\right)}{-18.3\ \text{J K}^{-1}\text{mol}^{-1}} = 962\ \text{K}$$

56. *Result/Explanation:* The raw materials used in the synthesis of sulfuric acid are sulfur, water, oxygen, and catalyst (Pt or VO_5)

$$S_8(s) + 8\ O_2(g) \longrightarrow 8\ SO_2(g)$$

$$2\ SO_2(g) + O_2(g) \longrightarrow 2\ SO_3(g)$$

$$SO_3(g) + H_2SO_4(\ell) \longrightarrow H_2S_2O_7(\ell)$$

$$H_2S_2O_7(\ell) + H_2O(\ell) \longrightarrow 2\ H_2SO_4(aq)$$

58. *Result:* **See resonance structures below**

Analyze and Plan: Use the methods described in Chapter 6. The N atom in nitric acid has one double bond, so we can draw three resonance forms for this molecule. The third of these has the highest formal charges; hence, it is considered to be the worst structure of the three.

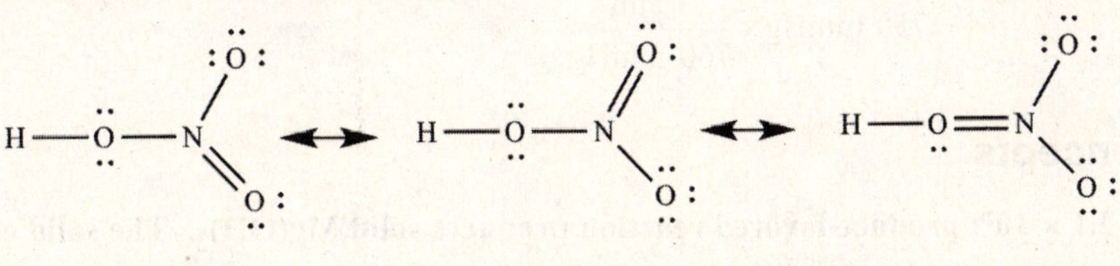

60. *Result:* $\mathbf{K_c = 2.7}$

Analyze and Plan: Use the methods described in Chapter 12.

Initially: (conc. SO_2) = 1.00 mol SO_2/1.00 L = 1.00 M

(conc. O_2) = 5.00 mol O_2/1.00 L = 5.00 M

Change:

77.8% SO_2 reacted, so (change conc. SO_2) = – 0.778 × (1.00 M) = – 0.778 M

Equilibrium: $[SO_2]$ = 1.00 M – 0.778 M = 0.22 M

	$2\ SO_2(g)$	+	$O_2(g)$	$\rightleftharpoons$	$2\ SO_3(g)$
initial conc. (M)	1.00		5.00		0
change as reaction occurs (M)	– 0.778 = – 2x		– x		+ 2x
equilibrium conc. (M)	0.22		5.00 – x		2x

We use SO_2 concentration change to find x: $0.778\ M = 2x = [SO_3]$

$$x = 0.389\ M$$

$$5.00 - x = 5.00\ M - 0.389\ M = 4.61\ M = [O_2]$$

$$K_c = \frac{[SO_3]^2}{[SO_2]^2[O_2]} = \frac{(2x)^2}{(1.00-2x)^2(5.00-x)} = \frac{(0.778)^2}{(0.22)^2(4.61)} = 2.7$$

62. *Result:* **3 L**

Analyze and Plan: Balance the equation and use the methods from Chapter 3 to determine how much of one reactant decomposes and the volume of a product gas using the ideal gas law, PV=nRT

Execute: Begin by writing a balanced equation for the decomposition of aqueous hydrogen peroxide (H_2O_2).

$$2\ H_2O_2(aq) \longrightarrow O_2(g) + 2\ H_2O(\ell)$$

To determine the amount of oxygen gas produced the amount of hydrogen peroxide must first be determined. We assume the dilute aqueous solution has a density the same as water, and that the percent provided in the question is a mass percent, the mass of hydrogen peroxide decomposed can be determined:

$$250.\ \text{mL soln} \times \frac{0.998\ \text{g soln}}{1\ \text{mL soln}} \times \frac{3\ \text{g}\ H_2O_2}{100\ \text{g soln}} = 7.5\ \text{g}\ H_2O_2 \approx 8\ \text{g}\ H_2O_2\ \text{decomposed}\ \textit{(keep 1 sig fig)}$$

The mass of hydrogen peroxide can then be converted to moles of oxygen using relevant molar masses and the stoichiometry of the balanced equation.

$$7.5\ \text{g}\ H_2O_2 \times \frac{1\ \text{mol}\ H_2O_2}{34.0146\ \text{g}\ H_2O_2} \times \frac{1\ \text{mol}\ O_2}{2\ \text{mol}\ H_2O_2} = 0.11\ \text{mol}\ O_2 \approx 0.1\ \text{mol}\ O_2\ \textit{(keep 1 sig fig)}$$

The moles of oxygen produced are then inserted into the ideal gas law, PV=nRT.

Remember to convert the temperature to Kelvin and the pressure to atmospheres and use the correct value of R.

$$V = \frac{(0.11\ \text{mol}\ O_2)\left(0.08206 \frac{\text{L}\cdot\text{atm}}{\text{mol}\cdot\text{K}}\right)(22 + 273.15)\ \text{K}}{750\ \text{mmHg} \times \frac{1\ \text{atm}}{760\ \text{mmHg}}} = 2.70\ \text{L} \approx 3\ \text{L}\ \textit{(round to 1 sig fig)}$$

Applying Concepts

64. *Result:* **K = 3.1 × 10^5; product-favored reaction produces solid $Mg(OH)_2$. The solid can be isolated after it settles.**

Analyze and Plan: Use the method described in Problem-Solving Example 12.2.

Execute:

$$Ca(OH)_2(s) \rightleftharpoons Ca^{2+}(aq) + 2\ OH^-(aq) \qquad K_{sp,1} = 5.5 \times 10^{-6}$$

$$Mg^{2+}(aq) + 2\ OH^-(aq) \rightleftharpoons Mg(OH)_2(s) \qquad K = 1/K_{sp,2} = 1/(1.8 \times 10^{-11})$$

$$Ca(OH)_2(s) + Mg^{2+}(aq) \rightleftharpoons Ca^{2+}(aq) + Mg(OH)_2(s) \qquad K_{net} = K_{sp,1} \times (1/K_{sp,2})$$

$$K_{net} = (5.5 \times 10^{-6}) \times [1/(1.8 \times 10^{-11})] = 3.1 \times 10^5$$

Putting sea water in the presence of $Ca(OH)_2$ will cause the product-favored precipitation of $Mg(OH)_2$. The solid can be isolated after it settles.

66. *Result:* 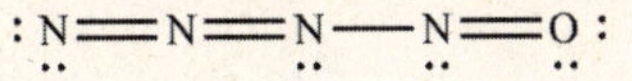 **(Other plausible structures are possible.)**

Analyze and Plan: Use method described in the solution to Question 6.15. Using 26 electrons, follow the octet rule on each atom.

Execute: The name helps confirm this structure: "Azide" is formed when three N atoms are bonded together, seen in this structure between the left-most three N atoms. "Nitrosyl" is the –NO functional group seen on the right side of the structure:

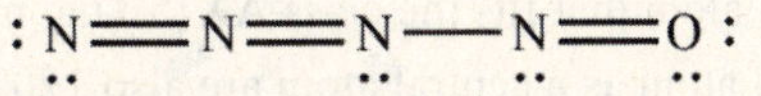

Other plausible resonance structures are also valid.

68. *Result:* **(a) See structures below (b) +140. kJ/mol with single and triple bond; +410. kJ/mol with two double bonds**

(a) *Analyze and Plan:* Use methods described in Chapter 6. Follow the octet rule.

Execute: HN_3 has 16 valence electrons:

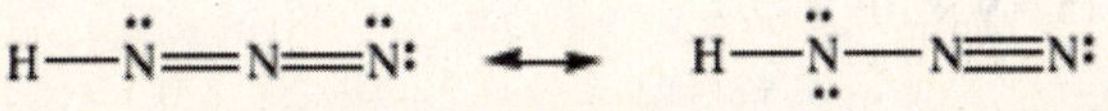

(b) *Analyze and Plan:* Adapt the method described in the solution to Question 6.34.

Formation is described as the production of one mol of a compound from its standard state elements. This compound is formed from $N_2(g)$ and $H_2(g)$. *Therefore, we must not use the given $\Delta_f H°$ for N or O.*

Execute: $\frac{3}{2}\,N_2(g) + \frac{1}{2}\,H_2(g) \longrightarrow HN_3$

Formation equation for the first structure:

$$\frac{3}{2}(\text{:N}\equiv\text{N:}) + \frac{1}{2}(\text{H—H}) \longrightarrow \text{H—}\ddot{\text{N}}\text{=N=}\ddot{\text{N}}\text{:}$$

$$\Delta_f H \cong \frac{3}{2}\,D_{N\equiv N} + \frac{1}{2}\,D_{H\text{–}H} - D_{N\text{–}H} - 2\,D_{N=N}$$

Use the average bond enthalpies (D) in Table 6.2:

$$\Delta_f H \cong \frac{3}{2}\,(946\ \text{kJ/mol}) + \frac{1}{2}\,(436\ \text{kJ/mol}) - (391\ \text{kJ/mol}) - 2(418\ \text{kJ/mol}) = 410.\ \text{kJ/mol}$$

Formation equation for the second structure:

$$\frac{3}{2}(\text{:N}\equiv\text{N:}) + \frac{1}{2}(\text{H—H}) \longrightarrow \text{H—}\ddot{\underset{..}{\text{N}}}\text{—N}\equiv\text{N:}$$

$$\Delta_f H \cong \frac{3}{2}\,D_{N\equiv N} + \frac{1}{2}\,D_{H\text{–}H} - D_{N\text{–}H} - D_{N\text{–}N} - D_{N\equiv N}$$

Use the average bond enthalpies (D) in Table 6.2:

$$\Delta_f H \cong \frac{3}{2}\,(946\ \text{kJ/mol}) + \frac{1}{2}\,(436\ \text{kJ/mol}) - (391\ \text{kJ/mol}) - (160\ \text{kJ/mol}) - (946\ \text{kJ/mol}) = 140\ \text{kJ/mol}$$

70. *Result:* **(a) $NO_2(g) + NO(g) \longrightarrow N_2O_3(g)$ (b) see resonance structures, below (c) O–N–O angle is 120°; N–N–O angle is slightly less than 120°.**

Analyze and Plan: Given the description of a reaction, write a chemical equation and Lewis structure with resonance structures and predict the bond angles. Use the techniques described in Chapters 3, 6, and 7.

Execute:

(a) $NO_2(g) + NO(g) \longrightarrow N_2O_3(g)$ 2 N and 3 O, on each side.

(b) There are 28 valence electrons. Follow the octet rule.

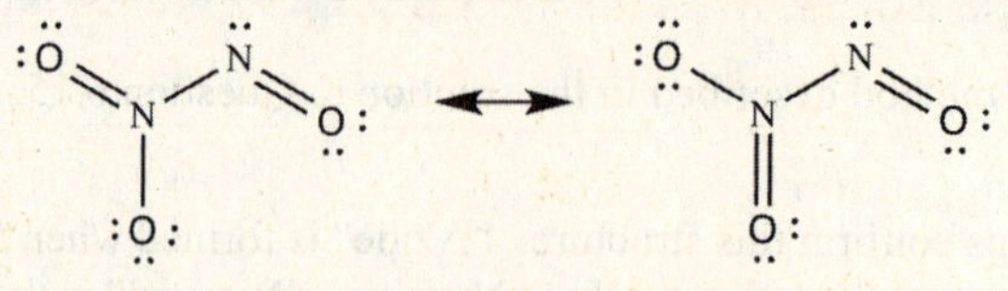

(Notice: A third structure was eliminated because it has two adjacent atoms have +1 formal charges.)

(c) The O–N–O bond is on an N atom that fits the class AX_3. That means the bond angle is 120°. The N–N–O angles involving that same N atom as a central atom are also 120°. The N–N–O angle on the second N atom is slightly less than 120°, due to the repulsion of the lone pair on the central N atom.

73. *Result:* **(a) 7 × 10² kJ (b) 3 V must be used to overcome the cell potential.**

Analyze and Plan: Adapt the methods described in Chapter 16 and Chapter 17 to get the reaction Gibbs energy and the cell potential from the reaction enthalpy and the reaction entropy.

Execute:

(a) $T = 600\ °C + 273.15 = 9 \times 10^2\ K$

$$\Delta_r G° = \Delta_r H° - T\Delta_r S° = (820\ kJ) - (9 \times 10^2\ K) \times (180\ J\ K^{-1}) \times \frac{1\ kJ}{1000\ J} = 7 \times 10^2\ kJ$$

(b) $$E°_{cell} = -\frac{\Delta G}{nF} = -\frac{7 \times 10^2\ kJ}{(2)(96485\ C)} \times \frac{1000\ J}{1\ kJ} \times \frac{1\ C \times 1\ V}{1\ J} = -3\ V$$

3 volts must be used to overcome this cell potential.

75. *Result:* **(a) three; 140°C, 10³ atm; 95°C,10⁻⁵ atm; 10⁻⁴ atm and 120°C (b) (i) Solid (rhombic) (ii) liquid (iii) solid (monoclinic) (iv) gaseous**

Explanation:

(a) The triple point of a phase diagram is a point at which more than two phases are in equilibrium. Three point satisfies this definition: points, B, C, and E on the diagram. The approximate pressure and temperatures are 140°C, 10^3 atm; 95°C,10^{-5} atm; 10^{-4} atm and 120°C.

Notice that the terms monoclinic and rhombic are specific unit cell configurations used to describe how the sulfur atoms pack in a three dimensional array in a solid. We learned about the cubic unit cells (primitive cubic, fcc, and bcc) that are adapted by many crystalline solids in Section 9.6. Rhombic and monoclinic are two possible variations.

(b) To determine the phase described by the given pressure and temperature, we simply draw a horizontal line intersecting the given pressure and a vertical line intersecting the given temperature. The area in which these two lines intersect identifies the phase of the material.

(i) Solid (rhombic) sulfur

(ii) Liquid sulfur

(iii) Solid (monoclinic) sulfur

(iv) Gaseous sulfur

77. *Result:* **(a) $4\ H_3O^+ + MnO_2 + 2\ Br^- \longrightarrow Mn^{2+} + 6\ H_2O + Br_2$ (b) 0.0813 mol (c) 3.54 g MnO_2**

Analyze and Plan: Balance an equation using the half-reaction method of balancing redox reactions introduced in Section 17.2 then use methods from Chapter 3 to determine the amount of reactant consumed and product produced.

Execute:

(a) Since (several) elements in the reaction change oxidation numbers, this is a redox reaction. The provides the most straightforward route to determining the balanced overall reaction.

$$2\ e^- + 4\ H_3O^+(aq) + MnO_2(s) \longrightarrow Mn^{2+}(aq) + 6\ H_2O(\ell)$$

$$2\ Br^-(aq) \longrightarrow Br_2(\ell) + 2\ e^-$$

$$4\ H_3O^+(aq) + MnO_2(s) + 2\ Br^-(aq) \longrightarrow Mn^{2+}(aq) + 6\ H_2O(\ell) + Br_2(\ell)$$

(b) Use the stoichiometry of the balanced equation to determine the amount of bromide ions consumed.

$$6.50\ g\ Br_2 \times \frac{1\ mol\ Br_2}{159.808\ g\ Br_2} \times \frac{2\ mol\ Br^-}{1\ mol\ Br_2} = 0.0813\ mol\ Br^-$$

(c) Use the stoichiometry to determine the mass of manganese(II) oxide needed to produce 6.50 g of bromine.

$$6.50\ g\ Br_2 \times \frac{1\ mol\ Br_2}{159.808\ g\ Br_2} \times \frac{1\ mol\ MnO_2}{1\ mol\ Br_2} \times \frac{86.9368\ g\ MnO_2}{1\ mol\ MnO_2} = 3.54\ g\ MnO_2$$

More Challenging Questions

85. *Result:* **6.3×10^5 J/mol**

Analyze and Plan: The Clausius-Clapeyron equation is discussed in Section 9-3a. Two pressure-temperature points can be roughly estimated from the phase diagram vapor-pressure curve (line BE) provided in Question 75.

Execute:

$$P_1 = 0.0254\ mm\ Hg$$

$$T_1 = 20.°C + 273.15 = 293\ K$$

$$P_2 = 0.133\ mm\ Hg$$

$$T_2 = 40.°C + 273.15 = 313\ K$$

Inserting these values into the two-set form of the Clausius-Clapeyron equation provides the final answer as shown in the Problem-Solving Example 11.1:

$$\ln\left(\frac{0.133\ mmHg}{0.0254\ mmHg}\right) = \frac{-\Delta_r H}{8.31\ JK^{-1}mol^{-1}}\left(\frac{1}{313\ K} - \frac{1}{293\ K}\right)$$

$$\ln(5.24) = \frac{-\Delta_r H}{8.31\ JK^{-1}mol^{-1}}\left(-0.00022\ K^{-1}\right)$$

$$\Delta_r H = \frac{\left(8.31\ J\,mol^{-1}\right)\ln(5.24)}{0.00022} = 6.3 \times 10^5\ J/mol$$

$$\Delta_r H = 6.3 \times 10^5\ J/mol$$

88. *Result:* **(a)** $P_4(s) + 5\ O_2(g) \longrightarrow P_4O_{10}(s)$ **(b) 2.25**
(c) $3\ Ca(NO_3)_2(aq) + 2\ H_3PO_4(aq) \longrightarrow Ca_3(PO_4)_2(s) + 6\ HNO_3(aq)$; **25.0 g $Ca_3(PO_4)_2$** **(d) 5.42 L**

Analyze and Plan: Write appropriate chemical equations for the formation of a nonmetal oxide and its reaction with water to make an acid, then calculate the molarity and pH of the acid solution, the mass of precipitate when a metal ion is added, and the volume of a gas produced when it reacts with a metal. Apply techniques from Chapter 14, 15, 3, and 9.

Execute:

(a) When a nonmetal burns in excess oxygen it typically attains its highest possible oxidation state.

$$P_4(s) + 5\,O_2(g) \longrightarrow 2\,P_2O_5(s) \quad \text{or}$$

$$P_4(s) + 5\,O_2(g) \longrightarrow P_4O_{10}(s)$$

(b) All nonmetal oxides are acidic. In many cases this means they will react with water to form a common aqueous acid.

$$P_2O_5(s) + 3\,H_2O(\ell) \longrightarrow 2\,H_3PO_4(aq) \quad \text{or}$$

$$P_4O_{10}(s) + 6\,H_2O(\ell) \longrightarrow 4\,H_3PO_4(aq)$$

First determine the molarity of the acid produced using the balanced equations provide in (a).

$$\frac{5.00\text{ g }P_4 \times \dfrac{1\text{ mol }P_4}{123.8952\text{ g }P_4} \times \dfrac{2\text{ mol }P_2O_5}{1\text{ mol }P_4} \times \dfrac{2\text{ mol }H_3PO_4}{1\text{ mol }P_2O_5}}{0.250\text{ L}} = 0.646\text{ M }H_3PO_4$$

$$\frac{5.00\text{ g }P_4 \times \dfrac{1\text{ mol }P_4}{123.8952\text{ g }P_4} \times \dfrac{1\text{ mol }P_4O_{10}}{1\text{ mol }P_4} \times \dfrac{4\text{ mol }H_3PO_4}{1\text{ mol }P_4O_{10}}}{0.250\text{ L}} = 0.646\text{ M }H_3PO_4$$

Since phosphoric acid is a weak acid, the pH will be affected by the first ionization of H_3PO_4 to form $H_2PO_4^-$ (other equilibria can be ignored).

The equation for the equilibrium and the equilibrium expression are:

$$H_3PO_4(aq) + H_2O(\ell) \rightleftharpoons H_3O^+(aq) + H_2PO_4^-(aq) \qquad K_{a1} = \frac{[H_3O^+][H_2PO_4^-]}{[H_3PO_4]}$$

As the reactants react, the concentrations of the products increase stoichiometrically, until they reach equilibrium concentrations.

	$H_3PO_4(aq)$	$H_3O^+(aq)$	$H_2PO_4^-(aq)$
initial conc. (M)	0.646	0	0
change as the reaction occurs (M)	– x	+ x	+ x
equilibrium conc. (M)	0.646 – x	x	x

K_{a1}, provided in the text (Appendix F) is 7.2×10^{-3}.

At equilibrium: $$K_{a1} = \frac{(x)(x)}{(0.010 - x)} = 7.2 \times 10^{-3}$$

$$x^2 = (7.2 \times 10^{-3})(0.010 - x)$$

$$x^2 + (7.2 \times 10^{-3})x - 7.2 \times 10^{-5} = 0$$

Using the quadratic equation, determine the value of x, then use that to determine the pH:

$$x = 0.0056\text{ M} = [H_3O^+]$$

$$pH = -\log[H_3O^+] = -\log(0.0056) = 2.25$$

(b) Precipitation of calcium phosphate occurs in this reaction, as shown in balanced form below. The limiting reagent, as indicated in the question, is phosphoric acid.

$$3\,Ca(NO_3)_2(aq) + 2\,H_3PO_4(aq) \longrightarrow Ca_3(PO_4)_2(s) + 6\,HNO_3(aq)$$

$$0.1614\text{ mol }H_3PO_4 \times \frac{1\text{ mol }Ca_3(PO_4)_2}{2\text{ mol }H_3PO_4} \times \frac{310.1768\text{ g }Ca_3(PO_4)_2}{1\text{ mol }Ca_3(PO_4)_2} = 25.0\text{ g }Ca_3(PO_4)_2$$

(c) Zinc, an active metal, reacts with acid to form hydrogen gas and a metal salt (ionic compound). Notice that all (active) metals react in a similar manner.

$$Zn(s) + 2\ HNO_3(aq) \longrightarrow Zn(NO_3)_2(aq) + 6\ H_2(g)$$

$$0.1614 \text{ mol } H_3PO_4 \times \frac{6 \text{ mol } HNO_3}{2 \text{ mol } H_3PO_4} \times \frac{1 \text{ mol } H_2}{2 \text{ mol } HNO_3} = 0.242 \text{ mol } H_2 \text{ generated}$$

Use the molar volume of a gas at STP (1 atm and 0°C) to determine the volume of hydrogen gas.

$$0.242 \text{ mol } H_2 \times \frac{22.414 \text{ L}}{1 \text{ mol } H_2} = 5.42 \text{ L}$$

90. *Result:* **(a) Al and N are oxidized; Cl is reduced (b) –2674 kJ/mol**

Analyze and Plan: Determine which elements are oxidized and reduced and the reaction enthalpy for a redox reaction.

Execute:

(a) Aluminum metal is oxidized (from Ox. # = zero to Al^{+3}).

Nitrogen is oxidized (from Ox. # = +3, in the ammonium ion of NH_4ClO_4 to Ox. # = +2, in NO).

Chlorine is reduced (from Ox. # = +7 in the perchlorate ion of NH_4ClO_4 to Ox. # = –1 in chloride Cl^-).

(b) The enthalpy change is the difference between the sum of the enthalpies of formation of the products and the sum of the enthalpies of formation of the reactants. Enthalpies of formation are found in Appendix J.

$$\Delta_r H° = \sum\left[(\text{coefficient of product}) \times \Delta H_f^\circ(\text{product})\right] - \sum\left[(\text{coefficient of reactant}) \times \Delta H_f^\circ(\text{reactant})\right]$$

$$= [\Delta_f H°\{Al_2O_3(s)\} + \Delta_f H°\{AlCl_3(s)\} + 6\ \Delta_f H°\{\{H_2O(g)\} + 3\ \Delta_f H°\{(NO(g)\}]$$
$$- [3\ \Delta_f H°\{NH_4ClO_4(s)\} + 3\ \Delta_f H°\{Al(s)\}]$$

$$= [(-1657.7 \text{ kJ/mol}) + (-704.2 \text{ kJ/mol}) + 6(-241.818 \text{ kJ/mol}) + 3(90.25 \text{ kJ/mol})]$$
$$- [3(-295 \text{ kJ/mol}) + 3(0 \text{ kJ/mol})] = -2674 \text{ kJ/mol}$$

92. *Result:* **–2708 kJ/mol**

Analyze and Plan: Adapt the methods used in the solution to Question 90(b). We will assume that the form of carbon in the reaction is graphite.

Execute: First, we will use the enthalpy of sublimation of phosphorus to determine the enthalpy of formation of $P_4(g)$. The standard enthalpy of formation of white phosphorus, $P_4(s) = 0$, according to Appendix J.

$$\Delta_r H° = \sum\left[(\text{moles of product}) \times \Delta H_f^\circ(\text{product})\right] - \sum\left[(\text{moles of reactant}) \times \Delta H_f^\circ(\text{reactant})\right]$$

Sublimation: $P_4(s) \longrightarrow P_4(g)$

$\Delta_r H° = 13.06 \text{ kJ/mol} = \Delta_f H°\{P_4(g)\} - \Delta_f H°\{P_4(s)\} = \Delta_f H°\{P_4(s)\} - 0$

$\Delta_f H°\{P_4(g)\} = 13.06 \text{ kJ/mol}$

$$\Delta_r H° = [6\ \Delta_f H°\{CaSiO_3(s)\} + 10\ \Delta_f H°\{CO(g)\} + \Delta_f H°\{P_4(g)\}]$$
$$- [2\ \Delta_f H°\{Ca_3(PO_4)_2(s)\} + 6\ \Delta_f H°\{SiO_2(s)\} + 3\ \Delta_f H°\{C(s)\}]$$

$$\Delta_r H° = [6\ \Delta_f H°\{CaSiO_3(s)\} + 10(-110.525 \text{ kJ/mol}) + (13.06 \text{ kJ/mol})]$$
$$- [2(-4138 \text{ kJ/mol}) + 6(-910.94 \text{ kJ/mol}) + 10(0 \text{ kJ/mol})] = -3060 \text{ kJ/mol}$$

Solve for $\Delta_f H°\{CaSiO_3(s)\}$:

$$6\ \Delta_f H°(CaSiO_3) = -3060 \text{ kJ} + (1105.25 \text{ kJ}) - (13.06 \text{ kJ}) - (8276 \text{ kJ}) - (5465.64 \text{ kJ})$$

$$\Delta_f H°(CaSiO_3) = -2618 \text{ kJ/mol}$$

94. *Result:* **$K_p = 1.985$**

Analyze and Plan: Use the methods described in Section 16.7 and Equation 16.9 to solve for the equilibrium constant. The equilibrium constant requested in the question is for the reverse of the equation provided. On reversing a reaction's equation the sign of the Gibbs free energy changes.

Execute: 25°C + 273.15 = 289.15 K.

$$\Delta G° = -RT\ln K_p$$

$$-1700 \text{J/mol} = -(8.314 \text{ J mol}^{-1}\text{K}^{-1})(298.15\text{K}) \ln K$$

$$K_p = 1.985$$

96. *Result:* **$\Delta H_{Lattice}$ = 2961 kJ; Mg^{2+} is smaller and has a greater charge density than Sr^{2+}, so the Mg^{2+} ions and the F^- ions are closer together and the ionic bond is much stronger.**

Analyze and Plan: Lattice enthalpy is defined as the energy necessary to convert a solid salt to its constituent ions in the gas phase, as illustrated for strontium fluoride below.

$$SrF_2(s) \longrightarrow Sr^{2+} + 2\,F^- \qquad \Delta H_{Lattice} = 2496 \text{ kJ}$$

An application of Hess' Law, called a Born-Haber cycle, can be used to determine the lattice enthalpy of an ionic compound. The equations needed are provided in the question. As always, pay close attention to the reaction stoichiometry and sign.

Execute: For the enthalpies listed below the abbreviations are as follows:

sub = sublimation (conversion of a solid to the gas phase);

IE = ionization energy (the energy required to remove an electron)—notice in this case the magnesium metal requires two separate ionizations;

BE = bond enthalpy—the energy required to break a bond between two elements;

EA = electron affinity;

f = formation

Notice in the cycle below, the equation used is the opposite of a formation reaction, thus the sign of the enthalpy must be reversed.

$$Mg(s) \longrightarrow Mg(g) \qquad \Delta_{sub}H = 146 \text{ kJ}$$

$$Mg(g) \longrightarrow Mg^+(g) + e^- \qquad \Delta_{IE_1}H = 738 \text{ kJ}$$

$$Mg^+(g) \longrightarrow Mg^{2+}(g) + e^- \qquad \Delta_{IE_2}H = 1451 \text{ kJ}$$

$$F_2(g) \longrightarrow 2\,F(g) \qquad \Delta_{BE}H = 158 \text{ kJ}$$

$$2\,F(g) + 2\,e^- \longrightarrow 2\,F^- \qquad 2 \times \Delta H_{EA} = -656 \text{ kJ}$$

$$+\ MgF_2(s) \longrightarrow Mg(s) + F_2(g) \qquad -1 \times \Delta_f H = 1124 \text{ kJ}$$

$$MgF_2(s) \longrightarrow Mg^{2+} + 2\,F^- \qquad \Delta_{lattice}H = 2961 \text{ kJ}$$

Comparison of strontium fluoride and magnesium fluoride lattice enthalpies: Ionic interactions dominate the bonding in most salts (ionic compounds). This effect is further aided by large differences in electronegativity between the cation and anions that form a given salt. In general, the greater the electronegativity difference between the elements in the salt, the greater the strength of the ionic bond.

For cations of the same charge, as is the case for Mg^{2+} and Sr^{2+}, electrostatic attractions to F^- are affected by the distance between the anion and the cation (electrostatic attractions fall off as $1/r^2$ where r = the radius of the charged particle). The ionic radii for Mg^{2+} and Sr^{2+} are 0.72 and 1.16 angstroms respectively. Because Mg^{2+} is significantly smaller, it can more closely approach the F^- resulting in a stronger interaction. In addition, because of its smaller size (and fewer electrons) the charge of the magnesium ion is distributed over a smaller volume in comparison to the strontium ion, resulting in a greater charge density. The combination of these two factors, both dependent on the ion size, result in a stronger ionic bond and stronger lattice interactions for Mg^{2+} relative to Sr^{2+}.

98. *Result:* $\mathbf{B_6H_{12}}$

Analyze and Plan: The mass of boron and hydrogen in a 1 mol of sample (76.7 g) can be determined by multiplying percent by the molecular formula. Converting these masses to moles give rations for the molecular formula.

Execute:
$$\frac{76.7 \text{ g compound}}{1 \text{ mol compound}} \times \frac{84.2 \text{ g B}}{100 \text{ g compound}} \times \frac{1 \text{ mol B}}{10.811 \text{ g B}} = 6.00 \text{ mol B per mol}$$

$$\frac{76.7 \text{ g compound}}{1 \text{ mol compound}} \times \frac{15.7 \text{ g H}}{100 \text{ g compound}} \times \frac{1 \text{ mol H}}{1.0079 \text{ g H}} = 12.0 \text{ mol H per mol}$$

This gives a whole number ratio per mol of compound: 6.00 mol B : 12.0 mol H

Molecular Formula = B_6H_{12}

Chapter 20: Chemistry of Selected Transition Elements and Coordination Compounds

Solutions for Red-Numbered Questions for Review and Thought

Topical Questions

Transition (d-block) Elements (Section 20-1)

13. *Result:* **(a) Ag [Kr]$4d^{10}5s^1$ and Ag^+ [Kr]$4d^{10}$ (b) Au [Xe]$4f^{14}5d^{10}6s^1$, Au^+ [Xe]$4f^{14}5d^{10}$, and Au^{3+} [Xe]$4f^{14}5d^8$**

Analyze and Plan: Follow the method described in Sections 5-7, 5-8 and 20-1.

Execute:

(a) Silver is Ag, in Group 1B. The most common oxidation states of silver are zero and +1.

The $_{47}Ag$ electron configuration: [Kr]$4d^{10}5s^1$ or $1s^22s^22p^63s^23p^63d^{10}4s^24p^64d^{10}5s^1$

The $_{47}Ag^+$ electron configuration: [Kr]$4d^{10}$

(b) Gold is Au, in Group IB. The most common oxidation state of gold is zero. The $_{79}Au$ electron configuration:

[Xe]$4f^{14}5d^{10}6s^1$ or $1s^22s^22p^63s^23p^63d^{10}4s^24p^64d^{10}4f^{14}5s^25p^65d^{10}6s^1$

Other oxidation states of gold are +1 and +3.

The $_{79}Au^+$ electron configuration: [Xe]$4f^{14}5d^{10}$

The $_{79}Au^{3+}$ electron configuration: [Xe]$4f^{14}5d^8$

15. *Result:* **Cr^{2+} and Cr^{3+}**

Analyze, Plan, and Execute:

Chromium is Cr. The neutral Cr atom has the following orbital diagram, with six unpaired electrons:

	3d	4s
[Ar]	[↑] [↑] [↑] [↑] [↑]	[↑]

The two more common oxidation states of chromium are Cr^{2+} and Cr^{3+}:

		3d
Cr^{2+}	[Ar]	[↑] [↑] [↑] [↑] [↑]

		3d
Cr^{3+}	[Ar]	[↑] [↑] [↑] [↑] [↑]

Both of these ions are paramagnetic.

17. *Result:* **$Cr_2O_7^{2-}$ (in acid) > Cr^{3+} > Cr^{2+}, the species with a more positive oxidation state of Cr has the greater tendency to be reduced.**

Explanation: The species with the more positive oxidation state of Cr has the greater tendency to be reduced, hence acting as a stronger oxidizing agent. $Cr_2O_7^{2-}$(in acid) has Ox. # Cr = +6, Cr^{3+} has Ox. # Cr = +3, and Cr^{2+} has Ox. # Cr = +2. So, $Cr_2O_7^{2-}$(in acid) is the best oxidizing agent, followed by Cr^{3+}, then Cr^{2+}.

19. *Result:* **(a) $Fe_2O_3 + 3\,CO \longrightarrow 2\,Fe + 3\,CO_2$ (b) $Fe(s) + 2\,H^+(aq) \longrightarrow Fe^{2+}(aq) + H_2(g)$ or $2\,Fe + 6\,H^+ \longrightarrow 2\,Fe^{3+} + 3\,H_2$**

Explanation: Roasting is described in Section 20-3a:

(a) $$Fe_2O_3(s) + 3\,CO(g) \longrightarrow 2\,Fe(s) + 3\,CO_2(g)$$

(b) Balancing the first of these equations can be done using standard balancing methods. The second may require the use of redox balancing methods shown in Appendix F Problem-Solving Example F.1

If Fe^{2+} is produced: $Fe(s) + 2\,H^+(aq) \longrightarrow Fe^{2+}(aq) + H_2(g)$

If Fe^{3+} is produced: $Fe(s) \longrightarrow Fe^{3+}(aq) + 3\,e^-$

$+ \quad 2\,H^+(aq) \longrightarrow H_2(g)$

$$2\,Fe(s) + 6\,H^+(aq) \longrightarrow 2\,Fe^{3+}(aq) + 3\,H_2(g)$$

Fe^{3+} is the more likely product.

21. *Result:* **$3\,NO_3^-(aq) + 6\,H^+(aq) + Fe(aq) \longrightarrow Fe^{3+}(aq) + 3\,NO_2(g) + 3\,H_2O(\ell)$**

Analyze and Plan: Use standard redox balancing methods shown in Appendix F Problem-Solving Example F.1

Execute:

$$3\,NO_3^-(aq) + 6\,H^+(aq) + 3\,e^- \longrightarrow 3\,NO_2(g) + 3\,H_2O(\ell)$$

$$+ \quad Fe(aq) \longrightarrow Fe^{3+}(aq) + 3\,e^-$$

$$3\,NO_3^-(aq) + 6\,H^+(aq) + Fe(aq) \longrightarrow Fe^{3+}(aq) + 3\,NO_2(g) + 3\,H_2O(\ell)$$

(Check: 3 N, 9 O, 6 H, 1 Fe, +3 net charge)

23. *Result:* **(a) Ox. # V = +5 (b) Ox. # Cr = +6 (c) Ox. # Mn = +4 (d) Ox. # Os = +8**

Analyze and Plan: Given the several formulas, determine the oxidation state of the transition metal in each. Use the rules given in Section 3-5c. Several elements in compounds have predictable oxidation numbers (Rules 1-3). The oxidation number (which identifies the oxidation state) of the transition metal can be determined using the sum rule (Rule 4). The term "oxidation number" is abbreviated below as "Ox. #".

Execute:

(a) V_2O_5 is a binary compound containing oxygen. Rule 3 gives us Ox. # O = –2. For a compound, the sum of the oxidation numbers is equal to zero.

$$0 = 2\,(\text{Ox. \# V}) + 5\,(-2)$$

Therefore, Ox. # V = +5. The transition metal vanadium is in the +5 oxidation state.

(b) $K_2Cr_2O_7$ contains potassium ion, K^+, and dichromate ion, $Cr_2O_7^{2-}$, which contains oxygen. Rule 2 gives us the Ox. # K = +1. Rule 3 gives us Ox. # O = –2. We use the sum rule to find the Ox. # Cr. For a compound, the sum of the oxidation numbers is equal to zero.

$$0 = 2(+1) + 2(\text{Ox. \# Cr}) + 7(-2)$$

Therefore, 2 (Ox. # Cr) = +12, so Ox. # Cr = +6. The transition metal chromium is in the +6 oxidation state.

(c) MnO_2 is a neutral compound. Rule 3 gives us Ox. # O = –2. We use the sum rule to find the Ox. # Mn. For a neutral compound, the sum of the oxidation numbers is equal to zero.

$$0 = 1(\text{Ox. \# Mn}) + 2(-2)$$

$$0 = \text{Ox. \# Mn} - 4$$

Therefore, Ox. # Mn = +4. The transition metal manganese is in the +4 oxidation state.

(d) OsO_2 is a neutral compound. Rule 3 gives us Ox. # O = –2. We use the sum rule to find the Ox. # Os. For a neutral compound, the sum of the oxidation numbers is equal to zero.

$$0 = (\text{Ox. \# Os}) + 4(-2)$$

$$0 = \text{Ox. \# Os} - 8$$

Therefore, Ox. # Os = +8. The transition metal osmium is in the +8 oxidation state.

☑ *Reasonable Result Check:* We expect transition metals to have positive oxidation states.

25. *Result/Explanation:* Atomic size depends on the largest orbitals. In main-group elements, the electrons are filling the largest orbitals. In contrast, all the transition elements have a fixed number of electrons in large (*n*)s-orbitals, and they differ only by the number of electrons in the (*n-1*) d-shell, which has smaller orbitals than the (*n*)s-orbitals. (See Section 20-1b)

27. *Result:* **(a) Ag (b) Ti (c) Hg**

Analyze and Plan: Given pairs of atoms, predict which is larger. Use the trends described in Question 25 and Section 20-1b

Execute:

(a) Cu and Ag are in the same group, Group 1B. Cu is in Period 4 and Ag is in Period 5. So, Ag is larger.

(b) Ti and Cr are in the same period, Period 4. The general trend for most of Period 4 is a decrease in atomic radius from left to right. Because Ti is to the left of Cr, Ti is larger.

(c) W and Hg are in the same period, Period 6. The first and last transition elements (d^1 and d^{10}) in a period are the largest. Because Hg is at the end (to the right) of Period 6, it is larger than W. Hg is larger.

Iron and Steel (Section 20-2)

29. *Result/Explanation:* Iron is produced by reduction because it is found in nature in compounds containing Fe^{2+} and Fe^{3+}, such as FeO and Fe_2O_3. To make steel, oxidation is used to remove carbon impurities from iron. (See Section 20-2)

31. *Result/Explanation:* The basic oxygen process has pure oxygen blown though a water-cooled tube inserted in the surface of molten, impure iron. At 1900°C, dissolved carbon reacts with oxygen to form gaseous CO and CO_2, which are vented.

Copper (Section 20-3)

33. *Result/Explanation:* These two metals differ by which metal is alloyed with copper. Brass is copper alloyed with zinc (Zn). Bronze is copper alloyed with tin (Sn).

35. *Result/Explanation:* In electro-refining, pure copper metal is set up as the cathode and blister copper metal is set up as the anode. The electrolyte solution contains $CuSO_4$ and H_2SO_4. Copper from the blister copper is oxidized into aqueous Cu^{2+} ions at the anode. The Cu^{2+} ion migrants to the pure Cu cathode, where they are reduced to pure copper metal. (See Figure 20.7)

Silver, Gold, and Chromium (Section 20-4, 20-5)

37. *Result/Explanation:* As described in Section 20-4, Ag reacts with concentrated nitric acid to form Ag+. Au does not react with concentrated nitric acid unless HCl is also present.

39. *Result/Explanation:* Silver and gold are soft metals. Their hardness and durability is increased by alloying them with other metals.

41. *Result/Explanation:* Adapt the method described in Question 23.

(a) $FeCr_2O_4$ contains iron(II) ion, Fe^{2+}, and the ion, $Cr_2O_4{}^{2-}$, which contains oxygen. Rule 2 gives us the Ox. # Fe = +2. Rule 3 gives us Ox. # O = –2. We use the sum rule to find the Ox. # Cr. For a compound, the sum of the oxidation numbers is equal to zero.

$$0 = (+2) + 2(\text{Ox. \# Cr}) + 4(-2)$$

Therefore, 2 Ox. # Cr = +6, and Ox. # Cr = +3. Chromium is in the +3 oxidation state.

(b) Cr_2O_5 is a binary compound containing oxygen. Rule 3 gives us Ox. # O = –2. For a compound, the sum of the oxidation numbers is equal to zero.

$$0 = 2\,(\text{Ox. \# Cr}) + 5\,(-2)$$

Therefore, Ox. # Cr = +5. Chromium is in the +5 oxidation state.

(c) $K_2[CrF_6]$ contains potassium ion, K^+, and a complex ion, CrF_6^{2-}, which contains fluorine. Rule 2 gives us the Ox. # K = +1. Rule 3 gives us Ox. # F = –1. We use the sum rule to find the Ox. # Cr. For a compound, the sum of the oxidation numbers is equal to zero.

$$0 = (+2) + (\text{Ox. \# Cr}) + 6(-1)$$

Therefore, Ox. # Cr = +4. Chromium is in the +4 oxidation state.

(d) $[Cr(en)_3]Cl_2$ contains a complex ion, $Cr(en)_3^{2+}$ and chloride ion, Cl^-. The ethylene diamine ligand (en) is neutral. Rule 2 gives us the Ox. # Cl = –1. We use the sum rule to find the Ox. # Cr. For a compound, the sum of the oxidation numbers is equal to zero.

$$0 = (\text{Ox. \# Cr}) + 3(0) + 2(-1)$$

Therefore, Ox. # Cr = +2. Chromium is in the +2 oxidation state.

☑ *Reasonable Result Check:* We expect transition metals to have positive oxidation states.

Coordination Compounds (Section 20-6)

43. *Result:* **(a) $[Cr(NH_3)_2(H_2O)_3(OH)]^{2+}$; 2+ (b) Counter ion would be a doubly charged anion, like SO_4^{2-}.**

Explanation: Cr^{3+} is bonded to two NH_3 ligands, three H_2O ligands, and one OH^-.

(a) The complex ion has a formula that looks like this: $[Cr(NH_3)_2(H_2O)_3(OH)]^{2+}$. The net charge is +2.

(b) The complex ion formed is a cation, so it will need a counter ion that is a doubly charged anion, such as SO_4^{2-}.

45. *Result:* **(a) $C_2O_4^{2-}$ ligands with –2 charge; Cl^- ligands with –1 charge (b) +3 charge (c) $[Co(NH_3)_4Cl_2]^+$ with +1 charge**

Explanation:

(a) The complex ion $[Co(C_2O_4)_2Cl_2]^{3-}$ has two $C_2O_4^{2-}$ ligands, each with 2– charge and two Cl^- ligands, each with 1– charge.

(b) The net charge is 3–. The sum of the individual charges must add up to this number:

$$-3 = (\text{cobalt charge}) + 2(-2) + 2(-1)$$

The cobalt is in the form of Co^{3+} with a charge of 3+.

(c) $C_2O_4^{2-}$ is a bidentate ligand and NH_3 is a monodentate, so two $C_2O_4^{2-}$ must be replaced by four NH_3. Replacing two –2 ions with four molecules, reduces the net charge of the ion by 4 also. Therefore, the new complex ion has a formula that looks like this: $[Co(NH_3)_4Cl_2]^+$. It has a 1+ charge.

48. *Result:* **For $Na_3[IrCl_6]$: (a) Six Cl^- (b) Ir^{3+} with +3 charge (c) $[IrCl_6]^{3-}$ with –3 charge (d) Na^+**
For $[Mo(CO)_4]Br_2$: (a) Four CO (b) Mo^{2+}; +2 charge (c) $[Mo(CO)_4]^{2+}$ with 2+ charge (d) Br^-

Explanation:

Consider $Na_3[IrCl_6]$

(a) The complex ion has six Cl^- ions as ligands.

(b) $0 = +3(+1) + (\text{iridium charge}) + 6(-1)$ The iridium is in the form of Ir^{3+} with a 3+ charge.

(c) The formula of the complex ion is $[IrCl_6]^{3-}$ with a 3– charge.

(d) The ions that are not in the complex ion are the three Na^+ cations.

Consider $[Mo(CO)_4]Br_2$

(a) The complex ion has four CO molecules as ligands.

(b) 0 = (molybdenum charge) + 4(0) + 2(−1) The molybdenum is in the form of Mo^{2+} with a +2 charge.

(c) The formula of the complex ion is $[Mo(CO)_4]^{2+}$ with a 2+ charge.

(d) The ions that are not in the complex ion are the two Br^- anions.

49. *Result:* **(a) 4 (b) 4**

Analyze and Plan: Monodentates need one coordination site. Bidentates need two coordination sites. Thus, we need to add the number of monodentate ligands to 2(number of bidentate ligands) to find the coordination number.

Execute:

(a) In $[Pt(en)_2]^{2+}$, we have two bidentate ligands (2 en), so the coordination number is 2(2) = 4.

(b) In $[Cu(ox)_2]^{2-}$, we have two bidentate ligands (2 ox), so the coordination number is 2(2) = 4.

52. *Result:* **(a) $[Pt(NH_3)_2Br_2]$ (b) $[Pt(en)(NO_2)_2]$ (c) $[Pt(NH_3)_2BrCl]$**

Analyze and Plan: Use the method described in the solutions to Questions 44 to 51.

(a) Br^- and NH_3 are monodentate ligands, making $[Pt(NH_3)_2Br_2]$.

(b) The ligand en is a bidentate ligand. NO_2^- is a monodentate ligand. The complex is $[Pt(en)(NO_2)_2]$.

(c) Br^-, Cl^-, and NH_3 are monodentate ligands, making $[Pt(NH_3)_2BrCl]$.

54. *Result:* **(a) 4 (b) 4 (c) 6 (d) 6**

Explanation: Use the method described in the solution to Question 49.

(a) In $[FeCl_4]^-$, we have four monodentate ligands (4 Cl^-), so the coordination number is 4(1) = 4.

(b) In $[PtBr_4]^{2-}$, we have four monodentate ligands (4 Br^-), so the coordination number is 4(1) = 4.

(c) In $[Mn(en)_3]^{2+}$, we have three bidentate ligands (3 en), so the coordination number is 3(2) = 6.

(d) In $[Cr(NH_3)_5H_2O]^{3+}$, we have six monodentate ligands (5 NH_3 and 1 H_2O), so the coordination number is 6(1) = 6.

56. *Result:* **(a) monodentate (b) tetradentate (c) tridentate (d) monodentate**

Explanation:

(a) $(CH_3)_3P$ has only one lone pair of electrons on the P atom, so it is a monodentate ligand.

(b) (^-OOC)–CH_2–$N(COO^-)_2$ has lone pairs of electrons on every O atom as well as the N atom, so in theory it could be heptadentate. In practice, only one of the two O atoms on the carboxyl anion groups and the N atom probably have simultaneous access to the metal atom, due to geometric constraints, so this ligand will usually be tetradentate.

(c) H_2N–CH_2–CH_2–NH–CH_2–CH_2–NH_2 has lone pairs of electrons on every N atom, so is a tridentate ligand.

(d) H_2O has electron pairs on only one atom, so is a monodentate ligand.

58. *Result:* **$FeCl_3$ (Other correct answers are possible.)**

Explanation: We need a compound that is composed of a 3+ cation, analogous to the $[Rh(en)]^{3+}$ ion, combined with three simple −1 anions. One example is $FeCl_3$ made with Fe^{3+} and three Cl^- ions.

Naming Complex Ions and Coordination Compounds (Section 20-6)

59. *Result:* **(a) $K[Co(H_2O)_2(ox)_2]$ (b) $[Cr(NH_3)_2(H_2O)_3OH]NO_3$ (c) $(NH_4)_2[CuCl_4]$**

Analyze and Plan: Determine the charges of the cations and anions using Figure 20.10 and adapting the method described in the solution to Question 43(b). Then balance the ions in a neutral compound.

Execute:

(a) This compound is composed of K^+ ions and $[Co(H_2O)_2(ox)_2]^-$. The neutral compound formula looks like this: $K[Co(H_2O)_2(ox)_2]$.

(b) This compound is composed of $[Cr(NH_3)_2(H_2O)_3OH]^+$ ions and NO_3^- anions. The neutral compound formula looks like this: $[Cr(NH_3)_2(H_2O)_3OH]NO_3$.

(c) This compound composed of two NH_4^+ ions and a complex anion $[CuCl_4]^{2-}$. The neutral compound formula looks like this: $(NH_4)_2[CuCl_4]$.

(d) This anion is composed of $[CoCl_4(en)]^-$.

(e) This compound is composed of $[Co(H_2O)_3F_3]$.

61. *Result:* **(a) tetrachloromanganate(II) (b) potassium trioxalatoferrate(III) (c) diamminedicyanoplatinum(II)**

Explanation:

(a) Four chloride ligands coordinate-covalently bonded to a Mn^{2+} ion to make a complex anion with a 2– charge, called tetrachloromanganate(II) ion.

(b) Three potassium counter ions combine with a complex anion composed of three oxalate anion ligands coordinate-covalently bonded to Fe^{3+} to form a neutral salt called potassium trioxalatoferrate(III) ion.

(c) Two ammonia ligands (NH_3) and two cyanide ligands (CN^-) are coordinate-covalently bonded to a Pt^{2+} metal ion forming a neutral compound called diamminedicyanoplatinum(II).

Geometry of Coordination Complexes (Section 20-6)

63. *Result:* **See sketches, below**

Explanation:

(a) *cis*-$[Pt(H_2O)_2Cl_2]^{2-}$ has the H_2O molecules adjacent to each other in the square planar coordination arrangement:

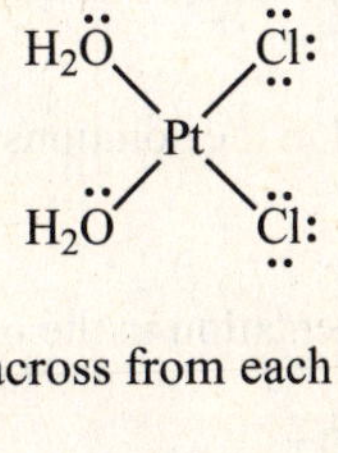

(b) *trans*-$[Cr(H_2O)_4Cl_2]^+$ has the Cl^- ligands across from each other in the octahedral coordination arrangement:

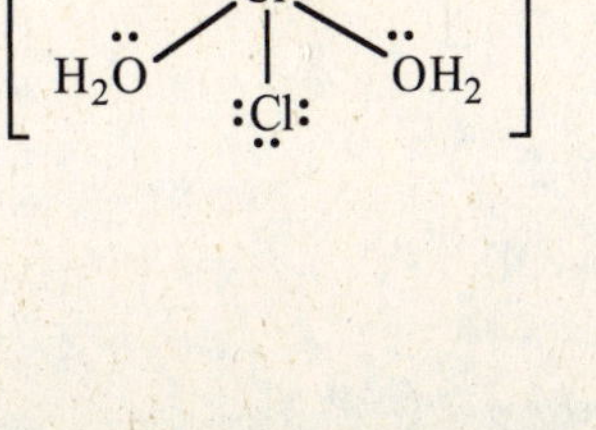

65. *Result:* **See sketch, below**

Analyze and Plan: Co(acac)$_3$ has the acetylacetonate ions in an octahedral coordination arrangement:

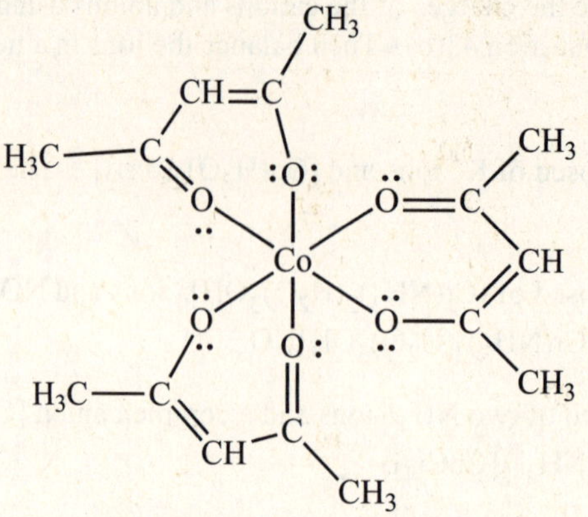

67. *Result:* **Both (a) and (b)**

Explanation:

(a) The triples could have an all-*cis*-orientation, or a pair of each triple could be *trans* while both are *cis* to the third:

(b) The Cl$^-$ ligands can be *cis*- or *trans*-orientation:

Both (a) and (b) can exhibit geometric isomerism.

69. *Result:* **(a), (c), and (d)**

Analyze and Plan: Adapt the methods used in the solutions to Questions 63 to 68.

Execute:

(a) The Cl$^-$ ligands can be *cis*- or *trans*-orientation in the octahedral coordination arrangement:

(b) This square planar structure has no isomers:

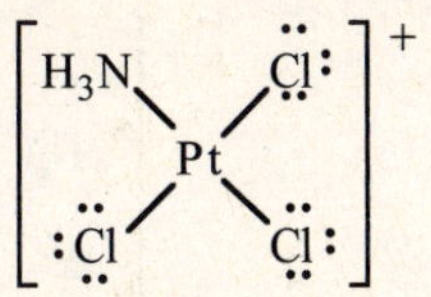

(c) The triples could have an all-*cis*-orientation, or a pair of each triple could be *trans* while both *cis* to the third in the octahedral coordination arrangement:

(d) The NH_3 ligands can be *cis*- or *trans*-orientation in the octahedral coordination arrangement:

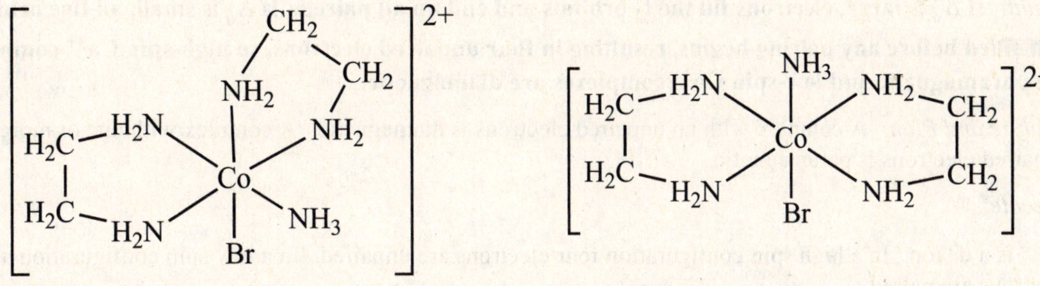

Crystal-Field Theory and Magnetic Properties of Complex Ions (Section 20-7)

71. *Result:* **See crystal-field splitting diagrams, below**

Explanation:

(a) Cr^{+2} is a d^4 ion

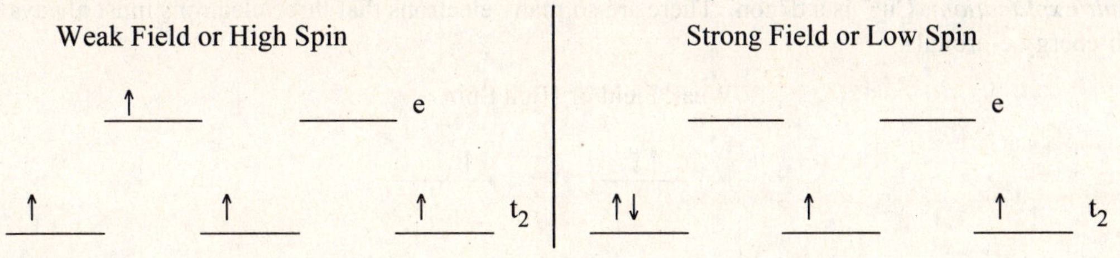

(b) Mn^{+2} is a d^5 ion

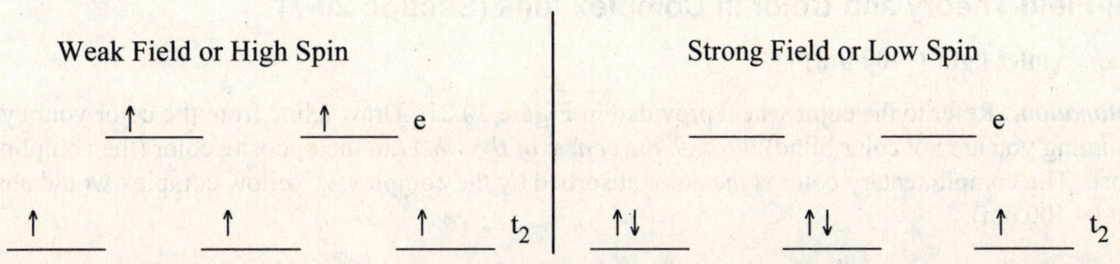

(c) Fe^{3+} is a d^5 ion

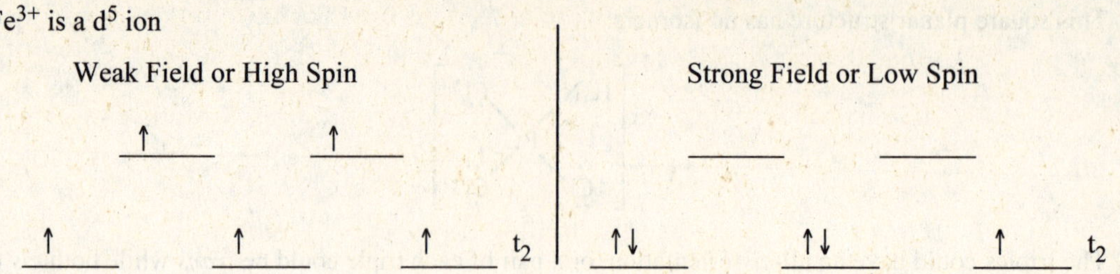

(d) Cr^{3+} is a d^3 ion

Only one configuration is possible because the lower-energy orbitals must each be singly filled first.

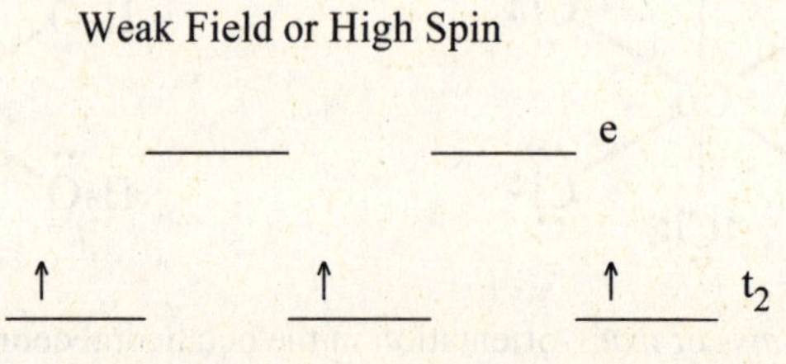

76. *Result:* **If Δ_o is large, electrons fill the t_2 orbitals and end up all paired. If Δ_o is small, all five orbitals are half-filled before any pairing begins, resulting in four unpaired electrons, so high-spin Co^{3+} complexes are paramagnetic and low-spin Co^{3+} complexes are diamagnetic.**

Analyze and Plan: A complex with no unpaired electrons is diamagnetic. A complex with one or more unpaired electrons is paramagnetic.

Execute:

Co^{3+} is a d^6 ion. In a high spin configuration four electrons are unpaired. In a low spin configuration all electrons are paired.

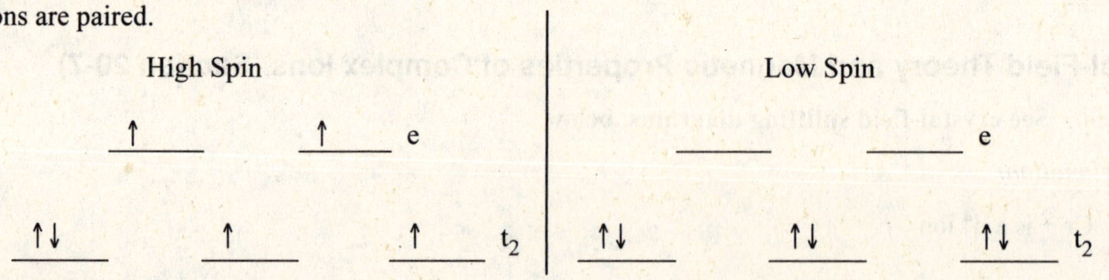

78. *Result/Explanation:* Cu^{2+} is a d^9 ion. There are so many electrons that three electrons must always go in the high-energy e-orbitals.

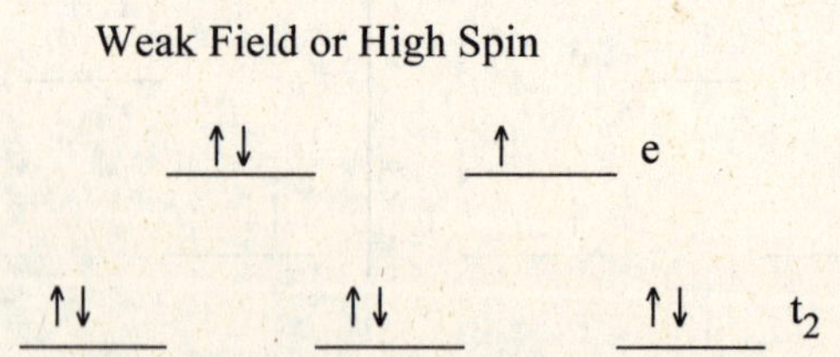

Crystal-Field Theory and Color in Complex Ions (Section 20-7)

79. *Result:* **violet light (~400 nm)**

Explanation: Refer to the color wheel provided in Figure 20.21. Draw a line from the color your eyes detect (assuming you are not color blind) *through the center of the wheel* to the opposite color (the complimentary color). The complimentary color is the color absorbed by the complex. A yellow complex would absorb violet light (~ 400 nm).

81. *Result:* **With cyanide ligand, Δ_o should increase, color changes from purple to yellow-orange; with chloride ligand, Δ_o should decrease, color changes purple to a blue.**

Explanation: Cyanide ion is a strong field ligand, thus Δ_o should increase. The color of the solution is expected to change to a yellow-orange shade (assuming approximately 450 nm or blue light is absorbed). Chloride ion is a weak field ligand (weaker than water), thus Δ_o should decrease. The color of the solution is expected to change to a blue shade (assuming approximately 600 nm or yellow light is absorbed).

General Questions

84. *Result:* **(a) [Ar]3d^4 (b) [Ar]3d^{10} (c) [Ar]3d^7 (d) [Ar]3d^3**

Analyze and Plan: Follow the method described in Sections 5-8 and 20-1 and in the solution to Question 14.

Execute:

(a) $_{24}Cr^{2+}$ electron configuration: [Ar]3d^4 or $1s^22s^22p^63s^23p^63d^4$

(b) $_{30}Zn^{2+}$ electron configuration: [Ar]3d^{10} or $1s^22s^22p^63s^23p^63d^{10}$

(c) $_{27}Co^{2+}$ electron configuration: [Ar]3d^7 or $1s^22s^22p^63s^23p^63d^7$

(d) $_{25}Mn^{4+}$ electron configuration: [Ar]3d^3 or $1s^22s^22p^63s^23p^63d^3$

86. *Result:* **See orbital box diagrams, below (a) 4 (b) 0 (c) 3 (d) 3**

Analyze and Plan: Follow the method described in the solution to Question 5.64 using the solution from Question 84.

Execute:

(a) 3d

$_{24}Cr^{2+}$ orbital box diagram [Ar] [↑] [↑] [↑] [↑] [↑]

The ion has 4 unpaired electrons.

(b) 3d

$_{30}Zn^{2+}$ orbital box diagram [Ar] [↑↓] [↑↓] [↑↓] [↑↓] [↑↓]

The ion has zero unpaired electrons.

(c) 3d

$_{27}Co^{2+}$ orbital box diagram [Ar] [↑↓] [↑↓] [↑] [↑] [↑]

The ion has 3 unpaired electrons.

(d) 3d

$_{25}Mn^{4+}$ orbital box diagram [Ar] [↑↓] [↑↓] [↑] [↑] [↑]

The ion has 3 unpaired electrons.

89. *Result:* **14.8 g Cu**

Analyze and Plan: Follow the method described in the solutions to Question 17.65.

Execute: $Cu^{2+}(aq) + 2\ e^- \longrightarrow Cu(s)$

$$5.00\text{ hr} \times 2.50\text{ A} \times \frac{3600\text{ s}}{1\text{ hr}} \times \frac{1\text{ C}}{1\text{ A}\cdot 1\text{ s}} \times \frac{1\text{ mol e}^-}{96485\text{ C}} \times \frac{1\text{ mol Cu}^{2+}}{2\text{ mol e}^-} \times \frac{63.546\text{ g Cu}}{1\text{ mol Cu}^{2+}} = 14.8\text{ g Cu}$$

91. *Result:* **0.40 ton SO_2**

Analyze and Plan: See Chapter 3 for assistance with this Question:

Execute: $2\ Cu_2S(s) + 3\ O_2(g) \longrightarrow 2\ Cu_2O(s) + 2\ SO_2(g)$

$$1.0 \text{ ton CuS} \times \frac{2000 \text{ lb}}{1 \text{ ton}} \times \frac{453.6 \text{ g}}{1 \text{ lb}} \times \frac{1 \text{ mol } Cu_2S}{159.157 \text{ g } Cu_2S} \times \frac{2 \text{ mol } SO_2}{2 \text{ mol } Cu_2S}$$

$$\times \frac{64.064 \text{ g } SO_2}{1 \text{ mol } SO_2} \times \frac{1 \text{ lb}}{453.6 \text{ g}} \times \frac{1 \text{ ton}}{2000 \text{ lb}} = 0.40 \text{ ton } SO_2$$

92. *Result:* **(a) 4 (b) 6 (c) 4 (d) 6**

Analyze and Plan: Use the methods described in the solutions to Question 49.

Execute:

(a) In $[Ni(en)Cl_2]^{2+}$, we have 1 bidentate ligands (en) and two monodentate ligands (2 Cl^-), so the coordination number is 2 + 2(1) = 4.

(b) In $[Mo(CO)_4Br_2]$, we have 6 monodentate ligands (4 CO and 2 Br^-), so the coordination number is 6.

(c) In $[Cd(CN)_4]^{2+}$, we have 4 monodentate ligands (4 CN^-), so the coordination number is 4.

(d) In $[Co(CN)_4(OH)]^{3-}$, we have 6 monodentate ligands (5 CN^- and OH^-), so the coordination number is 6.

94. *Result:* **See sketches below**

Analyze and Plan: Be systematic and remember that the complex ion has six ligands, so one of the ions is an uncomplexed counter ion. (Orientations and locations of ligands may vary, but four distinct combinations can be found: two chlorides in trans orientation with a bromide counter ion, two chlorides in *cis* orientation with a bromide counter ion, bromide and chloride in *trans* orientation with a chloride counter ion, and bromide and chloride in *cis* orientation with a chloride counter ion.)

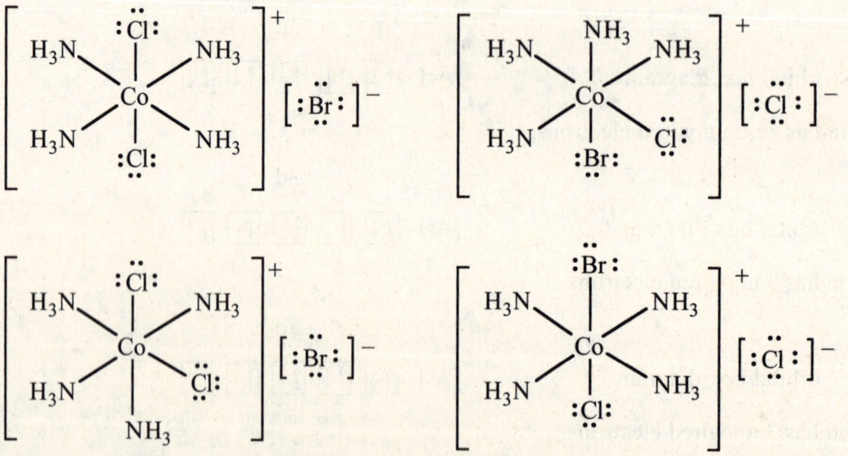

97. *Result:* **(a) False (b) False (c) False; see corrections, below**

Analyze and Plan: Use the methods described in the solutions to Questions 44 - 54.

Execute:

(a) The statement is false. The $C_2O_4^{2-}$ ligand is bidentate. The coordination number of the Fe^{3+} ion in $[Fe(H_2O)_4(C_2O_4)]^+$ is six.

(b) The statement is false. Cu^+ has no unpaired electrons.

(c) The statement is false. The net charge of a coordination complex of Cr^{3+} with two NH_3 and one en is +3.

99. *Result:* **(a) +2 (b) see structural formula below**

Explanation:

(a) The net charge = 0 = (Ox. # Pt) + 2(0) + (–2). So, Ox. # Pt is +2.

(b) The structural formula for $[Pt(NH_3)_2(C_2O_4)]$:

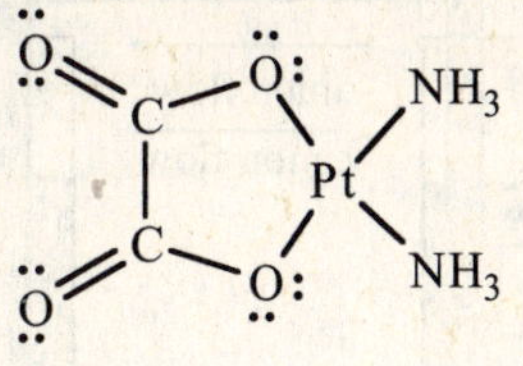

Applying Concepts

101. *Result/Explanation:* $Fe(s) + 2\ Fe^{3+}(aq) \longrightarrow 3\ Fe^{2+}(aq)$

103. *Result/Explanation:* Each N atom in the structure has a lone pair of electrons that can be used to make a coordinate covalent bond. Hence, this ligand can make **four** coordinate covalent bonds with a metal atom.

105. *Result:* **See three structures and explanation, below**

Explanation: Platinum(II) forms a d^8 complex ion with the square planar geometry. When all the ligands are monodentate, two geometrical isomers are possible.

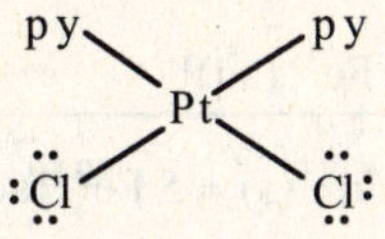

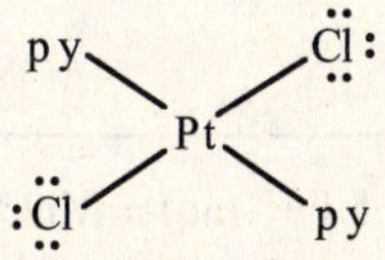

However, when one of the ligands is a bidentate ligand, such as 1,10-phenathroline (depicted in Question 104), the trans-isomer cannot exist.

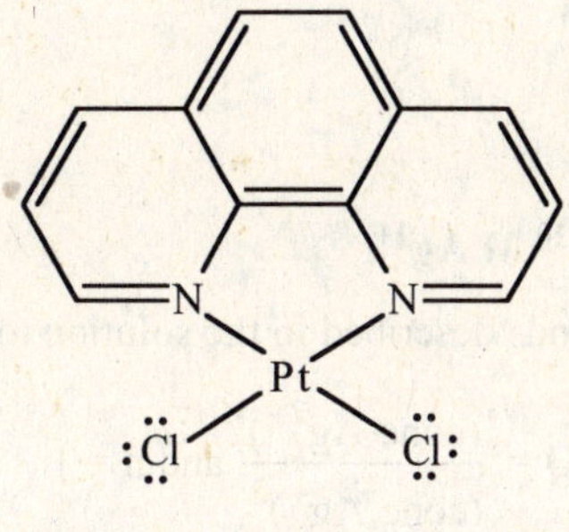

107. *Result:* **(a) $2\ Ag^+(aq) + Ni(s) \longrightarrow Ni^{2+}(aq) + 2\ Ag(s)$ (b) + 1.05 V (c) see sketch below**

Analyze and Plan: Use Appendix I or Table 17.1 and methods described in the solution to Question 17.21.

Execute:

(a) Find a positive $E°_{cell}$ for the product-favored reaction in a cell containing Ni/Ni^{2+} and Ag/Ag^+.

$$Ni(s) \longrightarrow Ni^{2+}(aq) + 2\ e^- \qquad E°_{anode} = +\ 0.25\ V$$

$$2\ Ag^+(aq) + 2\ e^- \longrightarrow 2\ Ag(s) \qquad E°_{cathode} = +\ 0.7991\ V$$

$$2\ Ag^+(aq) + Ni(s) \longrightarrow Ni^{2+}(aq) + 2\ Ag(s) \qquad E°_{cell} = +\ 1.05\ V$$

(b) The cell potential is + 1.05 V.

(c)

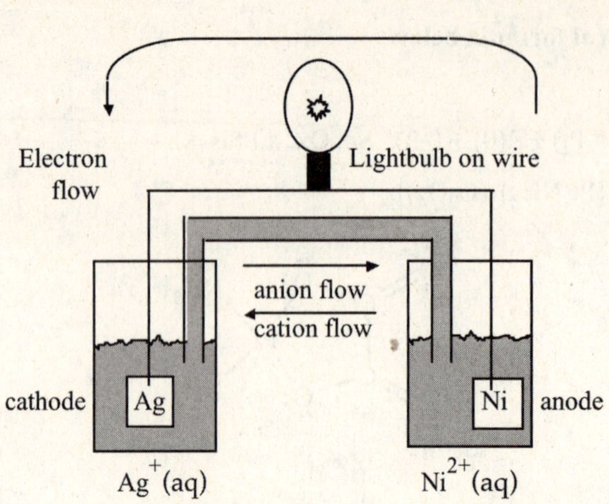

109. *Result:* **17.4% Fe**

Analyze and Plan: Balance the equation for the redox reaction as described in Appendix F Problem-Solving Example F.1, then use conversion factor methods from Chapters 2 and 3 to determine the percent iron in the ore.

Execute:

$$1 \times [5\ e^- + 8\ H^+(aq) + MnO_4^-(aq) \longrightarrow Mn^{2+}(aq) + 4\ H_2O(\ell)]$$

$$5 \times [Fe^{2+}(aq) + 1\ e^- \longrightarrow Fe^{3+}(aq)]$$

$$5\ Fe^{2+}(aq) + MnO_4^-(aq) + 8\ H^+(aq) \longrightarrow Mn^{2+}(s) + 5\ Fe^{3+}(aq) + 4\ H_2O(\ell)$$

$$\frac{18.6\text{ mL}}{1.500\text{ g ore}} \times \frac{1\text{ L}}{1000\text{ mL}} \times \frac{0.05012\text{ mol } MnO_4^-}{1\text{ L}} \times \frac{5\text{ mol } Fe^{2+}}{1\text{ mol } MnO_4^-}$$

$$\times \frac{1\text{ mol Fe}}{1\text{ mol } Fe^{2+}} \times \frac{55.847\text{ g Fe}}{1\text{ mol Fe}} \times 100\% = 17.4\%$$

111. *Result:* **(a) – 1.38 V (b) 1 × 10⁻²⁰ M Ag²⁺**

Analyze and Plan: Use the methods described in the solution to Question 17.46.

Execute: For this reaction, $Q = \dfrac{(\text{conc } Ag^{2+})}{(\text{conc } Ag^+)^2}$ and n = 1

(a) $$E_{cell} = E°_{cell} - \frac{0.0592\text{ V}}{n}\log\left(\frac{(\text{conc } Ag^{2+})}{(\text{conc } Ag^+)^2}\right)$$

$$= -1.18\text{ V} - \frac{0.0592\text{ V}}{1}\log\left(\frac{\frac{1}{5}(1\times10^{-4}\text{ M})}{(1\times10^{-4}\text{ M})^2}\right)$$

$$= -1.18\text{ V} - 0.0592\text{ V} \times \log(2\times10^3) = -1.18\text{ V} - 0.0592\text{ V} \times 3.3$$

$$E_{cell} = -1.18\text{ V} - 0.20\text{ V} = -1.38\text{ V}$$

(b)
$$\frac{n}{0.0592\ V}(E^\circ_{cell} - E_{cell}) = \log\left(\frac{(\text{conc } Ag^{2+})}{(\text{conc } Ag^+)^2}\right)$$

$$\frac{1}{0.0592\ V} \times (-1.18\ V - 0.00\ V) = \frac{1}{0.0592\ V} \times (-1.18\ V) = -19.9$$

(When $E_{cell} = 0$ V, the reaction is at equilibrium, so the (conc. Ag^{2+}) = [Ag^{2+}].)

$$-19.9 = \log\left(\frac{[Ag^{2+}]}{(1.0\ M)^2}\right)$$

$$1 \times 10^{-20} = \frac{[Ag^{2+}]}{(1.0\ M)^2}$$

$$[Ag^{2+}] = 1 \times 10^{-20}\ M$$

113. *Result:* **(a) no ions (b) 2.00 mol ions (c) 4.00 mol ions (d) 3.00 mol ions**

Analyze and Plan: Ions separate in aqueous solutions of ionic compounds, unless they are part of stable complex ions.

Execute:

(a) [Pt(en)Cl_2] is a tetracoordinated coordination complex. When it is dissolved in water, the coordination complex becomes aqueous, but no ions are formed:

$$[Pt(en)Cl_2](s) \longrightarrow [Pt(en)Cl_2](aq)$$

(b) Na[Cr(en)$_2$(SO_4)$_2$] is a hexacoordinated complex anion with a sodium counter ion. When it is dissolved in water, it ionizes:

$$Na[Cr(en)_2(SO_4)_2](s) \longrightarrow Na^+(aq) + [Cr(en)_2(SO_4)_2]^-(aq)$$

According to this equation, when 1.00 mol of the solid dissolves, 2.00 mol of ions form.

(c) K_3[Au(CN)$_4$] is a tetracoordinated complex anion with a potassium counter ion. When it is dissolved in water, it ionizes:

$$K_3[Au(CN)_4](s) \longrightarrow 3\ K^+(aq) + [Au(CN)_4]^{3-}(aq)$$

According to this equation, when 1.00 mol of the solid dissolves, 4.00 mol of ions form.

(d) [Ni(H_2O)$_2$(NH_3)$_4$]Cl_2 is a hexacoordinated complex cation with a chloride counter ion. When it is dissolved in water, it ionizes:

$$[Ni(H_2O)_2(NH_3)_4]Cl_2(s) \longrightarrow [Ni(H_2O)_2(NH_3)_4]^{2+}(aq) + 2\ Cl^-(aq)$$

According to this equation, when 1.00 mol of the solid dissolves, 3.00 mol of ions form.

116. *Result:* **See structural formula, below**

Analyze and Plan: A coordination compound contains one Co^{3+}, one SO_4^{2-}, one Cl^-, and four NH_3. It is clear that Co^{3+} and the four NH_3 molecules are part of the complex ion. We can deduce which of the ions is in the complex ion by observing predicted precipitation reactions.

If the ions are in the solution they are free to react; however, they can be protected from participating in a predicted precipitation reaction if they are part of the complex ion.

If you need a refresher on solubility rules, refer to Table 3.1.

Execute: The $BaCl_2$ was added to see if SO_4^{2-} would precipitate, since $BaSO_4$ is insoluble. The $AgNO_3$ was added to see if Cl^- would precipitate, since AgCl is insoluble. Because there was not a precipitate when the $BaCl_2$ was added, the SO_4^{2-} is a part of the complex ion. Because there was a precipitate when the $AgNO_3$

was added, the Cl^- is not part of the complex ion.

$$[Co(NH_3)_4SO_4]Cl$$

Cobalt complexes are hexacoordinated, so the sulfate must be a bidentate, to take up the other two bonding positions that the four ammonia molecules are not occupying. The chloride ion is ionically bonded to the complex cation, shown in structural formula as a separate charged anion, nearby the cation:

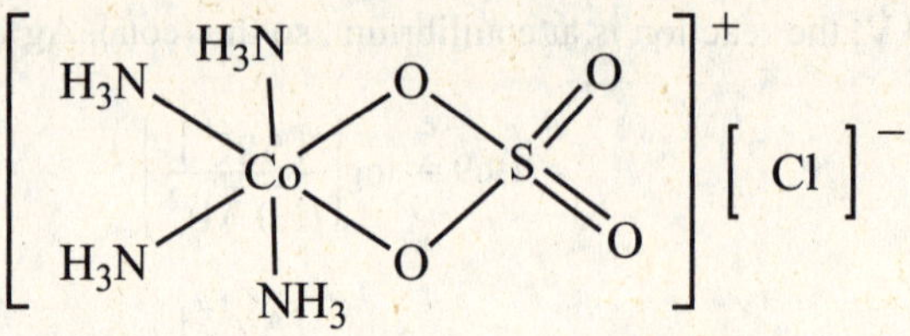

More Challenging Questions

118. *Result:* **(a) $[Co(NH_3)_6]Cl_3$ (b) $[Co(NH_3)_6]Cl_3(s) \longrightarrow [Co(NH_3)_6]^{3+}(aq) + 3\ Cl^-(aq)$**

Analyze and Plan: Use the percent by mass to determine the formula, using methods from Chapter 2. Arrange the atoms to form a complex ion with the cobalt such that there are four ions when one formula unit is dissolved in water. For this step, adapt the method described in the solution to Question 100.

Execute:

(a) In 100.00 g sample of compound, we have 22.0 g Co, 31.4 g N, 6.78 g H, and 39.8 g Cl.

$$22.0\text{ g Co} \times \frac{1\text{ mol Co}}{58.9332\text{ g Co}} = 0.373\text{ mol Co} \qquad 31.4\text{ g N} \times \frac{1\text{ mol N}}{14.0067\text{ g N}} = 2.24\text{ mol N}$$

$$6.78\text{ g H} \times \frac{1\text{ mol H}}{1.0079\text{ g H}} = 6.73\text{ mol H} \qquad 39.8\text{ g Cl} \times \frac{1\text{ mol Cl}}{35.453\text{ g Cl}} = 1.12\text{ mol Cl}$$

$$0.373\text{ mol Co} : 2.24\text{ mol N} : 6.73\text{ mol H} : 1.12\text{ mol Cl}$$

$$1\text{ Co} : 6\text{ N} : 18\text{ H}: 3\text{ Cl}$$

$$CoN_6H_{18}Cl_3$$

A typical ligand in complex ions is NH_3, so let's use the N atoms and H atoms to build NH_3 molecules. $Co(NH_3)_6Cl_3$. We can now arrange the components so that there is one hexacoordinated complex cation, $[Co(NH_3)_6]^{3+}$, with three Cl^- counter ions to make four ions. The formula is $[Co(NH_3)_6]Cl_3$.

(b) The dissociation equation to make an aqueous solution looks like this:

$$[Co(NH_3)_6]Cl_3(s) \longrightarrow [Co(NH_3)_6]^{3+}(aq) + 3\ Cl^-(aq)$$

120. *Result:* **See structural formula below**

Analyze and Plan: Adapt the methods described in the solutions to Questions 100 and Chapter 2.

Execute: The simplest formula, $PtN_2H_6Cl_2$, has a formula mass of 300.05 g/mol. The molar mass is twice that number, so the molecular formula must be $Pt_2N_4H_{12}Cl_4$. A typical ligand in complex ions is NH_3, so let's use the N atoms and H atoms to build NH_3 molecules. $Pt_2(NH_3)_4Cl_4$.

We can now arrange the components so that there is one tetracoordinated platinum(II) complex cation and one tetracoordinated platinum(II) complex anion. We do this by putting the neutral NH_3 ligands on Pt^{2+} to form the cation, and the negative Cl^- ligands on Pt^{2+} to form the anion. Therefore, a logical formula for the compound is $[Pt(NH_3)_4]^{2+}[PtCl_4]^{2-}$. Its structural formula looks like this:

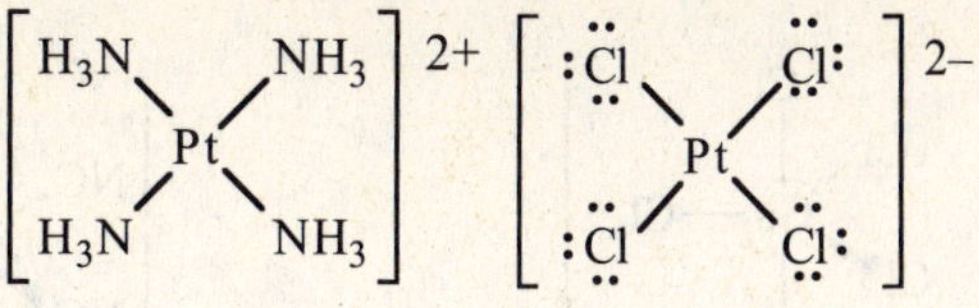

One other formula would also fit the given information: $[Pt(NH_3)_3Cl]^+[Pt(NH_3)Cl_4]^-$.

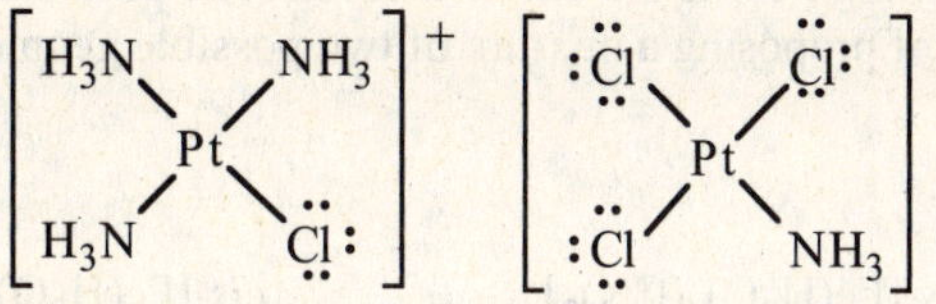

121. *Result:* **See structural formula below**

Analyze and Plan: Adapt the method described in the solutions to Questions 63 to 70.

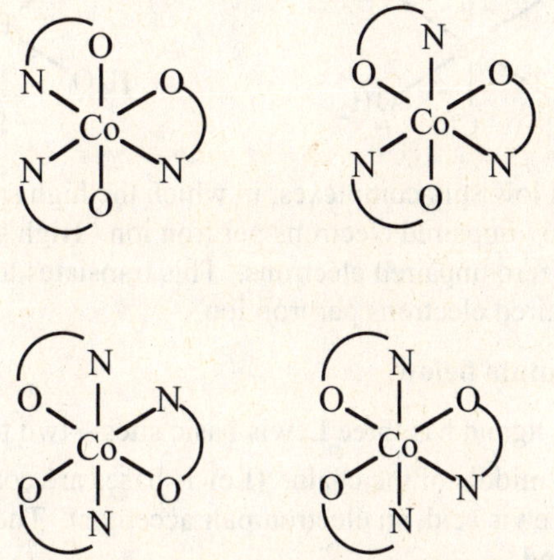

The difference between the top two structures is simply a reverse of the N and O ends of the top ligand.

The difference between the two structures on the right is simply a reverse of the N and O ends of the bottom ligand.

The difference between the bottom two structures is simply a reverse of the N and O ends of the right ligand.

122. *Result:* **See structural formula below; there are two possible geometries: the copper complex is shown in a trigonal bipyramidal geometry while the nickel complex is shown in a square pyramidal geometry**

Analyze and Plan: Five-coordinated complexes can adapt two possible geometries, either square pyramidal or trigonal bipyramidal.

The geometry of a given complex depends on a variety of factors which influence the electronic structure of the complex, such as the types of ligands, the central transition metal oxidation number, the complex ion's surrounding environment (such as counter ion), and crystal packing.

In solution, both geometries are often in dynamic equilibrium (fluxional) because the energy difference between the two geometries is very small.

In the solid state, one often sees a structure that is a mixture of the two ideal geometries (i.e., it is neither square pyramidal nor trigonal bypyramidal but rather somewhere in between).

Only in cases where there is a clear energetic preference for one structure over another will one geometry be favored.

Execute: In this case the copper complex is shown in a trigonal bipyramidal geometry while the nickel complex is shown in a square pyramidal geometry.

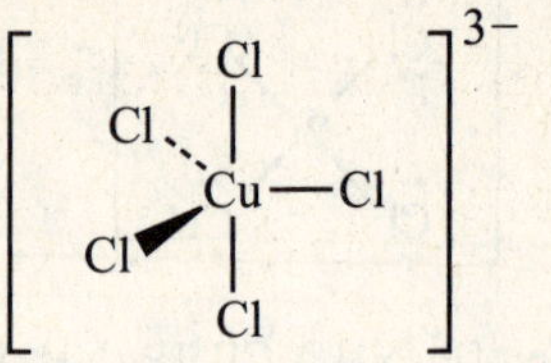

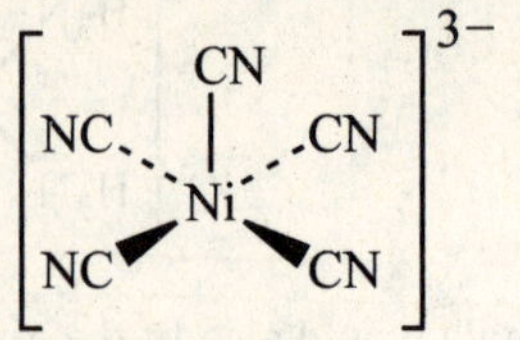

124. *Result/Explanation:* Assuming Fe^{+2} is the only oxidation state possible and all complexes are octahedral limits us to the possibility of proposing a mixture of two possible geometrical isomers (the *trans*- and *cis*-cyano complexes):

trans-$[Fe(H_2O)_4(CN)_2]$

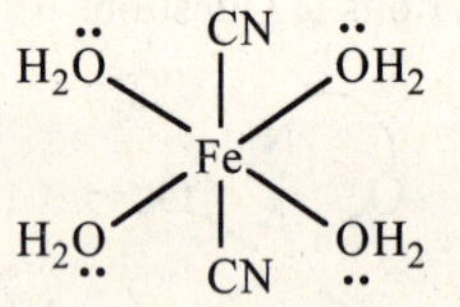

cis-$[Fe(H_2O)_4(CN)_2]$

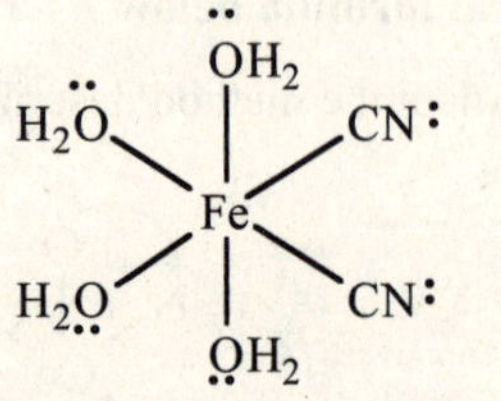

A mixture of high spin and low spin complexes, in which the high spin species is favored by a 2:1 ratio, would give the observed 2.67 unpaired electrons per iron ion. High spin iron(II) has four unpaired electrons, while low spin iron(II) has zero unpaired electrons. This translates to 8 unpaired electrons for every three iron ions or 8/3 = 2.67 unpaired electrons per iron ion.

126. *Result:* **See structural formula below**

Analyze and Plan: The det ligand has three Lewis basic sites—two terminal amine nitrogens (NH_2), and another amine –NH– in the middle of the chain. (Lewis bases are compounds that can donate one or more lone pair of electrons to a Lewis acid, an electron pair acceptor). The Lewis acid-Lewis base bond is known as a coordinate covalent bond.

Execute: In this case two det ligands form a total of 6 coordinate covalent bonds with iron(III) to give an octahedral coordination geometry.

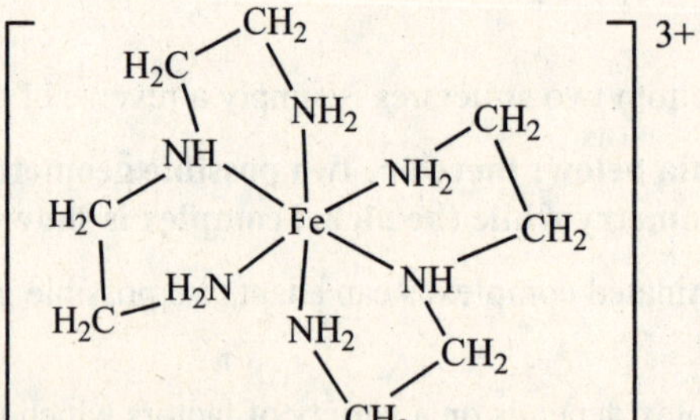

It is also possible for the terminal N atoms in each det ligand to be *trans* to each other.

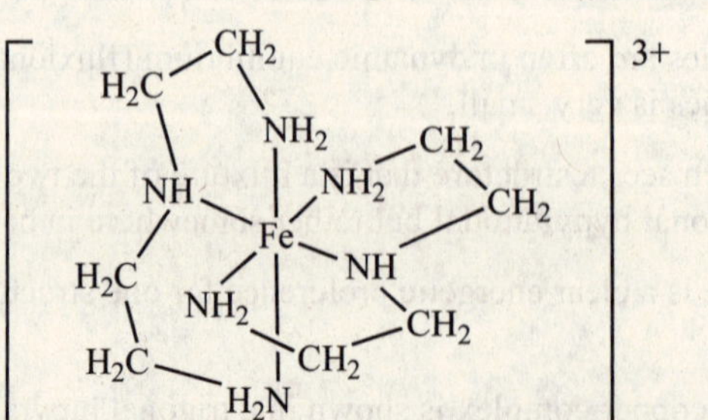

Appendix A: Problem Solving and Mathematical Operations

Solutions for Red-Numbered Questions

Numbers, Units, and Quantities

4. *Result/Explanation:* If the units for the answer to a calculation do not make sense, then we must conclude that the solution to the problem contains one or more errors. Faced with a situation like this, we must review the plan and its execution to find out where errors were made.

6. *Result:* **(c); it has the proper units (in) and precision; the others do not**

Analyze: Given the description of a specific device for measuring length and a list of results for a measurement, determine which choice would be a suitable record of the observation and why the others are unsuitable.

Plan: Eliminate choices that are incomplete or have incorrect units. Examine the remaining choices to see that they have proper significant figures.

Execute: We can immediately eliminate choice (d) since it has no units at all and the result must be reported with proper units. We can eliminate (a) and (b), since the units of this measured quantity will not be feet or meters. The answer provided in (c) looks like a suitable way to record the measurement of 8 inches and 6.1 sixteenths, because $6.1 \times 0.0625 = 0.38$, the decimal part of the reported quantity. Choice (d) could be used to describe the length of the pencil, but it should not be used to describe the result of this measurement, because the measuring device is calibrated in inches and fractions of inches, not in feet. The answer to (d) is the result of a conversion calculation, not the observed measurement.

☑ *Reasonable Result Check:* The result selected has the proper units (in) and precision.

8. *Result:* **(a) Units should not be squared when doing addition; 9.95 g (b) The density conversion factor is upside-down; 1.07 mL (c) The conversion factor needs to be cubed and the units of the answer should be m^3; 3.57×10^{-6} m^3**

Analyze: Given flawed calculations, determine what was done incorrectly and show the correct calculation and result for each.

Plan: Review the calculation and determine the flaws. Develop a revised plan and execute the calculation to get the correct result.

Execute:

(a) The sum of two quantities with the same units will also have those units. The calculation shown indicates that two masses added together have units of mass squared. The units are incorrect. The corrected calculation looks like this:

$$4.32 \text{ g} + 5.63 \text{ g} = 9.95 \text{ g}$$

(b) The conversion shown looks like it was intended to calculate the volume of a sample from the mass and density. If that is the case, then the density conversion factor is upside-down and the resulting units should be mL. The corrected calculation looks like this:

$$5.23 \text{ g} \times \frac{1.00 \text{ mL}}{4.87 \text{ g}} = 1.07 \text{ mL}$$

(c) The conversion shown looks like it was intended to convert cubic centimeters to cubic meters. If that is the case, then the conversion factor needs to be cubed. The corrected calculation looks like this:

$$3.57 \text{ cm}^3 \times \left(\frac{1 \text{ m}}{100 \text{ cm}}\right)^3 = 3.57 \times 10^{-6} \text{ m}^3$$

☑ *Reasonable Result Check:* In (a), when mass is added, the resulting mass will be larger and in units of mass. In (b), when the density is greater than 1, the mass quantity will have a larger numerical value than the volume quantity. In (c), the number of cubic meters will be much smaller than the number of cubic centimeters, since a meter is much longer than a centimeter.

Precision, Accuracy, and Significant Figures

10. *Result:* **(a) 95.9 ±0.59 in (b) Three (c) It is accurate.**

Analyze: Given several measured quantities for the length of a pole, determine what should be reported as the length and determine the number of significant figures to be reported. Given the actual length, determine whether the result is accurate.

Plan and Execute:

(a) Add up the five individual measurements and divide by five to get the average:

$$\text{Average length of the pole} = \frac{95.31\text{ in} + 96.44\text{ in} + 96.02\text{ in} + 95.78\text{ in} + 95.94\text{ in}}{5} = \frac{479.49\text{ in}}{5} = 95.898\text{ in}$$

The scatter includes measurements lower than the average by as much as (95.898 – 95.31 =) 0.59 inches and higher than the average by as much as (96.44 – 95.898 =) 0.54 inches. So, the pole is reported to be 95.9 ± 0.59 inches long because the uncertainty is the tenths place.

(b) The result should be reported with three significant figures, since the uncertainty is found in the first decimal place.

(c) 95.9 ± 0.59 inches. The pole's actual height is given to be exactly 8 feet, which is 96 inches, since 12 inches = 1 foot. The actual height is within the range of the uncertainty (between 95.31 inches and 96.49 inches) of the calculated height, so the result is accurate, though not very precise.

☑ *Reasonable Result Check:* All the original measured lengths are within the range of 95.9 ± 0.59 inches.

12. *Result:* **(a) Four (b) Two (c) Two (d) Four**

Analyze: Given several measured quantities, determine the number of significant figures.

Plan: Use rules given in Sections 1-5 and A-3, summarized here: All non-zeros are significant. Zeros that precede (sit to the left of) non-zeros are never significant (e.g., 0.003). Zeros trapped between non-zeros are always significant (e.g., 3.003). Zeros that follow (sit to the right of non-zeros) may be significant or may not be significant. They are not significant, if a decimal point is not specified (e.g., 3300) and they are significant if a decimal point is explicitly given (e.g., 3000. Or 0.30000).

Execute:

(a) 3.274 has **four** significant figures (The 3, 2, 7, and 4 digits are each significant.)

(b) 0.0034 L has **two** significant figures (The 3 and the 4 digits are each significant. The zeros are all before the first non-zero-digit, 3, and therefore they are not significant.)

(c) 43,000 m has **two** significant figures (The 4 and the 3 digits are significant. The zeros after the 3 are not significant because the number does not have an explicit decimal point specified.)

(d) 6200. mL has **four** significant figures (The 6 and 2 digits are each significant. The zeros after the 2 are also significant because the number has an explicit decimal point specified.)

☑ *Reasonable Result Check:* The significant figures rules have been properly applied.

14. *Result:* **(a) 43.32 (b) 43.32 (c) 43.32 (d) 43.32 (e) 43.32 (f) 43.32**

Analyze: Given several measured quantities, round them to four significant figures.

Plan: Use rules for rounding given inSections 1-5 and A-3, as summarized here. If the last digit is below 5, then rounding does not change the digit before it. If the last digit is above a five, the digit before it is made one larger. If the last digit is exactly five, round the digit before it to an even number, up if odd and down if even.

Execute:

(a) 43.3250 has six significant figures. To round it to four significant figures, we need to examine the fifth significant figure, the 5. Since the digit we're rounding is a 5, we look at the number to the left of the 5, which is 2. Since 2 is even, we will round down and leave the 2 unchanged: **43.32.**

(b) 43.3165 has six significant figures. To round it to four significant figures, we need to look at the fifth significant figures, the 6. The removal of the fifth digit 6 rounds up the 1 next to it to a 2. The result is **43.32**.

(c) 43.3237 has six significant figures. To round it to four significant figures, we need to look at the fifth significant figures, the 3. The removal of 3 does not change the 2 next to it. The result is **43.32**.

(d) 43.32499 has seven significant figures. To round it to four significant figures, we need to look at the fifth significant figures, the 4. The removal of 4 doesn't change the 2 next to it. The result is **43.32**.

Note: 0.00499 is *less* than 0.05, so it is makes sense that we're rounding down. It is inappropriate here to use the 9 to the right of the 4 to round it to a 5 before rounding to the second decimal place.

(e) 43.3150 has six significant figures. To round it to four significant figures, we need to look at the fifth significant figures, the 5. Since the digit we're rounding is a 5, we look at the number to the left of the 5, which is 1. Since 1 is odd, we will round up and change the 1 to a 2: **43.32**.

(f) 43.32501 has seven significant figures. To round it to four significant figures, we need to examine the fifth significant figure, the 5. Since the digit we're rounding is a 5, we look at the number to the left of the 5, which is 2. Since 2 is even, we will round down and leave the 2 unchanged: **43.32**.

☑ *Reasonable Result Check:* Each result has four significant figures.

16. *Result:* **(a) 13.7 (b) 0.247 (c) 12.0**

Analyze: Given some numbers combined using calculations, determine the result with proper significant figures.

Plan: Perform the mathematical steps according to order of operations, applying the proper significant figures (addition and subtraction retains the least number of decimal places in the result; multiplication and division retain the least number of significant figures in the result). Decimal places are the digits that follow the decimal point to the right)

If operations are combined that use different rules, it is important to stop and determine the intermediate result any time the rule switches.

It is important to keep the unrounded number in your calculator so you can use it in subsequent calculations to prevent excessive round-off errors.

Execute:

(a) Divide the two numbers and use the division rule:

$$\frac{4.47}{0.3260} = 13.71166\ldots \simeq \mathbf{13.7}$$

The numerator has three significant figures and the denominator has four significant figures, so the result will have three significant figures. Therefore, we report the result: **13.7**.

(b) First, add the numbers in the numerator and use the addition rule:

$$4.03 + 3.325 = 7.355\ldots \simeq 7.36$$

The first number has two, the digits "0" and "3." The second number has three decimal places, the digits "3," "2," and "5", so the result of the addition is rounded to two decimal places; hence, the last five rounds the next five to a six: 7.36.

Second, divide the two numbers and use the division rule.

$$\frac{7.36}{29.75} = 0.24723\ldots \simeq \mathbf{0.247}$$

The numerator has three significant figures and the denominator has four significant figures, so the result will have three significant figures. Therefore, we report the result: **0.247**.

(c) First, subtract the numbers in the denominator: $5.673 - 4.987 = 0.686$

Uses the subtraction rule: The first number has three decimal places, the digits "6," "7," and "3." The second number has three decimal places, the "9," "8," and "7". So, the result of the subtraction has three decimal places.

Second, divide the two numbers and use the division rule.

$$\frac{8.234}{0.686} = 12.00292\ldots \simeq \mathbf{12.0}$$

The numerator has four significant figures and the denominator has three significant figures, so the result will have three significant figures. Therefore, we report the result: **12.0**.

☑ *Reasonable Result Check:* The proper significant figures rules were used. The size and units of the results are appropriate.

Exponential or Scientific Notation

18. *Result:* **(a) 7.6003×10^4 (b) 3.7×10^{-4} (c) 3.4×10^4**

Analyze: Given some numbers, express them in scientific notation.

Plan: Convert the number to scientific notation by writing it as a number between 1 and 9.999… multiplied by a size factor represented as ten to a whole-number power.

Execute:

(a) 76,003 is $7.6003 \times 10{,}000$ or 7.6003×10^4.

(b) 0.00037 is 3.7×0.0001 or 3.7×10^{-4}.

(c) 34,000 is $3.4 \times 10{,}000$ or 3.4×10^4.

20. *Result:* **(a) 2.415×10^{-3} (b) 2.70×10^8 (c) 3.236 (d) 116 cm^3**

Analyze: Given some numbers combined using calculations, determine the result with proper significant figures.

Plan: Do the mathematical steps using order of operations, applying the proper significant figures (addition and subtraction retains the least number of decimal places in the result, or rounds the result to the position of greatest uncertainty; multiplication and division retain the least number of significant figures in the result).

Execute:

(a) Divide the two numbers and use the division rule:

$$\frac{0.7346}{304.2} = 0.0024149\ldots \simeq 0.002415$$

The numerator and the denominator both have four significant figures, so the result will have four significant figures (2, 4,1, and 5). This is a small number, so change it to scientific notation:

$$0.002415 \text{ is } 2.415 \times 0.001 \text{ or } \mathbf{2.415 \times 10^{-3}}$$

Therefore, we report: **2.415×10^{-3}**.

(b) Multiply and divide the numbers and use the multiplication/division rule:

$$\frac{(3.45\times 10^{-3})(1.83\times 10^{12})}{23.4} = 269{,}807{,}692\ \ldots \simeq 27\underline{0}{,}000{,}000$$

All three numbers have three significant figures, so the result will have three significant figures; hence the eight rounds the 69 to 70, for 27<u>0</u>,000,000. This is a huge number, so change it to scientific notation:

$$27\underline{0}{,}000{,}000 \text{ is } 2.70 \times 100{,}000{,}000 \text{ or } \mathbf{2.70 \times 10^8}$$

Therefore, we report: **2.70×10^8**.

(c) Subtract the two numbers and use the subtraction rule. The first number has three decimal places (2, 4, and 0). Changing the second number from scientific notation to decimal notation: $4.33 \times 10^{-3} = 0.00433$ shows that it has five decimal places (0, 0, 4, 3, and 3). So, the result of the subtraction has three decimal places.

$$3.240 - 4.33 \times 10^{-3} = 3.240 - 0.00433 = 3.23567\ldots \simeq \mathbf{3.236}$$

Therefore, we report **3.236**.

(d) Multiply all the numbers and use the multiplication rule:

$$(4.87\text{ cm})^3 = 4.87\text{ cm} \times 4.87\text{ cm} \times 4.87\text{ cm} = 115.5013\ldots\text{cm}^3 \simeq \mathbf{116\ cm^3}$$

The numbers have three significant figures, so the result will have three significant figures; hence the right five rounds the left five to a six, and we report: **116 cm^3**.

☑ *Reasonable Result Check:* The proper significant figures rules were used. Size and units are appropriate.

Logarithms

22. *Result:* **(a) –0.1351 (b) 3.541 (c) 23.7797 (d) 54.7549 (e) –7.455**

Analyze: Given some numbers and calculations using logarithms, determine the results with appropriate significant figures.

Plan: Use the information in Section A.6 describing the operations of logarithms and their significant figures. When taking the log of a number, the mantissa (the digits to the right of the decimal point) should have as many significant figures as the numbers whose log was found. So, the number of significant figures in the number we start with gives the number of decimal places in the result.

Execute:

(a) The log of 4 significant figures, gives a result with 4 decimal places: log(0.7327) = –0.1351

(b) The ln of 3 significant figures, gives a result with 3 decimal places: ln(34.5) = 3.541

(c) The log of 4 significant figures, gives a result with 4 decimal places: $\log(6.022 \times 10^{23}) = 23.7797$

(d) The ln of 4 significant figures, gives a result with 4 decimal places: $\ln(6.022 \times 10^{23}) = 54.7549$

(e) The ratio of 3 significant figures gives a result with 3 significant figures. The log of 3 significant figures, gives a result with 3 decimal places:

$$\log\left(\frac{8.34\times10^{-5}}{2.38\times10^{3}}\right)=\log(3.50 \times 10^{-8}) = -7.455$$

☑ *Reasonable Result Check:* For log calculations, the power of ten is approximately reflected in the ordinate (the number to the right of the decimal point), as in (e), where we see –7.xx when the number was a little larger than 10^{-8}. For ln calculations, the power of ten will be approximately reflected as the ordinate times 2.303. The result in (d) is 2.303 times the result in (c).

24. *Result:* **(a) 5.404 (b) 1 × 10^{15} (c) 110. (d) 1.000320 (e) 3.755**

Analyze: Given some numbers and calculations using antilogarithms, determine the results with appropriate significant figures.

Plan: Use the information in Section A.6 describing the operations of antilogarithms and their significant figures. When taking the antilog of a number, the number of decimal places in the number whose antilog was taken should be the same as the number of significant figures in the result of the antilogarithm operation.

Execute:

(a) The antilog of 4 decimal places, gives a result with 4 significant figures: antilog(0.7327) = 5.404

(b) The antiln of 1 decimal place, gives a result with 1 significant figure: $\ln(34.5) = 1 \times 10^{15}$

(c) The 10^x of 3 decimal places, gives a result with 3 significant figures: $10^{2.043} = 110.$

(d) The number 3.20×10^{-4} is also 0.000320. Therefore, it has 6 decimal places. The e^x of 6 decimal place, gives a result with 6 significant figure: : $e^{0.000320} = 1.000320$

(e) The ratio of 4 significant figures gives a result with 4 significant figures. The exp of 4 decimal places, gives a result with 4 significant figures:

$$\exp\left(\frac{4.333}{3.275}\right) = \exp(1.323) = 3.755$$

☑ *Reasonable Result Check:* Antilogarithms of small numbers (i.e., numbers near zero) give results near one. Antilog on larger numbers, give results with large powers of ten.

Quadratic Equations

26. *Result:* **(a) 0.480 and –1.80 (b) 3.23 and 1.34**

Analyze: Given quadratic equations, find the roots.

Plan: Set up the quadratic equation in the form of: $ax^2 + bx + c = 0$, then plug into the quadratic equation:

$$x = \frac{-b \pm \sqrt{b^2 - 4ac}}{2a}$$

The two roots are generated when we use the "+" or the "–" of the "±" in this equation.

Execute:

(a) $3.27x^2 + 4.32x - 2.83 = 0$, so, $a = 3.27$, $b = 4.32$, and $c = -2.83$:

$$x = \frac{-4.32 \pm \sqrt{4.32^2 - 4(3.27)(-2.83)}}{2(3.27)}$$

$$x = \frac{-4.32 + 7.46}{6.54} = \frac{3.14}{6.54} = 0.480$$

$$x = \frac{-4.32 - 7.46}{6.54} = \frac{-11.78}{6.54} = -1.80$$

(b) Rearrange: $x^2 + 4.32 = 4.57x$

$x^2 - 4.57x + 4.32 = 0$, so, $a = 1$, $b = -4.57$, and $c = 4.32$:

$$x = \frac{-(-4.57) \pm \sqrt{(-4.57)^2 - 4(1)(4.32)}}{2(1)}$$

$$x = \frac{4.57 + 1.90}{2} = \frac{6.47}{2} = 3.23$$

$$x = \frac{4.57 - 1.90}{2} = \frac{2.67}{2} = 1.34$$

☑ *Reasonable Result Check:* Each root solves the equation:

(a) $3.27(0.480)^2 + 4.32(0.480) - 2.83 = 0.753 + 2.07 - 2.83 = 0.00$

$3.27(-1.80)^2 + 4.32(-1.80) - 2.83 = 10.6 - 7.78 - 2.83 = 0.0$

(b) $(3.23)^2 + 4.32 = 10.4 + 4.32 = 14.8$ matches $4.57(3.23) = 14.8$

$(1.34)^2 + 4.32 = 1.80 + 4.32 = 6.12$ matches $4.57(1.34) = 6.12$

Graphing

28. *Result:* **See graph below; yes**

Analyze: Given the masses of several volumes of a sample, draw the graph and determine if mass is directly proportional to volume.

Plan: Plot the graph and see if the data fits a linear equation.

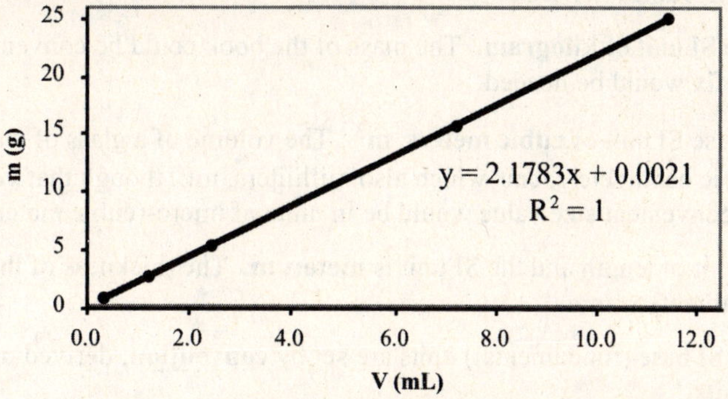

The graph is linear and the value of $R^2 = 1$, so the mass is directly proportional to volume.

☑ *Reasonable Result Check:* Mass and volume are extensive properties that are related to each other through a substance's fixed density, so it makes sense that the graph is linear.

Appendix B: Units, Equivalences, and Conversion Factors
Solutions for Red-Numbered Questions

Units of the International System

1. *Result/Explanation:* Determine which SI units and prefix would be used to measure a mass, a volume, and a thickness. We use the Tables B.1-B.3.

   (a) Mass would use SI unit of **kilogram**. The mass of the book could be conveniently measured in kilograms, so **no other prefix** would be needed.

   (b) Volume would use SI unit of **cubic meters, m^3**. The volume of a glass of water could be conveniently measured in cubic centimeters, cm^3 which also milliliters, mL, though that volume will still be hundreds of mL. The most convenient size value would be in units of micro-(cubic meters) or **nano**-(cubic meters).

   (c) Thickness is a unit of length and the SI unit is **meter, m**. The thickness of this page could be conveniently measured in **milli**meters, mm.

3. *Result/Explanation:* SI base (fundamental) units are set by **convention**; derived units are **derived from the SI** base fundamental units.

Conversion of Units for Physical Quantities

5. *Result:* **(a) 4.75×10^{-10} m (b) 5.6×10^{7} kg (c) 4.28×10^{-6} A**

   Analyze and Plan: Given several measured quantities, express it in SI base units. Determine the metric conversion factors required.

   Execute:

   (a) pico- is 10^{-12}, so 475 pm is 475 $\times 10^{-12}$ m or 4.75 $\times 10^{-10}$ m, the SI base unit of length.

   (b) giga- is 10^{9}, so 56 Gg is 56 $\times 10^{9}$ g or 5.6 $\times 10^{10}$ g, which is 5.6 $\times 10^{7}$ kg, the SI base unit of mass.

   (c) micro- is 10^{-6}, so 4.28 μA is 4.28 $\times 10^{-6}$ A, the SI base unit of electric charge.

7. *Result:* **(a) 8.7×10^{-18} m^2 (b) 2.73×10^{-17} J (c) 2.73×10^{-5} N**

   Analyze and Plan: Given several measured quantities, express it in SI base units. Determine the metric conversion factors required.

   Execute:

   (a) nano- is 10^{-9}, so 8.7 nm^2 is 8.7 $\times(10^{-9})^2$ m^2 or 8.7 $\times 10^{-18}$ m^2, area using the SI base unit of length.

   (b) atto- is 10^{-18}, so 27.3 aJ is 27.3 $\times 10^{-18}$ J or 2.73 $\times 10^{-17}$ J, the SI base unit of energy.

   (c) micro- is 10^{-6}, so 27.3 μN is 27.3 $\times 10^{-6}$ N or 2.73 $\times 10^{-5}$ N, the SI derived unit for force.

9. *Result:* **(a) 2.20 pounds (b) 2.22×10^{3} kg (c) 60 in^3 (d) 1×10^{5} pascal, 1 bar**

   Analyze and Plan: Express quantities in units given with scientific notation. Determine the metric conversion factors required.

   Execute:

   (a) Table B.4 gives: 1 pound = 0.45359 kg, so: $1.00\text{ kg} \times \dfrac{1\text{ pound}}{0.45359\text{ kg}} = 2.20\text{ pounds}$

   (b) Table B.4 gives: 1 short ton = 907.2 kg, so: $2.45\text{ ton} \times \dfrac{907.2\text{ kg}}{1\text{ ton}} = 2.22 \times 10^{3}\text{ kg}$

(c) Table B.4 gives: 1 L = 1000 cm³ and 1 inch = 2.54 cm so: $1\text{ L}\times\frac{1000\text{ cm}^3}{1\text{ L}}\times\left(\frac{1\text{ in}}{2.54\text{ cm}}\right)^3=6\times10^1\text{ in}^3$

(d) Table B.4 gives: 1 atm = 1.01325×10^5 pascals and 1 bar = 10^5 pascals, so:

$$1\text{ atm}\times\frac{1.01325\times10^5\text{ pascals}}{1\text{ atm}}=1\times10^5\text{ pascals}\quad\text{and}\quad 1\times10^5\text{ pascals}\times\frac{1\text{ bar}}{10^5\text{ pascals}}=1\text{ bar}$$

11. *Result:* **(a) 99° F (b) – 10.5 °F (c) – 40.0 °F**

Analyze and Plan: Use temperature conversion given in Table B.4: $t_F=\frac{9}{5}t_C+32$

Execute:

(a) $t_F=\frac{9}{5}(37\ ^\circ\text{C})+32=99\ ^\circ\text{F}$

(b) $t_F=\frac{9}{5}(-23.6\ ^\circ\text{C})+32=-10.5\ ^\circ\text{F}$

(c) $t_F=\frac{9}{5}(-40.0\ ^\circ\text{C})+32=-40.0\ ^\circ\text{F}$

Appendix F: Balancing Oxidation-Reduction Reactions

Solutions for Red-Numbered Questions

1. *Results:* **See half reactions below**

Analyze and Plan: Given descriptions of oxidation and reduction reactions, write their half reactions. Balance atoms and charge following the method described in Problem-Solving Example F.1 Step 3

Execute:

(a) Nickel is Ni(s). We balance Ni atoms, then balance charge with electrons:

$$Ni(s) \longrightarrow Ni^{2+}(aq) + 2\ e^-$$ ☑ Check: 1 Ni, zero net charge

(b) Hydrogen gas is $H_2(g)$. We balance H atoms, then balance charge with electrons:

$$2\ H^+(aq) + 2\ e^- \longrightarrow H_2(g)$$ ☑ Check: 2 H, zero net charge

(c) We balance Fe atoms, then balance charge with electrons:

$$Fe^{3+}(aq) + e^- \longrightarrow Fe^{2+}(aq)$$ ☑ Check: 1 Fe, +2 net charge

(d) Bromine is a diatomic liquid, $Br_2(\ell)$. We balance Br atoms, then balance charge with electrons:

$$Br_2(\ell) + 2\ e^- \longrightarrow 2\ Br^-(aq)$$ ☑ Check: 2 Br, –2 net charge

(e) Nitrogen dioxide is $NO_2(g)$ and nitrite ion is $NO_3^-(aq)$. We balance N atoms, then balance O atoms with H_2O, then balance H atoms with H^+, then balance charge with electrons:

$$H_2O(\ell) + NO_2(g) \longrightarrow NO_3^-(aq) + 2\ H^+(aq) + e^-$$ ☑ Check: 2 H, 3 O, 1 N, zero net charge

3. *Result:* $\mathbf{4\ Cd(s) + 7\ OH^-(aq) + 6\ H_2O(\ell) + NO_3^-(aq) \longrightarrow 4\ [Cd(OH)_4]^{2-}(aq) + NH_3(aq)}$

Analyze and Plan: Given an unbalanced redox reaction reactions, balance it. Adapt the method described in Problem-Solving Example F.2 for reactions in basic solutions.

Execute: Separate the primary components of the two half reactions:

$$Cd(s) \longrightarrow [Cd(OH)_4]^{2-}(aq)$$

$$NO_3^-(aq) \longrightarrow NH_3(aq)$$

Balance each of these half reactions.

Here, we'll add $OH^-(aq)$ to the first half reaction, since that is the ligand in the complex ion, then use electrons to balance the charge:

$$Cd(s) + 4\ OH^-(aq) \longrightarrow [Cd(OH)_4]^{2-}(aq) + 2\ e^-$$

$$8\ e^- + 9\ H^+(aq) + NO_3^-(aq) \longrightarrow NH_3(aq) + 3\ H_2O(\ell)$$

Multiply the first reaction by 4 to balance the number of electrons to 8:

$$4 \times \{\ Cd(s) + 4\ OH^-(aq) \longrightarrow [Cd(OH)_4]^{2-}(aq) + 2\ e^-\ \}$$

$$8\ e^- + 9\ H^+(aq) + NO_3^-(aq) \longrightarrow NH_3(aq) + 3\ H_2O(\ell)$$

Add the two half reactions and cancel the electrons:

$$4\ Cd(s) + 16\ OH^-(aq) \longrightarrow 4\ [Cd(OH)_4]^{2-}(aq) + 8\ e^-$$

$$+ \quad 8\ e^- + 9\ H^+(aq) + NO_3^-(aq) \longrightarrow NH_3(aq) + 3\ H_2O(\ell)$$

$$4\ Cd(s) + 16\ OH^-(aq) + 9\ H^+(aq) + NO_3^-(aq) \longrightarrow 4\ [Cd(OH)_4]^{2-}(aq) + NH_3(aq) + 3\ H_2O(\ell)$$

☑ Check: 25 H, 4 Cd, 19 O, 1 N, – 8 net charge)

Neutralize the nine H^+ (aq) with nine OH^-(aq) to make nine $H_2O(\ell)$:

$$4\ Cd(s) + 7\ OH^-(aq) + 9\ H_2O(\ell) + NO_3^-(aq) \longrightarrow 4\ [Cd(OH)_4]^{2-}(aq) + NH_3(aq) + 3\ H_2O(\ell)$$

Simplify by eliminating redundant water molecules and check:

$$\mathbf{4\ Cd(s) + 7\ OH^-(aq) + 6\ H_2O(\ell) + NO_3^-(aq) \longrightarrow 4\ [Cd(OH)_4]^{2-}(aq) + NH_3(aq)}$$

☑ Check: 4 Cd, 16 O, 19 H, 1 N, – 8 net charge

5. *Result:* **(a) $2\ NO(g) + O_2(g) \longrightarrow 2\ NO_2(g)$ (b) $3\ H_2(g) + P_2(g) \longrightarrow 2\ PH_3(g)$**
(c) $2\ Fe^{2+}(aq) + 2\ H^+(aq) + H_2O_2(aq) \longrightarrow 2\ Fe^{3+}(aq) + 2\ H_2O(\ell)$
(d) $4\ H_2O(\ell) + 2\ MnO_4^-(aq) + 6\ Br^-(aq) \longrightarrow 2\ MnO_2(s) + 3\ Br_2(g) + 8\ OH^-(aq)$
(e) $6\ CH_3OH(aq) + 8\ H^+(aq) + Cr_2O_7^{2-}(aq) \longrightarrow 3\ CH_2O(aq) + 2\ Cr^{3+}(aq) + 7\ H_2O(\ell)$
(f) $7\ H_2O(\ell) + 3\ As_2O_3(s) + 4\ H^+(aq) + 4\ NO_3^-(aq) \longrightarrow 6\ H_3AsO_4(aq) + 4\ NO(g)$

Analyze and Plan: Follow methods described in the solutions to Problem-Solving Examples F.1 and F.2.

Execute:

(a) This reaction can be balanced the simple way: **$2\ NO(g) + O_2(g) \longrightarrow 2\ NO_2(g)$** ☑ Check: 2N, 4 O

(b) This reaction can be balanced the simple way: **$3\ H_2(g) + P_2(g) \longrightarrow 2\ PH_3(g)$** ☑ Check: 6 H, 2 P

(c) Separate the primary components of the two half reactions:

$$H_2O_2(aq) \longrightarrow H_2O(\ell)$$

$$Fe^{2+}(aq) \longrightarrow Fe^{3+}(aq)$$

Balance each of these half reactions.

$$2\ e^- + 2\ H^+(aq) + H_2O_2(aq) \longrightarrow 2\ H_2O(\ell)$$

$$Fe^{2+}(aq) \longrightarrow Fe^{3+}(aq) + e^-$$

Multiply the second reaction by 2 to balance the number of electrons to 2:

$$2\ e^- + 2\ H^+(aq) + H_2O_2(aq) \longrightarrow 2\ H_2O(\ell)$$

$$[Fe^{2+}(aq) \longrightarrow Fe^{3+}(aq) + e^-] \times 2$$

Add the two half reactions and cancel the electrons:

$$2\ e^- + 2\ H^+(aq) + H_2O_2(aq) \longrightarrow 2\ H_2O(\ell)$$

$$+ \quad 2\ Fe^{2+}(aq) \longrightarrow 2\ Fe^{3+}(aq) + 2\ e^-$$

$$\mathbf{2\ Fe^{2+}(aq) + 2\ H^+(aq) + H_2O_2(aq) \longrightarrow 2\ Fe^{3+}(aq) + 2\ H_2O(\ell)}$$

☑ Check: 2 Fe, 4 H, 2 O, + 6 charge

(d) Separate the primary components of the two half reactions:

$$MnO_4^-(aq) \longrightarrow MnO_2(s)$$

$$Br^-(aq) \longrightarrow Br_2(g)$$

Balance each of these half reactions:

$$3\ e^- + MnO_4^-(aq) + 4\ H^+(aq) \longrightarrow MnO_2(s) + 2\ H_2O(\ell)$$

$$2\ Br^-(aq) \longrightarrow Br_2(g) + 2\ e^-$$

Multiply the first reaction by 2 and the second reaction by 3 to balance the number of electrons to 6:

$$2 \times [3\,e^- + 4\,H^+(aq) + MnO_4^-(aq) \longrightarrow MnO_2(s) + 2\,H_2O(\ell)]$$

$$[2\,Br^-(aq) \longrightarrow Br_2(g) + 2\,e^-] \times 3$$

Now add them:

$$6\,e^- + 8\,H^+(aq) + 2\,MnO_4^-(aq) \longrightarrow 2\,MnO_2(s) + 4\,H_2O(\ell)$$

$$+\ 6\,Br^-(aq) \longrightarrow 3\,Br_2(g) + 6\,e^-$$

$$8\,H^+(aq) + 2\,MnO_4^-(aq) + 6\,Br^-(aq) \longrightarrow 2\,MnO_2(s) + 3\,Br_2(g) + 4\,H_2O(\ell)$$

☑ Check: 2 Mn, 8 O, 8 H, zero net charge

Add eight $OH^-(aq)$ to neutralize the eight $H^+(aq)$ to form eight $H_2O(\ell)$

$$8\,H^+(aq) + 8\,OH^-(aq) + 2\,MnO_4^-(aq) + 6\,Br^-(aq) \longrightarrow 2\,MnO_2(s) + 3\,Br_2(g) + 4\,H_2O(\ell) + 8\,OH^-(aq)$$

$$2\,MnO_4^-(aq) + 6\,Br^-(aq) + 8\,H_2O(\ell) \longrightarrow 2\,MnO_2(s) + 3\,Br_2(g) + 4\,H_2O(\ell) + 8\,OH^-(aq)$$

Simplify and check:

$$\mathbf{4\,H_2O(\ell) + 2\,MnO_4^-(aq) + 6\,Br^-(aq) \longrightarrow 2\,MnO_2(s) + 3\,Br_2(g) + 8\,OH^-(aq)}$$

☑ Check: 2 Mn, 12 O, 8 H, – 8 charge

(e) Separate the primary components of the two half reactions:

$$CH_3OH(aq) \longrightarrow CH_2O(aq)$$

$$Cr_2O_7^{2-}(aq) \longrightarrow Cr^{3+}(aq)$$

Balance each of these half reactions:

$$CH_3OH(aq) \longrightarrow CH_2O(aq) + 2\,H^+(aq) + 2\,e^-$$

$$6\,e^- + 14\,H^+(aq) + Cr_2O_7^{2-}(aq) \longrightarrow 2\,Cr^{3+}(aq) + 7\,H_2O(\ell)$$

Multiply the first reaction by 3 to balance the number of electrons to six:

$$[CH_3OH(aq) \longrightarrow CH_2O(aq) + 2\,H^+(aq) + 2\,e^-] \times 3$$

$$6\,e^- + 14\,H^+(aq) + Cr_2O_7^{2-}(aq) \longrightarrow 2\,Cr^{3+}(aq) + 7\,H_2O(\ell)$$

Now add them and cancel the electrons:

$$3\,CH_3OH(aq) \longrightarrow 3\,CH_2O(aq) + 6\,H^+(aq) + 6\,e^-$$

$$+\ 6\,e^- + 14\,H^+(aq) + Cr_2O_7^{2-}(aq) \longrightarrow 2\,Cr^{3+}(aq) + 7\,H_2O(\ell)$$

$$\mathbf{6\,CH_3OH(aq) + 8\,H^+(aq) + Cr_2O_7^{2-}(aq) \longrightarrow 3\,CH_2O(aq) + 2\,Cr^{3+}(aq) + 7\,H_2O(\ell)}$$

☑ Check: 3 C, 20 H, 13 O, +12 charge

(f) Separate the primary components of the two half reactions:

$$As_2O_3(s) \longrightarrow H_3AsO_4(aq)$$

$$NO_3^-(aq) \longrightarrow NO(g)$$

Balance each of these half reactions:

$$5\ H_2O(\ell) + As_2O_3(s) \longrightarrow 2\ H_3AsO_4(aq) + 4\ H^+(aq) + 4\ e^-$$

$$3\ e^- + 4\ H^+(aq) + NO_3^-(aq) \longrightarrow NO(g) + 2\ H_2O(\ell)$$

Multiply the first reaction by 3 and the second reaction by 4 to balance the number of electrons to 12:

$$[\ 5\ H_2O(\ell) + As_2O_3(s) \longrightarrow 2\ H_3AsO_4(aq) + 4\ H^+(aq) + 4\ e^-] \times 3$$

$$4 \times [\ 3\ e^- + 4\ H^+(aq) + NO_3^-(aq) \longrightarrow NO(g) + 2\ H_2O(\ell)\ ]$$

Now add them and cancel the electrons, the excess $H_2O(\ell)$ and the excess $H^+(aq)$, :

$$15\ H_2O(\ell) + 3\ As_2O_3(s) \longrightarrow 6\ H_3AsO_4(aq) + 12\ H^+(aq) + 12\ e^-$$

$$+\quad 12\ e^- + 16\ H^+(aq) + 4\ NO_3^-(aq) \longrightarrow 4\ NO(g) + 8\ H_2O(\ell)$$

$$\mathbf{7\ H_2O(\ell) + 3\ As_2O_3(s) + 4\ H^+(aq) + 4\ NO_3^-(aq) \longrightarrow 6\ H_3AsO_4(aq) + 4\ NO(g)}$$

☑ Check: 18 H, 28 O, 6 As, 4 N, no net charge

6. *Result:* **(a)** $\mathbf{6\ FeO(s) + O_2(g) \longrightarrow 2\ Fe_3O_4(s)}$

(b) $\mathbf{3\ I^-(aq) + 2\ H^+(aq) + ClO^-(aq) \longrightarrow I_3^-(aq) + Cl^-(aq) + H_2O(\ell)}$

(c) $\mathbf{Pb(s) + 2\ H^+(aq) + PbO_2(s) + 2\ HSO_4^-(aq) \longrightarrow 2\ PbSO_4(s) + 2\ H_2O(\ell)}$

(d) $\mathbf{Al(s) + MnO_4^-(aq) + 2\ H_2O(\ell) \longrightarrow [Al(OH)_4]^-(aq) + MnO_2(s)}$

(f) $\mathbf{Zn(Hg)(amalgam) + HgO(g) \longrightarrow ZnO + 2\ Hg(\ell)}$

Analyze and Plan: Follow methods described in the solutions to Problem-Solving Examples F.1 and F.2.

Execute:

(a) This reaction can be balanced the simple way: $\mathbf{6\ FeO(s) + O_2(g) \longrightarrow 2\ Fe_3O_4(s)}$ ☑ Check: 6 Fe, 8 O

(b) Separate the primary components of the two half reactions:

$$I^-(aq) \longrightarrow I_3^-(aq)$$

$$ClO^-(aq) \longrightarrow Cl^-(aq)$$

Balance each of these half reactions:

$$3\ I^-(aq) \longrightarrow I_3^-(aq) + 2\ e^-$$

$$2\ e^- + 2\ H^+(aq) + ClO^-(aq) \longrightarrow Cl^-(aq) + H_2O(\ell)$$

The electrons are already balanced, so add the equations and cancel the electrons:

$$3\ I^-(aq) \longrightarrow I_3^-(aq) + 2\ e^-$$

$$+\quad 2\ e^- + 2\ H^+(aq) + ClO^-(aq) \longrightarrow Cl^-(aq) + H_2O(\ell)$$

$$\mathbf{3\ I^-(aq) + 2\ H^+(aq) + ClO^-(aq) \longrightarrow I_3^-(aq) + Cl^-(aq) + H_2O(\ell)}$$

☑ Check: 3 I, 2 H, 1 Cl, 1 O, –2 charge

(c) Separate the primary components of the two half reactions.

None of the atoms in $HSO_4^-(aq)$ are changing oxidation state. One Pb atom is changing from a zero oxidation state (in Pb) to a +2 oxidation state (in $PbSO_4$). Another Pb atom is going from a +4 oxidation state (in PbO_2) to a +2 oxidation state (in $PbSO_4$). $HSO_4^-(aq)$ needs to be added to both half reactions, to provide the sulfate ion for the $PbSO_4$ product.

$$Pb(s) + HSO_4^-(aq) \longrightarrow PbSO_4(s)$$

$$PbO_2(s) + HSO_4^-(aq) \longrightarrow PbSO_4(s)$$

Balance each of these half reactions:

$$Pb(s) + HSO_4^-(aq) \longrightarrow PbSO_4(s) + H^+(aq) + 2\ e^-$$

$$2\ e^- + 3\ H^+(aq) + PbO_2(s) + HSO_4^-(aq) \longrightarrow PbSO_4(s) + 2\ H_2O(\ell)$$

The electrons are already balanced, so add the equations and cancel the electrons:

$$Pb(s) + HSO_4^-(aq) \longrightarrow PbSO_4(s) + H^+(aq) + 2\ e^-$$

$$+\ 2\ e^- + 3\ H^+(aq) + PbO_2(s) + HSO_4^-(aq) \longrightarrow PbSO_4(s) + 2\ H_2O(\ell)$$

$$Pb(s) + HSO_4^-(aq) + 3\ H^+(aq) + PbO_2(s) + HSO_4^-(aq) \longrightarrow PbSO_4(s) + H^+(aq) + PbSO_4(s) + 2\ H_2O(\ell)$$

Consolidate and simplify:

$$\mathbf{Pb(s) + 2\ H^+(aq) + PbO_2(s) + 2\ HSO_4^-(aq) \longrightarrow 2\ PbSO_4(s) + 2\ H_2O(\ell)}$$

(☑ Check: 2 Pb, 4 H, 10 O, 2 S, zero net charge

(d) Separate the primary components of the two half reactions:

$$Al(s) \longrightarrow [Al(OH)_4]^-(aq)$$

$$MnO_4^-(aq) \longrightarrow MnO_2(aq)$$

Balance each of these half reactions.

Here, we'll add $OH^-(aq)$ to the first half reaction, since that is the ligand in the complex ion, then use electrons to balance the charge:

$$Al(s) + 4\ OH^-(aq) \longrightarrow [Al(OH)_4]^-(aq) + 3\ e^-$$

$$3\ e^- + MnO_4^-(aq) + 4\ H^+(aq) \longrightarrow MnO_2(s) + 2\ H_2O(\ell)$$

The electrons are already balanced, so add the equations and cancel the electrons:

$$Al(s) + 4\ OH^-(aq) \longrightarrow [Al(OH)_4]^-(aq) + 3\ e^-$$

$$+\quad 3\ e^- + MnO_4^-(aq) + 4\ H^+(aq) \longrightarrow MnO_2(s) + 2\ H_2O(\ell)$$

$$Al(s) + 4\ OH^-(aq) + MnO_4^-(aq) + 4\ H^+(aq) \longrightarrow [Al(OH)_4]^-(aq) + MnO_2(s) + 2\ H_2O(\ell)$$

☑ Check: 25 H, 4 Cd, 19 O, 1 N, – 8 net charge

Neutralize the four H^+ (aq) with the four OH^-(aq) to make four $H_2O(\ell)$:

$$Al(s) + MnO_4^-(aq) + 4\ H_2O(\ell) \longrightarrow [Al(OH)_4]^-(aq) + MnO_2(s) + 2\ H_2O(\ell)$$

Simplify by eliminating redundant water molecules and check:

$$\mathbf{Al(s) + MnO_4^-(aq) + 2\ H_2O(\ell) \longrightarrow [Al(OH)_4]^-(aq) + MnO_2(s)}$$

☑ Check: 1 Al, 1 Mn, 6 O, 4 H, –1 net charge

(e) Separate the primary components of the two half reactions:

$$Cr(s) \longrightarrow Cr(OH)_3(s)$$

$$CrO_4^{2-}(aq) \longrightarrow Cr(OH)_3(s)$$

Balance each of these half reactions. Here, we'll add $OH^-(aq)$ to each half reaction, since that is the anion of the product compound, then use electrons to balance the charge:

$$Cr(s) + 3\ OH^-(aq) \longrightarrow Cr(OH)_3(s) + 3\ e^-$$

$$3\ e^- + 5\ H^+(aq) + CrO_4^{2-}(aq) \longrightarrow Cr(OH)_3(s) + H_2O(\ell)$$

The electrons are already balanced, so add the equations and cancel the electrons:

$$Cr(s) + 3\ OH^-(aq) \longrightarrow Cr(OH)_3(s) + 3\ e^-$$

$$+ \quad 3\ e^- + 5\ H^+(aq) + CrO_4^{2-}(aq) \longrightarrow Cr(OH)_3(s) + H_2O(\ell)$$

$$Cr(s) + 3\ OH^-(aq) + 5\ H^+(aq) + CrO_4^{2-}(aq) \longrightarrow 2\ Cr(OH)_3(s) + H_2O(\ell)$$

☑ Check: 2 Cr, 7 O, 5 H, 19 O, 1 N, – 8 net charge

Neutralize the five H^+ (aq) with the three OH^-(aq) and add two more OH^-(aq) to each side to make five $H_2O(\ell)$:

$$Cr(s) + 3\ OH^-(aq) + 5\ H^+(aq) + 2\ OH^-(aq) + CrO_4^{2-}(aq) \longrightarrow 2\ Cr(OH)_3(s) + H_2O(\ell) + 2\ OH^-(aq)$$

$$Cr(s) + 5\ H_2O(\ell) + CrO_4^{2-}(aq) \longrightarrow 2\ Cr(OH)_3(s) + H_2O(\ell) + 2\ OH^-(aq)$$

Simplify by eliminating redundant water molecules and check:

$$Cr(s) + 4\ H_2O(\ell) + CrO_4^{2-}(aq) \longrightarrow 2\ Cr(OH)_3(s) + 2\ OH^-(aq)$$

☑ Check: 2 Cr, 8 H, 8 O, –2 net charge

(f) This reaction can be balanced the simple way: **Zn(Hg)(amalgam) + HgO(s) $\longrightarrow$ ZnO + 2 Hg(ℓ)**

☑ Check: 1 Zn, 2 Hg, 1 O